The FIDIC Forms of Contract

Third Edition

In September 1999, FIDIC introduced its new Suite of Contracts, which included a 'new' Red, Yellow, Silver and Green forms of contract. The 'new' Red Book was intended to replace the 1992 Fourth Edition of the Red Book, with the ambition that its use would cease with time. This ambition has not materialised and is unlikely to do so in the future.

Despite the importance of the 1999 forms, there has been very little published on the new concepts adopted in them and how they interact with the previous forms. This important work considers these aspects together with the many developments affecting the Fourth Edition of the Red Book that have taken place since 1997, when the second edition of this book was published, and relates them to key contracting issues. It is written by a chartered engineer, conciliator and international arbitrator with wide experience in the use of the FIDIC Forms and in the various dispute resolution mechanisms specified in them.

Important features of this book include:

— background and concepts of the various forms of contract;
— a detailed comparison of the wording of the 1999 three main forms, which although similar in nature nevertheless significantly differ in certain areas due to their intended purpose;
— analysis of the rights and obligations of the parties involved in the contract and the allocation of risks concerned;
— a range of 'decision tree' charts analysing the main features of the 1992 Red Book, including risks, indemnities and insurances, claims and counterclaims, variations, procedure for claims, programme and delay, suspension, payments and certificates, dispute resolution mechanisms, and Dispute Boards;
— a much enlarged discussion of the meaning of 'claim' and 'dispute' and the types of claim with a discussion of the notice provision in the 1999 forms of contract for the submittal of claims by a contractor and by an employer;
— the FIDIC scheme of indemnities and insurance requirements; and the methods of dispute resolution provided by the various forms of contract; and
— five new chapters in this third edition, four chapters dealing with each of the 1999 forms and the fifth chapter focusing on the topic of Dispute Boards.

The FIDIC Forms of Contract

Third Edition

**The Fourth Edition of the Red Book, 1992
The 1996 Supplement
The 1999 Red Book
The 1999 Yellow Book
The 1999 Silver Book**

Nael G. Bunni
*BSc, MSc, PhD, CEng, FIEI, FICE,
FIStructE, FCIArb, FIAE, MConsEI*

© 2005 Nael G. Bunni

Editorial offices:
Blackwell Publishing Ltd, 9600 Garsington Road, Oxford OX4 2DQ, UK
 Tel: +44 (0)1865 776868
Blackwell Publishing Inc., 350 Main Street, Malden, MA 02148–5020, USA
 Tel: +1 781 388 8250
Blackwell Publishing Asia Pty Ltd, 550 Swanston Street, Carlton, Victoria 3053, Australia
 Tel: +61 (0)3 8359 1011

The right of the Author to be identified as the Author of this Work has been asserted in accordance with the Copyright, Designs and Patents Act 1988.

All rights reserved. No part of this publication may be reproduced, stored in a retrieval system, or transmitted, in any form or by any means, electronic, mechanical, photocopying, recording or otherwise, except as permitted by the UK Copyright, Designs and Patents Act 1988, without the prior permission of the publisher.

First edition published 1991
Reprinted 1993, 1994, 1996
Second edition published 1997
Reprinted 1998, 2000, 2001, 2004
Third edition published 2005

Library of Congress Cataloging-in-Publication Data

Bunni, Nael G.
 The FIDIC forms of contract : the fourth edition of the Red Book, 1992, the 1996 Supplement, the 1999 Red Book, the 1999 Yellow Book, the 1999 Silver Book / Nael G. Bunni.—3rd ed.
  p. cm.
 Rev. ed. of: The FIDIC form of contract, 2nd ed. 1997.
 Includes bibliographical references and index.
 ISBN 13: 978-14051-2031-9 (alk. paper)
 ISBN 10: 1-4051-2031-2 (alk. paper)
 1. Engineering contracts. 2. Standardized terms of contract. I. Bunni, Nael G. FIDIC form of contract. II. International Federation of Consulting Engineers. III. Title.

K891.B8B86 2005
343'.07862—dc22

2005041189

ISBN 10: 1-4051-2031-2
ISBN 13: 978-14051-2031-9

A catalogue record for this title is available from the British Library

Set in 9.5/12.5pt Palatino
by Graphicraft Limited, Hong Kong
Printed and bound in Great Britain
by TJ International Ltd, Padstow, Cornwall

The publisher's policy is to use permanent paper from mills that operate a sustainable forestry policy, and which has been manufactured from pulp processed using acid-free and elementary chlorine-free practices. Furthermore, the publisher ensures that the text paper and cover board used have met acceptable environmental accreditation standards.

For further information on Blackwell Publishing, visit our website:
www.blackwellpublishing.com

Contents

List of Figures		xx
List of Tables		xxii
The Author		xxiii
Preface		xxv
Acknowledgements		xxvii

Part I	Background and Concepts of the Red Book	1
Chapter 1	**Background of the Red Book**	**3**
1.1	The ACE Form	4
1.2	The First Edition of the Red Book	6
1.3	The Second and Third Editions of the Red Book	7
1.4	The Fourth Edition of the Red Book	11
1.5	The 1996 Supplement to the Red Book	14
1.6	Concepts of the Red Book	15
1.7	The new suite of FIDIC contracts	15
Chapter 2	**The Red Book is Based on a Domestic Contract**	**17**
2.1	Introduction	17
2.2	Diversity of legal systems	17
2.3	The applicable law in international construction	19
2.4	The applicable law of the contract	20
2.5	Law governing procedure	22
2.6	Law governing enforcement of awards	23
2.7	Grouping of the contemporary legal systems	24
2.8	The Romano-Germanic group	25
	2.8.1 Sources of law in the Romano-Germanic group	28
	2.8.2 Legal authoritative writing	31
	2.8.3 Areas of the law affecting construction in the Romano-Germanic group	32
2.9	The common law group	32
	2.9.1 Sources of law in the common law group	35
	2.9.2 Areas of the law affecting construction in the common law group	40

2.10	The law in Islamic countries		41
	2.10.1	Sources of Islamic law	42
	2.10.2	Contracts in general and construction and engineering contracts in particular under Islamic law	44
	2.10.3	Some salient principles in construction and engineering contracts	44

Chapter 3 Legal Concepts Based on the Common Law System 49

3.1	The law applicable to the contract		49
	3.1.1	Clause 5	49
	3.1.2	Clause 26	50
	3.1.3	Clause 70	50
3.2	Conflict		51
3.3	Some specific concepts under the common law		52
	3.3.1	Substantive law and procedural law	52
	3.3.2	Legislation, common law and equity	52
3.4	Tort		52
3.5	Contract – general principles		54
	3.5.1	Prerequisites of a contract	55
	3.5.2	Limitation periods	57
3.6	Privity of contract		58
3.7	Performance of a contract		58
3.8	The contents of a contract		59
3.9	Remedies for breach of contract		62
3.10	Exclusion clauses		68
3.11	The responsibility to complete		68

Chapter 4 Drafting Principles 70

Chapter 5 The Concept of a Trusted Independent Engineer 73

5.1	Introduction		73
	5.1.1	FIDIC's Statutes and By-Laws and the independent engineer	73
	5.1.2	FIDIC's Code of Ethics	76
	5.1.3	FIDIC's Quality-Based Selection, 'QBS'	77
5.2	Other suppliers of consulting services		78
5.3	Services provided by the consulting engineer		79
	5.3.1	Counselling services	79
	5.3.2	Pre-investment studies	79
	5.3.3	Design, preparation of documents and supervision	80
	5.3.4	Specialised design and development services	80
	5.3.5	Project management	80
	5.3.6	Programme manager	81
5.4	Independence		81

Chapter 6	**A Traditional Re-measurement Contract**			**83**
	6.1	Factors governing choice of contract		83
	6.2	The allocation of essential functions		83
		6.2.1	The allocation of the function relating to finance	84
		6.2.2	The allocation of the functions of design and construction	86
		6.2.3	The allocation of risk, quality control and the method of pricing and payment	86
	6.3	Re-measurement contracts		87
		6.3.1	The Red Book is a re-measure contract	88
		6.3.2	Contracts with a bill of quantities	89
		6.3.3	Contracts with a schedule of rates	90
	6.4	Cost-reimbursable contracts		90
	6.5	Lump sum contracts		90
		6.5.1	The Supplement to the Fourth Edition of the Red Book, Section B	91
		6.5.2	Main features of FIDIC's Form for Payment on a lump sum basis	91
Chapter 7	**Sharing of Risks**			**93**
	7.1	Introduction		93
	7.2	The definition of 'risk'		94
	7.3	Measurement of risk		98
	7.4	Risk management		99
	7.5	Allocation of risks and their management		100
	7.6	Allocation of risks in the Red Book		105
	7.7	Responsibility and liability		108
	7.8	Indemnity and insurance		109
Chapter 8	**The Concepts in Practice**			**112**
	8.1	The Red Book in use		112
	8.2	Areas of conflict		114
		8.2.1	A relationship of trust	114
		8.2.2	The role of the engineer	114
		8.2.3	Avoidance of risk	115
		8.2.4	The design function	115
		8.2.5	Absence of a legal system	116
		8.2.6	Distrust of changes	117
		8.2.7	Legal questions	117
	8.3	EIC/FIDIC survey of 1996		120
	8.4	A brief summary of Part I		120
Part II	**The Fourth Edition: A Commentary**			**123**
Chapter 9	**The Revisions – Purposes and Consequences**			**125**
	9.1	Introduction		125
	9.2	Clause 1		129

	9.2.1	Definition of 'Engineer' under group (a)	129
	9.2.2	Definition of 'tests on completion'	130
	9.2.3	Definitions under group (e)	130
	9.2.4	Definitions under group (f)	130
	9.2.5	Definitions under group (g)	131
	9.2.6	Definition of 'approved'	134
9.3	Clause 2		134
	9.3.1	Requirement for consultation	135
	9.3.2	Responsibility for delegation	136
	9.3.3	Requirement for writing	136
	9.3.4	Express requirement for impartiality	136
9.4	Sub-clause 5.2		136
9.5	Sub-clauses 6.1, 6.4 and 6.5		137
9.6	Clause 7		137
9.7	Clause 8		138
9.8	Clause 10		138
9.9	Sub-clause 12.2		139
9.10	Clause 13		140
9.11	Sub-clauses 14.1 and 14.3		140
9.12	Clause 15		141
9.13	Clause 19		141
9.14	Clause 20		141
9.15	Clause 21		142
9.16	Clause 23		143
9.17	Clause 25		143
9.18	Clause 27		143
9.19	Clause 28		144
9.20	Clause 30		144
9.21	Clauses 34 and 35		144
9.22	Sub-clause 36.5		144
9.23	Clause 37		145
9.24	Clause 40		145
9.25	Clause 41		145
9.26	Sub-clause 42.3		146
9.27	Clause 44		146
9.28	Clause 46		147
9.29	Clause 51		147
9.30	Sub-clause 52.3		147
9.31	Clauses 53 and 54		148
9.32	Sub-clause 57.2		148
9.33	Clause 60		148
9.34	Sub-clauses 65.4 and 66.1		150
9.35	Clause 67		150
9.36	Clause 69		151
9.37	Other changes made in the 1992 Reprint		152
9.38	Concluding remarks		152

Part III	**The Fourth Edition in Practice**	**153**
Chapter 10	**Role of the Engineer**	**155**
	10.1 Introduction	155
	10.2 The engineer as a designer	157
	10.3 The engineer as the employer's agent	162
	10.3.1 Authority and duties of the engineer	163
	10.4 The engineer's proactive duties and authority	167
	10.5 The engineer's reactive duties and authority	170
	10.6 The engineer's passive duties and authorities	173
	10.7 The engineer as a supervisor	175
	10.8 The engineer as certifier	178
	10.9 The engineer as adjudicator or quasi-arbitrator	180
	10.10 Concluding remarks	183
Chapter 11	**Responsibility and Liability of the Engineer**	**184**
	11.1 Introduction	184
	11.2 Responsibility of the engineer towards the employer	187
	11.3 Responsibility of the engineer towards the contractor	192
	11.3.1 In the common law countries	193
	11.3.2 In the Romano-Germanic system	198
	11.4 The responsibility of the engineer towards third parties (other than the contractor)	199
	11.5 The responsibility of the engineer towards society; employees; and the engineer himself	200
	11.6 Liability in construction	200
	11.7 Levels of liability	201
Chapter 12	**The Employer's Obligations**	**208**
	12.1 Introduction	208
	12.2 Identification of specific elements of the project	209
	12.3 Appointment of engineer	209
	12.4 Possession of site	211
	12.5 To provide instructions as and when they are required	213
	12.6 The employer is to refrain from taking any action which would impede or interfere with the progress of the works	215
	12.7 The employer is to supply materials and carry out works if these form part of the work as defined in the contract	216
	12.8 The employer is to nominate specialist sub-contractors and suppliers as and when they are required	217
	12.9 To permit the contractor to carry out the whole of the works	218
	12.10 To make payments and to make them on time	218
	12.11 Additional obligations for the employer under the Fourth Edition of the Red Book	219

Chapter 13	**The Contractor's Obligations**		**222**
	13.1 Introduction		222
	13.2 The contractor's obligations during the tendering stage		223
	13.3 The contractor's obligations following the letter of acceptance and during the construction stage up to substantial completion		224
		13.3.1 Finalising documentation required prior to commencement of the works	224
		13.3.2 Construction and completion of the works with due diligence and within the time for completion	225
		13.3.3 Use of materials, plant and workmanship	230
		13.3.4 Provision of securities, indemnities and insurances	233
		13.3.5 Supply of information, notices or alerts	234
		13.3.6 Performance of certain administrative functions	236
	13.4 Contractor's obligations after substantial completion of the works		240
Chapter 14	**Risks, Liabilities, Indemnities and Insurances**		**241**
	14.1 Introduction		241
	14.2 The Red Book provisions relating to risk, responsibility, liability, indemnity and insurance		241
	14.3 Clause 20 of the Red Book – '20.1: care of the works'; '20.2: responsibility to rectify loss or damage'; '20.3: loss or damage due to employer's risks'; and '20.4: employer's risks'		245
		14.3.1 Sub-clause 20.1: care of the works	245
		14.3.2 Sub-clause 20.2: responsibility to rectify loss or damage; and sub-clause 20.3: loss or damage due to employer's risks	245
		14.3.3 Sub-clause 20.4: the employer's risks	246
	14.4 Clause 65 of the Red Book (sub-clauses 65.1 to 65.8) – special risks		248
	14.5 Clause 21 of the Red Book – insurance		252
		14.5.1 Importance of adequacy of cover	252
		14.5.2 Period of insurance and extent of cover	253
		14.5.3 Joint names	256
		14.5.4 Scope of insurance cover	257
		14.5.5 Provisions for payment in foreign currency	258
		14.5.6 Provision for deductibles	259
	14.6 Clause 22 of the Red Book – indemnity for damage to persons and property other than the works		259
	14.7 Clause 23 of the Red Book – third party insurance		260
	14.8 Clause 24 of the Red Book – injury to workmen and insurance		261
	14.9 Clause 25 of the Red Book – general insurance requirements		262

14.10	Part II of the Red Book – insurance arranged by the employer	263
14.11	Definitions	266

Chapter 15 Performance and other Securities 269

15.1	Introduction	269
15.2	The spectrum of securities	270
15.3	Types of securities	271
15.4	Characteristics of performance bonds and guarantees	272
	15.4.1 Payment guarantees	272
	15.4.2 Performance bonds	273
	15.4.3 Demand guarantees	276
15.5	The ICC Uniform Rules for Demand Guarantees	280
15.6	Uniform Rules for Contract Bonds	281
15.7	Insurance against unfair calling	284
15.8	Performance securities under the Red Book	284
15.9	Examples of securities provided	288
15.10	Other securities associated with a construction contract	290
	15.10.1 Bid bonds or guarantees	290
	15.10.2 Advance payment guarantees	290
	15.10.3 Retention money bonds	291
	15.10.4 Maintenance or defects liability bonds	291
	15.10.5 Company suretyship	291
15.11	Concluding remarks	292

Chapter 16 Claims and Counterclaims 293

16.1	Introduction	293
16.2	Definition and legal basis of claims and counterclaims	293
16.3	A claim under the contract and based on its provisions	298
	16.3.1 Variations	299
	16.3.2 Measurement changes	304
	16.3.3 Adverse physical obstructions or conditions	305
	16.3.4 Employer's risks	315
	16.3.5 Compliance with statutes, regulations, price fluctuations, currency and other economic causes	316
	16.3.6 Defects and unfulfilled obligations	317
	16.3.7 Failure to commence, delays, suspension of work, release from performance, default and termination	320
	16.3.8 Other miscellaneous specified events	325
16.4	A claim arising out of or in connection with the contract	327
16.5	Procedure for claims for additional payment – clause 53	328
	16.5.1 Procedural steps	329
	16.5.2 Records	332
16.6	The presentation of claims	333

	16.7	Quantum	333
		16.7.1 Heads of claim	333
		16.7.2 The global approach	336
	16.8	Failure to follow the claims procedure	341
	16.9	Concluding remarks	342
Chapter 17	**Delay in Completion and Claims for Extension of Time**		**343**
	17.1	Time is of fundamental importance	343
	17.2	Clauses 43, 44, 46, 47 and 48 of the Red Book	344
	17.3	Relevant clauses of the Fourth Edition of the Red Book to an extension of time under clause 44	353
	17.4	Programming	358
	17.5	Concurrent delays	364
	17.6	Claims for both extension of time and money	365
		17.6.1 Prolongation	366
		17.6.2 Disruption	368
	17.7	Liquidated damages	370
		17.7.1 Liquidated damages and penalties	371
	17.8	The Society of Construction Law *Delay and Disruption Protocol*	375
Chapter 18	**Certificates and Payments**		**377**
	18.1	Introduction	377
	18.2	Interim payment certificates	377
	18.3	Taking-over certificate	379
	18.4	Defects liability certificate	381
	18.5	Final payment certificate	381
	18.6	The engineer is to certify a valuation at date of termination	384
	18.7	Common requirements	384
	18.8	Late certification	384
Chapter 19	**Disputes Settlement by Arbitration**		**385**
	19.1	Introduction and background	385
	19.2	Advantages of arbitration	387
	19.3	What is a dispute?	389
	19.4	What is arbitration?	393
	19.5	The arbitration agreement	394
	19.6	Sources of law in arbitration	395
		19.6.1 General	396
		19.6.2 The arbitration agreement as a source of law	396
		19.6.3 Practice and custom	396
	19.7	The arbitrator	397
	19.8	The arbitration agreement under clause 67 of the Red Book	399
		19.8.1 Procedure under clause 67	403
	19.9	The 1996 supplement to the Fourth Edition of the Red Book	409

	19.10	The ICC Rules of Arbitration	414
		19.10.1 The procedure under the ICC Rules of Arbitration	416
		19.10.2 The advantages of the ICC Rules	422
		19.10.3 Some constructive criticism	424
	19.11	Why does arbitration in construction disputes continue to lose favour?	425
	19.12	Concluding remarks	437
Chapter 20	**Amicable Settlement Using Alternative Dispute Resolution**		**438**
	20.1	Introduction	438
	20.2	Methods of dispute settlement	439
	20.3	Direct negotiation	441
		20.3.1 Negotiators	441
		20.3.2 Distinguishing features of direct negotiation	442
		20.3.3 When should negotiation be used and what are the steps?	442
	20.4	Mediation	443
	20.5	Conciliation	445
		20.5.1 What is conciliation?	446
		20.5.2 Why conciliation?	446
		20.5.3 When should conciliation be used?	446
		20.5.4 Who should be a conciliator?	447
		20.5.5 Who should attend the conciliation?	448
		20.5.6 The conciliation process	448
	20.6	Mini-trial procedure	451
	20.7	Dispute Board, Dispute Review Board and Dispute Adjudication Board	452
	20.8	Adjudication	452
	20.9	The ICC Rules for Amicable Dispute Resolution	453
		20.9.1 The Rules	454
		20.9.2 Fees and costs	456
		20.9.3 General provisions	456
	20.10	Pre-arbitral referee procedure	457
	20.11	The ICC Rules for Expertise	457
		20.11.1 The Rules for Expertise	458
		20.11.2 Functions of the ICC Centre for Expertise	458
	20.12	Concluding remarks	460
Part IV	**Other Documents Related to the Red Book**		**461**
Chapter 21	**FIDIC's Other Forms of Contract (1993–1999)**		**463**
	21.1	Introduction	463
	21.2	The Yellow Book, third edition	463
		21.2.1 Background	463
		21.2.2 Differences in the nature of civil engineering and E & M engineering projects	464
		21.2.3 Essential features of the Yellow Book	465

	21.3	The Orange Book, first edition	471
		21.3.1 Background	473
		21.3.2 Differences in the nature of the Red and Yellow Books on the one hand and the Orange Book on the other	474
	21.4	The conditions of subcontract for works of civil engineering construction	479
		21.4.1 Format	479
		21.4.2 Clause 1, definitions and interpretation	479
		21.4.3 Clause 4, 'main contract'	480
		21.4.4 Clauses 13 and 15, 'indemnities; insurances'	480
		21.4.5 Clause 16, 'payment'	481
		21.4.6 Clause 19, 'settlement of disputes'	483
	21.5	Other publications of FIDIC	483
Part V		The 1999 Green Book; The 1999 Red Book; The 1999 Yellow Book; The 1999 Silver Book; Dispute Boards	485
Chapter 22		The 1999 FIDIC Suite of Contracts	487
	22.1	Introduction	487
	22.2	Differences in format	489
	22.3	Differences in concept	490
	22.4	The 1999 Green Book	494
Chapter 23		The 1999 Red Book	503
	23.1	Introduction	503
	23.2	The 1999 Red Book: concepts and content	504
	23.3	The 1999 Red Book: new concepts	519
		23.3.1 Sub-clause 1.12: 'Confidential Details'	519
		23.3.2 Sub-clause 2.4: 'Employer's Financial Arrangements'	520
		23.3.3 Sub-clause 2.5: 'Employer's Claims'	520
		23.3.4 Clause 3: 'The Engineer'	522
		23.3.5 Sub-clause 4.1(c): 'Fitness for Purpose'	526
		23.3.6 Sub-clause 4.2: 'Performance Security'	527
		23.3.7 Sub-clause 4.12: 'Unforeseeable Physical Conditions'	527
		23.3.8 Sub-clause 4.21: 'Progress Reports'	529
		23.3.9 Sub-clause 13.2: 'Value Engineering'	529
		23.3.10 Sub-clause 14.7: 'Payment'	529
		23.3.11 Sub-clause 14.8: 'Delayed Payment'	530
		23.3.12 Sub-clause 15.5: 'Employer's Entitlement to Termination'	530
		23.3.13 Clause 17: 'Risk and Responsibility'	530
		23.3.14 Sub-clause 17.6: 'Limitation of Liability'	532
		23.3.15 Clause 18: 'Insurance'	533

		23.3.16	Clause 19: 'Force Majeure'	534
		23.3.17	Clause 20: 'Claims'	536
		23.3.18	Sub-clauses 20.2 to 20.4 'Appointment of the Dispute Adjudication Board'; 'Failure to Agree Dispute Adjudication Board'; 'Obtaining Dispute Adjudication Board's Decision'	537
		23.3.19	Sub-clauses 20.5 to 20.8: 'Amicable Settlement'; 'Arbitration'; 'Failure to Comply with Dispute Adjudication Board's Decision'; 'Expiry of Dispute Adjudication Board's Appointment'	537
	23.4	Some highlights of the 1999 Red Book		539
		23.4.1	Sub-clause 1.1.4.3: 'Cost'	539
		23.4.2	Clause 2: 'The Employer'	539
		23.4.3	Clause 4: 'The contractor'	541
		23.4.4	Clause 6: 'Staff and Labour'	542
		23.4.5	Clause 7: 'Plant, Materials and Workmanship'	542
		23.4.6	Clause 8: 'Commencement, Delays and Suspension'	543
		23.4.7	Clause 9: 'Tests on Completion'	544
		23.4.8	Clause 10: 'Employer's Taking Over'	545
		23.4.9	Clause 11: 'Defects Liability'	546
		23.4.10	Clause 12: 'Measurement and Evaluation'	546
		23.4.11	Clause 13: 'Variations and Adjustments'	546
		23.4.12	Clause 14: 'Contract Price and Payment'	547
		23.4.13	Clause 15: 'Termination by Employer'	548
		23.4.14	Clause 16: 'Suspension and Termination by Contractor'	549
		23.4.15	Sub-clause 20.7: 'Failure to Comply with Dispute Adjudication Board's Decision'	549
		23.4.16	Guidance for the preparation of particular conditions	550
Chapter 24	The 1999 Yellow Book			551
	24.1	Introduction		551
	24.2	Tendering under, and using, the 1999 Yellow Book		552
	24.3	The 1999 Yellow Book: The employer's requirements		553
		24.3.1	Sub-clause 1.1.3.4 'Tests on Completion'	556
		24.3.2	Sub-clause 1.1.3.6 'Tests after Completion'	556
		24.3.3	Sub-clause 1.1.6.7 'Site'	556
		24.3.4	Sub-clause 1.1.6.9: Definitions – 'Variation'	556
		24.3.5	Sub-clause 1.5: General Provisions – 'Priority of Documents'	556
		24.3.6	Sub-clause 1.8: 'Care and Supply of Documents'	557
		24.3.7	Sub-clause 1.9: General Provisions – 'Errors in the Employer's Requirements'	557

24.3.8	Sub-clause 1.11: General Provisions – 'Contractor's Use of Employer's Documents'	558
24.3.9	Sub-clause 1.13: General Provisions – 'Compliance with Laws'	558
24.3.10	Sub-clause 2.1: The Employer – 'Right to Access to the Site'	558
24.3.11	Sub-clause 4.1: The Contractor – 'Contractor's General Obligations'	558
24.3.12	Sub-clause 4.6: The Contractor – 'Co-operation'	559
24.3.13	Sub-clause 4.7 'Setting Out'	559
24.3.14	Sub-clause 4.18: The Contractor – 'Protection of the Environment'	559
24.3.15	Sub-clause 4.19: The Contractor – 'Electricity, Water and Gas'	559
24.3.16	Sub-clause 4.20: The Contractor – 'Employer's Equipment and Free-Issue Material'	560
24.3.17	Sub-clause 5.1: Design – 'General Design Obligations'	560
24.3.18	Sub-clause 5.2: Design – 'Contractor's Documents'	560
24.3.19	Sub-clause 5.4: Design – 'Technical Standards and Regulations'	561
24.3.20	Sub-clause 5.5: Design – 'Training'	561
24.3.21	Sub-clause 5.6: Design – 'As-Built Documents'	562
24.3.22	Sub-clause 5.7: Design – 'Operation and Maintenance Manuals'	562
24.3.23	Sub-clause 6.1: Staff and Labour – 'Engagement of Staff and Labour'	562
24.3.24	Sub-clause 6.6: Staff and Labour – 'Facilities for Staff and Labour'	562
24.3.25	Sub-clause 7.4: 'Testing'	563
24.3.26	Sub-clause 7.8: Plant, Materials and Workmanship – 'Royalties'	563
24.3.27	Sub-clause 8.2: 'Time for Completion'	563
24.3.28	Sub-clause 8.3: Commencement, Delays and Suspension – 'Programme'	563
24.3.29	Sub-clause 9.1: Tests on Completion – 'Contractor's Obligations'	564
24.3.30	Sub-clause 10.2: 'Taking Over of Parts of the Works'	564
24.3.31	Sub-clause 11.1: 'Completion of Outstanding Works and Remedying Defects'	564
24.3.32	Sub-clause 17.5: Risk and Responsibility – 'Intellectual and Industrial Property Rights'	564
24.4	Comparison between the 1999 Yellow Book and the 1999 Red Book	565

24.4.1	Clause 1: General Provisions	565
24.4.2	Clause 3: The Engineer	566
24.4.3	Sub-clause 4.1: The Contractor – 'Contractor's General Obligations'	566
24.4.4	Sub-clause 4.4: 'Subcontractors', sub-clause 4.5: 'Nominated Subcontractors' and sub-clause 4.6: 'Co-operation'	567
24.4.5	Clause 5: Design	568
24.4.6	Sub-clause 7.5: Plant, Materials and Workmanship – 'Rejection'	572
24.4.7	Sub-clause 8.3: Commencement, Delays and Suspension – 'Programme'	572
24.4.8	Sub-clause 9.1: Tests on Completion – 'Contractor's Obligations'	573
24.4.9	Sub-clauses 11.2 and 11.6: Defects Liability – 'Cost of Remedying Defects and Further Tests' and 'Further Tests'	574
24.4.10	Clause 12: Tests after Completion	575
24.4.11	Sub-clauses 13.1, 13.2 and 13.3: Variations and Adjustments – 'Right to Vary', 'Value Engineering', and 'Variation Procedure'	576
24.4.12	Sub-clauses 14.1, 14.3 and 14.9: Contract Price and Payment – 'The Contract Price'; 'Application for Interim Payment Certificates' and 'Payment of Retention Money'	578
24.4.13	Sub-clause 17.5: Risk and Responsibility – 'Intellectual and Industrial Property Rights'	579
24.4.14	Clause 20.2: Claims, Disputes and Arbitration – 'Appointment of the Dispute Adjudication Board'	580

Chapter 25	The 1999 Silver Book	581
25.1	Introduction	581
25.2	The 1999 Silver Book: the shifted risks	582
25.2.1	Sub-clause 3.1: The Employer's Administration – 'The Employer's Representative'	582
25.2.2	Sub-clause 3.5: The Employer's Administration – 'Determinations'	583
25.2.3	Sub-clause 4.7: The Contractor – 'Setting Out'	584
25.2.4	Sub-clause 4.12: The Contractor – 'Unforeseeable Difficulties'	584
25.2.5	Sub-clause 5.1: Design – 'General Design Obligations'	585
25.2.6	Sub-clause 5.8: Design – 'Design Error'	585
25.2.7	Sub-clause 8.4: Commencement, Delay and Suspension – 'Extension of Time for Completion'	586

		25.2.8	Sub-clause 17.3: Risk and Responsibility – 'Employer's Risks'	586

25.2.8 Sub-clause 17.3: Risk and Responsibility – 'Employer's Risks' — 586
25.2.9 Sub-clause 20.1: Claims, Disputes and Arbitration – 'Contractor's Claims' — 587

25.3 The 1999 Silver Book: concepts and content — 587
- 25.3.1 Clause 1: General Provisions — 587
- 25.3.2 Clause 3: The Employer's Administration — 589
- 25.3.3 Clause 4: The Contractor — 590
- 25.3.4 Clause 5: Design — 591
- 25.3.5 Clause 6: Staff and Labour — 591
- 25.3.6 Clause 7: Plant, Materials and Workmanship — 592
- 25.3.7 Clause 8: Commencement, Delays and Suspension — 592
- 25.3.8 Clause 9: Tests on Completion — 593
- 25.3.9 Clause 10: Employer's Taking Over — 594
- 25.3.10 Clause 11: Defects Liability — 594
- 25.3.11 Clause 12: Tests after Completion — 595
- 25.3.12 Clause 13: Variations and Adjustments — 595
- 25.3.13 Clause 14: Contract Price and Payment — 595
- 25.3.14 Clause 15: Termination by Employer — 597
- 25.3.15 Clause 16: Suspension and Termination by Contractor — 598

Chapter 26 Dispute Boards — 599

26.1 Introduction — 599
26.2 Main advantages of the Dispute Board — 600
26.3 Background and evolution — 602
26.4 Types of Dispute Boards — 604
- 26.4.1 Dispute Review Board — 604
- 26.4.2 Dispute Adjudication Board — 606
- 26.4.3 Combined Dispute Board — 607

26.5 Varieties of Dispute Boards — 607
- 26.5.1 Varieties based on the number of Dispute Board members — 608
- 26.5.2 Varieties based on the date of appointment of the Dispute Board members — 609
- 26.5.3 Varieties peculiar to the type of Dispute Board — 610

26.6 Dispute Adjudication Boards under the FIDIC Contracts — 610
26.7 The role of the Dispute Adjudication Board — 612
26.8 Establishment of the Dispute Adjudication Board — 613
26.9 Obligations of the parties and the members of the Dispute Adjudication Board — 617
- 26.9.1 Obligations of the parties towards members of the Dispute Adjudication Board — 617
- 26.9.2 Obligations of the members of Dispute Adjudication Board towards the parties — 620

	26.10 Powers of the Dispute Adjudication Board	622
	26.11 Procedures relating to site visits and meetings	623
	26.12 Procedures relating to referral of a matter to the Board for its opinion	625
	26.13 Procedures relating to referral of a dispute to the Board for its decision	626
	26.14 Remuneration of the members of the Dispute Adjudication Board	629
	26.15 Cost of maintaining the members of the Dispute Adjudication Board	631
	26.16 The decision of the Dispute Adjudication Board	632
Part VI	**Comparison between text of the three 1999 Major Books: Red, Yellow and Silver Books**	**635**
Chapter 27	**A Precise Record of the Alterations, Omissions and Additions in the 1999 Yellow and Silver Books as compared with the 1999 Red Book**	**637**
Appendices		743
A	*Editorial Amendments in the 1988 Reprint of the Fourth Edition of the Red Book*	745
B	*Further Amendments in the 1992 Reprint of the Fourth Edition of the Red Book*	747
C	*Part II – Conditions of Particular Application*	753
References		787
Table of Cases		821
Index		827

List of Figures

Figure 2.1	Areas of the law affecting construction in the Romano-Germanic group.	33
Figure 2.2	Areas of the law affecting construction in legal systems based on common law.	40
Figure 3.1	Remedies for breach of contract in the common law system.	63
Figure 7.1	Risks in construction under the FIDIC Red Book, Fourth Edition.	106
Figure 7.2	Indemnities and possible insurance covers for a construction project.	110
Figure 13.1	Acceptance of tender and commencement of works.	228
Figure 14.1	Flow of risk into responsibility, liability, indemnity and insurance.	242
Figure 14.2	Indemnities and insurances relating to risks of injury and damage under the Fourth Edition of the FIDIC Red Book.	243
Figure 14.3	Indemnity and insurance relating to financial risks, etc.	244
Figure 14.4	Consequences of risks eventuating.	254
Figure 14.5	The insurance scheme as in the Fourth Edition of the Red Book.	264
Figure 15.1	The two alternatives of issuing a performance guarantee.	277
Figure 16.1	Claims and counterclaims.	297
Figure 16.2	Variation orders.	306
Figure 16.3	Suspension.	322

Figure 16.4	Procedure for claims.	330
Figure 17.1	Programme – time – delay – rate of progress.	356
Figure 17.2	An example of a network analysis.	360
Figure 17.3	Critical path diagram associated with the network analysis in Figure 17.2.	361
Figure 18.1	Certificates and payments.	382
Figure 19.1	Procedure under clause 67 of the Red Book.	410
Figure 19.2	Procedure under the provisions of the new alternative version of clause 67 contained in the 1996 Supplement.	412
Figure 23.1	Number of words in the FIDIC forms.	503
Figure 23.2	Number of sub-clauses in the 1999 Red Book.	504
Figure 26.1	Agenda for the first meeting and site visit, to be adjusted for other visits.	624

List of Tables

Table 9.1	Determination by the engineer in favour of the contractor.	132
Table 9.2	Determination by the engineer in favour of the employer.	133
Table 9.3	Determination by the employer against the contractor.	134
Table 10.1	The engineer's proactive duties and authority.	167
Table 10.2	The engineer's reactive duties and authority.	171
Table 10.3	The engineer's passive duties.	174
Table 12.1	Notices required to be given by the employer to the contractor under the contract.	214
Table 13.1	The phases of a construction project in chronological order.	226
Table 13.2	Notices required to be given by the contractor under the contract.	237
Table 23.1	The Fourth Edition of the Red Book and the corresponding conditions in the 1999 Red Book.	506
Table 23.2	The 1999 Red Book and the corresponding conditions in the Fourth Edition of the Red Book.	513
Table 23.3	Cost, with or without profit.	540
Table 23.4	Sub-clauses relating to extension of time.	544

The Author

Nael G. Bunni

Dr Bunni is a chartered engineer, conciliator/mediator and registered chartered arbitrator. He is Past President of the Association of Consulting Engineers of Ireland and Past President of the Chartered Institute of Arbitrators and Past Chairman of its Irish Branch. He received his MSc from Manchester University and his PhD from London University. He has extensive experience in civil and structural engineering design, supervision of construction, contract management, construction insurance, arbitration and other methods of dispute resolution. He has acted as an expert witness, dispute board member, conciliator/mediator or arbitrator in hundreds of domestic and international disputes (as a sole arbitrator, member or chairman of a tribunal in over 105 cases of dispute with values in excess of £1m, involving parties from over 45 different jurisdictions).

Dr Bunni is a member of various technical committees in Ireland and abroad, including: the Dispute Resolution Panel of the Institution of Engineers of Ireland; the Board of Directors of the London Court of International Arbitration; the Commission on International Arbitration of the ICC, Paris; the Standing Committee of the ICC International Centre for Expertise, Paris; and the Board of Trustees of the Dubai International Arbitration Centre. Dr Bunni is a member of the Panel of Arbitrators of a number of Arbitral Institutions and Organisations. He is Past Chairman of FIDIC's Standing Committee on Professional Liability; FIDIC's Task Committee on Construction, Insurance and Law; and its follow-up committee.

Dr Bunni was appointed Visiting Professor at Trinity College Dublin in December 1996, a position he continues to hold, and in March 2000 he was elected a member of the International Council for Commercial Arbitration, ICCA, which is a gathering by co-option of the foremost leaders in the field of dispute resolution.

Besides this book, Dr Bunni is the author of a large number of technical papers and two books on construction insurance, the latest of which is now in its second edition under the title *Risk & Insurance in Construction*, published by Spon Press, London, in 2003.

Dr Bunni has lectured extensively and has been invited to speak in many countries in Europe, Asia, Africa, North and South America and in New Zealand. He has organised and lectured at courses on various topics relating to construction contracts for FIDIC, the World Bank, the Munich Reinsurance Company, the Institution of Engineers of Ireland, the Chartered Institute of Arbitrators and other organisations.

Dr Bunni has been awarded a number of awards for: innovation; structural design for work he has done; and in November 1995 he was awarded the Institution Prize by the Institution of Engineers of Ireland (its premier prize) for his work and series of lectures on its Conciliation Procedure. He was awarded this prestigious prize once again in March 2004 for his 'contribution of outstanding merit . . . to the benefit of the members (of the Institution)'. Only very few have received this prize twice.

Preface

The Fourth Edition of the Red Book was published in 1987. It was first subjected to minor editorial amendments in 1988 and was later amended more significantly in 1992. The 1992 form was such a success that the World Bank adopted it in its Standard Bidding Documents in January 1995, albeit with some further mandatory amendments of its own. The most important of these mandatory requirements concerned the role of the engineer under clause 67 of the Red Book and the adoption by the Bank of the concept of a Dispute Review Board to replace the engineer in the process of dispute resolution. FIDIC responded in 1996 by introducing its Supplement to the Fourth Edition of the Red Book, which was published in November 1996. By that time, the Orange Book had already been published in 1995 as a standard form for design and build contracts. The second edition of this book, which was published in 1997, dealt with all these developments in contract forms.

Although the 1992 Fourth Edition of the Red Book with the 1996 Supplement formed an excellent combination in providing a standard form of contract that answered most of the criticisms at the time, and although its good and bad points were understood by its users, FIDIC decided in 1999 to replace it with a different form rather than introduce a fifth edition that could have simply tackled some of the issues that had developed in the meantime. The 'new' Red Book was intended to replace the 1992 Fourth Edition of the Red Book, with the ambition that the use of the Fourth Edition would cease with time and that new projects that were being contemplated at that time should immediately utilise the 1999 forms of contract. However, this ambition did not materialise as there appeared to be significant reservation on the part of users to adopt a form that was untested and untried within the industry. It is also unlikely to happen to any large extent in the future, for many reasons.

The main reason is perhaps that users are by now very familiar and comfortable with the provisions of the Fourth Edition of the Red Book, and particularly their version of it if they have changed the standard form to suit their own particular needs.

Furthermore, after 40 years of use, the meaning of most of the provisions of the Fourth Edition of the Red Book has become known through court decisions and some arbitral awards. Users were and are reluctant to start the process of adopting a new form of conditions of contract as they believe, maybe rightly so, that it is likely to cause further disputes and problems in an industry that is already overloaded with conflict.

The 1999 Red Book was accompanied by a new Yellow Book to replace the Orange Book, and a totally new Silver Book where most of the risks were allocated to the contractor. However, despite that spectrum of colour, the Fourth Edition of the Red Book stood the test of time and remained as a major source for conditions of contract for civil engineering projects worldwide, and particularly in certain regions of the world, such as the Middle East.

Hence, it became necessary for a third edition of this book to be published encapsulating these developments, but leaving the text of the Fourth Edition of the Red Book as the main feature and reference point. This decision was taken mainly because of the continued use and popularity of the Fourth Edition and because of the fact that the 1999 Red Book has retained most of the principles and many of the concepts of the Fourth Edition of the Red Book. However, despite retaining the 1992 Red Book as the main feature of this book, each of the 1999 forms has been allocated a separate chapter, with the anticipation that many of the problems that exist in these forms would demand Second Editions in the very near future for which more detailed commentaries would be appropriate.

The third edition of this book also takes into account the constructive comments that were received from reviewers and the changes and events that have taken place since 1997, for example: the changes which have taken place in the statutes and bye-laws of FIDIC and its Code of Ethics; the latest edition of the ICC Arbitration Rules, published in 1998; developments in claim and dispute procedures; the ADR Rules of the ICC; and developments in Dispute Boards and the very recent ICC Rules relating to them.

Chapters 7 'Sharing of Risks' and 14 'Risks, Liabilities, Indemnities and Insurances' were reconsidered in view of the Australian and New Zealand Standards relating to the topic of risk and risk management. Chapters 16 and 17 on claims and counterclaims were enlarged to consider new developments on the topic of delay and extension of time and to take into account the recent experiences in that field, including some of the new publications.

Chapter 19 'Dispute Settlement by Arbitration' was extended to incorporate various aspects of the topic of arbitration and material relevant to clause 67 of the Fourth Edition of the Red Book, including some problems encountered in arbitration as a method of dispute resolution; and finally some recommendations towards a more cost effective and speedy arbitration procedure.

A new part has been added to the book. Part V comprises five new chapters: Chapter 22 provides an overview of the 1999 FIDIC forms of contract; Chapter 23 deals with the 1999 Red Book; Chapter 24 deals with the 1999 Yellow Book; Chapter 25 deals with the 1999 Silver Book; and Chapter 26 provides a comprehensive description of Dispute Boards: their advantages and disadvantages; types; the procedures that should be followed; a typical agenda; and experiences of their use.

Part VI replaces the chapter in the previous edition that provided the alterations, omissions and additions made in producing the Fourth Edition, and in this edition of the book gives a comparative analysis of the text of the 1999 three main forms of contract. This analysis was carried out by denoting the differences that exist between the three forms using the Red Book as the reference text.

Finally, I have deviated from the convention of using 'he or she' and adopted the use of the masculine pronoun; this should be taken to refer to both male and female.

Nael G. Bunni
January 2005

Acknowledgements

I am indebted to the Fédération International Des Ingénieurs-Conseils (FIDIC) for the permission granted to me to quote and reproduce some material from the various FIDIC forms of contract published by FIDIC from 1992 to date. These publications are the copyright of FIDIC and can be purchased directly from FIDIC's offices at PO Box 311, CH-1215 Geneva 15, Switzerland, or from any of FIDIC's member associations in over 60 different countries around the world. My thanks are also due to FIDIC for permission to have this book available for purchase from their bookshop, at the above address and at www.fidic.org/bookshop

I would like to express my gratitude to those people who helped in producing this edition of the book. I am especially indebted to a number of my colleagues for their specialist commentary and input into certain passages of this edition of the book. Although not mentioned individually, they will recognise themselves.

I owe special gratitude to Mary Farrell, my secretary, for patiently word-processing the successive drafts of the new material in this edition. I also gratefully acknowledge the assistance of Ms Siobhan Fahey BA, BAI, LLB, CEng, MIEI, MCIArb, for reading and commenting on the manuscript for the new material in this edition, for her valued research that followed and her suggestions.

To my daughters Lara and Lydia, I owe a special tribute. To Lara, BSc, HDipAppSc, for her IT input. In particular, however, I gratefully acknowledge the help extended by my daughter Lydia, LLB, LLM, ACIArb, to whom I owe special gratitude for her valuable research, suggestions and secretarial assistance throughout the preparation of this edition of the book.

Finally, I wish to add a special word of thanks to Julia Burden of Blackwell Publishing for her continued encouragement, patience and support during the production of this edition of the book.

The Third Edition

To Anne
Without her own help, support and understanding
this third edition would not have been possible

Part I

Background and Concepts of the Red Book

Chapter 1

Background of the Red Book

In the commercial activities of today's highly complex society, standard forms of contract have become an essential part of the day-to-day transactions of most agreements. The majority of standard forms have been developed by commercial organisations for the purpose of efficiency, to build on the experience gained from the repeated use of these forms, but most of all for the optimum protection of one or both parties' interests. Standard forms of contract developed for construction activities, however, have mostly been drawn up by independent professional organisations, rather than by one or other of the parties to the contract, in order to establish or to consolidate a fair and just contract. Knowledge accumulated through experience and recurrent use over a long period of time has brought about revisions and modifications in construction standard forms with the aim either of achieving greater certainty in the intention of the wording or of providing a response to the needs of the parties and/or society. The use of a standard form in construction contracts where tendering is the conventional method of obtaining quotations has also ensured a common basis for the comparison and evaluation of tenders.

In Europe, and more particularly in the United Kingdom and in Ireland, such forms were produced as early as the nineteenth century. A standard form for building contracts was used under the aegis of the Royal Institute of British Architects some time towards the end of the nineteenth century. This led to what became known as the 'RIBA Form' which was published in successive editions between 1909 and 1957. It later developed into what became known as the JCT Form (Joint Contracts Tribunal) when the 1963 and the 1980 Editions were published. In Ireland, the RIBA Form was followed by the RIAI Articles of Agreement and Schedule of Conditions of Building Contract, issued by the Royal Institute of the Architects of Ireland.

In civil engineering contracts, various forms which were used by different employers prior to the Second World War were combined by the Institution of Civil Engineers and the Federation of Civil Engineering Contractors in the United Kingdom into an agreed standard document. This was published in December 1945, and the document was thereafter known as the General Conditions of Contract and Forms of Tender, Agreement and Bond for Use in Connection with Works of Civil Engineering Construction, in short the ICE Form. In January 1950 it was revised and issued with the added agreement of the Association of Consulting Engineers, London. Other revisions followed in March 1951 (Third Edition); in January 1955 (Fourth Edition which was later amended in 1969); in 1973 (Fifth Edition); and in 1991 (Sixth Edition). These revisions reflected some changes in the law and in the practice of civil engineering.

1.1 The ACE Form

To the credit of those responsible for drafting the ICE Form, many professional institutions all over the world modelled their own conditions of contract on its text, making only minor amendments to accommodate differences in local matters of law and nomenclature. The ICE Form was, however, drawn up mainly for the domestic scene in the United Kingdom. The obvious need for a similar form in the international construction field prompted the Association of Consulting Engineers in the United Kingdom, jointly with the Export Group for the Construction Industries in the United Kingdom, and with the approval of the Institution of Civil Engineers, to prepare a document for use in other parts of the world. It was published in August 1956 and became commonly known as the Overseas (Civil) Conditions of Contract (the ACE Form). Although in text and format this latter Form differed only slightly from the ICE Form, there were some minor changes in forty clauses as well as a small number of major alterations.

The most important of the minor changes were as follows:

(a) a definition of the word 'approved' was added;
(b) a clarification statement was added in clause 3 in relation to assignment;
(c) the words 'which shall not be unreasonably withheld' were added in respect of the consent of the engineer to the contractor to sub-let any part of the works;
(d) the words 'touching or concerning the Works' were added in clause 13 to describe the engineer's directions;
(e) clause 15 in relation to contractor's superintendence was expanded;
(f) the exception relating to damage to crops in clause 22 was re-worded;
(g) the words 'affecting the safety of the Works' were added in sub-clause 40(l)(b) to describe the weather conditions as a reason for suspension of the works; and
(h) the day as a unit of measurement of time replaced the week for the purpose of calculating liquidated damages in clause 47.

The major alterations were as follows:

(a) the document was published in two parts: Part I which incorporated 68 clauses as general conditions of contract; and Part II which included notes and a number of new clauses to be considered for inclusion in Part I. Part II was intended as 'a guide in the preparation of clauses (some of which are referred to in Part I)' but which were expected to 'vary as necessary to take account of the circumstances and locality of the works.' These additional clauses were intended to be drafted for each particular project to cover matters such as, definitions; labour; temporary reinstatement; material and plant; and certificates and payment. Part II was referred to as 'Conditions of Particular Application';
(b) a greater involvement and authority was given to the engineer's representative under a number of the clauses of the ACE Form as compared with the ICE Form;

(c) the explicit procedural provisions under clause 12 in the case of adverse physical conditions and artificial obstructions were deleted;
(d) allocation of the risk of damage due to unforeseen forces of nature was shifted in clause 20 from the contractor to the employer by including the following words into the excepted risks:

> 'any such operation of the forces of nature as reasonable foresight and ability on the part of the Contractor could not foresee or reasonably provide against'.

Whilst this shift in risk in respect of accidental damage to the works was implemented in clause 20, a similar shift in risk was not implemented in respect of financial loss resulting from suspension of work, under sub-clause 40(l)(b), due to weather conditions which are also a form of the forces of nature;
(e) the requirement that joint insurance for the employer and the contractor be provided against third party liability risks was deleted from clause 23 of the ACE Form;
(f) a change in clause 26 was made in connection with payment of fees under foreign statutes, ordinances and bylaws;
(g) a new sub-clause (4) was added to clause 30 in respect of water-borne traffic;
(h) the provision for labour under clause 34 was recommended to be drafted for each contract;
(i) temporary reinstatement as referred to in clause 49 of the ICE Form was omitted in the corresponding clause of the ACE Form;
(j) a condition was incorporated in the ACE Form requiring an amendment of the amount of the contract price in the case where the 'net effect of all variations' is found to result in a reduction or an addition greater than 15 per cent of the sum named in the tender;
(k) reference to the standard method of measurement was omitted from clause 57 of the ACE Form;
(l) failure by the contractor to proceed with the works with due diligence was deleted from the list of grounds entitling the employer to determine the contract under clause 63(l);
(m) a major revision was made to clause 65 which deals with special risks under which the employer was required to provide an indemnity to the contractor in respect of increased costs arising from these risks;
(n) a new clause was added under the title 'Default of Employer' entitling the contractor to determine the contract where no payment is made by the employer within a set period of time or where the employer interferes with or obstructs the issue of any certificate or where the employer becomes bankrupt; and
(o) a number of new clauses were included in Part II of the ACE Form to be considered for inclusion in Part I depending on the circumstances and locality of the works. These new clauses related to conditions of contract for price variations, customs duties and other dues, taxation, bribery and corruption, non-disclosure of information, other matters peculiar to the specific contract and finally, but most importantly, the law governing the contract.

1.2 The First Edition of the Red Book

The ACE Form as published in 1956 included a standard Form of Tender, an Appendix and a standard Form of Agreement. It was published in a blue cover which helped to distinguish it from the ICE Form. It was, perhaps, the first standard form of international conditions of contract for civil engineering works. In concept and style, however, it remained faithful to the original domestic form.

The ACE Form had only been used for a short period of time when the Conditions of Contract (International) for Works of Civil Engineering Construction was published in August 1957. This was based on the ACE Form, described above, and was also published in two parts. Perhaps because of its long title, in a very short time it became popularly known as the 'Red Book' (its cover was printed in red). It was prepared by the Fédération Internationale des Ingénieurs Conseils (the International Federation of Consulting Engineers, FIDIC) and the Fédération Internationale du Bâtiment et des Travaux Publics (the International Federation of Building and Public Works, now known as the International European Construction Federation, FIEC).

FIDIC is the international Federation of duly elected associations of consulting engineers representing the profession in their respective countries. Membership in the Federation is restricted to one association for each country. To qualify for membership, an association must demonstrate that its statutes, bylaws and regulations ensure that its members comply with the ethics and professional code of practice of a consulting engineer as outlined and according to the principles endorsed by FIDIC.[1.1] These principles have developed over the years and significant changes were recently introduced as explained later in Section 5.1 of this book.

In addition to some editing changes and a few minor revisions in clauses 1, 16, 31, 34, 40, 53, 60, 65 and 69(2), a number of important modifications were made to the ACE Form in the evolution of the first edition of the Red Book. These were:

(a) a reference to the ruling language of the contract was incorporated in clause 6(l);
(b) the reference to sureties in clause 10 was changed to performance bond;
(c) it was provided in clause 11 of the Red Book that the tender is to have been based on data supplied by the employer;
(d) the reference in the ACE Form to weather conditions or conditions due to weather conditions in clause 12 was deleted;
(e) a provision for the payment of a bonus under clause 47 for early completion of the works or any part thereof was added in Part II;
(f) in clause 52(1) of the ACE Form, the words 'If the Contract shall not contain any rates applicable to the extra or additional work then reasonable prices shall be fixed by the Engineer' were changed to:

> 'If the Contract shall not contain any rates applicable to the extra or additional work then suitable prices shall be agreed upon between the Engineer and the Contractor. In the event of disagreement the Engineer shall fix such prices as shall in his opinion be reasonable and proper.'

(g) it was provided in the Red Book that the appointment of the arbitrator (or arbitrators) for the settlement of disputes under the contract was to be under the Rules of Conciliation and Arbitration of the International Chamber of Commerce in Paris (ICC); and

(h) two clauses were added at the end of the general conditions: the first provided for any increase or decrease in the costs of labour and/or materials or any other matters affecting the cost of execution of the works; the second provided for currency restrictions or devaluation.

1.3 The Second and Third Editions of the Red Book

The Second Edition of the Red Book was published in July 1969, when the document was approved and ratified by the International Federation of Asian and Western Pacific Contractors' Associations. A supplementary section containing Conditions of Particular Application to Dredging and Reclamation Work was then added as Part III. The Second Edition, however, included no changes in the text. A reprint of the Second Edition in 1973 added the approval and ratification by the Associated General Contractors of America and the Inter-American Federation of the Construction Industry.

However, the publication of the controversial Fifth Edition of the ICE Form in June 1973 provided an impetus for a further revision of the Red Book. This Fifth Edition of the ICE Form provided the civil engineering industry in the United Kingdom with a document which included major departures from the practice followed in its Fourth Edition. Three main commentaries were published in July and November 1973, analysing the effect of these changes. Both Abrahamson and Akroyd criticised the style, language and lack of clarity of the document. Akroyd inquired as to 'whether this new document forms a contractor's charter to riches?'[1.2] Duncan Wallace, in an article published in November 1973 (with the title 'The Modest Revision which Became a Torrent of Change') called the Fifth Edition of the ICE Form 'a new and radically revised contract'.[1.3]

The Fifth Edition of the ICE Form obviously provided food for thought for those responsible for the Red Book, and so it was in March 1977 that the Third Edition of the Red Book was published, incorporating some significant changes. These changes, however, did not follow in all respects those made in the ICE Fifth Edition. In the preface to his book on the Fifth Edition of the ICE Form,[1.4] Duncan Wallace wrote:

> '. . . it is apparent that they [the draftsmen of the Third Edition of the Red Book] evidently considered and studiously avoided, all the principal difficulties (and indeed innovations) in the Fifth Edition, as well as a number of the anomalies in the Fourth Edition (with the single exception of the difficult Maintenance Certificate provisions in Clause 62, which have been swept away, quite rightly, in the Fifth Edition, but which are retained unchanged in the 1977 FIDIC Contract).'

As well as editing and other minor changes, a full list of the revisions is given in a supplement to a book by Duncan Wallace, dealing with the Red Book.[1.5] The most significant revisions made for the Third Edition of the Red Book were as follows:

(a) the definition of the word 'cost' was added as a new sub-clause 1(4) of the Third Edition, providing for cost 'to include overhead costs whether on or off the site'. It replaced the word 'expense' in some of the clauses of the previous edition which led to the inference that profit should not be paid to the contractor in a claims situation under these clauses of the contract. Such situations, for example, arose in:
   (i) sub-clause 5(2), under ambiguities or discrepancies in the contract documents,
   (ii) sub-clause 6(4), under a failure or inability of the engineer to issue drawings or orders within a reasonable time,
   (iii) clause 12, under conditions which could not have been reasonably foreseen by an experienced contractor;

(b) the duties and power of the engineer were defined under sub-clause 2(l) of the Third Edition and for the first time it was accepted that the engineer may be required under the terms of his appointment by the employer to obtain the latter's specific approval for the execution of any part of these duties in which case this information was to be set out in Part II of the Conditions.

In clause 2, it was clearly stated that the engineer's representative is responsible to the engineer;

(c) clause 5 of the Second Edition which specified the extent of the contract was enlarged and incorporated into a new clause 8 within the section dealing with the general obligations of the contractor. These obligations were explicitly stated. However, the obligation of the contractor to complete the works which is accepted in most, if not all, construction contracts was replaced by the term 'execute and maintain the Works'. Similarly under clauses 12 and 13, this obligation to complete was omitted. This omission was criticised: although some may argue that it is implicit in construction contracts that the contractor must complete, it would have been much clearer had this requirement remained in explicit terms. This is because the remaining clauses in the general conditions assume such an obligation and, perhaps more importantly, because such an obligation is not necessarily implied in all legal jurisdictions;

(d) sub-clause 6(1) of the Second Edition became sub-clause 5(1) of the Third Edition. It was expanded to include both the ruling language and the applicable[1.6] law of the contract;

(e) the concept of disruption to the progress of the works in the case of non-availability of information which is considered necessary for the implementation of the contract, was incorporated in the Third Edition by the inclusion of sub-clause 6(3). Sub-clause 6(4) was added to deal with extensions of time and payment to the contractor as a result of such disruption.

Under the Second Edition, delay in the receipt of information or instructions which resulted in extra cost to the contractor could have formed a valid claim for damages for breach of contract. This was because it is an implied term of the contract that the contractor is entitled to receive such information or instructions within a reasonable time before the date on which the relevant work is required to be executed.[1.7] See Section 3.8 for further commentary on implied and express terms. Reference should also be made to Section 17.4.2 where the question of calculating the cost of disruption is discussed;

(f) under clause 11 of the Third Edition, distinction was made between the data provided by the employer and the contractor's interpretation of such data. It is worth noting that the effect of the risks associated with the site is recognised and allocated to the contractor 'in so far as is practicable';
(g) clause 14 of the Third Edition was amended slightly and expanded making it necessary for the contractor to submit a programme within a certain period after the acceptance of his tender. The required programme in the Third Edition deals only with the proposed 'order of procedure and not the method of working previously stipulated'. If required, the contractor was also obliged under the latter edition to revise his programme;
(h) the provision in the Second Edition that the contractor is not liable to insure against the necessity for the repair or reconstruction of any work constructed with materials or workmanship not in accordance with the requirement of the contract, was omitted from the Third Edition. This change was criticised as there did not seem to be any reasonable explanation for the decision to omit this sentence and as drafted it contradicted the requirement in the clause itself that all loss or damage for which the contractor is responsible must be insured. Accordingly, unless the contractor was able to insure against defective materials and workmanship, he would automatically be in breach of contract;
(i) sub-clause 23(3) was added in the Third Edition in which it was stipulated that the terms of the third party insurance, required under sub-clause 23(l), should 'include a provision whereby, in the event of any claim in respect of which the contractor would be entitled to receive indemnity under the policy being brought or made against the employer, the insurer will indemnify the employer against such claims and any costs, charges and expenses in respect thereof'.

This sub-clause 23(3), which became known as the 'Principal Clause' was added to fill the gap formed originally in the First Edition when it followed the ACE Form by omitting the requirement for joint insurance. In filling this gap the Red Book followed the revision in the Fifth Edition of the ICE Form, published in 1973;
(j) the extra cost incurred by the contractor and borne by the employer in the case of suspension of work as described in clause 40 was detailed in more precise terms in the Third Edition;
(k) clause 46 of the Third Edition was re-drafted to accord with the Fifth Edition of the ICE Form. The noise and disturbance indemnity provided by the contractor in the Second Edition was omitted;
(l) clause 48 of the conditions was expanded in the Third Edition providing for more specific terms under which a certificate of completion of the works may be issued by the engineer to the contractor with a copy to the employer. The revision to this clause followed in broad terms that in the Fifth Edition of the ICE Form;
(m) time limits were added in the Third Edition for written confirmation of an oral order given by the engineer under sub-clause 51(2);
(n) the word 'rates' in clause 52 of the Second Edition was replaced in the Third Edition by 'rates and prices', thus permitting the valuation of variations to be based on both rates and prices. It was also envisaged in this clause in the Third Edition that a decrease as well as an increase may result in the valuation of variations;

(o) the percentage of variations which would trigger the operation of clause 52(3) was reduced to 10 per cent in the Third Edition and a more precise definition, although not precise enough, was given to the method of calculation of that percentage;
(p) the reference to prime cost sums in clause 58 of the Second Edition was omitted in the Third Edition. Clause 58 of the latter edition dealt only with provisional sums as defined therein, the power of the engineer to order such sums and the production of documentation related thereto;
(q) clause 59 dealing with nominated sub-contractors was expanded in the Third Edition with the inclusion of three new sub-clauses providing for:
 (i) an express statement of design requirements where such exist in a nominated sub-contract;
 (ii) the payments to be made in respect of actual price paid or to be paid by the contractor, labour to be supplied by the contractor and all other charges and profit; and
 (iii) the assignment of nominated sub-contractors' obligations;
(r) nuclear and pressure wave risks as described in sub-clause 20(2) were added to the list of special risks in clause 65 of the Third Edition;
(s) a provision was added under clause 66 of the Third Edition whereby the contract would be treated as frustrated should the parties to the contract be released from further performance under the law governing the contract;
(t) a provision was added under clause 67 of the Third Edition permitting the reference to arbitration to proceed during the progress of the works;
(u) three new clauses were added to Part I of the Third Edition under the headings of:
 (i) Changes in Costs and Legislation: Clause 70;
 (ii) Currency and Rates of Exchange: Clause 71;
 (iii) Rates of Exchange: Clause 72.

A set of 'Notes on Documents for Civil Engineering Contracts' was published by FIDIC in 1977 which referred to some selected aspects of the various clauses of the Third Edition of the Red Book. The status of these Notes was briefly alluded to by the statement that they 'must not be taken as representing in any way an interpretation of the text of these clauses' (the individual clauses of the Third Edition).[1.8]

Unfortunately, these Notes did not deal with the background or the reasons behind the extensive changes which were made in producing the Third Edition, especially when only some of these changes reflected the revisions which had taken place in producing the Fifth Edition of the ICE Form in 1973.

As in the ACE Form, the importance of the law governing a specific contract, the applicable law of the contract, was recognised in the Red Book but despite this, no attempt was made to depart from the principles of the common law under which the ACE and the ICE Forms were drafted. Neither was there any attempt to recognise that there could be a conflict of laws between the common law system and any other system of law to which the applicable law of the contract belonged.

Despite this lack of recognition, it seems that the provisions of the Red Book coped extremely well, and for a long time, with any conflict of laws which may have existed. The Second and Third Editions of the Red Book proved to be successful in many projects throughout the world. The Third Edition in particular coincided with the major economic growth which took place in developing countries towards the end of the 1970s and the major part of the 1980s and, particularly, in the Middle East and the Far East. The Third Edition was translated into French, German and Spanish.

Criticism came to the surface only in recent years when the number of disputes ending in arbitrations increased and every clause and term in the Red Book came under the scrutiny of lawyers experienced in discovering differing interpretations to a set of words. This problem and others were dealt with quite successfully when the revision of the Third Edition was undertaken by FIDIC.

1.4 The Fourth Edition of the Red Book

The Third Edition of the Red Book remained unaltered and no amendments were issued until the Fourth Edition was published in September 1987, when major revisions were made which extended even to the title of the document. The word 'international' was deleted, inviting parties from all over the world to use the Red Book not only in international contracts but also in domestic contracts.

Part II of the Red Book which is referred to as the 'Conditions of Particular Application' was expanded and produced in a separate booklet. It is linked to Part I by the corresponding numbering of the clauses, so that Parts I and II together comprise the conditions governing the rights and obligations of the parties. Part II must be specifically drafted to suit each individual contract. To assist in the preparation of Part II, explanatory material and example clauses are included providing the parties with options for their use where appropriate.

In 1988, the Fourth Edition of the Red Book was reprinted with a number of editorial amendments which were identified at the end of the document. These amendments were of a very minor nature and did not affect the meaning of the relevant clauses but simply clarified their intention. They are reproduced in Appendix A at the end of this book.

Later, in 1992, further amendments were introduced in a reprint of the Fourth Edition of the Red Book. The list is reproduced in Appendix B at the end of this book and the editorial amendments relating to punctuation and the use of 'and', 'or' or both are individually set out. Some of these 1992 amendments were directed towards a more uniform style of drafting but others were of a more significant nature, either adding to or changing the meaning of the relevant clauses of the Form. These significant amendments are as follows:

Page	Amendment
2	Sub-Clause 1.1, sub-paragraph (e): Definitions (iii) 'Interim Payment Certificate' and (iv) 'Final Payment Certificate' have been added.
6	Sub-Clause 8.1: Second paragraph has been added.
7	Sub-Clause 12.2, Marginal note: The word 'Adverse' has been changed to read 'Not Foreseeable' (also amended in the Contents and the Index).

Page	Amendment
8	Sub-Clause 13.1: Last sentence has been shortened by deleting the words 'or, subject to the provisions of Clause 2, from the Engineer's Representative.', and adding the words '(or his delegate).'
Sub-Clause 15.1, paragraph 1: Last sentence has been shortened by placing a full stop after the word 'Engineer', deleting the words 'or subject to the provisions of Clause 2, the Engineer's Representative.'
10	Sub-Clause 21.1, sub-paragraph (a): The words '(the term "cost" in this context shall include profit)' have been added.
11	Sub-Clause 21.4, sub-paragraph (a): The word 'where' has been corrected to read 'whether'.
18	Sub-Clause 40.3: The word 'written' has been deleted at the end of the first line.
19	Sub-Clause 42.3: The word 'wayleaves' has been changed to read 'rights of way' in the text and marginal note (also amended in the Contents and the Index).
29	Sub-Clause 60.1, sub-paragraph (e): The words 'or otherwise' have been added at the end.
Sub-Clause 60.2: The words 'certify to the Employer' have been changed to read 'deliver to the Employer an Interim Payment Certificate stating', the word 'thereof' has been changed to read 'of such statement' and the word 'he' has been changed to read 'the Engineer'. Sub-paragraph (b): The words 'Interim Certificates' have been changed to read 'Interim Payment Certificates'.
Sub-Clause 60.3, sub-paragraph (b): In the eighth line, the word 'ordered' has been changed to read 'instructed'.
Sub-Clause 60.4: The words 'interim certificate' in the first and fourth lines, and the word 'certificate' in the second line, have been changed to read 'Interim Payment Certificate'.
30	Sub-Clause 60.5: In the second line, after the word 'Engineer', the words 'six copies of' have been added.
Sub-Clause 60.6: In the second line, after the word 'consideration', the words 'six copies of' have been added.
Sub-paragraph (b): The words 'or otherwise' have been added at the end. At the end of the sub-clause, the final paragraph has been added.
Sub-Clause 60.7 and Sub-Clause 60.8 (text and marginal note): The words 'Final Certificate' have been changed to read 'Final Payment Certificate' (also amended in the Contents and the Index).
Sub-Clause 60.8 (a): The words 'or otherwise' have been added.

Page	Amendment
	Sub-Clause 60.8 (b): The words 'under the Contract other than Clause 47' have been changed to read 'other than under Clause 47'.
31	Sub-Clause 60.10: In the first and fourth lines, the words 'interim certificate' have been changed to read 'Interim Payment Certificate'. In the fifth and sixth lines, the words 'Final Certificate' have been changed to read 'Final Payment Certificate'. The words 'or otherwise' have been added at the end.
33	Sub-Clause 65.6: In the ninth line, the words 'and to the operation of Clause 67' have been changed to read 'and Clause 67'.
34	Sub-Clause 66.1: In the second line the word 'party' has been changed to read 'or both parties', in the third line between the words 'his' and 'contractual' the words 'or their' have been added. In the fourth line after the word 'then', the words 'the parties shall be discharged from the Contract, except as to their rights under this Clause and Clause 67 and without prejudice to the rights of either party in respect of any antecedent breach of the Contract, and' have been added.
35	Sub-Clause 67.2: The words 'arbitration of such dispute shall not be commenced unless an attempt has first been made by the parties to settle such dispute amicably' have been changed to read 'the parties shall attempt to settle such dispute amicably before the commencement of arbitration.' The words 'whether or not any attempt at amicable settlement thereof has been made' have been changed to read 'even if no attempt at amicable settlement thereof has been made'.
37	Sub-Clause 69.1, Sub-paragraph (d): The words 'unforeseen reasons, due to economic dislocation' have been changed to read 'unforeseen economic reasons'.
	Sub-Clause 69.4: In the second line of the second paragraph, the word 'cost' has been changed to read 'costs'.
38	REFERENCE TO PART II: In the third line, the words '5.1 part' have been changed to read '5.1 (part)'.
TENDER	Paragraph 1: In the last line, the word 'sums' has been changed to read 'sum'.
Appendix	In the ninth line, the words 'and Plant' have been added.
	In the twelfth line, the word 'Payment' has been added.
	In the thirteenth line, the words 'per annum' have been added.
EDITORIAL AMENDMENTS	For page 35, after the words 'Sub-Clause 67.1', the first sentence has been inserted.

1.5 The 1996 Supplement to the Red Book

In November 1996, FIDIC published a document entitled 'Supplement to Fourth Edition 1987 – Conditions of Contract for Works of Civil Engineering Construction – Reprinted 1992 with Further Amendments'. It is intended to provide the user with alternative arrangements in three controversial areas of the Red Book, thus giving him a choice in the method to be used for: settlement of disputes; payment; and preventing delay in certification for the purpose of payments. The Supplement comprises three sections, referred to as follows:

(a) Section A, entitled 'Dispute Adjudication Board'. This section provides an alternative wording to clause 67 of the Fourth Edition of the Red Book, *'Settlement of Disputes'*. The new wording was drafted in response to mounting criticism of the role of the engineer as an adjudicator or quasi-arbitrator in the resolution of disputes under clause 67 of the Red Book (see Chapter 9 later). Various alternative methods of dispute resolution were considered in the past few years, both domestically and internationally. The method finally chosen by FIDIC in its new supplement is that based on the use of an adjudication board composed of one or three experts who can render a decision in respect of a dispute without having to resort to the engineer for a final determination under the present clause 67 (see Chapter 19 later). This new method requires an expert or experts to be appointed at the beginning of a contract, who must keep in touch with the work in progress on the site which is achieved by visiting the site at regular intervals. The appointed expert, or experts, are to be available to act in the resolution of any disputes that may arise. In the supplement, FIDIC embraces this alternative method of dispute resolution as an acceptable substitute to the engineer's traditional role in dispute settlement.

The new supplement provides the necessary amendment to the wording of clause 67 of the Red Book and also contains a guide to this amended wording. Moreover, it contains Model Terms of Appointment and Procedural Rules for the Dispute Adjudicaton Expert or Board and the necessary amendments required to the Appendix to Tender which correspond to the amended wording of clause 67.

(b) Section B, entitled 'Payment on Lump Sum Basis'. This section provides the necessary amendments to the relevant clauses in Part I, General Conditions of Contract, of the Red Book which enable payment to be made to the contractor on a lump sum basis instead of using bills of quantities. Section B contains an introductory note followed by the amendments, together with amended forms of Tender and Agreement corresponding to the lump sum basis. See Chapter 6 later in this book for further discussion of this topic.

(c) Section C, entitled 'Late Certification'. This section provides alternative wording to safeguard the interests of the contractor where the engineer is late in certifying interim payments. See Chapter 18 later in this book for further discussion on this part of the supplement.

A detailed analysis and commentary on the Fourth Edition of the Red Book as a whole including the above mentioned amendments is provided later in Chapter 9. The text of the whole Form including the amendments of 1988 and 1992 is provided in Chapter 23 of the second edition of this book where it is compared with the Third Edition.

1.6 Concepts of the Red Book

Despite the universal use of the Red Book, its Fourth Edition retained some essential features and concepts which formed the foundation of its previous editions. In order to understand the provisions of the Red Book and the implications of the changes made in the Fourth Edition and its most recent amendments, it is essential to elaborate on these concepts as a background to the form of contract itself, on the reasoning underlying the revisions it was necessary to make, and underlying further changes which could have been made but were not. These concepts are set out below and are dealt with in detail in the following six chapters.

— Apart from a few revisions which have been made, it is based on a domestic contract: Chapter 2.
— Its legal concepts are based on the common law system: Chapter 3.
— Its wording is based on English legal drafting principles: Chapter 4.
— Its concept, in relation to the design and supervision of construction of the project, is based on the appointment of a consulting engineer trusted by both parties to the contract and referred to as the 'Engineer': Chapter 5.
— Its concept of remuneration is based on a re-measurement contract with a provisional bill of quantities which serves as a basis for final remeasurement and payment, under certificates from the engineer: Chapter 6.
— Its concept of responsibility and liability is based on the sharing of risks: Chapter 7.

1.7 The new suite of FIDIC contracts

In October 1999, FIDIC produced a totally new set of standard forms of contract alongside those that were in use at that time. The new set comprises the following four contract forms:

(a) The Construction Contract (Conditions of Contract for Building and Engineering Works, Designed by the Employer) – General Conditions, Guidance for the Preparation of the Particular Conditions, Forms of Tender, Contract Agreement, and Dispute Adjudication Agreement, referred to in this text as the '1999 Red Book';
(b) The Plant and Design-Build Contract (Conditions of Contract for Electrical and Mechanical Plant, and for Building and Engineering Works, Designed by the Contractor) – General Conditions, Guidance for the Preparation of the Particular Conditions, Forms of Tender, Contract Agreement and Dispute Adjudication Agreement, referred to in this text as the '1999 Yellow Book';
(c) The EPC and Turnkey Contract (Conditions of Contract for EPC Turnkey Projects) – General Conditions, Guidance for the Preparation of the Particular Conditions,

Forms of Tender, Contract Agreement and Dispute Adjudication Agreement, referred to in this text as the '1999 Silver Book'; and

(d) The Short Form of Contract – Agreement, General Conditions, Rules for Adjudication and Notes for Guidance, referred to in this text as the '1999 Green Book'.

The old set is divided into the 'Red' and 'Yellow' forms on the basis of the type of the project to be constructed. The new forms are divided on the basis of who designs the project. Whilst the 1999 Red, Yellow and Silver books are similar in many areas, they are in fact three separate and distinct forms of contract. A comparative analysis of all three books can be seen in Chapter 27 in Part VI.

Chapter 2

The Red Book is Based on a Domestic Contract

2.1 Introduction

As discussed earlier, the Red Book was modelled on the ACE Form which originated from the ICE Standard Form of Contract. The ICE Form was drafted for use as a domestic contract in the United Kingdom. The changes made to transform that domestic form to an international one were minimal.

A brief knowledge of the legal systems around the world and how these relate to the legal concepts on which the FIDIC Red Book is based is therefore helpful, if not essential, to the understanding of the problems which may result from its use.

This task is attempted in the present chapter and in Chapter 3. It is a difficult but necessary task. Comments on the law are of necessity generalised, and statements of general rules and principles are not intended to mean that they are without exception or qualification. These rules and principles may differ from one jurisdiction to another and even within one legal group. The present chapter and Chapter 3 are intended to expose the non-lawyer reader to legal topics which are considered important in the international construction scene. These topics include: the diversity of legal systems, the applicable law in international construction, the applicable law of the contract, the law governing procedure, the law governing enforcement of awards and the various groups of contemporary legal systems.

2.2 Diversity of legal systems

As long as there is human endeavour, there will always be conflict. The idea of law was born and developed independently in communities around the world at very early stages of civilisation in order to provide an instrument to regulate the various aspects of human behaviour and relationships between one individual and another and thus achieve a balance between the freedom of choice of the individual and the control of this freedom for the protection of others. Societies aspired to have laws that mirrored justice so as to eliminate the necessity to resort to force except for the purpose of upholding the supremacy of the law itself.

To that end, rules were written in the form of legal codes as early as 1700 BC,[2.1] and as communities sprang up, grew and later declined, the legal rules changed from a few basic, simple codes to sophisticated, complex and voluminous systems of law.

By that time, legal concepts had developed from what was earlier considered to be the law of the gods to a three-tier hierarchy. At the top, the law of the gods changed to the laws of God due to the evolution of religion. The second tier represented natural law or the law of reason and common sense, and the third tier represented man-made law. The latter has been subject to evolution from time to time and from place to place while endeavouring to respect the boundaries laid down by the divine and natural laws.

Whilst this aim and concept prevailed in the West, Near East, and parts of the Middle East where law was held supreme, it was quite different in the Far East. In China and indeed in some communities in the Middle East, law has been and is still considered an instrument of arbitrary action rather than necessarily one of justice. Thus, it is expected that the good citizen would not resort to the law or the courts. Rather, the conduct of individuals should be animated by the search for harmony and peace through methods other than the law. In these parts of the world, conciliation and mediation have been considered to be superior and of greater value in resolving conflicts. Countries in the Far East have traditionally held the view that law is for barbarians and that one should avoid ending with a winner and a loser.

As these developments in the law took different directions in different cultures around the world, it is practically impossible today to achieve any degree of international standardisation of the law dealing with legal relationships of individuals from different states. Even worse, it is now impossible to unify the national laws of all political entities and states of the world. The contemporary legal systems of the world have evolved across societies and cultures and across political systems differently. Because they are rooted in different cultures, they are written in different languages, influenced by different religious beliefs and formed under different customs. Given such socio-cultural variety, it may, therefore, be easier and more acceptable to reach international agreement on the basic concepts and rules which should govern contractual relationships in an international field, such as, construction. In a simple form, this is what is attempted by a standard form of contract for civil engineering works. However, the fact remains that as soon as one specifies in an international contract the applicable law of the contract as the law of the locality of the project, one automatically introduces a diversity of laws governing the relationship between the parties to the contract.

At first glance, the task of understanding the differences between two or more legal systems and realising their implications seems insurmountable. However, at closer scrutiny one may simplify the task by grouping these legal systems into a small number of categories. Each category should include systems with recognisable fundamental similarities and criteria upon which a classification can be successfully made. These groups can then be compared and contrasted using as a base matters such as legal principles, sources of law, influences on form, drafting and substance.

Such an analysis comes within the boundaries of comparative law, a complex and specialised area of the law, but it is essential in international contracts where a mix of legal systems is used. The problem for the analyst is that there is no agreement on the matters which should form the essential fundamental criteria for such an analysis and for establishing these groups. However, a certain minimum basic knowledge of the law governing the areas of professional activity is necessary for the engineer. Furthermore, for the engineer in international contracts, it is essential for him to understand the implications of the applicable law of the contract in a particular project since it is accepted that ignorance of the law is no excuse for mistakes.

From the point of view of construction, it is suggested that a simple classification should suffice in any study of comparative law, at least as a beginning, since a large number of the relevant basic concepts are similar in many potential subdivisions of legal systems. However, before dealing with the contemporary legal systems, it is important to consider the topic of the applicable law in international construction contracts.

2.3 The applicable law in international construction

In an international construction contract, the law under which the parties' rights and obligations are determined may be one of many. It could be the law of the country where the contract is made or where the project is constructed or that of the domicile of one of the parties to the contract. It could also be the law of the state where a significant part of the contract works are manufactured, or that where the contract is financed or simply the law which the parties regard as well-suited to govern the particular contractual relationship. In this connection, it has been stated that 'a contract is only international if the parties have either their places of business, or habitual residence, in different states'.[2.2] In the context of international commercial arbitration, however, the word 'international' has a secondary criterion in respect of the arbitration process itself. Thus, it is defined by the United Nations Commission on International Trade Law (UNCITRAL) Model Law as follows:

'An Arbitration is international if:
(a) the parties to an arbitration agreement have, at the time of the conclusion of that agreement, their places of business in different States; or
(b) one of the following places is situated outside the State in which the parties have their places of business:
  (i) the place of arbitration if determined in, or pursuant to, the arbitration agreement;
  (ii) any place where a substantial part of the obligations of the commercial relationship is to be performed or the place with which the subject-matter of the dispute is most closely connected;
or
(c) the parties have expressly agreed that the subject-matter of the arbitration agreement relates to more than one country.'[2.3]

In general, there is broad international acceptance that subject to few limitations, the parties are free to choose for themselves the law applicable to their contract. This freedom to choose is referred to as the principle of autonomy of the parties and although it was developed first by jurists under various systems of law, it was adopted by national courts and has been accepted by most legal systems.[2.4] This principle has also been adopted in international conventions.[2.5] In international commercial arbitration, the European

Convention of 1961 provides in Article VII that the parties are to be free to determine, by agreement, the law to be applied by the arbitrators to the substance of the dispute. Similarly, Article 33.1 of the UNCITRAL Arbitration Rules provides that '(T)he arbitral tribunal shall apply the law designated by the parties as applicable to the substance of the dispute.'

2.4 The applicable law of the contract

The law which governs a contract between certain parties and by which questions as to the validity, application and interpretation of its terms are addressed, is referred to as the 'applicable law of the contract'. In some jurisdictions, the terms 'proper law of the contract' or 'governing law of the contract' are used instead. However, as stated earlier in Section 1.3, the term applicable law is preferred by this writer. It is essential to recognise, however, that in certain circumstances it is possible to have different contractual matters of a contract governed by different systems of law, by agreement either of the parties or of a competent authority such as a court or an arbitrator, should the contract be silent on the point or if the parties disagree. In such a case, the applicable law of these specific matters could be different from the applicable law of the contract as a whole.

Where there is no certainty as to the applicable law of the contract, it would have to be selected in accordance with the principles of a branch of law known as 'private international law' sometimes called 'the conflict of laws'. This is a body of principles which attempts to provide answers as to what law is the most appropriate to apply and which forum is appropriate to determine a particular issue with an international dimension. This branch of the law forms part of the legal system of every jurisdiction and therefore, there are as many systems of conflict of laws as there are jurisdictions.[2.6]

There are three alternatives in respect of the determination of the applicable law of the contract:

(a) where there is an express choice of the applicable law;
(b) where there is an inferred choice of the applicable law; and
(c) where there is no choice of the applicable law.

These alternatives are now considered separately.

Where there is an express choice of the applicable law

At the time of making the contract, the parties may expressly choose the law which they wish to apply to their contract. As in clause 5 of the FIDIC Form, this choice may be expressed by a simple statement naming the country to which the chosen law belongs. In such a case, the only question which remains to be answered is whether there are any limits to the freedom of that choice and if so what are these limits.

Where litigation is concerned, there is a great deal of controversy over this question. At one end of the scale there are those who advocate that the parties are free to submit the validity of their contract to any law of their own choosing. Article 3(l) of the EEC Convention on the Law Applicable to Contractual Obligations (see Section 2.8 later) provides as follows:

> 'A contract shall be governed by the law chosen by the parties. The choice must be expressed or demonstrated with reasonable certainty by the terms of the contract or the circumstances of the case. By their choice the parties select the law applicable to the whole or a part only of a contract.'

Others prefer the view that a court should not necessarily regard an express choice of law by the parties as being the governing consideration in cases where:

(a) a system of law is chosen which has little or nothing to do with either party or with the contract; and/or
(b) the chosen law may frustrate the mandatory provisions of the law which has in fact the closest connection with the contract.

In contrast to litigation, the situation is different in international commercial arbitration where, unlike a judge of a national court, an arbitral tribunal is not bound to follow the rules of conflict of laws of the country in which it has its seat.[2.7] An arbitral tribunal may refuse to recognise and apply a chosen law only in the unlikely event of the effect of that chosen law violating international public policy. Subject to this limitation, party autonomy is unlimited and an arbitrator is obliged to recognise and give effect to a choice of law by the parties.[2.8]

Accordingly, subject to the above consideration, it is wise for parties to a construction contract to exercise their discretion carefully and to choose an appropriate system of law as the applicable law of the contract. Of course, what may be 'appropriate' for one party may not be so for another. In construction contracts, the choice of the applicable law of the contract is generally made by the promoter of the project or the 'employer'.

Where there is an inferred choice of the applicable law

In some cases where there is no express and clear choice of the applicable law of the contract, it may be possible to infer a choice of law from the other provisions of the contract and the relevant surrounding circumstances. The most important provision from which such inference may be taken is the arbitration clause where the place for arbitration has been selected. Other factors from which a choice of the applicable law of the contract may be inferred include the form of the conditions of contract adopted by the parties, the residence of the parties, the currency in which payment is to be made, and the nature and location of the subject matter of the contractual obligations.

Where there is no choice of the applicable law

Where there is no express choice of the applicable law of the contract and no inference can be made to establish such law, the principles of conflict of laws are used to select the legal system which should apply. In international commercial arbitration, however, the tribunal is faced firstly with the question of whether it has a free choice of the applicable law of the contract or whether it must follow the private international law of the place of arbitration. As this law differs from one jurisdiction to another, the result may be different. In this connection, it is interesting to note the provisions of Article 4(l) of the EEC Convention, discussed later in Section 2.8.1. It provides that to the extent that the law applicable to the contract has not been specified in accordance with Article 3 of the Convention, the contract is to be governed by the law with which it is most closely connected.

If the applicable law of the contract is not selected and specified when the contract is formed, extremely complex and difficult problems could arise should it be found necessary at a later stage to make such a selection. Further consideration of the topic is outside the scope of this book, and especially so in view of the fact that the Fourth Edition of the Red Book provides, in clause 5, the opportunity to select and specify the applicable law of the contract.[2.9]

However, it is not sufficient simply to choose the applicable law of the contract. It is equally important to understand the implications of such choice. Unfortunately, despite its obvious importance, clause 5 of the Fourth Edition of the Red Book is often given no more than a passing reference by the parties when international construction contracts are initiated. Indeed, in a large number of these contracts, the characteristics of the applicable law of the contract are unscrutinised or even ignored. Where disputes arise but are not settled under the provisions of the contract, the applicable law of the contract can form a leading section of the dispute resolution process. The special characteristics of the chosen applicable law of the contract then become the focal consideration of the parties to the contract and their legal advisers.

In this connection, it is essential to appreciate that when the applicable law of the contract is being selected or considered, not only its prevailing characteristics should be scrutinised but one ought also to consider the likelihood of any changes which may be enacted and the nature of such changes. This is because such changes may bring matching effects on the contract itself, thus altering its character and causing an imbalance in the relationship between the parties. Of course, the probability of such a change may have to be considered by one of the parties or all as part of the risk undertaken in pursuing and executing the contract. The risk is of greater severity if the other party to the contract, for example, a government ministry, has some or total control over such changes.

2.5 Law governing procedure

Distinction must be made between the law applicable to procedure and that applicable to the substance of a dispute or the applicable law of the contract discussed above. This distinction is important because it is generally accepted that the law applicable to procedure is the law of the forum where the litigation or arbitration takes place; whereas the law applicable to substance is the law governing the matters in principle: in a contract it is the applicable law of the contract or the particular term in question.

Accordingly, this distinction assumes greater importance where international contracts are concerned because one may find that the applicable law of the contract and that applicable to procedure belong to two different jurisdictions or even two different systems of law.

Although this distinction is not an easy one, it may be taken that in general the concept of procedure includes, amongst others, the areas of evidence, assistance of the courts and in some cases the rules on limitation. An arbitrator appointed to determine a dispute in an international commercial arbitration may find that the applicable law of the contract is different from that which regulates the internal arbitration proceedings. In general and subject to those arbitrations which are delocalised,[2.10] an arbitration is governed in respect of procedural law by the law of the jurisdiction in which the arbitration proceedings are held – the place of arbitration – which is technically referred to as the Seat of arbitration. In many cases, the choice of the Seat of arbitration is left to the arbitrator or to a third party, such as the authority named as responsible for the appointment of the arbitrator. In the latter case, should this appointing authority be an institution, such as the International Court of the International Chamber of Commerce (ICC) in Paris or the London Court of International Arbitration, the Seat of arbitration is fixed in accordance with the institution's rules.

In the case of the International Court of the ICC, whose Rules are named in clause 67 of the Fourth Edition of the Red Book, the Seat of arbitration (referred to in French as 'siège' and in the English version as 'place'), if not agreed by the parties, is fixed by the Court under Article 12 of the Rules. It is usually chosen on the basis of its neutrality and as a place other than that to which either of the parties is connected.

In the case of the London Court of International Arbitration, Article 7 of the Court's Rules provides that 'The parties may choose the place of arbitration. Failing such a choice, the place of Arbitration shall be London, unless the Tribunal determines in view of all the circumstances of the case that another place is more appropriate'.

Accordingly, should the parties wish to determine the law applicable to the arbitration proceedings, a choice of the place of arbitration should be made by them at the time of the formation of the contract.

2.6 Law governing enforcement of awards

Besides the applicable law of the contract and the law applicable to the procedure, the parties in an international construction contract may be involved in yet another system of law: the law of the country where a decision or an arbitral award is to be enforced.

A party seeking to enforce an award may have a choice of jurisdiction where to do so. The selected location will depend on where the assets of the losing party are situated and on the ease with which the award will be enforced. Such enforcement usually means that some legal proceedings would have to be taken in the jurisdiction where the assets are located, a location usually different from the place of arbitration. One of the major considerations is whether the country where the assets are located is a signatory to the 1958 New York Convention, or to some other treaty for the recognition and enforcement of foreign awards.[2.11] Another major consideration is the legal system of the place of intended enforcement and its provisions. It is perhaps worthwhile for the parties to a contract to look carefully at the question of asset location and enforceability at the time of, or before, the formation of the contract.

2.7 Grouping of the contemporary legal systems

Where international construction is concerned, there are four major groups of legal systems which apply today. These are:

(a) the Romano-Germanic group;
(b) the common law group;
(c) the Islamic law group, including those with origins from the first two groups;
(d) the socialist laws group.

There are of course other minor groups in existence, some of which are totally distinct from those mentioned above whilst others share some of their concepts. A detailed analysis of such groups is, however, outside the scope of this work.

The Romano-Germanic group owes its origin to Roman law during the times of Julius Caesar and the Emperor Augustus (63 BC to 14 AD) and subsequently to compilations of the Emperor Justinian in the sixth century, promulgated from 529 to 534. In later times, between the thirteenth and nineteenth centuries, and in a tradition of promoting through teaching the search for just laws and the pursuit of a model law, different universities in Latin and Germanic countries in Europe developed this group of legal systems. The term 'civil law' is often used in the English-speaking world to denote the Romano-Germanic group of legal systems and to indicate that the origin of these laws is the Roman law.

However, because the term 'civil law' is used in other contexts in legal terminology, it is more appropriate here to refer to the above group as the Romano-Germanic group. Romano-Germanic laws are, in principle, based on the judicial application of a certain legal code to a particular case by learned jurists and theorists, in conformity with logical and systematic deduction. Under this group, the rules of law have developed as rules of conduct linked to ideas of 'justice and morality'. They are usually formulated by legal scholars who are not involved in the practical administration and application of these rules.

The common law group, on the other hand, originated as the law common to all England after the Norman Conquest in 1066. It developed from a body of law which is almost entirely the product of judicial decisions by courts which applied custom and reason to everyday disputes, aided by only a few formal enactments of law. These decisions are embodied in an extensive series of reports extending back to the end of the thirteenth century. The common law continued to be developed in England by judges, rather than by legislators, through the accumulation of tradition expressed by upholding certain principles. The resultant case law continues at present to fill in gaps in the law, or to change the law in a certain direction, or to interpret the meaning of the large number of legislative enactments, by declaring precedents which impose authority on future judicial decisions. Such authority is not granted to precedent in the Romano-Germanic systems: a volume of reports of judicial decisions in common law has a similar authority to that of an authoritative legal text book in Romano-Germanic systems.

The common law as a system of law, however, incorporates two other closely linked elements besides the accumulated body of precedents. These are equity and statute law, see Chapter 3.

Both the common law and the Romano-Germanic systems spread into other parts of the world either through colonisation or by voluntary adoption where there was a need to modernise or an attempt to inject certain ideas to fill gaps into an already developed civilisation. In the latter case, they formed new groups which have certain common principles with either the Romano-Germanic or the common law groups or with both.

Islamic law is based on the Qur'an, the sacred book of Islam, which is a collection of God's revelations to the Prophet Muhammed just before the year 622 AD, the year of his flight from Mecca to Medina which is called the Hijra. The Islamic calendar starts from that year.[2.12] However, whilst the Qur'an is the primary source of Islamic law, it contains only some of the principles needed to form the law in a secular society. Islamic law draws its rules from three other sources: the Sunna, which is the traditional and model behaviour of the Prophet Muhammed; the Ijma' or consensus of scholars of the Islamic communities; and the Qiyas which is the juristic interpretation by the process of reasoning. Collectively, they are called the Shari'ah which can be literally translated as 'the way to follow'. But as modern society is much more complex than the times when the Shari'ah was used, other legal rules in the form of legislation have been added by various Islamic states to form their own law.

Therefore, in countries where Islamic law applies, one finds that in some, legislation is based on the Romano-Germanic group of laws, whereas in others, legislation is based on the common law group.

The socialist laws group is a distinct development from a base line of a Romano-Germanic concept. It originated in the former Union of Soviet Socialist Republics, where under the 1917 Revolution, law was treated as strictly subordinate to the task of creating a new economic structure, and the field of private law was limited and curtailed in favour of public law.

Taking these groups individually and analysing their principal concepts should lead to some understanding of the variations under which one would have to operate in international construction.

2.8 The Romano-Germanic group

As they did in all the provinces they conquered, the Romans established their law in Gaul, an ancient name for the region in Western Europe roughly corresponding to present France, Belgium, Western Germany and Northern Italy. Following the collapse of the Western Roman Empire in 476, Roman law survived in the southern part of France. In the north of France, with the incursion of the Franks, Roman law was replaced, though not entirely, by customary laws of Germanic origin.

Unlike the common law in England which was supported and developed by strong centralised royal courts in Westminster, the revival of the Romano-Germanic group in the thirteenth century took place without any reference to any European political system, but only to a community of culture and endeavour by various universities. This work was led by the University of Bologna in Italy and was followed by other universities such as Montpellier and Toulouse in France where law was considered as a means for bringing about social organisation and justice. A number of law books were written at that time by famous authors.

Two principles were ultimately enunciated, which perhaps constitute the two main differences between the Romano-Germanic and the common law systems. The first principle determined the manner in which a division between public and private law was achieved. In the Romano-Germanic system, the decision is based on the idea that the relationship between those who govern and those who are governed requires a different approach from that required for the relationship between individuals, and that the interests of the individual cannot be given the same weight as those of society. Thus, all the fundamental branches of the law were compiled individually for both divisions. Within public law lies administrative law which denotes a whole section of law whose rules deal with two aspects in the relationship between the administration and the public. The first aspect is the problem of encroachment on the rights and duties of the individual. The second aspect is the requirement of an effective operation of the public service. The concepts of an administrative contract are different from those of one under private law. The administrative contract is important from the point of view of construction law because it includes a public works contract.

The second principle which developed was that the legal writing and work done by the universities gave way to enacted law and thus a change evolved from compilation of the law to its codification, fusing the theoretical and practical aspects of the law of the time. Codification became a success under the Napoleonic regime in France immediately after the French Revolution of 1789 which established the newly adopted ideas of justice, freedom and dignity of the individual. Napoleon devoted considerable energy to the creation of the *Code Civil*. It is recorded that at the meetings of the *Conseil d'Etat*, he constantly focused attention on the realities of life rather than the technicalities of law. He immediately saw the relevance of abstract rules and insisted on a style of drafting which was clear and comprehensive to a non-lawyer. It has been said that the *Code Civil* owes the clarity and comprehensibility of its language to the fact that its draftsman had constantly to ask himself whether the words he had chosen would withstand the criticisms of a highly intelligent layman like Napoleon, unfamiliar with the jargon of the law.[2.13]

The French Code of 1804 was adopted by many European countries to varying degrees including Belgium, Italy, Luxembourg, the Netherlands and Poland, but it was rejected in Germany. The Civil Code in Germany dates back only to 1896. Therefore, it should be emphasised that within the Romano-Germanic group of laws, there are differences as well as similarities and that these differences may even extend to the precise meaning of certain legal terms, such as good faith or unjust enrichment or movables and immovables.

Codification spread from the European countries to vast territories in the remaining continents of the world through colonisation or voluntary adoption, and these codes and legal systems are considered as part of the Romano-Germanic group.

An important concept which binds the whole Romano-Germanic group is that of the legal rule. The legal rule which is the most basic element of the Code is not considered as merely a rule appropriate to the solution of a specific case.

'It is viewed as a rule of conduct endowed with a certain generality and situated above the specific application which courts or practitioners may make of it in any concrete case ... According to the Romano-Germanic notion, a code should not attempt to provide rules that are immediately applicable to every conceivable concrete case, but rather an organised system of general rules from which a solution for any given problem may be easily deduced by as simple a process as possible.'[2.14]

The legal rules applicable to administrative contracts can be different from those which apply to private contracts. Accordingly, the act of the jurist in this group consists in finding and formulating a rule so that it must not be too general, for then it would no longer be a sufficiently reliable practical guide; on the other hand, the rule must be general enough to cover a series of cases rather than merely apply to some particular situation as does a judicial decision.

The task of the lawyer in the Romano-Germanic group is conceived as essentially one of interpreting legislative provisions, and is thus unlike that in common law countries where the legal technique is characterised generally and in broad terms by the process of distinguishing applicable judicial decisions. The applicable legal rule in a given situation is not considered in the same manner across the legal systems: in common law countries, it is expected that the rule which provides a solution to a dispute will be framed as precisely as possible; whereas in the Romano-Germanic group, it is the contrary. This is because of the function of the latter, which is simply to establish the framework of the law and to furnish the judge with guidelines for decision making. Thus, in the Romano-Germanic group, it is considered desirable that the legal rule leaves a certain margin of discretion in its application.

It follows, therefore, that there are fewer actual rules of law in the Romano-Germanic group than in the common law groups where the legal rule, because it is less abstract, enters into greater detail for specific situations.

Accordingly, where civil engineering contracts are concerned, it is generally surprising to a lawyer from the Romano-Germanic group of countries to find that general conditions of contract, like those in the Fourth Edition of the Red Book, are in such detail. For example, he would be surprised to find that, despite the fact that clause 5 does specify the applicable law of the contract, the conditions continue to specify the rights and obligations of the parties, which is in effect a repetition of the provisions of the specified law.

Some observers speculate that the reason for this repetition is the mistrust of the technologists in the decisions of the judiciary. Others give the reason that the Red Book is based on the ICE Form whose legal concepts are rooted in the common law system, whilst some say that the reason lies simply in the lack of knowledge of engineers as to the legal concepts of a contract and more so when such a contract is an international one.

2.8.1 Sources of law in the Romano-Germanic group

The sources of law in the Romano-Germanic group are legislation, the creative role of the judge, judicial decisions, custom and international conventions and treaties.

Legislation

The primary source of law in the Romano-Germanic jurisdictions is legislation, but one must not confuse 'laws' and 'legislation' since the latter is not the exclusive source of law. Nearly all the countries within the Romano-Germanic group have written constitutions alongside their codes. The written constitution is held at the highest point of authority. All legislation is subordinate to the constitution, where it exists. The constitution usually comprises a series of rules which guarantee fundamental rights and freedoms to the individual and limit any arbitrary exercise of power by the state or government.

A statute, on the one hand, is intended to be an expression of the intention of the legislature and, on the other hand, is intended to be a guide to the lawyer and his client in determining what is the law within its field. Some believe that the intention of the legislature is best expressed as broadly but as concisely as possible even at the cost of obscurity and that obscurity can be interpreted and resolved by the judiciary. Others believe that the reverse is a more appropriate style and that statutes should be drafted in comprehensible and precise terms so that the ordinary practising lawyer, rather than the specialist, and perhaps even his client, might be able to understand its provisions readily.

Therefore, in drafting legislation, a balance has to be achieved between brevity and precision. Legislators in all countries have this problem and are divided between the two extremes. The division also extends to jurisdictions within the common law group although a distinction is sometimes drawn between the two groups as expressed in the following quotation by Lord Wilberforce:

> '... the English system and the French. These are thought – perhaps there is some mythology about it – but they are generally thought to be typical of two extreme methods. The English of elaborate, detailed drafting covering every individual case: the French of elegant generalities from which applications are deduced. It is the belief in this country, possibly ill founded, that the French method derives from your Revolution which decided, breaking with the old system, that laws should be drafted in understandable yet precise terms, the job of judges being merely that of applying the words to a situation. The English method is supposed to be due to English pragmatism, English dislike of principle.'[2.15]

The creative role of the judge

The creative role of the judge goes hand in hand with the logical process of interpretation of the codified law. It is reported that in a famous speech in 1904 on the occasion of the centennial celebrations of the French *Code Civil*, Ballot-Beaupré, the then President of the French Supreme Court (*Cour de Cassation*), explained how the judges must approach the interpretation of legislation:

'When a text expressed in imperative language is clear, precise, unambiguous, the judge must apply its literal meaning . . . But when the text is ambiguous, when there are doubts as to its meaning and intent, when it can be either restrained or extended or even contradicted by some other text, then in my opinion, the judge has the widest power of interpretation; he must not then stubbornly attempt to ascertain what the original thought of the draftsmen of the civil code was 100 years ago; he must rather ask himself what their intention would be were that provision to be drafted by them today – in the face of all the changes which have come about in the last century in ideas, social manners, institutions, the economic and social conditions in France, he must say to himself that justice and reason require that the text be liberally and humanly adapted to the realities and requirements of modern life.'[2.16]

However, unlike the common law system, once judgment is rendered in a particular case in the Romano-Germanic system, it is not then used as a precedent for another. In fact, most codes prohibit judges from laying down general and regulatory rules, as provided, for example, in Article 5 of the French *Code Civil*.[2.17]

There are, however, exceptions to this principle. For example, in Germany the decisions of the Federal Constitutional Court are binding and for this reason are published in the official *Federal Journal*. Other exceptions exist although they are basically related to the decisions of the Constitutional Court or a court of similar status. In most of the Romano-Germanic jurisdictions there is judicial control of the constitutionality of legislation and in some countries the courts are permitted to set aside laws that violate any provision of the constitution whilst in others this role is allocated to a specially formed council or supreme court.

Judicial decisions

In judicial decisions in the Romano-Germanic group, written reasons must be included in the judgments and this requirement is sometimes embodied in the constitution, as in Italy, to ensure that a well thought-out solution is given. Two styles of formulating the decision exist in the Romano-Germanic group: the first is the French style which is very concise and formulated on the basis of a series of conclusions, one following the other. This style is followed in many other European countries, such as Belgium, the Netherlands, Spain, Portugal and Luxembourg. The second style is practised in Germany, Greece, Italy and others and is based on a dissertation which sometimes includes references to previous decisions and legal doctrine.

Whilst one cannot emphatically deny that judicial decisions in Romano-Germanic jurisdictions have any role in the formulation of a source of law, they do not in fact have the same role as in the common law countries. This difference between the two groups of law is perhaps one of the earliest and most distinguishing factors between them. However, in a certain way, decided cases do advance the law in a certain direction in the Romano-Germanic group and, furthermore, cases decided by important courts, such as the French Supreme Court (*Cour de Cassation*) or those of the French Council of State (*Conseil d'Etat*) are studied in, and exercise an influence on, various neighbouring or distant French-speaking countries. In the main, judicial decisions do not create rules of law to be followed in future cases, because this role is left to the legislators. A judicial decision may therefore be rejected or modified in a subsequent decision of a new case. If followed, an earlier decision in a previous case is not quoted, but the legal rule underlying both decisions is quoted.

Therefore whilst it could be incorrect or imprecise to say that judicial decisions are not a source of law in the Romano-Germanic group, it is correct to say that they are not a source of legal rules.

Judges in the Romano-Germanic group are obedient to enacted law but, as in the case of judges elsewhere, they may have to decide a case where there is a lacuna in the legislation. Unlike the judges in Roman times, a judge today cannot refuse to adjudicate when the law is uncertain. In such a case the judge must give a decision as if he were the legislator. The Swiss Civil Code has an express provision for such cases (Article 1, paragraph 2). In this respect, the situation is the same in common law countries. Judges in the Romano-Germanic countries, in general, have formal legal training and in a number of countries they are appointed for life and are thus irremovable, a status intended to guarantee their independence.

Custom

Attitudes towards the role of custom as a source of law differ from one country to another in the Romano-Germanic Group. French, Italian and Austrian jurists subscribe to the view that custom may be applied when the law itself expressly states so. On the other hand, in Germany, Switzerland and Greece, custom is recognised at a higher level approaching legislation. In practice, however, the role of custom has diminished with the growth of codification and whilst it had, in the past, an important influence upon the development of law, it exerts a minimal effect at present.

International conventions and treaties

International conventions and treaties are a source of law only in the countries which have both signed and ratified them. Upon ratification, signatories to conventions become bound by the provisions of the convention or treaty, and in fact many conventions and treaties become part of the domestic law of the country. In some cases, the treaty or convention is sovereign to domestic law and even to constitutional law. It is, however, open to a contracting state, where a convention or treaty allows, to derogate from or to make a reservation in respect of certain provisions, for example, the New York Convention of 1958.

Two examples of such international conventions, referred to earlier in this chapter, are discussed below.

(i) The EEC Convention on the Law Applicable to Contractual Obligations. This convention which is limited to Member States of the European Community has the objective of unifying within the European Community the 'choice of law' rules where contract is concerned. It was opened for signature on 19 June 1980 and has already been signed and ratified by 10 Member States. Although it has a very wide scope, nine types of contractual agreements are excluded from it, as specified in Articles 1(2) and (3). Amongst those excluded which are relevant here are arbitration agreements and agreements on the choice of court, and contracts of insurance which cover risks situated in the territories of the member states.

The articles which deal with the determination of the applicable law are Articles 3 and 4 to which reference was made earlier in this chapter.

(ii) The New York Arbitration Convention of 1958. This is the most significant international convention governing the recognition and enforcement of arbitral awards which have been made in a territory of a state foreign to the country where recognition and enforcement is sought.

The convention was established in New York on 10 June 1958 as a result of an initiative by the International Chamber of Commerce in Paris, which, in 1953, promoted a new treaty to govern international commercial arbitration. The proposals made were later adopted by the United Nations Economic and Social Council (UNECOSOC) and developed into the New York Convention of 1958.

The New York Convention of 1958 represented an improvement on the Geneva Convention of 1927, which it replaced, as it provided for a much simpler and effective method of obtaining recognition and enforcement of arbitral awards. The Convention has two main objectives. The first is the recognition and enforcement of foreign arbitral awards. The second is to require the courts of contracting states to recognise and give effect to arbitration agreements made in writing within the meaning of Article 2 of the Convention and to refer to arbitration disputes arising in respect of which the parties have made such an agreement to arbitration.

At present, there are over 130 contracting states, making it a truly international convention.

A contracting state is permitted to make two reservations from the provisions of the Convention. The first, referred to as the reciprocity reservation, applies to the words: 'Of awards made only in the territory of another contracting state.' This reservation has the effect of limiting the field of application of the New York Convention to awards made in a state which is a contracting state. Such awards are sometimes referred to as 'convention awards'. It is applied by about two-thirds of the contracting states.

The second reservation is referred to as the commercial reservation which entitles a contracting state to declare that it will apply the Convention only to disputes arising 'out of legal relationships, whether contractual or not, which are considered as commercial under the national law of the State making such declaration'. (See Article 1.3 of the Convention.) Of course, under the national law of a contracting state, relationships which are considered commercial may not be so under the law of another state.

2.8.2 *Legal authoritative writing*

Works of legal scholarship nurtured the development of the Romano-Germanic laws and, particularly so, through the work of universities. At present, however, the role of legal writing is complex and delicate. It is considered by some merely as a source for stimulating and creating the ideas which are subsequently used by legislators to enact laws. Others consider it as a living source of interpretation from which methods may be established to understand and interpret the law.

2.8.3 Areas of the law affecting construction in the Romano-Germanic group

At a glance, the areas of the law affecting construction in the Romano-Germanic group are shown in Figure 2.1.

2.9 The common law group

The common law system originated in England after the Norman Conquest of 1066 but it was mainly created by the judges appointed by the Crown in the twelfth and thirteenth centuries. It is therefore judge-made law, sometimes referred to as case law. Disputes were brought before the various royal courts of justice which developed the idea that their intervention was justified in the interests of the Crown even if private interests were in question.

All cases brought before the English royal courts, which by the thirteenth century had established their seat at Westminster, concerned matters of public law. This originated historically from the idea that to bring a matter before the royal courts was not a right but a favour, which might or might not be granted. The claimant, having obtained a writ, had in effect an order given by the king to his officers to order the defendant to act in accordance with the law and thereby satisfy the claimant. If the defendant refused to act accordingly, the claimant could then proceed against him not only because his claim was not satisfied but also because of the defendant's disobedience of an order of the royal administration. The trial was therefore a matter of public law.

In the nineteenth century, the royal courts became automatically accessible to the citizens, but the law remained that of public law. No separation between public and private law exists in the common law system. But the essential feature of the common law is that where there is no legislation directly on a particular point, it would then be legally based on evolving precedent. Thus it has been said that to the lawyer from the Continent of Europe, English law has always been, 'something rich and strange'.[2.18]

Another important feature of the common law group is that it developed in a manner in which procedural considerations gained extreme importance. The way in which litigation was initiated led legal practitioners to think not so much in terms of rights as in terms of types of action and to focus more on the facts within the various actions or writs rather than on elaboration of the substantive law into a system based on a specific method. In law, 'writ' meant a command of the king directed to the relevant official, judge or magistrate, containing a brief indication of the matter under dispute and instructing the addressee to call the defendant into his court and to resolve the dispute in the presence of the parties. In this context, it has been stated that the procedure observed before the royal courts at Westminster varied according to the manner in which the suit was begun. To each writ there corresponded in effect a fixed procedure which laid down the other steps to be followed, the handling of incidental questions, the admissibility of evidence and the means of enforcing the decision. In any given procedure, the plaintiff and the defendant had to be styled by a specific wording; their inappropriate use in another procedure would be fatal to the proceeding.[2.19]

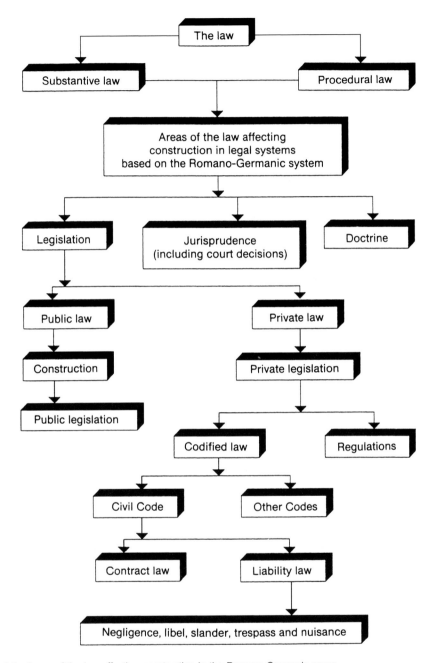

Fig. 2.1 Areas of the law affecting construction in the Romano-Germanic group.

In many respects this procedure caused difficulties and problems, with the result that the courts of law failed to administer justice. Aggrieved litigants turned to the king and petitioned him for an order to compel the opponent to do what justice, morality and good conscience required. Petitions were at first transmitted by the king to his chancellor who was the highest administrative officer and a prominent churchman. With time, however, the chancellor began to receive these petitions directly. The decisions made by the chancellor developed into a body of special rules and principles which are referred to as 'equity'.

It is extremely difficult, if not impossible, to define the rules and principles of equity in a short paragraph but the following statement is probably the best attempt that has been made in that direction:

'In its broad popular sense equity is practically equivalent to natural justice or morality; yet it would be a mistake to suppose that the principles of equity as administered in the courts . . . are coextensive with the principles of natural justice. Owing to the difficulty and doubtful wisdom of framing and enforcing any general rules to cover them, many matters of natural justice are not subject to legal sanctions but are left to the dictates of public opinion or to the conscience of each individual. Thus, in dismissing a claim against a company director who has perpetrated a sharp but not illegal practice, Fry LJ was constrained to say, "if we were sitting in a court of honour, our decision might be different".'[2.20]

From that beginning, where the chancellor was not bound by precedent, the Court of Chancery developed which by the eighteenth century administered equity as a recognised part of the law and in a similar manner to the common law.

A construction related example of the remedies developed under equity is that of injunction. It developed against the background in common law where a party could not protect itself against a recurring illegal act of another but had to claim damages in respect of each incident. Still less could a party protect itself against an impending illegal act of another but had to wait until the harm was inflicted and then sue for damages. The chancellor, when petitioned, did not accept this rule under common law and granted what became known as an injunction to restrain a party from acting in a manner which would cause harm to another. Eventually, as more and more specific cases came before him, answers to many related questions were provided. See later in Section 3.8.

A confrontation was inevitable between common law and equity and this happened in 1615 when a dispute between Chief Justice Coke and Lord Chancellor Ellesmere was submitted to King James 1. The King decided in favour of the Chancellor.[2.21] Since then, where there is conflict between the common law and equity, the rules of equity prevail. A consolidation of the two areas of common law and equity was achieved by the enactment of the Judicature Acts 1873 and 1875. The consolidation meant that all areas of the law including equity could be applied by the superior court. The Judicature Act section 25(11) provided that where there is conflict, the rules of equity are to prevail. However, it would be wrong to view equity as a contradiction or replacement of the common law as it is simply a set of supplementary rules. The Judicature Acts of 1873 and 1875 were repealed and replaced by the Supreme Court of Judicature Act 1925 which was itself replaced by the Supreme Court Act 1981.

An achievement of the reform in 1873 was the transformation of the courts system so that they were brought within a single Supreme Court of Judicature, consisting of the High Court of Justice and the Court of Appeal. Procedure in England today is governed by statutory rules, orders or instruments. The rules which apply in the Supreme Court are known as the Rules of the Supreme Court. These rules are published in a document extending over two volumes with extensive notes and annotation describing the fundamental steps to be followed in procedural matters from the inception of an action through the various stages to the trial. It is worth noting that, because of the basic difference in the structure between the common law group and other groups, similar legal terms may convey different meanings. English equity is different from the French *equité* and the division in French law between public law and private law does not exist in England. Even at a fundamental level, the equivalent of the French legal rule *règle de droit* is very different in that it evolved through judicial decisions.

Although it is true that there have always been statutes affecting various areas of the law in England and that during certain periods of its development a large number were enacted, it was the end of the nineteenth century that marked a huge increase in legislation. Comprehensive statutes were enacted codifying existing rules which covered specific areas of commercial law.

In recent times, there has been a tremendous increase in the number and volume of statutory enactments which make legislation a major source of law. Legislation is enacted by the sovereign with the advice and consent of the House of Lords and the House of Commons and by their authority. The rule contained in the statute is in the final analysis subject to interpretation by the courts.

2.9.1 Sources of law in the common law group

Whilst there is no written constitution in England, some of the countries within the common law group do have a written constitution which is held supreme as to authority within that state.

In England, the situation is perhaps best described in the following words of Sir George Engle:

'Under our unwritten constitution there are no constitutional or legal restrictions on Parliament's power to legislate. Parliament can pass any law it pleases on any subject, and no court in the United Kingdom has the right to declare an Act of Parliament unlawful or to set it aside as unconstitutional.'[2.22]

Sources of law other than the constitution, where it exists in a written form, may be stated as follows:

(a) judicial decisions when they are recognised as precedents in the common law;
(b) equity;
(c) legislation or statute law;
(d) regulations and delegated or subordinate legislation;
(e) international treaties; and
(f) custom.

Considering these in turn:

Judicial decisions

One must appreciate that the role of judicial decisions is not only to apply the legal rules but also to define them. Furthermore, the rules already defined by decided cases must be followed by lower courts – the rule of precedent. The hierarchy of the courts establishes a distinction between the courts within the Supreme Court of Judicature and the 'lower' courts and it is only a decision of a court within the Supreme Court that can be established as a precedent. In England, there are three levels in the hierarchy of the courts. These start with the High Court as the first level followed; at the second level, by the Court of Appeal, where cases are usually heard by a panel of three judges; and finally, at the top lies the Appellate Committee of the House of Lords where cases are usually heard by a panel of five judges but never by fewer than three. This Committee does not deliver judgments but opinions which are then adopted by the full House of Lords as its judgment.

In respect of the rule of precedent, the decisions rendered by the House of Lords are considered binding on all other courts. Until 1966, the House of Lords considered itself strictly bound by its own previous decisions, but in a Practice Statement in that year it announced that in future it might consider itself no longer bound if the circumstances indicated this to be necessary in the interest of justice. The first decision by the House of Lords where they refused to follow their own previous decision was in 1968.[2.23]

The decisions of the Court of Appeal are binding precedents for all courts of a lower level in the hierarchy and also for the Court of Appeal itself except in criminal cases. The decisions of the High Court are binding on the lower courts and persuasive for subsequent cases in the High Court.

A system of hierarchy is followed in other countries of the common law group with perhaps different names given to the various courts in the levels of hierarchy. The top level generally ends with the Supreme Court except in some of the Commonwealth and British overseas territories where the final court of Appeal is the Privy Council. The Privy Council is composed of members of the English House of Lords and others from the countries concerned and in theory it gives advice to the sovereign as the head of state of the country concerned.

When a judgment is rendered in the superior courts of a country within the common law group, the judge sets out the reasoning and the logic he has followed in reaching his decision in the particular case before him. Besides the necessary basis for his decision, the judge may comment on other matters or indeed on the decision itself. The rule of precedent applies only to the necessary reasoning. Any commentary which for all intents and purposes could have been omitted does not form part of the precedent. It is, however, considered persuasive in deciding subsequent cases if it is rendered by a judge considered to be authoritative. To distinguish between what is and what is not part of a precedent, a judge in a later case must analyse previous decisions in cases dealing with similar issues and must first distinguish between the essential reasoning and the commentary and then on whether or not the former applies in the case before him.

Some of the judgments rendered in cases where a precedent is set are published, but others remain unreported. The reports, where published records of judicial decisions are contained, are referred to as law reports. Figures quoted for the number of cases selected for publication from the various levels of the courts in England are as follows: 75 per cent of the decisions of the House of Lords, 25 per cent of the decisions of the Court of Appeal and 10 per cent of the High Court.

There are different series of law reports, each of which is given a title which is abbreviated for ease of reference. Therefore the 'All England Law Reports' is abbreviated as All ER and the 'Appeal Cases Law Report' is designated as AC and so forth.

The report of a decision contains the title of the case, a statement of the facts giving rise to the litigation, its history in the courts and a reproduction of the judgment of the court and of the decision. In England, until 1865, reporting was not organised and was by private enterprise, often overlapping and irreconcilable.

In the United States of America, law reports were also privately published at first. Today, however, reported opinions are almost invariably written by the court and are officially published. However, a private publishing organisation which began unofficial publication in the nineteenth century still continues today to publish its National Reporter System.

Equity

Equity, as already defined earlier (Section 2.9), forms part of the substantive law developed by the decisions of the Court of Chancery. There are certain general principles on which the court exercised its jurisdiction, many of which are embodied in what is called the twelve maxims of equity. A number of these maxims have an effect on construction contracts. An example of an application of one of these maxims is the area of penalties. The maxim concerned provides that 'equity looks to the intent rather than to the form'.

In *Parkin* v. *Thorold* (1852) 16 Beav 59 at 66 Romilly MR said:

'Courts of equity make a distinction in all cases between that which is matter of substance and that which is matter of form; and if it finds that by insisting on the form, the substance will be defeated, it holds it to be inequitable to allow a person to insist on such form, and thereby defeat the substance.'

Since equity regards penalties as inequitable, it will not therefore be bound by a description, such as 'liquidated damages', if in truth the sum specified is a penalty. See later in Section 17.5.

Although equitable remedies are discretionary and are viewed as an expression of fairness, the rule of precedent does apply nowadays to matters of equity.

Legislation

The third most important source of law is legislation and although it developed as a secondary source of law to the common law and the rules of equity, it has grown to a considerable volume embodying the legal principles of many areas of human activity. Besides developing the law in a certain area or direction, legislation has an important role in altering an existing flawed legal rule or restating a legal principle which had been developed through precedent in the courts. In doing so, legislation can overrule common law. This is extremely important where the error in the law is of a serious nature and cannot be left until a case may be brought before the courts permitting such correction. Because of the rule of precedent, the route through the courts, which is the only alternative for correcting or restating the law, requires a plaintiff who has the courage of his convictions and who is prepared to challenge a previous decision of a higher court.

Legislation is also an important vehicle to fill any gaps in the law or to augment it in matters relating to new concepts or changes in the organisation of modern society without having to establish the law through judgments in the courts. However, as explained earlier, it is worth noting that statutory law may come under the scrutiny of the courts if and when the interpretation of some wording is at issue. In the event of inadequate drafting, the courts may be called upon to resolve the ambiguity.

Primary and subordinate legislation

Legislation is divided into two forms: primary and subordinate. Primary legislation means an Act of Parliament, sometimes referred to as a statute. Subordinate legislation refers to rules, regulations or bylaws and are matters of detail relevant to a parent Act.

Primary legislation is drafted first in the form of a Bill by full time professional draftsmen which is presented, usually by the government, to Parliament for debate. If and when Parliament ratifies the Bill with or without amendment, it becomes an Act or a statute. Subordinate legislation is drafted by qualified lawyers who are experts in the field of law under specific powers conferred by Acts of Parliament on a minister of the government or some other official of an organisation. Primary legislation in England is drafted by full time professional draftsmen who are lawyers by profession. Subordinate legislation on the other hand is drafted by legal advisers to the governmental department concerned. These are also lawyers by profession and are generally experts in the field of law in question.

In England, the general belief of the draftsmen is that the ordinary citizen does not concern himself with legislation, as can be surmised from the following two quotations:

> 'I think that most citizens hardly ever read an Act of Parliament. I hardly ever met anyone who has read a whole Act of Parliament. Most people get their idea of what the law is from newspapers, or from the Citizens Advice Bureau, which gives them leaflets if they want to know what the position is about taking a tenant in their house, or getting rid of a tenant in a house which is controlled. The last thing they would do would be to go and read the Act of Parliament. They would go to a solicitor if they needed legal advice or they would go to some less exalted source of rather general legal advice.'[2.24]

and

> 'I believe that the clients whose interests we must primarily consider, in the process of legislative drafting, are, on the one hand, the legislator, and on the other hand, the

lawyer and his clients (in which term I include our biggest client, the government)...
It is too often suggested on the other hand, that the intelligibility of statute to the general public is a prime consideration in drafting it. I do not believe that this is so.

I am firmly of the opinion that however we draft our statutes, the British public, except perhaps for the remote lunatic fringe, never read the statutes nor are likely to. (I believe that this has more to do with the British temperament than with the state of the statute book: I am told that the Frenchman-in-the-street does read his statutes, and is not slow to assert his rights, whether to his employer or his landlord, by quoting the *Journal Officiel* at him.) The British citizen is much more likely to read the departmental booklet which is often issued to explain the law, or to go to his Citizens Advice Bureau or his lawyer, if necessary.'[2.25]

These statements may explain the difficulties experienced by lay people in understanding the legal language of statutes.

Furthermore, once an Act comes into force, the meaning of its wording may become a matter of a dispute in which case interpretation of the Act becomes the function of the courts. In specific cases, certain provisions have been left to the courts to interpret and thus statutory provisions have become burdened by a large number of precedents under common law, the text of which replaces that of the statute itself.

International conventions and treaties

As in the Romano-Germanic group, international conventions and treaties are a source of law only to the states which are signatories to them and which have ratified the provisions contained therein. They are then bound by the provisions of the convention or treaty.

Custom

Custom was an important source of law in earlier times. Now, however, as legal systems grow, the importance of custom as a source of law has diminished.

Custom can only attain legal status when it is recognised by the courts as being reasonable and in conformity with statute law. Thus, the courts have a great deal of discretion in determining whether custom is given legal validity. Should custom attain this legal validity, it then has the force of law but it remains subordinate to the written law and, as such, has lost much of its importance as a source of law in modern society.

The role of textbooks and legal writing

As the common law is based on reason, it implies that certain general principles deriving from already existing rules must be sought out and applied. Thus, textbooks by distinguished authors have a role in analysing, commenting and elaborating on the law at a particular point in time. In such legal writing, reasoning, ideas and interpretations are of the essence in establishing the tools for argument. Passages from such established material are often quoted in litigation or arbitration in support or in explanation of a point of view or a decision. As an example, reference may be made to the case of *Morgan Grenfell (Local Authority Finance) Ltd* v. *Sunderland Borough Council and Seven Seas Dredging Ltd*,[2.26] where the opinions of 'two distinguished text book writers' with regard to the ICE Conditions of Contract were quoted.

2.9.2 *Areas of the law affecting construction in the common law group*

At a glance, the areas of the law affecting construction in the common law group are shown in Figure 2.2.

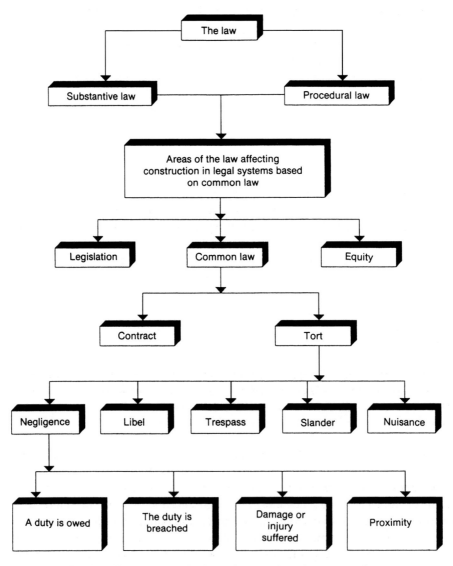

Fig. 2.2 Areas of the law affecting construction in legal systems based on common law.

2.10 The law in Islamic countries

Islamic law forms only a part of the law in most Islamic countries and it must not be confused with the whole body of law that applies in a particular jurisdiction or state. Like canon law in the Christian religion, it is the law of the religion of Islam and although its authority is stipulated in the constitution of many Islamic countries, none of them is exclusively governed by it. Legislation, codification or custom have been used to augment it and complement its provisions in order to regulate the activities of society and to provide public order. The direction followed differs depending on whether the Romano-Germanic, common law or, in a few cases, the socialist system is followed. This has not contravened Islamic law which provides that civil authorities have the power to regulate society.

Codification was used, for example, in Egypt, Syria and Iraq, inspired by the French Code. Other countries in North Africa have also formulated their own codes inspired by the Italian and French Codes. Pakistan, Malaysia and Northern Nigeria are examples of countries that were influenced by the common law group.

Therefore, the law in Islamic countries differs greatly from one jurisdiction to another because of the different directions they followed to augment the principles of Islamic law and because these countries extend over large areas with many basic differences in social and political traditions.

Where contract law is concerned, contractual principles from other legal systems in the West have for many years been integrated into trade agreements between parties from Western and Islamic regions. This integration, together with the influence through colonisation and the existence of legal authorities and literature in English and French, resulted in the applicable law in various Islamic jurisdictions being a combination of principles from various legal systems to varying degrees. Some critics believe that this is because the contract law that was developed by the founders of Islam envisaged only few conflicts, and thus there were no contractual provisions that dealt with general principles of contract law. As such, they believed that Islamic law alone was insufficient to deal with modern day contractual problems. These critics contend that:

> 'since the Shari'ah as a system of law had attained formal perfection centuries before modern technology, finance and marketing were dreamed of, it is incapable of recognizing or dealing with modern economic development or complex agreements'.[2.27]

It is argued that such a point of view can be clearly seen in Lord Asquith's Arbitral Award in the case of *Petroleum Development (Trucial Coasts) Ltd* v. *Shaikh of Abu Dhabi*, where he was of the opinion that the law, as it was applied in Abu Dhabi, was not capable of sufficiently managing a modern commercial agreement. His alternative was to apply in the Arbitral Award legal principles that were 'rooted in good sense and common practice of the generality of civilized nations – a sort of modern law of nature'.[2.28] It is noted that the assumption made that there was no general body of Islamic law governing contractual agreements was incorrect, since the vast Arabic legal scholarship that had for many years communicated principles of contracts and contractual obligations in Islamic law had been completely disregarded, most probably because of lack of knowledge of that legal scholarship or of the sources of Islamic law that they should have delved into.

2.10.1 Sources of Islamic law

Besides legislation and binding decisions there are two fundamental sources of Islamic law and two subsidiary ones. The two fundamental sources are:

(a) The Qur'an, which is the sacred book of Islam containing God's revelations to the Prophet Muhammed and recognised as the primary source of Islamic law. It is divided into 114 Surahs (chapters) of unequal length. However, the Qur'an is in no sense a comprehensive legal code. No more than 80 verses deal with strictly legal matters; while these verses cover a wide variety of topics and introduce many novel rules, their general effect was mainly to modify the then existing customary law in certain important areas; and
(b) The Sunna, which refers to the exemplary conduct or the model behaviour of the Prophet Muhammed in what he said, did or approved, serving as an example to be followed by all Muslims and a guide to jurists. The collection of his statements is called the *Hadieth*. The *Hadieth*, a report, or collection of sayings, approvals and practice attributed to the Prophet, provides the means by which the Prophet's words and deeds are made known, to be followed by the Muslim community.

The combination of these two sources is called the Shari'ah, which literally translated means 'the way to follow'.

The two subsidiary sources of Islamic law are:

(a) The Ijma', which is the consensus of Muslim scholars (Fuqaha') in studies and reflections called Fiqh meaning jurisprudence. The rules or consensus of Ijma' are drawn from the Shari'ah and are interpretations by the scholars of the wording of the Qur'an and the Sunna. Ijma' was made an important source of Islamic law through a verse of the Qur'an[2.29] and two statements of the Hadieth in the form of:

> 'My community will never agree upon an error'; and
> 'What Muslims find to be just is just by God'.

The doctrine of Ijma' was introduced in the second century AH (eighth century AD) in order to standardise legal theory and practice and to overcome individual and regional differences of opinion. Though conceived as a 'consensus of scholars', in actual practice Ijma' was a more fundamental operative factor. In terms of actual efficacy, consensus is the most important factor of all, for in effect, even the validity of the Sunnah itself is dependent on, amongst other things, the consensus or general acceptance by the community of its authority.

(b) The Qiyas, which is the juristic reasoning by strict analogy, is a formal procedure of deduction used only for the interpretation and application of Shari'ah and not to form a new rule. Some sects of Islam reject the applicability of Qiyas as a source of law and recognise it simply as a method of analogy.

In the early period of Islam, because *ijtihad*, which means 'to endeavour' or 'to exert oneself', took the form of individual opinion (*ra'y*), there was a wealth of conflicting and chaotic opinions. Qiyas came into being in the second century AH, as a method of *ijtihad*. It developed as the need grew to find a legal or doctrinal solution to a new problem. Thus, if a new case or issue arises that is not textually covered by the Qur'an and the Sunnah, and there was no Ijma on it, the issue may be settled by arguing analytically from such texts or cases that resemble the new issue, allowing for differences.

Although Ijma' is the consensus of the Muslim scholars, the degree of agreement required allows for some differences of opinion. In one of the traditional statements widely used by Muslims it is stated: 'The differences of opinion existing in my community are a manifestation of the grace of God'.

Differences have in fact given birth to a number of rites of Islamic religion based on differences of detail in the interpretation of Islamic law. Principally, there are four orthodox rites which form the following groups: the Sunnite, the Shi'ite, the Zahiri and the Ibadi.

Four schools of jurisprudence have survived in the *Sunni* rite of Islam. The *Hanafi* school (after *Abu Hanifah*, eighth century) holds sway in India, Pakistan, Afghanistan, Central Asia, Turkey and Lower Egypt; the *Maliki* school (after *Malik ibn Anas*, late eighth century) is operative in North and West Africa and in Upper Egypt; the *Shafi'i* school (after *ash-Shafi'i*, ninth century) is popular in South-east Asia; and the *Hanbali* school (after *Ahmad ibn Hanbal*, ninth century) is officially accepted in Saudi Arabia, and widely followed in certain parts of the Middle East.

The Shi'ites are so called because they are partisans of Ali, the son-in-law of the prophet Muhammed. The word Shi'a means partisan. They differ from the Sunnites mainly on constitutional law and the position of the caliphates who followed prophet Muhammed in ruling the communities of Islam. The *Shi'ah* school of jurisprudence is ascendant in Iran and parts of Iraq and its *Zaydi* version is in the Yemen.

The development of Shari'ah as a body of law was halted in the tenth century when political problems led to the disintegration of the Islamic Empire of that time. Thus, Islamic law established then is today immutable. However, this immunity to change is complemented by a great flexibility and resourcefulness for development through the operation of custom and the needs of society, the right to formulate administrative regulations and the right of the individual to contract, rights which are held sacred in Islamic law.

Many rules can thus be modified through the use of contract without violation of Islamic law. Loans bearing interest, which are forbidden, can be made by means of a double sale or by giving the creditor the use of some property as security and the enjoyment of any revenue that might accrue. It is also possible to hold that the ban on interest applies to persons and not organisations such as banks and companies. Contracts dependent on contingencies and insurance are forbidden but only the person collecting the premium commits a sin. Insurance contracts with companies or with a non-Muslim can therefore be made.

2.10.2 Contracts in general and construction and engineering contracts in particular under Islamic law

Contracts in general are divided under Islamic law into two groups: those that are specifically named and referred to in Shari'ah, such as contracts for sale, for labour, for agency, of marriage, etc; and those that are not specifically named. Unnamed contracts are those that are not specifically referred to in Shari'ah and for which there are no provisions. They were introduced as need arose and as society developed and continued to develop, and include the contract for manufacture, discussed below.[2.30]

Contractual arrangements under Islamic law prohibit dealing with matters that are not in existence at the time of making the agreement, with two specific exceptions of unnamed contracts that developed through Ijtihad.[2.31] The reason for this prohibition is to avoid what is referred to as *gharar*, see section 2.10.3 below. The first exception is a specific contract for sale known as *A'qd Al Salam*, which is defined as a contract for the sale of an item which is not in existence at the time of the agreement, but which will be in existence and will be delivered within a specific period of time. The nature, quantity, specification and price must be known at the time of agreement. The second exception is a contract of manufacture, known as *A'qd Istisna'*, which is defined as an agreement for the delivery of future, non-existent physical goods, but their nature and type are specified, for a certain price that is generally not due until delivery.[2.32] These two forms of contract were and are permitted under Islamic law despite the fact that the subject matter does not exist at the time of contracting.

As we are concerned here with construction and engineering contracts, it is worthy of note that the construction contract or contract for work, known as *A'qd Muqawala*, draws its origin in Islamic law from the A'qd Istisna' and thus it is accepted by all jurists in Islamic law as a valid form of contract on the grounds of custom and need of society, which has prevailed from the time of the Prophet Muhammed and is therefore justified by the necessities of business and equity. This is despite the belief held that it constitutes an element of uncertainty in the future performance.

There are certain conditions that apply to A'qd Istisna'. There should be:

- A clear description of the subject matter of the contract agreement, including scope, type and specification;
- A known price or method of calculating the price; and
- Generally, a date for the delivery of the subject matter and if delivered in accordance with specified characteristics, then its delivery cannot be refused by the buyer, but if not in accordance with the specification he has the option of applying damages. This provision forms part of *Al-Majalla to Al Ahkam Al Adliyah*, which is the Ottoman civil code, drawn up in accordance with the Hanafi School of Islamic Law.[2.33]

2.10.3 Some salient principles in construction and engineering contracts

There are some principles of Islamic law relevant to construction contracts. Although, as stated earlier, the Qur'an is not a comprehensive code of law, it does contain statements requiring the believer to honour agreements and to observe good faith in commercial dealings. However, it must first be emphasised that a contract, a'qd, in Islamic jurisprudence means no more and no less than a legal undertaking, the essentials of which are very different from the binding promise which constitutes a contract in Western systems of law. For example, a contract or a'qd does not necessarily involve 'consideration'. In the Shari'ah, a contract originates from agreement and consent and not necessarily from the observance of any specific form or procedure. Once the provisions of the treaty are agreed upon, the treaty becomes binding on both parties. The writing as well as the signing and the dating, and in some cases the witnessing of the treaty, are not necessarily legal prerequisites; they are merely to indicate that an agreement has been reached, as well as to record the actual terms of the treaty and its duration.[2.34]

The following four principles are of particular importance to construction contracts: *riba*, *gharar*, good faith and due process. These are explained below.

Riba

Riba means 'usury' in some verses of the Qur'an, but in others it seems to express the broad notion of illicit gain or unjustified profit and enrichment. There is also a warning in the Qur'an in the following terms:

> 'O you who believe, fear God and give up what remains of your demand for usury, if you were believers.[2.35] And if you do not, then be warned of war from God and his messenger, but if you repent, you shall have your capital sums; deal not unjustly and you shall not be dealt with unjustly.[2.36] And if the debtor is in difficulty, grant him time until it is easy for him to repay; and if you remit it as charity, it is best for you, if you knew.'[2.37]

Further, in Surah 4, it is also stated:

> 'And of their usury when they were forbidden it, and of their devouring people's wealth by false pretences. We have prepared for those of them, who disbelieve, a painful doom.'[2.38]

Riba is also linked in the Qur'an to the specific prohibition of gambling and other activities such as extortion, bribery and corruption.

Gharar

Etymologically, *gharar* means an act with three characteristics: cheating, danger and unwariness.[2.39] It is said that *gharar* is an arrangement that is alluring on the outside surface, but unknown in its inner content. In jurisprudence, the definition of *gharar* differs from one school of law to another. However, the definitions are closely related, as they all stem from the linguistic meaning and differ from one another on the basis of the following four criteria: the emphasis placed by the jurists on the above-mentioned three characteristics of *gharar* of cheating, danger and unwariness; on how strictly they apply them; on their interpretation of the specific wording of Shari'ah, referring to it expressly or implicitly; and on the intensity of the *gharar* being dealt with.[2.40]

An agreement containing *gharar* is not permitted under Islamic law. The Qur'an disallowed not only injustice in all its forms, but also inflicting harm, as apparent in many verses of the Qur'an, some of which were interpreted to imply *gharar*. One of the verses is quoted as follows: 'O you who believe, do not squander your wealth illicitly, but let there be amongst you trade by mutual consent and kill not one another. God is ever merciful unto you'.[2.41] In another verse, it is ordered 'you shall not take each other's money illicitly, nor shall you bribe officials to deprive others of some of their rights illicitly, while you know'.[2.42]

Whilst these two verses do not mention the word *gharar*, Hadieth did so in many quotations from the Prophet Muhammed. It is noteworthy that *gharar* was also forbidden by Ijma', the consensus of jurists.

As *gharar* is not mentioned expressly in the Qur'an, jurists in the different schools of Islamic law divide it into three groups based on its intensity: minor, major and gross. In some sects minor *gharar* is permitted in contractual agreements with certain conditions. For example, a contract of insurance is treated by jurists of the Hanbali schools of law as a forbidden type of transaction, whilst it is permitted by the Ja'afari school and some jurists of the Hanafi school. Some jurists of the Ja'afari school go further and declare that it is *gharar* not to insure.[2.43]

Hanafi jurisprudence allows the sale of goods where there is only minor *gharar*, for example items where the content is hidden by a shell, such as melons, walnuts, etc., as the buyer has the ability to see what he is buying.

Some authors translate *gharar* in English as 'risk', which is incorrect and widely off the mark; as explained above and in Chapter 7, *gharar* is very different from risk, and although risk contains the possibility of danger, one of the characteristics of *gharar*, it does not contain the other two, cheating and unwariness. It is worth noting that all construction and engineering contracts, which are obviously permitted under Islamic law, involve risk. According to Ibn Rushd,[2.44] recognised jurist in Islamic law:

> '*Gharar* in transactions of sale causes the buyer to suffer a loss and is the result of lack of knowledge concerning either the price or the subject matter. *Gharar* is averted if both the price and the subject matter are proved to be in existence at the time the transaction is concluded, if their qualities are known and their quantities determined, if the parties have control over them so as to ensure that the exchange takes place, and, finally, if any term of time involved is precisely determined.'

Traditional Shari'ah authorities consistently emphasise that the essential certainty in contractual obligations is to be achieved by exact definition of the rights and obligations of the parties. This can also be clearly seen in the common law where one of the general principles of the law of contract is the certainty of terms. For example, the contract price is so essential a term in a contract that if it has not been agreed, or the method by which it should be ascertained is not agreed, then there is no binding contract.[2.45] The fact that A'qd Istisna' is permitted under Islamic law proves that the existence of uncertainty alone does not constitute *gharar*.

Good faith

The theory of good faith in Islamic law can be traced as far back as the Ottoman Empire where it took the form of the Al-Majalla Code.[2.46] It has been said that this codification is proof that general principles of Islamic contract law do exist. The Al-Majalla clearly states that one of the fundamental principles of Islamic contract law is the duty to act in good faith.[2.47] It should also be borne in mind that the various customs of the Islamic world play a significant role in the concept of good faith in contract law. The reason for this can be attributed to the fact that agreements were mostly oral. If an agreement was in writing, such words were merely evidence of the oral agreement, not the actual contract itself. Therefore, all private agreements between parties are enforceable under Islamic law, once they comply with Islamic norms. The reality of such a principle is clear in most modern Islamic legislation. For example, under the Egyptian Civil Code of 1948, article 89 states that 'a contract is created, subject to any special formalities that may be required by law for its conclusion, from the moment that two persons have exchanged two concordant intentions'. Al-Sanhuri, the distinguished Egyptian jurist, in his book *Masadar al haqq* identified five events of contract law where the principle of good faith would apply:

(i)   good faith in the conclusion of a contract;
(ii) good faith in the performance of the contract;
(iii) good faith in the termination of the contract;
(iv) the notion of *riba* or usury; and
(v)   the notion of *gharar*.[2.48]

Due process

Due process is a fundamental principle of Islamic law. It is of such vital importance since it is prescribed in the Qur'an in the following text:

> 'O you who believe. Stand out firmly for justice, as witnesses to God, even as against yourselves, or your parents, or your kin, and whether it be [against] rich or poor: for God can best protect both. Follow not the lusts [of your hearts], lest you swerve, and if you distort [justice] or decline to do justice, verily God is well-acquainted with all that you do.'[2.49]

One of the components of the principle of due process is the idea that both parties should be heard and have a chance to present their case. There are very few situations where the Qadhi, the judge, can dismiss a case without first hearing both parties, and these are where the subject matter of the disagreement concerns something that is forbidden by the Shari'ah. Having said this, however, if one of the parties does not turn up before the Qadhi to present its case, then the Qadhi may either dismiss their case or pass judgment on an *ex parte* basis.

Alongside the right of the parties to be heard, there is also the right of the parties to submit evidence and to call witnesses to prove their case. The parties, however, must carry out and conduct the proceedings in accordance with the principle of good faith. Under Islamic law, the use of witnesses is a more popular and prudent option since it is a legal system where the opinion or evidence given by an upstanding member of society would be more effective than something that is written down on paper. Under Islamic law, the witness owes a duty to the court and not to the parties for whom he is appearing, thereby making the whole process more impartial, and therefore ensuring due process.[2.50]

The relevance of these four principles of *riba*, *gharar*, good faith and due process to construction and engineering contracts must be obvious, and in particular to such matters as variations, unforeseeable underground conditions, and other risks to which the parties to such contracts are exposed.

The necessity for making known the meaning of the words *riba* and *gharar* and integrating them into modern jurisprudence is also clear since very few parties are aware of the precise meaning and relevance of *gharar* to modern construction and engineering contracts.[2.51]

As Islamic law is rooted in the Islamic religion, it is interesting to note the following:

(a) For a Muslim not to obey Islamic law is a sin punishable in the next world after death; and
(b) Islamic law is applicable only to dealings between Muslims, unless non-Muslims accept to be subjected to it.

This chapter has briefly considered the diversity of legal systems around the world, the applicable law in international construction, the applicable law of the contract, the laws governing both procedure and enforcement of awards and the groupings of contemporary legal systems. The latter includes the Romano-Germanic group, the common law group and the Islamic group. Since the Red Book is based on legal concepts rooted in the common law system, the next chapter specifically considers the concepts of that system.

Chapter 3

Legal Concepts Based on the Common Law System

Once again it is worthwhile repeating the caveat given at the beginning of Chapter 2 that the comments on the law are of necessity generalised and that statements of general rules and principles are not intended to mean that they are without exception or qualification. These rules and principles may differ from one jurisdiction to another, even within one legal group, and therefore there is no alternative to a thorough study of the applicable law in all its relevant parts. In this chapter, the law of torts and the law of contract will be considered in some detail.

3.1 The law applicable to the contract

As explained earlier, the Red Book, which is based on legal concepts rooted in the common law system, originated from the ACE form which in turn originated from the ICE Form. The legal framework of the contract is basically set out in three clauses in the Red Book. These are clauses 5, 26 and 70.

3.1.1 Clause 5

Clause 5 of the Red Book provides for the law of the contract, usually referred to as the 'applicable law of the contract', to be specified in Part II of the conditions. Where that law follows the common law system there would be little, if any, conflict of laws between the concepts of the Red Book and those under which the contract might have to be construed because of the fact that the Red Book is based on the common law concepts. In fact, in the early years after the Red Book was first introduced, it was customary to stipulate English law to govern these contracts whenever the engineer or the contractor was of British nationality. The general rule under private international law of most countries follows in general terms the statement made by Lord McNair and quoted below:

> 'It is often said that the parties to a contract make their own law, and it is, of course, true that, subject to the rules of public policy and *ordre public*, the parties are free to agree upon such terms as they may choose. Nevertheless, agreements that are intended to have a legal operation (as opposed to a merely social operation) create legal rights and duties, and legal rights and duties cannot exist in a vacuum but must have a place within a legal system which is available for dealing with such questions

as the validity, application and interpretation of contracts, and generally, for supplementing their express provisions.'[3.1]

It has become more usual and appropriate in recent years to specify as the applicable law of the contract the law of the country where the project is constructed. Restrictions upon this freedom of choice of the applicable law are taking place through increasing intervention by modern-day governments intent on regulating their own economic activities. In such cases or where the contract follows a system of law other than the common law, it is inevitable that some conflict would occur between the applicable law of the contract and some of the legal concepts incorporated in the Red Book. However, as the Red Book was devised for use anywhere in the world, the extent and type of conflict of laws could not be predicted and thus the draftsman of the Red Book made no attempt to deal with this problem. Furthermore, the general conditions, of which there are over 70, were considered to be sufficient for most contracts.

In certain countries, the rule of law that property rights are subject to the law of the country where the immovable property, 'the works', are located has been extended so that it applies to personal obligations arising in connection with an 'immovable'. In these countries, the applicable law of the contract must be that of the country where the project is located. If this is not done, it may happen that, should a dispute arise, the legal problems arising from that dispute supersede the technical problems and should an arbitration award be rendered, it might be unenforceable on the grounds of public policy.

3.1.2 Clause 26

The second clause forming the legal framework of the Red Book is clause 26. It requires the contractor to conform in all respects with all national statutes, ordinances, laws, regulations and by-laws of any local or other duly constituted authority in connection with the execution and completion of the works and the remedying of any defects therein. It therefore requires conformity not only with the laws of the country where the project is located but also with the laws of any other country where part of the work is executed. This is required irrespective of what law is chosen as the applicable law of the contract. It is a far-reaching requirement as, in effect, it means that if the work or the supply of materials for such work is carried out in more than one country, then the laws of all these countries involved would have to be observed.

It should be noted that the branches of law which are related to construction are many and in a complex project could extend across the whole spectrum of legal systems.

3.1.3 Clause 70

Clause 70 of the Red Book is the third clause in the legal framework. It provides for the possibility of either introducing new, or making changes to already existing, national or state statutes, ordinances, decrees or other laws or any regulation or by-law of any local or other duly constituted authority in the country in which the works are being, or are to be, executed.

If such laws are introduced or if such changes are implemented in already existing laws causing additional or reduced costs to the contractor, then such additional or reduced costs must be determined by the engineer and accordingly, an adjustment should be made to the contract price. The matters covered by this clause are referred to in Part II of the Form.

In this connection, reference should also be made to clauses 71 and 72 whose provisions relate indirectly to those of clause 70. They deal with changes imposed by the government of the country in which the works are being or are to be executed in respect to currency restrictions or the rates of currency exchange.

The parties entering into a contract where the Red Book is used as a standard form of general conditions are therefore advised in the foreword to the conditions as follows:

'In the preparation of the Conditions it was recognised that while there are numerous Clauses which will be generally applicable there are some Clauses which must necessarily vary to take account of the circumstances and locality of the Works. The Clauses of general application have been grouped together and are referred to as Part I – General Conditions. They have been printed in a form which will facilitate their inclusion as printed in the contract documents normally prepared.'

The general conditions are linked with the conditions of particular application, referred to as Part II, by the corresponding numbering of the clauses, so that Parts I and II together comprise the conditions governing the rights and obligations of the parties.

The clauses in Part II must be specifically drafted to suit each particular contract. Some notes and explanatory material have been included in Part II of the document and are intended as an *aide-mémoire* in relation to the matters which should be covered by the various clauses. These notes should be detached from the document when inviting tenders.

3.2 Conflict

Depending on the choice of the applicable law of the contract as determined under clause 5 of the Red Book, conflict between its rules and the conditions in Part I of the Red Book may arise. If such conflict arises it may take the form of any of the following possibilities:

(a) that there is no legal rule in the applicable law of the contract similar to that assumed to exist by the draftsmen of the Red Book;
(b) that a legal rule similar to that assumed to exist by the draftsmen of the Red Book does exist in the applicable law of the contract, but only in a very basic and underdeveloped form; and/or
(c) that a legal rule similar to that assumed to exist by the draftsmen of the Red Book does exist in the applicable law of the contract, but its effect is different from, if not directly opposite to, that assumed.

3.3 Some specific concepts under the common law

Because of the possibility of conflict outlined above, it is necessary to consider some specific legal concepts under the common law system to establish the features upon which the Red Book is based. However, the purpose of the present section is not to give a comprehensive description of contract law under the common law system but simply to highlight the essential features. Those who wish to investigate this subject in greater depth should study standard reference books on contract law or specialised works on engineering contracts.

In a construction contract, in addition to criminal negligence, one may encounter all the areas of law dealing with civil wrongs as distinct from those which are liable to result in criminal proceedings. The area of civil wrongs can be divided in two different ways. The first is by dividing the law into substantive law and procedural law. The second is by dividing the law into legislation, common law and equity. (See Figure 2.2.)

3.3.1 Substantive law and procedural law

Figure 2.2 in the previous chapter illustrates both methods of categorisation of the law, where construction contracts are concerned. As to the first method, substantive law refers to all areas which define and deal with the rights and duties of individuals, groups, organisations and the state. Contract law comes under this heading.

Procedural law deals with the legal rules through which the process of law is set in motion to enforce some substantive right or remedy. These rules are often complex and numerous.

3.3.2 Legislation, common law and equity

As already noted, another method of categorisation of areas of the law which may be encountered in construction contracts divides the law into legislation, common law and equity. Legislation and equity are discussed in Section 2.9 above. The third area is that of the common law from which contract law and the law of torts emanate. (See Figure 2.2.)

The remainder of the present chapter considers the law of torts and the law of contract in more detail.

3.4 Tort

Tort is a legal term used in both the common law and the Romano-Germanic systems of law to describe various wrongs which may give rise to civil proceedings mainly in the form of an action for damages. The word tort has its origin in the Latin 'tortus', which means 'twisted or crooked'. The French word 'tort' means 'wrong'. It is defined as 'a breach of a legal duty which affects the interests of an individual to a degree which the law regards as sufficient to allow that individual to complain on his or her own account rather than as a representative of society as a whole'.[3.2]

A tort must be distinguished from a breach of contract. If there is a breach of contract, the obligations and rights of the parties to an agreement emanate from the terms of that agreement whereas under the law of tort they arise from the operation of the law without any need for consent from the parties. Thus, unlike the situation under contract, a person has no choice in the obligations placed upon him by the law in his behaviour towards others.

The same facts of a particular situation may give rise to an action in contract and in tort under the common law system. Concurrent liability in contract and in tort is not, however, universally recognised.

The law of torts regulates a wide variety of unlawful behaviour. Those which are related to construction are shown in Figure 2.2 as: nuisance, slander, libel, trespass and negligence.

Nuisance

The enjoyment of land and any right over, interest in, or premises thereon is protected by the tort of nuisance. Excessive dust, noise, vibration, fumes, seepage, gases, smoke etc. produced by someone may expose him to liability for nuisance. The court may order the nuisance to be stopped by an injunction or may award damages or both.

Libel and slander

A defamatory statement is one which injures the reputation of someone by its tendency to lower him in the estimation of right-thinking members of society or to cause him to be shunned or avoided.

A defamatory statement is libellous if the defamation is in a permanent form, for example, in writing or in a televised interview. It is slanderous if it is in a non-permanent form. A libellous statement is actionable without proof of damage, and may be considered criminal. On the other hand, a slanderous statement is actionable only on proof of damages capable of being expressed in terms of money (with some exceptions) and it cannot be a crime.

Trespass

There are three forms of trespass: to land, to chattels and to the person.

A person is liable for trespass when he intentionally enters or remains on land in possession of another without that person's permission. An action for trespass to chattels arises from any intentional, physical interference with a chattel in the possession of another. The loss suffered forms the basis of the liability.

Any conduct which causes some physical contact with another person without that person's consent may form the basis of an action of trespass to the person in tort. Consent may be either explicit, as in the case of a surgical operation, or implicit, as when one participates in a sport.

Negligence

The most important tort where construction is concerned is the tort of negligence. It is also the branch of the law which has been the subject of intensive development and change in the twentieth century. Negligence may be broadly defined as failure to comply with that standard of care which would correspond to the conduct expected of a reasonable person of ordinary prudence under similar circumstances, which results in damage.

The tort of negligence is concerned with breach of the duty to take care. However, the mere failure to comply with the standard of care is not sufficient to constitute the tort of negligence. In order to succeed in an action for negligence, a plaintiff must prove:

(i) a duty of care is owed, requiring conformity to a certain standard of conduct for the protection of others against exposure to risk;
(ii) a breach of that duty has been committed;
(iii) damage, injury or actual loss is suffered as a result of that breach; and
(iv) a proximate connection exists between the conduct in question and the resulting damage. This condition of proximity or remoteness is sometimes linked to either (i) or (iii) above. However, it is of sufficient importance to be considered on its own.

Contribution to the loss and assumption of the risk

Liability for negligence may be negated or lessened if the injured party was also at fault. Closely related is the situation where a party expressly or implicitly agrees to assume the risk of negligence which may be committed by another. The liability in this case may also be excluded.

In contrast to contributory negligence between two negligent parties causing damage or injury to one of them, are the contrasting rights of contribution which exist between persons who are liable in respect of the same injury or damage to a third innocent party.

It should also be noted that the law imposes a duty on the injured party to take all reasonable steps to mitigate the loss and debars him from claiming compensation for any part of the loss which is due to his own negligence.

Breach of statutory duty and crime

In English law some statutes which create criminal offences may also give the right to any person who has been injured or who has suffered damage by the offence to recover damages by a civil action.[3.3] This is a separate tort, known as 'breach of statutory duty'.

3.5 Contract – general principles

As stated earlier, the Red Book is based on a background rooted in the common law legal system. In the following sections important characteristics of this legal system with regard to contract are discussed.

The simplest definition of a contract is: 'a legally binding agreement'. A slightly more elaborate definition is 'an agreement between two or more parties in which each party binds himself to do or forebear to do some act and each acquires the right to what the other promises'.

In a valid contract, one may have to establish in certain circumstances the precise meaning of what the parties have undertaken to perform and how, where, when and by whom that performance was to be expected.

The contents of a contract may not necessarily be confined to the words which appear within its written boundary. Under English law, however, an agreement is an act and not a mental exercise, and its meaning is therefore to be inferred from what the parties have said or written or done and not from what is in their minds. This is a long established principle going back to 1478 when Chief Justice Brian proclaimed, in the context of a sale agreement, that 'the intent of a man cannot be tried, for the Devil himself knows not the intent of a man' (Anon (1477), YB 17 Edn. 4 fo 1, pl 2).

This subject is quite complex and embraces areas from the law of contract and beyond. Whilst the engineer, whether a designer or a constructor, is not expected to have detailed knowledge of the law, he must have general knowledge. Furthermore, where construction contracts are concerned, administration of the contract requires a certain minimum amount of knowledge of the law. As already stated, the discussion of the legal aspects presented here is by necessity a very brief one and it is neither intended to be without exception or qualification nor is it in any way to be a substitute for expert legal knowledge or advice.

3.5.1 Prerequisites of a contract

In many day-to-day activities, most people enter into contracts without giving much thought to their actions. Except for the very few, these contracts require no special forms and can be oral or in writing. They are called 'simple contracts'.

For simple contracts to be legally binding and enforceable under the common law system, the prerequisites of intent, agreement, consideration, legal capacity and legality of objectives must be satisfied.

(a) Intent

The intention to create legal relations and to be legally bound by the contents of a contract must be clearly apparent. There is a very strong presumption in construction and other commercial contracts that such intention does exist. This prerequisite is therefore of greater importance in other branches of the law.

(b) Agreement

There must be an agreement between the parties to the contract based on a definite offer by one of the parties and an unqualified acceptance of the offer by the other party. A conditional acceptance or a counter-offer may supersede the original offer and cancel its effect. Once an offer is accepted, a binding contract is formed, subject to the remaining prerequisites, even if a provision is included in the contract for the subsequent execution of formal documentation, as provided in clause 9 of the Red Book.

When seeking to determine what the parties consented to do by their contract, one should not look into the parties' minds but should ascertain their intentions by the outward expression as conveyed by the written or spoken words of that contract. It is also worth noting that where a person signs a written or a printed standard form of contract, he is bound by it, whether or not he has read it, understood it or accepted it.

(c) Consideration

There must be an exchange of an act or a promise of an act, referred to as a 'consideration'. Consideration has been defined by Sir Francis Pollock as 'An act of forebearance of the one party or a promise thereof, is the price for which the promise of the other is brought, and the promise thus given for value is enforceable'.[3.4] Consideration must be real, clear, definite, possible to perform and legal. The most usual form of consideration in construction contracts is the promise to perform work or the payment of money.

As stated later in this section, contracts under seal are enforceable without consideration.

(d) Legal capacity

In order for a legal contract to come into existence, the parties must have the capacity to contract. Any sane adult person can enter into a contract and such capacity is referred to as that of a 'natural person'.[3.5] Corporations, on the other hand, have a judicial capacity referred to as that of an 'artificial person' and created by law through a separate identity distinct from their members. The capacity to contract is defined and regulated by the 'objects clause' of a corporation's articles and memorandum of association. A corporation is entitled to do only those things set out in these articles and an act outside these objects is *ultra vires*, that is, exceeding the powers granted to the company by law.

It is important to note, however, that where the European Union is concerned, the first EU Companies Directive required Member States to abolish the *ultra vires* rule as it affected outsiders. Therefore, in EU Member States, a corporation can now be bound by contracts which are *ultra vires*. The Directive, however, in an important proviso, permits Member States to maintain the rule provided the company could prove the outsider knew that the act was beyond the authority of the company.[3.6]

(e) Legality of objectives

The objectives of a contract must be lawful, otherwise it is void. Legally, an agreement may not be given its required effect if it involves the commission of a legal wrong or is in some other way contrary to public policy or public safety.

In addition to simple contracts, however, another type which is used by English lawyers is that of a 'contract under seal' where a person undertakes an obligation by expressing his intention in a formally drawn up document, on paper or parchment, and signs it and has it delivered as his deed. Such a contract does not have to include a consideration and could be a gratuitous promise. It has the advantage of a limitation period of twelve years from the date on which the cause of action accrues instead of the six year limit for simple contracts.[3.7] For this reason, it is extensively used in construction contracts and almost exclusively in international construction contracts.

3.5.2 Limitation periods

In contract, the cause of action accrues on the date of the breach of contract and it is independent of the date of any damage which may have occurred. In construction contracts, however, the date of the breach may or may not be that date when the act causing the breach was committed, since it could be extendible to the date of practical completion of the works. This extension depends on the nature of the breach and on the circumstances of the case.

By comparison, under the law of tort, the cause of action accrues only when damage is suffered, so that the cause of action may not arise until long after the relevant work was carried out. The periods of time which limit an action under the law of tort vary from one country to another, even within the common law group. In England, the Latent Damage Act 1986 stipulates the following periods of limitation in a case of tortious negligence (other than for personal injury or death), as either:

(a) six years from the date on which the cause of action accrued; or
(b) three years from the starting date, if that period expires later than the period set out in (a) above.

The starting date is the earliest date on which the plaintiff or any person in whom the cause of action was vested before him, first had both the knowledge required for bringing an action for damages in respect of the relevant damage and a right to bring such an action.[3.8]

However, it is important to note that both under contract and tort, an exception to the normal periods of limitation exists in certain situations. For example, under English law, fraud, deliberate concealment and mistake extend the period of limitation so that time does not begin to run against the plaintiff until he discovers or ought to have discovered the fraud. Section 32 of the Limitation Act 1980 which replaced section 26 of the 1939 Limitation Act deals with this exception.[3.9] The operation of this exception is well illustrated in the case of *Applegate* v. *Moss* (1971). Briefly the circumstances were as follows:

> 'By a contract made in February 1957, the defendant agreed to build two houses for the plaintiffs and to support them on a raft foundation reinforced with a steel network of a specified type. He employed a Mr. Piper, an independent contractor, to do the work. The plaintiffs went into occupation of the houses when they were completed towards the end of 1957. In 1965, it was observed that, owing to the defective manner in which the foundation had been constructed, the houses were irreparable and unsafe for habitation. There was no raft, the reinforcement was grossly inferior to that specified and wide cracks had appeared beneath the houses.

The plaintiffs claimed damages for breach of contract. Despite the fact that their action was brought more than six years after the breach of contract they succeeded on the ground that there had been concealment within the meaning of section 26 of the Limitation Act 1939. "The builder put in rubbishy foundations and then covered them up".'[3.10]

Although section 32 of the Limitation Act 1980 involves some changes to section 26 of the Limitation Act 1939 referred to in the case of *Applegate* v. *Moss*, its effect is substantially unchanged.

Attention is drawn to the fact that, under English law, the right to claim beyond the end of the limitation period is not extinguished. It is simply barred by the relevant statutory provisions if the defendant successfully pleads that the cause of action arose outside the limitation period. If he does not plead this defence, no court or arbitrator can take the point for him. On the other hand, the situation is totally different outside the common law group where the right to sue is extinguished once the period of limitation is passed.

3.6 Privity of contract

With few exceptions, no one may be entitled to the benefits or be bound by the obligations of a contract to which he is not a party. An example of an exception to this rule which may apply to a construction contract, is an assignment, although of course only the benefits can be assigned, not the obligations unless the parties agree otherwise.

The contract between a promoter and a contractor cannot therefore extend to the engineer. Similarly, the contract between a promoter and an engineer cannot extend to the contractor. However, the terms of these two separate contracts would obviously be of interest to these three parties.

3.7 Performance of a contract

When the parties to a contract perform the obligations undertaken by its content, the contract is then discharged through performance. In a construction contract, performance of the contract means, on the one hand, completion of the work as well as any matter relating to the obligations of the contractor for maintenance of defects and, on the other hand, payment by the employer. The employer, however, does in general retain his rights to sue for 'breach of contract' should a latent defect be discovered within the period of limitation specified under the relevant statute, unless the contract provides otherwise.

There are three other ways to discharge a contract and these are:

(a) By agreement of the parties, a contract may be varied or discharged through a release by a subsequent contract under seal, or by accord or through a waiver, or rescission.

(b) Through the doctrine of frustration in circumstances where an event of a fundamental nature and beyond the control of either of the parties occurs which renders the contract's performance impossible or of a totally different nature than what was originally intended by the parties, then the contract is regarded as having been frustrated. The event cannot be one which has been catered for in the contract and neither can it be one which simply makes the performance of the contract more difficult or expensive.

In a recent case heard by the Judicial Committee of the Privy Council on appeal from the Court of Appeal of Hong Kong, a landslip occurred taking with it a block of flats of 13 storeys as well as hundreds of tonnes of earth. They all landed on the site of partly completed buildings, totally obliterating them. It was accepted that the landslip was an unforeseen natural disaster and it was uncertain as to whether the partly completed contract could ever be completed, and even if it could, it was uncertain when it could be completed. This was held to be a frustrating event.[3.11]

(c) By a breach of a primary obligation by one party, the contract may become discharged if the injured party decides to treat the breach as a repudiation of the contract, although damages may be sought by that injured party. This is an extensive and complex topic which cannot be given its due consideration in the space available here. Reference should be made to Chapters 18 and 19 of *Law of Contract* (see Reference 3.9).

3.8 The contents of a contract

The contents of a contract depend on the intention of the parties at the time the contract was entered into, or to be more accurate, what a reasonable man would judge on the evidence available to have been the intention of the parties.

It is important to bear in mind the distinction between what the parties intended and the evidence available to prove what it was they intended. If it is clear that a contract actually exists then the following questions must be answered before the intentions of the parties and, hence the scope of the contract, can be determined:

(a) What did the parties say or write?
(b) In making such oral or written statements, did the parties intend to create obligations and rights?
(c) If the parties intended to create rights and obligations, what is their relative importance?
(d) Are there any matters which constitute terms of the contract by reason of their being implied as such by statute, custom or the courts?

Not all statements (in whatever form) made by the parties become terms of the contract as some are what are called 'mere representations'. The question as to which of these applies can only be answered by repeating the question: 'What was the intention of the parties?'

For a representation or statement to become a term of the contract, the statement must have been made either simultaneously with the close of the bargain or at some earlier moment, and the parties must have intended that it would become a part of the bargain. There are tests that may be useful when trying to decipher whether or not a term is part of a contract. These are:

(a) At what stage of the transaction was the crucial statement made?
(b) Was the oral statement followed by a reduction of the terms to writing?
(c) Had the person who made the statement special knowledge or skill as compared with the other party?

It is important to note that the above criteria are mere aids to find out the exact terms of the contract.[3.12]

Again attention must be drawn to the difference between the situation which the parties actually intended and the evidence available to prove what their intention was.

Implied terms constitute an exception to what has just been said since, by their very nature, they do not depend on the intention of the parties but are imported into the contract. It is sometimes possible to identify such implications from the nature or wording of the terms of the contract, but more often than not specialist knowledge is required to identify situations where they arise. Such situations include:

(a) Any prevailing *trade or professional or local custom*, unless a contrary intention is shown in the contract. However, custom is not imported into a contract to contradict an express term therein, but to serve to reinforce it and assist its general purpose. Thus:

> 'An alleged custom can be incorporated into a contract only if there is nothing in the express or necessarily implied terms of the contract to prevent such inclusion and, further, a custom will only be imported into a contract where it can be so imported consistently with the tenor of the document as a whole.'[3.13]

(b) *Statute* which increasingly plays an important role in consumer protection.
(c) The *courts*, where there is no customary or statutory authority for an implied term. In this case, the courts may imply a term which is needed to give business efficacy to the contract. However, it is important to note that the courts would not import a term into a contract unless it is inevitable and absolutely essential in order to make the contract effective.

In this connection, reference is made to the case of *Tai Hing* v. *Liu Chong Hing Bank* (1986).[3.14] The courts will not make a contract workable or improve its terms. Many cases may be quoted in this connection but two are chosen here to provide an easy understanding of the principles: the *Moorcock* case in 1889[3.15] and *Trollope & Colls Ltd* v. *North-West Metropolitan Regional Hospital Board* in 1973.[3.16]

In the *Moorcock* case, the defendants owned a jetty and wharf and agreed to allow the plaintiff, a shipowner, to discharge his vessel at their jetty which extended into the River Thames. The ship was to be moored alongside the jetty and to be discharged and loaded at the wharf. It was understood by the parties that at low tide the ship would rest on the riverbed, but the defendants had no control over the bed of the river, and had taken no steps to ascertain whether it was or was not a safe place for the vessel to lie. They gave no warranty as to the safety of the area for ships. At low tide, the ship settled on a ridge of hard ground beneath the mud and was damaged. The plaintiff sued for the resultant damage.

The court held that there was an implied warranty that the area was suitable for mooring and, therefore, the plaintiff could recover damages. The test which was applied to settle whether a term should be implied is known as the 'officious bystander test'. This means that if at the time when two parties to a contract are negotiating its terms, a bystander asks: 'What will happen in such and such a case?', then both parties would reply: 'Of course so and so will happen; it is too clear to say that.'[3.17]

In the *Moorcock* case, Lord Justice Bowen explained the nature of the implied term as follows:

'I believe if one were to take all the cases, and they are many, of implied warranties or covenants in law, it will be found that in all of them the law is raising an implication from the presumed intention of the parties with the object of giving to the transaction such efficacy as both parties must have intended that at all events it should have. In business transactions such as this, what the law desires to effect by the implication is to give such business efficacy to the transaction as must have been intended at all events by both parties who are business men ... The question is what inference is to be drawn where the parties are dealing with each other on the assumption that the negotiations are to have some fruit, and where they say nothing about the burden of this unseen peril, leaving the law to raise such inferences as are reasonable from the very nature of the transaction.'

Since 1889, the authority of the *Moorcock* case has been invoked on many occasions, but where construction is concerned, the words of Lord Pearson quoted below from the case of *Trollope & Colls Ltd* v. *North-West Metropolitan Hospital Board* (1973), are most apt. In that case, the House of Lords refused to imply a term to make the building contract more workable. The work was to be carried out in three phases, governed by different conditions.

The contract provided that the work in phase III was to be completed by a set date, but it did not include a term for the consequence of the work in phase I being delayed. In the event, the work in phase I was delayed and there was a dispute about whether or not the date for completion of phase III should be amended. Lord Pearson stated:

'The court does not make a contract for the parties. The court will not even improve the contract which the parties have made for themselves, however desirable the improvement might be. The court's function is to interpret and apply the contract which the parties have made for themselves. If the express terms are perfectly clear and free from ambiguity, there is no choice to be made between different possible meanings: the clear terms must be applied even if the court thinks some other terms would have been more suitable. An unexpressed term can be implied if and only if the court finds that the parties must have intended that term to form part of their contract. It is not enough for the court to find that such a term would have been adopted by the parties as reasonable men if it had been suggested to them: it must have been a term that went without saying, a term necessary to give business efficacy to the contract, a term which, though tacit, formed part of the contract which the parties made for themselves.'

In the context of standard forms of contract, it is important to realise that terms which have been implied in decided cases may be considered as precedents for similar contracts, thus establishing a contractual duty under the operation of the common law. Such implied contractual duties in construction contracts under the common law include, but are not limited to, the following:

(a) the employer must give possession of the site within a reasonable time;
(b) the employer or his agents must give any necessary instruction and information and have the design completed to enable the contractor to achieve completion of the works by the contract completion date;
(c) the contractor must carry out his work with proper skill and care and with materials of good quality and fit for this purpose.

Similarly, where it has been decided that certain duties or obligations cannot be implied in a particular type of contract, a precedent would be established for the non-existence of an implied duty in that context. For instance, the owner does not impliedly undertake an obligation with regard to the nature or suitability of the site or subsoil or the design.

3.9 Remedies for breach of contract

The concept of law, where a contract is concerned, is embodied in what ought or ought not to be done rather than in what will or will not be done. Whilst what will or will not be done cannot be made certain by the application of law, what ought or ought not to be done is secured through a back-up of sanctions.[3.18] Therefore, if an obligation under the contract is not fulfilled, sanctions in the form of damages or equitable remedies will be applied (see below).

This approach is different from that under some jurisdictions within the Romano-Germanic system of law where the performance of the contract has to be actually done, that is, there is a legal requirement that what was promised to be done will be done.[3.19]

The chart in Figure 3.1 provides a schematic analysis of the subject of remedies for breach of contract.

Under English law, where there is a breach of contract, an injured party may seek any or a combination of the following remedies:

(a) Remedies in the form of damages.
(b) Equitable remedies in the form of:

Legal Concepts Based on the Common Law System

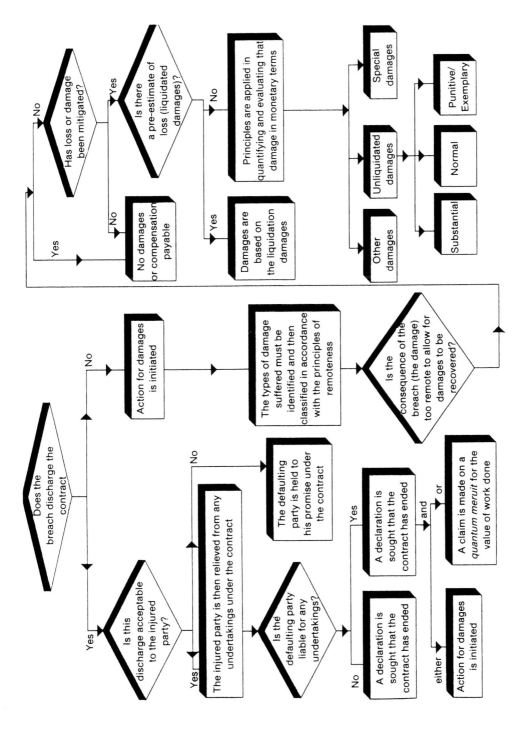

Fig. 3.1 Remedies for breach of contract in the common law system.

(i) *Specific performance*

Specific performance is defined in *Law of Contract* (see Reference 3.9), page 661, as a decree issued by the court which constrains a contracting party to do that which he has promised to do. This remedy is rarely available in construction disputes because courts will not enforce a contract requiring constant supervision.

(ii) *Injunction*

Injunctions are court orders compelling or restraining a specific act.

Like all equitable remedies injunctions are discretionary and the courts usually grant them only if they are satisfied, following certain established principles, that justice or convenience requires it. Injunctions may be of three kinds: interim, interlocutory and perpetual.

(a) Interim injunctions

In cases of real emergency, application can be made to the court for injunction on an *ex parte* basis which means that the person against whom injunction is sought is not on notice of or aware of the application. In granting such an injunction, the court will invariably require an undertaking from the applicant to pay any loss or damage suffered by the party against whom the injunction is granted should the applicant's claim ultimately fail. An interim injunction is normally granted merely to maintain the *status quo* and will only be valid for such time as would be necessary to obtain an interlocutory injunction.

(b) Interlocutory injunctions

Where there is no real emergency but some interim relief is required pending the full trial of the issue, an interlocutory injunction may be obtained which essentially gives the same relief as an interim injunction, except that the party against whom the relief is sought is put on notice of the claim and his *prima facie* defence to the claim is heard. As in the case of an interim injunction a court, in granting an interlocutory injunction, will generally require an undertaking as to damages from the applicant.

(c) Perpetual injunctions

Following the full trial of the issue, the court may make absolute the interlocutory injunction already granted. In doing so, it may grant one of the following equitable remedies:

(a) *Rectification*

Rectification is a remedy in the case of a mistake resulting in the terms of contract being incorrectly expressed.

(b) *Rescission*

Rescission is:

'the cutting down or terminating of a contract by the parties or one of them. It may be done by agreement, or by one party who is entitled to do so by reason of the repudiation or material default of another or by reason of the contract having been induced by fraud or misrepresentation by the other party. It is effected by taking proceedings to have the contract judicially set aside, or by giving notice to the other party of intention to treat the contract as at an end.'[3.20]

(c) *Remedies in contract and under quasi-contractual relationships* through a claim on the basis of a *'quantum meruit'*.

Quantum meruit is the basis for a claim or a remedy sought for the payment of a reasonable remuneration for work done or services rendered. It arises where:

(a) a contractual agreement exists but either:
    (i) a precise price has not been agreed; or
    (ii) conditions agreed as a basis for not charging for certain work do not eventually materialise; or
    (iii) an original contract has been replaced by another and payment is sought for work done under this new contract.
(b) for one reason or another, no contract exists. This situation could arise where:
    (i) there has never been a contract; or
    (ii) a contract could have existed but is discharged and no new contract is substituted; or
    (iii) there is a presumption of the existence of a contract but this is later found to be without legal validity.

Where there is a breach of the contract, the type of action to be followed by an injured party would depend on whether or not the contract becomes discharged as a result of the breach (see Figure 3.1). If the contract becomes discharged, then the injured party has the choice of either holding the defaulting party to his promise under the contract or accepting the contract as discharged. In the former case, the injured party remains liable for his own undertakings under the contract. If the latter course is chosen, then the consequences are:

(a) The injured party is relieved from the undertaking to perform his obligations and may seek to have a declaration that the contract is terminated.
(b) The injured party may either sue for the recovery of damages or claim on a *quantum meruit* basis for the value of any work already carried out.

If the breach does not discharge the contract, then an action for damages would raise two issues – the first is the extent of the loss suffered and the second is the measure of the damages (see Figure 3.1). In this connection, the difference between the two words 'damages' and 'damage' should be fully recognised to avoid any confusion in what is intended. The fundamental principle of establishing the extent of loss for which damages may be claimed must be distinguished from the assessment of monetary compensation. The type and extent of loss which may be compensated was first defined in the case of *Hadley* v. *Baxendale* by Baron Alderson as follows:

> 'Where two parties have made a contract, which one of them has broken, the damages which the other party ought to receive in respect of such a breach should be such as may fairly and reasonably be considered either:
>
> (a) arising naturally, i.e. according to the usual course of things from such a breach itself, or
> (b) such as may be supposed to have been in the contemplation of both parties at the time they made the contract as the probable result of the breach of it.'[3.21]

The facts of the case were summarised in *Law of Contract* as follows:[3.22]

> 'The mill of the plaintiffs at Gloucester was brought to a standstill by a broken crank shaft and it became necessary to send the shaft to the makers at Greenwich as a pattern for a new one. The defendant, a common carrier, promised to deliver it at Greenwich on the following day. Owing to his neglect, it was unduly delayed in transit, with the result that the mill remained idle for longer than it would have done had there been no breach of the contract of carriage. The plaintiffs, therefore, claimed to recover damages for the loss of profit caused by the delay.
>
> The evidence of the parties was conflicting, but the Court of Exchequer considered the case on the footing that the only information given to the carrier was that the article to be carried was the broken shaft of a mill and that the plaintiffs were the millers of that mill.
>
> It was obvious that the failure of the carrier to perform the contract punctually was the direct cause of the stoppage of the mill for an unnecessarily long time, and, if the plaintiffs were entitled to an indemnity against all the consequences of the breach, they should have been awarded damages for the loss of profit. At the trial the jury did indeed allow the claim, but on appeal the court ordered a new trial. Alderson B demonstrated that, in accordance with the principle that he had just expressed, there were only two possible grounds upon which the plaintiffs could sustain their claim. Firstly, that in the usual course of things the work of the mill would cease altogether for the want of the shaft. This, he said, would not be the normal occurrence, for, to take only one reasonable possibility, the plaintiffs might well have had a spare shaft in reserve. Secondly, that the special circumstances were so fully disclosed that the inevitable loss of profit was made apparent to the defendant. This, however, was not the case since the only communication proved was that the article to be carried was the shaft of a mill and that the plaintiffs were the owners of the mill. The jury, therefore, should not have taken the loss of profit into consideration in their assessment of damages.
>
> The words "either", and "or", used in the formulation of the rule as explained by Alderson B, shows that it contains two branches. The first deals with the normal damage that occurs in the usual course of things; the second with abnormal damage that arises because of special or exceptional circumstances. The defendant is taken

to have contemplated both kinds of damage, but where it is abnormal only if he knew of the special circumstances at the time of the contract.'

Many cases have come before the courts since then where the above two rules were debated, explained and reiterated. The words 'in the contemplation of both parties' were considered in these cases in the context of reasonable foreseeability and probability of occurrence but the principle of remoteness still holds[3.23] and the two important questions in a claim for damages are still:

(a) for what extent of loss is the claimant entitled to recover compensation? and
(b) what principle should be applied in evaluating or quantifying that damage in monetary terms?

In respect of the second question, the courts in England have tended to apply the principle that 'If the plaintiff has suffered damage that is not too remote, he must, as far as money can do it, be restored to the position he would have been in had that particular damage not occurred.'[3.24]

In this connection, it is necessary to remember that English law imposes a duty on all parties to mitigate the loss or damage resulting from the breach. The courts, therefore, in assessing the damages to be awarded, would examine the measures, if any, which could or ought reasonably to have been implemented to reduce the loss or damage. If the injured person is considered to have failed to mitigate the loss, such failure would be taken into consideration in the assessment of damages to be awarded.

The measure of damages is based on whether or not an assessment of the damages to be paid has been included in the contract in case of a breach. Thus, the parties may foresee the possibility of a particular breach and accordingly may stipulate a genuine pre-estimate of a certain sum of money to be paid as compensation if a breach actually takes place. This is called liquidated damages and, provided it is a true and genuine pre-estimate, the claimant is entitled to it without having to prove that he has in fact sustained a loss or damage; see later in Section 17.5.

If, on the other hand, the sum stipulated is not a true and genuine pre-estimate of the loss or damage which would result from the breach, the damages would then be considered as a penalty. Under English law, a penalty is subject to the rules of equity.[3.25]

If the damages are not fixed in the contract, then they are referred to as unliquidated damages and an arbitrator or the court would be responsible for assessing their value as compensation for actual loss suffered subject to the principle of remoteness discussed earlier in this chapter. They may be in the form of:

(a) substantial damages which put the claimant in the same situation as he would have been if the contract had been performed; or
(b) nominal damages which apply if no loss is suffered and represent only a token sum awarded for the infringement of a contractual right.

Apart from the remedies for breach of contract, it is worth mentioning exemplary or punitive damages which are awarded in certain tort cases. The purpose of awarding this type of damages is not to compensate the plaintiff, nor even to strip the defendant of his profit, but to express the court's disapproval of the defendant's conduct, for example, where he has deliberately committed a wrong (such as defamation) with a view to profit.[3.26]

Damages may also be classified as *general damages* which are payable as compensation for the loss presumed to follow from a breach of contract or from a tort; and *special damages* which are payable as compensation for particular losses not presumed, but which in fact have followed in a particular case.

Special damages have to be specifically claimed and strictly proven.

3.10 Exclusion clauses

Exclusion clauses are usually, though not exclusively, found in standard forms of contract. They are used by one party to a contract in an attempt to exclude or limit a certain liability or liabilities which would otherwise be attached to that party. Because of their nature and due to the conflict between commercial and consumer interests, exclusion clauses have produced a great number of cases under English law, but only a few rules. Specialist knowledge and advice are particularly required in this area.

Where construction contracts are concerned, it is generally accepted that where there may be any doubt about the meaning and scope of the exclusion clause, the ambiguity is resolved against the party who has inserted the clause into the contract and who is relying on it.

An exclusion clause must therefore be carefully drafted and clearly expressed with precise phraseology. This is more stringently applied where the clause is attempting to exclude liability rather than limit it.

3.11 The responsibility to complete

Under common law, there is a duty on the contractor to complete a construction contract whatever happens. This duty was defined as early as 1867 in the case of *Appleby v. Myers*,[3.27] where it was held that the plaintiff contractor was not entitled to recover anything from the employer in respect of any portion of the machinery he had erected which was destroyed in a fire prior to the completion of the work.

More recently, in the case of *Charon v. Singer Sewing Machines Ltd* (1968), the defendants employed the plaintiffs to convert a shop and living accommodation. The contract included the words 'allow for covering up and protecting the works during frosty and inclement weather or from damage from any other cause and reinstating any work so damaged'. Vandals broke into the shop one day before completion of the work and caused damage which had to be repaired by the plaintiff by repeating work already carried out and paid for. The court held that the contractor had to bear the cost of repair.[3.28]

In this case, the following statement from *Hudson's Building and Engineering Contracts*, 10th Edition, was quoted with approval:[3.29]

'Indeed, by virtue of the express undertaking to complete (and in some contracts to maintain for a fixed period after completion) the contractor would be liable to carry out his work again free of charge in the event of some accidental damage occurring before completion even in the absence of any express provisions for protection of the work.'

This chapter has considered the main legal concepts of the common law system on which the Red Book is based. The next chapter briefly discusses drafting principles in relation to contract documents.

Chapter 4

Drafting Principles

'And the Lord said, "behold they are one people and they have all one language; and this is only the beginning of what they will do; and nothing that they propose to do will now be impossible for them. Come let us go down and then confuse their language that they may not understand one another's speech."'

Revised Standard Version
Common Bible, Genesis, Chapter 11

The ICE and ACE Standard Conditions of Contract upon which the Red Book was based were drafted using legal language which is expected to be so precise as to remain unequivocal even when subjected to detailed and hostile legal scrutiny. Unfortunately, the very reasons assumed to give precision tend to promote complexity and incomprehensibility, thus leading to ambiguity. The principles of English legal drafting were discussed in Chapter 2, and reference is specifically made here to the quotations given in Section 2.9.1, References 2.22, 2.24 and 2.25. Where such drafting is applied to a standard form of conditions of contract, one may look at the words of an authoritative lawyer in this field:

'As to the risks on our list that are created by lawyers and law, take legal language to start with. Lawyers are lawyers, and the lawyer/draftsmen and the lawyer/interpreter between them should make it possible to obtain satisfactory forms of construction contract with reasonable effort. The construction industry may take the view that lawyers as technicians have a job to do in producing contracts which, as interpreted by the courts, will be satisfactory in construction matters (complicated as they are by their nature and not merely because the parties make them complicated to annoy the judges) and that they are not doing this job very well. Rules of interpretation are maintained which require the draftsmen of standard forms to do the impossible – to visualise all conceivable eventualities for many thousands of contracts and spell out terms to deal with them in language which is so clear that it cannot be distorted even by a party who finds himself with an interest in distortion, and without making the document so cumbersome as to be useless to those for whose benefit it is presumably intended. When a case comes before the courts many days may be spent in analysing a sentence in a conditions of contract or specification or bill of quantities without allowing for the fact that the draftsman had to do his work with reasonable speed. The courts also must find the game more fun when it is played in the dark, since they restrict the information that may be used to clarify the meaning of the words used. One must wonder precisely what function the courts are fulfilling by interpreting a standard form document of great importance to a complex industry in a legalistic way.

The result is that when a standard form or a special term for a contract has to be drafted, much time and money is spent employing lawyers to translate it into legal language, and then employing other lawyers to translate it back again when the users want to know what it means. The serious risk, often realised, is that much is lost, distorted or overlooked in the process of translation and retranslation.'[4.1]

However, even at its best, language as a means of communication has its limitations. For example, it is accepted that comprehension of sentences read for the first time, depends mainly on the length of a sentence, the number of ideas expressed in it, the complexity of these ideas and the intelligence quotient (IQ), of the reader. Long sentences often contain complex ideas that would be more clearly expressed in several sentences. Whilst sentences of around 18 words are direct and can be readily understood, informative and technical text may require sentences of 35 to 45 words for the expression of a complex idea or a number of ideas. It is calculated that sentences with 28 words or more, representing a multiplicity of statements read for the first time can only be easily and readily understood by four per cent of the population (equivalent to an IQ of 130 and over).[4.2]

The length of paragraphs and the connection between them are also very important in presenting to the reader a simple, logical and clear text.[4.3]

With repetition, this limit in the number of words in a sentence increases of course but is still far below the average number of words per sentence in the two standard forms of contract referred to above. Thus, complaints about the complexity of the legal language used in these forms have been made even by lawyers and judges who are repeatedly exposed to them.[4.4]

In international contracts, other difficulties arise in the use of languages as a means of communication, and these are:

(a) Various languages have their own unique peculiarities, for example: vowels are not written in Arabic; definite and indefinite articles are not used in Russian; some words in English are not spelt as pronounced and can be equivocal in their meaning; and so on.
(b) Words used in normal circumstances may have different meanings when used in the legal sense, for example, the words 'error' and 'mistake'.
(c) Some words in English, when used in the legal sense, impose obligations unfamiliar to the layman, for example, the use of 'may' versus 'will' versus 'shall'.[4.5]
(d) The sequence of words in some languages is a determinative factor in their meaning. For instance the words: 'I wish him to survive' have a totally different meaning from 'I wish to survive him' despite the fact that the same words are used in both statements.

The relevance of the provision of language has been recognised from the early days of civilisation. Confucius said:

'If language is not correct, then what is said is not what is meant, if what is said is not what is meant, then what ought to be done remains undone. If this remains undone, morals and arts will deteriorate. If morals and arts deteriorate, justice will go astray. If justice goes astray the people stand about in helpless confusion. Hence there must be no arbitrariness in what is said. This matters above everything.'

But, in order to formulate a fair contract, it is not sufficient to use simple, logical and clear wording. What is equally important, is to recognise the different interpretations which might be placed on that wording, not only by the average person but also by a clever advocate who would attempt to argue for an interpretation suitable to his case.

History has demonstrated the ingenuity of mankind to produce different interpretations to a particular set of words. A masterful demonstration of how useful a strict construction of a set of words can be was made in Shakespeare's *Merchant of Venice*. Mr. Justice Niall McCarthy referred to it in the following manner:

'Portia opted for the strict construction –
"Therefore prepare thee to cut off the flesh,
Shed thou no blood; nor cut thou less, nor more,
But just a pound of flesh: if thou tak'st more
Or less than a just pound – be it but so much
As makes it light or heavy in the substance,
Or the division of the twentieth part
Of one poor scruple: nay, if the scale do turn
But in the estimation of a hair –
Thou diest and all thy goods are confiscate."

(Act IV, Scene 1)

The arbitrator, the Duke, never made any ruling on the true construction of the bond; the unfortunate Shylock, an early victim of the lawyers' art, not merely lost his claim for interest – for principal – but was damnified for having dared to sue.'[4.6]

However, Shakespeare's plot is capable of a different interpretation as was demonstrated in the following statement:

'Shakespeare illustrated in fiction that the terms of a contract which are repugnant to nature can be interpreted, with sufficient ingenuity, to produce a just result even to the point that the party seeking to impose and enforce oppressive terms might be deprived of all benefit.'[4.7]

Faced with such drafting difficulties as described above, there have been various calls to simplify the language and the wording of the Red Book.[4.8] Although considerable change in that direction has been achieved in the most recent revision and its reprints, the original concept was retained.

Chapter 5

The Concept of a Trusted Independent Engineer

5.1 Introduction

This chapter considers the concept of a trusted, independent engineer which is central to a contract using the Red Book, under the headings of the qualities of a consulting engineer, the services provided by him and his independence. The definition, qualities and independence of a consulting engineer are outlined in the Statutes and By-Laws of FIDIC which were first published in 1955.

The initiator of a civil engineering project may be a government agency, municipal authority, public corporation, semi-state organisation, incorporated company, a group of individuals or a private person.

As in the ICE Form of Contract for civil engineering works, the Red Book is drafted on the basis that when a project is initiated by a developer or a promoter, the duties related to feasibility, design and supervision during construction of the project are entrusted to an independent consulting engineer who is referred to as the 'Engineer' in the contractual arrangements between the developer and the contractor. The developer of such a project is referred to in these forms as the 'Employer'.

5.1.1 FIDIC's Statutes and By-Laws and the independent engineer

FIDIC's Statutes and By-Laws have been kept under constant scrutiny and review in order to react promptly to, and sometimes be proactive in, international developments in the construction industry and the manner in which construction projects are procured and executed. Thus, the definitions of independence and of a consulting engineer have developed over the years.

Under Article 2 of *FIDIC's Statutes and By-Laws* the objectives of the FIDIC organisation as a whole have been set out.[5.1]

'The objectives of the Federation are to:
(1) Represent consulting engineering firms globally
(2) Enhance the image of consulting engineers
(3) Be the authority on issues relating to business practice
(4) Promote the development of a global and viable consulting engineering industry
(5) Promote quality
(6) Actively promote conformance to a code of ethics and to business integrity
(7) Promote commitment to sustainable development.'[5.2]

FIDIC's main objective is, therefore, to guarantee that it is the worldwide representative of the construction industry, and in acting as such it can ensure that the position of the consulting engineer is understood to be the primary provider of the design of infrastructure. FIDIC has continuously developed its different forms of contract in order to strengthen the business interests of its member associations.

The third edition of the *Guide to the Use of Independent Consultants for Engineering Services* published by FIDIC in 1980 defined a consulting engineer as follows:

> 'The consulting engineer is a professional engineer in private practice. He maintains his own engineering office either alone or in association with other engineers. He employs the necessary staff to assist in carrying out the services which he provides. His organisation may be that of a sole proprietorship, a partnership or a company. This depends on the type and magnitude of his operations and the conditions of practice set by his national association. He must carry out his practice on a highly ethical professional basis. The technical knowledge, experience and ability of the consultant, his associates and assistants must be fully adequate for the projects undertaken.'[5.3]

Until October 1996, Article 3(2) of FIDIC's Statutes and By-laws set out the following three principles relating to consulting engineers. These are professional status, independence and competence.

> '(1) Professional Status
> In the exercise of his Profession a Consulting Engineer must act in the legitimate interest of his Client. He must discharge his duties with complete fidelity and conduct himself in such a manner as faithfully to serve the best interest of Society and to uphold the standing and reputation of the profession.
>
> (2) Independence
> A Consulting Engineer's professional advice, judgement or decision must not be influenced in any way by a connection with another person or organisation.
> Control of the policies and management of a Consulting Engineering Firm which is partly or wholly owned by entities other than Consulting Engineers shall be vested in Consulting Engineers.
> The remuneration of a Consulting Engineer for professional services should be derived only from fees paid by Clients, with the exception that a Consulting Engineer may nevertheless also benefit from equity participation in ad hoc groups formed with others to undertake design and construct, project management or similar activities.
>
> (3) Competence
> A Consulting Engineer must have the necessary knowledge and experience to enable him to fulfil his mission.'[5.4]

Article 25 of these Statutes and By-Laws provided that the words 'Consulting Engineer(s)' are deemed to include individual members or member firms of Member Associations of FIDIC.

In October 1996 significant changes were introduced into FIDIC's Statutes and By-laws leading to the following definition of a consulting engineer being incorporated in Article 25 of the 1996 Edition:

> 'The words "Consulting Engineer" shall define persons who are individual members of a Member Association, or are qualified practising professionals principally occupied as advisers on engineering and related matters, with the responsibility for the management of member firms of a Member Association, or persons who have recently held such positions before their professional retirement.'

Article 3(2) was revised again in October 2002 by the following text which represents a fundamental development of the concepts of professional conduct, status and independence of a consulting engineer:

> 'To qualify for membership of the Federation an Association must demonstrate that its Statutes, By-Laws and regulations ensure that its members comply with the FIDIC Statutes, By-Laws and Code of Ethics.

In accordance with Article 3 of FIDIC's Statutes and By-Laws, published in 2002, the following criteria should be met to be eligible for FIDIC membership:

(a) Members of FIDIC shall only be Associations representing suitably qualified and experienced individuals and firms who derive a substantial portion of their income from the provision of impartial consulting services to a client for a fee.
(b) The members of the Member Associations should be faced with the pressures and discipline of the relevant consultancy market. Members shall:
  (i) be properly constituted as business enterprises or employees of such enterprises;
  (ii) operate as financially independent entities, free from subsidies or preferences; and
  (iii) subscribe to the FIDIC principles and policies.
(c) FIDIC strives for a high standard of competence and professional performance of its membership. Member Associations in their statutes shall define the professional qualifications and experience which are required as a prerequisite for membership.
(d) Member Associations may require stricter membership criteria which exceed the basic requirements set out by FIDIC.
(e) In addition to the full (voting) membership of the Federation, there shall be other non-voting classifications of membership as follows:
  (i) Honorary membership: A person who has rendered outstanding service to the Federation may be elected as an Honorary Member by the General Assembly.
  (ii) Sustaining membership: Any firm (or group of firms) from a country not having a Member Association, which meet(s) all of the Federation's membership criteria.
  (iii) Affiliate membership: Any association, organisation or group which supports the objectives of the Federation may be approved as an Affiliate Member by the General Assembly.

5.1.2 FIDIC's Code of Ethics

The Statutes and By-laws of FIDIC also refer to a Code of Ethics.[5.5] This code focuses on the principles which are fundamental to the behaviour of a consulting engineer if society is to have the necessary confidence in its advisers. It specifies that the consulting engineer shall:

'Responsibility to Society and the Consulting Industry
1. Accept the Responsibility of the consulting industry to society.
2. Seek solutions that are compatible with the principles of sustainable development.
3. At all times uphold the dignity, standing and reputation of the consulting industry.

Competence
4. Maintain knowledge and skills at levels consistent with development in technology, legislation and management, and apply due skill, care and diligence in the services rendered to the client.
5. Perform services only when competent to perform them.

Integrity
6. Act at all times in the legitimate interest of the client and provide all services with integrity and faithfulness.

Impartiality
7. Be impartial in the provision of professional advice, judgement or decision.
8. Inform the client of any potential conflict of interest that might arise in the performance of services to the client.
9. Not accept remuneration which prejudices independent judgement.

Fairness to others
10. Promote the concept of "Quality-Based Selection" (QBS).
11. Neither carelessly nor intentionally do anything to injure the reputation or business of others.
12. Neither directly nor indirectly attempt to take the place of another consulting engineer, already appointed for a specific work.
13. Not take over the work of another consulting engineer before notifying the consulting engineer in question, and without being advised in writing by the client of the termination of the prior appointment for that work.
14. In the event of being asked to review the work of another, behave in accordance with appropriate conduct and courtesy.

Corruption
15. Neither offer nor accept remuneration of any kind which in perception or in effect either a) seeks to influence the process of selection or compensation of consulting engineers and/or their clients or b) seeks to affect the consulting engineer's impartial judgement. In this connection it is worth mentioning that FIDIC publishes a *Policy Statement on Business Integrity* which provides the view of the profession on this topic.[5.6]
16. Co-operate fully with any legitimately constituted investigative body which makes inquiry into the administration of any contract for services or construction.'

5.1.3 FIDIC's Quality-Based Selection, 'QBS'

In 1997, FIDIC published its document on Quality-Based Selection, referred to in item 10 of its Code of Ethics, in the form of a brochure. It sets out to help all those responsible for the selection of consultants to achieve that objective on the basis of qualifications, experience, ability and integrity.[5.7]

This document was replaced in 2003 by a more thorough statement, when FIDIC published the first edition of its *Guidelines for the Selection of Consultants*. The new Guidelines examine quality selection in a more detailed manner and propose various methods of selection that are appropriate to the developing scene of the procurement of engineering services. The selection of a consultant is one of the most important decisions an owner or client will make, and the success of any project will largely depend on the professional qualities of the consultant.[5.8] The tendency has been in the recent past to select consultants on the basis of the simplistic criterion of price alone with disastrous consequences to the project, in terms of quality, cost of the project, maintenance cost and time of construction. The compulsion towards price competition seems to have stemmed from a number of factors, which should be rejected if the aim is a successful project.[5.9] It is worthwhile in this connection to quote from a paper read at the 1983 FIDIC Conference in Florence, by Mr Norman Westerberg, a Finnish consulting engineer, in which he quoted a statement by Rear-Admiral D.C. Iselin, the Commander of the US Navy's Naval Facilities Engineering Command, to the US Sub-Committee on Military Construction in 1978.[5.10]

> 'In a typical project, we find that outfitting, operating, maintenance, and repair costs represent 56% of the life cycle cost, construction costs represent 42%, and the design cost represents approximately 2%. This relatively modest cost, notwithstanding the A–E design effort, has critical influence on both the 42% for construction costs, and especially the 56% for operations, maintenance, and repair. It is vitally important that we get the highest possible technical quality in the design effort. In my professional opinion, any proposal which seeks to reap a near term saving by reduction in design costs, but which increases the risk of diminished technical quality of the design effort, is shortsighted in the extreme. We will live with the cost of impacts of that diminished technical quality for the full economic life of the facility: this concern is a cornerstone of my opposition to price competition.'

FIDIC lays down seven criteria that they recommend be used when an owner/client is selecting a consultant. These are:

(a) professional competence;
(b) managerial ability;
(c) availability of resources;
(d) impartiality;
(e) fairness of fee structure;
(f) professional integrity; and
(g) quality assurance system.

The Guidelines set down a procedure that may be followed during the selection process, which includes drafting a terms of reference document that details the extent and the scope of the project and any resource requirements. The owner/client is also advised to make a list of consultancy firms that they feel would be suitably qualified to carry out the required services, and then make a short list of those firms that are best qualified for the task. The owner/client should then invite the consultancy firms on the short list to submit a proposal for the project, and once these are received, they should be reviewed with the utmost care and attention. FIDIC discourages owners/clients basing their selection of a consultant purely on price. They state that:

> 'it is neither in the best interest of the client nor of the project itself that consultants be selected on the basis of a procurement system which includes a price comparison of their professional services. Once a price is introduced the selection process becomes biased in favour of lowest fee rather than quality.'[5.11]

They suggest five basic methods of selection where price will only play a part in the decision-making process for selection. These are the two-envelope method,[5.12] the cost-weighted method,[5.13] the budget method,[5.14] design competition,[5.15] or price negotiation.[5.16] The system of QBS is followed by numerous institutions throughout the world, in Australia, Asia, Canada, Germany, Japan and the United States.[5.17]

The QBS system will have great influence over the success of a project. The system delivers the best quality for money, in that the selection of a consultant based on his capabilities and professional experience will ultimately prove to be best value for the client. The owner/client and consultant relationship is strongly successful from the very beginning of the project, and the fact that there is a mutually agreed scope of services from the outset will also ease any tensions that may arise later on during the project.

5.2 Other suppliers of consulting services

The Statutes and By-Laws of FIDIC exclude some other suppliers of services unless they are individual members of a member association of FIDIC, or are qualified practising professionals principally occupied as advisers on engineering and related matters, with the responsibility for the management of member firms of a member association, or persons who have recently held such a position before professional retirement. The following are examples of these excluded suppliers:

— In-house departments within the employer's organisation;
— Government agencies or organisations;
— Contracting firms;
— Supply firms;
— Public corporations;
— Non-profit making organisations;
— Universities and research institutions;

— Any non-independent enterprise with an interest in or possible benefit drawn from matters which include amongst others, the determination of design standards; choice and acceptability of materials; and the adoption of methods of ensuring a saving in essential costs.

5.3 Services provided by the consulting engineer

The services provided by the consulting engineer may be one or a combination of any of the following: counselling, pre-investment studies, design and preparation of documents, supervision, specialised design and development services and project management.

5.3.1 Counselling services

These services are very wide in scope and can be anything from the provision of advice on the identification of resources, project identification and quantification of objectives to advising on a particular case or situation. Such services may require the involvement of an individual consulting engineer, or a small group of specialists qualified to deal with the particular task.

5.3.2 Pre-investment studies

Pre-investment studies are perhaps the most critical of all the services provided by the consulting engineer as they determine the investment policy of the promoter as well as the feasibility and basic features of individual projects which could be suitable for the particular situation. The promoter usually initiates the process of procurement of a particular project knowing in general the objectives for which it is required.

However, it is the pre-investment study that determines whether this project is feasible or not, economically viable or not, environmentally acceptable or not, politically constrained or not, physically workable in one detail or another or not at all and, finally, sustainable or not and at what cost.

At this stage, the base of knowledge required to provide such service is extremely wide and the team composition may include engineers, planners, economists, sociologists, ecologists and other specialists. Various alternative solutions are normally presented to the particular problem under study with commentary on advantages and disadvantages of each option. It is clear that such a study has to be totally independent and free from any bias or influence, be it commercial, technical, political, financial or of any other kind.

In some cases, this service may be divided into two stages. The first entails the pre-feasibility study during which the consulting engineer investigates whether or not it is worthwhile to proceed to an advanced stage of a feasibility study. It is extremely difficult to define the boundary between these two stages without reference to a particular project.

5.3.3 Design, preparation of documents and supervision

Engineering and detailed design, preparation of contract documents and supervision services during project implementation are all services provided by the consulting engineer. Engineering and detailed design services should normally be followed by supervision services during the implementation period. The same consulting engineer should normally provide both services as it is a mistake to employ different professionals for these two interrelated services. There are many reasons for this conclusion, the most important of which is perhaps the avoidance of any doubt about the responsibility for the engineering and other services provided.[5.18] The documents showing the essential engineering input into a project are usually based on an admeasurement contract. An admeasurement contract is one where bills of quantities are used to show the quantities of work to be carried out for the price quoted and whilst the contractor remains liable to complete all the work shown on the drawings and specifications, the quantities are remeasured at the end of the contract to recalculate the contract price, whether or not variations had been ordered. In addition to any instructions given to tenderers, the documentation supplied to them usually comprises the following parts:

(a) form of tender;
(b) general conditions of contract;
(c) specification and description of the works prepared in conjunction with the design;
(d) bill of quantities;
(e) drawings; and
(f) form of agreement.

Although the design of the permanent works is expected to be carried out by the consulting engineer, the contractor may be expected to perform some design function.

5.3.4 Specialised design and development services

The consulting engineer is responsible for work of a specialist nature. Some consulting engineers provide specialised services which encompass the provision of original designs of certain processes or items of machinery or equipment or construction artefacts. Services may extend to research and development of innovative work and assistance in registration of patents for inventions.

5.3.5 Project management

This service is identified in Reference 5.3 and is defined as:

> 'The mobilisation of multiple resources and the coordination of their activities so that work performed by each accumulates into a multi-discipline team effort to achieve the owner's objective within an agreed schedule, budget and quality.
> A Project Management team, in addition to managing engineering, procurement, construction and commissioning, will also be required to cover other significant

elements of a project, such as securing project financing, maintaining public awareness and obtaining government approvals.'

5.3.6 *Programme manager*

This function differs from project management in that it involves elements other than those relating to construction, such as financing and operation.

Other details of these services are given in Reference 5.3. Standard Forms of Agreement between the promoter (employer or owner or client) and consulting engineers are published by FIDIC.[5.19]

5.4 Independence

The appointed engineer is expected to provide his services in a strictly fiduciary manner, one of trust, which is essential not only to the whole relationship between the employer and the consulting engineer but also to that between the employer and the contractor. Although the concept of independence is developing at a fast pace and especially in respect to the role of the engineer in dispute resolution, it is worthwhile to illustrate this point by the following quotation which is taken from a paper presented by Mr J. J. de Greef, then President of the European International Contractors, at a seminar held in Jakarta in October 1984:[5.20]

> 'Let us assume that the Employer with the advice of the Engineer has selected a number of Contractors qualified to bid for the job and that the chosen ones have received their tender documents.
> The first two questions the Contractor will ask himself are:
> "Who are my competitors and
> who is the Consulting Engineer?"
> . . .
> I come now to an awkward part in my lecture but I should be less than frank if I did not mention one point which strongly influences the price of the contract to be concluded between the Employer and the successful Tenderer: the Employer's choice of the Consulting Engineer.
> . . .
> Why am I putting so much stress on the quality of the Engineer or rather his suitability for that particular project? The answer is simple: because every experienced Contractor includes in his price calculation a factor for the way he expects the Engineer to function. The weight of this factor is determined by a number of considerations, e.g. how good (or how bad) is an Engineer in maintaining his time schedule for drawings? How efficiently does he deal with monthly payments? What is his attitude towards variation orders? Does he decide on them quickly or not? There are at least a dozen qualities in the Engineer's Organisation which will be weighed by the Contractor and which sometimes may lead to awarding to him a considerable nuisance value and this will increase the price the knowledgeable Contractor will

quote. The conclusion is therefore: "the better the Engineer, the more realistic the Contractor's price".'

This point is particularly valid in view of the fact that under the General Conditions of Contract between the employer and the contractor, the engineer is entrusted during the construction period with administrative duties which might alter the time for completion of the works and vary the cost of such works. These duties include, *inter alia*, measuring and valuing of the executed work, fixing rates and prices of additional and varied work and deciding all matters which are within the engineer's role under the contract. The engineer in carrying out his duties within the role allocated to him must do so with unfettered judgement. This is essential since in performing these administrative duties, the engineer is regarded as having a dual role. On the one hand, he is considered as acting in the role of an agent of the employer in ensuring through supervision that the work is executed by the contractor in accordance with the provisions of the contract and that the contractor is in possession of all required documentation for such execution. On the other hand, the engineer is regarded as a certifier or an adjudicator as the case may be.[5.21]

As in the ICE Form, the Red Book provides for arbitration when and if either the employer or the contractor or both are dissatisfied with the decision made by the engineer in performing his duties, see Chapter 19, clause 67.

In performing his duties, the engineer is under an implied duty in common law, where it applies, and an express duty under the Code of Ethics of FIDIC to act with complete impartiality of judgement or decision in applying the terms of the contract between the employer and the contractor. It should be noted that the 1996 Supplement to the 1992 Edition of the Red Book and the 1999 Red Book have removed the requirement of impartiality from the engineer, and his role as a quasi-arbitrator or adjudicator has been allocated to a Dispute Adjudication Board, see Chapter 26.

The previous codes of practice refer also to another requirement, that of independence of ownership. The independence of the engineer is a topic which has rarely been surpassed in producing a heated debate between those involved in the construction industry. Employers, contractors, engineers, lawyers and others quote different incidents to support one or the other of the conflicting views of the debate from which they draw divergent conclusions. See Chapters 10 and 11.

In such debates on independence, some argue that the engineer must be independent in both judgement and ownership. Some argue that he need only be impartial. Others argue that in performing his administrative duties, it is not humanly possible to act with independent judgement when there are questions relating to his own duties. Others contend that whilst the engineer should be and is capable of acting with independent judgement, even on matters which concern him, he does not necessarily need to be of independent ownership. These matters should be examined in each case and the conclusions may form specific provisions in Part II of the Red Book.

Chapter 6

A Traditional Re-measurement Contract

The previous chapter considered the concept of a trusted, independent engineer which is central to the contract under the Red Book. The present chapter examines the contractual arrangements available to an employer or a promoter in the construction field and the basis upon which a choice is made. In particular, four criteria are considered: the identity of the parties undertaking the major roles in construction; the identity of the designer; the manner in which the risks are shared between the parties concerned; and the method of management and of remuneration in respect of the works to be executed. As the Red Book is a traditional re-measurement contract, this type of contractual arrangement is considered in more detail.

6.1 Factors governing choice of contract

Before embarking on the implementation of a construction project, a number of steps would have to be successfully completed. It would be necessary to ascertain that the proposed project would be technically feasible, economically viable, financially supportable and operationally resourceable. Technical feasibility involves both design and construction. Economic viability and the provision of finance go hand in hand. Operation, expertise, transfer of technology and training are other aspects to be considered if a project is to be successfully implemented.

For such projects, the promoter and/or the employer have a wide variety of contractual arrangements to choose from for the various contracts required to initiate and implement the construction project. Before making such a choice, they would have to consider, carefully analyse and determine three main components of the contractual arrangement which govern that choice, namely:

— To whom is the essential function of finance allocated?
— To whom are the essential functions of design and construction allocated?
— To whom are the risks allocated and in particular those of quality control and valuation?

6.2 The allocation of essential functions

In addition to the fundamental role of the initiator of a construction project, the essential functions set out below must be allocated to or taken by certain parties who would then become involved in the implementation of the project. The initiator could be either the eventual owner of the project, known also as the employer, or an outsider in the role of a promoter who initiates and develops the idea of the project.

— initiation and development of the project;
— finance;
— design;
— management and co-ordination of the design functions;
— risk management;
— construction;
— safety and health;
— management of construction activities;
— supervision and quality control;
— certification;
— dispute resolution;
— other specialist advice, supply and construction;
— operation.

The promoter and/or the owner/employer have various options in allocating the above functions to one party or to a number of parties in different possible combinations. For a number of these options, there is a suitable standard form of contract within the family of standard forms published by FIDIC as enumerated below.

— Client/Consultant – model services agreement (the White Book), Parts I & II, Second Edition, issued in 1991;
— Conditions of Contract for Works of Civil Engineering Construction (the Red Book), Parts I & II, Fourth Edition, issued in 1987 and reprinted in 1988 with editorial amendments, and subsequently reprinted in 1992 with further amendments;
— Conditions of Contract for Electrical & Mechanical Works including erection on site (the Yellow Book), Parts I & II, Third Edition, issued in 1987;
— Conditions of Contract for Design-Build and Turnkey (the Orange Book), Parts I & II, First Edition, issued in 1995; and
— Conditions of Sub-contract for Works of Civil Engineering Construction (compatible with the Red Book), Parts I & II, First Edition, issued in 1994.

The Yellow Book, the Orange Book and the Sub-contract form are briefly discussed in Chapter 21, later in this book.

6.2.1 The allocation of the function relating to finance

A construction project can be financed in a number of ways as follows:

— self financing through the owner's own resources;
— borrowing from a lending agency; and
— financing through involvement of the public and/or the constructor of the project.

In the latter type of arrangement, it is more often than not that a concession agreement would be negotiated to own or operate or own and operate the project for a period of time for the benefit of all the parties involved before handing over the project to the owner. Thus, the concession agreement may be one for any of the following:

— Build-Own agreement, known by the acronym 'BO';
— Build-Operate-Transfer agreement, known by the acronym 'BOT';
— Build-Own-Operate-Transfer agreement, known by the acronym 'BOOT', which incorporates in the arrangements property developments rights;
— Build-Own-Operate agreement, known by the acronym 'BOO';
— Build-Rent-Transfer agreement, known by the acronym 'BRT'; and
— Build-Lease-Transfer agreement, known by the acronym 'BLT'.

In some cases, the financier may dictate the terms of contract for the project to be financed, such as in a case where the World Bank is involved, certain amendments to the provisions of the standard form of contract used are mandatory; see Chapter 22 of the second edition of this book.

The 'BOT' and 'BOOT' concepts

The concepts of Build-Operate-Transfer or 'BOT' and of Build-Own-Operate-Transfer or 'BOOT', are the most popular amongst those mentioned above. They require the presence of three elements: first, a feasible and viable project; secondly, a willing government to grant a concession agreement which empowers a concessionaire the right to operate and benefit from the constructed project by that concession; and thirdly, funders who are willing to take the financial risk of undertaking the project. On the expiry of the concession period, the project reverts to the authority which had granted the concession.

These concepts are not new although the acronym is.[6.1] The first construction project financed through such an arrangement was the Suez Canal which was completed and opened for international navigation in 1869. The concession agreement in that case was for a period of 99 years. A similar approach was later adopted for the construction of the Panama Canal project.

Although a period of 99 years proved to be too long in the case of the Suez Canal and was cut short in 1956 due to a political motivation, the project financing arrangements were successful. A more reasonable period nowadays for a concession agreement would be 10 to 40 years. However, despite the success of these arrangements, the traditional methods of financing infrastructure projects around the world remained, until recently, unchanged. These traditional methods entailed using either fiscal resources of the respective countries or sovereign borrowings from lending agencies, such as the World Bank, the International Monetary Fund, the African Development Bank and similar organisations.

In recent years, however, there have been major global shortcomings in funding the ever growing need for infrastructure projects around the world. This led many governments to reconsider their policy of self-financing such projects. Many resurrected the concepts of BOT and BOOT rather than seeking to impose higher taxes. This approach was encouraged by the fact that in many countries around the world, the national economic growth and the annual income per capita have increased to a level that permitted the participation of the general public in such financial activities for the benefit of the whole society.

In summary, although the traditional methods of financing a project, i.e. self-financing and borrowing, represent the main methods used today, the concept of involving private finance in the construction of infrastructure projects is becoming more popular. The use of private finance is expected to rise to an estimated 15% of infrastructure projects implemented world-wide.

6.2.2 The allocation of the functions of design and construction

The functions of sketch design, preliminary design and final detailed design, management of design, supervision, quality control, certification and dispute resolution may be allocated to one independent professional as is traditionally done in conjunction with the use of the Red Book, as described in the previous chapter. On the other hand, these functions may be split and allocated individually so that, for example, the design function may be entrusted to the contractor in a design and construct contract. The function of quality control or dispute resolution may be allocated to a third party.

The functions of construction and of the management of construction and supply activities may be entrusted to one main contractor as envisaged in the Red Book. These two functions may, however, be separated to create what is known as a management contract where a number of contractors are appointed, each of whom is entrusted with a separate part or a different aspect of the project. The whole project is then managed by an overall management contractor to whom the responsibility for coordinating the activities of the various contractors is allocated.

Other combinations of function allocation are also possible. However, clearly where the Red Book is concerned, the above functions are divided between the engineer and the contractor. The functions of design, management of design, supervision, certification, adjudication and dispute resolution are allocated to the engineer. The functions of construction and management of related activities are allocated to the contractor. Whilst modifications may be made to that concept, they must be specifically and explicitly stated. Thus, for example, sub-clause 8.2 of the Red Book permits the allocation of some design function to the contractor, provided that it is 'expressly' provided for in the contract.

6.2.3 The allocation of risk, quality control and the method of pricing and payment

Having determined the allocation of the essential functions to the various contracting parties, the employer, and the promoter if applicable, should then consider the matrix of risks generated. They should first consider and determine whether or not it is acceptable, in principle, to have the risks that are inherent in the contemplated contracts shared between the parties involved. This is particularly important in the construction contract where the risks are numerous and diverse. Furthermore, they should consider the risks inherent in the method of valuing the work done and the method of paying for it within the context of the form of contract chosen by them. If the concept of risk sharing is acceptable, its extent and manner should then be decided and made explicit in the contract documentation. Chapter 7 in this book deals with the topic of 'Risk'.

As to evaluating the work done and paying for it, there are, in broad terms, three categories of valuation depending on the extent of risk accepted by the employer and/or the promoter. These are:

(a) re-measurement contracts based on unit rates and prices;
(b) cost-reimbursable contracts; and
(c) lump sum contracts.

Clauses 55, 56 and 57 of the Red Book clearly identify the Red Book as a re-measurement contract. However, in some cases the use of a lump sum contract is attractive to the employer and technically suitable; for example, in a project where the design has been developed to a sufficiently complete stage that, from the information supplied in the tender documents, the contractor can prepare all drawings and details necessary for the construction without having to refer to the engineer for clarification or for further information and no design changes of significance are likely. Typically, this may be the case in building work rather than in civil engineering construction. In such circumstances, some employers used to modify certain provisions of the Red Book so that the contract becomes a lump sum type. Such modification was not standardised by FIDIC until the introduction of the Supplement to the Fourth Edition in November 1996.

In the context of the Yellow Book, sub-clause 33.8, *'Payment by Measurement'*, provides that *'For any part of the Works which is to be paid according to quantity supplied or work done, the provisions for measurement shall be stated in Part II.'* Thus, the option is left open for the most appropriate choice to be made not only for each particular project, but also for each section thereof.

The Orange Book, on the other hand, is a fixed price lump sum form of contract as provided in its clause 13.1.

6.3 Re-measurement contracts

In this category of contract the employer accepts the risk of variation in the quantities originally estimated and in some of the rates and prices tendered. This is particularly relevant in civil engineering works with a high content of work below the ground surface where the quantities are inherently unpredictable.

The Red Book, like its ICE model, is based on this category of contract where a bill of quantities is utilised. The bill of quantities is used for the final re-measurement of all the items of work executed and also for valuing the final contract price.

The section entitled 'Measurement' in the Red Book which includes clauses 55, 56 and 57 establishes the principle of re-measurement in the contract. It provides for the contractor to be paid pursuant to clause 60 for the value of the works executed in accordance with the contract. Sub-clause 60.1 provides for payment in respect of, amongst other things, the value of the permanent works executed together with any other sum to which the contractor may be entitled, whether under the contract or otherwise. The latter provision refers to all the clauses under which adjustment to the contract price may be made, such as variations, adverse physical conditions and other specified events. (See Chapter 16 later in this book.)

In addition to measurement changes, therefore, attention should be directed to variations and to events which lead to changes in the contract and, in particular, to clause 52 that deals with the valuation of variations. In sub-clause 52.2, the employer accepts that not only quantities may vary, but also *'the rate or price contained in the Contract for any item of the Works'* may be itself varied. Such variation in the rate or price would occur by reason of any variation in the nature or amount of any work done. In sub-clause 52.3, an adjustment to the contract price may be made where *'there have been additions or deductions from the Contract Price which taken together are in excess of 15 per cent of the "Effective Contract Price".'*

6.3.1 The Red Book is a re-measure contract

It is perhaps necessary to emphasise that under the Red Book, there is no warranty by the employer that the quantities measured in the bill of quantities are accurate. It follows that compensation for any unexpected differences in quantities should be paid for on precisely the same basis as that already used in the tender and not as a variation ordered by the employer. This emphasis is necessary in view of the confusion which has occurred as a result of some decisions in cases where similar conditions of contract to the Red Book were used, but with certain additional clauses, resulting in controversial decisions relating to the principle of re-measurement.[6.2] Some commentators believe as a result that increases in quantities under the Red Book can be taken as automatic variations.[6.3]

The true position under the Red Book has been explained clearly in the following terms:

'By clause 56.1, the engineer is required to ascertain and determine by measurement the value of the works in accordance with the contract and thus would be able to compensate the contractor for changes in the quantities and, if it were necessary and appropriate, be able to adjust any relevant items in the preliminary bills which could be shown to be affected by time or quantities, without any need for the assistance of the variations valuation clause or of wholly new prices under that clause.'[6.4]

In this connection, it is of importance to point out that a properly advised employer would require a detailed make-up of prices, to be supplied by the tenderers, against which future claims for preliminaries or other rate adjustments can be accurately and fairly measured in the light of the as-built final quantities. Furthermore, if any major change is implemented during the currency of the contract which affects that make-up, then a new make-up must be submitted by the contractor and verified by the engineer.

The question of preliminary items has become increasingly significant in quantifying proper adjustments to changes in quantities in international construction contracts. This is because some contractors now include into the preliminary bills not only the fixed items of expenditure such as mobilisation costs, but also certain variable time-related items, such as supervision costs; quantities-related items, such as temporary works, taxes or insurance premiums; and other items which should more properly be included in the construction bills. From the experience of the writer, the range in percentage terms for the value of preliminary items as part of the cost of the whole project is as wide as 8% to 58% in international construction contracts.

Of course, if the total contract price is built up in such a way that the preliminary items only represent fixed items of expenditure which will not be affected by changes in quantities or time, then even significant changes in the quantities would require no more than a recalculation of the contract price using the final quantities mulitplied by the unit rates of the respective construction items. However, if the preliminary items form a major part of the contract price, then a recalculation taking into account only the construction items, without an appropriate adjustment of the preliminary items, would not be accurate. On the one hand it would produce a contract price which might be too high if the quantities were reduced without a reduction in the preliminary items, or on the other hand too low if the quantities were increased without an increase in preliminaries.

6.3.2 Contracts with a bill of quantities

From a detailed analysis of the design calculations, specification and drawings, the project is divided into its constituent trades and types of activities to be performed. Each constituent trade or activity is subdivided into discrete items, compiled into a bill, with a brief description attached to each item of work to be done. Quantities for each item are calculated from the drawings either precisely, where possible, or estimated, where not.

Whatever method is chosen for the selection of a contractor, he would have entered, when tendering, a unit rate or price against each of the items in the bill of quantities. The price of the contract is then calculated by adding the extended priced bills. Accordingly, the purposes of the bill of quantities document are as follows:

(a) Where tenders are sought by employers, the cost of tendering is reduced by having these bills prepared only once on behalf of the employer. The alternative would be for each of the tenderers to compute his own quantities, a task which can be very costly in complicated projects and which would have to be borne in the end by employers as an overhead cost of contracting firms.

   Some also argue that if tenderers are expected to compute their own quantities, the time allocated for tendering would have to be increased considerably.
(b) Where tenders are sought and received, the bill of quantities is used as a basis for analysis of and comparison between the tenders submitted. Such analysis is essential since, for the experienced engineer, it can be quite revealing to scrutinise the rates and prices inserted by the tenderers and thus establish the intentions and the capabilities of the tenderers.
(c) The bill of quantities is used as a basis for valuation of variations, if they occur, during the construction period. This benefit, however, may often depend on a

schedule of make-up rates and prices being submitted by the successful tenderer after appointment but before he is permitted to commence work on site. The make-up of rates and prices should identify the cost of mobilisation, labour, material, plant, construction equipment, overheads and demobilisation. It should also include the make-up of any preliminary items which are incorporated in the bills.[6.5]

(d) The document is also used as a basis for valuing the work executed by the contractor during the construction period for the purpose of certification and interim payments.

(e) The bill of quantities is used at the end of the contract as a basis for re-measurement of the various items of work executed and for valuing the final contract price.[6.6]

6.3.3 Contracts with a schedule of rates

Where the design of the permanent work is not sufficiently developed for a bill of quantities to be compiled and where the work is urgently required, it may be possible to have a contract based on a schedule of rates. This type of contract leaves to the end of the contract period the task of establishing which of the items of work was carried out and the respective quantities. No quantities are given in the schedule of rates. Instead, the tenderers are invited either to quote a percentage to be added or deducted from rates previously entered by the engineer, or to enter their own rates to the various items of the schedule.

6.4 Cost-reimbursable contracts

Another type of contract that the employer may choose is the cost-reimbursable contract. In this category of contract, the employer essentially accepts the whole risk of carrying out the works. The contractor is reimbursed for the actual cost of carrying out the work plus an additional amount of money in respect of profit. This may be done on the basis of the following:

(a) cost plus percentage fee contracts;
(b) cost plus fixed fee contracts;
(c) cost plus fluctuating fee contracts; and/or
(d) target price contracts.

In all of these contracts, very detailed and extensive day-to-day professional administrative services are required to check the quality and quantity of work done.[6.7]

6.5 Lump sum contracts

The third category of contract is the lump sum contract, where the employer wishes to accept the least amount of risk with respect to quantities. It is sometimes referred to as a fixed price contract, since a duty is imposed on the contractor to carry out all the work included in the contract documents for a fixed specified tendered sum. A typical example of this type of contract is the Orange Book where clause 13.1 provides that:

'...
(b) the Contract Price shall not be adjusted for changes in the cost of labour, materials or other matters;
(c) ...
(d) any quantities which may be set out in a Schedule are only estimated quantities and are not to be taken as the actual and correct quantities of the Works to be executed by the Contractor in fulfilment of his obligations under the Contract; and
(e) ...'

Fixed lump sum contracts are used extensively in the United States of America and in other parts of the world for building work. They are useful where the quantities are expected to remain unchanged. Bills of quantities, however, are sometimes included in this type of construction contract for the reasons detailed above in paragraphs (a) to (d) of Section 6.3.2 above, but not that of paragraph (e). Thus, the contractor is made responsible for all the costs necessary for the completion of the work. Accordingly, despite the fixed price nature of this contract, there are a number of grounds on which a contractor can claim extra payment. If the drawings and contract documents do not accurately describe the work, or if they make provision for specific alterations, the contractor would be entitled to be paid additional sums. Payment is usually made at pre-determined stages related to the extent of progress achieved on site.

6.5.1 The Supplement to the Fourth Edition of the Red Book, Section B

It is worth noting that in certain countries and for one reason or another, employers have found the lump sum contract more attractive and have thus modified the Red Book and changed it to a lump sum fixed price contract leaving only a few specific provisions for price changes. Therefore, there was a considerable demand for FIDIC to standardise the necessary changes to the Red Book in order to transform it from a re-measurement contract to a lump sum contract. This was successfully accomplished by FIDIC in November 1996 with the publication of its Supplement to the Fourth Edition of the Red Book. Section B of that document gives guidance on the preparation of certain clauses of Part II of the Red Book which are necessary to amend Part I in order to produce a lump sum contract.

As is explained in the introductory part of this section of the Supplement, it is intended that this lump sum form of contract should be used for works which are simple and straightforward, of relatively low value and short duration. Moreover, it is recommended that for larger works FIDIC Conditions for Design-Build and Turnkey should be used.

6.5.2 Main features of FIDIC's Form for Payment on a lump sum basis

In order to adopt a lump sum basis for payment to the contractor in a contract based on the Red Book, a number of amendments are necessary to its Part I. These amendments should be incorporated into Part II of the Red Book. Furthermore, the wording of the Tender and of the Form of Agreement should also be amended to suit the amended provisions of the contract.

Section B of the Supplement recommends that the following clauses of the Red Book should be amended: 1(b), 7.1 to 7.3, 12.1, 18.1, 38.1, 51.2, 52.1, 52.2, 57.1, 58.1, 59.4 and 60.1. Clauses 52.3, 55 and 56 of the Red Book are recommended for deletion.

The lump sum form maintains the essential traditions of the Red Book philosophy including the authority of the engineer. However, the responsibility of the construction drawings is shifted to the contractor who is required to prepare them and submit them for the engineer's approval. Such approval does not relieve the contractor of any of his responsibilities under the contract. This allocation of responsibility necessitates that the tender documents must be brought to such a stage of completion prior to the invitation to tender that they properly reflect the employer's requirements.

As there is neither a bill of quantities nor a re-measurement of the contract price, interim and final payments should be made on the basis of percentage completion of the various components of the project. Accordingly, a schedule containing a breakdown of the lump sum is annexed to the Tender form for completion by the contractor. The term *'breakdown of the lump sum'* is a defined term under the amended clause 1 and is intended only for the purposes of the monthly interim payments.

Although the final cost of the project is fixed in respect of any variations in quantities, it is adjustable in accordance with a certain number of the contract provisions.

Chapter 7

Sharing of Risks

'It is only by risking our persons from one hour to another that we live at all.'

William James, 1897

7.1 Introduction

Based on the statistics gathered in the past two decades on topics such as disputes in the construction industry and international arbitration,[7.1] accidents at work[7.2] and exposure to natural hazards around the world,[7.3] it can be concluded that construction projects are sensitive to an extremely large matrix of hazards and risks. This sensitivity is due to some of the inherent characteristics of construction projects which are summarised as follows:

(a) The time required to plan, investigate, design, construct and complete a construction project spans such a lengthy period that it is often greater than the period of cyclical recurrence, known as the 'return period', of many of the hazards to which such projects are exposed. For example, the hazard of rainfall has usually a return period of one year depending on the time for the rainy season. Therefore, the risks associated with rainfall on a particular project would have to be assessed and managed for the number of years taken to complete it. Any reduction in the period of construction introduces its own risks.
(b) The number of people required to initiate, visualise, plan, finance, design, supply materials and plant, construct, administer, supervise, commission and repair any defects in a construction project is enormous. Such people usually come from different social classes and in international contracts, from different countries and cultures.
(c) Many civil engineering projects are constructed in isolated regions of difficult terrain, sometimes stretching over extensive areas and exposed to natural hazards of unpredictable intensity, frequency and return period.
(d) The materials selected for use generally include a number of new products of unproved performance or strength. Advanced and complex technology is also necessary in some construction projects.
(e) Extensive interaction is required between many of the firms involved in construction, including those engaged as suppliers, manufacturers, sub-contractors and contractors, each with its own different commitments and goals.

(f) Construction projects are susceptible to risk cultivation by the parties themselves or by others associated with them or advising them.

It is therefore extremely relevant for the construction industry and those involved in it to understand the concept of risk and to know how to properly manage the risk matrix generated when a construction project is initiated.

The subject of risk, its assessment, allocation and management in construction projects, has been developed and applied on an increasing scale over the last 20 years. The Health and Safety at Work Regulations introduced in a number of countries and in particular those recently imposed in the European Union gave the subject of risk in construction an even greater significance.[7.4] Amongst the requirements introduced in the European Union through the Construction (Design and Management) Regulations 1994, (CDM), there is a requirement to carry out risk assessment of planned work and to take reasonable measures to deal effectively with any significant risk.

However, there is little uniformity of approach to the topic of risk by those involved in the construction industry and, surprisingly, only a few useful general applications of the topic of risk have been developed in the area of planning and management of construction projects. The lack of uniformity relating to risk extends even to the definition of 'risk' and what is meant by it.

Etymologically, the origin of the word 'risk' in English, 'risqué' in French and 'rischis' in Italian is uncertain. The Latin word 'resecum' meaning 'danger' or 'rock' may throw some light on its origin, but the Chinese 'wej-ji' with the characters representing 'chance' and 'danger,' is more illustrative of the concept of risk as it applies to the construction industry. This concept has evolved with these two notions embodied in it: a positive chance of gain and a negative danger of a loss.[7.5]

Decision making often, if not always, involves risk taking. However, the well-informed decision maker will be aware of the risks associated with any decision he takes and will endeavour to reduce all foreseeable negative risks and their consequences to an acceptable minimum.

In making a decision, an individual may face any one of the following possibilities:

(a) 'pure risk' where only negative deviations from the desired outcome are possible and therefore danger of loss is predominant; or
(b) 'speculative risk' where both negative and positive deviations are possible and therefore there is a danger of loss as well as a chance of gain; or
(c) only positive deviations are possible and therefore only a chance of gain exists. Such events do not form part of the notion of risk.[7.6]

7.2 The definition of 'risk'

Risk is defined in British Standard No. 4778: Section 3.1: 1991,[7.7] as 'A combination of the probability, or frequency, of occurrence of a defined hazard and the magnitude of the consequences of the occurrence'. In the same British Standard, the definition of hazard is given as 'A situation that could occur during the lifetime of a product, system or plant that has the potential for human injury, damage to property, damage to the environment, or economic loss'. However, this concept reflects only pure risk and thus expresses only the negative result and ignores the positive chance of gain. To this extent, it is an inadequate definition that should be corrected, as was done in the New Zealand and Australian Standards by replacing the word 'hazard' with 'event'.[7,8] Based on the definition as contained in the New Zealand/Australian Standard, risk may be expressed in the form of a mathematical equation, as follows:

Risk = Probability, or frequency, of the occurrence of a defined event × Consequences of the occurrence of that event; or $R = P \times C$.

There are a number of points which flow from the above mathematical expression, as follows:

(a) An event may have a number of different causative factors, of which one or any combination could lead to its occurrence. For example, if the event is, say, the undesirable collapse of a cofferdam at a construction site, such collapse may have been caused as a result of bad ground conditions, material failure, defective design, or a combination of some or all of these factors. All these factors are referred to as hazards. If the event results in a positive consequence in the form of a profitable outcome, say better than expected ground conditions, then the event would represent a chance to gain.
(b) An event can, therefore, be construed as a dormant potential for, on the one hand, inconvenience, loss, damage to property, damage to the environment, moral damage, injury or loss of life, and on the other for gain. To eventuate, it is triggered by a particular incident, which may be referred to as a 'triggering incident'. A triggering incident is usually necessary for an event to occur causing either desirable or undesirable consequences. For example, if the cofferdam collapse mentioned in (a) above was due to a defective section in its wall, the triggering incident could be the imposition of an additional loading beyond the limit sustainable by that defective section;
(c) The desirable or undesirable consequences may result in different levels of severity or magnitude of consequences depending on the particular circumstances and timing of the event. For example, the consequence of the collapse of the cofferdam in (a) above may be a financial loss in the form of the cost of the repair to the cofferdam, or it may extend to a critical delay in the completion of the project, or it may go beyond the financial loss and delay into personal injury and death; and
(d) Expressing risk in a mathematical formula permits a comparison of the magnitude of the various risks to which a project is exposed. Such a comparative analysis may then be used to decide whether to accept a particular risk or take measures to eliminate it, mitigate its effect or, indeed, enhance it.

In summary, when an event occurs, the sequence can best be expressed as follows:

hazard or potential for gain → triggering incident → event → consequences → assessment and evaluation.

Any assessment and evaluation carried out of these consequences may be used subsequently in the measurement of risk relating to that event.

Two illustrative examples of the above concept and terminology are given here. The first is the disastrous collapse of the walkway bridge of the Hyatt Regency Hotel in Kansas City in the United States in 1981. The events were as follows.

On a Friday evening in July 1981, over 1,000 people were crowded onto the main floor of the lobby of the hotel and on three walkway bridges spanning it, to watch and participate in a dancing contest. Shortly after 7 pm a loud cracking noise was heard and the fourth-level walkway was seen to buckle and fall onto the second-level walkway two-storeys below, causing it to collapse and dump some 60 tonnes of debris, along with the spectators from both walkways, onto the crowded dance area. The death toll was 111, and 188 were injured.

Lawsuits were quickly launched, seeking compensation damages exceeding US$1 billion and punitive damages of more than US$500 million. Several technical investigations were commissioned which showed that the failure was triggered in the hanger rod connections to the floor beams. These connections were defectively constructed as a result of a change in their structural detail during the construction period. The change in detail of the connections resulted in doubling of the load applied against the lower flange of the upper floor beams. The walkway failure cycle began when one of the upper hanger rods pulled through the bottom flange of its supporting floor beam. The as-built connection reached its capacity under dead loading alone and was ready to fail when any significant live loading was imposed.[7.9]

In this example, the event, or the hazard in this case, was the defectively constructed connections and the triggering incident was the dancing contest.

A more recent and complex example occurred on 28 January 1986 when the space shuttle *Challenger* was given the clearance for ignition.[7.10] The space-shuttle-rocket-booster exploded after lift-off and all seven crew-members perished. The flight began in the late morning at 11.28 am and ended 73 seconds later in an explosive burn of hydrogen and oxygen propellants that destroyed the external fuel tank and exposed the space shuttle to severe aerodynamic forces that caused complete structural failure. Although the technical cause of the *Challenger's* explosion was the result of a faulty design of the O-ring seal, which failed at the launch, the Presidential Commission which was established to investigate and enquire into the cause of the disaster, found that the underlying cause 'was rooted in organisational failures and poor communication'.

The explosion was found to have been caused by hot combustion gases that escaped from a booster via a failed field-joint seal. The design of the joint included two O-rings that did not function correctly at launch due to the low ambient temperature that prevented them from responding correctly to the rising pressure after ignition and rotational movement within the joint.

For a number of years prior to the tragedy, engineers had been concerned about the behaviour of the seals at low temperatures and such temperatures were forecast for the morning of the launch. Analysis of the records showed that of the previous 23 launches in which the field-joints had been examined following booster recovery and where data was held, seven showed damage to the O-ring seals.[7.11] This damage had only occurred at ambient temperatures below 24°C and it occurred in all cases where the temperature was below 18°C. The lowest recorded temperature was 12°C. However, various factors, including the management structure of the project, and ultimately, time pressures to maintain the space shuttle programme, created a situation where launch proceeded despite technical advice to the contrary and at an ambient temperature near to freezing, where seal damage was likely to occur.

The Mission Management Team (MMT) had postponed the launch scheduled for 27 January due to high crosswinds. The MMT met again at 14.00 hours on that day and concerns were raised about the effect of the forecast low temperatures on such facilities as drains, eye wash and shower water, and fire suppression systems, but no concerns were raised about the O-rings. When the situation was relayed to the engineers at Morton Thiokol, they were adamant about their concerns over the low temperature: '... way below our database and we were way below what we qualified for...'. They contacted Morton Thiokol's liaison officer at the Kennedy Space Center, expressed their concern, and requested more forecast temperature data. He recognised the significance of the concerns and ensured that a teleconference was set up. This was in turn followed by a second teleconference.

At the second teleconference, Morton Thiokol engineers presented the history of O-ring erosion and blow-by. Their recommendation was not to launch until the O-ring temperature reached 53°F (12°C). A long, detailed, and reportedly, not acrimonious discussion followed. Thiokol's Vice-President of Engineering was asked for a recommendation and he replied that he could not recommend launch. The Deputy Director, Science and Engineering at Marshall was reported to have said he was 'appalled' at the recommendation not to launch. The Manager SRB (Solid Rocket Booster) Project at Marshall was said to have asked, 'My God, Thiokol when do you want me to launch, next April?' Under this pressure, Thiokol management asked for a recess to consider their recommendation further and a Thiokol management-level discussion took place. One of the managers is said to have remarked that he 'took off his engineering hat and put on his management hat'. The Thiokol managers seem to have concluded that, although blow-by and erosion were to be expected, there was not sufficient evidence to predict joint failure. In the absence of such evidence, Thiokol engineers described it: 'This was a meeting where the determination was to launch and it was up to us to prove beyond a shadow of a doubt that it was not safe to do so. This is in total reverse to what the position usually is in a pre-flight conversation or a flight readiness review.' The launch subsequently took place, with fatal results.

The Presidential Commission Report (Bermingham 1999, pers. comm.) traced the technical cause of the accident to hot gas escaping, known as blow-by, following the failure of the O-ring pressure seal in a joint of the casing of the booster. The failure was due to a faulty design, which was unacceptably sensitive to a number of factors, including the effects of temperature, physical dimensions and the character of the seal materials, as well as the reaction of the joint to dynamic loading.

The event, or once again the hazard in this case, was the defective design of the O-rings, but what was the triggering incident? Some might think that it was the low temperature, others might blame human error or judgment in yielding to unjustifiable pressure, whilst others would be tempted to focus on both.

7.3 Measurement of risk

When considering the acceptability of a particular negative risk or when comparing negative risks in a certain project or a number of projects, it is useful to use the formula referred to in the previous section. Considering the first variable, the consequences of an undesirable event, British Standard No. 4778 classifies the consequences of hazards into four categories of severity:

(a) negligible;
(b) controlled or marginal;
(c) critical; and
(d) catastrophic.

This classification is based on the effect produced once an undesirable event is triggered and may include any one or a combination of, loss of life; personal injury; damage to reputation; material damage; financial loss; loss of time; and mental distress. Therefore, the consequences of a hazard resulting in an undesirable event range from those which are negligible and can be disregarded to those at the other end of the scale which would include loss of life, personal injury, or material loss of various types.

In order to express this variable, the consequences of hazards, as numerical data to insert in the above formula of $R = P \times C$, further categories can be extrapolated from the BS 4778 classification. A range of numerical values can be assigned to the variable based on the expected financial loss expressed as a percentage of the overall cost of the project. For example:

Category of loss	*The consequences of a hazard eventuating*	*Percentage of cost of project*
0	No loss	Nil–0.09
1	Nuisance type small losses	0.1–0.99
3	Medium losses which can be borne by the individual concerned	1–4
5	Large losses	10–19
6	Probable maximum loss in the range of the largest previous losses of similar projects	20–40
7	Serious and exceeding any previous events	41–50
10	Catastrophic – total loss	81–100

Other sets of categories can be worked out for different ventures or projects. Values for the second variable, probability of occurrence, range from 0 to 1, depending on a schedule specifically calculated for the project.

For example:

Probability of occurrence	The event
0	It is certain that no loss would occur – loss is not possible
0.3	Slight probability of occurrence
0.5	Equal chance of occurrence or non-occurrence
0.7	Strong probability of occurrence
1.0	Certain to occur

Probability of occurrence of many desirable or undesirable events that are predicted in construction projects can be calculated from data available from various sources, such as health and safety reports on accidents; meteorological, climatic and seismic reports; geological data; insurance statistics; previous experience; and other technical reports. Where such data is unavailable, a best estimate based on experience and common sense could be used. In this regard, it is important to note that the probability of occurrence refers only to occurrences during the period of the relevant contract.

Various methods and formulae are available for the calculation of probability. Indeed, within the field of statistics, probability is a major field of study and the literature indicates the development of an ever increasing range of computerised statistical modelling techniques in the calculation of risk. Consequently, further consideration of the topic is outside the scope of this book.[7.12]

Finally, it is notable that in using the above formula to determine the acceptability of a particular risk, both the severity of the consequences of the event and its probability of occurrence contribute to the risk factor obtained. Thus, a high severity combined with a low probability of occurrence may lead to an acceptable risk, but a low severity combined with a high probability of occurrence may lead to an unacceptable risk. An example of the first combination is the matrix of risks which exist during the commissioning of a nuclear power station (although in some countries this risk has been rejected as being unacceptable). An example of the second combination is that which could result from the use of a low quality roofing system, such as the material known as mineral felt, a risk which is considered unacceptable in permanent construction in many jurisdictions.

7.4 Risk management

It is advocated by some commentators that for efficient and effective construction and completion of a project, risk management is essential. Risk management is defined in BS 4778 as 'the process whereby decisions are made to accept a known or assessed risk and/or the implementation of actions to reduce the consequences or probability of occurrence'. Risk management is also concerned with the mitigation of those risks deriving from unavoidable events through the optimum specification of warning and safety devices combined with risk control procedures, such as contingency plans and emergency actions.

If a decision is made to accept a risk, a further decision must be made on whether or not the risk should be shared between the contracting parties. Before such decisions can be made, it is necessary to go through a systematic process which involves analysis of the possible events to which the project may be exposed and evaluation of their intensity, frequency and return period. In this regard, the following terms and definitions from BS 4778 are relevant (see Reference 7.7):

Hazard analysis	The identification of hazards and the consequences of the credible accident sequences of each hazard.
Risk quantification	The estimation of a given risk by a statistical and/or analytical modelling process.
Risk evaluation	The appraisal of the significance of a given quantitative (or, when acceptable, qualitative) measure of risk.
Risk assessment	The integrated analysis of the risks inherent in a product, system or plant and their significance in an appropriate context.
Risk criteria	A qualitative and quantitative statement of the acceptable standard of risk with which the assessed risk needs to be compared.

7.5 Allocation of risks and their management

When desirable or undesirable events are identified, assessed and analysed, their management must be allocated to the various parties in order to keep these events under control, prevent the occurrence of harmful consequences or mitigate any undesirable consequences should they eventuate. By such management, it is intended to reduce the risk of harm and enhance the risk of gain. Allocation of the management role is part of the risk management process, where the party to whom a certain event and the associated risks are allocated should be selected in accordance with certain rules rather than haphazardly.

The conclusions of one of the earliest papers on the rules for allocation of risks in a construction project remains valid today[7.13], and may simply revolve around the ability of a party to:

(a) control any arrangements which might be required to deal with the hazard or any triggering incident relating to it;
(b) control the risk or to influence any of its resultant effects;
(c) perform a task relating to the project, such as obtaining and maintaining insurance cover; and
(d) benefit from the project.[7.14]

On the other hand, the rules for allocation of risks may revolve around an already established policy in a large organisation or a governmental agency, such as that contemplated by the following terms:[7.15]

> '... while an event may be foreseen ... employers may see advantages in a contract which requires him (the contractor) to assume that risk, and to include for the cost

of dealing with that situation in his tendered contract price. Where the risk is uncertain, this logically requires that a contingent element will have to be included in the original price, which, in the event, may possibly not be required. If so, the employer will have agreed to an unnecessarily high price but may regard that as preferable . . . than a lower price subject to post-contract upward adjustment at [a] late stage should the risk materialise . . . whether or not a particular risk should be so included in the price is in essence a question of policy and not of "fairness", "morality", or "justice.",'

and

'This (American) system . . . also seeks to preserve a large pool of competent contractors and obtain low contract bids by absorbing particular risks and seeking to assure the contractor he will be treated fairly.'

If a risk is not allocated in a contract and a dispute arises between the contracting parties as to whom it should be allocated, then an arbitrator or a judge would most likely examine the following criteria for risk allocation and determine which is the most appropriate to apply in the circumstances of the event in order to resolve the dispute:

(i) which party could best foresee that risk?
(ii) which party could best control that risk and/or its associated consequences?
(iii) which party could best bear that risk?
(iv) which party most benefits or suffers when that risk eventuates?

Although these rules for the allocation of risks were contemplated in 1983, their application was very recently referred to in a major report of a study commissioned by the Government of Hong Kong Special Administrative Region on the allocation and management of risk in the procurement of construction projects. In his report, Mr Jesse B. Grove stated that one aspect of the philosophy of risk allocation must be that:

[T]he ultimate goal of optimal risk allocation is to promote project implementation on time and on budget without sacrifice in quality, that is, to obtain the greatest value for money. The goal for a repeat employer should be to minimise the total cost of risk on a project, not necessarily the cost of either party. A study in the USA has shown that 5 percent of project cost may be saved by choice of the most appropriate terms of contract alone. The question is therefore what is 'most appropriate' and how can it be recognized? There is a variety of answers.'[7.16]

This review was followed in November 2000 by a timely conference held in Hong Kong.[7.17] A number of the papers presented at the conference were subsequently published in the *International Construction Law Review*.[7.18]

In a section entitled 'Application of Philosophy', Mr Grove referred to four criteria for allocation of risks:

(a) The fault standard: cost and time impacts of risks caused (or not avoided) through the fault of a party should be borne by that party;

(b) The foreseeability standard: he who is best able to foresee the risk is allocated that risk;
(c) The management standard: he who is best able to control and manage the risk is allocated that risk;
(d) The incentive standard: risks should be placed on the party most in need of incentive (presumably already with the ability) to prevent and control them.

He drew on conclusions made in an earlier report on the topic by Thompson and Perry.[7.19] He stated the following:

'9   Application of Philosophy
9.1 It is not enough to say that there should be a "balance of risk" or "efficiency in risk allocation" because all of us will never agree on what is a fair and reasonable balance between the contractor and the employer or which terms are most efficient for either of them.
9.2 When studying the views of the proponents of, and commentators on, the various philosophies of risk allocation, one is tempted to conclude that the same principles underlie them all. Certainly there seem to be the following common considerations:
   • Which party can best control the events that may lead to the risk occurring?
   • Which party can best manage the risk if it occurs?
   • Whether or not it is preferable for the employer to retain an involvement in the management of the risk.
   • Which party should carry the risk if it cannot be controlled?
   • Whether the premium charged by the transferee is likely to be reasonable and acceptable.
   • Whether the transferee is likely to be able to sustain the consequences if the risk occurs.
   • Whether, if the risk is transferred, it leads to the possibility of risks of different nature being transferred back to the employer.
   If these considerations are applied, it should be possible to achieve clear and realistic terms that are acceptable to the employer and on which contractors are prepared to tender at prices which do not contain contingencies for unclear terms or for significant risks which are not possible to estimate with some certainty or which are unlikely to materialise.
9.3 In my opinion, Max Abrahamson has come the closest to laying down an acceptable "formula" for risk allocation, as follows:
   [A] party should bear a construction risk where:
   1. It is in his control, i.e. if it comes about it will be due to willful misconduct or lack of reasonable efficiency or care; or
   2. He can transfer the risk by insurance and allow for the premium in settling his charges to the other party . . . and it is most economically beneficial and practicable for the risk to be dealt with in that way; or
   3. The preponderant economic benefit of running the risk accrues to him; or

4. To place the risk on him is in the interests of efficiency (which includes planning, incentive, innovation) and the long-term health of the construction industry on which that depends; or
5. If the risk eventuates, the loss falls on him in the first instance, and it is not practicable or there is no reason under the above four principles to cause expense and uncertainty, and possibly make mistakes in trying to transfer the loss to another.

The job of trying to balance the five principles in practice is the hard one ... But at least it is best to work from declared principles rather than undeclared and perhaps unconscious prejudices.'[7.20]

In the risk management process, an extensive matrix of general risks can be identified for most, if not all, construction work to which one can add further risk matrices for specific projects. There are examples and legal judgments in respect of events, where the liability was allocated in accordance with each of the above criteria.[7.21] These risks are traditionally shared between the parties involved, in accordance with the provisions of two contracts usually agreed; first, between the developer and the design professionals involved, and secondly, between the developer and the main contractor. From the latter agreement, where the developer is referred to as the employer, flows another line of risk sharing, between the main contractor on the one hand, and sub-contractors, suppliers, manufacturers, insurers and others, on the other hand.

If these risks are analysed on the basis of the effect they generate once they eventuate, two basic types of risk can be identified. The first type incorporates the risks which could lead to damage, physical loss or injury, and the second type incorporates risks which could lead to lack or non-performance of the contract, delay in completion of the works and/or cost over-run of the constructed project.

Examples of the first type of risk which involves damage, physical loss or injury include defective design, defective material, defective workmanship, Acts of God, fire, human error and failure to take adequate precautions. Examples of the second type include late possession of the site, delay in receipt of information necessary for timely construction, changes in design, and variations to the original contract.[7.22]

The treatment of these two types of risk in construction contracts differs in that the first type encompasses risks that might be insurable, whereas the second type involves, in principle, uninsurable risks. It is important to appreciate that the treatment of the identified risks in most of the forms of construction contracts is dealt with in two different ways.

First, there are the risks that lead to death, bodily injury and/or physical loss or damage, which are specified separately in certain parts of the contract and which might be insurable. So, in the Fourth edition of FIDIC's Red Book, 1987 to 1992; the Seventh Edition of the ICE Form, Measurement version; and the second edition of the ICE Form 'Design and Construct', these risks and the respective insurance provisions are dealt with in clauses 20 to 25.

Second, there are the risks that lead to economic and/or time loss, which are dealt with throughout the remaining part of the contract conditions, but whilst the employer's risks are explicitly specified, only some of the others that are allocated to the contractor are specified. If the above divisions and allocations of risks are not understood, many problems could arise.[7.23]

Irrespective of the method chosen to allocate risks in a construction contract, a most important question arises in relation to any unidentified risk, if and when it eventuates: to whom should the consequences be allocated? Under the applicable law, the contractor is liable for the consequences of all the risks that are not specifically allocated to the employer. This is the approach taken in FIDIC's standard forms of contract and in the Engineering and Construction Contract. Against that view, the contractor would argue that if and when such risks eventuate, they should best be borne by the party who gains in the long run the benefit of the project, namely the employer. The American Institute of Architects adopts that view and accepts that all risks belong to the employer when no other party can either control the risks or prevent the loss.[7.24] In other words, they adopt the principle that unidentified risks can neither be controlled nor can the resulting loss be prevented. Perhaps the least confusing from the above point of view is the Engineering and Construction Contract, where the phrase used in Clause 81.1 is '... the risks *which are not carried by the Employer* are carried by the Contractor'.[7.25]

Examples of problems resulting from lack of or incorrect allocation of risks can be very instructive. An owner may increase risk exposure by applying unreasonable monetary and time restraints, or by not implementing appropriate maintenance or operating procedures once the project is completed.

Unfortunately, the division of risk as referred to above is not clearly and explicitly explained in a number of the well-known standard forms of contract, a problem that has resulted in major misunderstandings. As an example, the wording of sub-clause 17.1 of the 1999 suite of FIDIC's forms of contract should have started by explaining that the risks included under clause 17 of the Conditions are only those risks of loss and damage and not the whole matrix of the risks to which the project and the contracting parties are exposed. The term 'employer's risks' in the context of this clause should have been replaced by 'employer's risks of loss and damage', since these risks are confined to those which lead to some form of accidental loss or damage to physical property or personal injury, which in turn may lead to economic and/or time loss risks, directly or through the other clauses of the contract.

As this explanation is not stated in the Conditions, the mistake of referring to the risks under clause 17 as 'employer's risks' could lead to serious error in that the reader, and of course the user, would conclude that having identified in clause 17 the employer's risks, all the other risks belong to the contractor, including the economic and/or time risks in the remaining provisions of the contract. This problem can be highlighted by reference to clause 17 of the Orange Book,[7.26] where the draftsman fell into that trap and stated expressly in sub-clause 17.5 that 'The Contractor's risks are all risks other than the Employer's Risks listed in sub-clause 17.3'. This mistake has led to many instances of misunderstanding, conflict and at least one serious arbitral proceedings, where the employer pointed out that by sub-clause 17.5 he bears no risks under the contract other than those specified in sub-clause 17.3.[7.27]

When the risks are allocated, the consequential flow of responsibility, liability, indemnity and insurance would apply.

7.6 Allocation of risks in the Red Book

The Red Book is drafted on the basis of sharing of risks between the employer and the contractor. The principles of sharing of risks in the Red Book are similar to those in the ICE Form which can be best stated by reference to the draftsman's reply to criticisms made after the publication of the Fifth Edition of that Form.[7.28] The principles were stated as follows:[7.29]

> 'It is a function of a contract to define upon whom the various risks of an enterprise shall fall, and it was decided that the Contractor should only price for those risks which an experienced contractor could reasonably be expected to foresee at the time of tender.... It is the right and the duty of the Employer to decide, and by his engineer to design and specify that which is to be done, and it is the Employer's duty to allow the contractor to do that which is to be done without hindrance. It is the duty of the contractor to do what the Contract requires to be done, as designed and specified by the engineer, but, subject to any specific requirement in the contract, it is his right and duty to decide the manner in which he will do it. If there are to be exceptional cases where the Contractor is to decide what to do or to design what is to be done, or where the Employer or the Engineer is to decide how the work is to be done, the contract must expressly provide for this and for the necessary financial consequences for the protection of the Contractor.'

Whilst these principles were stated in connection with the Fifth Edition of the ICE Form, they are in essence the same for both the Red Book and the Seventh Edition of the ICE Form,[7.30] although they may differ in detail. Figure 7.1 shows the categories of risk under the Red Book.

In allocating these risks between the employer and the contractor three concepts must be borne in mind. These are:

(a) the meaning and the significance of 'risk';
(b) not all undesirable events can be foreseen and identified; and
(c) whilst risks generally imply undesirable consequences, in certain circumstances undesirable consequences to one of the parties would be generally desirable to the other.

Taking these concepts in turn, the first which deals with the meaning and significance of risk has been considered earlier in this chapter under Section 7.2.

The second concept referred to above relates to foreseeability of risk. Not all undesirable events, or hazards, in a particular construction project can be foreseen and identified. Whereas it is possible to plan for those hazards which can be foreseen and identified, it is extremely difficult, if not impossible, to do so for the unidentified hazards. Accordingly, where risk management is concerned, the decision on what is practicable or acceptable risk exposure is generally made without a complete identification of the hazards and consequently their probability of occurrence. The danger then lies in the possibility that an individual's conduct may be judged to have been negligent, retrospectively, with abundant legal and expert help, ample time and full hindsight.

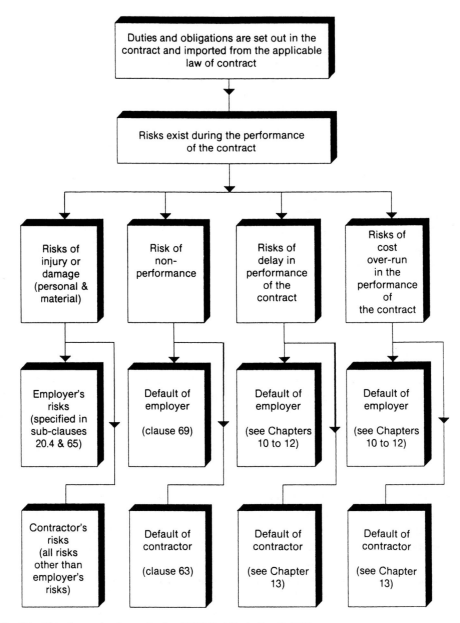

Fig. 7.1 Risks in construction under the FIDIC Red Book, Fourth Edition.

Risks involving unidentified hazards are, therefore, extremely difficult to allocate or apportion between the parties. By definition, they cannot be avoided and their consequences cannot be accurately assessed. However, it should be noted that notwithstanding the above, there are accepted statistical methods available for taking into consideration the possibility of their occurrence and thus dealing with them on the basis of a contingency provision. Such risks are usually allocated to one or the other of the parties involved either as a whole or in precisely defined categories.

The third concept referred to above is that whilst risks generally imply undesirable consequences, in certain circumstances desirable as well as undesirable consequences may occur. Where this happens, both negative and positive aspects must be assessed and allocated. If the negative aspects are identified in the contract and allocated to one party, the positive aspects, should they occur, must also be identified and allocated. It would be an error to ignore these positive aspects and have them considered as a windfall to the other party. For example, if the risk of unforeseen conditions is allocated to the employer and if during the execution of the works, the contractor encounters adverse unforeseen conditions, which cause additional cost to be incurred, then the additional cost of encountering such conditions is borne by the employer. However, should unforeseen conditions produce a more favourable and cost-saving operation for the contractor, then logically, the effect must be credited to the employer rather than to the contractor, unless explicitly stated otherwise in the conditions of contract. The change in the marginal note of sub-clause 12.2 of the Red Book (replacement of the word 'adverse' by the words 'not foreseeable'), which was implemented in the 1992 reprint, shows a recognition of this principle. However, whilst the effect of an adverse set of unforeseen conditions is dealt with clearly in clause 12, that of desirable unforeseen conditions is left without an express solution. Presumably, the engineer can deal with such circumstances in his due-consultation role.[7.31] (See also Chapter 9 of this book.)

As risks associated with foreseen and identifiable hazards are calculated, it should be generally a matter of policy to determine to whom each of the risks is allocated. The most cost-effective method of allocation from the point of view of controlling the occurrence of the risk and mitigating or eliminating its adverse effects is based on the ability to exercise such control. However, risks allocated to the contractor on the basis of this method would have a cost implication if they are not his own fault, since it would be prudent for the contractor to include in his original price an element relating to this additional risk he is asked to carry. If the risk does not eventuate, the employer would have paid a larger sum than necessary. On the other hand, this may be more advantageous to the employer than to assume the risk himself and be exposed to the possibility of having to make an additional payment should the risk eventuate. This is particularly so where there are strict budget restrictions or where the financial considerations of the project are such that the project would not be economically viable beyond a certain limit.

Should the employer decide, for financial or other reasons, that he is not prepared to assume certain of the identified risks, the tender documents must reflect that policy decision. Similarly, tenderers who participate in the tendering process must show that they are prepared to accept that policy decision and may reflect the acceptance of the risks allocated to them in their price.

In certain circumstances, the contractor's price for accepting a particular risk may be extremely high in comparison with the adverse effect on the employer should the risk eventuate. In such a case, the wisdom of allocating this type of risk to the contractor should be carefully examined. An example of such a situation is the 'on-demand' or 'unconditional' bond or guarantee requirement which has recently become popular in international contracts.

An on-demand bond creates an obligation on the party providing such a bond to pay a specified sum to the beneficiary as soon as he makes a simple demand. The demand may have to contain the description of an event which entitles the beneficiary to make such demand. However, specifying the event in itself provides no protection to the bonded party since the occurrence of the event does not have to be proven by the beneficiary, but may simply be alleged. (See later in Chapter 15 and in particular Section 15.3.) What does provide some protection, however, is the fact that once the bonded party is given notice of the call on the bond, he may attempt to injunct the proceedings to prevent the bonding bank from effecting payment.

7.7 Responsibility and liability

Whatever the rules or the reasons for allocation of risks, the responsibility and liability attaching to these risks when they eventuate, follow and flow from that allocation. Accordingly, the simplicity and clarity of the wording where such allocation is made is of paramount importance, since after all the purpose of the contract is to allocate the risks between the parties to that contract. This means that the contractual arrangements, the legal rules of the applicable law of the contract and the technical documentation, such as any specification and drawings, must be clearly stated and fully understood.

Once liabilities are assigned through the contract documents, the parties involved have the following options to finance the consequences of risks should these risks eventuate:

(a) To retain the responsibility for financing the costs of loss or damage or injury by providing any one or a combination of the following arrangements:
  (i)   an element of their cash flow;
  (ii) reserves created specifically for the purpose;
  (iii) funds assigned;
  (iv) creating captive societies.
(b) To transfer the responsibility for financing the costs of loss or damage or injury or non-performance to:
  (i) another party to the contract by agreement, thus creating a sharing of risks;
  (ii) an insurer through an insurance contract which in turn becomes transferred to reinsurers through reinsurance arrangements. An insurer may impose his own risk management conditions, thus creating another cycle of transfer.

  This second cycle of transfer which is enforced through either an incentive in premium reductions or conditions attached to the insurance policies may result in:
  — the insured having to take measures to eliminate or mitigate a certain risk;
  — the insured having to retain part of the responsibility by the imposition of a deductible or excess at the lower end of the scale, or a limitation of the part insured at the upper end of the scale;
  — the insured having to retain certain risks through exclusion clauses in the insurance policy; or
  — the insured having to seek another insurance cover from a different insurer.

7.8 Indemnity and insurance

Due to a number of inherent characteristics of construction projects, liabilities arising from the duties and obligations of the parties to the construction contract should be covered by indemnities given by one party to the other, or provided in the form of construction insurance policies as can be seen in Figure 7.2. These inherent characteristics may be summarised as follows:

(a) except in few specified circumstances, the contractor must complete the contract irrespective of whatever happens including accidents and other deterrents; see Chapter 13;
(b) construction projects involve vast sums of money, frequently provided by banks, financial institutions and insurance companies which require some form of guarantee as to the safety of the capital they provide for financing the project;
(c) construction projects are unique and no two projects are alike;
(d) numerous hazards and risks exist in a variety of categories; and
(e) analysis of tenders is simplified if a figure is included in the Bill of Quantities to cover the insurance premium in respect of providing the necessary indemnities in respect of the liabilities which might be incurred by the contractor.[7.32]

In this chapter, the concept of risk sharing in construction contracts is explained and developed to show the relationship between event, risk, liability, indemnity and insurance. Later in this book, Chapter 14 further develops this topic and the interrelationship between these areas to show how they are treated under the Red Book.

As mentioned at the end of the first chapter of this book, the essential features and concepts which remain to be at the core of the Red Book have been discussed and explored in the previous six chapters. The next chapter considers how suitable these concepts had been in practice and the extent to which they had satisfied the construction industry requirements in the international field.

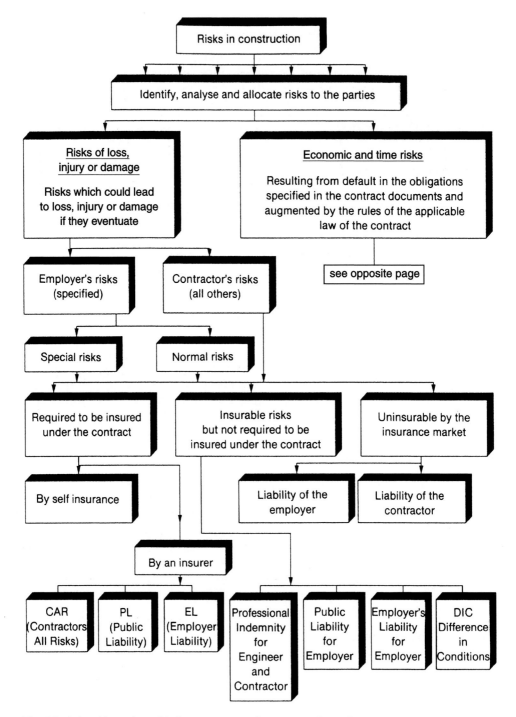

Fig. 7.2 Indemnities and possible insurance covers for a construction project.

Sharing of Risks

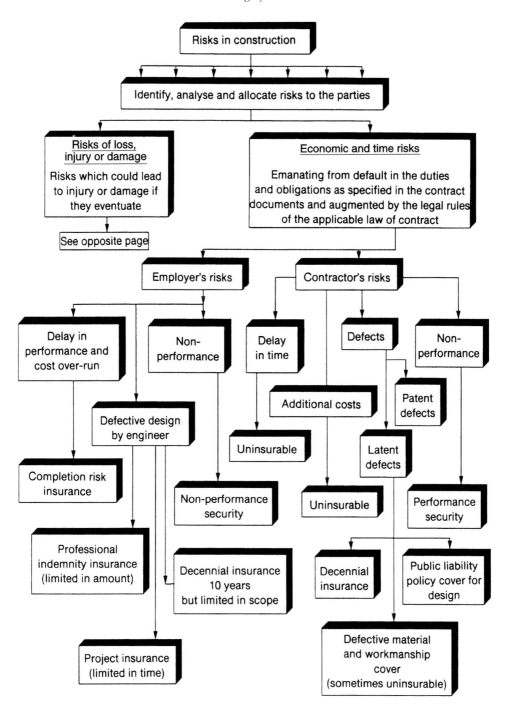

(Fig. 7.2 Contd.)

Chapter 8

The Concepts in Practice

The preceding chapters of this book have considered the main concepts underlying the Red Book. In summary, these concepts are:

— The Red Book is mainly based on a domestic contract;
— Its legal concepts are rooted in the common law system;
— Its wording emerged from a contract written using English legal drafting principles;
— The appointment of a trusted and independent 'Engineer' for design and supervision is central to the whole contract;
— Remuneration is based on re-measurement with a provisional bill of quantities;
— Responsibility and liability is based on the sharing of risks between the contracting parties.

The present chapter focuses on how some of these concepts operate in practice.

8.1 The Red Book in use

The Red Book is a balanced document providing an excellent Standard Form of Conditions of Contract with sufficient rules for a construction contract. Should disputes arise between the parties to a contract, and some disputes are perhaps inevitable in a complex contract, then the Red Book provides a three-tier mechanism for their resolution: the engineer, as a quasi-arbitrator; amicable dispute settlement procedure; and ultimately arbitration is used as a last resort. Whilst many contracts do end in arbitration, it must be noted that for every such contract, there are many others where disputes are resolved without the need to resort to arbitration. Unfortunately, only those contracts which end in arbitration are noted and become part of the statistics.[8.1]

Over the years, use of the Red Book has contributed to the successful completion of many projects around the world. The Red Book has, in fact, been used continuously more than any other form in the international scene for over 40 years. Its use has been endorsed by most international development and financial institutions including the World Bank. However, the last two to three decades have seen the erosion of trust between parties involved in construction which has led to many problems, often resulting in disputes ending in arbitration.

In arbitration, every detail of the construction contract comes under the magnifying glass of lawyers engaged by the parties and, consequently, their zealousness to protect the rights of their clients results in every detail of the contract being minutely scrutinised with hostility.

Employers are often tempted to make changes in the Red Book before using it as the Conditions of Contract for projects they promote. They do so in an attempt to reduce their exposure to some of the financial risks inherent in a construction contract. Consulting engineers have also sometimes attempted to amend the Red Book in response to a bad experience in a previous case. Contractors facing financial losses in a specific contract have sometimes attempted to eliminate similar losses in the future but have often adopted a wrong approach, thereby creating an even more difficult and complex situation.

Following the oil price explosion of the early 1970s, large resources became available to a number of new organisations and state agencies which resulted in the construction of some daringly large and complex projects.

In an article published in 1983, the following statistics were given:[8.2]

'... the lending operations of the development finance institutions increased sharply after the mid 1970s ... Their lending is limited to financially profitable projects and is based on economic criteria ... The cumulative commitments of these institutions (excluding the Iraq Fund for which detailed data are not available) rose from US$1.5 billion in 1975 to over US$11.0 billion in 1981 ...'

In 1978, the World Bank, which is an international financing agency that lends money for projects in developing countries, approved loans and provided disbursements of US$8.8 billion, in 46 countries. This figure increased steadily to US$25.57 billion in 1987, to US$29.04 billion in 1990 and is further increasing, especially in the area of infrastructure as can be seen from the figures quoted in the Bank's annual reports.[8.3]

The new developers depended to a large extent on foreign consulting engineers and contractors whose capacity to provide or retain experienced engineers and technicians was strained to the limits. Personnel with limited experience had to be engaged and were given responsibilities for which they were not well suited.[8.4]

There was a shift in the location of work, and engineers were not only working in developing countries but were also training there. The understanding of the new development agencies and organisations of the essential concepts of the Red Book, or the conviction that these were essential for their proper use, did not prevent them from varying these concepts. In some cases, the lack of proper understanding of the less-experienced engineers and contractors or their advisers of the possible risks did not prevent them from making changes. Whilst some changes were made to resolve a particular, real or expected conflict, others were made simply to shift the balance of risk sharing. In certain circumstances, some of the essential features of the Red Book, as described in the previous chapters, were sacrificed, resulting in conflict between the intention of the document and its changed wording. Some specific areas of conflict will now be examined.

8.2 Areas of conflict

Areas of potential conflict include: the erosion of the relationship of trust between the employer and the engineer, the role of the engineer and his independence in relation to both the employer and the contractor, attempts by a party to offload risk to others and in particular in matters relating to design, absence of a developed legal system, distrust and conflicts between the concepts embodied in the Red Book and the applicable law of the contract.

8.2.1 A relationship of trust

In many cases, the relationship of trust which should have existed between the parties in a construction contract either did not exist or was sacrificed. This relationship of trust must be based on the confidence which the employer has in the engineer's ability and professional integrity. Complete confidence is necessary for a successful delegation of authority from the employer to the engineer and in particular where such authority relates to monetary matters. However, such confidence can only develop with time after the completion of many successful assignments. Where the selection of the 'Engineer' is not based on ability and professional integrity, but on fee competition or other financial considerations, particularly in a field where identification and precise definition of a set of 'Terms of Reference' is almost an impossible task, then such trust even where it originally existed, often disappears.

Some engineers have contributed to the erosion of trust by behaviour which was unacceptable to certain employers resulting in the profession as a whole being branded as a business rather than a profession.[8.5]

In such situations, the employer has often resorted to allocate the authority of the 'Engineer' to himself or to an employee of his without considering the effect which this assumption of authority has on the independent role of the engineer under the Conditions of Contract. Some optimistic contractors did not foresee the risks involved in such a contract and found no reason to be wary of the possible adverse conditions or possible abuse of authority inherent in such a situation.[8.6]

8.2.2 The role of the engineer

The role of the 'Engineer' in the international field has come under attack from both employer and contractor on the question of bias. On the one hand, the 'Engineer' has been accused by the contractor of being biased in favour of the employer because:

— His fee is paid by the owner or developer, referred to as employer under the Red Book;
— He has acted as adviser to the employer prior to construction and may wish to continue this role upon completion of the construction stage;
— He is required to consult with the employer prior to making certain decisions.

Therefore, the engineer has been suspected of colluding with the employer. The case is even stronger against the engineer where he is an employee of the employer, as his future employment might well depend on his actions.

On the other hand, he has been accused by the employer of being biased towards the contractor during the administration and execution of the contract in areas such as awarding extensions of time and in determining amounts of claims, etc., in favour of the contractor. In general, it is sometimes claimed that he is too lenient towards the contractor thereby relieving him of certain of his obligations under the contract.

8.2.3 Avoidance of risk

Sometimes a party to a construction activity seeks to avoid the risks inherent in, and best allocated to, its own function, by attempting to offload them onto others or onto an insurer, even where insurance does not provide the answer.[8.7]

Such attempts have been made through the rewriting of some of the provisions of the Red Book. Inevitably, this has led to conflict and disputes between the parties as there are established principles of allocation of risks which ought to be followed whenever possible.[8.8] These principles are generally based on control over the occurrence of the risks and/or influence over the effect they cause if and when they eventuate (see Chapter 14). Only in certain circumstances can the allocation of risks be successfully based on the optimum ability of a party to perform a specific task related to the project or the inability of any party other than the promoter or the employer who initiated the project to accept a certain risk. There are exceptions to the freedom to allocate risks in accordance with the above principles and they depend on the prevalent legal rules, such as those which apply to exclusion clauses and limitation clauses.

8.2.4 The design function

For technical reasons or for cost-cutting purposes in order to compete on the basis of fees rather than ability or for many other reasons, the designer sometimes seeks to offload some of his design functions to the contractor. Such a shift in the responsibility for design functions invariably results in disputes as to liability for financial loss as a result of any of the following:

(a) subsequent defects;
(b) delays due to late availability of information;
(c) additional work being carried out which was not shown on the drawings but is necessary for the completion of the project;
(d) varied work which has to be implemented to a different standard or in a different form to complete the project; and
(e) many other circumstances.

These disputes, if not resolved amicably, end in arbitration, which then grow from technical into legal issues and arguments never intended in the first place. The cost thus multiplies and increases to limits disproportionate to the cost of the project.

This shift in the function of design may in some cases be totally justified for technical reasons or where the contractor is in control of elements which form part of the decision-making process relating to design. It may, however, be for reasons which in many cases could be considered unwise. As an example of where a shift in the design function may be justified, the contractor on a road project may be asked to design the asphalt mix for an asphaltic concrete layer. This is usually done for the simple reason that the design of such a mix would depend on parameters within the contractor's control, such as the type of aggregate he chooses, its grading, shape and crushing characteristics, and on the characteristics of the asphaltic material and the plant selected by the contractor. An example of where a shift in the design responsibility may be unwise is where the contractor may be asked to design the steel reinforcement bars for a reinforced concrete structure whose dimensions are already designed and finalised by the consulting engineer. Another common example of a shift in the design responsibility is where the contractor is asked to design the mechanical and electrical services in a project which has already been architecturally and structurally designed by the professional design team.

Where the design function is shifted to the contractor, a new matrix of risks is generated and unless these risks are precisely allocated and a definite procedure is adopted to differentiate between the design risks of the professional team and those of the contractor, problems will arise. This is not a new idea or conclusion. In 1964, the Report of the Banwell Committee on the Placing and Management of Contracts for Building and Civil Engineering Work published in the United Kingdom stated that:[8,9]

'The more that is known of what is involved in any project, the less will be the degree of uncertainty against which tenderers will be obliged to make provision; this applies to all construction work but is of particular importance where contractors are being asked to quote firm prices. Only if the work has been settled in all its critical details is it reasonable to expect a contractor to tender for a firm price and a fixed period.'

Time and money being the two elements behind most disputes, such shift in design responsibility in a standard form of conditions of contract which does not recognise the consequences of such a shift in clear, unambiguous and precise terms results in major problems in practice. Shared responsibility of design, whether expressed in the Conditions of Contract or in the specification, is the cause of some major disputes between the construction trinity in practice.

8.2.5 Absence of a legal system

The absence of a developed legal system relating to construction either in some of the countries where the Red Book has been used or in the international context, has resulted in a vacuum where different approaches have been contested. The result has not always been certain. Indeed, in some cases a number of the established principles under the Red Book which require absolute trust in the engineer were rejected by some employers in a country far away from that of the engineer or the contractor.

8.2.6 Distrust of changes

In some cases, the employer has objected to the changes made to the ICE Form in devising the international contract as in the Red Book. Questions have been asked as to 'why the overseas employer should be treated on a different footing and why sauce for the goose is not also sauce for the gander. The cynic might see in this the influence of contractors' associations.'[8.10] The critical analysis made by Ian Duncan Wallace QC, in *The International Civil Engineering Contract* was used by some employers to argue their case that a lot of imbalance exists in the Red Book against the employer's interests.[8.11]

8.2.7 Legal questions

As soon as disputes arise between the parties and come under the scrutiny of lawyers, any conflict which might exist between the common law system, upon which the Red Book is based, and the applicable law of the contract assumes paramount importance. Irrespective of the scale of the dispute, the legal conflict amplifies what could in essence be a technical problem to a complex and long drawn out set of legal arguments, the resolution of which is usually extremely costly. The conflict could involve mandatory provisions, envisaged solutions to contractual problems in the Red Book which may be different from those under the applicable law of the contract, imported provisions from the applicable law of the contract and some specific problems due to these imported provisions.

Mandatory provisions

There are certain imperative and mandatory provisions of the applicable law of the contract. In certain legal systems the concept of public law, as against private law, makes mandatory provisions in some specific areas. The concept of an administrative contract under public law is, for example, different in many substantial aspects from that of an ordinary contract under private law (see Chapter 2 above).

Envisaged solutions

Solutions envisaged in the Red Book may be different from those under the applicable law of the contract. For example, the term 'liquidated damages' which is a legal term rooted in the common law system is the solution envisaged in the Red Book for any default by the contractor in completing the works on time. Clause 47 of the Red Book provides that 'if the contractor shall fail to achieve completion of the works within the time prescribed', then the contractor is to pay to the employer the sum stated in the contract as liquidated damages for such default and not as a penalty. Systems other than the common law do not have an equivalent term for 'liquidated damages'. The problem is multiplied when the language of the contract is not English and where a translation of that term further shrouds the intention of the Form. In Arabic, for example, the term 'liquidated damages' when literally translated means 'fines'.

For both employer and contractor, it would have been essential to establish in the Third Edition of the Red Book a clear statement as to whether the amount of damages stated in the contract is the only remedy for delay, or whether in addition to such amount higher damages can be successfully claimed by the employer.

Furthermore, one must consider that in an administrative contract under certain jurisdictions, where the term 'unliquidated damages' is not known and the accepted solution to a problem of delay is the imposition of a penalty, the absence of a clause on penalties does not waive the right of the administration to impose penalties. The right is an imperative provision of administrative law which an administrative authority is not permitted to waive.[8.12] In this connection, it should be noted that penalties in an administrative contract are not subject to proof of damages.

Another example of such possible incompatibility between the Red Book and the applicable law of the contract is the term 'frustration' as used in the Third Edition. This term is also virtually unknown outside the common law system. (In the Fourth Edition this term has been replaced by the words 'release from performance'.)

Imported provisions

Provisions can be imported from the applicable law of the contract, whether through additional obligations or through deletions from the Red Book. In certain circumstances, the applicable law of the contract imports into the provisions of an agreement between the parties some additional rights and obligations on which the contract is silent. Such silence could be due either to a deletion of a specific condition from the standard form of contract or to a simple omission from that form.

Should these additional rights and obligations form the basis of claims between the parties to the contract, they would do so through the operation of the applicable law and not through a breach of contract. For example, in certain countries forming part of the Romano-Germanic group, unexpected changes in the laws, regulations and bylaws, and unforeseeable economic and political events come under the provisions of certain articles of the civil Code.[8.13] These events are collectively dealt with in what is known as the theory of 'exceptional circumstances'. Should the law applicable to a contract include legal rules which deal with such events, it is usual that any agreement between the parties outside the meaning of the provisions of the legal rule is considered null and void. Therefore, in general it is the civil Code that specifies what happens in such a situation and not the contract between the parties.

Thus, for events under clauses 70 and 71 of the 1992 Red Book and also for some of the events under clause 20, the rules of the applicable law of the contract may stipulate different solutions from those envisaged under the Red Book. Under the law of various Arab countries whose law is rooted in the Egyptian Code, events would fall within the theory of exceptional circumstances if the following three conditions are satisfied:

(a) the existence of an exceptional circumstance;
(b) that circumstance must be unforeseeable; and
(c) that circumstance must be so onerous as to result in great loss if the task is performed.[8.14]

Specific problems due to imported provisions

It is interesting to note in the above connection, that certain legal rules in some jurisdictions coincide in their provisions with the solutions adopted by the Red Book. For example, the provisions under clause 12 are to a large extent echoed in the theory of unforeseen exceptional physical obstructions.[8.15]

In such a case, any amendments or changes made to the standard form could be doubly dangerous as they not only affect the interpretation of the remaining part of the Conditions of Contract, but may also have serious consequences on the interpretation of other clauses in the light of the legal rule. Thus, if clause 12 is deleted, the intentions of the provisions of clause 11 may suffer when that clause alone is interpreted under the legal rule. In an extreme situation, release from performance of the contract may result, an equivalent to the term 'frustration' used in the Third Edition of the Red Book.

Another prominent example of a possible imported provision is that of joint and several liability of the engineer (as a designer and supervisor) and of the contractor (as the constructor) for latent defects or damage resulting to the works. Under the common law, this is a very complex subject where both contract law and tort law must be considered and where the contributory negligence criteria have to be invoked to establish a remedy involving more than one party.

Joint and several liability of the designer/supervisor and the contractor relieves the employer from the necessity of establishing the precise cause of the particular latent damage or defect before instituting legal action. It also avoids the problem of delay in reaching agreement on the appropriate method of repair or restitution. Furthermore, it establishes legal rights of recourse between the designer/supervisor and the contractor. Thus, the questions of who is really liable, for what, and in what proportion, are left to them to resolve, relieving the employer from facing lengthy and costly legal proceedings.

The principle of joint and several liability exists in many countries which follow the Romano-Germanic system. It exists under Swiss law, German law, French law and in a number of other countries whose law is rooted in the French legal system. Where they differ, however, is in the concept of the limitation period for legal actions connected with latent defects. This concept of the period of limitation which is in itself another imported provision from the applicable law to the Red Book, exists in most legal systems.[8.16] A survey of the period of limitation under different jurisdictions was made by the Standing Committee on Professional Liability of FIDIC. The results were published in the Committee's annual report of 1984 and updated subsequently in 1985 and 1986. This period was found to vary in contract from as short a period as one year in Norway to 12 years for contracts under seal in countries within the common law group. The period of limitation for liability in tort was found to vary, however, between one year to infinity as it depends on the date on which the damage or defect is discovered or ought to have been discovered.[8.17]

Under the French legal system and many others that are rooted in it, there is a further requirement in the case of building structures. It is called decennial liability and, as its name indicates, it operates for a period of 10 years. The provisions of decennial liability are usually mandatory and cannot be excluded by agreement of the parties. It has been adopted in, amongst others, Belgium, Italy, Saudi Arabia, Egypt, Kuwait and the United Arab Emirates.[8.18] See also Chapter 11 later.

Another important example of imported provisions into a construction contract is the element of termination which is treated differently under different systems of law. Thus, although the Third Edition of the Red Book includes a clause which deals with 'frustration', this term is unknown in Romano-Germanic systems of law.[8.19] The doctrine of frustration which is rooted in the common law system stipulates that a contract can be annulled if the parties to it cannot achieve their purpose because of unforeseen events. The source of this principle is relatively recent as it dates back to the beginning of this century when a man rented a room on Pall Mall in London to watch the planned coronation of Queen Victoria's son, Edward VII. However, when the coronation was cancelled, he demanded some of his money back, and won. Although this concept of termination in a construction contract is known in legal systems other than the common law, it is embodied in terms such as rescission and release from performance which are the subject of specific legal rules under different jurisdictions. For example, Article 1794 of the French Civil Code gives the employer the right to terminate the contract at any time before completion. This right to terminate which is provided simply for the convenience of the employer, exists independently of the right of termination in case of a breach of contract as provided in the Red Book.

8.3 EIC/FIDIC survey of 1996

A survey of the Red Book users' feelings about contract policy in general and about the Red Book in particular was undertaken by Reading University in England at the request of the European International Contractors and FIDIC. The final survey report was published in June 1996, which provided some interesting data on the following:[8.20]

(a) General contractual issues such as the use of specific clauses for individual projects; the importance of simplicity of expression; powers of the engineer and his impartiality; and payment clauses.
(b) Specific contractual issues relating to the Red Book including: risk distribution; settlement of disputes; applicability to civil law jurisdictions; provisions for money claims; best features of the Red Book; and worst features of the Red Book with specific reference to the clauses which are frequently amended.

8.4 A brief summary of Part I

This chapter concludes Part I. In summary, Part I considered the background of the Red Book; it explained that the Red Book is based on the ICE domestic contract; it demonstrated that the Red Book is rooted in legal concepts based on the common law system; it briefly discussed drafting principles in relation to international construction contracts; it considered the concept of the trusted, independent engineer as embodied in the Red Book; it explained that the contract under the Red Book is a traditional re-measurement contract; and it considered the concept of sharing of risk in construction contracts in general. Finally, the present chapter discussed how these concepts affected the use of the Red Book in practice and the resultant conflicts. The conflicts and the problems resulting therefrom prompted an extensive revision of the Third Edition of the Red Book. Part II gives a detailed commentary on these revisions which resulted in the Fourth Edition.

Part II
The Fourth Edition: A Commentary

Chapter 9

The Revisions – Purposes and Consequences

9.1 Introduction

In 1983, FIDIC decided that the problems facing the user of the Red Book necessitated a revision of the document. As indicated in Chapter 8, the Third Edition was being criticised and changes to the text of the document were regularly made by promoters, developers and employers. Consequently, FIDIC considered that it was desirable to reconcile the document with its practice and use.[9.1]

The drafting committee appointed in 1983 by FIDIC was requested to prepare a Fourth Edition of the document. Its Terms of Reference were briefly summarised as follows:

'— Change only where change is necessary.
— Maintain the basic role of the Engineer.
— Pay close attention to some specific topics such as Bonds and Guarantees, Apportionment of Risk, Insurance, Claims Procedures, Certificates and Payments and Dispute Procedures.
— Endeavour to update the language so that it is more understandable to those charged with administering the Conditions on site.'

Initially, all FIDIC National Member Associations were invited to submit their comments and suggestions drawn from their own experience with the operation of the Third Edition of the Form. At the same time, the European International Contractors (EIC), who were instructed by the Confederation of International Contractors Associations (CICA) to represent them in the consultative role allocated to them by FIDIC, sought the views of their members on the Third Edition. The EIC representatives were assisted by two others from the Associated General Contractors of America (AGC). The outcome of these enquiries was made available to the drafting committee. Contributions from the employers' point of view were put forward, amongst others, by various representatives of the World Bank and by officials from the various Arab Funds, through a number of seminars organised by FIDIC and other organisations. Specialist advice and help was sought and obtained by the drafting committee from an insurance broker on the insurance provisions of the document and from numerous organisations, firms, institutions and individuals who contributed in various ways to the revision of the document. Legal advice was also obtained on the final wording of the Fourth Edition prior to its publication.[9.2]

These comments and suggestions, received by FIDIC from all over the world, were assembled and analysed and much of this material was used in redrafting the Fourth Edition. The subsequent draft of the new clauses was then studied by the various interested parties, such as the EIC liaison group, representatives of the World Bank and of the various Arab Funds, and others. During the drafting process, every one of the 72 clauses was changed. Every clause was carefully drafted and redrafted numerous times, with one particular clause being the subject of seventeen redrafts.[9.3]

Following the above consultation and redrafting process, extensive revisions were made to produce the Fourth Edition of the Red Book. In some clauses there are changes in wording to improve the language, making it simpler and clearer to understand; in other clauses there are substantive and fundamental changes in the intention of the draftsman, in order to tackle and, thus, attempt to resolve some of the problems which faced the user of the Third Edition, as described earlier in Chapter 8.

Whilst the drafting committee remained faithful to the overall guiding principles for the revision, it was admitted by the chairman of the committee, Mr Helge Sorensen, that 'both the sheer number of changes and the depth of the whole exercise goes far beyond what was originally envisaged'. He explained the principles achieved in general as follows:[9.4]

— The basic role of the engineer has been maintained;
— The role of the employer has been made more visible;
— Every endeavour has been made to maintain the overall balance of rights and obligations between the two parties to the contract;
— Current practice has been reflected in the new conditions;
— Procedures have been spelled out in greater detail in an action oriented way;
— The conditions cater for a larger degree of plant (i.e. electrical and mechanical work);
— Greater recognition has been given to the fact that contractors are sometimes called upon to do some design; and
— Efforts have been made to harmonise the Red Book with the Third Edition of FIDIC's Form of Contract for Electrical and Mechanical Works, the Yellow Book.[9.5]

The above were obviously in addition to the principles contained in the Terms of Reference quoted earlier.

Overall, the Fourth Edition adhered to the clause numbering system and the headings of the different clauses of the Third Edition. Whilst there are those who will argue that the amendments do not go far enough or that some are not in themselves an improvement on the former edition, the essential features of the Form as described in the previous chapters have been retained in the Fourth Edition and all the guiding principles are reflected in the new text.

As an example, the role of the employer has been emphasised without diminishing the basic role of the engineer. However, some changes were made to ensure that the Form would be more compatible with legal principles other than those of the common law. Despite these changes, it is certain that arguments and counter-arguments will continue to be voiced as to whether or not the balance of risk has been maintained between the parties to the contract. Such arguments will be in respect of both the changes made and also that particular part of the Form which was not altered.

A new feature is added in some clauses placing an obligation on the engineer to consult with the employer or with both the employer and the contractor prior to reaching his determination on various issues. However, regardless of the obligation placed on the engineer to consult with the employer, it is the engineer who must determine the issues. But, according to the Fourth Edition, the employer must be provided with copies of certain items of correspondence or documents and is thereby kept informed of various issues during construction. For example, under clause 8.1, the contractor is required to give prompt notice to the engineer, 'with a copy to the Employer, of any error, omission, fault or other defect in the design of or Specification for the Works which he discovers when reviewing the Contract or executing the Works'. Furthermore, where the engineer determines a certain issue, the Fourth Edition requires the employer to be kept informed as, for example, in clause 6.4 where the engineer determines any extension of time or costs to which the contractor may be entitled as a result of the engineer's failure or inability to issue certain drawings or instructions.

There is generally a greater demand for communication and adherence to set procedures between the employer, the contractor and the engineer in their dealings with each other under the provisions of the Fourth Edition. Some will argue that this in itself only regulates and reflects what was already implied in the Third Edition and what was already happening to the industry.

The objective of achieving harmony with the FIDIC *Conditions of Contract for Electrical and Mechanical Works* (E & M, Yellow Book) (see Reference 9.5) did not progress sufficiently, but some progress was made in this area.[9.6] In fact, this objective was in conflict with that of 'change only when change is necessary' since the language of the E & M Conditions was dramatically changed to a simpler and clearer language, more comprehensible to the layman. This objective, however desirable, cannot ever be completely attained due to the different nature of the work involved. However, there is now an increased similarity in definitions between the two Forms of Conditions of Contract, and some of the provisions of the Red Book closely follow those of the Yellow Book, sub-clause 37.4 being one such example. Clear differences also exist, as will be seen later in this chapter where both Forms are compared.

Procedures are set out in greater detail in the Fourth Edition and greater emphasis is placed on the time limits that are required for certain actions to be initiated or completed. If followed, these constraints should introduce a greater degree of discipline into the administration of a project than that which formerly existed under the Third Edition. Accordingly, better project management ought to be expected from the provisions under the Fourth Edition.

The sequence of the clauses and the content of the Third Edition has essentially been preserved in the new document, the main exception being a change in the topic covered in clause 53 which is now confined to the procedure for claims. The text on claims was formerly included as part of clause 52 in the Third Edition. In numbering the clauses, numerals replaced alphabetic symbols.

Whilst important amendments have been made to the majority of the clauses, not all of them are discussed in this chapter, as some of the changes are obvious and their effect can be clearly established from a simple comparison of the two texts, as set out later in Part V. A selective approach is adopted here in order to highlight some of the more complex and significant changes. It is, however, important for those who are familiar with the Third Edition and who now intend to use the Fourth Edition of the Form, to be fully acquainted with all the amendments. To those who are interested only in the Fourth Edition, a comparison with the previous edition is perhaps of only academic interest.

At the outset, it is worth noting that the title of the document has been revised in that the word (International) has been removed. The intention of the committee in deleting this word was to emphasise that the document forms a sound basis for both international and domestic contracts, since it has also been successfully used in the latter situation in many places. A reference to this effect is included in the first paragraph of the Foreword to Part I of the document. A further point of note regarding the title of the document is that the Fourth Edition is now issued solely by FIDIC and there is no mention of other organisations which had previously ratified the document (see Section 1.3 earlier). This is probably due to two main reasons. The first reason is that the organisations which previously approved of the Third Edition could not approve all of the amendments proposed by FIDIC. Perhaps this led FIDIC to decide that where differences could not be reconciled, they would adopt a form of wording which in FIDIC's opinion best preserved the balance of risks, rights and obligations of both parties to the contract. The second main reason is that approval of such organisations may have given the erroneous impression to employers that the FIDIC Form is a contractor's orientated document. (See Section 8.2.6.)

In order to explain the changes, FIDIC published a Guide to the use of the revised Form referred to in Reference 9.1. However, despite its considerable length, the Guide does not provide a complete commentary on all the changes. The commentary was in the main confined to the fundamental changes implemented but, as stated earlier, for those who are familiar with the wording of the Third Edition and its intentions and who are now adopting the Fourth Edition, it is essential, if not imperative, that each change should be carefully considered and its consequences established. In Part V of this book, the text of the clauses of both editions is set out with the intention of highlighting all the changes by adopting the following method:

(a) The text of each sub-clause in the Third Edition is set out with the text of the Fourth Edition, as later amended in 1988 and 1992, alongside.
(b) Altered wording is shown in bold type, i.e. both the wording which originally appeared in the Third Edition and its replacement in the Fourth Edition (1992).
(c) Any words in the Third Edition which were omitted in the Fourth Edition (1992) are shown underlined and the location of that omission is marked by superimposing the symbol ∧ in the text of the Fourth Edition (1992).
(d) New words added to the original text of the Third Edition are shown in italics.

No commentary is given in Part V of the Book. Instead, the reader should refer to the commentary in the remaining part of the present chapter as well as the commentary provided in the *Guide to Use of FIDIC Conditions of Contract for Works of Civil Engineering Construction, Fourth Edition*, and referred to hereinafter as the 'Guide'. Commentary now follows in the present chapter on the more important and complex aspects of some of the changes made in producing the Fourth Edition in its 1987 format and on the subsequent amendments in 1988 and 1992. Reference is also made here to the 1996 Supplement to the Fourth Edition and to the impact it has on the relevant clauses of the Red Book.

9.2 Clause 1

Clause 1 of the Fourth Edition deals with definitions and provides a greater number of definitions than its predecessor. In order to assist the reader to understand the document more readily and more clearly, definitions are listed in groups in the present chapter, rather than in alphabetical order. Whilst the full list of definitions needs to be closely studied, the following are deemed worthy of particular mention.

9.2.1 Definition of 'Engineer' under group (a)

Under group (a), the 'Engineer' is defined as follows:

> '1.1(a)(iv) "Engineer" means the person appointed by the Employer to act as Engineer for the purposes of the Contract and named as such in Part II of these Conditions.'

In the Fourth Edition, the definition of the engineer is confined to a person, but sub-clause 1.3 provides that 'words importing persons or parties shall include firms and corporations and any organisation having legal capacity'. Therefore, whilst the engineer will be named usually as an individual, the firm he represents will in effect be involved in that capacity. The definition, however, is no longer qualified by the expression 'from time to time . . .', as in the Third Edition. The intention of this omission, as explained in the Guide, is to prevent the employer from replacing the engineer without the consent of the contractor, and the reason given is contained in the statement that 'The identity of the Engineer (and his reputation) has been a factor in the calculation of the Contractor's tender'. Consequently, the employer would have to discuss his intentions with the contractor if he were contemplating such replacement and presumably he would also have to discuss the identity of the intended replacement as he may not take a unilateral decision on this matter. However, while this may be in the intention of the definition as stated in the Guide, this intention is not repeated in the text of the document itself. If compared with the similar provision in the FIDIC Yellow Book, (see Reference 9.5) a clearer statement is found as can be seen in sub-clauses 2.8 and 46.1 of the Yellow Book, quoted below.

Sub-clause 2.8 of the Yellow Book: replacement of engineer

> 'The Employer shall not appoint any person to act in replacement of the Engineer without the consent of the Contractor.'

Sub-clause 46.1 of the Yellow Book: Employer's default

> 'The Contractor may, by giving 14 days notice to the Employer and the Engineer, terminate the Contract if the Employer:
>
> . . .
>
> (e) appoints a person to act with or in replacement of the Engineer without the Contractor's consent . . .'

Some critics have commented on this unfortunate difference between the two documents in view of the stated guiding principle of achieving harmony between the two texts.

9.2.2 Definition of 'tests on completion'

The Fourth Edition includes a group of new definitions as part of an exercise in providing definitions in respect of the full range of the terms used throughout the FIDIC document. The most worthy of mention is, perhaps, the definition of 'Tests on Completion' stated under group (d), paragraph (i) as:

> 'the tests specified in the Contract or otherwise agreed by the Engineer and the Contractor which are to be made by the Contractor before the Works or any Section or part thereof are taken over by the Employer.'

This specific provision for undertaking tests on completion is a development which will be welcomed by employers as it will constitute a very definite yardstick by which they can measure whether or not a project is satisfactorily completed. The concept is also closely related to the provision in the *Conditions of Contract for Electrical and Mechanical Works*, see Reference 9.5. Detailed provisions for the conduct of such tests are made under clause 37 of the E & M Conditions.

9.2.3 Definitions under group (e)

Two new definitions were introduced in the 1992 Reprint of the Form. These are interim payment certificate and final payment certificate which relate to the provisions of clause 60. The word 'payment' has been added to the two original terms used in clause 60: interim certificate and final certificate.

9.2.4 Definitions under group (f)

Under group (f), terms relating to the works are now defined with two important amendments. The term 'Section' is now a defined term requiring specific identification in the contract in order to operate the provisions of clauses 43 and 46. 'Constructional Plant' is replaced by 'Contractor's Equipment' and such equipment is separated from plant by introducing a new definition for 'Plant'. Plant in the Fourth Edition signifies machinery and apparatus 'intended to form or forming part of the Permanent Works'.

The permanent works are defined in paragraph (f)(ii) as the permanent works to be executed (including plant) in accordance with the contract.

There may be a problem of lack of clarity with the above definitions relating to the words (including plant) in the definition of permanent works, as it is not clear when 'plant' becomes permanent works. Is it part of the permanent works, for instance, when it is being manufactured and stored abroad or does it only become permanent works when it is delivered to the site or only when it is finally installed? From the definition of 'plant' given under paragraph (f)(iv), it would appear that plant is to be deemed part of the permanent works once it is 'intended to form part of the Permanent Works', regardless of where it is situated or stored. On the other hand, one may interpret the phrase 'to be executed (including plant) in accordance with the contract' to mean actually installed and commissioned. This is due to the fact that under sub-clause 60.1 of the Red Book, the contractor's monthly statements may include 'the percentage of the invoice value of listed . . . Plant delivered by the contractor on the site for incorporation in the Permanent Works but not incorporated in such works'.

It is interesting to note that this definition of plant is closely related to the definition under the Third Edition of the E & M Form although it differs from it in one important aspect. In the E & M Form, plant includes materials whereas in the Red Book it does not.

9.2.5 Definitions under group (g)

Under group (g) of the definitions in the Fourth Edition, the term 'cost' has been more clearly defined as exclusive of profit. Where the word cost is to include profit, it is expressly stated so, as in sub-Clause 21.1 where 'full replacement cost' is defined in the context of the sum to be insured by the contractor as including profit. New definitions have also been added for 'day', 'foreign currency', and 'writing'.

Now that 'cost' is clearly defined in the Fourth Edition to be exclusive of profit, it is interesting to distinguish the clauses which refer to cost from those which refer to an alternative form of calculation of payment. Tables 9.1 to 9.3 list the clauses where payments are to be made on the basis of cost or alternatively, on some other method described therein. These clauses have been considered under the following headings:

— where the engineer is to determine in favour of the contractor: Table 9.1;
— where the engineer is to determine in favour of the employer: Table 9.2; and
— where the employer can determine in favour of himself directly against the contractor: Table 9.3.

It is very important to distinguish the circumstances where 'cost' forms the basis of payment from others. For example, sub-clause 12.2(b) states that the engineer shall determine:

> '(b) the amount of any costs which may have been incurred by the Contractor by reason of such obstructions or conditions having been encountered, which shall be added to the Contract Price,'

Table 9.1 Determination by the engineer in favour of the contractor.

Clause description	Where the word 'cost' is used	Where an alternative form of wording is used
Delays and cost of delays	6.4(b)	
Adverse physical obstruction of conditions	12.2(b)	
Setting out		17.1(c), 'an Addition to the Contract Price in accordance with Clause 52'
Loss or damage due to employer's risks		20.3, 'an Addition to the Contract Price in accordance with Clause 52'
Fossils	27.1(b)	
Facilities for other contractors		31.2, 'an Addition to the Contract Price in accordance with Clause 52'
Engineer's determination where tests not provided for uncovering and making openings	36.5(b) 38.2	
Engineer's determination following suspension	40.2(b), 'the amount which shall be added to the Contract Price in respect of the cost incurred . . .'	
Failure to give possession	42.2(b)	
Cost of remedying defects		49.3, 'an Addition to the Contract Price in accordance with Clause 52'
Contract to search	50.1, 'amount in respect of the costs'	
Valuations of variations		52.1, 'Values at the rates and prices etc.'
Variations exceeding 15%		52.3, 'further sum'
Daywork		52.4, 'The Engineer . . . shall . . . be entitled to authorise payment . . . as daywork . . . or at such value therefor as shall in his opinion be fair and reasonable'
Failure to comply		53.4, 'such amount as . . . verified by contemporary records'
Payment of claims		53.5, 'such amount as the Engineer shall determine'
Use of provisional sums		58.2, 'an amount equal to the amount thereof determined in accordance with Clause 52'
Payment to nominated sub-contractors		5.4, 'actual price; in respect of labour . . . the sum entered in the BOQ or as Clause 58.2; and in respect of all other charges and profit, a sum being a percentage rate . . .'
Monthly payments		60.2, 'an amount which he (the Engineer) considers due . . .'
Final certificate		60.8, 'the amount, which, in the opinion of the Engineer, is finally due . . .'

(Table 9.1 Contd.)

Clause description	Where the word 'cost' is used	Where an alternative form of wording is used
Valuation at date of termination		63.2, 'the amount reasonably earned . . . ; and the value of any unused or partially used, any Contractor's Equipment and any Temporary Works'
Damage to works by special risks		65.3, 'an addition to the Contract Price in accordance with Clause 52'
Increased costs arising from special risks	65.5	
Payment if contract is terminated	65.8(b), (e) and (f)	65.8(a), (c) and (d) 'amounts payable in respect of any preliminary items . . .' 'a sum being the amount . . .' and 'additional sum . . . Sub-clauses 65.3 and 65.5'
Payment in event of release from performance		66.1, 'sum payable . . . same as . . . under Clause 65'
Payment on termination		69.3, 'the amount of any loss or damage . . .'
Contractor's entitlement to suspend work	69.4(b)	
Increase or decrease of cost	70.1	
Subsequent legislation	70.2	
Currency restrictions		71.1, 'any loss or damage'

Table 9.2 Determination by the engineer in favour of the employer.

Clause description	Where the word 'cost' is used	Where an alternative form of wording is inserted
Transport of materials or plant		30.3, 'the amount determined by the Engineer . . .'
Rejection (due to testing, inspection of material plant etc.)	37.4	
Default of contractor in compliance	39.2	
Rate of progress	46.1	
Contractor's failure to carry out instructions	49.4	
Payment after termination		63.3, 'the cost . . . and all other expenses incurred by the Employer . . .'
Urgent remedial work	64.1	
Payment if contract terminated (possibly)		65.8, 'any other sums which . . . were recoverable . . .'
Payment in event of release from performance (possibly)		66.1, 'the sum payable . . . the same . . . under Clause 65 . . .'
Increase or decrease of cost	70.1	
Subsequent legislation	70.2	

Table 3.3 Determination by the employer against the contractor.

Clause description	Where the word 'cost' is used	Where an alternative form of wording is inserted
Remedy on contractor's failure to insure		25.3, 'the amount so paid'
Liquidated damages for delay		47.1, 'the relevant sum stated in the Appendix'
Costs for the purposes of clauses 63 (i.e. remedies)	54.6	54.6, 'all sums properly paid . . .'
Certification of payments to nominated sub-contractors		59.5, 'the amount so paid . . .'

This means that the contractor is entitled to recover only 'costs' which he may have incurred as a result of encountering physical obstructions or conditions. Sub-clause 12.2 goes on to state that this determination by the engineer as to the amount of any costs must also take account of any instruction which the engineer may issue to the contractor in connection therewith. However, such instructions may involve the operation of sub-clause 52.4 on dayworks which provides that:

> 'The Engineer may, if in his opinion it is necessary or desirable, issue an instruction that any varied work should be executed on a daywork basis.'

Under sub-clause 52.4 the engineer is 'entitled to authorise payment for such work, either as daywork,... or at such value therefor as shall, in his opinion, be fair and reasonable', that is, the contractor is entitled to a fair and reasonable price for work carried out at the engineer's instruction.

In contrast, under clause 27 on fossils, the contractor is entitled to recover only his costs whether this is due to encountering fossils or to carrying out an instruction issued by the engineer for the purpose of dealing with such an encounter. In general, however, where the contractor is required to act and thereby actually do additional work, he is entitled to profit by virtue of the application of clause 52.

9.2.6 Definition of 'approved'

The definition of approved in sub-clause 1(1)(n) of the Third Edition, which necessitated that approval to be in writing, has been incorporated in sub-clause 1.5 of the Fourth Edition as part of the requirement for notices, etc. to be in writing. An additional requirement has also been included in sub-clause 1.5 to the effect that any consent, approval, certificate or determination must not be unreasonably withheld or delayed.

9.3 Clause 2

Clause 2 deals with the engineer and the engineer's representative. Whilst there are many clauses which regulate the role and authority of the engineer to act as the employer's representative within the contract between the employer and the contractor, clause 2 in conjunction with clause 67 contains the main provisions by which this role and authority are regulated.

These two revised clauses have been greatly expanded from their predecessors in the Third Edition, but the traditional role of the engineer has been preserved. The Red Book is still based on the requirement that the engineer is to act impartially. Accordingly, where circumstances dictate that the employer should restrict the engineer's role by vesting the whole of the decision-making process in the employer or an employee of the employer, then an alternative set of conditions of contract should be adopted. If decision making is totally removed from the engineer's role and allocated to the employer, then the credibility of the impartial adjudication role presently given to the engineer throughout the document, and specifically so under sub-clause 2.6, could no longer apply.

A further aspect which must be considered in any modification of the engineer's role with respect to vesting the decision-making process in the employer is that the employer may well have to consider the need to obtain professional indemnity insurance for himself, if he were to assume, as he must, the responsibility for a decision-making role on technical and professional matters. As every decision carries the concomitant risk of it being the wrong one, the responsibility for the consequences of a wrong decision would lie with the decision maker. Such responsibility may lead to the necessity for providing indemnity in respect of the cost of remedying the adverse effects. Where the indemnity is not specifically incorporated in the wording of the contract, it is generally imported into it in most legal systems through the applicable law of the contract between the parties. If such indemnity is not shifted to an insurer, the liability is self retained. (See also Chapter 10.)

Bearing in mind the impartial role of the engineer, the new text of clause 2 includes many important changes from its predecessor. These are discussed below.

9.3.1 Requirement for consultation

Sub-clause 2.1, through the wording in Part II of the Conditions, identifies the precise role of the engineer by indicating the particular clause of the contract where the engineer is required to obtain the specific approval of the employer before carrying out his duties. A new sentence is added in the Fourth Edition to avoid any doubt about whether or not approval had been sought and obtained by the engineer from the employer prior to his undertaking any of his duties. The new wording deems any requisite approval to have been given once authority is exercised by the engineer, thus placing the responsibility on him to ensure that approvals are sought and obtained whenever necessary. The Fourth Edition introduces a new term, 'due consultation', in a number of clauses. This term appears in clauses: 6.4, 12.2, 27.1, 30.3, 36.5, 37.4, 38.2, 39.2, 40.2, 42.2, 44.1, 44.3, 46.1, 49.4, 50.1, 52.1, 52.2, 52.3, 53.5, 64.1, 65.5, 65.8, 69.4 and 70.2. It requires the engineer to consult with the employer and the contractor before taking his decision as to what action is appropriate in the circumstances of the particular clause. Whilst this requirement for consultation between the engineer and the employer is an attempt to involve the employer more actively with the contract administration, the consultation with the constructor is in fulfilment of his investigative role.

9.3.2 *Responsibility for delegation*

Sub-clause 2(2) of the Third Edition is now replaced in the Fourth Edition by sub-clauses 2.2, 2.3 and 2.4 which deal not only with the engineer's representative and the engineer's authority to delegate, but also with a new person referred to as 'assistant to the engineer's representative'. It is now clearly stated that the appointment of the engineer's representative is the sole responsibility of the engineer to whom the engineer's representative is responsible. The appointment of assistants is made by either the engineer or the engineer's representative. The engineer or his representative are required to notify the contractor of the names, duties and scope of authority of such assistants.

9.3.3 *Requirement for writing*

A new sub-clause has been added to the Fourth Edition requiring that instructions given by the engineer must be in writing, or if oral, must be subsequently confirmed in writing. This requirement which was part of clause 51 of the Third Edition has now been transferred to clause 2 of the Fourth to give the principle of record keeping a wider application. However, it should be noted that some reference to 'instructions in writing' remains in the Fourth Edition, for example sub-clause 48.1. Accordingly, it may be taken that when the word 'instruction' appears on its own, it may be given either in writing or orally followed by confirmation in writing. On the other hand, when the words 'in writing' are added to the text, oral instructions may not be given. An example of the former is now in Clause 40.3 where the word 'written' has been omitted in the 1992 Reprint of the Fourth Edition.

9.3.4 *Express requirement for impartiality*

Finally and, perhaps, most importantly, sub-clause 2.6 of the Fourth Edition is a new provision with the express requirement that the engineer must act impartially whenever he gives a decision, opinion or consent; or expresses his satisfaction or approval; or determines value; or takes any action which may affect the rights and obligations of the parties. This express requirement replaces what was an implied term in previous editions of the Red Book. It is notable that the new provision incorporates a reference to clause 67 indicating that any such decision, opinion, consent, etc., may be opened up, reviewed or revised by an arbitrator.

9.4 Sub-clause 5.2

Sub-clause 5.2 deals with priority of contract documents, providing a hierarchy between the various documents forming the contract which must be taken to be mutually explanatory of one another. Whilst any discrepancies or ambiguities which arose in a contract under the Third Edition were to be 'explained and adjusted' by the engineer, under the Fourth Edition they must be explained and adjusted in accordance with a priority listing of the contract documents specified in sub-clause 5.2. It is to be noted that Part II of the Fourth Edition provides alternative forms of wording where other orders of precedence might be preferred or, alternatively, where it is decided that no order of precedence should be included.

One further change which should be noted is the deletion of the last sentence of sub-clause 5(2) of the Third Edition. This sentence catered for the situation where ambiguity or discrepancy in the contract documents involved the contractor in costs which could not reasonably have been foreseen by him. The deletion of this provision makes it clear that if the engineer issues an instruction under sub-clause 5.2, it constitutes a variation, the valuation of which will have to be made under clause 52.

9.5 Sub-clauses 6.1, 6.4 and 6.5

Sub-clauses 6.1, 6.4 and 6.5 deal with delays, the cost of the delay of drawings and the failure by the contractor to submit drawings.

Sub-clause 6.1 includes a new provision dealing with drawings to be supplied by the contractor. It also provides that unless it is strictly necessary, no document provided by the employer or the engineer is to be communicated to a third party by the contractor or used by a third party.

Sub-clause 6.4 is the first provision in the Red Book where the employer's role in the contract has been made more visible. This is achieved through two new provisions in this sub-clause: the first is through the inclusion of the process of due consultation, and the second is through the requirement that the engineer, after determination of any extension of time and additional costs, should give the employer a copy of his notification to the contractor of such determination.

Sub-clause 6.5 is a new provision added in the Fourth Edition to cater for the situation where the contractor contributes to a delay being the subject of sub-clause 6.4. The engineer is required to take any failure by the contractor to submit drawings, specifications or other documents which the contractor is required to submit, into account when making his determination pursuant to sub-clause 6.4. The inclusion of sub-clause 6.5 emphasises the fact that provision is made under the Fourth Edition for certain elements of design to be carried out by the contractor (see clause 7 of the Fourth Edition).

9.6 Clause 7

Clause 7 in the Fourth Edition has been extended to provide for the situation where the contractor is entrusted with the design of part of the permanent works. The procedure for submission of drawings, specifications, calculations and other information is detailed in the new provisions of sub-clauses 6.1 and 7.2. Paragraph (b) of sub-clause 7.2 caters for projects where elements of mechanical and electrical plant forming part of the contract are to be designed by the contractor. It is worth noting that, under this new provision, the works will not be considered to be completed for the purposes of taking over until the operation and maintenance manuals, together with drawings on completion, have been submitted to and approved by the engineer. Sub-clause 7.3 is a new clause dealing with responsibility for design when carried out by the contractor expressly stating that approval by the engineer of such design in accordance with sub-clause 7.2 does not relieve the contractor of any of his responsibilities under the contract.

9.7 Clause 8

A second paragraph was introduced in clause 8 of the 1992 Reprint of the Fourth Edition. In effect, it requires the contractor to give prompt notice of any defect he discovers in the design of or the specification for the works when 'reviewing the Contract or executing the Works'. The words quoted indicate that this duty is imposed on the contractor as soon as the contract comes into existence and not before. However, as soon as a contract is formed, the contractor is required to notify the employer or the engineer of any defects discovered during the tendering period.

A number of important aspects flow from the addition of this duty to the contractor's obligations. First, the principle is sound from a theoretical point of view since it would be much better and less costly to deal with any defect as soon as it is discovered. In fact, this added duty is taken from a similar clause recommended by the World Bank to its borrowers in its Standard Bidding Documents for major works. If the existence of a duty on the contractor to inform the engineer and employer of any defect discovered by him had hitherto been in doubt, it is now clear that such discovery must be notified. However, from a practical point of view, it would be extremely difficult to enforce such a duty, except in a few rare cases, because of the difficulty in proving that the contractor had known of the defect but failed to notify the engineer and the employer of its existence. If such knowledge is not proven, no sanction can be imposed.

Secondly, this new wording of clause 8 introduces the question of whether there is an implied duty on the contractor to check the design of or specification for the works. Such a potential shared responsibility between the contractor and the engineer would lead to an even greater number of, and more bitter, disputes than exist at present in the construction industry. Up to this new wording, the courts in some jurisdictions have wisely, as a matter of policy, resisted imposing such an implied obligation on the contractor.

In addition to the above new wording, a statement similar to that in sub-clause 7.3 has been added to sub-clause 8.2 in relation to the responsibility of the contractor for design carried out by him notwithstanding any approval by the engineer.

9.8 Clause 10

Clause 10 in the Fourth Edition has been substantially amended and expanded to incorporate two new sub-clauses. The former term 'performance bond' is now replaced by 'performance security', and the wording of sub-clause 10.1 is now in respect of 'proper' performance instead of the previously imprecise 'due' performance. Many argued that 'due' merely allocated an obligation to complete, whereas 'proper' has the added obligation of doing so with satisfactory material and workmanship and in accordance with the specification and drawings.

Sub-clauses 10.2 and 10.3 are new. The first deals with the period of validity of the performance security which is now expressly stated as ending with the issue of the defects liability certificate. The second stipulates that prior to making a claim under the performance security, the employer should notify the contractor of the nature of the default in respect of which the claim is to be made.

It should be noted in this connection that whilst such notification is a condition precedent to making a claim under the performance security, the condition is only slightly restrictive in that the default does not have to be proven at that stage (see later in Chapter 15).

9.9 Sub-clause 12.2

While the overall provisions of this sub-clause have not been altered in the Fourth Edition, a number of significant changes have been incorporated into the new wording. It is now evident that in order to warrant consideration under this clause in the Fourth Edition, both obstructions and conditions must be physical, as opposed, for example, to being administrative.

Furthermore, the role of the employer is again emphasised in that the engineer must 'duly consult' with the employer prior to reaching his determination on any entitlement due to the contractor. When the engineer determines any such entitlement, the employer should receive a copy of the determination, as notified to the contractor.

The new wording of sub-clause 12.2 also includes a provision for direct determination of an extension of time in favour of the contractor pursuant to the provision of clause 44. This express reference to extension of time clarifies what some commentators considered to be a vague area under the Third Edition.

Another important change in the wording of this sub-clause is the shift from 'approval' to 'acceptance' by the engineer of any proper and reasonable measure taken by the contractor pursuant to this clause.

The marginal note of this sub-clause was changed in the 1992 Reprint of the Fourth Edition by substituting the words 'Not Foreseeable' for the word 'Adverse' in the title of the sub-clause: 'Adverse Physical Obstructions or Conditions'. This change could be interpreted as suggesting that unforeseen physical obstructions or conditions may produce an advantageous as well as an adverse effect. However, the sub-clause deals only with an effect which entitles the contractor to an extension of time and/or additional payment and is a presumption of only an adverse effect. It is not clear as to whether the contractor is expected to give notice under this clause irrespective of whether or not the conditions were of an adverse nature. In any case, as stated in sub-clause 1.2

of the Form, headings and marginal notes should not be taken into consideration in the interpretation of the contract.

9.10   Clause 13

The wording of the last sentence of this clause was shortened in the 1992 Reprint by the deletion of both the express reference to the engineer's representative and to clause 2. However, the words 'or his delegate' were added in brackets to that sentence. The deletion of the reference to the engineer's representative was also made in clause 15 'Contractor's Superintendence', see Section 9.12 below. This deletion was obviously made in view of the power which is given to the engineer under sub-clause 2.3 'Engineer's Authority to Delegate' but it is puzzling as to why it was considered necessary to add the words 'or his delegate' in clause 13, since no such reference is made elsewhere. If the reason for adding these words is to counteract the forceful effect of the word 'only' in that sentence, then it would have been more logical to delete it leaving the text much clearer.

9.11   Sub-clauses 14.1 and 14.3

Although sub-clause 14.1 of the Fourth Edition closely follows its predecessor in the Third Edition, there are two important changes in its wording. Both of these changes occur in the provision that the contractor must:

> '... submit to the Engineer for his approval a programme showing the order of procedure in which he proposes to carry out the Works.'

which has changed to:

> '... submit to the Engineer for his consent a programme, in such form and detail as the Engineer shall reasonably prescribe, for the execution of the Works. The Contractor shall, whenever required by the Engineer, also provide in writing for his information a general description of the arrangements and methods which the Contractor proposes to adopt for the execution of the Works.'

The first change is from the word 'approval' to 'consent'. The second is the change in the requirements of what the programme should show: 'the order of procedure' is now changed to 'such form and detail as the Engineer shall reasonably prescribe, for the execution of the Works'. This change could have far-reaching consequences if the engineer reasonably prescribes that, for example, the contractor should submit in support of his programme a critical path analysis in such detail so as to show the resources he intends to employ in carrying out the works.

Sub-clause 14.3 now requires the contractor to provide a detailed cash flow estimate, in quarterly periods, of all payments to which the contractor will be entitled under the contract.

9.12 Clause 15

As with clause 13 referred to above, the wording here has also been altered by the deletion of the reference to the engineer's representative and to clause 2, but unlike clause 13, there is no reference to delegation by the engineer.

9.13 Clause 19

Clause 19 deals with safety, security and the protection of the environment. The text of this clause has been expanded to incorporate new provisions placing greater responsibility on the contractor for the safety of all persons entitled to be on the site and also for the protection of the environment on and off the site. A new sub-clause is also added to deal with the employer's responsibilities when he is expected to carry out work on the site with his own workmen.

9.14 Clause 20

Clause 20 deals with: care of works; responsibility to rectify loss or damage; and loss or damage due to employer's risks. The wording of this clause has been considerably altered in order to provide a simpler, clearer and more logical language. The important changes effected by this alteration can be summarised as follows:

(a) The clause is now divided into three sub-clauses providing a logical flow in the conditions, starting with the care of the works and continuing to the responsibility for rectifying any loss or damage which may occur and how that responsibility is allocated depending on the type of risk encountered.
(b) The excepted risks are now referred to as the employer's risks, probably in an attempt, although not admitted, to clarify that the responsibility for these risks is the employer's.
(c) The employer's risks are now listed, making the language of the clause simpler and clearer to comprehend. The definition of two of these risks is altered in the Fourth Edition. These are the design risk and the risk of the forces of nature. As to the design risk, the word 'solely' is now deleted from the previous phrase: 'a cause solely due to the Engineer's design of the Works'. The Guide to the use of FIDIC Conditions referred to earlier in Reference 9.1 includes the following reasoning for this deletion:

> 'It has been decided to change this so that the contractor is only responsible for errors in the designs for which he has been responsible. This is considered to be more balanced than was the Third Edition. It is now possible to establish design responsibility from the outset, rather than applying a test after the event (i.e., was there contribution to the error by another party or was it solely caused by the Engineer?). In practice that question was exceedingly difficult to answer.

Should there be damage to the works in circumstances where an error in design is a contributory factor, the Contractor is only relieved of responsibility to the extent that the damage was caused by a design not provided by him.'

The definition of the risk of forces of nature has also been altered in clause 20, thus removing two difficulties. The Third Edition referred to risks which could not be foreseen and to risks from the forces of nature which could not be insured against. Both of these references have been removed for the following reasons given in the Guide: The words 'could not be foreseen . . . could be argued that all risks of the forces of nature can be foreseen' and the words 'Could not insure against have been removed as this test is difficult to apply . . .'.

In this regard, the Fourth Edition provides that it is an employer's risk if the works are damaged by an operation of the forces of nature against which the contractor could not reasonably have been expected to take precautions. Unfortunately, the test of reasonableness here is also difficult to apply (see Chapter 14 later).

9.15 Clause 21

Clause 21 deals with: insurance of works and contractor's equipment; scope of cover; responsibility for amounts not recovered; and exclusions. As in clause 20, the wording of this clause has also been considerably altered to clarify the insurance provisions under the Red Book. The important changes which emerge from this alteration may be summarised as follows:

(a) The definition of the sum insured has been altered with an additional sum of 15 per cent of the 'replacement' cost now being required to be included in the insurance cover, unless otherwise specified in Part II of the Conditions. This additional sum is added to take into account additional costs of and incidental to the rectification of the loss or damage.

(b) A new sub-clause is added to clause 21 in the Fourth Edition, to identify any permitted exclusions from the insurance cover to be provided under the contract. This additional sub-clause is in recognition of the fact that insurance policies do, in general, contain exclusions. The permitted exclusions as identified in the Fourth Edition refer to some of the employer's risks listed in sub-clause 20.4. It is important to recognise that the contractor would be in breach of contract if the insurance cover he provides contains exclusions not permitted by this clause. Furthermore, as the employer's risks (e) to (h) of sub-clause 20.4 are not included in the list of permitted exclusions under sub-clause 21.4, the insurance policy required under clause 21 must provide cover against these risks which include, amongst others, defective design; defective material; and defective workmanship for the whole project. These risks are usually not insurable, see Section 14.5.4 of this book. This is obviously a mistake which should have been corrected in the 1992 Reprint, but no such correction was made. Perhaps a similar position was adopted to that outlined by one insurance broker who stated publicly that he saw no difficulty in a contractor being in breach of contract in such circumstances! Such an attitude is erroneous and contractors in practice seek and obtain letters from their insurers confirming that view in an attempt to excuse this breach.

The 1992 Reprint of the Form did, however, resolve two minor problems in clause 21. In sub-clause 21.1, the term 'replacement cost' as qualified by defining cost is inclusive of profit within the context of this clause. Further, in sub-clause 21.4, the word 'where' in sub-paragraph (a) was replaced by 'whether'. This was obviously a typographical error as the word 'whether' appears in the matching sub-clause 65.6.

9.16 Clause 23

This clause reinstates joint insurance in respect of third party liability which had been abandoned in the Third Edition. This third party insurance is required to be effected in the joint names of the contractor and the employer. Furthermore, a new sub-clause is added to provide for the insurance to apply to the contractor and the employer as separate insureds, thus acting as third parties to each other.

9.17 Clause 25

This clause has been expanded to incorporate standard insurance requirements which usually apply to the insurance clauses 21, 23 and 24. Accordingly, the contractor is now required under sub-clause 25.2 of the Fourth Edition to notify the insurers of any 'changes in the nature, extent or programme for the execution of the works and ensure the adequacy of the insurances at all times in accordance with the terms of the Contract . . .'.

A new provision is also added under sub-clause 25.4 to deal with any failure to comply with the insurance conditions as stipulated by the policies by requiring the parties to indemnify each other against all losses and claims arising from such failure.

9.18 Clause 27

Clause 27 deals with fossils. The provisions of this clause have been expanded to allow for the process of 'due consultation' with the employer and the contractor to take place. In this regard, the provision in the Third Edition that the contractor is to carry out the engineer's representative's orders at the expense of the employer has now been altered to a provision whereby it is for the engineer to determine after due consultation, both time extension and additional payment in respect of costs. The change from 'expense' to 'cost' incurred by the contractor as a result, mentioned in Section 9.2.5 above, should also be noted.

9.19 Clause 28

Sub-clauses 28.1 and 28.2 reiterate what was formerly provided by clause 28 of the Third Edition, except that an added sensible protection has been included in favour of the contractor. This protection is in respect of the indemnity which the contractor must provide in favour of the employer for any infringement of patent rights. The contractor is not now required to provide indemnity where such infringement results from his compliance with the design or specification provided by the engineer.

9.20 Clause 30

Sub-clauses 30.2 and 30.3 have been redrafted to differentiate between the responsibility for the transportation of the contractor's equipment and temporary works, and that for the transportation of materials and plant as specified under the terms of the contract. The former relates to responsibility for matters selected by the contractor and, thus, his responsibility, whereas the latter is for elements outside his control and, accordingly, the responsibility is allocated to the employer in this regard.

It is interesting to note that sub-clause 30.3 provides for the only situation in the Red Book where the employer is expected to consult directly with the contractor without the presence of the engineer.

9.21 Clauses 34 and 35

These clauses have been reduced in Part I of the Fourth Edition to a general statement in respect of the engagement of staff and labour, in addition to the returns of labour and the contractor's equipment. However, in Part II of the Fourth Edition, it is noted that a number of sub-clauses should be added to these two clauses to take account of the circumstances and locality of the particular works. Various examples of such sub-clauses are given in Part II and the employer is invited to add these as relevant.

9.22 Sub-clause 36.5

This sub-clause has been added in the Fourth Edition to provide for the situation where a test required by the engineer in accordance with sub-clause 36.4 shows the materials, plant or workmanship to be in accordance with the provisions of the contract. In this clause, 'due consultation' with the employer and the contractor is also specified as a pre-condition to the determination by the engineer for any extension of time and for additional payment of costs.

9.23 Clause 37

Clause 37 has been greatly expanded from the version in the Third Edition. The provision of the new sub-clauses 37.2 to 37.5 describe in detail the various procedures which should be followed for the testing of materials and plant in addition to the consequences of both satisfactory and unsatisfactory testing.

9.24 Clause 40

A number of amendments have been made to clause 40 in an attempt to clarify its intent. These are:

(a) Paragraph 40.1(b) has been expanded to include reference to 'breach of contract' and the words 'or for which he (the Contractor) is responsible' have been added. It is understood that these additions were included on the basis of legal advice obtained which indicated their necessity to avoid complications which might arise in instances where sub-contracts are involved. (Similar additions have also been made to sub-clause 44.1(e) and to the last sentence of sub-clause 51.1.)
(b) The important provisions regarding the engineer's determination of extensions of time and any extra amount to be added to the contract price are now given under a new sub-clause 40.2. Such determination is once again subject to 'due consultation' by the engineer with the employer and the contractor. It should also be noted that the requirement in the Third Edition that the contractor gives written notice of his intention to claim within twenty-eight days of the engineer's order has been deleted.
(c) The new wording of sub-clause 40.3 of the Fourth Edition has an express reference to clause 69, default of employer, in case of a suspension lasting more than 84 days which does not fall within the definition in sub-clause 40.1. In such cases, the suspension is to be treated as an event of default by the employer and the provisions of sub-clauses 69.2 and 69.3 are to apply. Furthermore, in the 1992 Reprint of the Fourth Edition, the word 'written' at the end of the first line was deleted, providing an option for an instruction given by the engineer to be either in writing as required in sub-clause 2.5 or given orally followed by a confirmation in writing.

9.25 Clause 41

Unlike the Third Edition, there is no period for commencement in the Fourth Edition. The contractor's obligation is to commence as soon as is reasonably possible after receipt of a notice to this effect from the engineer. This notice, however, must be issued within a specific period of time after the date of the letter of acceptance, which is stated in the Appendix to Tender. The commencement date is now a defined term, in sub-clause 1.1(c)(i), and is established by the engineer when he issues a notice to the contractor to commence the work.

In many cases, this notice will probably be issued by the engineer as soon as the contractor has received and acknowledged the letter of acceptance from the employer. However, there may be occasions due to the site not being immediately available, or the employer having to take some procedural steps prior to commencement of the works, when the engineer may be obliged to delay the issue of the notice to commence the work. In any case, the contractor will know at the tender stage the permitted extent of any delay, as the time within which the notice to commence must be issued will have been stated in the Appendix to Tender.

9.26 Sub-clause 42.3

The word 'wayleaves' was replaced in the 1992 Reprint of the Fourth Edition by 'rights of way' in the text of the clause and in the marginal note. This is presumably due to the fact that the former term is relatively unknown internationally.

9.27 Clause 44

The text and format of clause 44 have been improved and expanded from those of its predecessor in the Third Edition. In sub-clause 44.1 'Extension of Time for Completion,' the list of events giving rise to entitlements to extension of time are presented in tabular form which improves the clarity of the wording.

The number of events themselves has also been expanded to include what was originally taken in some cases as implied. For example:

(a) The 'nature' of extra or additional work is now added alongside the amount of such work;
(b) 'Any delay, impediment or prevention by the Employer' is now explicitly referred to as an event giving rise to an entitlement to extensions of time. By deliberately inserting such an event, the risk of the contract being invalidated by such shortcomings on the part of the employer has been greatly reduced. While in the past, any such shortcomings were dealt with by mutual consent of the parties, as entitling the contractor to an extension of time, it is now clear that such entitlement is the remedy to be adopted;
(c) Paragraph (e) of sub-clause 44.1 now contains reference to 'breach of contract', and the words 'or for which he (the Contractor) is responsible' have been added to include reference to sub-contractors; and
(d) The role of the employer is again emphasised in sub-clause 44.1 through both the 'due consultation' provision and the requirement that the employer is to be provided with copies of whatever determination is made.

The procedure for notification of the occurrence of such events referred to above has also been amended. Consequently, under sub-clause 44.2, the contractor must now provide:

(a) A written notification of the delaying event to the engineer with a copy to the employer; and
(b) Within 28 days, or such other reasonable time, the contractor must supply detailed particulars of any extension of time to which he may consider himself entitled in order that such submission may be investigated at the time.

Where an event has a continuing effect and where it is not practicable for the contractor to submit detailed particulars within the stated time, a new provision has been added under sub-clause 44.3 to cater for the procedure in such a situation. Once again, this procedure includes the 'due consultation' provision by the engineer with the employer and the contractor. Attention should be given to the final sentence in this sub-clause where it is stated that: 'No final review shall result in a decrease of any extension of time already determined by the Engineer'.

9.28 Clause 46

Clause 46 deals with rate of progress. As in clause 14, 'approval' of the engineer is now replaced by his 'consent'.

An important provision is now added in the Fourth Edition which empowers the employer, through a determination by the engineer after 'due consultation', to recover extra supervision costs resulting from any steps taken by the contractor in meeting his obligations under this clause.

9.29 Clause 51

Clause 51 deals with variations; and instructions for variations. In addition to the general changes in the wording, a number of important changes were introduced into the text of this clause. These are:

(a) The engineer's authority to instruct the contractor to omit work does not extend to work to be carried out by the employer or by another contractor;
(b) The engineer's authority is now extended to include the issue of an instruction to change any specified sequence or timing of construction of any part of the works; and
(c) A new sentence is added expressly providing that where a variation is necessitated by a default of, or breach of contract by the contractor or any sub-contractor for which he is responsible, any additional cost incurred is to be borne by the contractor.

9.30 Sub-clause 52.3

In addition to the change from 10 per cent to 15 per cent as the limit which triggers the operation of this sub-clause, its wording has been expanded to elaborate on many of its provisions which were ambiguous in the Third Edition. The most important of these changes is the new last sentence which expressly provides that the sums determined by the engineer under this sub-clause are based only on the amounts by which the additions or deductions are in excess of 15 per cent of the effective contract price.

9.31 Clauses 53 and 54

The provisions of sub-clause 52(5) of the Third Edition have been modified and transferred to sub-clauses 53.3 and 53.5 of a new clause 53 in the Fourth Edition. The new clause 53 also incorporates provisions relating to the manner of dealing with claims; their substantiation within specific time limits and by contemporary records; and their payment. In addition to the claims made under the provisions of the contract, this clause encompasses claims made under the applicable law of the contract.

Clauses 53 and 54 of the Third Edition have now been combined in the Fourth Edition into one clause, clause 54, under the title 'Contractor's Equipment, Temporary Works and Materials'. In clause 54 of the Fourth Edition, there are new provisions to deal with the conditions of hire of contractor's equipment.

9.32 Sub-clause 57.2

This is a new sub-clause providing for the contractor to submit to the engineer within certain time limits a breakdown for each of the lump sum items contained in the tender. It also provides that such breakdown is subject to the approval of the engineer.

9.33 Clause 60

Clause 60 deals with certificates and payments. The provisions of clause 60 of the Third Edition were mainly given in Part II of the document, as it was considered that procedures for payments differed so much between one organisation and another that it was not practicable to draft a clause for general application. This view changed when the Fourth Edition was drafted, and a specific procedure is now set out for:

— the contractor when making applications for payment or when submitting a draft final statement;
— the engineer in certifying; and
— the employer in making payment.

It should be noted that sub-clause 60.10 of the Fourth Edition provides for payment of interest at a fixed rate upon all sums unpaid on the appointed date, but without prejudice to the contractor's entitlement under clause 69 (default of the employer).

Additional sub-clauses are also included in Part II of the Fourth Edition to cover certain matters relating to payments, such as when payments are to be made in various currencies, or where the place of payment is defined. These sub-clauses are to be added to the main document as relevant.

A number of amendments were made in the 1992 Reprint of the Fourth Edition, as listed below:

(a) The words 'or otherwise' were added after the words 'under the Contract' in each of sub-clauses 60.1(e), 60.6(b) and 60.8(a).
(b) In sub-clause 60.8(b), the words 'under the Contract other than Clause 47' have been changed to read 'other than under Clause 47'.
(c) The words 'or otherwise' were added in sub-clause 60.10 after the words 'under Clause 69' at the end of the sub-clause.
(d) In sub-clause 60.2(b), 60.4, 60.7 and 60.10, changes were made to take effect of the new definitions added in the 1992 Reprint, i.e. 'Interim Payment Certificate' and 'Final Payment Certificate'.
(e) In sub-clause 60.6, 'Final Statement', a new paragraph has been added requiring the engineer to deliver to the employer an interim payment certificate for those parts of the draft final statement which are not in dispute.
(f) In sub-clause 60.2, the words 'certify to the Employer' have been changed to read 'deliver to the Employer an Interim Payment Certificate stating', the word 'thereof' has been changed to read 'of such statement' and the word 'he' has been changed to read 'the Engineer'.
(g) In the eighth line of sub-clause 60.3, sub-para (b), the word 'ordered' left from the relics of previous Editions to read 'instructed'.
(h) In the second line of sub-clause 60.5, after the word 'Engineer', the words 'six copies of' have been added.
(i) In the second line of sub-clause 60.6, after the word 'consideration', the words 'six copies of' have been added.
(j) The final paragraph of sub-clause 60.6 described in item (e) above ends with the provision that any dispute in relation to the final statement may be settled in accordance with clause 67 of the Form.

The changes referred to in items (d) to (i) above are self-explanatory and need no comment. However, the changes indicated in items (a), (b), (c) and (j) above are far-reaching and require special attention. They provide an end to the debate as to whether damages for breach of contract should form part of the contractor's monthly statement showing the amounts to which he considers himself entitled. The new wording of sub-clause 60.1(e) confirms that the contractor may include in such a statement 'any other sum to which the Contractor may be entitled under the Contract or otherwise.'

More importantly, the changes mentioned above to sub-clauses 60.6(b) and 60.8(a) confirm that when submitting his final statement for the engineer's consideration, the contractor may include such damages and that the engineer may include these damages in the final payment certificate which he must issue to the employer within the set 28 days.

Furthermore, and as a consequence of the newly added final paragraph of sub-clause 60.6, any dispute regarding the final statement which must now include a dispute relating to damages in breach of contract is to be resolved under clause 67 of the Form. This clearly means that the mechanism of clause 67 including the decision of the engineer under that clause does apply to a claim or counterclaim for breach of contract.

Finally, as discussed earlier in Chapter 1, FIDIC published in November 1996 a Supplement to the Fourth Edition of the Red Book in three sections. Section C deals with sub-clause 60.10 and is intended to provide some safeguard for the contractor in the event that the engineer is late in certifying interim payments. This is achieved by deleting in lines 3 and 4 of the sub-clause the words 'within 28 days after such Interim Payment Certificate has been delivered to the Employer' and substituting the words 'within 56 days after the Contractor's statement has been received by the Engineer'.

9.34 Sub-clauses 65.4 and 66.1

In the 1992 Reprint of the Fourth Edition the following changes were made:

— In the ninth line of sub-clause 65.4, the words 'and to the operation of Clause 67' have been changed to read 'and Clause 67'; and
— In the second line sub-clause 66.1, the word 'party' has been changed to read 'or both parties', in the third line between the words 'his' and 'contractual' the words 'or their' have been added. In the fourth line after the word 'then', the words 'the parties shall be discharged from the Contract, except as to their rights under this Clause and Clause 67 and without prejudice to the rights of either party in respect of any antecedent breach of the Contract, and' have been added.

These changes were presumably made to highlight the difference between termination of the employment of the contractor and termination of the contract itself.

9.35 Clause 67

Clause 67 deals with the settlement of disputes. The wording of this clause has been improved, clarifying many of the provisions of the Third Edition which had been a source of jurisdictional challenges in many arbitration cases. The clause now clearly states that the reference to the engineer should state that it is made pursuant to clause 67.

The decision of the engineer should also state that it is made pursuant to clause 67. The consequences of the engineer's failure to give notice of his decision has now also been clarified in that it is treated in the same manner as if either the employer or the contractor are dissatisfied with the decision of the engineer. Furthermore, it has now been clearly established that a notice given by either party to the other of his intention to commence arbitration entitles that party to arbitration. It is also established that without such notice, there is no entitlement to arbitration.

A further important change has been made by deleting the word 'difference' from the first sentence of clause 67 restricting its application to a dispute situation. Clause 67 has also been expanded to include:

(a) A new provision for an amicable settlement of disputes prior to the commencement of arbitration; and
(b) A new sub-clause to provide for the situation where either of the parties fails to comply with a decision of the engineer which has become binding. In any such case, the other party may refer such failure to arbitration, thus transforming the engineer's decision into an arbitral award which may be enforceable under international conventions.

In the November 1996 Supplement to the Fourth Edition of the Red Book, referred to above, Section A provides an alternative wording to clause 67. The new wording was drafted in response to mounting criticism of the role of the engineer as an adjudicator or quasi-arbitrator in the resolution of disputes under clause 67 of the Red Book. Various alternative methods of dispute resolution were considered in the past few years, both domestically and internationally. The method finally chosen by FIDIC in its new Supplement is that based on the use of an adjudication board composed of one or three experts who can render a decision in respect of a dispute without having to resort to the engineer for a final determination under the present clause 67 (see Chapter 19 later). This new method requires an expert or experts to be appointed at the beginning of a contract, who must keep in touch with the work in progress on the site which is achieved by visiting the site at regular intervals. The appointed expert, or experts, must be available to act in the resolution of any disputes that may arise. In the supplement, FIDIC embraces this alternative method of dispute resolution as an acceptable substitute to the engineer's traditional role in dispute settlement.

The new supplement provides the necessary amendment to the wording of clause 67 of the Red Book and also contains a guide to this amended wording. Moreover, it contains model terms of appointment and procedural rules for the dispute adjudication expert or board and the necessary amendments required to the Appendix to Tender which correspond to the amended wording of clause 67.

The effect of the changes to clause 67 are discussed in more detail in Chapters 19 and 20.

9.36 Clause 69

There are two new sub-clauses within clause 69: sub-clauses 69.4 and 69.5. They provide a possible temporary solution to the problem of non-payment by the employer under a certificate. According to the new provisions of sub-clause 69.4, the contractor can give notice to the employer entitling him to suspend the work or to reduce the rate of production after a period of 28 days. Sub-clause 69.5 provides for resumption of work when the employer subsequently effects payment.

The 1992 Reprint included two amendments: the first in sub-clause 69.1 where in sub-paragraph (d), the words 'unforeseen reasons, due to economic dislocation' were replaced with 'unforeseen economic reasons'; and the second in sub-clause 69.4 where the word 'cost' was replaced with 'costs'.

9.37 Other changes made in the 1992 Reprint

Besides the changes made in the 1992 Reprint of the Fourth Edition of the Red Book to Part I of the 1988 Reprint as discussed in the present chapter, a number of minor changes were also made to the Tender, Appendix, Reference to Part II and Editorial Amendments. These are set out in Chapter 1 and also in Appendix B.

Furthermore, there were amendments made to Part II of the 1988 Reprint of the Red Book which are set out in Appendix B.

9.38 Concluding remarks

In this chapter, the revisions made in producing the Fourth Edition of the Red Book, its 1988 and 1992 Reprints and its 1996 Supplement were highlighted and discussed.

Part III considers the Fourth Edition of the Red Book in practice, under the headings: the role of the engineer; responsibility and liability of the engineer; the employer's obligations; the contractor's obligations; risks, liabilities, indemnities and insurances; performance and other securities; claims and counterclaims; delay in completion and claims for extension of time; certificates and payments; disputes settlement by arbitration and finally, amicable settlement using alternative dispute resolution.

Part IV deals with related documents to the Red Book; namely: other standard forms of contract published by FIDIC, e.g. the Yellow Book, the Orange Book and the Sub-contract Form; and the World Bank Standard Bidding Documents for major works which is based on the Fourth Edition of the Red Book as reprinted in 1992.

Part III
The Fourth Edition in Practice

Chapter 10

Role of the Engineer

10.1 Introduction

It is evident from the previous chapters that the role of the engineer is central to the contract under the Red Book. Indeed, the engineer has many, sometimes what appear to be conflicting roles. This chapter deals with the engineer in his roles of designer, employer's agent, supervisor, certifier, adjudicator or quasi-arbitrator, and also with his proactive, reactive and passive duties and authority under the contract.

The concept of engaging a consulting engineer for any of the tasks detailed in Chapter 5 stems from the idea that when a promoter initiates a construction project he is faced with a multiplicity of technical, commercial and legal considerations with which he is not familiar, or at least in which he is not an expert. In civil engineering construction, in order to transform the promoter's ideas into reality, the traditional method, and also the method adopted by FIDIC, has been to engage the services of a consulting engineer to carry out the following functions:

(a) To complete a skilful design of the project sought by the promoter. Such design includes, but is not limited to, the preparation of drawings which should express and communicate the details of every aspect of the project to be constructed; to draft a specification of the materials to be used and of the standard of the workmanship to be achieved; and to prepare the bill of quantities;
(b) To prepare all documents necessary for obtaining a competitive price for carrying out the work by a competent contractor and to advise the promoter on the tenders received and on the selection of the contractor;
(c) Once work starts on the project, to supervise or to inspect the work carried out by the contractor in order to ensure conformity with the design requirements; and
(d) To administer the contract, to deal with situations as they arise, to certify and to act as an adjudicator of disputes.

In carrying out the duties set out in steps (a) and (b), the consulting engineer acts as an adviser and consultant to the promoter. Once a contract is placed with a contractor, the developer is called an employer under the Red Book and the consulting engineer is referred to as the engineer for the purposes of steps (c) and (d). During these steps the engineer acts as agent of the employer. For the purposes of the construction of a civil engineering project, two contracts are traditionally formed, one between the consulting engineer and the employer and another between the employer and the contractor.[10.1]

In the case of the first contract, the contractor is not a party. Similarly, the engineer is not a party to the contract between the employer and the contractor.

This traditional method has distinct advantages over others. First, it permits the completion or near completion of the design of the project before the appointment of a contractor and thus allows for competitive tendering in respect of its construction. Secondly, it eliminates or at least minimises the problem of professional judgement being influenced by any commercial aspirations which might influence the designer. Thirdly, it permits a higher degree of control over the quality of the constructed works, such control being exercised by the designer who would subsequently, normally act as the supervisor. Fourthly, it embodies greater flexibility of the final design of the project in that it incorporates procedures which allow for modification of requirements and of detail which might be required during the construction period.

As already discussed in Part I, the role of the engineer under the Red Book is based on, and developed from, the English tradition and in particular from the ICE Form of Contract. Although similar contractual arrangements, where the design and specification for a project are prescribed by an employer and expected to be strictly followed by a contractor, are traditionally also used in countries outside the English influence, the difference between the latter type and the Red Book in the position and the role of the engineer seems to focus on his role as described in step (d) above.[10.2]

Many employers and commentators unfamiliar with either the ICE Form of Contract or the Red Book find the position and the role of the engineer under these two forms of contract confusing and abnormal. This is particularly so when they find that the engineer may act, on the one hand, as an agent of the employer during the construction of the works, and on the other hand, as an adjudicator of disputes between that employer and the contractor. They consider it even more unusual when some of these disputes relate to the engineer's own design, specification or instructions.

Others find it peculiar that the engineer, whilst not a party to the contract between the employer and the contractor, is in fact empowered under various clauses of the FIDIC Conditions of Contract to give instructions, to make decisions, determinations or variations, which are binding on the parties to the contract unless and until they are subsequently rescinded or varied by an arbitrator. Furthermore, those unfamiliar with the Red Book also consider it peculiar that because of the principle of privity of contract under the common law system (on which the Red Book is based, see Chapter 3 above), the engineer's actions or inaction do not expose him to any liability under the contract between the employer and the contractor. In addition, any liability towards the employer for lack of care or skill exists only under the engineer's contract with the employer. On the other hand, should the contractor suffer any loss or damage under his contract with the employer due to the engineer's negligence he can obtain a remedy against the employer only through the contract with that employer. The employer may then, if appropriate, initiate a claim against the engineer for the loss or damage sustained. For a detailed discussion on this aspect of the relationship between the engineer, the employer and the contractor, see Chapter 11 later.

Therefore, whilst the functions set out in steps (a), (b) and (c) above are known almost worldwide, it is step (d) which is controversial and requires detailed discussion. To summarise, in steps (a) and (b), the engineer's role is that of a designer; in step (c) as a supervisor and as an employer's agent; and in step (d) as administrator, certifier and adjudicator. The engineer's proactive, reactive and passive duties are performed throughout the construction and completion of the works. These roles are discussed below.

10.2 The engineer as a designer

The role of a designer is the first and primary role of the engineer and should be completed, or almost completed, by the time the Conditions of Contract come into operation. The more complete the design of the project is at the time of inviting tenders for its construction, the lower the risk of an unsatisfactory outcome. To ensure the completeness of the design, the design brief must first be defined and its parameters decided, so that the best person can be selected to undertake the design. It is difficult, if not impossible, however, to define in general terms the concept of 'Design' because each civil engineering project is a unique artefact and the design parameters and boundaries are therefore different in every case.[10.3] Nevertheless, it may be useful to clarify the path followed in performing the design function and the expected results.

The design of a particular project involves decision making on the form and shape of the project; the constituent elements of that form and shape; their precise dimensional characteristics; the specification of the material and workmanship of each module of these elements; the quality to be achieved; and how these elements relate to and interact with each other.

The design process described above applies equally to temporary works, with the additional requirement of ensuring that the temporary works are capable of performing a specific function for a certain period of time to facilitate the construction of the permanent works.

The nature of the project for which the design is required is identified in the contract between the employer and the consultant. Whereas the temporary works must be fit for their purpose and, thus, are normally left to the contractor to design and to be liable for (see Chapter 13 later), the liability for the design of the permanent works extends only to skill and care, unless further extension of liability is specifically stipulated in the contract. When the entire design has been completed the following aspects of the project would normally have been determined:

(a) The shape, form and dimensions of the project.
(b) The function which the project is expected to perform and the level and quality of such performance.
(c) The selection of materials and workmanship to produce the shape, form and dimensions of the project; and the production of documents necessary to express and communicate the design precisely and clearly to the employer and to prospective contractors.

   Some aspects of the method of construction which would be necessary for the implementation of the project may have to be included depending on the project and its design. Other aspects remain within the control of the contractor and the division between the two is an important aspect which cannot be sufficiently emphasised.

(d) The projected cost of the project on completion, based on an accurate Bill of Quantities, and also the projected maintenance cost of the project's life span.

It is generally accepted that there is a relationship between the quality of the materials and workmanship selected in the design and the necessity for maintenance and its frequency.

(e) In certain aspects, the timing or sequence of construction of any part of the works.

A check list prepared by an advisory committee to the Institution of Civil Engineers in London provides a valuable reminder of what should be done by the engineer at the pre-contract and the post-contract stages.[10.4] Although drafted with reference to the Fifth Edition of the ICE Form, most of its items apply to the FIDIC Form and accordingly it is quoted below.

'Pre-contract

1. Ensure the employer is aware that he carries the financial risk for unforeseen events and of the financial managerial and advisory resources required for the contract.
2. Warn the employer of the decision and actions required of him giving programme dates of finalisation of designs, provision of access, construction and taking over the works.
3. Design and detail the contract works and as far as possible prepare clear working drawings and a concise specification.
4. Prepare accurate bills of quantities, detailing the works required and complying with the standard method of measurement where possible. Keep provisional items to a minimum.
5. Ensure the employer and his staff understand the role of the Engineer under the *ICE Conditions of Contract*, to ensure fair dealings between the contractor and the employer.
6. Adopt the *ICE Conditions of Contract* or for ancillary works, national and well understood *Conditions of Contract*, in full, without any variations or deletions, and draw the employer's attention to the powers and duties of the Engineer under these contracts.
7. Ensure the employer and his auditors accept the ICE/CIPFA joint statement on the Engineer's and auditor's relationship and accept that the Engineer has the quasi-judicial powers to make decisions that are final and binding on the employer and contractor subject only to reference to arbitration.[10.5]
8. Ensure the Engineer has a defined and readily understood method of selecting contractors and recommend the number invited should be limited.
9. Ensure all tenderers receive the same tendering information and are given a sufficient period for the preparation of tenders.
10. Make all site and service information in the employer's and Engineer's possession available to those invited to tender.
11. Ensure tenders are delivered in specifically marked envelopes to the employer or Engineer, by a fixed date and time, and are opened with witnesses at a declared fixed time.

12. Check tenders carefully and correct any errors in extension of item rates, times or quantity. Notify tenderer of any resulting change in the total of the priced bills of quantities and tender sum. Review tenders received with particular regard to the proposed construction methods and degree of risk involved and with the implications of sectional completion dates on the employer's and contractor's cash flow, as well as the anticipated final contract price. Submit a report to the employer pointing out any rate that is less than the known cost of carrying out the work and giving a recommendation of a tender acceptance with reasons. If rates are in doubt recommend a tenderer be invited to stand by his rates or withdraw.
13. Advise the employer to give tenderers the name of the successful tenderer at the earliest opportunity. It is also recommended that the list of values of tenders received should be circulated.

Post contract

14. On the appointment of a contractor, confirm the appointment by letter.
15. State by letter to the contractor and Engineer's representative details of the delegated powers and responsibilities and names of the Engineer's representative and project team, and give a date for commencement of the works.
16. Agree the extent and methods of payment for variations, extras and supervision and recording of dayworks, preferably before work is commenced, and confirm in writing.
17. Do not exceed the powers granted by the employer, e.g. do not take on responsibility for redesign or significant variations and extra works without the employer's agreement to the works and to provide finance.
18. Make decisions on extensions of time at stages and times required under the contract.
19. Ensure that a site diary and site records are properly kept and agreed where appropriate with the contractor and arrange for regular progress photographs to be taken.
20. Ensure site meetings are held at least monthly and that minutes are kept and agreed.
21. Issue certificates for payments after interim measurements promptly.
22. Visit the site regularly, at least monthly. Inspect works in progress and review compliance with the contract programme.
23. Ensure nominated subcontractors are properly appointed by the main contractor and that appropriate *subcontract conditions of contract* are applied.
24. Agree measurements of quantities for completed works as the work proceeds and agree with the contractor that they are to be carried to the final account, unaltered.
25. Ensure claims are detailed and the sums due settled as soon as possible.
26. Ensure certification of maintenance and completion are issued to the contractor on time.
27. Ensure the employer is aware of his new insurance liability when the maintenance certificate is issued.

28. On clause 66 decisions, review all the evidence available and if possible, arrange for the Engineer's representative to put the employer's case, and the contractor his, to enable a clear judgment to be made on the issues.
29. In the event of arbitration, keep the dispute to the areas of the clause 66 determination and present your evidence fairly and concisely. The guiding principles for the Engineer and his staff are that the contract is a joint enterprise for the benefit of both parties and the employer is entitled to a project well executed and the contractor to fair dealing and a fair profit. Remember always the contractor could only price and resource for the works which were defined at the time he tendered.'

Although it is accepted that the whole design, including the drawings, the specification and any other documentation, should ideally be completed before tenders are invited, the Fourth Edition of the Red Book allows for the possibility of not achieving this ideal situation. In sub-clause 7.1, the engineer is given the authority to issue to the contractor during the currency of the contract 'supplementary Drawings and instructions as shall be necessary for the purpose of the proper and adequate execution and completion of the Works . . .'.

Sub-clause 7.1 should, however, be considered in conjunction with sub-clauses 6.3 and 6.4, 'Disruption of Progress' and 'Delays and Cost of Delay of Drawings'. The flexibility provided by sub-clause 7.1, in allowing for finalising the design during the construction period, is matched by a possible cost implication which can sometimes form the basis of substantial claims by the contractor against the employer. In this regard, sub-clause 6.3 states in essence that 'the Contractor shall give notice . . . whenever planning or execution of the Works is likely to be delayed or disrupted unless any further drawing or instruction is issued by the Engineer within a reasonable time . . .'.

Sub-clause 6.4 states briefly that:

'if, by reason of any failure or inability of the Engineer to issue, within a time reasonable in all the circumstances, any drawing or instruction for which notice has been given . . . , the Contractor suffers delay and/or incurs costs then the Engineer shall, . . . , determine:
a) any extension of time . . .
b) the amount of such costs . . .'

The Fourth Edition of the Red Book also provides, in sub-clauses 7.2 and 8.2, for the possibility of part of the permanent works being designed by the contractor. These two sub-clauses have been drafted with mechanical and electrical plant in mind (see sub-clause 7.2(b)), where operation and maintenance manuals are required to be submitted by the contractor to the engineer. In such an event, where the contractor is required to design part of the permanent works, the specification must expressly provide for such a requirement. In this connection, it is notable that sub-clause 7.3 provides that approval by the engineer of drawings, specifications, calculations, operation and maintenance manuals and other information is not to relieve the contractor of any of his responsibilities.

The responsibility for design is stipulated in sub-clause 20.4(g) where the risk of loss or damage due to the design of the works is allocated to the employer. Where any part of the design is provided by the contractor or the contractor is responsible for it, the risk of any loss or damage is listed as a contractor's risk under sub-clause 20.2.

In the latter connection, it should be noted that clause 38(ii) of the 'Model Services Agreement between Client and Consultant', published by FIDIC in 1990 (see Reference 10.1), states that 'Neither the Client nor the Consultant shall assign obligations under the Agreement without the written consent of the other party'. Therefore, should there be a need for the contractor to design any part of the permanent works, it would be necessary to alter clause 38(ii) of the Model Services Agreement, if used, unless consent of the employer is first obtained. If another form of agreement is used where no reference to assignment is made, then it is important to consider the necessity of incorporating an express term into the agreement authorising the designer to delegate to the contractor the relevant design obligations relating to that part of the permanent works. This would certainly be prudent under English law since it was decided in *Moresk v. Hicks*,[10.6] that there is no implied authority for the designer to delegate any of his duties to others. In that case, an architect sub-contracted the design of the structural frame of the building to a contractor who supplied and erected the frame. The frame, however, proved to be defective in two major design areas. The architect sought to argue that there was an implied authority to delegate the duty of the design of the frame. The court disagreed with this submission and held that:

'... if a building owner entrusts the task of designing a building to an architect he is entitled to look to that architect to see that the building is properly designed. The architect has no power whatever to delegate his duty to anybody else. Certainly not to a contractor who would in fact have an interest which was entirely opposed to that of the building owner.'

In this connection, it is also important to refer to clause 59.3 of the Red Book which provides for the situation where the services to be provided in respect of a provisional sum include a matter of design or specification of any part of the permanent works or of any plant to be incorporated therein. Under this clause, such design requirement should be expressly stated in the contract and should be included in any nominated sub-contract. Accordingly, similar express terms should be incorporated into the agreement between the promoter and the consulting engineer authorising the latter to delegate the relevant design duties.

As the designer is under a continuing duty during the construction period to check that his design will work in practice and to correct any errors which may emerge, a reference to clause 51 of the Red Book is necessary in this context.[10.7] Sub-clause 51.1(c) authorises the engineer to change the character or quality or kind of work included in the contract. Such change falls within the definition of 'Design' as propounded above, as does any change in levels, lines, position and dimensions of any part of the works which is referred to in sub-clause 51.1(d) as being within the authority of the engineer. Change of any specified sequence or timing of construction of any part of the work may also be part of the 'Design', a change provided for under sub-clause 51.1(f).

Finally, 'Design' may include additional work of any kind necessary for the completion of the works as provided in sub-clause 51.1(e), whether such additional work is required to rectify the original design or simply to represent an addition to it. Sub-clause 7.1 provides the authority for the engineer to issue to the contractor, from time to time, supplementary drawings and instructions which would be necessary for the proper and adequate execution and completion of the works and for remedying of any defects therein.

Apart from the permanent works, it is usual to expect the contractor to design the temporary works, and reference in this connection is made to sub-clause 8.2 of the Red Book. This sub-clause provides that the 'Contractor shall take full responsibility for the adequacy, stability and safety of all Site operations and methods of construction'. The sub-clause further provides that the contractor is not to be responsible for the design or specification of any temporary works not prepared by the contractor, suggesting that it is normally expected that the temporary works would be designed and specified by the contractor and would fall within his responsibility.[10.8]

10.3 The engineer as the employer's agent

When the employer appoints a contractor to construct a project, certain duties are generally created which must be performed by the employer, or on his behalf, in order to ensure that the project is completed on time, within the budget and, perhaps more importantly, with qualities as designed and specified. Whoever performs these duties is called the employer's agent. These duties are, in general, concerned with the following functions:

— *Design*: As explained earlier, the process of design continues during the construction period and many questions could arise during this period which would relate to matters of design. The philosophy and practical calculations of the design process are best known by the engineer. Design, being more of an art than a science, could best be modified or changed, should it be required, by the original designer.
— *Quality control*: The design is expressed through drawings and specifications which set out the required quality to be achieved. Quality must be monitored by someone familiar with the original concepts and parameters of the design.
— *Administration and management*: Progress on site depends to a large extent on the availability of information required by the contractor. This information may involve queries as to how a certain provision of the contract should be interpreted or how problems encountered on the site should be dealt with promptly and properly. It may involve questions relating to whether or not a certain item is necessary for the completion of the work; whether or not an item should form a variation to the contract; whether or not a specific contractual obligation has been fulfilled; whether or not a certain item of the work is properly measured; etc.

   When such information is required, it is necessary that the employer's agent be familiar with the details of the project in order to deal with it promptly.
— *Cost accountancy and certification*: The contract is based on interim payments being made periodically, mostly on a monthly basis, by the employer to the contractor.

The employer's agent must possess the necessary knowledge to evaluate the work carried out by the contractor, periodically and ultimately in a final certificate.

The functions set out above are detailed in 51 out of the 72 clauses of Part I of the Conditions. They are clauses: 1, 2, 4, 5, 6, 7, 12, 13, 14, 15, 16, 17, 18, 19, 20, 27, 30, 31, 33, 34 (as worded in Part II), 35, 36, 37, 38, 39, 40, 41, 42, 44, 45, 46, 48, 49, 50, 51, 52, 53, 54, 56, 57, 58, 59, 60, 62, 63, 64, 65, 67, 68, 69 and 70.

Of the above clauses, however, clauses 1 and 2 should be recognised as the principal clauses with respect to setting out the duties and authority of the engineer and in particular his role as the employer's agent. It is important, therefore, to consider the provisions of clauses 1 and 2 as they relate to the engineer.

10.3.1 Authority and duties of the engineer

Under the Red Book, the role of the employer's agent is performed by the engineer who is defined in sub-clause 1.1(a)(iv) as the person appointed by the employer to act as engineer for the purposes of the contract and named as such in Part II of the Conditions. There are many reasons why this role is usually taken by the engineer, the most important one being that he is perhaps the most competent person to assume that position and to perform these duties.

Previous editions of the Red Book defined the engineer as follows:

' "Engineer" means the Engineer designated as such in Part II, or another Engineer appointed from time to time by the Employer and notified in writing to the Contractor to act as Engineer for the purposes of the Contract in place of the Engineer so designated.'

It is significant to note that the words 'from time to time' in the above definition have been deleted from the Fourth Edition. The definition without this phrase may imply that the employer cannot replace the engineer without clear bona fide reasons and without the consent of the contractor since the latter may have attached some importance to the reputation and expertise of the named engineer when submitting his tender. As a minimum requirement, it would therefore be necessary for the employer, if he were intending to replace the named engineer with another, to discuss his intentions with the contractor and to attempt to obtain his consent.

Consequently, so far as the duties and the authority of the engineer in the context of his role as the employer's agent are concerned, it is important to note that whilst he is not a party to the contract, both the employer and the contractor have, in signing the contract, recognised and accepted the duties and authority of the engineer as specified. Sub-clause 2.1(a) of the Red Book provides that the 'Engineer shall carry out the duties specified in the Contract'. The term 'Contract' is defined in sub-clause 1.1(b)(i) as the:

'Conditions (Parts I and II), the Specification, the Drawings, the Bill of Quantities, the Tender, the Letter of Acceptance, the Contract Agreement (if completed) and such further documents as may be expressly incorporated in the Letter of Acceptance or Contract Agreement (if completed).'

Each of the documents enumerated in this definition except for the Conditions (Parts I and II) is subsequently defined in clause 1. Collectively, they are exclusive to the particular project for which they have been drafted. On the other hand, the duties which are common to all projects where the Red Book is used are those which are incorporated in the Form itself. In order to carry out these duties, the engineer is given certain authority either specified in, or necessarily to be implied from, the provisions of the contract, which he may exercise under the provisions of sub-clause 2.1(b).

The duties and authority of the engineer may be divided into proactive, reactive and passive categories. The proactive category includes any duty performed or action taken where the initiative lies with the engineer in administering the contract. The reactive category includes any duty performed or action taken in response to a request by the contractor or the employer. The passive category includes rules and regulations which must be observed. Tables 10.1 to 10.3 set out check lists for these three categories.

Furthermore, sub-clause 2.1(b) provides that if the engineer is required under the terms of his appointment by the employer to obtain the specific approval of the employer before exercising the authority delegated to him, the particulars of such requirements should be precisely set out in Part II of the Conditions. It is most important for all to recognise that any such requirement must be capable of implementation. In this connection, page 4 of Part II of the Conditions contains a recommendation that if the obligation on the engineer to obtain the approval of the employer could result in his being unable to take action in an emergency where matters of safety are involved, an additional paragraph may be necessary and an example wording is given therein as follows:

'Notwithstanding the obligations, as set out above, to obtain approval, if, in the opinion of the Engineer, an emergency occurs affecting the safety of life or of the Works or of adjoining property, he may, without relieving the Contractor of any of his duties and responsibilities under the Contract, instruct the Contractor to execute all such work or to do all such things as may, in the opinion of the Engineer, be necessary to abate or reduce the risk. The Contractor shall forthwith comply despite the absence of approval of the Employer, with any such instruction of the Engineer. The Engineer shall determine an addition to the Contract Price, in respect of such instruction, in accordance with Clause 52 and shall notify the Contractor accordingly, with a copy to the Employer.'

The proviso to paragraph (b) of sub-clause 2.1 states that 'any requisite approval shall be deemed to have been given by the Employer for any such authority exercised by the Engineer' (i.e., for any of the items identified in Part II which require the employer's approval). The contractor can therefore proceed with the works without questioning the validity of the engineer's authority irrespective of whether or not the engineer has actually obtained such approval and the contractor would be entitled to payment for any work carried out should the employer be unwilling to give retrospective approval for any action by the engineer.

Furthermore, if it subsequently transpires that the engineer was seeking approval from the employer before taking action on matters not specifically identified in Part II, the contractor would have legitimate grounds for complaint having regard to the provisions of sub-clause 2.6. Under sub-clause 2.6, the engineer is required to 'exercise his discretion' impartially whenever he is:

(a) Giving his decision, opinion or consent; or
(b) Expressing his satisfaction or approval; or
(c) Determining value; or
(d) Otherwise taking action which may affect the rights and obligations of the employer or the contractor.

It is essential to recognise that the engineer can exercise his discretion only if his authority is not curtailed in accordance with sub-clause 2.1 of Part II of the Conditions. If the engineer's liberty to exercise his discretion is restricted in respect of certain duties which are specifically stated in Part II of the Conditions, then the question of acting impartially does not arise. That is, of course, if the engineer understands the true meaning of impartiality.

> 'Impartiality does not consist in not having prejudices . . . we all have them. The art consists in the ability to discard prejudices [which] largely depends on a sufficient measure of self-awareness to be conscious of this defect of character, in so far as it is a defect, and to learn to recognise it when the devil comes and tempts one. My advice is: be aware of your prejudices, and hang them upon the peg with your raincoat before you come into court.'[10.9]

Accordingly, it is clear that sub-clause 2.1 provides the main safeguard for the contractor in the FIDIC Conditions where the actions of the engineer are concerned. Whilst it can be argued that the provisions of this sub-clause could always be inferred from previous editions of the Red Book, the fact that they are now expressly stated is to be welcomed and the parties can no longer plead ignorance of this fact. Should such an impartial stance by the engineer be unacceptable by an employer, it would be necessary for him to make express provision in Part II of the Conditions to alter sub-clause 2.1 or strike out sub-clause 2.6. In such circumstances, tenderers would be alerted to this situation and consequently would have an opportunity to adjust their tender prices accordingly.

In a number of the clauses of the Fourth Edition a condition is imposed on the engineer prior to his exercising his discretion which requires him to determine any of his duties only after 'due consultation with the Employer and the Contractor' (see Section 9.3). This combined with the wording of sub-clauses 10.1 and 25.1 where the employer is required to liaise directly with the contractor, gives more prominence to the role of the employer under the Fourth Edition of the Red Book.

In addition to the above, clause 2 of the Red Book deals with three other aspects of the engineer's duties and authority as follows:

(a) The duty to appoint an engineer's representative;
(b) The authority to appoint assistants to the engineer's representative; and
(c) The duty to put his instructions into writing.

The first of these, the duty to appoint an engineer's representative, is expressed in sub-clause 2.2. The engineer appointed by the employer is not normally resident at close proximity to the site, hence the necessity arises for the engineer to appoint a representative to whom he can delegate certain duties and authority. The engineer's representative is appointed by, and is responsible to, the engineer who in turn is responsible for the action or inaction of the engineer's representative. The delegation of duties and authority vested in the engineer to the engineer's representative must, therefore, be in writing and the Fourth Edition now expressly states, in sub-clause 2.3, that any such delegation (or revocation) of duties and authority 'shall not take effect until a copy thereof has been delivered to the Employer and the Contractor'. As there is no reference to the engineer's representative in the Fourth Edition, except in clause 2, the engineer's representative is totally dependent on the letter of delegation for his own duties and authority. The employer may in fact place certain restrictions on the engineer, in the contract between them, as to exactly what can or cannot be delegated to the engineer's representative. In such circumstances, it would be prudent to state such limitations in Part II of the Conditions.

Clearly, certain functions should not be delegated by the engineer and the *Guide to Use of FIDIC Conditions of Contract for Works of Civil Engineering Construction, Fourth Edition*, (referred to in Reference 9.1), specifically excludes delegation in respect of clauses 41, 44, 48, 62, 63, 65, 67 and 69. It is, however, stated in the Guide that in order to avoid unnecessary interruption of work at site, delegation of authority (in whole or in part), is common in respect of the following clauses: 4, 5, 7, 12, 14, 16, 17, 18, 19, 20, 27, 30, 31, 33, 34, 35, 36, 37, 38, 39, 40, 42, 45, 49, 50, 53, 54, 56, 59, 60 (except 60.6 and 60.8), 64 and 70.

The engineer's representative has the authority to appoint persons to assist him in carrying out his duties. This authority, which is new under the Fourth Edition, comes under the provisions of sub-clause 2.4. Although limited in scope, it can be particularly useful in situations where the engineer's representative is required to attend various meetings, inspections, tests, etc., some of which may take place simultaneously or far away from the site.

It is important to note that the contractor is to be notified of the precise duties which such assistants may carry out and, more importantly, the scope of authority of such persons in carrying out their duties. It is necessary to retain a clear distinction between the authority of the engineer's representative and that of his assistants.

In this connection, reference must also be made to sub-clause 37.5 of the Fourth Edition entitled 'Independent Inspection' which empowers the engineer to delegate the duty of inspection and testing of materials or plant to an independent inspector. This delegation which should be read in conjunction with sub-clause 37.2 of the Conditions is to be effected in accordance with sub-clause 2.4 and, accordingly, any person so appointed is to be 'considered as an assistant of the Engineer'.

It is notable that some ambiguity exists between sub-clause 2.4 and sub-clause 37.5 regarding the appointment of persons to assist the engineer's representative and the engineer, respectively. Specifically, sub-clause 2.4 deals with the appointment of persons to assist the engineer's representative while in sub-clause 37.5, the appointment is that of an assistant of the engineer. Of course, the chain of authority created through delegation in clause 2 would mean that an authorised instruction given by an assistant of the engineer's representative could be deemed to have been given by the engineer's representative (see sub-clause 2.4). In this connection, reference to the wording of clause 13 should be made where it is stated that 'the Contractor shall take instructions only from the Engineer, or, subject to the provisions of Clause 2 from the Engineer's Representative'. There is no mention of the assistants, thus creating a further ambiguity. However, if the engineer's representative is authorised through delegation to give an instruction, then it would 'have the same effect as though it has been given by the Engineer' (sub-clause 2.3). Consequently, the engineer should monitor very closely the actions of appointed inspectors and assistants as they would be acting with some specified authority of the engineer himself.

The obligation placed on the engineer in sub-clause 2.5 to give instructions in writing is extremely important with regard to keeping contemporary records of all matters related to the contract as work proceeds. Such records would provide important evidence when payment for work executed is considered, or when claims are made for additional work. The provisions of clause 53, for example, highlights the necessity of such records.

10.4 The engineer's proactive duties and authority

The engineer's proactive duties and authority, defined earlier in Section 10.3.1, are set out in Table 10.1, arranged in sequence corresponding to the clauses of the Conditions of Contract. Where a duty includes the obligation to give notice to the employer or the contractor or to both, the word 'Notify' in bold type is used. The proactive duty or authority of the engineer is shown in italics with the first letter capitalised.

Table 10.1 The engineer's proactive duties and authority.

Sub-clause No.	Duties and authorities where an action is initiated by the engineer
2.1(b)	*Obtain* the approval of the employer before exercising any authority for which he is required to obtain such approval, as set out in Part II of the Form.
2.2	*Appoint* the engineer's representative.
2.3	*Delegate* to the engineer's representative, in writing, any of the duties and authorities vested in the engineer, with a copy to the employer and the contractor.
2.4	*Appoint* assistants to the engineer's representative. **Notify** the contractor of the names, duties and scope of authority of assistants.
6.1	*Provide* two copies of the drawings to the contractor, free of charge.
6.4	*Determine*, after due consultation, extension of time and amount of costs, if any, due to failure or inability of the engineer to issue any drawing or instruction for which notice has been given under sub-clause and **Notify** the contractor accordingly with copy to the employer.
6.5	(The engineer shall take into account any failure by the contractor to submit drawings, specifications or other documents which he is required to submit under the contract.)

(Table 10.1 Contd.)

Sub-clause No.	Duties and authorities where an action is initiated by the engineer
7.1	*Issue* to the contractor any supplementary drawings and instructions as necessary for completion of the works and remedying of any defects.
13.1	*Consider* whether or not satisfied with the contractor's execution and completion of the works including remedying defects and *Issue* instructions where necessary (see clauses 37, 49 and 50).
14.1	*Prescribe* the form and the detail of the programme to be submitted by the contractor within the stated period in Part II. If appropriate give *Consent* to such programme.
14.1	*Consider* and, if appropriate, *Ask* the contractor to submit a general description of the arrangements and methods he proposes to adopt for the execution of the works.
14.2	If appropriate, *Request* the contractor to produce a revised programme.
15.1	Where appropriate, *Approve* the authorised representative of the contractor and keep under review such approval. *Determine* the period of superintendence required after completion of the works. If the engineer's approval of the contractor's authorised representative is withdrawn, **Notify** the contractor accordingly.
16.2	If applicable, *Object to and Require* the contractor to remove any person who is incompetent or negligent or who misconducts himself or whose presence on site is undesirable.
17.1	*Inform* the contractor of the original points, lines and levels of reference and in case of error in the position, levels, dimensions or alignment of the setting out of any part of the works, *Require* the contractor to rectify such error either at his own cost or against an addition to the contract price *Determined* by the engineer. **Notify** the contractor accordingly with a copy to the employer.
20.2	*Consider* whether or not he is satisfied with the contractor's rectification of any loss or damage to the permanent works due to any of the contractor's risks. If not, *Refer* to clauses 37.4, 49 and 50 for action to be taken.
20.3	In case of loss or damage due to any of the employer's risks, *Indicate* the extent of repair required and *Determine* an addition to the contract price in accordance with clause 52. **Notify** the contractor accordingly, with a copy to the employer.
31.2	Where the contractor provides facilities for other contractors, *Determine* an addition to the contract price and **Notify** the contractor accordingly with a copy to the employer.
33.1	*Consider* whether or not he is satisfied with the clearance of site on completion and inform the contractor accordingly.
35.1	*Consider* and if required, *Obtain* from the contractor a return of staff, labour and contractor's equipment, as prescribed by the engineer.
36.1	*Specify* any tests which may be required and *Select* any materials or plant for testing.
36.5	*Determine*, after due consultation, any extension of time and the amount of costs as a result of tests not intended by, or provided for, under the contract. **Notify** the contractor accordingly with a copy to the employer.
37.3	**Notify** the contractor when intending to inspect or to attend tests on materials and plant to be supplied under the contract.
37.4	*Consider* and if required, *Request* that the tests of rejected materials or plant be made or repeated. *Determine*, after due consultation, the costs recoverable from the contractor and **Notify** the contractor accordingly, with a copy to the employer.
37.5	*Consider* and if required *Delegate* inspection and testing of materials and plant to an independent inspector in accordance with sub-clause 2.4. **Notify** such delegation and appointment to the contractor with a notice not less than 14 days.

(Table 10.1 Contd.)

Sub-clause No.	Duties and authorities where an action is initiated by the engineer
40.2	Where supervision of the progress of the works is not in accordance with sub-clause 40.1, *Determine*, after due consultation, extension of time and amount of costs, if any, and **Notify** the contractor with a copy to the employer.
41.1	**Notify** to the contractor to commence the works within the time stated in the Appendix to Tender.
46.1	*Consider* whether or not the rate of progress of the works is too slow to comply with the time for completion and if so, whether this is for a reason which does not entitle the contractor to an extension of time, and if so, **Notify** the contractor accordingly. If the employer is involved, as a result, in additional costs, *Determine*, after due consultation, the amount recoverable from the contractor and **Notify** him accordingly with a copy to the employer.
49.2	*Inspect* works and *Instruct* the contractor to execute amendments; reconstruction; and to remedy defects, shrinkages or other faults.
49.3	*Determine* an addition to the contract price in accordance with clause 52 should an instruction be given under sub-clause 49.2(b) for reasons other than in sub-clause 49.3. **Notify** the contractor accordingly with a copy to the employer.
49.4	*Determine*, after due consultation, all costs incurred by the employer in an instruction under sub-clause 49.2(b) for which the contractor is liable and which shall be recoverable from the contractor by the employer. **Notify** the contractor accordingly with a copy to the employer.
50.1	Where the cause of any defect, shrinkage or other fault which appears prior to the end of the defects liability period is not known, *Consider* the necessity for conducting a search for such a cause and if such a search is considered necessary *Instruct* the contractor, with a copy to the employer, to search.
50.1	Where the defect, shrinkage or other fault is the responsibility of the employer, *Determine* after due consultation the amount in respect of such search incurred by the contractor which should be added to the contract price. **Notify** the contractor accordingly with a copy to the employer.
51.1	Where necessary, *Vary* the form, quality or quantity of the works, or any part thereof and for that purpose, or if appropriate, *Instruct* the contractor in accordance with clause 51.1 with the effect to be valued in accordance with clause 52.
52.2	*Consider* whether or not any rate or price should be varied as a result of any instruction under clause 51, and if so, **Notify** the contractor within 14 days of the date of such instruction of the intention to vary a rate or a price.
52.3	If, on issue of the taking-over certificate for the whole of the works, all varied work and all adjustments, as defined in sub-clause 52.3, exceed 15% of the effective contract price, *Consult* with employer and contractor and agree such sum to be added or deducted from the contract price. Failing agreement, *Determine* such sum and **Notify** the contractor accordingly with a copy to the employer.
52.4	If necessary or desirable, *Issue* an instruction that any varied work shall be executed on a daywork basis in accordance with the procedures set out in sub-clause 52.4.
56.1	*Ascertain and Determine* by measurement the value of the works. **Notify** the contractor giving reasonable time when any part of the work is to be measured and follow the procedure under clause 56.
57.2	*Receive* within 28 days after receipt of the letter of acceptance and, if appropriate, *Approve* the breakdown for each of the lump sum items contained in the tender.

(Table 10.1 Contd.)

Sub-clause No.	Duties and authorities where an action is initiated by the engineer
58.1	*Determine* the provisional sum to be used, in whole or in part, and *Instruct* and **Notify** the contractor accordingly with a copy of the notification to the employer.
59.4	*Consider* and, if appropriate, *Instruct* the contractor on the matter of payments to sub-contractors.
59.5	*Consider* and, if appropriate, *Demand* from the contractor reasonable proof that all payments less retentions included in previous certificates for sub-contractors have been discharged. If proof is not supplied, *Follow* the procedure laid down in sub-clause 59.5.
60.1 & 60.2	*Prescribe* from time to time the form of statement to be submitted by the contractor under sub-clause 60.1. *Certify*, within 28 days of receiving such statement, to the employer the amount to be paid to the contractor.
60.3	Upon the issue of the taking-over certificate for a section or part of the permanent works or for the whole of the works, *Certify* for payment one half of the relevant retention money. Upon the expiration of the defects liability period, *Certify* the second half of the relevant retention money.
60.8	Within 28 days after receipt from the contractor of the final statement and the written discharge in accordance with sub-clause 60.7, *Issue* a final certificate to the employer with a copy to the contractor.
62.1	Within the period, and the procedure set out in sub-clause 62.1, *Sign and Deliver* to the employer a defects liability certificate with a copy to the contractor.
64.1	*Inform* the employer of any remedial or other work which is urgently necessary for the safety of the works and which the contractor is unable or unwilling to do at once, having been previously notified. If work so done by the employer is the responsibility of the contractor, *Determine*, after due consultation, all costs consequent thereon or incidental thereto which are recoverable from the contractor by the employer. **Notify** the contractor accordingly with a copy to the employer.
68	*Comply* with procedures laid down in sub-clauses 68.1, 68.2 and 68.3 regarding notices and change of address.
70.2	If, after 28 days prior to the latest date for submission of tenders for the contract, changes in legislation occur which cause additional or reduced costs to the contractor other than due to rise or fall in the cost of labour and/or materials and other matters under sub-clause 70.1, *Determine* after due consultation such additional or reduced costs from the contract price and **Notify** the contractor accordingly with a copy to the employer.

10.5 The engineer's reactive duties and authority

The engineer's reactive duties and authorities defined earlier in Section 10.3.1, are set out in Table 10.2, arranged in sequence corresponding to the clauses of the Conditions of Contract. Where the duty includes the obligation to give notice to the employer or the contractor or to both, the word 'Notify' in bold type is used. The reactive duty or authority of the engineer is shown in italics with the first letter capitalised.

Role of the Engineer

Table 10.2 The engineer's reactive duties and authority.

Sub-clause No.	Duties and authorities which are in response to an action initiated by the employer or the contractor
4.1	If appropriate, *Give consent* to the contractor to sub-contract any part of the work.
5.2	*Issue* instructions to explain and *Adjust* any ambiguities or discrepancies which may be present in the contract documents.
6.1	If appropriate, and only when strictly necessary for the purposes of the contract, *Give consent* to the contractor to use or communicate the drawings, specifications and other contract documents to a third party.
6.3	*Check* that any notice given by the contractor under sub-clause 6.3 contains all the details required. *Avoid*, or if not possible, *Minimise*, likely delay or disruption.
7.2	Where drawings, specification, calculations, maintenance manuals, and other information is submitted by the contractor, where appropriate, *Approve* such documents. (Such approval does not relieve the contractor of any of his responsibilities.)
10.1	*Receive* notification from the contractor regarding the provision of performance security.
12.2	When the contractor gives notice of encountering physical obstructions or physical conditions under sub-clause 12.2 and if such obstructions or conditions could not have been reasonably foreseen by an experienced contractor, *Determine*, after due consultation, extension of time and amount of any costs incurred by the Contractor. **Notify** the contractor accordingly, with a copy to the employer.
14.3	*Receive*, within the specified time, a detailed cash flow estimate. If required, *Ask* the contractor to submit revised estimates at quarterly intervals.
27.1	If and when a notice is received from the contractor of any discovery of fossils, etc., *Instruct* the contractor on how to deal with them. *Determine*, after due consultation, any extension of time and the amount of costs incurred and **Notify** the contractor accordingly with a copy to the employer.
30.3 & 30.4	If and when notification from the contractor of any damage to any bridge or road is received and if a claim is made successfully as a result of a failure by the contractor to observe and perform any of his obligations, *Determine*, after due consultation, the amount recoverable from the contractor. **Notify** the contractor with a copy to the employer.
31.1	If applicable, *Require* the contractor to afford all reasonable opportunities for other contractors and workmen to carry out their work.
38.1	*Receive* notice in respect of any part of the works or foundations which are ready for examinations and without unreasonable delay *Attend* to examine and measure.
39.2	In case of default by the contractor in carrying out instructions in accordance with sub-clause 39.1, *Determine* after due consultation all costs recoverable from the contractor by the employer and **Notify** the contractor accordingly with a copy to the employer.
40.3	When a notice is received within sub-clause 40.3, *Consider* and, if appropriate, *Grant* permission within 28 days from receipt of that notice to proceed with the suspended works.
42.2	If the employer fails to give possession of site or access thereto, *Determine* after due consultation any extension of time and amount of any costs incurred and **Notify** the contractor accordingly, with a copy to the employer.
44.1	In the event of certain circumstances which entitle the contractor to an extension of the time for completion, *Determine* after due consultation the extent of such extension and **Notify** the contractor accordingly, with a copy to the employer.
44.2	(Note that the engineer is not bound to make any determination unless the contractor within 28 days after such event notifies the engineer with a copy to the employer; and within 28 days or as may be agreed by the engineer after such notification, submits detailed particulars of any extension of time claimed.)

(Table 10.2 Contd.)

Sub-clause No.	Duties and authorities which are in response to an action initiated by the employer or the contractor
44.3	Where an event which fairly entitles the contractor to an extension of time for completion has a continuing effect such that it is not practicable for the contractor to submit detailed particulars within sub-clause 44.2(b), and provided that interim details have been submitted, *Make* without undue delay, and after due consultation, an interim determination and finally review and *Determine* an overall extension, after due consultation. (Such final review shall not result in a decrease of the extension of time already determined.) **Notify** the contractor accordingly, with a copy to the employer.
46.1	If and when notified by the contractor under sub-clause 46.1, *Consider* and if appropriate, *Give consent* to the steps proposed by the contractor to expedite progress of the works. If any steps taken by the contractor under this clause involve the employer in additional supervision costs, *Determine*, after due consultation, such costs and **Notify** the contractor accordingly, with a copy to the employer.
48.1	When requested by the contractor under sub-clause 48.1, and within 21 days of the date of delivery of the notice, either *Issue* a taking-over certificate stating the date of substantial completion or *Give* instructions specifying outstanding work required to be done before the issue of such certificate. **Notify** the contractor of any defects in the works affecting substantial completion.
48.2 & 48.3	As in the last sub-clause, but in respect of taking-over of sections or parts of the works.
48, 20 & 21	Before issuing a taking-over certificate in respect of the works or any section or part thereof, the engineer should *Inform* the employer of his intention to do so, giving him sufficient time to arrange for the transfer of the responsibility for the care of that section or part and for any insurances.
52.2	If notice is given either by the contractor to the engineer of his intention to claim extra payment or a varied rate or price; or by the engineer to the contractor of his intention to vary a rate or a price in accordance with sub-clause 52.2, and within the time stipulated therein, then:
52.1	i If the contract does not contain any rates or prices applicable to any varied work and failing the use of the contract rates or prices as a basis for valuation, *Consult* with the employer and the contractor to agree on suitable rates or prices with the contractor. In the event of disagreement, *Fix* such rates or prices as appropriate and **Notify** the contractor accordingly with a copy to the employer;
52.1	ii Until such time as rates or prices for varied work are agreed or fixed, *Determine* provisional rates or prices to enable on-account payments to be included in interim certificates;
52.2	iii If the nature or amount of any varied work relative to the nature or amount of the whole works or to any part thereof renders the rates or prices in the contract inappropriate, *Consult* with the employer and the contractor to agree suitable rate or price with the contractor. In the event of disagreement, *Fix* such other appropriate rate or price and **Notify** the contractor accordingly with a copy to the employer;
52.2	iv Until such time as rates or prices under the contract which are rendered inappropriate due to varied work are agreed or fixed, *Determine* provisional rates or prices to enable on-account payments to be included in interim certificates.
53.2 & 53.3	If notified by the contractor that he intends to claim under sub-clause 53.1, *Inspect* such contemporary records and, if appropriate, *Instruct* the contractor to keep further records and/or supply copies of such records. Furthermore, *Consider* and, if appropriate, *Follow* the procedure set out in sub-clause 53.3 regarding interim accounts.
53.5	Where the contractor has given notice under sub-clause 52.1, kept records under sub-clause 53.2 and supplied particulars under sub-clause 52.3, *Consult* with the employer and the contractor and *Determine* amount due to the contractor for inclusion in any interim certificate and **Notify** the contractor accordingly with a copy to the employer.

(Table 10.2 Contd.)

Sub-clause No.	Duties and authorities which are in response to an action initiated by the employer or the contractor
54.1	If requested and if appropriate, *Consent* to the contractor's removal of equipment, temporary works and material from the site.
60.5	When the contractor submits a statement at completion (within 84 days after the issue of the taking-over certificate) with supporting documents, and if appropriate, *Approve* the form of details of these documents and *Certify* payment in accordance with sub-clause 60.2.
60.6	When the contractor submits a draft final statement (within 56 days after the issue of the defects liability certificate) with supporting documents, if appropriate, *Approve* the form of details of these documents. If not, *Require* the contractor to submit further information.
63.4	Where the employer enters upon the site and the works and terminates the employment of the contractor, *Fix, Determine and Certify* as soon thereafter as may be practicable a valuation of the work done under the contract and unused or partially used materials, any contractor's equipment and any temporary works. Furthermore, within 14 days of such entry and termination *Consider* and, if appropriate, *Instruct* the contractor to assign to the employer the benefits of agreements for the supply of goods or materials or services and/or for the execution of any work for the purposes of the contract, which the contractor may have entered into.
65.3	If the works or any materials or plant on or near or in transit to the site or any of the contractor's equipment sustain destruction or damage by reason of any of the special risks and when notified in accordance with clause 52, *Determine* an addition to the contract price and **Notify** the contractor accordingly with a copy to the employer.
65.5	If and when notified by the contractor of increased costs arising from any of the special risks, *Determine* after due consultation the amount of such costs which shall be added to the contract price and **Notify** the contractor accordingly with a copy to the employer.
65.8	If the contract is terminated under the provisions of sub-clause 65.6, *Determine* after due consultation, any sums payable under sub-clause 65.8 and **Notify** the contractor with a copy to the employer.
67.1	If a dispute is referred to the engineer in writing under clause 67, **Notify** within 84 days of receipt of such reference, the employer and the contractor of the decision stating that such decision is made pursuant to clause 67.
69.4	If notified by the contractor under sub-clause 69.4 that work will be suspended or the rate of work reduced, and following such suspension or reduction the contractor suffers delay or incurs costs, *Determine* after due consultation extension of time and amount of costs to be added to the contract price and **Notify** the contractor accordingly with a copy to the employer.

10.6 The engineer's passive duties and authorities

The engineer's passive duties and authorities, defined earlier in Section 10.3.1 are set out in Table 10.3, arranged in sequence corresponding to the clauses of the Conditions of Contract. Where the duty includes the obligation to give notice to the employer or the contractor or to both, the word 'Notify' in bold type is used. The passive duties and authorities are shown in italics with the first letter capitalised.

Table 10.3 The engineer's passive duties.

Sub-clause No.	Duties and authorities expressed as rules to be followed
1.5	*Ensure* that any notice, consent, approval, certificate or determination is in writing.
2.1(c)	*Ensure* that no action is taken which would relieve the contractor of his obligations under the contract, unless expressly stated therein.
2.5	When giving instructions, *Do so* in writing or if oral, confirm in writing.
2.6	*Exercise* impartially any discretion in giving a decision, opinion, or consent; or in expressing satisfaction or approval; or in determining value; or in otherwise taking action which may affect the rights and obligations, either of the employer or the contractor.
6.1	*Keep* the drawings in the sole custody of the engineer.
16.2	(The engineer has the proactive authority to object and require the contractor to remove any person who is incompetent or negligent or who misconducts himself or whose presence on site is undesirable.) The engineer may consent to allow such a person to return.
18.1	If boreholes or exploratory excavations are required, such requirement shall be the subject of an instruction in accordance with clause 51, unless provided for in the Bill of Quantities.
19.1	The engineer has the authority to require the contractor to provide and maintain lights, guards, fencing, warning signs and watching, at the latter's cost.
20.2	In case of loss or damage to the works due to any of the contractor's risks, *Consider* whether or not satisfied with the rectification and if so give instructions as necessary.
25.1	*Receive* notification from the contractor of providing evidence of insurance as required under the contract.
36.1	All materials, plant and workmanship shall be as described in the contract and in accordance with the engineer's instructions.
37.1	The engineer, or any person authorised by him, shall at all reasonable times have access to the site, workshops and place of manufacture.
37.2	The engineer is authorised to inspect and test the materials and plant to be supplied.
37.4	The engineer, as a rule, has the authority, when necessary, to reject materials or plant if defective or not in accordance with the contract and notify the contractor stating his objections with reasons.
38.2	The engineer, as a rule, has the authority, when necessary, to instruct the contractor to uncover or make openings in or through any part of the works. *Determine*, after due consultation, the amount of costs, if work is satisfactory, and **Notify** the contractor accordingly, with a copy to the employer.
39.1	The engineer, as a rule, has authority when necessary to instruct the contractor to remove any material or plant not in accordance with the contract.
40.1	The engineer, as a rule, has the authority, when necessary, to *Issue instructions* to suspend the progress of the works or any part thereof should he consider it necessary and to have the works properly protected and secured.
42.1	If a programme under clause 14 is not submitted, *Receive* notice from the contractor of reasonable proposals for possession of site.
45.1	None of the works, unless otherwise provided (see sub-clause 46.1), shall be carded out during the night or on locally recognised days of rest without the consent of the engineer.
52.1	All variations under clause 51 shall be valued at the rates and prices set out in the contract, if these rates and prices are applicable. If not applicable, they shall be used as a basis for valuation.

(Table 10.3 Contd.)

Sub-clause No.	Duties and authorities expressed as rules to be followed
54.8	The operation of clause 54 does not imply approval by the engineer nor does it prevent him from rejecting such material or other matters referred to in the clause.
58.2 & 58.3	The engineer, as a rule, has the authority when necessary, to issue instructions in respect of the use of the provisional sums included under the contract.
60.4	The engineer, as a rule, has the authority, when necessary, to correct or modify any interim certificate and to reduce or omit the value of any work not being carried out to his satisfaction.
63.1	The engineer, as a rule, has the authority, when necessary, to certify, if appropriate, to the employer, with a copy to the contractor, any default by the contractor.
67.3	The decision of the engineer under clause 67 does not disqualify him from being called as a witness and giving evidence before an arbitration on any matter relevant to the dispute.

10.7 The engineer as a supervisor

It must be stated at the outset that the task of achieving the quality specified under a construction contract is the responsibility of the contractor, as also is the method of performing that task. There are many legal cases which have dealt with this topic, but the most pertinent from the point of view of the engineer acting as a supervisor is, perhaps, the case of *Oldschool* v. *Gleeson*.[10.10] The plaintiffs in that case were the owners of two adjoining properties, Nos. 30 and 31 Islington Green, London N1, which they wished to redevelop. The first defendants were the building contractors engaged by the plaintiffs and the second defendants were the consulting engineers. The works necessitated the total demolition of No. 31 and the partial demolition of No. 30. Work commenced in November 1971, and on 14 March 1972 when the demolition and excavation at No. 31 was almost complete, the party wall separating No. 31 and No. 32 collapsed.

The plaintiffs sued the first defendants and the second defendants seeking to be indemnified against all damages which might be awarded against them in respect of the damage suffered by the owners of No. 32 or their tenants.

The first defendants admitted their liability to indemnify the plaintiffs and instituted third party proceedings against the second defendants, the consulting engineers, claiming an indemnity or contribution in respect of the plaintiffs' claim and damages in respect of the loss suffered by them in that their contract works had been delayed. In those third party proceedings, the first defendants alleged that the second defendants owed them a duty of care in relation to the design and/or supervision of the works and that such duty had been breached by the second defendants in producing a design which could not have been constructed without causing the collapse of the party wall and/or in failing to provide adequate supervision. In his judgment, Judge W. Stabb, QC stated:

'Plainly it is the consulting engineer's duty to produce a suitable design for the works which will achieve what the building owner requires, and it is further his duty to ensure that design is carried out. The difference of opinion between the experts – Mr H. on behalf of the first defendants and Mr M. on behalf of the second defendants – is as to the extent of the consulting engineer's duty in regard to the manner in which the contractors execute the work in order to achieve the required results. Here may I pay tribute to the two experts, both of whom are consulting engineers of high qualification and considerable experience, and both of whom in my estimation gave their evidence in support of their respective opinions in a manner deserving of the highest praise.

Mr H. obviously is, if he will forgive me saying so, of the older school. Although he was disposed to agree at one stage that a consulting engineer's duty was to design and see that the design was properly carried out, but otherwise to leave the contractors to get on with the job and not give instructions as to how the work was to be done, he nevertheless maintained that it was still the consulting engineer's duty to see that the contractors executed the work in a competent manner, particularly where the safety of the works was involved. He regards the consulting engineer as what he described as being 'the father and mother of the job', whose duty it is to direct the contractors as to the manner in which the work is to be done, if he sees that the method which they are employing might endanger the safety of the works, and to stop the work if necessary. He considers it to be the consulting engineer's duty to ensure that the contractors carry out the work in a manner which will not endanger the safety of the works and thereby to assume responsibility for insisting that the contractors should undertake the work, if necessary, in a manner and sequence different from that which they have planned or may be proposing to follow.

Mr M, on the other hand, was equally insistent that the manner of execution of the works is a matter for the contractors. He considered that the consulting engineer is in no position, for instance, to require the contractors to comply with any particular sequence of works; he has no right, let alone duty, to involve himself in the work of the contractors. Of course he would interest himself in their work, would offer advice to assist the job to go better and would certainly not turn his back on a situation that he could see was likely to give rise to danger to life. Equally he would intervene if he could see imminent damage to property. Those are matters of common sense; but that is a very different matter from assuming responsibility for the method of work to be adopted by the contractors.

In my judgment, Mr M's view is the right one. I do not think that the consulting engineer has any duty to tell the contractors how to do their work. He can and no doubt will offer advice to contractors as to various aspects of the work, but the ultimate responsibility for achieving the consulting engineer's design remains with the contractors. To take the present case as an example, I have no doubt that it was the contractors' duty to set whatever shoring might have been necessary. It was also for them to decide upon the sequence of excavation that was to be adopted and how such excavation was to be temporarily supported if required ... It was the responsibility of the contractors to decide upon the method and sequence of excavation so as to achieve the consulting engineer's design ...'

Dismissing the first defendants' claim, the court held that:

(a) The second defendants' design was not fundamentally unsound and the collapse of the party wall was not due to any fault of that design.
(b) The second defendants were not under a duty and had no right to instruct the first defendants as to the manner of execution of the contract works.
(c) Even if it is the law that when a consulting engineer knows or ought to know that the contractors are failing to take proper precautions in the absence of which there is a risk of damage to property, he owes those contractors a duty to take care to prevent such damage occurring then:
    (i) his duty does not extend beyond warning the contractors to take the necessary precautions; and
    (ii) the second defendants in fact gave to the first defendants ample warning and were not in breach of any such duty.

It must be recognised, however, that a contract is a promise of what ought or ought not to be done rather than a guarantee of what will or will not be done and that a breach of that promise can lead only to sanctions. The form of sanction to be applied would depend on the applicable law of the contract. In common law, it is in the form of damages (see Chapter 3). To minimise the number of events where such sanctions might apply, it is necessary to discover any irregularities and defects in the performance of the contract as soon as they occur or at the earliest possible time thereafter. This is particularly important in construction contracts because of some peculiar characteristics which are recognised in these contracts, such as:

(a) Most construction contracts take a long time to complete during which time various parts of the work become covered or concealed by other parts and, thus, become practically inaccessible.
(b) The work is carried out on the employer's site, this being a fixed piece of land or waterway. A completed but defective project cannot be rejected and replaced in practice by another, as is normally done in manufactured products. Instead it has to be repaired and accepted in a condition which might be less than perfect. Such repair may not in fact be acceptable and even if acceptable, may not be feasible if the contractor becomes insolvent.
(c) As the construction process includes a great number of activities, some of which cannot start before the completion of others, the earlier a defect is discovered the greater the possibility of finding a possible or feasible solution. Furthermore, the earlier the discovery of a defect, the lower the cost of implementing a solution and the shorter the period of disruption and delay to the completion of the project.

Accordingly, the supervisory role of the engineer is an essential part of any construction contract, but such a role must be viewed as a supportive role to the contractor's objective of quality rather than a primary one for its achievement. This view is supported by other judicial decisions which include the decisions of the Court of Appeal in *Clayton* v. *Woodman*[10.11] and the House of Lords in *East Ham* v. *Bernard Sunley & Sons*.[10.12]

It follows, therefore, that unless the contractor intends to fulfil this objective of quality control by allowing in his tender for personnel to carry out his own supervision; planning his construction methods to include for such an objective; and controlling and executing his work to achieve the required quality within the prescribed time, it would be extremely difficult for the contractor and the engineer to have a smooth and satisfactory relationship on site. The engineer will simply be left to act as a policeman busy in discovering mistakes and devising methods to remedy the effect of such mistakes.

The question which must therefore be asked is, what is the extent of the supervisory role of the engineer? A policy statement by FIDIC on the role of the consulting engineer during construction provides some insight into supervision, but unfortunately it deals mainly with the recommendation that the engineer should be appointed to provide a full professional service inclusive of pre-contract and post-contract activities.[10.13] This policy statement still leaves unanswered the question of whether supervision means inspection or overseeing or something beyond those functions. Obviously, it does not mean one supervisor for each worker. In the scheme of the Red Book, the engineer as a supervisor is expected to monitor, through inspection and testing, the work being carried out and to make sure that on completion the employer, for whom the project has been constructed, has a project completed in accordance with the contract and with any supplementary instructions which may have been given. (See Section 10.1 above, for a definition of design.) What then is the engineer to monitor? The main aspects which need to be monitored by the engineer are:

(a) compliance with the specified quality;
(b) progress in accordance with the planned programme;
(c) budget control in accordance with the cost plan; and
(d) compliance with other matters specified, such as safety, environmental controls, etc.

The relevant clauses of the Red Book dealing with these four aspects include clauses 2, 17, 19, 20, 33, 36, 37, 38, 39, 46, 49, 50, 51, 56, 57 and 65. In considering these clauses, it should be borne in mind that the contractor must employ supervisory staff who should also ensure the quality of the work being executed through direct contact with the site personnel. Such contact includes giving direct instructions which no one else, including the engineer, can give.

10.8 The engineer as certifier

Sub-clauses 2.6(c) and (d) make it clear that wherever the engineer is required to exercise his discretion by 'determining value' or 'otherwise taking action which may affect the rights and obligations of the Employer or the Contractor', he is to exercise such discretion impartially. When combined with sub-clause 60.2 which provides that the engineer shall '... certify to the Employer the amount of payment to the Contractor which he considers due and payable ...', the role of the engineer in respect of payment is clear. It is also a powerful role, since on the strength of an interim certificate issued by the engineer under sub-clause 60.10, the contractor must be paid by the employer any amount due to him within 28 days after the certificate has been delivered to the employer. In the case of the final certificate, the period for making payment is 56 days.

Sub-clause 60.10 further provides that failure to pay within the stated period entitles the contractor to be paid interest at the rate stated in the Appendix to Tender. This is without prejudice to the contractor's other entitlements under clause 69 in case of non-payment within the specified period.

Under sub-clause 69.1, the contractor is entitled to terminate his employment under the contract if the employer fails to pay the amount due under any certificate issued by the engineer within 28 days after the expiry of the time stated in sub-clause 60.10. Furthermore, under sub-clause 69.4, the contractor is entitled, after giving 28 days notice to the employer, with a copy to the engineer, to suspend work or reduce the rate of work should the employer fail to pay within the specified time.

Besides issuing certificates under clause 60, the engineer's role as a certifier includes other certification duties under the contract. These duties include:

(a) issuing a taking-over certificate under clause 48;
(b) certifying the date of completion of the works under clause 49;
(c) certifying payments to nominated sub-contractors under clause 59;
(d) issuing a final certificate under sub-clause 60.8;
(e) signing and delivering a defects liability certificate under sub-clause 62.1; and
(f) valuation of work done, materials supplied and other equipment on site in case of termination under clause 63.2.

Despite the diversity of the duties of the engineer as a certifier, the most important of these is, perhaps, his role under clause 60 enabling him to certify monthly payments to be made to the contractor. These payments are essential for the smooth performance of the contract as they represent to the contractor the lifeline with respect to the contract. In essence, under the Red Book the contractor is responsible for financing the cost of construction for the period stated in sub-clause 60.10, whilst it is the employer who must ensure the continued financing of the contract works for the whole period of construction.

Certification under clause 60 is also important in an indirect manner to show:

(a) A mode of expressing interim satisfaction with the work in relation to quality notwithstanding that such satisfaction may be subsequently reversed and that interim certificates are to be taken only as on-account payments and can be corrected under sub-clause 60.4; and
(b) Evidence, in the form of a sum included in the certificate, of any determination or assessment made by the engineer in respect of any claim which may have been made by the contractor. Such a claim could be in respect of either a matter of fact or a matter of interpretation of the wording of the technical and legal aspects of the contract documents.

These two aspects and particularly the latter reveal the actions of the engineer in performing his role as the employer's agent in determining additional payments and extensions of time to which the contractor may be entitled. Accordingly, they provide the link to the final role of the engineer, that of an adjudicator or quasi-arbitrator.

10.9 The engineer as adjudicator or quasi-arbitrator

As stated above, the Red Book provides for yet another role for the engineer: the role of an adjudicator which is sometimes referred to as the role of a quasi-arbitrator. This role is embodied in clause 67 which provides that any dispute may be referred to the engineer for his decision and such reference is a condition precedent to arbitration. The term 'quasi-arbitrator' as applied to this role was rejected by Mr Justice Purchas in the case of *Pacific Associates Inc* v. *Baxter*.[10.14] He stated in that respect:

> '... I am quite unable to accept the proposition that the role played by the engineer under GC67 was anything other than a review of an earlier executive decision made in the course of its function as supervising engineer and subject to arbitration under the latter provisions of GC67. To adopt the words of Viscount Dilhorne in a similar position in *Sutcliffe* v. *Thackrah*,[10.15] to view the function of the engineer under GC67 as an arbitral function would be to construct an arbitration on an arbitration.'

It appears from the above judgment that the engineer's role under clause 67 is an adjudicator's role rather than a role similar to that of an arbitrator. Whichever it is, this role is of particular importance to those contracts where the engineer is required, under the terms of his appointment by the employer, to obtain the specific approval of the employer before exercising certain authorities delegated to him under the Conditions of Contract. As discussed earlier, sub-clause 2.1(b) provides that should there be such a requirement, it should be set out in Part II of the Conditions so that tenderers may recognise the framework within which the engineer is permitted to operate without possible interference or hindrance.

However, whilst it may be accepted that the employer might wish to curtail the engineer's authority in respect of any of the provisions of the contract by requiring the engineer to obtain his prior approval, it is not logical to curtail the authority provided under clause 67 by such a requirement. Should this be done, the two-tier system of dispute settlement, devised in the Red Book, for a primary determination by the engineer followed by arbitration where either or both of the parties involved are dissatisfied with the engineer's determination, would be meaningless (see Chapter 19).

Accordingly, the engineer must be free to decide the issues referred to him under clause 67 without obtaining the approval of the employer. Similarly, the contractor by entering into the contract must be deemed to have accepted the procedure laid down by the two-tier system of dispute resolution and its time allowances and constraints. However, it must also be recognised that only disputes can be referred to the engineer under clause 67 for his decision. Furthermore, before a dispute comes into being there must be a claim or an assertion made by a claimant party and that claim or assertion is subsequently either denied or rejected by the other party and then that denial or rejection is rejected by the claimant party. This assertion could be in the form of a claim for money; a claim for an extension of time; an asserting statement as to quality or quantity or an interpretation of a particular clause whether technical or legal; or a declaration or other similar form. When the assertion is a claim for money or time, the engineer must be allowed sufficient time to investigate and process the claim in accordance with the relevant provisions of the Conditions of Contract and in particular clause 53. If this is not done, the employer would be deprived of a service to which he is entitled under the contract.

The two-tier system of dispute resolution in the Red Book begins, therefore, with an assertion or a claim by one or both of the parties to the contract under any of its clauses, other than clause 67, whether or not such an assertion or claim is due to an opinion, instruction, determination, certificate or valuation of the engineer. If such an assertion as made by the claimant party is then rejected or denied by the other party, the rejection or denial might be accepted or rejected by the claimant party. If accepted, whether by persuasion or for commercial or other reasons, that is the end of the matter. If the rejection is unacceptable to the claimant party, the situation would only then develop into a dispute between the parties. It is relevant here to draw attention to the fact that the word 'difference' has been deleted from the text of clause 67 of the Fourth Edition of the Red Book (see section 9.35 above). Thus, the second tier of the system begins when the dispute is referred by the claimant party to the engineer under clause 67 for his decision as an adjudicator.

The engineer fulfils his role as an adjudicator through the following framework:

(a) The contract is between the employer and the contractor and only a dispute between the employer and the contractor can be referred to the engineer under clause 67.
(b) The word 'dispute' means: a verbal controversy; an argument; a debate; a quarrel. A dispute can result only from a previous assertion made by one party which is rejected by another party, and the rejection is finally pronounced to be unacceptable by the asserting party.
(c) A dispute may emanate from an action or inaction by the engineer or the employer in response to an assertion made by the contractor, or vice versa.
(d) After a dispute has arisen, there is no time limit to the reference by either party to the engineer for his decision under clause 67. A time bar may, however, exist under the provisions of the applicable law of the contract or, of course, by a final settlement of the account. However, once a dispute is referred to the engineer for his decision under clause 67, then a strict timetable applies with which both the engineer and the parties must comply.
(e) In reaching his decision, the engineer should not only be impartial, but must also do so in a logical and reasoned manner which may have to be explained later should he be called upon to give evidence in any subsequent arbitration proceedings (see sub-clause 67.3). Although the engineer is not obliged to give reasons with his decision, he is not precluded from doing so, and in some cases it would be advisable for him to consider such action. In this connection, it is worthwhile noting that the decision of the engineer may subsequently be re-opened, reviewed and revised by an arbitrator. Therefore, any reasons given by the engineer with his decision would be extremely helpful during arbitration proceedings.

(f) Many of the matters which may be referred to the engineer for a decision under clause 67 are matters relating to the interpretation and application of the terms of the contract documents. As such, the engineer would require both technical knowledge as well as a clear understanding of various aspects of construction law.

(g) In deciding disputes under clause 67, the engineer may even find himself having to rule on matters which may concern his own work and professional activities with respect to the project and may thus be of relevance to his own responsibility and liability under his contract with the employer. The engineer must deal with these matters in an impartial manner. It must be pointed out that the possibility of the engineer having to decide claims, the cause of which may be attributed to his own actions, has brought the role of the engineer as adjudicator under attack on the basis of bias.[10.16] In such circumstances, the engineer is open to the suspicion that he may not be willing to disclose or admit any of his failings or shortcomings and, therefore, that not only he can not act independently and impartially, but also that he would not appear to be acting as such.

The role of the engineer has recently been subjected to further attack emanating from both parties to the contract. On the one side, the engineer is accused by the contractor of being biased in favour of the employer because:

— his fee is paid by the employer;
— he has acted as adviser to the employer prior to construction and may wish to continue this role upon completion of the construction stage;
— he is expected to go through 'due consultation' with the employer prior to making certain decisions;
— he may have to obtain the specific approval of the employer prior to taking certain actions or decisions.

Therefore, he is sometimes assumed by those who attack the engineer's role as adjudicator, to act in collusion with the employer. The case is even stronger against the engineer who is in the employment of the employer, as his future employment may depend on his actions.

On the other side, he is sometimes accused by the employer of being biased towards the contractor during the administration and execution of the contract in such areas as awarding extensions of time and in determining amounts of claims, and giving instructions in favour of the contractor. He is also sometimes accused of being too lenient on the contractor, thereby relieving the contractor of some of his obligations under the contract.

It seems inevitable that where there is substance in such attack, and in situations where either party to the construction contract is dissatisfied with the impartiality of the engineer, the only answer would lie in arbitration proceedings where the arbitrator is empowered to re-open, review and revise the engineer's decision.

As discussed earlier, the above mentioned criticism of the role of the engineer under clause 67 of the Red Book led to the introduction of an alternative wording to that clause in the new supplement published by FIDIC. As described later in Chapter 19, the new wording is based on the use of an adjudication board composed of one or three experts who can render a decision in respect of a dispute without having to resort to the engineer. Thus, if the new wording of clause 67 is chosen in preference to that presently in use, the traditional role of the engineer as an adjudicator will be transferred to the dispute adjudication expert or board.

10.10 Concluding remarks

In this chapter, the duties of the engineer and his traditionally complex role under the Red Book were discussed in detail. The weaknesses and strengths of that role were explored. Should the engineer fail to perform any of his functions properly, the liability towards the employer for any lack of care or skill which causes loss or damage to the employer is discussed in Chapter 11. Furthermore, should the contractor suffer any loss or damage under his contract with the employer due to the engineer's negligence, the contractor might obtain a remedy against the employer through the contract between the contractor and the employer, leaving the employer with the option of obtaining redress against the engineer. This matter is also discussed in Chapter 11.

Chapter 11

Responsibility and Liability of the Engineer

11.1 Introduction

It is appropriate first to define both responsibility and liability with some precision, owing to the relevance of these two terms to the contents of this chapter. The word 'responsibility' is defined in *The Oxford Companion to Law* as follows:[11.1]

> 'Responsibility. A word used in several senses. A person may be said (1) to be responsible if he generally displays care and forethought and considers the possible results of his actions. He may also be said (2) to be responsible for certain events if his conduct has been a material factor in bringing them about; thus a reckless driver may be said to be responsible for an accident. In this sense the word means little more than that he has caused the events and does not necessarily imply accountability. An animal or a snowfall may be said to be responsible for causing a happening. A person may also be said (3) to be legally responsible when of such an age and in such a state of mind and body that he is deemed to be capable of controlling his conduct rationally and such that he can fairly be held accountable and legally liable for the consequences of what he does. Conversely, a person mentally ill or under the influence of drink or drugs may be held legally irresponsible. Responsibility in this sense is fundamental to liability to punishment. Responsibility in the third sense has a substantial moral flavour, but moral responsibility or blameworthiness and legal responsibility are not wholly equivalent. A person may by law be held responsible in cases where he has not been personally blameworthy at all. Thus under the principle of vicarious liability a person is held responsible for, and legally liable for the wrongs of, his employees, though not personally in fault at all. . . .'

Similarly, the word 'liability' is defined in the same dictionary as follows:

> 'Liability. The legal concept of being subject to the power of another, to a rule of law requiring something to be done or not done. Thus a person who contracts to sell goods is liable to deliver them and the buyer is liable to pay the price. Each is required by law to do something, and can be compelled by legal process at other's instance to do it; the other is empowered to exact the performance or payment. It is sometimes called subjection. The correlative concept is power.
>
> A person is said to be under a liability when he is, or at least may be, legally obliged to do so or suffer something. Thus, one may be said to be liable to perform, to pay,

to be sued, to be imprisoned, or otherwise to be subject to some legal duty or legal consequence. In general, liability attaches only to persons who are legally responsible; an insane person does not generally incur any liability.

Liability may arise either from voluntary act or by force of some rule of law. Thus, a person who enters into a contract thereby becomes liable to perform what he has undertaken, or to pay for the counterpart performance, or otherwise to implement his part of the contract. If he acts in breach of contract, he becomes liable by law to pay damages in compensation for the breach. Similarly, if a man acts in breach of any of the general duties made incumbent on him by statute or common law, such as to refrain from injuring his neighbour, or to maintain his tenant's house in reasonable repair, or to exercise diligence in administering property of which he is trustee, he incurs legal liability to make good his omission or default.

Liability is commonly distinguished according to its legal grounds into civil liability, whereby one is subject to the requirement to pay or perform something by virtue of rules of civil law, and criminal liability, whereby one is subject to being fined, imprisoned, or otherwise treated by virtue of rules of criminal law. Civil liability may arise from many grounds, from the natural relations of the family, from undertaking or contract, the commission of a harm, from trust, statute, or decree of court. In respect of liability arising from harm done, it arises from intentional harm, harm brought about in breach of duty, and in some cases there is strict liability if harm befalls despite care taken. Criminal liability arises from the admitted or proved commission of some kind of conduct declared by the rules of criminal law to be a crime inferring punishment. At common law criminal liability normally also requires that the conduct has been done intentionally or recklessly but not merely negligently or accidentally, and sometimes proof of a particular intent is a necessary ingredient of a crime, but many cases under statute have been held to impose strict or absolute liability, i.e. liability irrespective of the actor's state of mind. . . .'

Whilst these two notions of responsibility and liability are expressed in two different words in the English language, in other languages one may find them to be combined in one expression. The difference between the two concepts is that of obligation under the law, for example, in the French language where the word *'responsabilité'* is used to mean legal responsibility.

Dealing first with the responsibility of the engineer, his responsibility/accountability applies to the manner in which he discharges the duties allocated to him under the contract as discussed in Chapter 10. Thus, the question which must be answered is: to whom, and in respect of what, is responsibility owed by the engineer?

In essence, the general principles which apply to the first part of this question can be summarised by listing the parties to whom the engineer owes a duty:

(a) the employer, as client;
(b) the contractor, as the party responsible for construction;
(c) third parties;
(d) society in general;
(e) his employees; and
(f) himself.

The answer to the second part of the question (i.e. in respect of what is responsibility owed?) is extremely complex and depends not only on the contractual arrangements, but also on the applicable law of the contract or contracts between the parties involved. As stated earlier in Chapter 2, the applicable law imports into the general conditions, certain principles of law which govern the contract. However, the situation becomes even more complex in the international scene where a contract incorporating the Red Book is executed in a country where the accepted role of the engineer is different from that described in the previous chapter. In such cases and because of the similarities between the Red Book and the ICE Form, legal cases which apply to the ICE Form are of value in establishing what legal principles are relevant in the application of the science of comparative law.

For example, in construction projects in France, the nearest corresponding position to the engineer under FIDIC is that occupied by the *maître d'œuvre*. The role of the *maître d'œuvre* normally includes the design of the works and the administration and supervision during the construction period. Under French law, a *maître d'œuvre* carrying out his duties during the construction period is neither regarded as an agent of the employer, nor as an adjudicator or a 'quasi-arbitrator', unless he is specifically requested and empowered to do so under the contract. He is recognised as an independent professional bound by contract for the hire of work under Articles 1710 and 1779 of the French Civil Code (*contrat de louage d'ouvrage*). Thus, his principal duty is to advise his client, but not to act on his behalf.

A comparison between the standard form of contract for private work in France (*'Norme Afnor – Marchés – Cahiers Types'* – NI P03 – 001, April 1982) and the Red Book would conclude that the powers of the *maître d'œuvre* under a construction contract are generally much more limited than those of the engineer under the Red Book. He states that under the French form:[11.2]

(a) Variations in the works are ordered by the owner (art. 07.1.4.1) not the engineer. They are implemented by written orders signed by the engineer and countersigned by the owner.
(b) Completion of the works is established by minutes of completion drawn up by the engineer but signed by the owner (art. 14.4).
(c) The engineer is accorded no power to approve or disapprove the contractor's work programme (art. 01.4.1.5).
(d) The engineer has no power to stop the works, although the owner can do so for periods of up to six months without being deemed in breach of contract (art. 19.1.3).
(e) The engineer has no formal role in the settlement of disputes between the owner and the contractor. Moreover, the provisions of FIDIC, which might permit the engineer to render decisions that in certain circumstances may become 'final and binding' on the contractor, may be unenforceable against the contractor under French law (*Cass. Comm.* 9 March 1965).

Thus, when dealing with a FIDIC-type contract, lawyers, engineers and others from countries using the Romano-Germanic system of law should acquaint themselves with the role of the engineer under the Red Book before dealing with his responsibility and liability.

In this connection, it should be realised that other systems of law may have a different role for the equivalent position to the 'FIDIC Engineer' and, furthermore, that not all countries within the Romano-Germanic system of law have exactly the same legal principles.

Returning then to the second part of the question posed above: 'what is the responsibility owed by the engineer?', one would have to accept the role of the engineer under the Red Book as a basis and add to it whatever is imported by the applicable law of the contract.

Sections 11.2 to 11.5 below will consider the six elements of the topic of 'to whom the engineer owes a duty'.

11.2 Responsibility of the engineer towards the employer

Under the conditions of the contract with his employer, the engineer is responsible for the duties he undertakes. If these duties are related to a contract between the employer and the contractor under the Red Book, the engineer is responsible in his role as a designer, and during the construction of the project, as employer's agent, supervisor, certifier and adjudicator.

Failure properly to perform these duties properly in accordance with the provisions of the contract would result in the engineer being liable to the employer under the provisions of the law of the contract applicable to his agreement with the employer. However, in certain jurisdictions, and in some circumstances, the engineer may be found to be concurrently liable in respect of a particular event and in respect of the resultant damage both in contract and in tort.

In countries within the common law system, the law on whether or not concurrent liability in contract and in tort applies is still evolving. In England, the construction case of *Bagot* v. *Stevens Scanlan & Co.* in 1964 marks, perhaps, a starting point for the development of this area of the law. The court held then that duties owed by a professional person arose only in contract and not in tort.[11.3]

In 1976, however, the case of *Esso Petroleum* v. *Mardon* paved the way for the latter concept to be abandoned.[11.4] In his judgment in that case, Lord Denning remarked that the court in the *Bagot* case was not referred to two earlier decisions of higher authority where it was held: first, in 1844, that 'the plaintiff may either recover in tort or in contract'[11.5]; and secondly, in 1914, that 'for failure to perform his obligations, he may be liable at law in contract or even in tort, for negligence in breach of a duty imposed on him'.[11.6]

These remarks led in 1978 to the adoption of the concept of concurrent liability in the case of *Midland Bank Trust Co. Ltd* v. *Hett, Stubbs and Kemp*.[11.7] Later decisions in other cases confirmed this position. While some courts outside England and Wales were willing to adopt the concept of concurrent liability,[11.8] others were not.[11.9]

Shortly afterwards, the trend towards concurrent liability was halted and the Court of Appeal in England held that a claim in tort could be no wider than the claim in contract.[11.10] In the mid 1980s, the courts began to look more favourably at the dissenting judgments in previous decisions to restrict the right to impose concurrent liability. The Privy Council in *Tai Hing Cotton Mill Ltd* v. *Liu Chong Hing Bank Ltd*, in 1986, decided to:

> 'adhere to the contractual analysis [firstly] on principle because it is a relationship in which the parties have, subject to a few exceptions, the right to determine their obligations to each other; and [secondly] for the avoidance of confusion [in the law] because different consequences do flow according to whether the liability arises from contract or tort, e.g. in limitation of action'.[11.11]

Two passages from the judgment in that case are worthy of mention as they have been cited in subsequent cases: 'Their Lordships do not believe that there is anything to the advantage of the law's development in searching for a liability in tort where the parties are in contractual relationship', and '. . . Their Lordships do not accept that the parties' mutual obligations in tort can be any greater than those to be found expressly or by necessary implication in their contract.'

This trend continued with the case of *Ernst and Whinney* v. *Willard* (1987) where the assignees of a long lease of an office development brought a claim against engineers, nominated mechanical sub-contractors and ductwork sub-contractors for alleged deficiencies in the air-conditioning system.[11.12] It was held that it would not be just and reasonable to impose liability in negligence on the defendants for various reasons, amongst which was reliance on the decision in *Tai Hing*. The general trend continued towards denying concurrent liability, at least in certain circumstances, culminating in the decision in *Greater Nottingham Co-operative Society* v. *Cementation Piling & Foundations Ltd* (1988).[11.13] In this case, the plaintiffs were building owners, and the defendants piling sub-contractors, who had executed a collateral warranty agreement. Damage occurred to an adjoining property due to certain negligent piling. The employers claimed the additional cost of executing a revised piling scheme which had been paid to the main contractor. The employers' claim was brought in tort notwithstanding the direct contract since it was not relevant to deal with the particular issues of the case. In deciding whether there should be a concurrent but more extensive liability in tort as between the two parties arising out of the execution of the contract, Lord Justice Purchas in the Court of Appeal considered as relevant the fact that the 'parties had an actual opportunity to define their partnership by means of a contract, and took it'. The Court of Appeal went on to decide that there was no liability in tort in this case. This decision may be taken to indicate that a contract can have the effect of limiting, rather than excluding, tortious liability.

However, the recent trend in construction in England seems to be developing in a way that each case will have to be considered on its own merits as to whether or not there is concurrent liability in contract and in tort. Furthermore, it seems that a consideration of the particular facts and circumstances of each case would determine whether a duty of care in tort, which might be imposed by the general law, was of wider scope than any contract in which the same parties had agreed during the same course of dealing. In *Barclays Bank plc* v. *Fairclough Building Limited* (1993),[11.14] the plaintiff, citing the two passages quoted above from the *Tai Hing* case, contended that if there were liability for breach of contract, there was no liability in tort for negligence. The case arose out of asbestos contamination of a building following the cleaning down of corrugated asbestos roof sheets by pressure jetting with water. Water and slurry which contained asbestos entered the buildings in substantial quantities. The slurry then dried out leaving asbestos dust which polluted the atmosphere, causing serious health problems. The court held in that case that there was a claim both under contract and in tort.

In *Lancashire & Cheshire Association of Baptist Churches Incorporated* v. *Howard & Seddon Partnership* (1993),[11.15] a claim arose in contract against the architects where it was alleged that there was condensation and inadequate ventilation in the church sanctuary designed by them and completed in 1980. But, the claim was statute barred in contract under the Limitation Act and a preliminary issue was ordered to be tried concerning whether the basis of the claim could be in tort where time was still within the limitation period. It was held that a professional person owed duties in both contract and tort and that 'unless the contracting parties have expressly excluded the duty in the law of tort, which they are free to do, it is reasonable to conclude that the duties under both heads survive concurrently' and thus the longer limitation period in tort could be relied upon. Thus, the claim succeeded and the judge went on to categorise *Tai Hing* as a case which provides authority for the proposition that where there is a contract, the tortious liability cannot be more extensive than that in contract.

In 1995, two cases followed the same trend. In *Gable House Estates* v. *The Halpern Partnership and Bovis Construction Ltd*[11.16] the court held that a concurrent duty in tort can exist where the parties are in a contractual relationship if the terms of the contract do not preclude it. In the case of *Holt* v. *Payne Skillington*, the court held that duties in contract and in tort can be concurrent, but are not co-extensive.[11.17] Furthermore, the duty in tort for negligence may be greater than for breach of contract and it is imposed by the general law, whereas the duty in contract comes about only by the obligation assumed by the parties.

In the Romano-Germanic system, whilst the contract between the employer and the engineer governs the relationship between them, many jurisdictions differ in their interpretation and application of the law in respect of concurrent liability in both contract and tort. In France, for example, an act or omission by either party to a contract which is considered to be in breach of that contract, is a basis for an action under contract law (Articles 1146 to 1155 of the Civil Code), but does not concurrently serve as a basis for an action in tort.

The Belgian courts also reject the idea of concurrent liability under both contract and tort. In German law, except in specific cases of loss or damage, the Code, in paragraph 823 (BGB), bars the recognition of a tort action in a contractual situation. However, under Dutch law, concurrent liability is permitted with the restriction that the requirements of a tort action should be met independently of those required for a breach of contract.

The relevance of whether liability is in contract or in tort stems from two major differences. The first is in respect of the time bar under the relevant Limitation Act. Although the time bar differs from one jurisdiction to another, in general, actions under tort have longer periods of time in that they start to run later than those under contract. The second relevant difference is in respect of the damages which may be awarded under tortious and contractual liability. There may be other differences which could conceivably alter the balance as to whether to have an action in tort or in contract. These are: the rules of proximity, the proof of negligence and the level of liability which may apply under the contract in question.

As the law in this area is obviously developing and since the law differs from one jurisdiction to another, it is necessary to treat the above discussion of the topic as a matter of general rather than specialist knowledge. It is also necessary when considering this question of concurrent liability to obtain expert knowledge and advice in the relevant jurisdiction.

Before leaving this topic, it is helpful to discuss the two standards of duties in construction. The first is the duty of reasonable skill and care and the second, which is of a higher standard, is the fitness for purpose. It is important to distinguish between these two standards, because in the first, negligence of the defendant has to be proven by the claimant party, whereas for fitness for purpose there is a strict obligation to achieve the objective irrespect of negligence.[11.18]

The standard of duty of care

In general terms, a professional person is under a duty of reasonable skill and care, whereas a contractor is under a duty of fitness for purpose. But, it is also generally accepted that a designer has to take into account the purpose of his design if that is specifically made known to him. If he does not, he would be lacking in reasonable care. For instance, it would be a lack of care if a pumping station is designed to discharge a certain volume of water per hour when it is known that the actual volume required to be handled is twice that designed.

With regard to his responsibility to the employer, the standard of duty of care and skill owed by the engineer as a designer is perhaps best described in the decision in the case of *Bolam* v. *Friern Hospital Management Committee*, in 1957[11.19] which was approved in *Whitehouse* v. *Jordan* (1981) by the House of Lords:[11.20]

> 'How do you test whether this act or failure is negligence? In an ordinary case it is generally said that you judge that by the action of the man in the street. He is the ordinary man. In one case it has been said that you judge it by the conduct of the man on the top of the Clapham omnibus. He is the ordinary man. But where you get a situation which involves the use of some special skill or competence, then the test as to whether there has been negligence or not is not the test of the man on top of the Clapham omnibus, because he has not got this special skill. The test is the standard of the ordinary skilled man exercising and professing to have that special skill. A man need not possess the highest expert skill; it is well established law that it is sufficient if he exercised the ordinary skill of an ordinary competent man exercising that particular art.'

The ordinary skill of an ordinary competent person would normally be judged by reference to the state of the art of the design when it was carried out and not at a later date.[11.21]

It is accepted that the precise scope of the duties undertaken by the engineer in the contract with his client, the employer, would determine his responsibilities and, consequently, his liability. Obviously, should the terms of the agreement between the promoter and the engineer include detailed expectations of the design, the standard of care would be elevated to that of fitness of purpose as was concluded in a Canadian case in 1980.[11.22] In that case, it was concluded that there was a 'common intention that the building should be fit for its purpose'. It was held that 'this gave rise to a term implied in fact that if the structure was completed in accordance with the design it would be reasonably fit for use as a . . . store'.

The designer is also under a continuing duty to check during the construction period that his design will work in practice and to correct any errors which may emerge.[11.23] In some cases, the continuing duty would extend not only to the specific matters in the design itself, but to any new developing knowledge emerging during the construction period. However, such duties can be exercised only if the design professional is involved in the supervision as well as the design of the project. The risk of something going wrong increases, therefore, when the designer is not engaged in supervision either because another professional is employed instead of him, or because no one is engaged to carry out the supervisory duties. There is one further matter which might have to be considered in connection with the continuing duty of the designer and that is, whether the duty continues beyond the practical completion of the project. The law in this area is developing in some jurisdictions and specialist knowledge would be required to deal with this topic.

In respect of delegation of responsibilities, there is, generally, no implied term in the contract between the employer and the engineer permitting the delegation of the duty of design: in *Moresk* v. *Hicks* (1966), an architect sub-contracted the design of the structural frame of the building to a contractor who supplied and erected the frame.[11.24] The frame, however, proved to be defective in two major design aspects and the architect was therefore held liable to his client. (See the quotation from the judgment in that case referred to in Section 10.2 above.) In order for the engineer to be relieved of the responsibility of design of a specific part, a contract must be established between the employer and the specialist designer.

Before leaving the subject of the liability of the engineer towards the employer, consideration must be given to two further aspects. First, knock-on liability; and secondly, decennial liability which exists in a large number of jurisdictions around the world.

Knock-on liability arises where an individual may be held liable for more than his actual share of blame in respect of a loss for which he is only partly to blame. This happens when other parties, also partly responsible for the loss, cease to remain solvent due to one reason or another. The engineer may, therefore, find himself not only partly responsible, but also totally liable to the employer for a loss event for which an insolvent contractor is also liable. Such liability may include the full cost of repair of the damage or replacement of the damaged item. This legal principle applies both in the common law and the Romano-Germanic systems of law. Further consideration of this principle or that of contribution between several wrongdoers is beyond the scope of this book, but for further reference the publication, *The Liability of Contractors* is informative.[11.25]

Decennial liability forms part of the law relating to construction projects in many jurisdictions, including France, Belgium, Spain, Latin America, some parts of Africa and the Middle East. It applies as soon as the works are completed and accepted by the owner without any reservation and continues for ten years thereafter. In France, Articles 1792 and 2270 of the Civil Code legislate for this liability.[11.26] Article 1792 provides as follows:

> 'Every constructor of a structure is legally responsible to the owner, or those deriving title from him, for any damage (including damage resulting from sub-soil conditions) which jeopardises the integrity of the structure or which by affecting one of its component elements, or one of the equipment elements, render the structure unfit for its intended purpose.
>
> Such responsibility will not be imposed where the builder demonstrates that the said damage results from causation outside his authority and control.'

Accordingly, under French law, every constructor of a structure is presumed to be liable to the owner for damage which jeopardises the integrity of the structure or which, by affecting one of its component elements, or one of the equipment elements, renders the structure unfit for its intended purpose. The constructor is defined as including, among other persons: an architect, a contractor, a technician or other person bound to the owner by a contract for the hire of work.

The defect giving rise to the damage defined in Article 1792 must be latent. Visible defects at the date of completion, (referred to in French as the equivalent of 'reception') do not give rise to a presumption of decennial liability.

The provisions of Article 1792 impose strict liability on all constructors and, therefore, the employer or owner does not have to prove either negligence or any causal association between the negligence and the damage. Under French law, these provisions are a matter of public policy and cannot be restricted or excluded in any agreement. In other jurisdictions it is permissible to modify the legal provisions regarding decennial liability, for example, in Holland.

Article 2270 of the French Civil Code deals with the ten-year period of this liability and Article 2820 provides for decennial insurance.

11.3 Responsibility of the engineer towards the contractor

In the traditional contractual arrangement where the Red Book is used, there is no contract between the contractor and the engineer. The contractor, therefore, is precluded from bringing an action against the engineer in contract. Instead, he must rely on the law of tort within the legal framework of the applicable law of the contract.

Clause 2 of the Red Book is of particular relevance in the above connection. Sub-clause 2.1 (engineer's duties and authority) states: (a) 'The Engineer shall carry out the duties specified in the Contract'; and (b) 'The Engineer may exercise the authority specified in, or necessarily to be implied from, the Contract, . . .' Therefore, the engineer, while not being a party to the contract between the employer and the contractor, is empowered thereunder to issue instructions, consent, approve, certify or determine all of which are binding on both the employer and the contractor unless and until they are challenged and/or subsequently amended or rescinded in arbitration.

Under sub-clause 2.6, the engineer is required to exercise his discretion impartially within the terms of the contract and having regard to all the circumstances.

It is clear from the above that clause 2 of the Red Book establishes that a duty of care is owed to the contractor by the engineer in exercising his discretion in an impartial manner within the terms of the contract, and having regard to all the circumstances. This duty of care exists alongside the other duties which may be imposed in tort under the applicable law of the contract in order to avoid causing physical loss or damage or, in some cases, economic loss, with or without physical damage.

The above duties might well include exercising skill and care in, amongst others, designing and issuing design drawings and specifications; in making valuations and in certifying; and in making statements and giving instructions. The precise nature of which duty is owed and which is not will depend not only on the applicable law of the contract, but also on the contractual structure and the circumstances of the event leading to loss or damage.

In the following parts of this section, the state of the law of negligence in the common law and the Romano-Germanic jurisdictions is examined in relation to the responsibility and liability of the engineer.

11.3.1 In the common law countries

The state of the law of negligence in the common law countries has yet to become clear in this regard, and recent judicial decisions raise more questions than provide answers. This leads to the conclusion that further developments must be anticipated. In connection with construction contracts based on conditions such as the Red Book, there are various considerations which must be taken into account wherever the case is considered and whatever jurisdiction applies. These are:

(1) A claim in negligence

The contractor may either claim against the engineer in tort if all the prerequisites of a successful tortious claim are established, or he may claim against the employer. If the first course of action is taken, then the requirements for the tort of negligence must be established. These requirements have been the subject of argument in many cases, and in *Portsea Island Mutual Co-operative Society* v. *Michael Brasher Associates*, they were set out as follows:[11.27]

(a) The defendant must have owed a duty of care to the plaintiff. For such a duty to have existed, the following requirements must have been fulfilled:
 (i) that the defendant foresaw or ought reasonably to have foreseen that failure to exercise reasonable care and skill in the performance of a specified conduct would likely result in injury or damage to the plaintiff;
 (ii) that there was at the material time a sufficient relationship of 'neighbourhood' or 'proximity' between the parties; and
 (iii) that it was just and reasonable that the defendant should have owed a duty of care to the plaintiff;

(b) The defendant should have acted in breach of duty by failing to take reasonable care; and
(c) As the result of the defendant's breach of duty, the plaintiff suffered damage which was of the type which the defendant foresaw or ought reasonably to have foreseen as likely to result.

However, even if causation and foreseeability of damage can be proven, it may still be contrary to public policy for a particular loss to be recovered.

By following this course of action and claiming in tort against the engineer, the contractor acts as any third party claiming under the tort of negligence. The only difference is, perhaps, in respect of the foreseeability or proximity requirements as set out above which would be easier for him to establish in the context of a construction contract.

(2) A claim in respect of design and supervision

The engineer owes a duty of care in respect of the design and supervision of the works and would incur a liability to the contractor if his design is such that a competent contractor could not have avoided any resultant damage. It was stated in the case of *Oldschool v. Gleeson (Construction) Ltd* by Judge Stabb QC that:[11.28]

> 'I take the view that the duty of care which an architect or a consulting engineer owes to a third party is limited by the assumption that the contractor who executes the works acts at all times as a competent contractor. The contractor cannot seek to pass the blame for incompetent work on to the consulting engineer on the grounds that he failed to intervene to prevent it ... the responsibility of the consulting engineer is for the design of the engineering components of the works and his supervisory responsibility is to his client to ensure that the works are carried out in accordance with that design. But if, as was suggested here, the design was so faulty that a competent contractor in the course of executing the works could not have avoided the resulting damage, then on principle it seems to me that the consulting engineer responsible for that design should bear the loss.'

In that case, the engineer's design was not fundamentally unsound and the contractor's claim failed, but the principle of responsibility is clearly set out where such design is faulty.

(3) A claim for negligent under-certification

> 'In a trade in which cash flow is especially important, this [negligent certification] might have caused the contractor serious damage for which the architect [or the engineer] could have been successfully sued.'

This statement from the decision in *Arenson v. Casson, Beckman, Rutley & Co.* (1975) sets out clearly that where the engineer has negligently certified, and the employer is unable to meet his obligations, for one reason or another, the contractor may be in a position to recover the loss from the engineer.[11.29]

However, in the case of *Pacific Associates* v. *Baxter* (1989),[11.30] to which reference is made later, a different set of circumstances presented themselves leading the English courts to hold that an architect or engineer owes no duty to the contractor in tort provided that the contractor is able to challenge the certificate in arbitration proceedings and provided that the employer is not insolvent.

(4) Is there a claim where loss or damage is anticipated?

What should happen in a case where a defect is discovered before any damage results and money is spent to avert existing or future danger? Can such loss be recovered? In this regard, Lord Bridge in the case of *D & F Estates* v. *Church Commissioners for England* (1989) stated:[11.31]

> 'If the hidden defect in the chattel is the cause of personal injury or of damage to property other than the chattel itself, the manufacturer is liable. But if the hidden defect is discovered before any such damage is caused, there is no longer any room for the application of the *Donoghue* v. *Stevenson* principle.[11.32] The chattel is now defective in quality, but it is no longer dangerous. It may be valueless or it may be capable of economic repair. In either case the economic loss is recoverable in contract by a buyer or hirer of the chattel entitled to the benefit of relevant warranty of quality, but is not recoverable in tort by a remote buyer or hirer of the chattel.
>
> If the same principle applies in the field of real property to the liability of the builder of a permanent structure which is dangerously defective that liability can only arise if the defect remains hidden until the defective structure causes personal injury or damage to property other than the structure itself. If the defect is discovered before any damage is done, the loss sustained by the owner of the structure who has to repair or demolish it to avoid a potential source of danger to third parties, would seem to be purely economic. Thus, if I acquire a property with a dangerously defective garden wall which is attributable to the bad workmanship of the original builder, it is difficult to see any basis in principle on which I can sustain an action in tort against the builder for the cost of repairing or demolishing the wall. No physical damage has been caused. All that has happened is that the defect in the wall has been discovered in time to prevent damage occurring . . .'

The above passage seems to settle the position regarding whether or not a claim exists in tort where loss or damage is anticipated but has not yet actually occurred.

(5) A claim against the employer

If the contractor follows the second course of action, referred to in (1) above, by claiming against the employer, then in general terms, and except in cases of fraud or dishonesty, the employer would be liable for the acts of the engineer as his agent, subject to any right of recourse he might have against his agent for indemnity. An action of this type is usually defended jointly by the employer and the engineer.

(6) The arbitration clause

The Red Book includes an arbitration clause which can be used to have any determination of the engineer which is unacceptable to the contractor examined by an arbitrator and, if appropriate, corrected. This remedy which is available to the contractor can be used to correct the results of economic loss or damage resulting from lack of care of the engineer.

(7) Concurrency of liability in contract and tort

There is a strong tendency to deny any duty of care in tort if at the material time the defendant owed a duty in contract either to the plaintiff or to someone else.

In *Greater Nottingham Co-operative Society* v. *Cementation Piling & Foundations Ltd* (1988), referred to earlier, the defendant piling sub-contractors were sued in negligence in respect of workmanship by building owners to whom they had given warranties in design and a choice of materials.[11.33] The Court of Appeal decided that there was no liability in tort in this case.

In the case of *Pacific Associates* v. *Baxter* (1989), the engineer owed a contractual duty to the employer based on conditions similar to those of the Red Book and the question was asked: based on the contractual structure into which the contractor was prepared to enter with the employer, did the contractor look to the engineer by way of reliance for the proper execution of his (the engineer's) duties under the contract in extension of the rights which would accrue to the contractor under the contract against the employer?

In other words, it may be argued that it is not just and reasonable to impose on the engineer by way of liability in tort, rights in favour of the contractor in excess of these rights which the latter was content to acquire against the employer under the contract.

A competent authority such as a court or an arbitrator may not impose a duty of care on the engineer if it is considered to be neither just nor reasonable to do so.

(8) The parties have the right to determine their own obligation

A verdict of neither just nor reasonable might be reached by a court or an arbitrator if the employer and the contractor in a set of circumstances had the opportunity to define their contractual relationship by means of a contract and they took it. In this connection, see the categorisation of the *Tai Hing* case by the court in the case of *Lancashire and Cheshire Association of Baptist Churches* referred to above in Section 11.2.

In that case, the contractor, by accepting the invitation to tender upon the terms in the 'Instructions to Tenderers' and the contractual documents, was assumed to have accepted the role of the engineer in the contract.

A similar conclusion may be reached if the answer to the following question is given in the negative: should the law intervene in a commercial relationship to impose on the agent of one party to a contract the duty of care towards the other party in respect of administrative acts for the consequences of which his principal would in any case be liable under the contract?

(9) Deviating from a standard form of contract

The *Pacific Associates* case referred to above serves as a good example of the importance of the contractual framework in determining the respective duties of the contracting parties. In that case, the plaintiff contractor was the successful tenderer for dredging and reclamation work in Dubai Creek Lagoon. The contract was based on the third edition of the Red Book with some additional clauses inserted in the 'General Conditions'. The most relevant to this case was clause 86, which provided as follows:

> '86. Neither any member of the Employer's staff nor the Engineer nor any of his staff, nor the Engineer's Representative shall be in any way personally liable for the acts or obligations under the contract, or answerable for any default or omission on the part of the Employer in the observance or performance of any of the acts, matters or things which are herein contained . . .'

In that case, disputes arose between the contractor and the employer which were subject to arbitration under clause 67 of the contract. However, a settlement was agreed in the arbitration and the contractor claimed in negligence against the engineer for an alleged breach of duty to act impartially in certifying payment in respect of encountered physical conditions which, the contractor claimed, could not have been foreseen by an experienced contractor.

The engineer applied to the court to have the claim struck out on the grounds that it did not disclose a reasonable cause of action and was an abuse of the process of the court. The court held that the contractor could not recover damages from the engineer. The contractor appealed to the Court of Appeal, which upheld the lower court's decision.

In the Court of Appeal, Lord Justice Purchas reasoned that in view of the contractor's acceptance of the invitation to tender with the complete contractual framework, including the disclaimer in clause 86, it would:

> 'be impossible to support the contention that either the engineer was holding itself out to accept a duty of care with the consequential liability for pecuniary loss outside the provisions afforded to the contractor under the contract, or to support the contention that the contractor relied in any way on such an assumption of responsibility on the part of the engineer in any way to bolster or extend its rights'.

The existence of the arbitration clause enabling the contractor to challenge certificates of the engineer was also considered to be important by the court, if not decisive. In the words of Lord Justice Purchas: even if clause 86 'were not included in the contract in this case, the provisions of GC67 would, in my view, be effective to exclude the creation of any direct duty on the engineer towards the contractor.'

Thus, this case serves to show that any modifications to a standard form must be carefully examined and its consequences cautiously considered before acceptance.

The facts and the law in the *Pacific Associates* case has been dealt with in detail here as it seems to have settled the law in the referred legal system in respect of the duty, if any, of the engineer towards the contractor under similar circumstances. Two decisions in 1991 in the British Columbia Court of Appeal followed the conclusions reached in the *Pacific Associates* case by holding that the engineer owed the contractor no duty of care.[11.34] However, it is notable that under a different set of circumstances in Canada, the engineer was held liable where he was found to have negligently misrepresented relevant information relating to tender documents, on which he knew or ought to have known that the contractor would rely.[11.35]

11.3.2 In the Romano-Germanic system

In contrast to the situation in the common law countries, the position under the Romano-Germanic system of law may be influenced by the existence of some express provisions in the Civil Code of the particular jurisdiction. A legal principle may be imported into the legal framework of the relationship in the construction contract.

For instance, Articles 1382 to 1386 of the French Civil Code contain the general principles of tort law which briefly provide that any act of man which causes damage to another shall oblige the person by whose fault the damage occurred to make it good.

In some jurisdictions, however, it is necessary for a contractor to bring into operation the principle of 'benefit' in order to bring an action against the designer in tort. A typical example of this principle is Article 1029 of the Civil Code of Quebec in Canada which provides:

'A party in like manner may stipulate for the benefit of a third party, when such is the condition of the contract which he makes for himself, or of a gift which he makes to another; and he who makes the stipulation cannot revoke it if the third party has signified his assent to it.'

This particular article came into focus in the Canadian case of *Demers* v. *Dufresne* (1979).[11.36]

In some jurisdictions, decennial liability has been extended from its French origin, where it refers to damage jeopardising the integrity of the structure or its fitness for its intended purpose, to a reference to any defect that might threaten the soundness and safety of the structure, as in the United Arab Emirates where Article 880 of the Civil Code states:

'Article 880
(1) If the intention of the contract was to construct buildings or other permanent construction works which were designed by the Engineer and to be executed by the Contractor under the Engineer's supervision, they both jointly and severally warrant, to indemnify the Employer in respect of all occurrences of partial or total failure of what they had constructed or built and for all defects that might threaten the soundness and safety of the structure for a period of ten years unless a longer period is stipulated in the Contract. All this applies unless the period for which both parties of the Contract require these construction works to last is less than ten years.

(2) The obligation to indemnify remains even if the damage or failure was due to a fault in the ground itself or if the faulty construction was carried out with the approval of the Employer.

(3) The period of Ten Years starts from the date of handing over of the Works.'

The reference in item 2 of this Article to the 'ground' is based on the theory that if the fault was discoverable then it must be the responsibility of the constructor, but if it could not be discovered then it would come within the definition of *force majeure* which absolves the constructor from liability.[11.37]

11.4 The responsibility of the engineer towards third parties (other than the contractor)

The responsibility of the engineer and the liability which follows towards third parties originates in tort under the applicable law of the contract. However, liability towards third parties can also arise under statute or under contracts with third parties in what is known as collateral warranties. It is a matter of law as to whether or not a duty of care arises in a particular situation. Thus, it is beyond the scope of this book to consider in detail the nature of liability of the engineer towards third parties to the construction contract. This is particularly so in view of the fact that in a number of jurisdictions the state of the law of negligence is far from settled. Recent judicial decisions have reinterpreted even the most celebrated statements of previous important decisions.[11.38]

New concepts are continually evolving and the liability of the engineer would have to be established in accordance with the prevailing legal principles through specialist knowledge.[11.39] For instance, in the recent past, the European Community issued its directives 89/391/EEC which is referred to as the Framework Directive for safety and health at the work place; and Directive 92/57/EEC on the implementation of minimum safety and health requirements at temporary or mobile construction sites. Each member state is expected to implement these two Directives by enacting Regulations to transpose their provisions to the local law. The purpose of the Regulations would be to improve site safety on any building, civil engineering or engineering construction work. A number of states have already issued their Regulations.

Amongst those included in the liability net in these Directives is the engineer in both capacities, as a designer and as a supervisor. As a matter of interest, the employer and the contractor are also included.

The duties of the designer include a long list of elements which must be taken into account for the purpose of prevention of accidents. The list starts with elements of risk avoidance and evaluation, goes on to technical matters of design, material and systems of work, and ends with the requirement of providing appropriate training and instructions to employees.

Perhaps the most remarkable result of the Regulations implementing these Directives is the nature of the liability of the parties to a construction contract arising from a breach of the statutory duties imposed. The liability arising from such a breach could change from civil to criminal liability and the penalty which might be imposed may reach a fine of unlimited amount and/or imprisonment for a number of years.

11.5 The responsibility of the engineer towards society; employees; and the engineer himself

Construction work is closely related to the environment and society. In most cases the results of designs prepared by the engineer, financed by the employer and implemented by the contractor, can be seen and felt for a long time, thereby shaping the environment and, to some extent, the society. In some cases, it involves the 'act of directing the great source of power in nature for the use and convenience of man . . .'.[11.40] The responsibility in this area tends to belong to the owner and the engineer. Some therefore argue that the engineer's first duty and responsibility is towards the environment and society in which the project is located. To fail society is to fail one's self. It is not very difficult to succeed in this objective if there is in existence legislation to control development, in which case it is the duty of the engineer to abide by the rules set out in such legislation. Should he fail to do so, he will be in breach of a statutory duty and/or will have committed a crime. A problem arises, however, where legislation is either non-existent or deficient, in which case a conflict of interest will almost certainly arise between the duty and responsibility towards the owner and that towards society. Should such a conflict arise, the responsibility would tend acutely towards the employer. It must be recognised here that the first responsibility must lie with the legislative authority in not specifying what might and might not happen with society and its environment. Professionals in the forefront of technology in their appropriate fields must be given the task of drafting such legislation.

The engineer will also owe a duty to his own employees. However, although the responsibility in this area is a moral and contractual one, it is now to a large extent part of legislation also. Thus, to fail in the latter area would be to breach statutory requirements.

Finally, the engineer owes himself a duty and interestingly, it is this area of responsibility that is perhaps the most ignored by the engineer. There are basically two demands: the first is for the engineer to remain in existence, i.e. to work at a profit, and the second is for the engineer to be able to uphold his own standards and reputation. The responsibility of the engineer towards himself is a very real one. Profitability must also be accompanied by a high standard, otherwise it would be short-lived. The responsibility of maintaining such a reputation of high standard is fundamental to construction.

11.6 Liability in construction

Generally, responsibility leads to liability, but as can be seen from the definitions in Section 11.1, in certain circumstances one could be responsible for an act or omission but not liable for the resultant damages; and in other circumstances one could be liable for the act or omission but not responsible. These situations can be illustrated by the example of the employer who is held liable for the negligent acts of his employee, even though the employer himself may have been completely without blame and thus not responsible for the negligent act. Another example, which may help to clarify this point, is that of an insurance contract that imposes on the insurer a liability to indemnify the insured in respect of an occurrence for which the insured, but not the insurer, is responsible.

Contracting parties generally define in their contract the responsibility and the liability attached to each of them in respect of certain events. However, in case of dispute it is left to arbitral tribunals and/or the courts, and sometimes only to the courts, to establish who is liable and in what proportion. The parties must therefore fully understand their rights and obligations and the legal implications that would follow.

In this connection, one of the fundamental precepts of law is that ignorance of the law is no defence; but then one may ask, how much of the law is a construction professional expected to know?

The situation in common law jurisdictions is particularly difficult because one has to know the most up-to-date position at any particular time. As construction is a subject bridging most national borders, this is true not only within the jurisdiction in which one practises but also in all other common law jurisdictions. The reason is that, although a legal principle setting a precedent in one jurisdiction is not binding on courts in another, it is nevertheless understood to be persuasive. In practical terms this means that, unless there is a good reason to ignore that principle, it will be used in reaching a decision in other jurisdictions.

Where professional liability is concerned, there is a certain minimum amount of knowledge of the law that it is usually necessary to know, and more is preferable. This minimum level includes an understanding of the levels of liability that may apply in different circumstances, as explained below.

11.7 Levels of liability

The levels of liability under the law are rather similar in the various jurisdictions around the world. In general, there are three standards under the law of tort:

(a) liability based on lack of care and negligence;
(b) strict liability; and
(c) absolute liability.

Liability based on lack of care and negligence is the most usual, and for one to be liable in negligence, it is essential for the claimant to establish a proof of the respondent's negligence. A higher and more stringent standard of liability exists in the case of strict liability, which is the standard sometimes set by statute where liability arises if the harm to be prevented takes place irrespective of whether or not care and precautions have been taken. In such a case the onus of proof shifts from the claimant having to prove negligence, to the respondent having to prove non-negligence and even then liability may attach. In common law jurisdictions, strict liability also occurs under the principle of *Rylands* v. *Fletcher* where harm or damage is caused by the escape of a danger from one's own land.[11.41]

An even more stringent level of liability is the absolute liability. This is imposed under certain statutory provisions and is incurred by reason of the intentional occurrence of an event of a kind deemed prohibited, without regard to care or precautions taken and without need for proof of negligence or fault.

The standard of liability under the civil law system is generally based on whether negligence is a criterion, but there one has also to consider whether gross negligence has been committed.

In construction contracts, contractual liability is basically generated either pursuant to the contractual provisions or in breach of them. There are two levels of legal liability attached to the work and services supplied by contractors and professionals. These are:

(a) reasonable skill, care and diligence; and
(b) fitness for purpose.

Fitness for purpose is a greater obligation than that of reasonable skill, care and diligence. It is an absolute obligation independent of negligence. Therefore, negligence does not have to be proved where there is an obligation to provide fitness for purpose. On the other hand, where the duty is simply to use reasonable skill and care, in order to establish liability the employer or any claimant against a designer must show that the designer has been negligent. This aspect of onus of proof forms an important element in the cost incurred during dispute settlement, if and when disputes arise.

Dealing with the first level of liability, it generally applies to professional persons providing services to an employer, whereas the higher level of fitness for purpose applies to contractors. Therefore, designers are under a duty to use reasonable skill and care, but the word 'reasonable' has to be interpreted as appropriate in circumstances where the level of skill varies.

The statement, quoted below, made in the American case of *Cagne* v. *Bertran*, has remained as the criterion in the United States since 1954:[11.42]

> '... those who sell their services for the guidance of others in their economic, financial, and personal affairs are not liable in the absence of negligence or intentional misconduct.'

That judgment was used as a basis and was followed in the case of *Allied Properties* v. *Blume* and also in others.[11.43] Further, in *Xerox Corporation* v. *Turner Construction Co. and Others*, the court noted:[11.44]

> 'In the absence of an express agreement to the contrary, the duty of an architect, in performing his duties to his employer, is to exercise reasonable care and diligence, to use ordinary and reasonable skill usually exercised by one in that profession.'

However, this duty of an architect is to his client. In *Gravely* v. *The Providence Partnership*, in 1977, the plaintiff was a guest in a hotel suite which had a spiral staircase connecting the bathroom at an upper level to the lower level.[11.45] During the night, he went upstairs and opened the door of the bathroom but discovered that it opened towards him. As he stepped back, he fell and was seriously injured. The architect who designed the facility was sued in negligence and in breach of warranty. The court decision was in favour of the architect. The plaintiff then appealed. The Federal Appellate Court assessed the duty as follows:

'An architect, in the preparation of plans and drawings, owes to his employer the duty to exercise his skill and ability, his judgement and task reasonably and without neglect.... Even if there were a warranty of expertness, it would not be actionable by the plaintiff, for it would not run to the public generally but only to the architect's employer. This absence of mutuality we know as want of privity of contract.'

It should be noted that an error of judgement may not necessarily be regarded as negligence. This was highlighted in the medical negligence case in England, *Whitehouse v. Jordan*.[11.46]

A fundamental distinction between the use of reasonable skill and care and an obligation of fitness for purpose in connection with design activities is that in the former case negligence has to be shown, whereas in the latter case there is an absolute obligation, which is independent of negligence. Therefore, it is not necessary to prove negligence where an obligation of fitness for purpose exists. However, as can be seen from the case of *Greaves v. Baynham*, referred to in the next section, where a particular purpose is made known to a designer, that designer has an obligation to produce a design that is fit for that particular purpose, and this is so whether or not there is negligence. Moreover, a designer may be in breach of his obligation to use *only* reasonable skill and care if he knew of a particular purpose to which a specific part of his design was to be put, and he did not sufficiently take into account the relevant particulars, as then this could amount to negligence.

The skill of an ordinary competent person would normally be judged by reference to the state of the art of the design when it was carried out and not at a later date.

It is accepted that the precise scope of the duties undertaken by the engineer in the contract with his client would determine his responsibilities and, consequently, his liability. Obviously, should the terms of the agreement between the employer and the engineer include detailed expectations of the design, the standard of care would be elevated to that of fitness for purpose as was concluded in the Canadian case *Medjuck & Budovitch Ltd v. Adi Ltd* in 1980.[11.47] In that case, it was concluded that there was a 'common intention that the building should be fit for its purpose'. It was held that 'this gave rise to a term implied in fact that if the structure was completed in accordance with the design it would be reasonably fit for use as a ... store'.

Therefore, the liability of the engineer can become higher than simply having to use skill and care when he is made aware of the purpose for which the design is intended.[11.48]

It is also worth noting that the designer is under a continuing duty to check during the construction period that his design will work in practice, and to correct any errors that may emerge.[11.49]

Fitness for purpose in construction

The obligation for fitness for purpose in construction contracts has to be viewed from three separate angles: fitness for purpose for the supply of materials; for workmanship; and for design.

In common law jurisdictions, there are generally two main sources for the legal obligation of fitness for purpose in a construction contract: statutory and common law sources. From a statutory point of view, the applicable Sale of Goods and Supply of Services Act is relevant to the application of the principles concerned.[11.50]

The Sale of Goods and Supply of Services Act generally sets out the legal principles that apply in this area. Where the supply of materials is concerned, there is usually an implied warranty on the part of contractors that the materials they supply for use in their work will be reasonably fit for the intended purpose and are of good quality. The leading case in that connection in England is *Young & Marten Limited* v. *McManus Childs* in 1969.[11.51]

On this topic, if an owner, relying on a manufacturer's information, specifies an item in his contract with a contractor, then a separate and distinct contract between the owner and the manufacturer is created. This latter contract also implies fitness for purpose, see *Shanklin Pier Ltd.* v. *Detel Products Ltd* (1951).[11.52]

However, where the building owner relies on his own judgment rather than on that of the contractor, the warranty as to fitness for purpose will not be implied and the contractor will have no responsibility to supply materials which are fit for their purpose. In such circumstances, the warranty of quality will still usually be implied. A designer who selects materials and specifies them, so that the contractor has no choice as to what materials to buy, will remove the employer's cause of action against the contractor in respect of an implied warranty for fitness for purpose. The employer can still usually rely on the implied warranty as to good quality, at least in cases where the defect complained of is one of quality rather than fitness for purpose.

A similar implication would apply to workmanship where the contractor is obliged to ensure that the finished work is reasonably fit for the purpose intended.

Where a professional is responsible for design, the Sale of Goods and Supply of Services Act generally provides that there will be an implied term for the supplier to use due skill, care and diligence. The use of the word 'due' instead of 'reasonable' should be noted. However, it is important to recognise that this provision does not prejudice any rule of law, which imposes a duty stricter than that of skill and care, and therefore the Act does not of itself prevent the implication of a fitness for purpose term, where appropriate. Four cases are relevant and helpful to the understanding of the difference between that obligation and the higher obligation of fitness for purpose. These are:

(a) *Greaves (Contractors) Limited* v. *Baynham Meikle & Partners*;[11.53]
(b) *IBA* v. *EMI and BICC*;[11.54]
(c) *Norta Wallpapers (Ireland)* v. *Sisk and Sons (Dublin)*;[11.55] and
(d) *George Hawkins* v. *Chrysler (UK) Limited and Burne Associates*.[11.56]

However, it must be noted that all these cases were on particular facts and were not cases that arose out of the commonly found relationship between designers, contractors and employers. However, some quotations from the first two cases could be instructive.[11.57]

Where the contractor is responsible for the design of a construction project, the employer relies on the contractor, not only in respect of the selection of the materials and the proper workmanship, but also in respect of design. In the *Greaves* case referred to above, which related to a package deal contract, Lord Denning MR said:

> 'Now, as between the building owners and the contractors, it is plain that the owners made known to the contractors the purpose for which the building was required, so as to show that they relied on the contractors' skill and judgment. It was, therefore, the duty of the contractors to see that the finished work was reasonably fit for the purpose for which they knew it was required. It was not merely an obligation to use reasonable care, the contractors were obliged to ensure that the finished work was reasonably fit for the purpose. That appears from the recent cases in which a man employs a contractor to build a house: *Miller* v. *Cannon Hill Estates Limited* (1931);[11.58] *Hancock* v. *B. W. Brazier (Anerley) Limited* (1966).[11.59] It is a term implied by law that the builder will do his work in a good and workmanlike manner; that he will supply good and proper materials; and that it will be reasonably fit for human habitation.'

Lord Denning also made the following statement:

> 'The law does not usually imply a warranty that he (the designer) will achieve the desired result but only a term that he will use reasonable skill and care. The surgeon does not warrant that he will cure the patient. Nor does the solicitor warrant that he will win the case. But, when a dentist agrees to make a set of false teeth for a patient, there is an implied warranty that they will fit his gums.'

That statement of Lord Denning was made where the liability of the package deal contractor to the building owner had been admitted and it is not, therefore, binding, although it is of great persuasive authority. The same applied to statements made in the House of Lords in England in *IBA* v. *EMI and BICC* (1981), referred to above, where Lord Scarman said:

> 'In the absence of any term (express or to be implied) negativing the obligation, one who contracts to design an article for a purpose made known to him undertakes that the design is reasonably fit for the purpose.'

Lord Scarman had equated the position to that of a dentist making a set of false teeth where it has been held that there is an implied term that the false teeth will be reasonably fit for their intended purpose: *Samuels* v. *Davis* (1943). In *Samuels*, Lord Justice Du Parcq said:[11.60]

'If someone goes to a professional man . . . and says: "Will you make me something which will fit a particular part of my body?" and the professional gentleman says "Yes", without qualification, he is then warranting that when he has made this article, it will fit the part of the body in question . . . If a dentist takes out a tooth or a surgeon removes an appendix, he is bound to take reasonable care and to show skill as may be expected from a qualified practitioner. The case is entirely different where a chattel is ultimately to be delivered.'

Lord Scarman also distinguished between the dentist, using reasonable care in taking out a tooth, and the more onerous task of providing false teeth, but he went one step further in allocating a duty of fitness for purpose to a designer who contracts to design 'an article' for a purpose made known to him: 'Such a design obligation is consistent with the statutory law regulating the sale of goods'.

The above cases were followed in 1985 by the case of *Viking Grain Storage Limited v. T. H. White Installations Limited and Another*,[11.61] where there is now little doubt on the applicability of fitness for purpose on design and build projects. This case concerned a preliminary issue before the court as to whether the following terms were to be implied:

(a) that the design and build contractor would use materials of good quality and reasonably fit for their purpose; and
(b) that the completed works be reasonably fit for their purpose.

The works in this case were related to a grain drying and storage installation. The Official Referee had little difficulty in concluding that there was reliance by the plaintiff owners on the skill and judgment of the defendant contractors and it followed that those two implied terms contended for were implied.

The question remains as to whether an employer might consider restricting the contractor's liability for design from the standard of 'fitness for purpose' to the lower standard of 'due skill and care'. However, whatever the acceptability of such an action from a legal point of view, the practical aspect of its advisability should be carefully considered. This is in the context that by restricting the liability to a narrower application, one of the main advantages of such a method of procurement of works, the single source of liability in design and build contracts, disappears. The employer is told that in design and build contracts he does not have to enquire into distinctions between defects arising from material, workmanship or design. Neither would he have to show that negligence is the source of a complaint when problems arise.

The role of insurance in fitness for purpose

It is accepted that where supply of material is concerned, the party responsible for selection is to some extent covered through that selection by a manufacturer's warranty or by the insurance cover provided by that manufacturer. Where design is concerned, however, the designer does not usually provide a warranty for fitness for purpose in certain jurisdictions and the professional indemnity insurer does not generally extend the insurance cover to include for fitness for purpose.

In such circumstances, and in these jurisdictions, contractors are therefore exposed to liability for which they are not indemnified, whether through suppliers of such service or through insurance. The suppliers of design services do not usually have assets to provide any real comfort to a contractor undertaking design and build contracts. Furthermore, the insurers of professional indemnity resist the provision of a cover against fitness for purpose for reasons of insurability. They argue that such a cover would be against the principles of insurance. Similarly, contractors argue that their liability should be restricted to the lower standard of due skill and care.

Fitness for purpose in standard construction contracts

The above problem relating to contractors' exposure to liability exists all over the world. FIDIC, which consults widely throughout its over 70 national associations worldwide, concluded when publishing the Orange Book in 1995 that it is necessary to impose a fitness for purpose liability on contractors in respect of design in addition to that usually imposed in respect of material and workmanship.[11.62] Fitness for purpose is also imposed in the recently published Silver Book, *Conditions of Contract for EPC Turnkey Projects*, and the 1999 Yellow Book *Conditions of Contract for Plant and Design-Build for E & M Plant, and for Building and Engineering Works designed by the Contractor*. For both of these forms of contract, the relevant provisions are in sub-clause 4.1.[11.63]

Against that background, it is necessary to mention that contractors in the UK have secured a restriction in the design liability imposed under the ICE 1992 Design and Construct Conditions of Contract, where sub-clause 8(2)(a) states:

> '(2)(a) In carrying out all his design obligations under the Contract . . . (and including the selection of materials and plant to the extent that these are not specified in the Employer's Requirements) the Contractor shall exercise all reasonable skill care and diligence.'

However, in considering the precise meaning of the words quoted above, it could be argued that the contractor has in fact a lower standard of design liability than that which would attach to a professional designer, if the latter were made aware of the intended purpose of the design he has been asked to undertake.

It may therefore be concluded that unless expressly restricted in the contract, the implied liability of a contractor for design, including any duty of selection of materials and methods of workmanship, is one of fitness for purpose.

On the other hand, if a design professional is made aware of the intended purpose of a particular aspect of a project, the liability could be elevated from one of due care and skill to one of fitness for purpose.

Chapter 12

The Employer's Obligations

12.1 Introduction

The employer has obligations under both the applicable law of the contract and under the Conditions of Contract. The applicable law of the contract is specified in Part II of the Conditions with reference to sub-clause 5.1(b). In general, the employer would be concerned with those branches of the law as described earlier in Chapter 3.

Under the FIDIC Conditions of Contract, the developer of a project will have taken a number of specific actions before reaching the stage where he is involved in the role of employer. He will already have received professional advice and have taken a number of decisions in connection with the following:

(a) The choice of professional advisers for the planning, engineering and other aspects of the construction project;
(b) The most suitable contractual arrangements for the employment of these professional advisers;
(c) The design of the project, the site, assessment of budget costs, choice of financial arrangements and any services which must be obtained to permit construction to start on site;
(d) The most appropriate procurement and contractual arrangements for the purpose of constructing the project;
(e) The criteria to be adopted for selecting the contractor who will be responsible for construction of the project;
(f) The most appropriate arrangements for the day-to-day control over the quality of the construction and final completion of the project;
(g) The manner in which any legislative or governmental approvals are to be obtained; and
(h) The relevant information required in respect of the various clauses of the conditions of contract chosen to govern the construction of the works. In particular, a decision must be made in respect of the details to be inserted against the items listed in the Appendix to Tender and also in connection with the various clauses of the Conditions of Particular Application as contained in Part II of the document.

Compared with previous editions of the Red Book, under the Fourth Edition, additional obligations have been allocated to the employer in order that he may have more direct control in the day-to-day progress and administration of the project. Together with those traditionally recognised, these new obligations will obviously lead the employer towards accepting greater responsibilities and liabilities. He must be particularly aware of these additional duties and should familiarise himself with the obligations assigned to him, especially when these involve him in adherence with time schedules. The employer's obligations under the Fourth Edition are discussed below, mainly in the order in which they appear in the standard Form.[12.1] These are to:

(a) identify specific elements of the project, such as the site and the personnel to be involved in it;
(b) appoint an engineer to administer the contract;
(c) give possession of the site;
(d) provide information, instructions, consents, approvals and give notices as and when they are required;
(e) refrain from taking any action which would impede or interfere with the progress of the works;
(f) supply materials and carry out works if these form part of the works as defined in the contract;
(g) nominate specialist sub-contractors and suppliers as and when they are required;
(h) permit the contractor to carry out the whole of the works; and
(i) make payments and to make them on time.

12.2 Identification of specific elements of the project

Part II of the Red Book is specific to each particular project. Thus, the employer is required to insert certain information regarding the project either in Part II of the Form or in the drawings and the specification. Definitions of these requirements are given in Clause 1 of the conditions. Sub-clause 1.1(a)(i) defines the employer as the person named as such in Part II which once given cannot be changed through assignment without the consent of the contractor. Sub-clause 1.1(a)(iv) defines the engineer as the person appointed by the employer to act as engineer for the purposes of the contract and named as such in Part II of the Form. Although not specifically stated, the employer by identifying the engineer in Part II of the Conditions is restrained from changing the engineer without the consent of the contractor (see earlier in Chapter 9 for a detailed comment on this matter).

The site is defined in sub-clause 1.1(f)(vii) and this definition must also be identified in Part II of the Form. The relevance of this declaration can be recognised when read against clause 42, *'Possession of Site and Access thereto'*, discussed later in Section 12.4.

12.3 Appointment of engineer

As discussed in Chapter 10, the role of the engineer during the construction of a project is extremely complex and involves:

(a) a continuing design role;
(b) an administrative role as the employer's agent;
(c) a supervisory role;
(d) a certifying role; and
(e) an adjudicating role.

Whilst it is recommended and highly desirable that the employer appoint the engineer from the beginning of the project so that he would be the person responsible for carrying out the pre-contract duties of design as defined in Chapter 10, this is not an absolute requirement. The obvious disadvantages of appointing someone different for these two distinct roles, i.e. pre-contract and post-contract roles, would have to be considered very carefully before adopting that course. In any case, the Red Book has been drafted in such a manner that subject to minor clarifications in connection with clause 6, whosoever is appointed could fulfil the duties and authority delegated to the engineer in both of the above roles.

Where the employer wishes to restrict the authority of the engineer in such a way that he is required to obtain the specific approval of the employer before exercising any such authority, the employer must set out such restrictions in the terms of appointment of the engineer and in the Articles of any agreement with him. Such restrictions must also appear in precise terms in Part II of the Red Book so that all tenderers are aware of the restricted roles of the engineer, prior to pricing and submitting their tenders.

Attention must be given by the employer to three aspects of the Red Book in the above connection: first, to sub-clause 2.6 where the engineer is required to exercise his discretion impartially in respect of any decision, opinion, consent, expression of satisfaction, approval, or determination of value or action; secondly, throughout the Red Book, the engineer is required to 'duly consult' with the employer and the contractor (see Chapter 9), but having so consulted, he is entitled to form his own opinion regardless of what the employer or the contractor may say to him, unless such authority is restricted under sub-clause 2.1(b); and thirdly, the provision at the end of sub-clause 2.1(b), where it is stated that any requisite approval shall be deemed to have been given by the employer for any authority exercised by the engineer, safeguards the contractor from having to establish that the engineer had in fact obtained the approval of the employer prior to exercising his authority.

The employer also has an indirect role in the delegation of authority by the engineer to the engineer's representative. Under sub-clause 2.3, any delegation or revocation of duties or authority to the engineer's representative must be delivered from time to time and in writing to the employer and the contractor. If the employer intends to have control over what can or cannot be delegated to the engineer's representative, he should have such control incorporated in the terms of appointment of the engineer.

The Fourth Edition of the Red Book incorporates two important changes from previous editions (see Chapter 9), which would create some difficulty if the person named as the engineer were to die or become seriously ill or resign. As the position of engineer is essential in the Red Book, and the contract becomes unworkable without an engineer, the employer is implicitly required to appoint another person within a reasonable time. However, he is not expressly empowered under the Fourth Edition to appoint from '*time to time*' and unless the contractor accepts the appointment of the new engineer, the situation could be deadlocked. To avoid that happening, it would be advisable for the employer to name the engineer and his firm in Part II giving himself the option of appointing another member of the firm as the engineer. However, that proposition does not solve the problem that would arise should both the engineer and his firm resign.

12.4 Possession of site

The duty of the employer to give possession of the site and to explain the process by which such possession is accomplished is expressly stated in sub-clause 42.1 where it is provided, *inter alia*, that:

'... the Employer will, with the Engineer's notice to commence the Works, give to the Contractor possession of
(c) so much of the Site, and
(d) such access as, in accordance with the Contract, is to be provided by the Employer ...'

Sub-clause 42.2 provides for the remedy in the case of failure by the employer to give possession of the site in accordance with the terms of sub-clause 42.1. Such remedy may take the form of an extension of time under clause 44 and/or payment of any costs incurred by the contractor.

Whilst the contractor is given possession of the site, an express right is given to the engineer, and to any person authorised by him, to have access to the site at all reasonable times (see sub-clause 37.1). The Red Book is, however, silent on whether the employer has such right of access. Of course, he could obtain authorisation through the engineer under sub-clause 37.1, but it remains ambiguous as to whether the employer has an implied right of access or that it is intended that the contractor should have exclusive possession. Whichever is the correct interpretation, the contractor's obligation to 'have full regard for the safety of all persons entitled to be upon the Site' under sub-clause 19.1 must not be affected. Similarly, the contractor's obligation to 'take full responsibility for the care of the Works and materials and Plant for incorporation therein' under sub-clause 20.1 must not be diminished. Of course, the employer may enter upon the site and the works and terminate the employment of the contractor under sub-clause 63.1, 'Default of Contractor', but this is a specific situation different from that of access during the normal execution of the works.

Under a separate heading, unless it is expressly stated to the contrary, the employer is under no duty to the contractor to make any alterations to the site which would make it easier for the works to be executed. A clause in the specification would normally be inserted to the effect that the contractor must accept the site as he finds it. However, whilst there is no duty owed by the employer to identify the subsoil conditions or, for that matter, topography of the site, it is essential for cost implications and also for the purposes of design that such surveys and investigations be conducted and the results obtained made known to tenderers in the contract documents. The provisions of clause 11 deal explicitly and precisely with this aspect of the site.

However, clause 11 provides that the contractor is responsible for his own interpretation of all the data made available to him by the employer. The clause also provides that the contractor is to be deemed to have inspected and examined the site and have satisfied himself with amongst other things, the means of access to the site.

Access to site

Two other aspects of the process of possession of the site require detailed explanation in the contract documents at the tender stage. The first is the possible provision already made by the employer for access to the site with the owners of properties forming part of the site or with the owners of neighbouring properties, or in respect of other properties owned or controlled by him.

The second is whether the whole of the site is to be made available to the contractor at the commencement date or whether possession of the site is to be given in stages as the works proceed. This is particularly important when a date for completion is specified in the contract documents. The following quotation is apt in this respect:

'If in the Contract one finds the time limited within which the builder is to do the work, that means, not only that he is to do it within that time, but it means also that he is to have that time within which to do it.'[12.2]

These two aspects should also form part of the specification and the drawings, thus clarifying a usual matter of contention in civil engineering contracts.

Finally, any discussion regarding the site is incomplete without reference to the precise definition of the word 'site'. In this regard it has been argued by many commentators that the definition of the word 'site' is far too vague as contained in most standard forms of Conditions of Contract. This argument is supported by the fact that engineering works frequently take place over wide areas which may or may not be already occupied by the employer and these areas very often do not have readily recognisable boundaries or limits. Examples of such works range from pipelaying under highways and through open country to others confined within the existing complex of a water treatment works. It is important to define exactly what is meant by the word 'site' because of the many references which are made to it throughout the contract documents.

The word 'site' appears in the following clauses of the Red Book: <u>1.1</u>, 6.2, 8.2, <u>11.1</u>, <u>12.2</u>, 16.1, 16.2, <u>19.1</u>, <u>19.2</u>, <u>21.1</u>, <u>21.2</u>, <u>25.1</u>, <u>30.1</u>, <u>30.2</u>, 30.3, <u>31.1</u>, <u>31.2</u>, <u>33.1</u>, 35.1, 36.1, 36.4, 37.1, 39.1, 40.1, <u>42.1</u>, <u>42.3</u>, <u>52.3</u>, 54.1, 54.5, 54.7, 60.1, 63.1, 65.3, 65.7 and 69.2. However, only those which are underlined above are relevant to the possession of, identification of and access to the site.

The precise details pertaining to the site are best defined in the specification and drawings. If it can be established at the outset where the exact description of the 'site' will be given in the contract documents, then this could be stated in the definition of sub-clause 1.1(f)(vii) or, alternatively, it could be given as an addition in the Appendix to Tender by reference to sub-clause 1.1(f)(vii) or sub-clause 42.1.

The precise definition of the site may also have implications in connection with the provisions of the liability and insurance clauses of the Conditions. In particular, the terms of the insurance policies provided under clauses 21 and 23 of the Conditions may be affected by the boundaries of the site, where within such boundaries the works are to be executed and where outside such boundaries there are neighbouring properties. Where neighbouring properties are owned by the employer, sub-clause 23.3 assumes particular importance, since such properties should to all intents and purposes be treated as properties of third parties to the contract between the employer and the contractor.

12.5 To provide instructions as and when they are required

Under a number of the clauses of the FIDIC Fourth Edition, the employer is required to provide directly to the contractor certain information, instructions, consents and approvals. He is also required to give notices to the contractor. These are in addition to information, instructions, consents, approvals and notices provided by the engineer to the contractor, in his role as the employer's agent.

The clauses where the employer is under the obligation to provide the above instructions, etc., and give notices, are listed below:

(a) Under clause 3, the contractor would require the prior consent of the employer should he wish to assign the contract or any part thereof, or any benefit or interest therein or thereunder.

The employer is, therefore, required to provide a reply to the contractor if such a request is made, and such reply, whether in granting consent or withholding it, should be in writing as provided in sub-clause 1.5.

(b) Should the contractor be called upon to enter into and execute a contract agreement in accordance with clause 9, the employer is responsible for the cost of the preparation and completion of this agreement.

(c) Under sub-clause 10.1, the choice of institution providing the security for the contractor's proper performance of the contract is subject to the approval of the employer. Such approval, if appropriate, should be given by the employer in writing. Although there is no time limit for a decision on providing or withholding such approval, there is a time limit imposed on the contractor under sub-clause 10.1 for obtaining and providing such security. This time limit is 28 days from the day of receipt by the contractor of the letter of acceptance.

(d) Should the employer decide to make a claim under the provisions of the performance security provided by the contractor under sub-clause 10.1, he must first notify the contractor stating the nature of the default in respect of which the claim is to be made. This obligation is provided under sub-clause 10.3. There are four other situations where the employer is required to give notice to the contractor. These are listed in Table 12.1.

Table 12.1 Notices required to be given by the employer to the contractor under the contract.

Sub-clause	Description of the event
10.3	When the employer is making a claim under the performance security, the nature of the default in respect of which the claim is to be made should be stated
30.3	Whenever a settlement is to be negotiated between the employer and the road authority in respect of damage to any bridge or road communicating with or on the routes to the site
63.1	When the employer wishes to terminate the employment of the contractor due to the latter's default
65.6	If and when the employer terminates the contract due to outbreak of war
67.1	Whenever the employer is dissatisfied with any decision of the engineer or if the engineer fails to give notice of his decision on or before the appointed date in accordance with clause 67.1, the employer should give notice to the contractor of his intention to commence arbitration as to the matter in dispute, with a copy to the engineer.

(e) Sub-clause 11.1 provides that the employer shall have made available to the contractor, before the submission by the contractor of the tender, such data on hydrological and sub-surface conditions as have been obtained by or on behalf of the employer.

Although there is no duty on the employer to carry out such investigations and surveys to provide the necessary data, it is nevertheless essential that whatever information is available to the employer should also be made available to the tenderers during the tendering period and hence to the contractor.

As stated earlier, it is usual practice to have such surveys and investigations done at the pre-investment stage of the project for the purposes of design, cost analyses and budgeting.

(f) The employer is required under sub-clause 22.3 to indemnify the contractor against all claims, proceedings, damages, costs, charges and expenses in respect of any of the following:
  (i) death or injury to any person; or
  (ii) loss of or damage to any property (other than the works), which may arise as a result of the 'exceptions' referred to in sub-clause 22.2. Essentially, these exceptions are risks which are allocated under the contract to the employer.

(g) In accordance with sub-clause 25.1, the insurance policies required to be effected by the contractor under the provisions of the contract should be submitted to the employer within 84 days of the commencement date. The terms of these policies are subject to the approval of the employer. Once again, although there is no time-limit imposed on the employer to grant his approval and/or comments, there is the overall limit of 84 days to resolve any questions which may arise.

(h) Under clause 26, the employer is responsible for obtaining any planning, zoning or other similar permission required for the works to proceed. The employer is further required under this clause to indemnify the contractor in respect of such responsibility in a similar manner to that stipulated under sub-clause 22.3.

(i) The employer is required under the provisions of clause 71 to reimburse the contractor for any loss or damage arising from currency restrictions imposed by the government or authorised agency of the government of the country in which the

The Employer's Obligations

works are being or are to be executed 'after the date 28 days prior to the latest date for submission of tenders for the Contract'. Similarly, the employer is required to reimburse the contractor for any loss or damage arising from restrictions on transfer of currency.

12.6 The employer is to refrain from taking any action which would impede or interfere with the progress of the works

As discussed in Section 12.5(h) above, it is an express obligation of the employer under clause 26 to obtain any planning, zoning or other similar permission required for the works to proceed. To fail in obtaining such permission, the employer would be in breach of a term of the Conditions of Contract. However, in addition to the obligation to obtain such permission, it is necessary for the employer to do so in a timely manner in order not to impede or interfere with the progress of the works.

Under sub-clause 44.1(d), it is expressly provided that:

'... in the event of any delay, impediment or prevention by the Employer ... being such as fairly to entitle the Contractor to an extension of the Time for Completion of the Works, or any Section or part thereof, the Engineer shall, after due consultation with the Employer and the Contractor determine the amount of such extension ...'

Besides this express provision entitling the contractor to an extension of time, it is an implied obligation of the employer not to hinder or prevent the contractor in the performance of his own obligations.

Under the Fourth Edition of the Red Book, the contractor's programme for the execution of the works is expected to be submitted to the engineer for his consent within a period of time prescribed in Part II of the Conditions. The form and detail of this programme which would have been prescribed by the engineer under sub-clause 14.1 would have had to include the date on which the contractor would require the employer to obtain such planning or other permissions.

Under sub-clause 14.1, the general description of the arrangements and methods which the contractor proposes to adopt for the execution of the works would be submitted only if required by the engineer. However, it must be remembered that such description is submitted only for the information of the engineer who, in the absence of express provisions to the contrary, has no right to instruct the contractor on how the latter should do his work nor to interfere with his methods or the order in which the work will be carried out.

In the words of Mr Justice Mocatta:

'It is the function and right of the builder to carry out his own building operation as he thinks fit.'[12.3]

Accordingly, should the employer or the engineer on his behalf interfere with the progress of the works, the contractor would be entitled to damages for breach of contract.

In connection with the employer's duty not to interfere with the progress of the works, clause 31 expressly provides for the possibility of the presence of other contractors or workmen on the site. It states that the contractor should 'afford all reasonable opportunities' for other contractors employed by the employer, for the workmen of the employer and for the workmen of any duly constituted authorities, in carrying out their own work.

There is no reference made, however, as to the duties and obligations of the employer on behalf of these third parties in respect of interfering with or impeding the progress of the works. Each situation would have to be considered on its merits, but in general terms it may be taken that 'an Employer will be liable to the Contractor if other contractors of the Employer disturb the Contractor in his work in circumstances which he could not reasonably have foreseen at the time of tendering.'[12.4]

One of the contentious issues in connection with interference with the progress of the works is the timing of any instructions given by or on behalf of the employer either for additional drawings or instructions under sub-clauses 6.3 and 6.4, or for alterations, additions and omissions under clauses 51 and 52. In addition to the provisions in the Red Book that alterations, additions and omissions may be ordered, there is a mechanism for these variations to be valued and for time for the completion of the contract to be extended should the instruction to vary the works necessitate such an action.

However, the following hypothetical questions are often asked in the situation where a contractor is substantially behind programme, but shortly before completion he receives either some new design information, which would not have been required to be given to him earlier, or an instruction for a change in a part of the works not yet executed. In such a situation can the contractor be excused the liability for the substantial antecedent culpable delay? Can such previous delays be ignored because of the subsequent additional requirements? Furthermore, can the assumption be made that design could not be completed until the latest design information or the final instruction be given to the contractor?

Whilst there is no judicial authority known to the author to help resolve such questions, it is illogical for the draftsman of a construction contract to contemplate any positive interpretation without an express term to that effect. This is particularly so in view of the provision in the Red Book which regulates the consequences of late design information, as in sub-clause 6.4, and of variations, as in clause 52.

12.7 The employer is to supply materials and carry out works if these form part of the work as defined in the contract

The possibility that the employer might provide workmen on the site is expressly referred to in sub-clauses 31.1(b) and 19.2. In respect of their work, there is an implied obligation to have it executed in a timely manner and in such a way so as not to impede the contractor in the performance of the whole of the works.

Of course, the employer is entitled under sub-clauses 39.2 and 49.4 to carry out work in respect of instructions with which the contractor has not complied, but this is an entitlement rather than an obligation. Under sub-clause 39.2, the instructions not complied with refer to the removal of improper work, materials or plant according to sub-clause 39.1, whereas those under sub-clause 49.4 refer to amendments, reconstruction and remedying defects, shrinkages or other faults during the defects liability period or within 14 days after its expiration.

12.8 The employer is to nominate specialist sub-contractors and suppliers as and when they are required

Nominated specialist sub-contractors are defined in sub-clause 59.1 to include all those 'specialist merchants, tradesmen and others executing any work or supplying any goods, materials, Plant or services for which Provisional Sums are included in the Contract'. A provisional sum is defined in sub-clause 58.1 as a sum included in the contract and so designated in the bill of quantities for the execution of any part of the works or for the supply of goods, materials, plant or services, or for contingencies. Sub-clause 58.2 provides that a provisional sum may be used at the instruction of the engineer in respect of the supply of goods, materials, plant or services by either the contractor or a nominated sub-contractor as defined in sub-clause 59.1. Furthermore, the sum may be used in whole or in part or not at all on the instruction of the engineer.

The decision as to whether the contractor or a nominated sub-contractor is to be used for the supply of specific goods, materials, plant or services is usually made by the employer on the advice of the engineer. The intention behind the appointment of nominated sub-contractors stems from two possible advantages. First, by nominating a particular sub-contractor, the employer can select a particular quality or type of goods, materials, plant or service without creating a direct contract with the sub-contractor. Secondly, by such selection, the employer can ensure through direct tendering that a competitive price is obtained for this specialist work.

Accordingly, for every provisional sum included in the bill of quantities there must be either a direction in the contract or an instruction given later as to who is to provide the specialist work involved and how it is to be provided. There must be, therefore, an implied duty under sub-clause 59.1 for the employer, or the engineer on his behalf, to nominate or select or approve such nominated sub-contractor in sufficient time to enable the contractor to perform his obligations under the contract.

In making a timely decision, account should be taken not only of the time required to obtain quotations from specialist sub-contractors, but also for any time necessary for the nominated sub-contractor or supplier to provide any information, specification or drawings required, under sub-clause 59.3, for the proper completion of the design of the project. In general terms, the work of any nominated sub-contractor whose duties include design of the elements to be supplied would affect the detailed design of the remaining part of the project. Some of the most complicated claims in construction projects result from problems related to this obligation of the employer. Arbitration cases contain numerous situations which could provide valuable lessons to employers, contractors, engineers and sub-contractors, but the proceedings are private and confidential.

Attention should also be given to sub-clause 59.2, where it is provided that the contractor is not required to be under any obligation to employ any nominated sub-contractor against whom he has reasonable objections or who declines to enter into a sub-contract with the contractor in the terms provided in sub-clause 59.2.

12.9 To permit the contractor to carry out the whole of the works

The employer is entitled under sub-clause 51.1(b) to omit any work included under the contract provided that such omitted work is not to be carried out by the employer or by another contractor. Furthermore, under sub-clause 8.1 the contractor is required to complete the works and accordingly, it can be deduced that the employer should permit the contractor to carry out the whole of the works and the only exception to that rule relates to any part which is genuinely not required and is thus omitted.

12.10 To make payments and to make them on time

The obligation of the employer to pay and to do so in stages specified under the contract and on time is expressly provided in clause 60. Details of the reasons for which payment is to be made, however, are distributed throughout the various clauses of the Red Book.

Taking first the obligation to pay, sub-clause 60.10 makes it clear that the 'amount due to the Contractor ... shall subject to Clause 47, be paid ... within 28 days after such interim certificate has been delivered to the Employer ...'. In the case of a final certificate, the payment shall be made within 56 days after the certificate has been delivered to the employer.

Sub-clause 60.10 further provides for interest to be paid in the event of failure by the employer to pay within the specified periods. The rate of interest is specified in the Appendix to Tender and it is applicable to all unpaid sums from the date on which these sums were due to the contractor.

The importance with which this obligation of the employer is viewed can be clearly seen from the provisions of clause 69. This clause provides that in the event of any failure by the employer to pay under sub-clause 60.10, or of any interference with or obstruction or refusal of any required approval to the issue of a payment certificate, or of any one of a number of financial problems stipulated therein, the contractor is entitled to terminate his employment under the contract.

The staging of payments is specified in sub-clause 60.2 and is dependent on the submission of monthly interim statements by the contractor to the engineer who should, within 28 days from the date of receipt of such statements, certify to the employer the amount of payment due to the contractor. A minimum amount for these interim certificates is specified in the Appendix to Tender.

Finally, the clauses relevant to the obligation of the employer to pay are as follows (titles are not precisely reproduced):

— Sub-clause 6.4   Delays and costs of delay of drawings;
— Sub-clause 12.2   Adverse physical obstructions or conditions;
— Sub-clause 27.1   Fossils;
— Sub-clause 36.5   Tests for which there is no provision;
— Sub-clause 38.2   Uncovering and making openings;
— Sub-clause 40.2   Costs following suspension;
— Sub-clause 42.2   Failure to give possession of site;
— Sub-clause 50.1   Contractor to search;
— Clause 52   Valuation of variations and daywork;
— Sub-clause 53.5   Payment of claims;
— Sub-clause 59.4   Payment to nominated sub-contractors;
— Clause 60   Certificates and payments;
— Sub-clause 63.3   Payment after termination;
— Sub-clause 65.5   Increased costs arising from special risks;
— Sub-clause 65.8   Payment if contract is terminated;
— Clause 66   Release from performance;
— Sub-clause 69.3   Payment on termination;
— Sub-clause 69.4   Contractor's entitlement to suspend work;
— Sub-clause 70.1   Increase or decrease of cost;
— Sub-clause 70.2   Subsequent legislation; and
— Clause 71   Currency and rates of exchange.

It is to be borne in mind that sub-clause 60.9 provides that the employer is not liable to the contractor 'for any matter or thing arising out of or in connection with the Contract or execution of the Works, unless the Contractor shall have included a claim in respect thereof in his Final Statement and ... in the Statement at Completion ...'.

A claim in respect of any matters which arise after the issue of the taking-over certificate would obviously have to be included in the final statement, if it is to be pursued.

12.11 Additional obligations for the employer under the Fourth Edition of the Red Book

As stated earlier in the introduction to this chapter, the employer takes a more active role in the administration of the contract under the Fourth Edition of the Red Book compared with earlier editions. This extended role means an extended boundary of his obligations under the contract. See Part IV of this book for a comparison between the Third and Fourth Editions of the Red Book. The most important of these additional obligations are:

(a) Due consultation: As discussed in Section 9.3.1 above, there is a new requirement imposed on the engineer to 'duly consult' both with the employer and the contractor before making assessment or determination of a number of specific matters should they come before him. This duty on the engineer must be matched by an obligation on both the employer and the contractor to provide the time and the intention to consult with the engineer. This consultation may have to be done at short notice or within a very short time. The duty to consult appears in 24 sub-clauses and with the exception of two, all are concerned with assessment of costs. The two exceptions cover the engineer's assessment of extension of time. 16 of the 22 references are in respect of payments to the contractor.

This new feature probably does not constitute any deviation from commonly accepted practice, since the engineer would normally have found it appropriate to consult both parties, even without the explicit provision of this new feature, if he considered that his assessment would have certain implications for them. The main point of the present section is that this procedure is now clearly and expressly stipulated in the text of the contract.

The term 'due consultation' is not defined in the Conditions and this in itself has led to some debate as to what exactly 'due consultation' should mean. Some commentators have expressed the view that it should be interpreted as consultation appropriate to the circumstances under which the particular action falls to be considered, while others view it as a definite restraint on the engineer by the employer. It is not specified whether this consultation should take place with both parties present or with each separately, but here again individual circumstances will dictate what can or should be achieved. However, it must be remembered that the engineer should 'duly consult' with the employer and not be dictated to by him, since he is required to act impartially, under the provisions of sub-clause 2.6.

(b) The Fourth Edition of the Red Book provides, in many of its terms, that the employer should receive copies of certain documents. The employer is therefore expected to be kept informed about certain events as they take place. Accordingly, he is expected to take action should he disagree with what is happening, and should provide the staff to pursue such action.

(c) The employer, when he is to carry out work on the site with his own workmen, is responsible under the new sub-clause 19.2 for the safety of all persons entitled to be on site, and to keep the site in an orderly state appropriate to the avoidance of danger to such persons.

(d) The employer has the obligation under sub-clause 21.3 to bear any amount not insured, or not recovered from the insurers, which falls within his responsibility under clause 20.

(e) Under the Fourth Edition of the Red Book, third party insurance is required to be effected in the joint names of the employer and the contractor (see sub-clause 23.1). Insurance of the works as required under the provisions of sub-clause 21.1 has always been required in the joint names of the employer and the contractor. As jointly insured, the employer has the obligation under the terms of the relevant insurance policy to observe and comply with the conditions stipulated in that policy. In the event that either the employer or the contractor fails to comply with the conditions imposed by any of the insurance policies effected pursuant to the contract, a new provision has been added under sub-clause 25.4 to the effect that 'each shall indemnify the other against all losses and claims arising from such failure'.

(f) Under sub-clause 25.1, the general terms of the insurance policies required under the contract are to be agreed between the employer and the contractor prior to the issue of the letter of acceptance. This is obviously a joint precondition of the contract between the employer and the contractor and an obligation on the employer to make a decision as to the insurance arrangements to be followed.

(g) Under sub-clause 30.3, the employer is now responsible for negotiating a settlement of and paying all sums due in respect of terms emanating from damage to any bridge or road inflicted by the haulier of materials or plant on a route to the site, unless the haulier is required by law or regulation to indemnify the road authority against such damage. In this connection, the employer has the obligation to inform the contractor whenever a settlement is to be negotiated. Where any amount may be due from the contractor, the employer is required to consult with the contractor before such settlement is reached. This, however, would necessarily involve the engineer as it is the engineer who, after due consultation with the employer and the contractor, determines the amount to be due to the employer.

(h) An obligation has been added to both the employer and the contractor to attempt to resolve disputes through amicable settlement under the provisions of sub-clause 67.2, before proceeding to arbitration.

Chapter 13

The Contractor's Obligations

13.1 Introduction

In a construction contract, the contractor's obligations may be divided into two main categories: first, the obligations which emanate from the agreement between the parties to the construction contract as prescribed in the Conditions of Contract; and secondly, a set of obligations which apply under the applicable law of the contract between the parties. This second category extends in legal terms beyond the contractor's obligations under the contract with the employer and in time beyond the completion of that contract.

In simple terms, the contractor's obligations under the Fourth Edition of the Red Book revolve around five main areas:

(1) the construction and completion of the works with due diligence and within the time for completion as specified in the contract;
(2) the use of materials, plant and workmanship as described in the contract and in accordance with the engineer's instructions;
(3) the provision of securities, indemnities and insurances in respect of such works and obligations during the contract period;
(4) the supply of information and notices required for the execution and completion of the works and also for alerting the employer whenever an event occurs which is likely to increase the cost of the work or the time for completion; and
(5) the performance of certain administrative and other functions, (including, if explicitly required, the design of certain elements) to facilitate the process of construction and its various activities.

Although the contractor's obligations under the Fourth Edition of the Red Book can be stated in summary as above, a detailed study would reveal that they are, in fact, spread throughout the whole document and the provisions of most of its clauses impose some obligations on the contractor. A number of these obligations start with the commencement of the project cycle and continue to the end of the contractor's involvement, such as the obligation to provide securities and indemnities. Others emerge and continue for only a part of the project cycle. The various clauses regulate in detail the timing, extent and procedure for the contractor's obligations in respect of most of the complex situations associated with civil engineering construction. In the remainder of the present chapter, the five main areas of the contractor's obligations, referred to above, are discussed with particular reference to the three main stages in a construction contract which are as follows:

(a) tendering stage up to the letter of acceptance;
(b) following the letter of acceptance and during construction up to substantial completion of the works;
(c) after substantial completion of the works.

13.2 The contractor's obligations during the tendering stage

Sub-clauses 11.1 and 12.1 impose obligations on the contractor retrospective of the date of the contract in relation to the site, its surroundings and any information available in connection with it. These two sub-clauses provide that prior to submitting his tender, the contractor shall be deemed to have:

(a) inspected the site and its surroundings;
(b) inspected any information available in connection with the site;
(c) received and interpreted data on hydrological and sub-surface conditions obtained by or on behalf of the employer;
(d) satisfied himself (so far as is practicable, within financial and time constraints) as to:
    — the form and nature of the site, including the sub-surface conditions,
    — the hydrological and climatic conditions,
    — the extent and nature of work and materials necessary for the execution and completion of the works and the remedying of any defects in these works, the means of access to the site and the accommodation he may require;
(e) obtained, in general, all necessary information as to risks, contingencies and all other circumstances which may influence or affect his tender;
(f) based his tender on the data made available by the employer and on his own inspection and examination; and
(g) satisfied himself as to the correctness and sufficiency of his tender to cover all his obligations and all matters for the proper execution and completion of the works and the remedying of any defects therein.

Sub-clause 25.1 states that the contractor is to provide evidence to the employer prior to the start of the work that the insurance policies required under the contract have been effected. Sub-clause 25.1 also imposes a retrospective duty on both the employer and the contractor to agree the general terms of the insurance policies required under the contract. Such agreement is to be reached prior to the issue of the letter of acceptance. Furthermore, sub-clause 70.1 refers to Part II of the Conditions in relation to what should be done in the case of a change in the cost of labour and/or materials or any other matters affecting the cost of the execution of the works. The contractor should reflect these arrangements in his tender rates and prices.

13.3 The contractor's obligations following the letter of acceptance and during the construction stage up to substantial completion

The letter of acceptance, defined in sub-clause 1.1(b)(vi) as the formal acceptance by the employer of the tender, forms the contract between the employer and the contractor. In some jurisdictions, the date of the contract is the date of receipt of this letter whilst in others it is the date of transmission. Within the specified period of time under sub-clause 41.1 in Part II of the Red Book, the engineer should issue to the contractor the notice to commence the works.

The time lag between the issue of the letter of acceptance and the notice to commence is an important period during which many activities should take place. For the employer, this period should be utilised to ensure that possession of the site and access to it can take place in accordance with clause 42 and that all the necessary legal and financial matters are processed. For the contractor, this period should be utilised in initiating mobilisation formalities; finalising commitments with suppliers and sub-contractors; finalising arrangements for securities and insurances; arranging the work programme; and finalising estimates of cash flow requirements and details of the breakdown of lump sum items. It is important therefore to specify in Part II of the FIDIC Form a realistic period between the date of the letter of acceptance and the notice to commence.

13.3.1 Finalising documentation required prior to commencement of the works

The obligations which must be fulfilled by the contractor within specific time periods from the date of receipt of the letter of acceptance or the commencement date are as follows:

(a) In accordance with sub-clause 10.1, and if the contract requires it, the contractor should obtain security for his proper performance of the contract within 28 days after receipt of the letter of acceptance, in the sum stated in the Appendix to Tender. The contractor should notify the engineer when he has provided such security to the employer.

(b) Within the time stated in Part II of the Conditions from the date of the letter of acceptance, the contractor should submit to the engineer, for his consent, a programme for the execution of the works as required in sub-clause 14.1. The programme should be in the form and detail reasonably prescribed by the engineer. The Red Book does not provide for the date when the engineer should prescribe the form and detail of the programme nor what is meant by the word 'reasonable' in describing such form and detail.

(c) Within the time stated in Part II of the Conditions from the date of the letter of acceptance, the contractor should submit to the engineer, for his information, a detailed cash flow estimate, in quarterly periods, of all payments to which the contractor will be entitled under the contract (see sub-clause 14.3).

(d) Prior to the start of work at the site and in accordance with sub-clause 25.1, the contractor should provide evidence to the employer that the insurances required under the contract have been effected and should notify the engineer of so doing.

(e) Within 84 days of the commencement date, the contractor should provide the insurance policies to the employer which should be consistent with the general terms agreed prior to the issue of the letter of acceptance. The contractor should also notify the engineer of so doing, as required by the provisions of sub-clause 25.1.

(f) Within 28 days after the receipt by the contractor of the letter of acceptance, he should submit to the engineer, for his approval, a breakdown of each of the lump sum items contained in the tender (see sub-clause 57.2).

Figure 13.1 shows a flow diagram of these obligations and their effect on commencement of work, and Table 13.1 sets out, in a chronological manner, the position of the contractor's obligations under the Red Book with respect to the other parties from the date of invitation of tenders to the date of the final certificate.

The contractor's obligations, set out in (a) to (f) above, form a prerequisite to the execution of the works as can be seen in the flow chart in Figure 13.1. As the contractor moves into the construction stage, all the five areas of obligations identified earlier in the introduction to this chapter become alive and relevant. A more comprehensive analysis of these obligations is given below.

13.3.2 Construction and completion of the works with due diligence and within the time for completion

The date of receipt by the contractor of the notice to commence is the commencement date as defined in sub-clause 1.1(c)(i) and accordingly it is also the commencement of the construction stage. From thereon, the contractor's first obligation is to proceed with the works with due expedition and without delay. The time for completion specified in Part II of the Red Book under sub-clause 43.1 is calculated from the commencement date. The contractor should complete the works within the period stated in the Appendix to Tender or such extended time as may be allowed under clause 44.

The contractor's duty to complete the whole of the works may be divided into two stages: first, the duty to complete the works substantially to the stage where a taking-over certificate can be issued by the engineer under sub-clause 48.1 stating the date on which the works were substantially completed; and, secondly, the duty to complete the works to the stage where a defects liability certificate can be issued by the engineer under sub-clause 62.1 stating the date on which the contractor 'shall have completed his obligations to execute and complete the Works and remedy any defects therein to the engineer's satisfaction'. It should be borne in mind, however, that the contractor is not considered to have completed the contract until a defects liability certificate is signed by the engineer and delivered to the employer with a copy to the contractor.

With very few exceptions, under the Red Book the contractor undertakes to execute and complete the whole of the works irrespective of any difficulties he may experience. This obligation is expressly stated in sub-clause 8.1 and supported throughout the Conditions by other sub-clauses either by express reference to 'completion', such as in sub-clauses 1.1(c), 7.1, 12.1, 13.1, 14.2, 16.1, 19.1, 22.1, 22.2, 26.1, 29.1, 43.1, 47.1, 49.2, 51.1(e), 62.1, 65.6 and/or by implication, such as in sub-clauses 14.1, 14.4, 20.1, 44.1, 47.2, 48.1, 48.2, 48.3, 60.3, 60.5 and 65.3.

Table 13.1 The phases of a construction project in chronological order.

Sub-clause No.	Emp.	Eng.	Con.	Description of main events from invitation to tender until the final certificate	Time schedule
	●	●	○	Invite Tenders and supply documents	Tendering period
11.1			●	— Inspect the site	
11.1			●	— Inspect any information	
11.1			●	— Satisfy himself as to form and nature of the site	
11.1	●		○	— Provide data	
11.1			●	— Interpret the data provided by the employer	
11.1			●	— Base tender on the data	
12.1			●	— Satisfy himself as to correctness and sufficiency of the tender and of the rates and prices	
70.1			●	— Consider the provisions in Part II regarding increase or decrease of cost	
				Tender Submissions, Analysis and Decision on Award	
25.1	○ ●		○ ●	— Agree general terms of insurance	
1.1(b)(vi)	●		○	Letter of Acceptance	28 days / 28 days / Part II Cl.14.1 / Part II Cl.14.3 / See Appendix to Tender
10.1	○		●	— Provide security for proper performance, if required, (shall notify engineer of so doing)	
57.2		○	●	— Provide breakdown for each of the lump sum items	
14.1		○	●	— Submit programme for the execution of the works	
14.3		○	●	— Submit detailed cash flow estimate (in quarterly periods)	
41.1		●	○	Notice to Commence	Post, Cable, Telex, Fax or by Hand
25.1	○		●	— Provide evidence prior to the start of work at the site that the insurances required have been effected	
1.1(c)(i)				Commencement Date (Date of receipt of Notice to Commence)	
41.1			●	— Commence the works as soon as is reasonably possible after the receipt of the Notice to Commence	84 days / Time for Completion as stated in Appendix
42.1	●		○	— Access to site and possession of site	
2.2		●		— Engineer's representative is to be appointed	
25.1	○		●	— Provide insurance policies	
8.1			●	— Carry out design (to the extent provided for in the contract), execute and complete the works	
48.1			●	— Tests on completion	
48.1	○	○	●	— Notice of Substantial Completion and written undertaking to finish with due expedition any outstanding work during the Defects Liability Period	21 days
48.1	○	●	○	Taking-Over Certificate Date stated in Taking-Over Certificate as...	

● Action by
○ Receipt by

(Table 13.1 Contd.)

Sub-clause No.	Emp.	Eng.	Con.	Description of main events from invitation to tender until the final certificate	Time schedule
48.1	○	●	○	Taking-Over Certificate — Date stated in Taking-Over Certificate as the date on which, in the opinion of the Engineer, the works were substantially completed in accordance with the Contract	84 days / Period as specified in the Appendix to Tender Cl.49.1
33.1			●	— Clear away site and leave clean	
60.3(a)		●		— One half of retention money certified for payment to the contractor	
60.5		○	●	— Submit Statement at completion	
49.1				End of Defects Liability Period	14 days / 28 days
49.2			●	— Complete the works, and execute any outstanding work and remedy defects	
60.3(b)		●		— Second half of retention money certified for payment to contractor	
62.1	○	●	○	Defects Liability Certificate	14 days / 56 days
10.2	●		○	— Return performance security to contractor (no claim to be made against security after issue of Defects Liability Cert.)	
60.6		○	●	— Submit Draft Final Statement	Agreement between Eng. & Contractor
60.6		○	●	Submit Final Statement	
60.7	○	○	●	— Give written discharge confirming that the total of the Final Statement represents full and final settlement of all monies (only effective after payment due under Final Certificate and return of performance security under sub-clause 10.1)	Upon submission / 28 days
60.8	○	●	○	Final Certificate	

The exceptional situations where the contractor would be excused from the obligation to complete the whole of the works and, thus, not be liable for breach of contract, are as follows:

(a) In the case of legal or physical impossibility as provided under clause 13, failure by the contractor to execute such work which is either legally or physically impossible would not lead him to be in breach of contract. Impossibility is not to be confused with either difficulty or complexity. Therefore, work which is more difficult or more complex does not fall within the provisions of clause 13; nor does work which requires more attention, labour, equipment, or cost. For instance, it would be physically impossible to install a prestressing duct in concrete work through a space between reinforcement bars smaller than the diameter of the duct. To eliminate the physical impossibility, the solution would be either to increase the space or reduce the diameter of the duct or re-route the duct within the unit, any of which

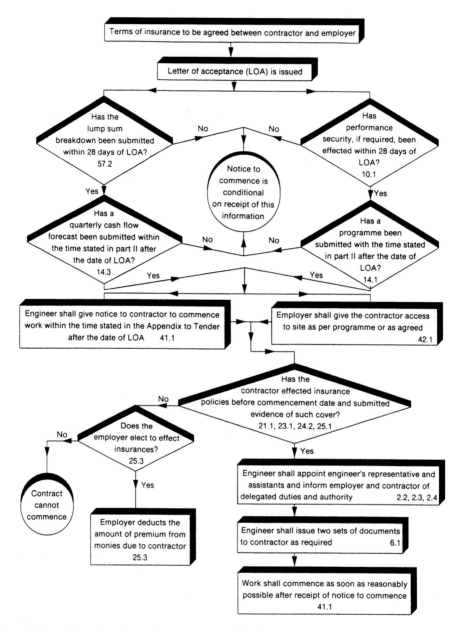

Fig. 13.1 Acceptance of tender and commencement of works.

should be ordered in an instruction varying the form of the works in accordance with clause 51. Unless such an instruction is issued, the contractor would be in default if he proceeded to install the duct in a manner different from that shown in the contract documents.

Legal impossibility would vary with the provision of the applicable law of the contract and an act may be legally sound in some jurisdictions but not in others. For instance, the use of certain materials may be barred by law. An example of legal impossibility is the position where machines or materials which are protected by a patent are specified, but the holder of the patent right will not supply or consent to the use of the protected item in any circumstances.[13.1] The use of this particular item is legally impossible, but it would not be an impossibility if the only obstruction to its use is that a larger payment than that anticipated by the contractor is demanded by the patent holder. This situation may also serve as an example of legal impossibility in one jurisdiction whilst not in another if the protected item is barred from being imported into one jurisdiction but not into the other.

(b) Where an employer's risk, as defined under sub-clause 20.4, eventuates causing loss or damage to the works, the contractor's obligation to complete would be restricted to exclude from the whole of the works the elements which had suffered such loss or damage, unless and until the engineer requires the contractor to rectify the loss or damage under the provisions of sub-clause 20.3. Where the engineer requires the contractor to rectify such loss or damage, he (the engineer) must determine an addition to the contract price in accordance with clause 52.

(c) Where the progress of the works or any part thereof is suspended on the instruction of the engineer under sub-clause 40.1, and where permission to resume work is not given by the engineer within 84 days from the date of suspension, the contractor's obligation to complete the whole of the works would be limited in accordance with the provisions of sub-clause 40.3.

(d) In the event of a default by the employer as provided in sub-clause 69.1, the contractor would be entitled to terminate his employment under the contract by giving 14 days notice of termination to the employer with a copy to the engineer.

Regarding the contractor's obligation to complete the works, a question often arises as to the extent of the work included in the obligation to complete. This question is usually followed by another: are there any limits on the variations to the form, quality or quantity of the works which the engineer can make? The answer is therefore relevant from the point of view of completion and the ordering of variations under clause 51. Two principles may help as a guideline:

(a) Construction work involves many processes, materials and methods of construction and if it is to be done properly, then no description, however comprehensive, would be sufficient to include every detail and description of the work to be done. In this regard, it has been submitted that:[13.2]

> 'in the absence of an expressed contrary intention, an obligation to do described work imports an obligation to do all the necessary ancillary work or processes, whether described or not, which are needed to produce the described work.'

For example, an express obligation to cast structural reinforced concrete elements would imply, if not specifically stated, that there is a quantity of steel reinforcement bars which has to be supplied, cut and bent to specific shapes and fixed in place before the concrete is poured. The question of whether or not this is to be paid for as an extra would depend on the contract documents, but the relevant principle in this example is that the contractor is under an obligation to carry out and complete the steel reinforcement work.[13.3]

(b) Construction work involves many risks. When such risks eventuate, they frequently result in cases where the contractor must carry out incidental work in order to complete the project. Such work may include the repair of damage, the replacement of an element in case of loss or the addition of some process in case of physical or legal impossibility to complete the works. It is submitted that such incidental work is covered by the contractor's obligation to complete.

The construction and completion of the works must not only be done but it must also be done with due diligence. This obligation is regulated in sub-clauses 8.1 and 41.1 in respect of the work during the construction period and in sub-clauses 48.1 and 49.2(a) for the period after the issue of the taking-over certificate. The obligation regarding the rate of progress in order to comply with the time for completion is repeated in sub-clause 46.1, which prescribes the procedure to be followed if in the opinion of the engineer, the rate of progress of the works or any section is at any time too slow to comply with the time for completion. According to the provisions of sub-clause 46.1, such procedure is initiated by the engineer by giving the contractor a notice to that effect. The relevant notice should be carefully worded in order not to be used as an instrument for an unjustifiable claim for acceleration. Should the contractor ignore or neglect to comply with the obligation to proceed with due diligence, the only sanction is the ultimate one of default which would be applied by the engineer when he would certify to the employer, under sub-clause 63.1(d), provided that the notice under sub-clause 46.1 had been properly given and could have served as a warning.

Finally under this heading, the contractor is under the obligation to execute and complete the works within the time for completion, as stated in the contract or as extended under clause 44 calculated from the commencement date. The time for completion is specified in the Appendix to Tender. However, a reason for an extension of time might arise under any of the following sub-clauses: 6.4, 12.2, 20.3, 27.1, 36.5, 40.2, 42.2, 44.1, 44.3, 65 and 69.4.

13.3.3 Use of materials, plant and workmanship

In addition to construction and completion of the works with due diligence and within time, the contractor is also obliged to use materials, plant and workmanship as described in the contract and in accordance with the engineer's instructions. The subject of quality of materials, plant and workmanship in construction work is vast. The Conditions of Contract are not usually the place to deal with such a subject except in prescribing the principles to be applied. In the Red Book, these principles are set out in sub-clauses 36.1, 39.1 and 37.4. In this regard, sub-clause 36.1 provides that the materials, plant and workmanship must be 'of the respective kinds described in the Contract and in accordance with the Engineer's instructions . . .'. The detailed description and specification of the 'respective kinds' of such materials, plant and workmanship is normally given in the other contract documents, particularly the specification. Describing and specifying the materials and plant to be used can generally be done with precision. Codes of practice and national standard specifications are available in many countries to help the designer specify with precise detail the type of materials and plant required. Where workmanship is concerned, however, it is more difficult to be precise in general terms. There is no standard specification describing the requirements of standard workmanship in construction and therefore one should be explicit in specifying the required standard.

In legal terms, it has long been established that the contractor has a very wide obligation with respect to the supply of materials and workmanship. Under English law, unless there are express terms to the contrary, there is an implied warranty in a construction contract for the contractor to:

— do the work with care and skill or, as sometimes expressed, in a proper and workmanlike manner;
— use materials of good and marketable quality; and
— ensure that both the work and the materials should be reasonably fit for the purposes for which they are required.

In a case in 1870, it was stated that:[13.4]

'I do not hesitate to say that I am clearly of the opinion, as a general proposition of law, that where one man engages with another to supply him with a particular article or thing, to be applied to a certain use or purpose, in consideration of a pecuniary payment, he enters into an implied contract that the article or thing shall be reasonably fit for the purpose for which it is to be used and to which it is to be applied.'

Later, in 1934,[13.5] the following was stated which was subsequently approved unanimously in the House of Lords in *Young & Marten Ltd* v. *McManus Childs*:

'I think the true view is that a person contracting to do work and supply materials warrants that the materials which he uses will be of good quality and reasonably fit for the purpose for which he is using them, unless the circumstances of the contract are such as to exclude any such warranty.'[13.6]

In cases where the work is precisely and sufficiently described in the design documents, however, the implied warranty amounts to no more than a warranty that reasonable care and skill will be taken in carrying out the specified work. There is usually a thin line separating workmanship and design; selection of materials through design and use of materials through construction; and between workmanship and use of materials. The latter is particularly relevant in non-homogeneous materials manufactured on site, such as concrete.

In the Romano-Germanic system of law, the obligation of the contractor and his liability is codified by statute and so, for example, the German statutory law governing contracts for work[13.7] provides that the contractor is obliged to perform the construction so that:

— it possesses the guaranteed quality; and
— there are no faults which either destroy or limit the purposes for which it was required, either as stipulated in the contract or for normal usage.

The Conditions of Contract may add a further requirement, such as that added by the VOB Conditions[13.8] in the form of 'observance of the professional standards of the construction industry'. Other jurisdictions within the Romano-Germanic system have similar rules in their Civil Codes providing for similar obligations of good quality materials and construction work of good workmanship, free from defects and in accordance with the contract documents.[13.9]

Sub-clause 39.1 of the Red Book provides a remedy in the case of improper work, materials or plant. It provides that the engineer is to have the authority to issue instructions for:

(a) the removal from the site of any materials or plant which, in the opinion of the engineer, are not in accordance with the contract;
(b) the substitution of proper and suitable materials or plant; and
(c) the removal and proper re-execution of any work where: either the material, plant, workmanship; or the design by the contractor is not, in the opinion of the engineer, in accordance with the contract.

In a similar vein, sub-clause 37.4 gives the engineer the authority to reject any materials or plant which are found to be defective or otherwise not in accordance with the contract.

The contractor is also under an obligation to ensure that the quality of materials, plant and workmanship is as prescribed in the contract. Thus, it is provided under sub-clauses 15.1 and 16.1 that the contractor should provide on the site: first, such skilled, semi-skilled and unskilled labour as is necessary for the proper and prompt fulfilment of the contractor's obligations under the contract; secondly, only such technical assistants as are skilled and experienced in their respective callings and such foremen and leading hands as are competent to give proper superintendence; thirdly, all necessary superintendence during the execution of the works and as long thereafter as the engineer may consider necessary; and, fourthly, an authorised representative who gives his whole time to the superintendence of the works.

An extensive set of provisions are included in the Red Book for sampling, testing and inspection of the quality of the materials, plant and workmanship. These provisions are included in clauses 36, 37, 38 and 39.

Where defects arise which, in the opinion of the engineer, are either due to the use of materials, plant or workmanship not in accordance with the contract, or due to neglect or failure on the part of the contractor to comply with any obligation under the contract, they should be repaired by the contractor at his own cost, as provided in sub-clause 49.3.

Where the contract expressly provides that part of the permanent works is to be designed by the contractor, an additional obligation is imposed on the contractor. This is done under sub-clauses 6.1, 7.2, 39.1(c)(ii) and 49(b). In relation to the elements designed by the contractor, the employer will rely on the contractor in the selection of the materials. The warranty for fitness for purpose will almost certainly extend from that in respect of material, plant and workmanship to include design, in most jurisdictions. As this is an absolute obligation independent of lack of care or negligence, it should be contrasted with the standard skill and care obligation of the engineer as a designer (see Section 11.2 above). Such comparison is particularly relevant from the point of view of onus of proof. Where the obligation is one based on negligence, the onus of proof is that of the claimant (usually the employer), whereas if the liability is independent of fault or negligence, the onus of proof shifts to the defendant contractor to prove no legal responsibility. Furthermore, from a practical point of view, the employer would not have to differentiate between defects arising out of design and those from a contractor's obligation as to materials, plant or workmanship, making it easier for him to substantiate a claim.

13.3.4 Provision of securities, indemnities and insurances

The FIDIC Fourth Edition also imposes on the contractor the obligation of providing the following securities, indemnities and insurances.

(a) If, having considered the matter of securities, the employer decides that security for proper performance should be obtained by the contractor, then in accordance with sub-clause 10.1, such security should be obtained and submitted to the employer within 28 days after the receipt of the letter of acceptance. The performance security should be in the sum stated in the Appendix to Tender. Chapter 15 below deals with the subject in more detail.

(b) Clauses 20 to 25, inclusive, deal with the subject of liabilities of the employer and the contractor, the indemnities which must be provided in that connection, and the insurance cover in respect of these indemnities. Chapter 14 below deals with the risks to which the employer and the contractor are exposed, the responsibility and liability triggered by the occurrence of such risks, and the indemnities and insurances to cover such events.

(c) Under sub-clause 26.1, the contractor is required to keep the employer indemnified against all penalties and liability for breach or non-compliance with national or state statute, ordinance or other law, or any regulation or bylaw of any local or other duly constituted authority in relation to the execution and completion of the works and the remedying of any defects therein. Such indemnity is also required in respect of any breach or non-compliance with the rules and regulations of all public bodies and companies whose property or rights are affected or may be affected in any way by the works.

(d) Under sub-clause 28.1, the contractor is required to provide indemnity from and against all claims and proceedings for or on account of infringement of any patent rights, design trademark or name or other protected rights in respect of any contractor's equipment, materials or plant used for or in connection with or for incorporation in the works, unless such infringement results from compliance with the design or specification of the engineer. The indemnity should extend to all damage, costs, charges and expenses whatsoever in respect of such claims and proceedings or in relation thereto.

(e) Under sub-clause 29.1, the contractor is required to provide indemnity to the employer against all claims, proceedings, damage, costs, charges and expenses whatsoever arising out of or in relation to interference with traffic and adjoining properties.

(f) The contractor is also required to provide indemnity, under sub-clause 30.2, to the employer against all claims for damage to any road or bridge communicating with or on the routes to the site used to facilitate the movement of the contractor's equipment or temporary works.

13.3.5 Supply of information, notices or alerts

It is a requirement throughout the Red Book that, in certain specified instances, each party to the contract undertakes to provide information, notices or alerts to the other party or to the engineer. Where such notices are submitted to the engineer, a copy should also be sent to the other party. Failure to give such information, notices or alerts would be a breach of the terms of the contract.

The information and alerts which should be provided by the contractor, during the construction period are set out below in items (a) to (n), while Table 13.2 sets out the notices which should be given by the contractor during this period.

Information and alerts required to be provided by the contractor during the construction period

(a) Design information: where the contractor is expressly required to design part of the permanent works, design information should be submitted by the contractor to the engineer for the latter's approval in accordance with sub-clauses 6.1 and 7.2.

(b) Programme: In accordance with sub-clauses 14.1 and 14.2, a programme should be submitted by the contractor to the engineer for the latter's consent, and whenever required by the engineer, the contractor should provide in writing a general description of the arrangements and methods which the contractor proposes to adopt for the execution of the works. A revised programme showing any modifications necessary to ensure completion of the works within the time for completion should be produced by the contractor for the engineer at the latter's request.

(c) Cash flow estimates: As provided in sub-clause 14.3, the contractor should provide to the engineer a detailed cash flow estimate, in quarterly periods, which should be subsequently revised if required by the engineer.

(d) Geological and archaeological finds: Immediately upon discovery of fossils, coins, articles of value or antiquity and structures and other remains or articles of geological or archaeological interest, and before removal, the contractor should acquaint the engineer of such discovery as provided under sub-clause 27.1.

(e) Details of labour and equipment: If required by the engineer, under sub-clause 35.1, the contractor should deliver to him a detailed return of labour in such a form and at such intervals as prescribed by the engineer. This return should show the staff and the number of the several classes of labour employed by the contractor from time to time on the site in addition to any other information in relation to the contractor's equipment as defined in sub-clause 1.1(f)(v).

(f) Particulars of extension of time: Where the contractor had notified the engineer, and sent a copy to the employer, of an event under the terms of sub-clause 44.1, the contractor should submit detailed particulars of any extension of time to which he may consider himself entitled. Such a submission should be made under the provisions of sub-clause 44.2(b) and within 28 days or such other reasonable time as may be agreed by the engineer.

Where an event has a continuing effect such that it is not practical for the contractor to submit detailed particulars within the 28 days referred to above, he should then provide such particulars in accordance with the terms of sub-clause 44.3.

(g) Emergency work: Whenever work is unavoidable or absolutely necessary for the saving of life or property or for the safety of the works, the contractor should immediately advise the engineer and carry out such work irrespective of the restrictions imposed under clause 45.

(h) Varied work: In the event of an instruction by the engineer that any varied work is to be executed on a daywork basis, under sub-clause 52.4, the contractor should furnish to the engineer such receipts or other vouchers as may be necessary to prove the amounts paid and, before ordering materials, the contractor should submit quotations for the same for the engineer's approval.

During the continuance of such work, the contractor should deliver each day to the engineer an exact list in duplicate of the names, occupation and time of all workmen employed on such work and a statement, also in duplicate, showing the description and quantity of all materials and contractor's equipment used thereon or therefor other than contractor's equipment which is included in the percentage addition in accordance with such daywork schedule. One copy of each list and statement should, if correct, or when agreed, be signed by the engineer and returned to the contractor.

At the end of each month, the contractor shall deliver to the engineer a priced statement of the labour, materials and contractor's equipment, except as aforesaid, used and the contractor shall not be entitled to any payment unless such lists and statements have been fully and punctually rendered.

(i) Contemporary records: Where the contractor has given a notice under sub-clause 53.1, he should keep such contemporary records as may reasonably be necessary to support any claim which he may subsequently wish to make. The contractor should supply the engineer with copies of such records as referred to in sub-clause 53.2 as and when the engineer so instructs.

The contractor should also send to the engineer an account giving detailed particulars of the amount claimed and the grounds upon which the claim is based. The period within which such account is to be sent is detailed in sub-clause 53.3.

(j) Measurement: The contractor should supply all particulars required by the engineer under the provisions of sub-clause 56.1 in connection with measurement of the value of the works in accordance with the contract.

(k) Lump sum items: A breakdown for each of the lump sum items contained in the tender should be submitted by the contractor to the engineer, for his approval, within the time specified in sub-clause 57.2.

(l) Provisional sums: The contractor should produce for the engineer in accordance with sub-clause 58.3 all quotations, invoices, vouchers and accounts of receipts in connection with expenditure in respect of provisional sums, except where work is valued in accordance with rates or prices set out in the tender.

(m) Payment to nominated sub-contractors: The contractor should provide the information required under sub-clause 59.5 in relation to payments to nominated sub-contractors.

(n) Monthly statements: The contractor should submit to the engineer after the end of each month, six copies of a statement in such form as the engineer may from time to time prescribe. The statement should contain the information detailed in sub-clause 60.1.

The notices which should be given by the contractor during the construction period in a contract under the FIDIC Form are identified in Table 13.2. These notices are required by the employer or the engineer and should be sent by post, cable, telex or facsimile transmission or delivered to the respective addresses nominated for that purpose in Part II of the Conditions pursuant to sub-clause 68.2. In this regard, it is notable that communication by cable and telex has become obsolete in the recent past with the rapid development of more modern electronic media. The relevant section in Part II of the Conditions should reflect these developments.

13.3.6 *Performance of certain administrative functions*

In the performance of the contract, the contractor undertakes to carry out many administrative obligations and functions which involve positive acts, such as the submission of drawings under sub-clause 6.1 and the payment of royalties under sub-clause 28.2. On the other hand, the contractor has obligations not to pursue certain lines of action, for instance, not to assign the contract without the prior consent of the employer as provided under sub-clause 3.1 and not to sub-contract the whole of the works as in sub-clause 4.1.

Details of the contractor's administrative obligations are as follows:

(a) Definition of terms: Whilst the definition of all the expressions included in clause 1 are relevant and significant to the contractor, it is generally necessary to scrutinise and pursue the implications of some of these definitions prior to the submissions of the tender. Others may have or continue to have an effect after the commencement date as they impose certain obligations on the contractor after that date. For instance, the meaning of 'sub-contractor' as given in sub-clause 1.1(a)(iii) raises two important points to be considered during the construction period should the contractor be expected to enter into a sub-contract with a named sub-contractor in the contract. First, it is a 'person' that has to be named as the sub-contractor; and secondly, the definition of a sub-contractor does not include an assignee of such person. (See below for the significance of assignment compared with sub-contracting.) Other definitions in this category include those for the expressions: tests on completion; temporary works; and contractor's equipment.

Table 13.2 Notices required to be given by the contractor under the contract.

Sub-clause	Notice to	Copy to	Description of the event
6.3	Engineer	Employer	Where planning or execution of the works is likely to be delayed or disrupted unless further drawing or instruction is issued by the engineer within a reasonable time. The notice should include details of the drawing or instruction required and of why and by when it is required and if any delay or disruption is likely to be suffered if such notice is late
10.1	Engineer		When providing performance security to the employer, to notify the engineer of so doing
12.2	Engineer	Employer	Whenever the contractor encounters physical obstructions or physical conditions, other than climatic conditions on the site, not foreseeable by an experienced contractor
25.1	Engineer		When providing evidence to the employer that the insurances required under the contract have been effected and also when providing the insurance policies
25.2	Insurers		Whenever there are changes in the nature, extent or programme of the works
26.1	Statutory authorities		Whenever notices are required in compliance with statutes or regulations or other laws or bylaws
30.3	Engineer	Employer	Whenever and as soon as any damage occurs to any bridge or road communicating with or on the routes to the site
38.1	Engineer		Whenever any part of the works is about to be covered up or put out of view or foundations are ready or about to be ready for examination
40.3	Engineer		When requiring permission to proceed with the works or that part thereof in regard to which progress had been suspended
40.3	Engineer		If and when the contractor elects to treat the suspension, where it affects only part of the works, as an omission of such part under clause 51; or where it affects the whole of the works, the contractor elects to treat the suspension as an event of default by the employer and terminate his employment under the contract in accordance with the provisions of sub-clause 69.1
42.1	Engineer	Employer	When the contractor makes proposals to commence and proceed with the execution of the works
44.2	Engineer	Employer	Within 28 days after an event has first arisen which fairly entitles the contractor to an extension of the time for completion of the works
48.1	Engineer	Employer	When the whole of the works have been substantially completed and have satisfactorily passed any tests on completion prescribed by the contract. Such notice should be accompanied by an undertaking to finish with due expedition any outstanding work during the defects liability period
52.2	Engineer		Whenever the contractor intends to claim extra payment or a varied rate or price in respect of varied work
53.1	Engineer	Employer	Within 28 days after an event for which the contractor intends to make a claim for additional payment pursuant to a clause under the Conditions
65.5	Engineer		As soon as any cost attributable to, or consequent on, or the result of, or in any way whatsoever connected with, any of the special risks comes to the contractor's knowledge
67.1	Employer	Engineer	Within 70 days of receipt of a notice of an engineer's decision under clause 67 with which the contractor is dissatisfied and the matters in respect of which he intends to commence arbitration
67.1	Employer	Engineer	If the engineer fails to give notice of his decision within 84 days after the day on which he received a reference to him of matters in dispute, then within 70 days after the expiry of the above-mentioned 84 days, a notice should be given by the contractor of his intention to commence arbitration in respect of the matters in dispute

(b) Assignment: Other than in specific circumstances referred to in sub-clause 3.1, the contractor is under the obligation not to assign the contract or any part thereof or any benefit or interest therein or thereunder without the prior consent of the employer.

As an assignment is a legal instrument which may have different meanings under different jurisdictions, it is essential to ascertain the precise position under the applicable law of the contract between the employer and the contractor before entering into or consenting to an assignment. In general, where a contract exists between two parties, A and B, assignment means the transfer of certain contractual rights by one of the parties, either A or B, to a third party, C. The assignor, A or B, who is originally entitled to those rights thus enables the assignee C to pursue the other party in the original contract and enforce those transferred rights directly against him.

It is essential in this context to distinguish between, on the one hand, the rights and, on the other, the obligations or liabilities of the parties involved. Under English law, for example, the assignment of contractual obligations or liabilities is not permitted without the permission of the other party to the contract, that is, agreement must be reached not only between the assignor and the assignee but also between them and the other party to the contract with the assignor. Accordingly, in a construction contract, the contractor cannot divest himself from the obligation to complete the works in the absence of an agreement with both the third party assignee and the employer. Such an agreement is called a 'Novation'.[13.10]

Whilst certain rights require specific formalities if they are to be assigned, not all rights are assignable. One of the rights where assignment may be questioned is the right of arbitration. An assignment of 'all monies due or to become due under the contract' by a contractor was held not to include the right to arbitration in the English case of *Cottage Club Estates Ltd* v. *Woodside Estates Co. (Amersham) Ltd*.[13.11] Although the decision in this case has been criticised in subsequent cases and by authoritative writers,[13.12] it is essential to ensure that the wording of an assignment reflects precisely and unequivocally what is intended by the parties and that such intentions are within the provisions of the applicable law of the contract.

(c) Sub-contracting: In similar terms, sub-clause 4.1 prohibits the contractor from sub-contracting the whole of the works. In the case of sub-contracting part of the works, the contractor is required to seek and obtain the prior consent of the engineer. This obligation does not, however, extend to situations where the sub-contractor is named in the contract.

As there is no contract between the employer and a sub-contractor, the contractor remains responsible and liable to the employer for the performance of the sub-contractor. The sub-contractor has no contractual rights or obligations towards the employer under the Red Book.

(d) Copies of drawings and other documents: The contractor is required under sub-clause 6.1 to make at his own cost any further copies of the drawings, beyond the two supplied to him free of charge. Where he is responsible for the design of part of the permanent works, the contractor is required to supply to the engineer, free of charge, four copies of all drawings, specifications and other documents, together with a reproducible copy of any material which cannot be reproduced to an equal standard by photocopying. Any further copies should be supplied at cost.

One copy of the drawings provided to or supplied by the contractor is to be kept by him on the site, as required under sub-clause 6.2.

(e) Contract agreement: Sub-clause 9.1 provides that the contractor should, if called upon to do so, enter into and execute the contract agreement.

(f) Other contractors on the site: The contractor is required to afford all reasonable opportunities for other contractors, the workmen of the employer and of any duly constituted authorities, in carrying out any work on or near the site, which is not included in the contract. This obligation is provided in sub-clause 31.1 and is subject to the requirements of the engineer.

(g) Engagement of staff and labour: Under the provisions of clauses 34 and 35, the contractor is required to make his own arrangements for the engagement of all staff and labour and, if required by the engineer, deliver a return of labour on the site and of the contractor's equipment.

(h) Access to the site: The contractor should afford to the engineer and to any person authorised by him, every facility for and every assistance in obtaining right of access to the site and to all workshops and places where materials or plant are being manufactured, fabricated or prepared for the works.

During such manufacture, fabrication or preparation, the engineer is entitled to inspect and test the materials and plant to be supplied under the contract. The contractor's obligation, described above, in respect of access is therefore extended to obtain permission for the engineer to carry out such inspection and testing, in accordance with clause 37.

(i) Monthly statements: In respect of certificates, the contractor is under an obligation to submit to the engineer, on a monthly basis, six copies of a statement showing the amounts to which he considers himself to be entitled. Such statements should be in the form and detail as provided in sub-clause 60.1.

(j) Assignment to the employer in case of termination: If the employment of the contractor is terminated in accordance with clause 63, he is required under sub-clause 63.4, if instructed by the engineer, to assign to the employer the benefit of any agreement which he (the contractor) may have entered into for the supply of any goods or materials or services and/or for the execution of any work for the purposes of the contract. This obligation is subject to the provisions of the applicable law of the contract in respect of assignment.

(k) Amicable settlement of disputes to be attempted: Where notice of intention to commence arbitration has been given under sub-clause 67.1, there is an obligation imposed on the contractor, as well as the employer, to attempt to settle the dispute amicably within limits stipulated in sub-clause 67.2.

(l) Making payments in certain circumstances: In a number of sub-clauses throughout the conditions, the contractor is under an obligation to make payments as detailed below:
   (i) sub-clause 28.2 in respect of royalties;
   (ii) sub-clause 30.2 in respect of the cost of strengthening any bridges or altering or improving any road communicating with or on the routes to the site;
   (iii) sub-clause 42.3 for special or temporary wayleaves required by him in connection with access to the site and for any additional facilities outside the site.

(m) Due-consultation: Finally, it is the duty of the engineer to 'duly consult' with the employer and the contractor in certain circumstances for which the contractor would have to provide the time and the desire to consult with the engineer in a prompt manner. See also Sections 9.3 and 12.11.

13.4 Contractor's obligations after substantial completion of the works

After substantial completion in accordance with sub-clause 48.1, the contractor has the obligation to complete the works and execute any outstanding work and remedy defects in accordance with sub-clauses 49.2 and 49.3. This obligation extends as can be seen from Table 13.2 to a number of provisions in the Conditions which are related to the defects liability period as well as to others which apply mainly to the construction period but also whenever work is required to be done or tested or inspected. These provisions are:

(a) Site clearance: Under sub-clause 33.1, the contractor, upon the issue of the taking-over certificate, is required to clear away and leave clean that part of the site to which such taking-over certificate relates. He is, however, permitted to retain on site certain items which are required by him for the purpose of fulfilling his obligations during the defects liability period.
(b) Defects: Under sub-clause 50.1, if any defect, shrinkage or other fault appears, the contractor is required, if instructed by the engineer, to search under the directions of the engineer for the cause. The contractor is also required to remedy any defects and faults for which he is found to be liable.
(c) Statement of completion: The contractor should submit a detailed statement at completion within 84 days after the issue of the taking-over certificate in respect of the whole of the works. This statement should contain all the information prescribed in sub-clause 60.5.
(d) Draft of final statement: The contractor, within 56 days after the issue of the defects liability certificate, should submit to the engineer, for his consideration, a draft final statement with supporting documents. The draft final statement should show the information required in sub-clause 60.6.
(e) Further information: In the event that the engineer disagrees with or cannot verify any part of the draft final statement, the contractor should submit such further information as the engineer may reasonably require as provided in sub-clause 60.6. He should also prepare and submit the final statement.
(f) Discharge: Upon submission of the final statement, the contractor should give to the employer, with a copy to the engineer, a written discharge confirming that the total of the final statement represents full and final settlement of all monies due. Such discharge is provided under sub-clause 60.7.

The above obligations are set out in the time schedule shown in Table 13.1.

Chapter 14

Risks, Liabilities, Indemnities and Insurances

'Measured on a scale of the needs for the tasks to come, mankind is a faulty design'

Unknown air force pilot instructor

14.1 Introduction

As already discussed in Chapter 7, the concept of responsibility and liability of both the employer and the contractor towards each other and towards third parties is based on the sharing of the risks in accordance with the obligations allocated to them under the contract. In Chapter 7, the definition and significance of the concept of 'risk' is explained. Chapters 10 to 13 set out in detail the duties and obligations of the employer and the engineer, as agent of the employer, on the one hand, and those of the contractor on the other, under the Fourth Edition of the Red Book.

As is evident from these chapters, such duties and obligations are extremely complex in any construction contract but particularly so as the size and cost of the project increases. Complexity also increases as the risk matrix becomes larger with the involvement of a greater number of parties and especially so when different nationalities and cultures are involved.

The present chapter considers the responsibilities, liabilities, indemnities and insurances as they relate to risks in construction projects under the provisions of the Fourth Edition of the Red Book. In this regard, a glossary of definitions is given in Section 14.11 below.

14.2 The Red Book provisions relating to risk, responsibility, liability, indemnity and insurance

The Red Book provisions relating to the topics of risk, responsibility, liability, indemnity and insurance are contained in clauses 20 to 25 and 65 of the general conditions of contract. A logical flow from the risks in construction (as illustrated in Figures 7.1 and 7.2) to their ensuing responsibilities and liabilities and how these risks are dealt with under the Red Book is shown in Figure 14.1.

Although the wording of these clauses has been much simplified in the Fourth Edition following many criticisms and some recommendations,[14.1] it remains complex to most engineers due to the fact that it covers legal and insurance topics in which engineers,

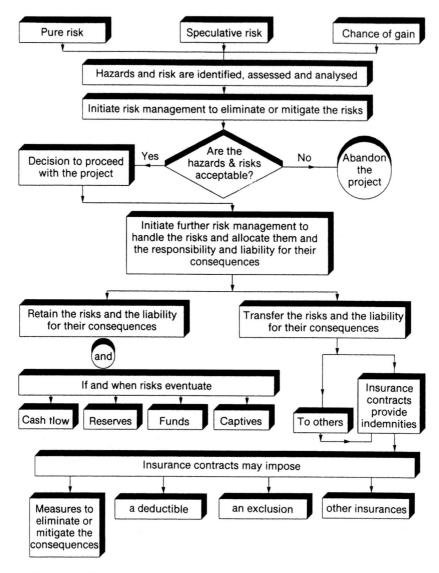

Fig. 14.1 Flow of risk into responsibility, liability, indemnity and insurance.

for whom the conditions are written, are not experts. In many respects, the wording of the Fourth Edition of the Red Book follows the Fifth Edition of the ICE Form which dates back to 1945, although some punctuation marks were added. The new punctuation resulted in a reduction of the average length of the sentences and therefore an improvement in the text. However, by following the ICE Form, the opportunity was missed of rewriting the whole section in a logically flowing sequence from the concept of risk sharing to responsibility to liability which leads to the indemnities in respect of which insurance is required. This limited improvement was such that the drafting committee of the Sixth Edition of the ICE Form adopted it without much change.

Risks, Liabilities, Indemnities and Insurances 243

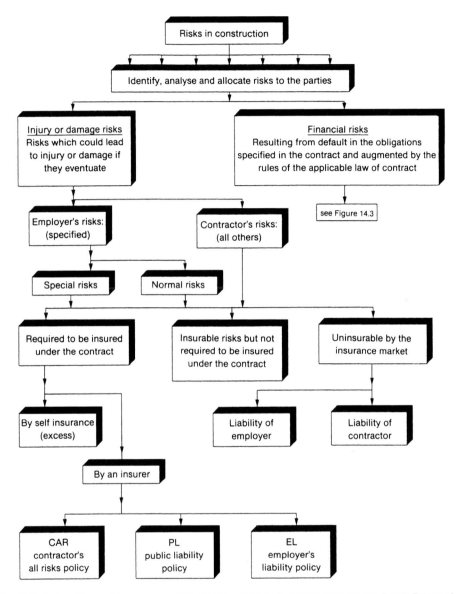

Fig. 14.2 Indemnities and insurances relating to risks of injury and damage under the Fourth Edition of the FIDIC Red Book.

Figures 14.2 and 14.3 show in a diagrammatic form the indemnities and insurances in respect of the risks allocated to the parties under the Fourth Edition of the Red Book. In addition, Figure 14.2 shows other insurance covers which are not required under the Red Book, but are either required under the contract between the employer and the engineer or voluntarily taken by the employer or by the contractor. (See also Section 13.3.4 for the indemnities and insurance covers required to be provided.) The

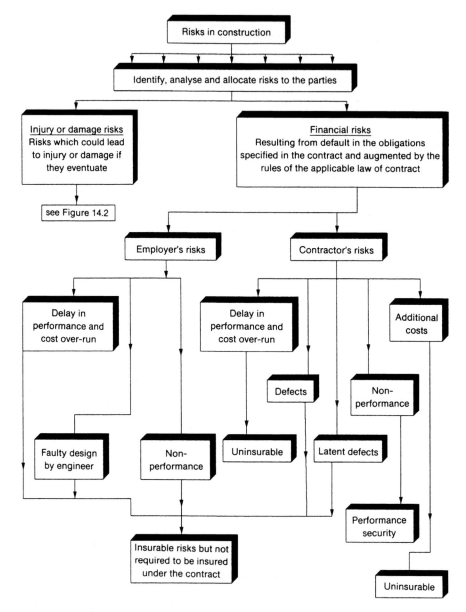

Fig. 14.3 Indemnities and insurance relating to financial risks under the Fourth Edition of the FIDIC Red Book.

indemnity provisions under the Red Book relating to the risks of damage, loss and injury are contained in clauses 20, 22, 24 and 65 while clause 25 deals with general insurance requirements. The remainder of the present chapter considers in detail clauses 20 to 25 and 65.

14.3 Clause 20 of the Red Book – '20.1: care of the works'; '20.2: responsibility to rectify loss or damage'; '20.3: loss or damage due to employer's risks'; and '20.4: employer's risks'

14.3.1 Sub-clause 20.1: care of the works

Clause 20 begins by placing the responsibility for care of the works, material and plant for incorporation in the works with the contractor. This responsibility continues until the date of issue of the taking-over certificate for the whole of the works except in the following two circumstances: the first exception is where the engineer issues a taking-over certificate for a section or part of the permanent works; and the second is in respect of any outstanding elements which the contractor undertakes to complete during the defects liability period. In the first case, the responsibility for the care of that section or part passes to the employer from the date of issue of the taking-over certificate in respect of such section or part of the permanent works. In the second case, the contractor remains responsible for the care of any outstanding elements until they are completed pursuant to clause 49.

It is unclear in the latter regard, as to whether completion of individual elements occurs when each element is completed or whether all outstanding elements have to be collectively completed before responsibility is passed on to the employer. It is also unclear as to whether there would have to be a completion certificate issued by the engineer for these outstanding elements before the responsibility passes to the employer. Clarification of these two points could have been made in sub-clauses 48.1 and 49.2(a). This lack of clarity is unsatisfactory from the point of view of the related insurance requirement which goes hand in hand with the duty of care and its transfer from the contractor to the employer on the date of issue of the taking-over certificate. In this connection, it should be noted that the relevant date is the date of issue of the certificate as distinct from the date of taking over as stated in that certificate.

14.3.2 Sub-clause 20.2: responsibility to rectify loss or damage; and sub-clause 20.3: loss or damage due to employer's risks

The remaining part of clause 20 consists of three sub-clauses which distribute and allocate the risks of injury, loss or damage between the employer and the contractor. They also describe the contractor's obligation to rectify such loss or damage when caused by either the risks allocated to him or to the employer.

Accordingly, these risks are divided into two bands, the first of which is allocated to the contractor and includes all risks 'whatsoever' other than the employer's risks as defined in sub-clause 20.4.

Sub-clauses 20.2 and 20.3 prescribe the extent of the contractor's obligation to rectify any loss or damage caused by the employer's and/or the contractor's risks. When the loss or damage occurs as a result of the contractor's risks during the period for which he is responsible for the care of the affected part, these clauses allocate the obligation to rectify such loss or damage and the liability for the cost to the contractor. It is further provided that the contractor should rectify the loss or damage so that the permanent works conform with the provisions of the contract to the satisfaction of the engineer.

The obligation to rectify any loss or damage which occurs as a result of any of the employer's risks or of a combination of the employer's risks with any of the contractor's risks is also placed on the contractor, but only if and to the extent required by the engineer. The word 'if' can be interpreted to indicate that the contractor may be relieved of his obligations to complete the project in respect of items lost or damaged due to an employer's risk, if the engineer does not require such loss or damage to be rectified. The liability for the cost of such rectification is allocated to the employer to the extent that it is necessitated by an employer's risk. The engineer is required under sub-clause 20.3 to determine an addition to the contract price in accordance with clause 52 taking into account the proportional responsibility of the employer and the contractor where the cause of the loss or damage is a result of a combination of employer's and contractor's risks. Reference in this connection should also be made to sub-clause 65.3 which deals with damage to the works due to any of the special risks.

It is important to note that in rare cases, there might not be an addition to the contract price but instead a reduction. This situation could arise as a result of a special risk eventuating causing destruction of or damage to a part of the works which had already been condemned under the provisions of clause 39.1 of the Red Book prior to occurrence of the special risk. Clause 39.1 deals with removal of improper work, material or plant. If the employer through the engineer does not require the contractor to rectify the loss or damage to the part of the works which had been already condemned, then the engineer must deduct the value of the condemned part from any payment due to the contractor. In these circumstances, the provisions of sub-clause 65.1(a), 65.5 and 65.8 in combination with sub-clauses 20.3 and 39.1 are relevant.

Where the contractor causes any loss or damage to the works during the defects liability period whilst carrying out any of his obligations under clauses 49 and 50 of the contract, the liability for such loss or damage is allocated to the contractor.

14.3.3 *Sub-clause 20.4: the employer's risks*

Sub-clause 20.4 deals with employer's risks and incorporates what are referred to as the special risks which are defined in clause 65. These special risks are discussed in Section 14.4 below. The present section considers the employer's risks as defined in sub-clause 20.4 and discusses their implications and whether or not they are properly allocated. These include the risk of war, the risk of pressure waves, risks relating to design and risks relating to the action of forces of nature. Definition of some of the relevant terms are given in the glossary in Section 14.11 at the end of this chapter.

(a) The Risk of War, etc.

Although the risk of war, which is an employer's risk under paragraph (a) of sub-clause 20.4, is expressly defined in sub-clause 65.6 as a special risk regardless of where it occurs in the world, by reference to sub-clause 65.2(b) there seems to be an intention that the combined risks of war, hostilities, invasion and acts of foreign enemies are also to be treated as employer's risks regardless of where they occur in the world. Similar intentions apply to the employer's risks as defined in paragraphs (c), (d) and (e) of sub-clause 20.4. However, it appears that this is intended to apply only in the context of the provisions of sub-clause 65.1 where destruction of or damage to the works and to other property and injury or loss of life arise due to the above risks. If that is the case, there is incompatibility with the provisions of sub-clause 65.2(b) where the definition of the risks of rebellion, revolution, insurrection, military or usurped power, and civil war are restricted to those relating to the country in which the works are to be executed. If destruction of an item covered by the description in sub-clause 65.1 happens due to war in the country where it is manufactured, it would be considered a special risk: why then would it not be a special risk under sub-clause 65.2(b) if it happens due to, for instance, civil war in that country?

(b) The risk of pressure waves

The risk of pressure waves caused by aircraft or other aerial devices travelling at sonic or supersonic speeds is included as an employer's risk under paragraph (d) of sub-clause 20.4. The inclusion of this risk under the list of employer's risks originates from the ICE Form where in the UK, the government undertook to pay compensation if damage resulted from the supersonic test flights originally made by Concorde. Consequently, insurers in the UK excluded this risk from their contractors' all risks insurance policy. Later when little or no damage was caused by supersonic test flights, the undertaking seems to have remained in respect of operational flights. In countries where the standard form of contract does not originate from the ICE Form, no specific reference is made to this risk. Therefore, there is no apparent reason for this risk to remain as one of the employer's risks and especially when insurance cover can be provided for it, if requested, under the contractors' all risks insurance policy.

(c) Risks relating to design

The employer's risk in respect of design as defined in paragraph (g) of sub-clause 20.4 is extremely wide as it refers to loss or damage due to the design of the works other than that provided by the contractor. This definition is not restricted to negligent design as it includes events where damage occurs without any fault, defect, error or omission committed by the designer. Such an event could occur simply because of the state of the art and knowledge in the field of engineering, such as the case of the bridge over the Pioneer river in Australia where three concrete piers collapsed due to the inadequacy of their design to withstand forces during an unprecedented flood. The design of the piers complied with the standard expected from professional bridge engineers at that time and there was no 'personal failure or noncompliance with standards which would be expected of designing engineers'.[14.2] The failure resulted from the effect of the prismatic shape of the piers, an effect which was not known at the time.

The employer's risk in terms of design does not include the word 'solely' previously included in the Third Edition. Accordingly, 'should damage occur to the Works in circumstances where an error in design is a contributory factor, the contractor is relieved of responsibility only to the extent that the damage was caused by a design not provided by him'.[14.3]

(d) Risks relating to forces of nature

The employer's risk in respect of the forces of nature as defined in paragraph (h) of sub-clause 20.4, although less contentious than its predecessor in the Third Edition, (see Chapter 9), remains imprecise. The wording of paragraph (h) is as follows: 'any operation of the forces of nature against which an experienced contractor could not reasonably have been expected to take precautions.' The ambiguity of this paragraph stems from the use of the following words, all in the same sentence: 'experienced', 'could not reasonably', and 'expected to take precautions'. There is no need to create such ambiguity in the contract since insurance is available for the risks of forces of nature. In fact, it can be argued that it is for this type of risk that insurance is generally available. It should be noted that with the present wording, these employer's risks are the responsibility of the employer and should the insurer decide to argue about whether or not any payment in respect of a particular event is covered under the terms of the contractors' all risks insurance policy, it is the employer who is at risk of not being covered. In this connection, reference should also be made to the mandatory amendment of this paragraph required by the World Bank in their Standard Bidding Document for Major Works (see Chapter 22 of the second edition of this book).

14.4 Clause 65 of the Red Book (sub-clauses 65.1 to 65.8) – special risks

It is appropriate now to deal with the provisions of clause 65 which come under the heading of 'special risks', as they are closely related to the provisions of clause 20 and, in particular, sub-clause 20.4 discussed above. In fact, it is clear from the wording of clause 65 that its purpose is to describe the contractual arrangements and to allocate the financial consequences when any of the special risks eventuates. To that extent, it is puzzling as to why it did not form part of clause 20, or alternatively follow immediately thereafter. As neither of these two alternatives was chosen, one would have expected some explanation to be given in the Guide, referred to in Reference 14.3 above, in order to remove the confusion caused by this separation. However no explanation is given.

Sub-clause 65.1 starts with the statement that the 'contractor shall be under no liability whatsoever in consequence of any of the Special Risks referred to in sub-clause 65.2 whether by way of indemnity or otherwise . . .'. The only restriction to this provision is in respect of improper work, material or plant by the contractor as stipulated under clause 39 of the Conditions.

Sub-clause 65.2 identifies the special risks in terms of the definition given under sub-clause 20.4. Thus, the special risks include the risks identified in paragraphs (a), (c), (d) and (e) of sub-clause 20.4 and also those in paragraph (b) of that sub-clause in so far as the risks relate to the country in which the works are to be executed. Thus, a distinction is made between the risks of war, hostilities and acts of foreign enemies which are defined as special risks regardless of where they occur, and the risks of rebellion, revolution, insurrection, military or usurped power, and civil war which are recognised as special risks only if they occur in the country in which the works are to be executed.

In broad terms, the remaining part of clause 65 deals with the consequences of the occurrence of a special risk.

Sub-clause 65.3 prescribes the contractual arrangements if the works, or any materials or plant on or near or in transit to the site, or any of the contractor's equipment, sustain destruction or damage by reason of any of the special risks. It provides that the contractor is entitled to payment in accordance with the contract for:

(a) any permanent works duly executed;
(b) any materials or plant destroyed or damaged; and
(c) in so far as the engineer requires or as may be necessary for the completion of the works, for
   (i) irectifying any destruction or damage to the works, and
   (ii) replacing or rectifying such materials or contractor's equipment.

With regard to (c) above, the phrase 'necessary for the completion of the Works' requires careful consideration. The intended meaning of this phrase is that the contractor is entitled to payment for materials and/or equipment under the provisions of sub-clauses 65.3 only if such material and/or equipment was present on the site for the purpose of completion of the works. Any materials or equipment present on the site for the convenience of the contractors but not required for the completion of the works at the time of occurrence of the special risk do not qualify for payment by the employer if it is destroyed or damaged as a consequence of the special risk. Furthermore, under sub-clause 65.3, payment for destroyed or damaged materials or equipment is not dependent on the resumption of construction work or completion of the works after the occurrence of the special risk. This is because the basis of clause 65 is the allocation of responsibility for these special risks to the employer.

Sub-clause 65.3 continues by requiring the engineer to determine:

'an addition to the Contract Price in accordance with Clause 52 (which shall in the case of the cost of replacement of Contractor's Equipment include the fair market value thereof as determined by the Engineer) and shall notify the Contractor accordingly, with a copy to the Employer.'

An important aspect of the wording in the above text is the use of the words 'the cost of replacement of the Contractor's Equipment [which shall] include the fair market value.' In this regard, the replacement cost or the fair market value must be reasonably and fairly determined without any subjective influence. This aspect assumes great importance in international contracts and particularly where expensive items of machinery and equipment are used.

Any payment to the contractor under clause 65 is determined by the engineer as an adjustment to the contract price in accordance with clause 52 of the conditions of contract. As discussed earlier, this determination may not always be an addition. For example, a reduction may result if the part destroyed includes defective work, materials or plant and the contract is terminated under the provisions of sub-clause 65.6 without the engineer requiring the replacement of such defective work.

The definition of destruction, damage, injury or loss of life caused by the special risks is extended in sub-clause 65.4 to include any 'destruction, damage, injury or loss of life caused by the explosion or impact whenever and wherever occurring of any mine, bomb, shell, grenade, or other projectile, missile, munition, or explosive of war'.

Sub-clause 65.5 deals with increased costs arising from the special risks and allocates the liability for any increase in such cost of the works to the employer. In this regard, the wording of sub-clause 65.5 is extremely wide in that any increase in cost 'consequent on or the result of or in any way whatsoever connected with the said special risks . . .' is allocated to the employer. However, the contractor is required to notify the engineer forthwith of such increase as soon as it comes to his knowledge. Furthermore, the engineer is required to 'duly consult' with the employer and the contractor before determining the amount of the contractor's costs. The engineer is also required to notify the contractor in writing of the amount he has determined providing a copy to the employer. Once again clause 39 is the only qualification to this provision.

The remaining parts of clause 65 deal with the risk of war and the contractual arrangements, whether war is declared or not, and irrespective of where such risks eventuate in any part of the world. Sub-clause 65.6 provides that the employer is entitled to terminate the contract at any time after the outbreak of war by giving notice to the contractor and the contract becomes terminated upon such notice being given subject to the operation of clause 67. Where the contractor is concerned, this sub-clause provides that if there is an outbreak of war which materially affects the execution of the works, whether financially or otherwise, he is required to use his best endeavours to complete the works, unless the contract is terminated by the employer. This is an extremely onerous provision where the safety of employees is threatened and the employer delays his decision to terminate the contract.

If a contract is terminated under sub-clause 65.6, then the mechanism and consequences of such termination are as follows:

(a) The contract is terminated only when a notice is given by the employer to the contractor. Upon such notice being given, the contract terminates except as to the rights of the parties under clauses 65, 'special risks'; and clause 67, 'settlement of disputes', but without prejudice to the rights of either party in respect of any antecedent breach of the contract. This means that once a contract is terminated under sub-clause 65.6, no provision of that contract survives other than the provisions of clauses 65 and 67. This wording of sub-clause 65.6 does not affect the rights and obligations of the parties to the contract in respect of a breach committed prior to the termination of the contract under that sub-clause.

(b) The rights of the parties following termination under sub-clause 65.6 are dealt with under the following sub-clauses:
— sub-clause 65.3, 'damage to works by special risks', discussed above;
— sub-clause 65.7, 'removal of contractor's equipment on termination'; and
— sub-clause 65.8, 'payment if contract terminated'.

In this regard, the provisions of sub-clause 65.3 have been discussed above, while the remainder of the present section considers sub-clauses 65.7 and 65.8. Specifically, sub-clause 65.7 provides that 'If the Contract is terminated under the provisions of Sub-Clause 65.6, the Contractor shall, with all reasonable dispatch, remove from the Site all Contractor's Equipment and shall give similar facilities to his Sub-contractors to do so.'

Accordingly, sub-clause 65.7 provides for the right of the employer to require the Contractor, 'with all reasonable dispatch' to 'remove from the Site all the Contractor's Equipment.' This provision is intended to ensure that the effects of the occurrence of any of the employer's special risks be reduced to a minimum by the removal from the site of such materials and equipment for which the employer would become liable if they remain on the site and sustain loss or damage.

Sub-clause 65.8 provides for the method and amount of payment to which the contractor is entitled if the contract is terminated under sub-clause 65.6. In such circumstances, the contractor is entitled to be paid by the employer for amounts not already paid in relation to all work executed prior to the date of termination at rates and prices provided in the contract. In addition, the contractor is entitled to be paid as follows under sub-clauses 65.8(c) to (f):

'(c) a sum being the amount of any expenditure reasonably incurred by the contractor in the expectation of completing the whole of the works insofar as such expenditure has not been covered by any other payments referred to in this Sub-Clause;
(d) any additional sum payable under the provisions of Sub-Clauses 65.3 and 65.5;
(e) such proportion of the cost as may be reasonable, taking into account payments made or to be made for work executed, of removal of Contractor's Equipment under Sub-Clause 65.7 and, if required by the Contractor, return thereof to the Contractor's main plant yard in his country of registration or to other destination, at no greater cost; and
(f) the reasonable cost of repatriation of all the Contractor's staff and workmen employed on or in connection with the Works at the time of such termination.'

There are various phrases used in the above provisions of sub-clause 65.8 which merit more detailed consideration. For example:

— The words 'expenditure reasonably incurred' in (c) above must mean an item of cost which has already been incurred. Otherwise, the term 'cost' would have been used instead of 'expenditure', 'cost' being defined as 'expenditure incurred or to be incurred'.
— Therefore, the expression 'expenditure reasonably incurred . . . in the expectation of completing the whole of the works' in (c) above, would mean an expenditure already incurred and which is essential for the project as a whole, such as payment for the contractor's insurance policies pursuant to clauses 21 and 23 of the general conditions of contract, or the expenditure incurred in obtaining the performance security required under clause 10 of the general conditions of contract. It would not include an item not already incurred and neither would it include consequential costs.

— The word 'proportion' in the expression 'such proportion of the cost as may be reasonable', which appears in (e) above, could mean anything from 0% to 100%. Thus, in certain circumstances, the word 'proportion' could mean zero if that is what is deemed to be the reasonable amount in the particular circumstances. Accordingly, when the latter meaning is incorporated in the remainder of sub-clause 65.8(c), the reasonable proportion of the cost of removal of the contractor's equipment may be very little if that equipment is simply removed and transported to a location nearer to the site than the 'main plant yard' in his (the contractor's) country of registration.

Figure 14.4 shows a flow chart of the consequences of risks when they eventuate during the contract period.

14.5 Clause 21 of the Red Book – insurance

The responsibility for insurance is based on the responsibilities and liabilities allocated not only to the contractor but also those allocated to the employer under clause 20. The gap which existed in the Third Edition where the contractor was required to insure the works only against those risks of loss or damage for which he was responsible, has been eliminated in the Fourth Edition (see *Construction Insurance* (Reference 14.1 above), pages 198 and 199).

In this regard, specific items are provided in the bill of quantities for such insurance to be priced by the contractor when tendering. Thus, the employer pays for such insurance to safeguard against the possibility of severe losses occurring during the currency of the contract which may render the contractor financially incapable of completing the project. The latter prospect is unacceptable not only to the contractor but also to the employer.

Clause 21 sets out the insurance requirements for which the contractor is responsible in respect of the works and the contractor's equipment. It deals with details of the scope and extent of the insurance cover required in respect of the works and contractor's equipment including the period of insurance, joint insurance, provision for payment in foreign currencies and other insurance features. These aspects are considered below.

14.5.1 Importance of adequacy of cover

The fact that insurance is in force as provided in clause 21 does not alter the obligations and duties of the parties under the contract. Sub-clause 21.1 provides that the works together with materials and plant for incorporation therein should be insured to the full replacement cost plus an additional sum of 15 per cent of such replacement cost, or as may be specified in Part II of the Conditions. This additional percentage is required to cover any extra costs of and incidental to the rectification of loss or damage including professional fees and the cost of demolition and removal of any part of the works and of removal of debris of whatsoever nature.

Sub-clause 21.1 also provides for insurance of the contractor's equipment and other belongings brought onto the site by the contractor, for a sum sufficient to provide for their replacement at the site. The replacement cost should be reflected in the sum insured or the monetary value of the insurance policy to be provided for the relevant elements. From an insurance point of view, the sum insured is usually defined in the Schedule section of the insurance policy in a precise manner so as to establish the basis of claim settlement if a risk covered under the policy eventuates. If there is under-insurance, the value of the claim is reduced in the proportion of the sum indicated in the policy to the sum representing the replacement cost of the damaged element. This insurance principle is referred to as 'the average clause' and is usually incorporated in most contractors' all risks policies.

Besides determining the insurer's liability, the sum insured must be in precise terms because it forms the basis of calculating the insurance premium and of any analysis of the performance of the insurance transaction. The calculation of the replacement cost is based on the contract price plus any subsequent adjustment made through variations, additions, omissions and normal inflation, such as increases in the cost of material, plant, labour and machinery. Such normal inflation referred to as the primary inflation, should be covered by the contractors' all risks policy in addition to two other elements of inflation in the event of loss or damage to any completed or partially completed part of the works. These are:

— The inflation between the time at which such part of the work is originally carried out and the time at which it is repaired or reinstated. Such inflation is referred to here as secondary inflation.
— The inflation which occurs during the delay in executing any uncompleted part of the works after such event of damage. This element is referred to here as the transitional inflation.

Sub-clause 21.3 provides for the situation where there is under-insurance of any item that suffers loss or damage. In such a case, sub-clause 21.3 allocates the responsibility for any amounts not insured or not recovered from the insurers in respect of the cost of the replacement or repair of the lost or damaged item to the party who is responsible for that item pursuant to clause 20.

14.5.2 Period of insurance and extent of cover

The period of insurance and the extent of cover to be provided during the period of construction and the defects liability period are set out in sub-clause 21.2. During the construction period, insurance cover is required from the start of work at the site until the date of issue of the relevant taking-over certificate in respect of the works or any section or part thereof. This can create a gap in the insurance cover between the commencement date and the date of start of work at the site, and particularly so where the contract contains a substantial plant element.

It is difficult to understand why this restriction has been incorporated in the Red Book since it does not affect any saving in premium and only makes a complicated subject

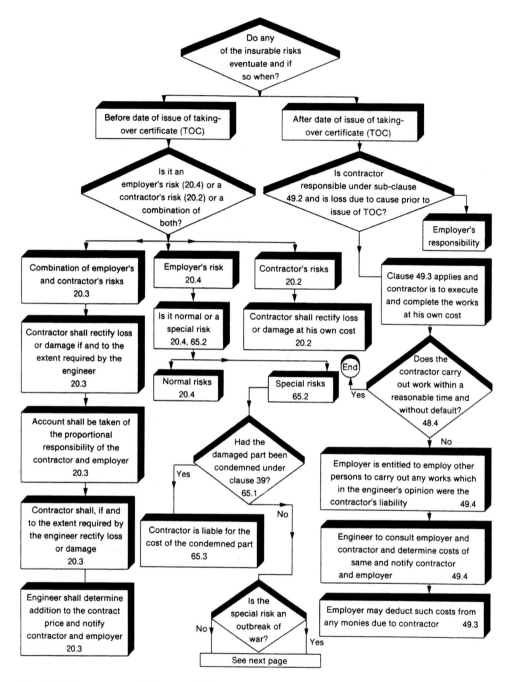

Fig. 14.4 Consequences of risks eventuating.

Risks, Liabilities, Indemnities and Insurances

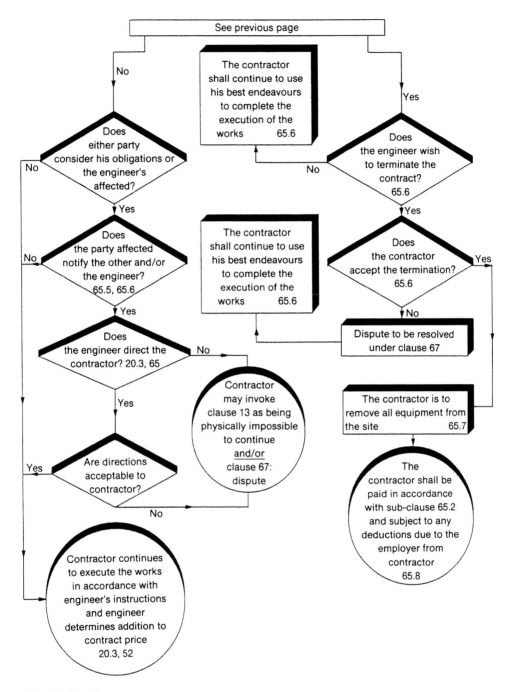

(Fig. 14.4 Contd.)

more complex. It is suggested in the Guide that one way of avoiding this gap might be to alter sub-clause 21.2 through replacing the words 'start of work at the Site' by the defined term 'Commencement Date'. However, a warning is given that such a change 'could lead to complications in arranging insurance' and that 'other wordings might be more appropriate in the context of a particular contract'. If this change is implemented in sub-clause 21.2, then it is suggested in the Guide, and rightly so, that a similar change should be made in sub-clause 25.1, in order to maintain consistency. It is, however, suggested here that no such complication should arise if the required procedures are followed properly and promptly (see Table 13.1).

During the defects liability period, the insurance cover is reduced to any loss or damage due to: first, an event which occurs during the construction period but is not discovered until the defects liability period; and secondly, an event which occurs during the defects liability period which is 'occasioned by the Contractor in the course of any operations carried out by him for the purpose of complying with his obligations under Clauses 49 and 50'. Clause 49 refers to any work that a contractor undertakes to complete pursuant to clause 48.

The division in the period of insurance into two parts is compatible with the period for care of the works as specified in clause 20, that division being based on the date of issue of the taking-over certificate. It is therefore essential that before issuing the taking-over certificate, the engineer should inform the employer of his intention to do so, leaving adequate time for the employer to arrange for whatever insurance cover he wishes to obtain. It is also essential for both the employer and the engineer to appreciate the difference in the scope of the contractors' all risks insurance policy between the cover provided before and after the issue of a taking-over certificate. The cover required and that provided by insurers is reduced in scope once the taking-over certificate is issued. For further details see *Construction Insurance* (Reference 14.1 above), pages 182 to 185.

14.5.3 Joint names

The insurance policy required pursuant to clause 21 must be issued in the joint names of the employer and the contractor in respect of the works, the materials, and plant for incorporation therein, together with the additional sum of 15 per cent stipulated under sub-clause 21.1(b). Sub-clause 21.2 provides that joint insurance is required in respect of the specified insurance covers for the whole period from the start of the work at the site until the end of the defects liability period. However, it should be noted that joint insurance is not required in respect of the contractor's equipment.

To provide joint insurance, the policy or policies must include in the relevant schedule an explicit statement to that effect, and it is not sufficient to name the employer in the policy as a principal or to note his interest in its provisions.

However, in connection with the topic of joint insurance, it is stated on page 73 of the Guide that:

> 'during the Defects Liability Period the insurance is only against that damage which the Contractor is required to repair under the terms of the Defects Liability Clause and so the Employer has no insurable interest in this part of the policy. This section of the insurance could, therefore be in the name of the Contractor alone.'

This statement is incorrect and misleading and is not in accordance with the requirements of sub-clause 21.2. As explained above, the required insurance cover extends beyond the repair of defects into loss or damage arising from a cause occurring prior to the commencement of the defects liability period. It would, therefore, be wise and would cost no more to have the full cover in the joint names of the employer and the contractor. The disadvantages of not having joint insurances are many and are enumerated in *Construction Insurance* (Reference 14.1 above), pages 197 and 198.

14.5.4 Scope of insurance cover

Sub-clause 21.2 also identifies the scope of the insurance cover required. It states that 'the insurance . . . shall cover . . . the Employer and the Contractor against all loss or damage from whatsoever cause arising, other than as provided in sub-clause 21.4 . . .'. In sub-clause 21.4, the permitted exclusions from the insurance cover are set out in the terms of the employer's risks (a) to (d). Thus, a loss or damage from whatsoever cause other than the employer's risks (a) to (d) ought to be covered by the insurance obtained under clause 21.

Regrettably, this means that faulty materials, faulty workmanship and faulty design must be insured if the requirements of sub-clause 21.2 are to be fulfilled. Faulty design is not the subject of a contractors' all risks policy, other than in respect of consequences of such faults. It is generally covered under professional indemnity insurance, but then the sum insured is only a fraction of the value of the construction project.

But, the risk of faulty materials and workmanship is usually uninsurable as such cover would be in conflict with the insurance principle of fortuity. Furthermore, employers should also uphold that principle, since if such a risk is insured, the use of defective materials and workmanship would in effect be encouraged. From an employer's point of view, even if such a risk were insurable, it should not be permitted to be insured.

The use of sub-standard material and workmanship is much less costly than specified material and workmanship. Hence, where contracts are awarded on the basis of competitive tendering, it is only a matter of time before the lowest and perhaps successful tenders are based on the supply of the cheapest possible materials and workmanship and, in many cases, at a standard lower than acceptable. In this connection, some may argue that it is not essential to have both 'if' and 'when' as unknown entities for the principle of fortuity to be satisfied in insurance and that the insurance prerequisites are satisfied if only one of those variables is unknown. Thus, for example, they may argue that if faulty material is used in a structure, fortuity would remain in respect of the time of the impending accident. Examples given from existing insurance practice in support of such arguments include life assurance where it is inevitable for a human to die and the only question is when. However, this comparison is erroneous since the question of benefit must be considered and whereas death does not provide financial gain to the person whose life is insured, the contrary is true in the case of faulty material and workmanship. Lessons should be learnt from the experience of tendering during times of recession. During such periods, competition for contracts is keenest and tenders are often at cost, if not below, with some contractors depending on claims made during the construction period to ensure a break-even situation.

It seems that these problems were recognised at the time of writing the Guide or when it was published.[14.5] In commenting on these two issues, it is suggested in the Guide that as 'insurance is not generally available for the costs of rectification of defects', the cover required 'relates to the costs of the repair of the damage which results from the defect'. It is then added that 'sub-clause 21.2 requires the insurance to be against 'loss or damage' and not for the part of the work which was itself defective'. Unfortunately, this is all wishful thinking because sub-clauses 21.2 and 21.4 will be interpreted only by their own text and not by importing some other wording into their text. Perhaps what was not appreciated by the draftsman is that in most cases when a defective or faulty part fails, it suffers loss or damage as well as causing loss or damage to surrounding parts. Thus, the loss or damage is not limited to the surrounding parts but also extends to the defective or faulty part itself.

It is essential, therefore, for anyone using the Fourth Edition to consider the risks of faulty material, workmanship and design and to decide on how they should be allocated and, if insurance is required, how they ought to be insured. In this connection, the experience of the past few years in using the Fourth Edition shows that prudent contractors have always written to the engineer and to the employer, as soon after the award of the contract as possible, informing them of the lack of availability of this type of insurance cover in the insurance market and requested guidance as to what action was expected of them in the circumstances.

In this connection it is worth noting that the 6th Edition of the ICE Conditions of Contract for Works of Civil Engineering Construction, published in 1991, has maintained the wording of its predecessor, the 5th Edition, in excluding the necessity for insuring against the risk of materials and workmanship incorporated in the works which are not in accordance with the requirements of the contract. This continued use of that exclusion was maintained despite the fact that the 6th Edition had borrowed a number of the provisions from the Fourth Edition of the Red Book.

A further problem exists in sub-clause 21.4 in that there are other risks which are not insurable and these should appear as part of the permitted exclusions, the most important of which are consequential losses and those losses due to wear and tear.

Finally, under the heading of 'Responsibility for Amounts not Recovered', sub-clause 21.3 provides that any amounts not insured or not recovered from insurers are to be borne by the parties in accordance with their responsibilities under clause 20.

14.5.5 *Provisions for payment in foreign currency*

It is suggested in the Guide that where there is provision in the contract for payments to be made to the contractor in foreign currency, it may be appropriate to add the following sentence to sub-clause 21.1:

> 'The insurance in paragraphs (a) and (b) shall provide for compensation to be payable in the types and proportions of currencies required to rectify the loss or damage incurred.'

This is an important provision, especially when the contractor is obliged to insure with a national insurance company which arranges for settlement of any claims in local currency, despite the usual reinsurance arrangements which are made by insurers with international reinsurance companies.

14.5.6 Provision for deductibles

A second useful suggestion is made in the Guide recommending the addition of the following phrase to paragraph (a) of sub-clause 21.1, in connection with deductibles:

'and with deductible limits for the Employer's Risks not exceeding ... (insert amounts)'.

It is not clear why this provision for deductibles is not extended to the contractor's risks or why the provision is not incorporated in the text of clause 21, as for example, was later incorporated in the 6th Edition of the ICE Form of Contract when published in 1991. The relevant item in the ICE 6th Edition is under sub-clause 21(2)(d) which states as follows:

'(d) Any amounts not insured or not recovered from insurers whether as excesses carried under the policy or otherwise shall be borne by the Contractor or the Employer in accordance with their respective responsibilities under Clause 20.'

14.6 Clause 22 of the Red Book – indemnity for damage to persons and property other than the works

Clause 22 sets out the indemnities required by the employer and the contractor towards each other in respect of:

(a) death or injury to any person; or
(b) loss of or damage to any property (other than the works) which may arise out of or in consequence of the execution and completion of the works and the remedying of any defects therein, and against all claims, proceedings, damages, costs, charges and expenses whatsoever in respect thereof or in relation thereto, subject to the exceptions defined in sub-clause 22.2.

These indemnities are unlimited in amount, and this fact should somehow be reflected in the tender evaluation procedure when the contractor is chosen.

In many respects, clause 22 complements clause 20 in that one deals with responsibilities and liabilities in respect of the works, while the other deals with those in respect of indemnities emanating from responsibilities and liabilities for matters other than the works. An immediate distinction between the works and the site becomes apparent and accordingly where there is damage to the site, the provisions of clause 22 and not clause 20 apply. This is important in projects involving the construction of, for example, tunnels or roads where the employer's decision to construct the works on a particular site results in certain risks being attached to that decision. Amongst these is loss or damage which emanates from the permanent use of, or occupation of, land by the works;

that which results from the right of the employer to execute the works; or that which is the unavoidable result of the construction of the works or the remedying of any defects therein in accordance with the contract. These events are listed in paragraphs (a) to (c) amongst the exceptions of sub-clause 22.2 which apply to the indemnity required by the employer from the contractor.

It should be noted, however, that the exception in sub-clause 22.2 relating to the 'unavoidable result of the execution and completion of the works and the remedying of any defects therein in accordance with the contract' is restricted to property and does not extend to personal injury as was the case in the Third Edition. This means that under the Fourth Edition the contractor is required to indemnify the employer in respect of the unavoidable personal injury as a result of the execution and completion of the works and the remedying of any defects therein. The reason for this extension in the indemnity required from the contractor is unclear, especially in view of the difficulty, if not unavailability, of insuring such a risk by the contractor under a public liability policy.[14.6]

The last exception is set out in paragraph (d) of that sub-clause and provides an exception to the indemnity required from the contractor for:

(a) death of or injury to persons or loss of or damage to property in so far as they result from any act or neglect of the employer, his agents, servants, or other contractors, not being employed by the contractor; or
(b) any claims, proceedings, damages, costs, charges and expenses in respect of the events in (a) above or in relation thereto,

with the proviso that where the injury or damage was contributed to by the contractor, his servants or agents, the exception to the indemnity extends only to the extent of the responsibility of the employer, his servants or agents or other contractors. The division of this responsibility is based on what is just and equitable in the circumstances of the case in question, since no generalisation could be made for all sorts of events and circumstances.

Sub-clause 22.3 sets out the requirements for the indemnity to be provided by the employer to the contractor for the exceptions specified in sub-clause 22.2.

14.7 Clause 23 of the Red Book – third party insurance

Clause 23 provides for the insurance requirements to underwrite the indemnities specified in clause 22 in respect of both damage to property and personal injury other than to workmen. Once again, the fact that insurance is effected does not alter the obligations of the parties as defined in clause 22. As in the insurance of the works, joint insurance is also required for third party liability to protect both the employer and the contractor against the liabilities specified in clause 22. While these liabilities are not limited in amount, the required insurance is limited to an amount stated in the Appendix to Tender under sub-clause 23.2. This fixed amount in the Appendix to Tender in respect of third party insurance is a minimum amount per occurrence. The number of occurrences per year should be unlimited to any one period of insurance.

The scope of insurance is also limited in so far as it may exclude the exceptions defined in paragraphs (a), (b) and (c) of sub-clause 22.2, but not paragraph (d). This means that the insurance cover provided under this clause would have to cover any act or negligence of the employer, resulting in death of or injury to persons or loss of or damage to property, other than the works.

Therefore, whilst the contractor is not required to indemnify the employer in case of a negligent act causing damage or injury to a third party, the insurance cover provided in compliance with sub-clause 23.1 should provide an indemnity to the limit specified under sub-clause 23.2.

The period of insurance is indirectly specified by reference to the loss or damage arising out of the performance of the contract.

Sub-clause 23.3 provides that the insurance policy should include a cross liability clause such that the insurance is to apply to the employer and the contractor as if they were separate insureds and accordingly each would be considered as a third party towards the other.

14.8 Clause 24 of the Red Book – injury to workmen and insurance

The liability in respect of accidents or injuries to employees of a contractor or a sub-contractor is dealt with differently in various parts of the world. In many countries, there are specific statutory provisions applying to this type of liability and to the insurance transactions connected with it, making it difficult to formulate a single clause which can apply to all circumstances of a construction contract.

Clause 24 stipulates that the employer is not to be liable for any damages or compensation payable to any workman or other person in the employment of the contractor or any sub-contractor unless as a result of an act or default for which he is responsible. That responsibility includes acts or defaults of the employer's agents or servants. The clause also requires the contractor to indemnify the employer against all such damages and compensation unless they are the responsibility of the employer.

Sub-clause 24.2 provides that the contractor must insure against his liability towards his workmen for the whole time at which any person is employed by him on the works. As the premium for this type of insurance is based on the pay-roll of the employer, sub-clause 24.2 provides that the contractor's obligation to insure would be satisfied if each sub-contractor insures against his own liabilities towards his own workmen in a similar manner to that done by the contractor so that the employer is indemnified under the policy. In this connection, the contractor must require his sub-contractors to present to the employer the policy of insurance and the receipt for the payment of the current premium.

The liability of the employer for death of or injury to workmen of the contractor and his sub-contractors as a result of an act or default of his own and arising out of the performance of the contract is generally covered under the insurance policy arranged in accordance with clause 23.1. However, this cover does not extend to the employer's servants or agents or to other contractors not being employed by the contractor as they are not named as joint insureds.

14.9 Clause 25 of the Red Book – general insurance requirements

There are certain insurance requirements which are common to the three insurance clauses 21, 23 and 24. These are set out in clause 25. Sub-clause 25.1 provides that evidence of compliance with the insurance requirements specified under the contract should be provided by the contractor to the employer prior to the start of work at the site. As discussed in Section 14.5 above, if the insurances are required to be in force from the commencement date, then the wording of this sub-clause should be altered accordingly. In any event, the contractor is required to produce, within 84 days of the commencement date, the insurance policies required under the contract and to deliver them to the employer so that they can be checked and approved by him. The contractor is also required to notify the engineer when he has provided these policies. Furthermore, the contractor is obliged under the provisions of sub-clause 25.2 to produce to the employer, when required to do so, the insurance policies in force and the receipts for payment of current premiums.

Some commentators have expressed doubt about the adequacy of the 84 day period for production of the insurance policies and especially so where complex projects have to be insured with national insurance companies. However, whilst this may have been the experience in the past, it is important to recognise that under sub-clause 25.1, the terms of the insurance policies must be discussed between the employer and the contractor before the letter of acceptance is issued. It is at that time that the important features of these policies are determined, and all that would be necessary after the issue of the letter of acceptance is simply to have these policies processed. These features should include all the critical elements of the insurance cover, such as:

— The sum insured in the case of the contractors' all risks policy and any required indexing to be applied on such sum during the contract period and which would be necessary to calculate the full replacement cost;
— The limit of indemnity in the case of the liability policies;
— The deductibles to be applied in case of claims;
— The general and special exclusions from the insurance cover provided;
— The conditions attached to the policies; and
— Most importantly, perhaps, the mechanism of claim settlement and whether or not there is agreement on the appointment of a specifically appointed loss adjuster by the insurer and the insureds. Where such agreement is not made, a loss adjuster appointed by the insurer should be professionally qualified and should act in a recognised professional manner. If he does not, the whole process would lead to disputes and disrepute.

Sub-clause 25.1 provides that the insurance policies when ultimately issued should be consistent with the general terms agreed prior to the issue of the letter of acceptance and that they should be placed with insurers approved by the employer. Some commentators[14.7] have suggested that these provisions would entail comprehensive discussions and negotiations prior to the award of the contract but after submission of the tenders. Otherwise, it would be difficult, if not impossible, for a contractor to quote a definite price for the provision of these insurances. This may be the case especially if the critical features of the insurance policies are not precisely specified in the tender documents or if they are required to be altered during the pre-award negotiations.

More importantly, it would be a speculative task for the contractor to quote a definite price if insurance is not available for some of the risks required to be covered under the contract, such as the faulty material, workmanship or design cover required under the strict interpretation of the wording of clauses 20 and 21.

Sub-clause 25.2 places an obligation on the contractor to notify the insurers of any change in the nature, extent or programme for the execution of the works. This provision is necessary as it is a usual condition of the contractors' all risks policy that any such change must be notified to the insurer in case there is a material change in the risks covered. The insurance cover may be invalidated if the insurers are not made aware of such changes or, indeed, any other condition imposed under the terms of the policy. The contractor is also required to ensure the adequacy of the insurances at all times in accordance with the terms of the contract. This obligation is a continuing one and the contractor would be wise to clarify the position regarding any risks for which insurance is not available, at the earliest possible time, but in all circumstances not later than the pre-award negotiations.

If the contractor fails to effect and keep in force any of the insurances required under the contract or if he fails to provide the policies within the 84 days specified in sub-clause 25.1, then the employer may effect and keep in force such insurances. This provision is under sub-clause 25.3 which also entitles the employer to pay the premium necessary for effecting and keeping in force such insurances and then deducting the amounts paid from any monies due or to become due or as a debt due from the contractor.

In the event that either the employer or the contractor fails to comply with the conditions of the insurance policy or policies, sub-clause 25.4 provides that each must indemnify the other against all losses or claims arising from such failure.

Figure 14.5 shows a summary of the insurance arrangements as required in the Fourth Edition of the Red Book.

14.10 Part II of the Red Book – insurance arranged by the employer

In certain circumstances, it may be preferable for the employer to arrange the insurances required under the contract. This possibility has been recognised in the Fourth Edition and Part II includes examples for such insurance arrangement.

Should this be the case, it is recommended that this course of action is decided upon prior to the tender stage and in such a case, the employer should specify in the tender documents the extent, type and duration of insurance to be provided by him. The contractor should then be given the opportunity to examine the details of the insurance cover provided by the employer and should be permitted to effect, at his own cost, any additional insurance cover he may require. In this regard, a clause should be inserted in the Conditions to the effect that:

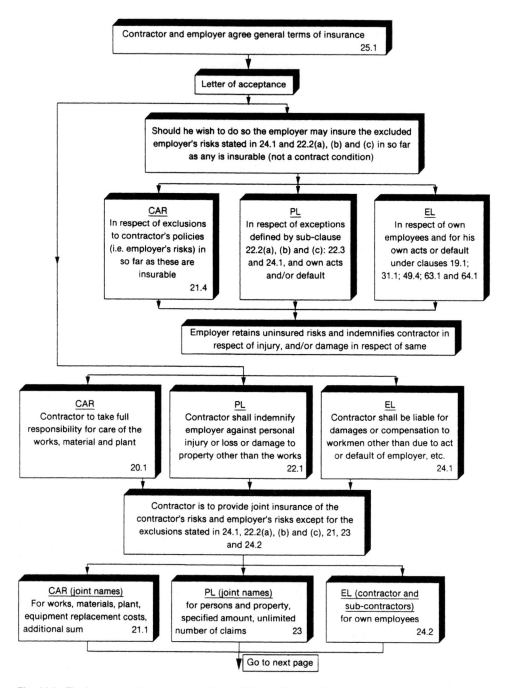

Fig. 14.5 The insurance scheme as in the Fourth Edition of the Red Book.

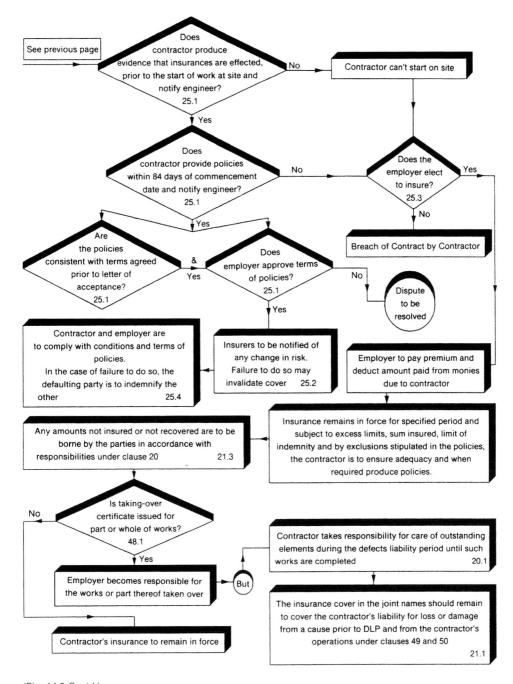

(Fig. 14.5 Contd.)

'The Employer shall, before commencing the Works and whenever required, produce to the Contractor for inspection any policy or policies of insurance required under these Conditions, together with the receipts in respect of premiums paid under such policy or policies.'

In drafting clauses 20 to 25 to replace those in Part I, the specific requirements of the project in question should be considered. Furthermore, consideration should be given to reorganising these clauses in a way that would reflect the logic in the flow from risk to responsibility to liability to indemnity to insurance.[14.8]

14.11   Definitions

The following is a glossary of terms used for the purpose of the present chapter, arranged in alphabetical order irrespective of their source or topic:[14.9]

Civil war

Civil war is defined as large scale and sustained hostilities (as distinct from mere revolt or rising) between the organised armed forces of two or more factions within one state, or between the government and a rebel or insurgent group within the same state.

Commotion

Commotion may be defined as public disorder and physical disturbance.

Hazard

Hazard is defined as a dormant potential for inconvenience, loss, damage, injury or loss of life. It is triggered by a certain incident referred to as a 'triggering incident' resulting in the occurrence of an undesirable event.

Hostilities (whether war declared or not)

Hostilities are acts of enmity and antagonism by persons acting as the agents of sovereign powers, or of such organised and considerable forces as are entitled to the rank of rebels as compared with mobs and rioters. This does not cover the act of a mere private individual acting entirely on his own initiative, however hostile his action may be.

Indemnity

Indemnity is an undertaking to compensate for loss, damage, or expense which a party has suffered in consequence of the act or default of another. Thus, a contract of indemnity is one where a party A undertakes to assume the legal liability to which another party B may be held. This liability may be either towards A or towards another party C.

Insurance

Insurance is defined as the equitable financial contribution of many for the benefit of an individual party which has suffered a loss. This contribution is made through payment of a premium for which an insurance policy is issued. The insurance policy takes the form of a contract of indemnity.

Insurrection

Insurrection is a term generally used to mean an uprising or mutiny against the constituted authority of a state. It is lesser in scope and purpose than that which is designated for revolution or rebellion.

Liability

Liability is the legal concept of one party being subject to the power of another, or to a rule of law requiring something to be done or not done. This requirement to do something or not to do it can be compelled by legal process at the other party's instance. It is sometimes called subjection.

Liability may arise either from voluntary act or by force of some rule of law. Thus, a person who enters into a contract becomes liable to perform what he has undertaken, or to pay for the counterpart performance, or otherwise to implement his part of the bargain. If he acts in breach of contract, he becomes liable by law to pay damages in compensation for that breach. Similarly, if a person acts in breach of any of the general duties imposed by law, such as the exercise of care and diligence in certain circumstances, that person incurs legal liability to make good any omission or default he commits.

Military or usurped power

Without using words of rigorous accuracy, military and usurped power suggests something more in the nature of war and civil war than riot and tumult. It is notable that the words 'military power' do not refer to military power of a government lawfully exercised. The disjunctive 'or' is used to denote contrast between the words 'military' and 'usurped'. Thus, these words should not be read as 'usurped, military power'.

Rebellion

Rebellion is a violent opposition by a substantial group of persons against the lawfully constituted authority in a state, so substantial as to amount to an attempt to overthrow that authority.

Responsibility

A person is said to be responsible for certain events if his conduct has been a material factor in bringing them about. In this sense, responsibility means little more than having caused the events to occur and does not necessarily imply accountability. Accountability for the consequences of an event entails legal responsibility which requires the person to be of such an age and in such a state of mind and body that he is deemed to be capable of controlling his conduct rationally and objectively.

Riot

Riot is a tumultuous disturbance of the peace by three or more persons assembled together without lawful authority, with intent to assist each other, if necessary by force, against anyone who opposes them in the execution of a common unlawful purpose, and who execute or begin to execute that purpose in a violent manner so as to alarm at least one person of reasonable firmness and courage. The usual consequence of riot is destruction of property.

Risk

Risk is a combination of the probability, or frequency, of occurrence of a defined hazard and the magnitude of the consequences of the occurrence.

Strike

Strike is a usual term given for a simultaneous and concerted cessation of work by an employer's employees, or a substantial group of them, normally related to an industrial dispute. It is a breach of contract by each workman, unless due notice of intent to terminate employment be given by each, in which case it is not considered a breach of contract, but is not illegal or criminal unless it involves committing criminal acts. Courts are generally forbidden to require an employee to attend for work by order of specific performance of a contract of employment or by an injunction restraining a breach of such contract.

War

War is a forcible contention between one state or group of states and another state or group of states through the application of armed force and other measures with the purpose of overpowering the other side and securing certain claims or demands. War is recognised and sought to be regulated by international law, even though it does involve breach of peaceful relations between the warring states. Its existence brings into play a great body of legal rules attempting to define parties' rights and duties.

Chapter 15

Performance and other Securities

15.1 Introduction

In the recent past, the need for securities has grown rapidly mainly due to world-wide economic growth in general terms. But intermediate periods of economic recession have served to emphasize the benefits, if not the necessity, for such securities from the point of view of sound business management. Furthermore, this field has emerged from its original, relatively restricted role of protecting the capital at risk into a credit and financial management role within the whole business decision-making process.

However, this development took place in different directions around the world and the terminology for similar concepts differs from one part of the world to another. Forms of performance security of various types are referred to as bonds, guarantees, demand guarantees, conditional guarantees, sureties and standby letters of credit. To a large extent, some of these securities are synonymous and their purpose only differs in detail. Some are issued by banks, others by certain insurance companies, and in some cases by both.[15.1] Except in a few cases, such as customs guarantees, securities rest on latent contractual obligations between two parties.[15.2] The surety or the guarantor acting for one party against a fee will, in the event of non-performance by that party, step in and fulfil a certain specified obligation. The responsibility passes to the surety or guarantor only if the party in question is unable or unwilling to fulfil its own obligation in the underlying contract. In this way, it differs from an insurance policy which is a contract of indemnity. Accordingly, a form of security does not demand the heavy duty of disclosure of all known risks by the guaranteed party as is the case for an insured party in an insurance contract.

Although there are fundamental differences between an insurance policy and a form of security, they both perform a similar function in providing protection against an undesirable event occurring during a particular contract. Reference to securities can be seen in Figure 14.3 in Chapter 14, where securities are marked as the appropriate route to protection against non-performance by either the employer or the contractor of certain aspects of their respective obligations and responsibilities towards the other. Where the employer is concerned, the main obligation covered by a security is the payment obligation, whereas for the contractor it is the execution of the works. Historically, securities were first required in the form of cash deposits and later developed into guarantees issued by a third party. The security for a payment protection developed in the shape of a bank guarantee in the same manner as a 'letter of credit' in an international sales contract, where the guarantee essentially depends on an event taking place or on a date reached.

A letter of credit is effectively a guarantee of an automatic payment by a purchaser in advance of receipt or acceptance of goods supplied to him by a supplier. Should the purchaser find a defect in the goods supplied after the letter of credit has been cashed, the only remedy available to him would be to pursue the matter in litigation. A less extreme case would be where under the letter of credit a certain event is made conditional upon the call for payment but usually such an event could be easily established. In construction, for instance, a letter of credit may be issued by an employer in one country to a manufacturer in another in respect of the supply of certain plant. The obligation to pay through the letter of credit can be activated once delivery is made on the evidence of a presentation of a bill of lading or some other form of receipt.

The security for protection against non-performance by the contractor developed in the form of a performance bond issued by a bank or a surety requiring proof of the contractor's default and of damage suffered. However, this protection was subsequently extended to provide for a type of bond where actual default or loss is irrelevant to the enforcement of the security provisions, thus resembling the employer's type of security described above. The latter is referred to as 'on demand' or 'unconditional' performance guarantees.

In this chapter, the differences between the various forms of security are highlighted, followed by the requirements of the Red Book which are then discussed in relation to the whole process of construction.

15.2 The spectrum of securities

There are three basic forms of securities. In all of these, the guaranteed party is referred to as the principal party; the counterparty in the underlying contract is referred to as the beneficiary; and the guarantor which issues the guarantee, the bond, the surety or the credit.

On one end of that spectrum of securities is the documentary credit form where the guaranteed party, the principal, arranges for a credit to be issued by a bank which would be directly called upon, by the beneficiary, to pay in the event of non-performance by the principal. The beneficiary may resort to the principal pursuant to the underlying contract only if this demand fails, without any fault on the part of the beneficiary.

On the other end of the spectrum is the suretyship guarantee. This guarantee is in the form of a bond where the beneficiary must establish default by the principal before he can demand from the guarantor payment or performance of the obligation which had not been performed by the principal. This latter option of performing the obligation can be exercised by the guarantor only if it is specifically permitted in the form of a bond. If default is established, the guarantor becomes liable for the amount of the loss suffered as a result of the default, within a maximum amount stated in the guarantee, but subject to any defences available to the principal. In this connection, a conditional guarantee is similar to the suretyship guarantee, but when default is established and the guarantee is called, the guarantor covers the beneficiary up to an agreed maximum sum.

The suretyship guarantee is accessory to the underlying contract and the entity giving the security is entitled to invoke against the beneficiary all defences which the principal debtor can raise in accordance with the underlying contract.

Between these two forms of securities lies the demand guarantee. Under this guarantee, performance is first required to be carried out by the principal through the underlying contract. The guarantee is intended to apply only if the principal fails to perform the obligations in the underlying contract, in which case the beneficiary can call on the guarantee. Such call on the guarantee should not be made unless there is default. Accordingly, the demand guarantee is in this respect dependent on the underlying contract. However, it is independent from the underlying contract to the extent that the obligations and liability of the guarantor are solely defined by the terms and conditions mentioned in the guarantee without the necessity of establishing default or the defences which the principal could raise.

There are a number of varieties of this form of guarantee contrasted by dependence on whether or not the beneficiary is required to present any documentation with his demand and also on the type of such documentation. For example, the guarantee may simply be payable on first written demand, or it may be that the demand be accompanied by a statement that a default has occurred, or it may further require that the default be identified. However, although default is an essential prerequisite to the call of the guarantee, this principle is sometimes abused intentionally and the guarantee is called in the knowledge that the principal had not been in default.

In this connection, it is worthy of note that standby letters of credit are a form of demand guarantees. They originated in the United States to circumvent the problem of inability of the banks there to issue guarantees. The different title does not, however, change their nature but the banking industry has tended to apply some characteristics of documentary credits thus creating differences in business practice. The confusion in terminology extends to Europe where in commercial practice both accessory suretyship and demand guarantees are referred to as guarantees. In some jurisdictions, including England, accessory suretyship is referred to sometimes as conditional guarantee and in construction contracts as performance bond, whereas demand guarantees are referred to as unconditional guarantees.

The above range of securities includes many types as the following section outlines.

15.3 Types of securities

Some of the main types of securities used in relation to construction contracts may be identified as follows:

(a) Customs, tax and/or similar bonds: These securities are issued to guarantee the payment of import duties, excise and related taxes in accordance with the legal requirements governing the entry of goods into a country.
(b) Contract bonds: These bonds secure due performance of a contract. The main categories of such bonds are bid, performance, payment and advance payment bonds. In connection with the performance bond, it is sometimes issued for the

'proper' execution and completion of the construction contract involving the proper performance in respect of each and every obligation in accordance with the terms of that contract. This may be interpreted to mean that the beneficiary is covered for latent defects in the works during the period of validity of the bond and, hence, careful consideration should be given to the date of release of the bond and whether it should extend beyond the date of the defects liability period.[15.3]

(c) Maintenance bonds: These bonds provide protection against defective materials and workmanship arising during the defects liability period of the contract.

(d) Bid bonds: These bonds are tendered by the party bidding for a supply or construction contract to guarantee that if awarded the contract, the bidder will in fact enter into the contract and furnish the prescribed performance and/or any other specified bond.

(e) Fidelity bonds: These bonds protect an employer against loss arising from fraudulent or dishonest acts of employees.

(f) Financial bonds: These bonds comprise guarantees for the payment of rent (lease bonds) and judicial bonds, including payments to a court in respect of legal proceedings (court bonds).

(g) Licence, concession or permit bonds: These bonds are usually required before a licence to engage in a particular business is granted by a public or statutory body.

In the next section of this chapter the three most important types of bonds from the point of view of the Red Book are discussed. They are: the payment guarantee, the performance bond, and the demand guarantee.

15.4 Characteristics of performance bonds and guarantees

It is necessary to consider in detail the characteristics of the performance securities relevant to the Red Book before examining the provisions of that document in later sections of this chapter for obtaining and providing such securities.

15.4.1 Payment guarantees

Payment guarantees are issued by a bank which undertakes by the guarantee to make, or to return, a certain payment should the guaranteed party, the principal, fail to do so in accordance with an underlying contract. For example, advance payments are sometimes made by an employer to a contractor for the purchase of certain machinery at the commencement of a contract. Similarly, advance payments are made by an employer to a contractor or by a contractor to a sub-contractor in respect of mobilisation or commissioning of some specific design duties.

These advance payments would have to be returned by deductions from monies that would become due at a later stage or in accordance with a schedule of repayments. Any default in the agreed programme for repayments would have to be safeguarded by the issue of such payment guarantees.

Accordingly, these guarantees are usually issued by a bank on the basis of an unconditional or on-demand call. In some cases, the amount secured by the guarantee is reduced progressively as the work covered by it is executed during the contract.

Another example of this type of security is a payment guarantee given by a bank at the request of an employer in respect of interim payment certificates to a contractor. In this case, the employer would be the principal under the guarantee and the contractor would be the beneficiary.

15.4.2 Performance bonds

As explained earlier, a performance bond may be defined as an undertaking to perform an obligation of a bonded party, called the principal, if and when that principal fails in its performance. The promised obligation can be a simple one or a complex combination of many duties expressly provided in an agreement. The consequences of a default by the bonded party under this type of bond can be either: the proper completion of the contract; or the payment of an amount of money required to complete the contract properly; or the payment of the whole amount of the bond. However, the wording of the bond must be specific in this respect and must also specify the circumstances or the conditions which would trigger these consequences.

Unfortunately, the standard form of bond used in many construction contracts has been the cause of considerable confusion and uncertainty, not only to those using it, but also to the courts whose responsibility it is to interpret the document in the event of litigation. In particular, there has been confusion as to the meaning of the exact nature of the obligation created by such a form in view of the archaic English wording generally used despite repeated complaints. Many construction court cases have resulted in the sureties being able to discharge themselves from liability they had undertaken by the bonds they had issued.

In a relevant case from this point of view, a bondsman in a construction contract sought unsuccessfully to argue that he had entered into an insurance contract. In this case, Lord Atkin referred to the traditional form of bond as follows:

'I entertain no doubt that this was a guarantee, and the rights of the parties should be regulated on that footing. I may be allowed to remark that it is difficult to understand why businessmen persist in entering upon considerable obligations in old-fashioned forms of contract which do not adequately express the true transaction... Why insurance of credits or contracts, if insurance is intended, or guarantees of the same, if guarantees are intended, should not be expressed in appropriate language, passes comprehension. It is certainly not the fault of lawyers.'

He did not offer his opinion on where the fault lay, but presumably he meant that it is the fault of those who had recommended and continued to recommend the use of such standard forms and of those who sign them; essentially both parties have been from within the construction industry.[15.4]

Adverse comments on the wording of the traditional standard form of bond were made on numerous occasions.[15.5] In 1992, in a case in the English Court of Appeal, *Mercers*

v. *New Hampshire*, it was commented that 'The construction of the bond is not assisted by its archaic language'.[15.6]

More recently, two cases are worthy of note and study. The first is the *Perar* case in the English Court of Appeal relating to a design and build contract under the 1981 JCT Form of Contract.[15.7] In the words of an authoritative author, a reasonably intelligent bystander would be likely to conclude that:

> 'the lawyers and judges had all danced happily on the head of a pin; that the person directly or indirectly paying for the premiums on the bond (i.e. the construction owner) had been cleverly and legally persuaded, as so often, into paying for a valueless bond; and that the bondsman had been permitted, as so often, to pocket his premiums and escape scot-free from his obligations without anyone expressing a word of regret and, so far as anybody could tell, with the blessing of the Court of Appeal.'[15.8]

However, due to the peculiar circumstances of the *Perar* case, its application in general is perhaps limited to highlighting the importance of having to secure a properly worded bond incorporating the essential characteristics of such a document. This is particularly relevant to the employer and his advisers who should pay attention to the requirements of a properly worded bond, as outlined below.

The second case was *Trafalgar House Construction (Regions) Limited* v. *General Surety and Guarantee* (1994) where the English Court of Appeal held, on an identical form of bond to that in the *Perar* case, that it imported an on-demand obligation. Attention was focused on the words 'or if, on default of the (principal), the surety shall satisfy the damages sustained by the (beneficiary) thereby up to the amount of the bond'. This wording, amongst other factors, led to the above decision which was later overruled in the House of Lords.[15.9] However, before it was overruled, it sent reverberating shocks throughout the construction industry, domestically and elsewhere, wherever similar traditional forms of bond had been used. In some instances, such a traditional form had been used for a very long time. Various organisations abandoned the standard form and commissioned legal advisers to rewrite the document. It is expected that some new forms will now emerge based on the experience gained from these and other previous cases, encompassing some, if not all, of the following suggested requirements for bonds and be adopted by the construction industry:

(a) The bond should remain in force until the obligations of the principal party, in whose name the bond is issued, are completely fulfilled. There should be no cancellation clause and no termination date within the performance period. In this regard, it is important to check the wording of the proposed form of bond to ascertain that it is worded in clear, understandable language and that it does not include any restrictions, sometimes fostered and promoted by the guarantor, which might permit its cancellation or invalidate its effect.

(b) In construction and particularly in international contracts, the traditional archaic bond wording should be abandoned, and instead there should be a simple undertaking in a clear wording to perform properly all the contractor's obligations under the relevant contract.

(c) The document should be legally sound under the provisions of the applicable law of the contract.
(d) There should be no reference to, or provision for, or release of this undertaking at any particular time during the validity period of the performance bond. The validity period should extend either to the date of the final certificate if the proper or due performance is required against a breach during the contract period or to the end of the limitation period, as specified under the applicable law of the contract, if proper or due performance is required to extend to latent defects.
(e) There should be no provision requiring notice to be given to the bondsman within a set period of time, whether of the contractor's default or of the employer's intention to claim, as this would impose an almost impossible burden on both the employer and the contractor to fulfil properly. An example of such an onerous provision is given hereunder:

> 'This bond is executed . . . upon the following express conditions, which shall be the conditions precedent to the right of the employers to recover hereunder. The surety shall be notified in writing of any non-performance or non-observance on the part of the Contractor of any of the stipulations or conditions contained in the contract and on their part be performed and observed, which may involve a loss for which the surety is responsible within one month after such non-performance or non-observance it shall have come to the knowledge of the Employer or his representatives having supervision of the contract . . . and the Employer shall insofar as it may be lawful permit the surety to perform the stipulations . . . of the said contract which the Contractor shall have failed to perform . . .'.

It is impossible to visualise how the draftsman of that provision could have been referring to a construction contract when formulating its text. The usual daily events on such a contract would have to result in continuous notifications to the surety. It is also difficult, if not impossible, to visualise the surety's reaction to such notifications.
(f) There should be no other condition or restriction imposed on enforcement (e.g., exhaustion of prior remedies against the contractor, or the obtaining of an award or judgment).
(g) There should be no reference to interest or legal costs in the context of any limit stipulated in the bond. The bond limit should ideally be expressed to be exclusive of interest and legal costs, whether due from the contractor or the bondsman.
(h) There should be express provision that the bond cannot be voided by reason of:
   (i) any alteration of the contract made between the employer and the contractor;
   (ii) any alteration in the nature or extent of the work to be carried out;
   (iii) any allowance of time to the contractor;
   (iv) any indulgence, or additional or advance payment, forbearance payment or concession to the contractor;
   (v) any compromise of any dispute with the contractor; and
   (vi) any failure of supervision or failure to detect or prevent any fault of the contractor.

It is suggested that those who seek and insist upon a bond in the wording they require will find that, in most circumstances, there will be no difficulty in obtaining one in the above terms.[15.10]

(i) In contrast with the above positive requirements, it is usual and perhaps necessary to incorporate the negative condition that the performance bond would be rendered unenforceable by a beneficiary who is himself in breach of contract. Furthermore, it may be necessary for the principal to insist that the beneficiary must not assign his rights under the bond.

Finally, it is notable that careful consideration should also be given to whether or not to accept the condition, generally imposed by standard forms of bond, that upon default by the principal, the guarantor may perform the principal's obligations under the contract and complete the works himself. The concern relating to the inclusion of such condition is due to the fear that the work will then be done at minimum cost and with inferior quality and the consequential difficulties in exercising control of the standards required under the contract.

15.4.3 Demand guarantees

As stated earlier, a demand guarantee is issued by a guarantor in such terms that should the guaranteed party, the principal, default in the performance of certain obligations, the guarantor undertakes to pay a certain amount to the beneficiary. Where the guarantor is a bank, it is referred to as the issuing bank, and if the beneficiary insists on the issuing bank being based in his own country and the guaranteed party is unwilling or unable to comply, then the demand guarantee is issued by a chain of banks.

In the first alternative, the issuing bank requires and obtains what is referred to as a counter indemnity from its customer, the principal or the guaranteed party, in the form of a debit to its account for all claims received and paid. Accordingly, as soon as the guarantee is called by the beneficiary and a payment is made, the system will simultaneously produce a debit in the principal's bank account.

In the second alternative where there is a chain of banks, the issuing bank requires and obtains a counter indemnity from either the instructing correspondent bank or directly from the principal, depending on the particular circumstances. The demand guarantee and the counter indemnity are usually worded in a careful manner so that one does not expire before the other.[15.11] Figure 15.1 shows these two alternatives. The accumulating costs involved in issuing the guarantee and the counter indemnities are payable by the principal.

To create the obligation to pay the guaranteed sum, the beneficiary needs only make the demand for payment from the issuing bank. In most, if not all cases, the only defence available to a bank is fraud by the beneficiary which must be known to the issuing bank at the time of presentation of the demand.

The unconditional nature of these securities has originated from other commercial contracts where the principle of an irrevocable letter of credit is used. The obligation to pay through the letter of credit can be activated once delivery is effected on the evidence of a presentation of a bill of lading or some other form of receipt.

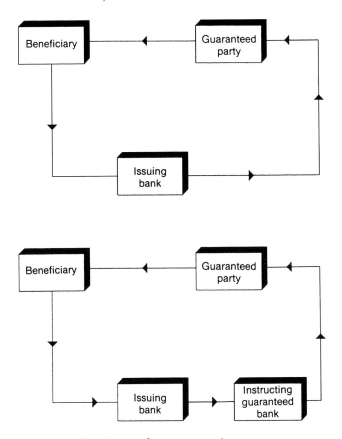

Fig. 15.1 The two alternatives of issuing a performance guarantee.

In a similar manner, a performance guarantee is usually unconditional but in the less extreme case, a condition can be inserted in the form of a necessary notification of a breach committed by the guaranteed party.

In construction, perhaps the most serious problem in this type of performance guarantee, whether unconditional or with simple and easily ascertainable conditions, is that it is extremely expensive and bad value for the money expended from the employer's perspective; and from a contractor's point of view, it could be ruinous if unjustifiably called.

In the well-known English case of *Edward Owen Ltd* v. *Barclays* between an engineering company and its instructing bank, the problems of such securities were highlighted and discussed.[15.12] Edward Owen Engineering, an English company, was awarded a contract which was subject to Libyan laws. Payment was to be by letter of credit issued by a Libyan bank on behalf of the Libyan buyers. The engineering company was to provide a performance guarantee in favour of the Libyan buyers. The bond was duly opened by Barclays Bank which gave the following undertaking as a counter indemnity to the Libyan bank: 'We confirm our guarantee payable on demand without proof or conditions'. The Libyan bank issued their guarantee to the Libyan beneficiary. The

Libyan bank, however, failed to issue the letter of credit in favour of Owen Engineering and the Libyan buyers claimed on the guarantee. The Libyan bank claimed from Barclays, the instructing bank, but an injunction was sought by Owen Engineering, to prevent them from making a payment. In a lengthy judgment the following comments summed up the judges' decision which went against the English company:

'So, as one takes instance after instance, these performance guarantees are virtually promissory notes payable on demand. So long as the Libyan customers make an honest demand, the banks are bound to pay: and the banks will rarely, if ever, be in a position to know whether the demand is honest or not. At any rate they will not be able to prove it to be dishonest. So they will have to pay.'

and

'The position of a Bank which has given a bond payable on demand is similar to that of a Bank which has opened a confirmed Irrevocable Letter of Credit. Such obligations are the life blood of International Commerce and it is only in exceptional cases, such as fraud known to the Bank, that the Courts will interfere.'

The Court further stated that the:

'Performance Bonds must be honoured to the letter as between Banks and that questions between the buyers and sellers must be dealt with between themselves – in this case presumably by Libyan Law.'

The judge enumerated the instances in which such a guarantee might be called and referred to the extreme case where mere allegations are made without any proof as one bearing the appearance of a discount. He stated:

'It is obvious that that course of action can be followed, not only when there are substantial breaches of contract, but also when the breaches are insubstantial or trivial, in which case they bear the colour of a penalty rather than liquidated damages: or even when the breaches are merely allegations by the customer without any proof at all: or even when the breaches are non-existent. The performance guarantee then bears the colour of a discount on the price of 10 per cent, or 5 per cent, or as the case may be. The customer can always enforce payment by making a claim on the guarantee and it will then be passed down the line to the English supplier. This possibility is so real that the English supplier, if he is wise, will have taken it into account when quoting his price for the contract.'

In a further case around the same time, *R. D. Harbottle (Mercantile) Ltd* v. *National Westminster Bank Ltd*, the English National Westminster Bank was taken to court in similar circumstances by R. D. Harbottle who claimed that a demand under a guarantee issued by National Westminster Bank on their behalf was unjustified or even fraudulent as the goods had been supplied by them as per the contract.[15.13]

Once again the court ruled in favour of the Bank stating that:

'Performance guarantees in such unqualified terms seemed astonishing but apparently they were not unusual particularly with customers in the Middle East. In effect, the sellers rely on the probity and reputation of their buyers and on their good relations with them. But this trust is inevitably sometimes abused, and I understand that such guarantees are sometimes drawn upon, partly or wholly, without any or any apparent justification, almost as though they represented a discount in favour of the buyers. In such cases the sellers are then left merely with claims for breaches of contract against their buyers and the difficulty of establishing and enforcing such claims.'

The judgment further stated that the:

'Courts were not concerned with enforcing claims and counterclaims between buyers and sellers – these were the risks merchants took. Harbottle had taken the risk on the unconditional wording of the guarantee. The commitments of a Bank were on a different level and must be allowed to be honoured free from interference by the courts, otherwise trust in International Commerce could be irreparably damaged.'

FIDIC does not advocate the use of on-demand guarantees. This is stated in the Guide (referred to in Reference 9.1) on page 58 and the reason given is that such guarantees can be called without justification, and their use is likely to increase the tender sum to reflect this risk. Authoritative writers have also condemned this type of securities. In *Construction Contracts, Principles and Policies in Tort and Contract*, Volume 2 (Reference 15.8 above), the following was said of these securities: 'The use of this type of bond is unwise, at least without a full understanding of the consequences'. Later, on page 314 of that book, the following comment was made:

'(a) Owners who have no intention of calling such a bond irresponsibly may well find that tendering contractors, influenced by adverse experience of the unwarranted calling of bonds elsewhere, may increase their prices excessively to cover the contingency. Indeed the mere requirement of such a bond inevitably suggests an unwillingness by the owner to accept, in a potential future dispute, the decisions of an independent tribunal, so that the psychological effect on pricing may be much more substantial than is realised.

'(b) In any event, the direct cost of obtaining such bonds (which must inevitably enter the contract price) is very substantial, and is increased still further by the fact that two, and sometimes three or more banks or bondsmen can involve themselves simultaneously in one project. Thus a frequently met arrangement is for the owner/buyer to require the bond or guarantee to be furnished to him by a bank in his own country (which is likely to be a docile payer on demand and may even be closely linked to the buyer) with an unconditional counter-indemnity provided to that bank by a trading bank in the seller's/contractor's country, with quite possibly a further counter-indemnity to that bank from the contractor's own domestic bank. All these banks will charge for these guarantees or indemnities (notwithstanding that only the contractor's domestic, or other last bank in the line, will be effectively at risk, since the remainder are only guaranteeing the solvency of other banks). Not only, however, must the contractor price directly for all these banks' direct charges, but since the string of indemnities will result in prompt debiting of the contractor's own bank account, should the bond be called by the owner, the contractor will need to price in addition for the contingent effect on his own finances of an irresponsible or unjustified call.'

An analysis carried out on 40 construction arbitration awards rendered under the Rules of Conciliation and Arbitration of the International Chamber of Commerce in Paris between 1988 and 1990 showed that from the legal point of view, conflict between the parties arose in matters relating to bank guarantees and performance bonds in 25 per cent of the cases. The comparable figure in relation to interpretation and qualification of the contract was 25 per cent; 17.5 per cent in relation to repudiation or termination of the contract; 12.5 per cent in respect of retention of monies due and 20 per cent in respect of other matters such as formation of the contract, the arbitrator's power to adopt the agreement, warranties, novation or monetary conversion. This is a very high percentage indeed for guarantees and bonds.[15.14]

15.5 The ICC Uniform Rules for Demand Guarantees

The abuse of the principles of demand guarantees, discussed above, by unfair calling led in 1992 to the publication of Uniform Rules for Demand Guarantees, URDG, by the International Chamber of Commerce in Paris.[15.15] These Rules sought to introduce uniformity of application and governing principles relating to demand guarantees. An important guide to these Rules was also published in the same year providing an important study of the operational and judicial issues on demand guarantees.[15.16]

The URDG Rules comprise 28 articles and provide in Article 2(a) a definition of a demand guarantee in the following terms:

'... any guarantee, bond or other payment undertaking, however named or described, by a bank, insurance company or other body or person (hereinafter called "the Guarantor") given in writing for the payment of money on presentation in conformity with the terms of the undertaking of a written demand for payment and such other document(s) (for example, a certificate by an architect or engineer, a judgment or an arbitral award) as may be specified in the Guarantee, such undertaking being given
  (i) at the request or on the instructions and under the liability of a party (hereinafter called "the Principal"); or
  (ii) at the request or on the instructions and under the liability of a bank, insurance company or any other body or person (hereinafter "the Instructing Party") acting on the instructions of a Principal to another party (hereinafter the "Beneficiary")'.

The main features of these Rules are as follows:

(a) Demand guarantees are invoked only if the principal has defaulted;
(b) The Rules do not apply to suretyship or conditional bonds or guarantees or other accessory undertakings;
(c) The principal can expect on the grounds of equity and good faith to be informed in writing of any claim made in which it is alleged that he is in breach of his obligations in the underlying contract and also in what respect such a claim is made; and
(d) The demand guarantees should not contain any condition other than the presentation of a written demand and other specified documents.

As explained in the introduction to these Rules, the specified documents required to be presented in demand guarantees vary widely. At one end is the guarantee which is payable on simple written demand, without a statement of default or other documentary requirements. At the other end is the guarantee which requires presentation of a judgment or arbitral award. Between these two extremes lie various intermediate forms of guarantee, such as guarantees requiring a statement of default by the beneficiary, with or without an indication of the nature of the default, or the presentation of a certificate by an engineer or another professional. All these fall within the scope of the Rules.

The documents presented must conform to the requirements of the guarantee, and when non-conformity is apparent, the beneficiary is not entitled to payment. Of course, these Rules do not affect principles or rules of national law concerning the fraudulent or manifest abuse or unfair calling of guarantees.

15.6 Uniform Rules for Contract Bonds

After the successful launch of the Uniform Rules for Demand Guarantees, the International Chamber of Commerce in Paris published its Uniform Rules for Contract Bonds, URCB, in 1993.[15.17] As explained in the introductory section of these Rules, they have been drawn up by an ICC working party of members representing the ICC Commission on insurance and the building and engineering industry for world-wide application. They came into effect on 1 January 1994. Essentially, they come from the insurance industry and are intended to apply to suretyship guarantees.[15.18]

The URCB relate to contract bonds which create obligations of an accessory nature, where the liability of the surety or guarantor arises and is conditional upon an established default by the contractor of an obligation set out in an underlying contract. The contractor is referred to as the principal. The Rules are intended therefore to apply where the intention of the parties is that the obligations of the guarantor will depend upon the duties or liabilities of the principal under the relevant contract. Furthermore, these Rules are intended to operate so as to confer upon the beneficiary in each case security for the performance or execution of the contract obligations or payment which may fall due to the beneficiary as a result of any breach of obligation or default by the principal under the underlying contract. Accordingly, subject to its financial limits, either the obligations set out in the underlying contract will be performed or executed, or upon default, the beneficiary will recover any sum properly due, notwithstanding the insolvency of the principal or the principal's failure for any other reason to satisfy or discharge his liability.

The relationship between the parties under a bond governed by the URCB differs from that arising under the URDG. In the URCB, the beneficiary obtains security for the obligations of the principal arising pursuant to the underlying contract but the guarantor's liability would only arise in case of an established default under that contract. Thus, in the event of a dispute arising as to the liability of a guarantor, the URCB contemplate that such dispute will be determined by reference to the underlying contract. The guarantor and the principal are protected in that liability will arise only when default is established. The beneficiary is protected by the assurance that any judgment or award will be discharged by the guarantor if the principal fails to do so. On the other hand the guarantee under the URDG should not stipulate any condition for payment other than the presentation of a written demand and other specified documents without requiring the guarantor to decide whether the beneficiary and principal have or have not fulfilled their obligations under the underlying contract.

Of course, both sets of Rules apply only where expressly incorporated by the parties in their underlying contract. By adopting these Rules, it is expected that some uniformity of practice in the operation and enforcement of bonds would be established in the face of the difficulties created by first, unconditional bonds and secondly, by the uncertainty in defining what is and what is not conditional.

The first of these two problems can be highlighted by reference to the case of *Edward Owen Engineering* v. *Barclays Bank International* which has been extensively discussed in the legal texts.[15.19] The second problem can be highlighted by reference to the two recent cases in the last few years mentioned above. The first is the *Trafalgar House* case, and the second is the *Perar* case which revolved around a design and build contract.[15.20]

New forms of bonds were hastily prepared by various organisations in the face of the criticism and the conflicting judgments in these cases. It is hoped that if both sets of Rules are extensively applied, such confusion would not arise.

The URCB contain eight articles, under the following headings: scope and application; definitions; form of bond and liability of the guarantor to the beneficiary; release and discharge of guarantor; return of the bond; amendments and variations to and of the contract and the bond and extensions of time; submission of claims and claims procedure; and jurisdiction and settlement of disputes.

Article 2 of these Rules defines the various forms of bond which fall within the application of these Rules. They are: advance payment bond; maintenance bond; performance bond; retention bond; and tender bond.

Article 3 of these Rules sets out first, the information which must be stipulated in a bond issued for a particular purpose under these Rules and secondly, the principles of liability of the guarantor to the beneficiary. Sub-paragraphs (ix), (xii) and (xiii) of Article 3 pose three important questions which must be carefully considered and answered for each specific bond. These are as follows:

'ix. Whether the Guarantor shall be entitled at its option to perform or execute the Contract or any Contractual Obligation.

xii. Whether sub-paragraph (i) of Article 7(j) is to apply and the name of the third party to be nominated thereunder for the purpose of Article 7 below (claims procedure).

xiii. How disputes or differences between the Beneficiary, the Principal and the Guarantor in relation to the Bond are to be settled.'

In answer to the first of these questions, it is submitted that it should be answered in the negative. In this connection, reference should be made to the arguments presented in Section 15.4.2 above.

The second question deals with the most interesting aspect of these Rules, Article 7(j), which, if answered in the affirmative, requires the appointment of a referee to settle the question of whether or not there is default by the contractor of any of his obligations under the underlying contract. Article 7(j) provides as follows:

'Notwithstanding any dispute or difference between the Principal and the Beneficiary in relation to the performance of the Contract or any Contractual Obligation, a Default shall be deemed to be established for the purposes of these Rules:

i. upon issue of a certificate of Default by a third party (who may without limitation be an independent architect or engineer or a Pre-Arbitral referee of the ICC) if the Bond so provides and the service of such certificate or a certified copy thereof upon the Guarantor; or

ii. if the Bond does not provide for the issue of a certificate by a third party, upon the issue of a certificate of Default by the Guarantor; or

iii. by the final judgment, order or award of a court or tribunal of competent jurisdiction, and the issue of a certificate of Default under paragraph (i) or (ii) shall not restrict the rights of the parties to seek or require the determination of any dispute or difference arising under the Contract or the Bond or the review of any certificate of Default or payment made pursuant thereto by a court or tribunal of competent jurisdiction.'

The appropriate answer to the question under sub-paragraph xii of Article 3 should therefore be in the affirmative and the name of the third party or an appointing authority should be inserted in the bond.

As to the question under sub-paragraph xiii of Article 3, it is submitted that any dispute or difference between the beneficiary, the principal and the guarantor in relation to the bond should be settled in the same manner as stipulated in the underlying contract. The reason for this proposal is to permit uniformity in the area of dispute resolution between the bond and the underlying contract, thus reducing the possibility of conflicting decisions being given under the same circumstances.

There is one other aspect of the URCB worthy of note which should be carefully considered once a surety guarantee is chosen as the appropriate method of obtaining security. This concerns Article 6(b) which relates to a tender bond and contains the words 'substantial or material variation of or amendment to the original tender'. This is an ambiguous wording which should be clarified for the sake of avoiding any doubt. If it is intended to be applied to a form of bond other than a tender bond, it should be clarified to mean a variation which transforms the project into one which is either beyond the financial capacity or the expertise of the contractor.

15.7 Insurance against unfair calling

In certain countries, it is possible to arrange some insurance cover against unfair calling of performance securities with some insurance companies for particular customers, but such insurance is difficult to obtain and expensive. The cost of such insurance varies from approximately 0.5 per cent upwards per annum of the value of the guarantee or bond.

In addition, certain governmental agencies in some countries provide a cover against unfair calling for their nationals who export goods and services. In the UK, the Export Credits Guarantee Department, ECGD, provides a bond support scheme which insures banks who issue performance bonds. Under this scheme, banks may claim from the ECGD which reserves the right to claim from the customer/supplier. If, however, the supplier has suffered an unfair call, then the ECGD will refund the supplier 100 per cent of the loss.

However, the eligibility for bond support is subject to sales contracts being a minimum of £250,000 and the terms must be cash or near cash terms.

A similar scheme exists in a number of other countries, for instance in France and in Germany where the relevant organisations are referred to as *Coface* and *Hermes* respectively.

15.8 Performance securities under the Red Book

Having discussed the principles and characteristics of the various forms of securities, it is appropriate to consider in the remaining part of this chapter how these relate to the Red Book. Under the Fourth Edition of the Red Book, the employer in his contract with the contractor is mostly concerned with the following non-performance risks:

(a) Improper execution of the works by the contractor involving material, plant, workmanship, or design leading to defective work or work not in accordance with the contract which is discovered during the contract period and ending with the issue of the final certificate;
(b) Improper execution of the works by the contractor involving material, plant, workmanship, or design leading to defective work or work not in accordance with the contract which is discovered after the issue of the final certificate;
(c) Delayed completion of the works beyond the stipulated date stated in the contract; and
(d) Failure to complete the works as a result of the contractor's inability to continue with the performance of his contractual obligations.

The above risks are unrelated and should be analysed and assessed separately. Should the risk described in (a) above eventuate, the contract includes various detailed provisions to deal with the situation. In particular, some of the terms of clauses 7, 8, 13, 36 to 39, 49, 50, 54, and 59 to 64, come into operation in such eventuality providing the employer with a number of remedies. Such remedies include interim payments due to the contractor; retention monies withheld under sub-clause 60.2 until such time as they become due under sub-clause 60.3; amounts which may be due under the final certificate; and the contractor's equipment which is required under the provisions of sub-clause 54.1 to be kept on site and not removed without the consent of the engineer.

If the risks under item (c) above eventuate, some of the terms of clauses 6, 14, 41 to 48 and 63 come into operation. In sub-clause 47.1, in particular, the employer has a certain amount of protection in the form of liquidated damages.

Although the percentage of retention and the amount of liquidated damages are limited in extent as may be stated in the Appendix to Tender, under sub-clauses 60.2 and 47.1 respectively, they can form a substantial security against the risks described in items (a) and (c).

However, if the employer is either dissatisfied with the extent of the above securities or not in a position to accept the risks in items (b) and (d) above without any security, then clause 10 of the Red Book enables him to require the contractor to provide the financial security or securities he chooses. Clause 10 is drafted in such a way that flexibility is provided for the employer to indicate the type of security required by appending its form to the Conditions.

Two example forms of performance security are provided on pages 7 to 9 of Part II of the Red Book. The first is an example of performance guarantee and the second is an example of surety bond for performance. However, neither of these two forms provides the security against the risk described under item (b), if required. The two forms are dealt with later in this chapter, in Section 15.9. The employer must therefore study his needs carefully and select the form of security which is most suitable to his needs. He must also ensure that the form selected is compatible with the applicable law of the contract; for example, it may be required under certain jurisdictions that a form of security must be executed under seal. The various forms available in the guarantee and surety market and their suitability to a construction contract are discussed later in Section 15.10.

When dealing with clause 10 of the Red Book which contains the provisions for obtaining the performance securities, two points should be noted. First, the clause applies only if the contract requires the contractor to obtain such security, in which case the form of security to be provided should be appended to the Conditions. Secondly, unlike previous editions of the Red Book, clause 10 does not refer specifically to the provision of a bond; instead, it refers to the provision of a performance security, thus allowing the employer to specify in Part II of the Conditions different types of security depending on the specific needs of the particular contract in question. Clause 10 does, however, require that if a performance security is required, it should be for the 'proper' performance of the contract.

In this connection, 'proper' must mean that it is not sufficient for the contractor to perform the contract, but that he must also perform it properly in accordance with its terms and provisions. This requirement for the performance security to be for the 'proper' performance of the contract is, however, restricted under sub-clause 10.2, in terms of period, to the date of issue of the defects liability certificate. The performance security should then be returned to the contractor within 14 days of the issue of that certificate. This condition is imposed on the employer to avoid the possibility of an extension of the period of validity of the performance security being imposed by the application of a rule of law in certain jurisdictions notwithstanding that the security has a termination date. For instance, in Turkey and Syria, the period of validity can be extended for up to 10 years and in India for up to 60 years.[15.21] Therefore, under sub-clause 10.2 there is a requirement for security for the risk described under item (a), while this sub-clause provides no such security for the risk of latent defects as described under item (b). Should this be unacceptable to the employer, the provisions of sub-clause 10.2 would have to be altered such that the performance security would be valid until the end of the limitation period, as stipulated under the applicable law of the contract.

The restriction in the period of validity of the performance security marks an important change from previous editions of the Red Book which did not address the question of the period of validity, thus implying unlimited time. Of course, allowing unlimited time is unacceptable to the surety, since liabilities will continue to accrue in respect of a multitude of construction contracts with vast sums of money involved. The change in the Fourth Edition has thus swung too far in the opposite direction.

Under the provisions of sub-clause 10.1, the contractor is also required to obtain and provide to the employer such performance security in the amount stated in the Appendix to Tender within 28 days after receipt by him of the letter of acceptance. Having obtained and provided the security to the employer, the contractor is then required to notify the engineer accordingly. It is interesting to note that the amount of the security is usually expressed as a percentage of the contract price which can vary from around ten per cent to much higher percentages depending on the practice adopted around the world.

Although the form of the performance security provided by the contractor should be as annexed to the Conditions, sub-clause 10.1 does refer to 'such other form as may be agreed between the Employer and the Contractor'. Presumably, such agreement would have to be reached during the period leading to the issue of the letter of acceptance, as any later discussions may involve addition to or reduction from the contract price depending on what is agreed between the employer and the contractor.

The institution providing the performance security is subject to the approval of the employer and the costs of complying with the requirements of clause 10 are to be borne by the contractor unless the contract provides otherwise. The third part of clause 10, sub-clause 10.3, provides that prior to making a claim under the performance security, the employer must, in every case, notify the contractor of the nature of the default in respect of which the claim would be made. However, this sub-clause stops short of stating what happens after such notification. This is unfortunate, since such notification can then be followed by the call upon the surety to execute the terms of the performance security depending on whether these terms specify the payment of a sum of money in the case of a performance guarantee or the completion of the contract in the case of a performance bond.

The Guide (see Reference 9.1) does, however, include the following statement in an attempt to explain this omission:

'During the drafting, the possibility of including a period of notice was examined but it was considered that such a provision might impede the use of the conditions. Normal practice is that when the Contractor is notified of any default, he is given an opportunity to remedy it prior to a claim under the security being made.'

However, it is not explained why explicit provisions as to the mechanics of reaching a claim situation were not incorporated in sub-clause 10.3 or even how such provisions would impede the use of the Conditions. If the only obligation that the employer has before making a claim is an obligation to notify the contractor of the default, then it is very easy in the case of a performance guarantee to make an unfair call upon such security. In construction, a search for a default in the form of a departure from the provisions of the contract is not very difficult. For example, a delay of one day in completing the project beyond the specified time for completion plus any extension granted under clause 44 is a default, the use of a substandard batch of concrete is also a default. Numerous other examples of such default can be cited.

By contrast, the Third Edition of the *FIDIC Form for Electrical and Mechanical Conditions of Contract* (the Yellow Book; see Reference 9.5), specified in detail the situations where the employer is entitled to make a claim under the performance security. These are:

(a) Where a technical or other breach of the contract is being asserted, the employer must give notice to the contractor stating his intention to claim under the security provided, and the amount claimed and the breach of contract upon which he is relying. If the contractor fails to remedy the breach within 42 days after receipt of that notice, the employer may make a claim. The obvious purpose of the 42 day period is to give the contractor a reasonable final opportunity to remedy the breach before a claim can be made upon the security. If the contractor's breach is incapable of being remedied, the employer will still have to wait for 42 days, but this is better than the risk of an unfair calling of the security. Of course, it is important to note that under many jurisdictions, the 42 day period would give the contractor an opportunity to seek an injunction from the courts to restrain the employer from making a call upon the security, should he decide to challenge the employer's assertion that a breach of contract was committed. However, if an injunction is granted, then it would undoubtedly be on the basis that the contractor would have to indemnify the employer in damages should it be subsequently found that the employer was entitled to make the assertion of breach.

(b) Where the employer has obtained an award in arbitration and the amount awarded has not been paid within 42 days after the date of the award. Of course this provision must interact with the arbitration clause and the enforcement of or challenge to an award under such a clause as well as with the arbitration law under the applicable law of the contract. Under most jurisdictions, the parties

are entitled to challenge or appeal an arbitrator's award, but under very restricted conditions.
(c) Where, rather than resort to arbitration, the employer and the contractor have agreed that a certain sum is due from one party to the other, but that the sum has not been paid.
(d) Where the employer calls upon the performance security after the contractor had gone into liquidation or had become bankrupt. Such a call may be made immediately after the above event and a copy of the claim made must be sent to the contractor.

Clause 10 of the Red Book is supplemented by two suggestions in Part II of the Conditions.

The first is for the situation where there is a provision in the contract for payments to be made in foreign currency. In that case, it is suggested that sub-clause 10.1 may be varied by inserting after the first sentence the following wording:

> 'The security shall be denominated in the types and proportions of currencies stated in the Appendix to Tender.'

Secondly, it is suggested that where the source of the performance security is to be restricted by the employer, an additional sub-clause may be added and two examples are given in Part II of the Conditions under the heading 'Source of Performance Security', as follows:

> '10.4 The performance security, submitted by the Contractor in accordance with Sub-Clause 10.1, shall be furnished by an institution registered in (insert the country where the Works are to be executed) or licensed to do business in such country.
> or
> 10.4 Where the performance security is in the form of a bank guarantee, it shall be issued by:
> (a) a bank located in the country of the Employer, or
> (b) a foreign bank through a correspondent bank located in the country of the Employer.'

15.9 Examples of securities provided

As previously mentioned, two examples of performance securities are given in Part II of the Red Book, the first, on page 7, in the form of a performance guarantee and the second, on pages 8 and 9, in the form of a performance bond. They will be referred to in this section as the First and the Second forms respectively. Both forms require modifications or additions depending on the particular circumstances of the contract before they can be used successfully. The following comments provide some of the points which may be considered:

(a) Although the wording and the text of both forms has been greatly simplified from the usual archaic language of such documents in response to criticism made, such as that by Lord Atkin, quoted earlier (Reference 15.4), there remains one expression which is unusual to the ordinary person associated with construction and particularly to someone whose mother tongue is not English. This is that the expression 'presents' is used in both example forms as a noun. The *Oxford Dictionary* gives the following meanings to that noun: 'the present time'; 'something freely given'; 'gift'. None of these render a suitable meaning. It should therefore be replaced.

(b) The First form includes the expression 'duly perform' and the Second form includes the expression 'promptly and faithfully perform'. Clause 10, with which these example forms have to comply, provides that the contractor is to obtain security for his 'proper performance'. Whilst it can be argued that the expression 'duly', 'promptly and faithfully' and 'properly' mean exactly the same thing, this is the sort of argument that would take lawyers a considerable time to advance and argue in an arbitration. Therefore, it is submitted that where that meaning is intended, the word 'duly' is used.

(c) Whilst it is provided in the Second form that 'any suit must be instituted before the issue of the Defects Liability Certificate' in compliance with the provision of clause 10 of the Red Book, no period of liability has been indicated in the First form.

(d) The First form provides that 'if on default by the Contractor the Guarantor shall satisfy and discharge the damages sustained by the Employer thereby up to the amount of the above written guarantee ...'. This suggests that the employer can call upon portions of the sum guaranteed or that the guarantor can satisfy the beneficiary's call upon the guarantee by making a payment less than the guaranteed sum. If this is correct, this condition should be stated in clearer and more precise language allocating to the employer the duty to account for any monies he receives, and also providing the exact mechanism for ascertaining the proper payment to be made and how. If not, then a different form of wording should be used to avoid any doubt or possible confusion.

(e) Whilst the First form provides specifically, and rightly so, that 'no alteration in terms ... or in the extent or nature of the Works ... and no allowance of time ... nor any forbearance or forgiveness ... shall in any way release the Guarantor from any liability ...', such provisions are not included in the Second form except by the curtailed phrase '(including any amendments thereto)'. As explained earlier, the above provisions are necessary to any properly drafted form of security.

(f) It is difficult to visualise how condition (a) of the First form can be operated in accordance with the provision of clause 10 of the Red Book whose only condition for the call upon the security is a simple notification by the employer to the contractor stating the nature of the default in respect of which a claim is to be made, in accordance with sub-clause 10.3.

Similarly, if condition (b) is to apply, then the provision of sub-clause 10.3 must be altered to allow for an arbitration mechanism to be invoked in advance of a claim under the security.

(g) The significance of the words 'well and truly to be made' in the recital part of the bond in the context of the payment of the guaranteed sum in the Second form is not clear. A payment is either to be made or not.
(h) It is a condition of the Second form, in its third paragraph, that the employer must perform his obligations under the contract with the contractor in order for the security to be valid. This condition is not included in the First form.
(i) The last paragraph of the Second form restricts the right of action under the bond to the employer or his heirs, executors, administrators or successors. Such a condition is not included in the First form.

The pitfalls of performance securities are many without complications from use of ambiguous language.

15.10 Other securities associated with a construction contract

In addition to the performance securities which may be required under clause 10 of the Red Book, there are others which have become necessary in the implementation of a construction project, mainly due to growing distrust between those involved in the construction industry. These are described below.

15.10.1 Bid bonds or guarantees

Bid bonds or guarantees are securities given by tenderers accompanying their tenders in which the guarantor undertakes to pay a sum or sums of money not exceeding a certain amount upon receipt of a demand stating that a particular event has taken place. They are intended, first, to avoid irresponsible tendering and, secondly, to ensure that the employer is protected if the successful bidder fails for any reason to proceed with the formation of a contract.

Bid bonds or guarantees are typically required for between three to five per cent of the tender amount. They would normally be available to the beneficiary either against his on-demand or unconditional call for payment, usually indicating that the tenderer has refused to proceed with the contract; or on the condition that an independent authority should certify that the tender has been withdrawn before an award has been made or before the end of the bidding period plus any agreed extension. They have a short life span which is the length of time between the date of issue and the date of any performance security which may be required and which normally follows the issue of the letter of acceptance. They should only be capable of extension if the tenderer agrees.

Bid bonds or guarantees should include a condition that they will be returned to the unsuccessful tenderer either as soon as a tender has been accepted or a decision has been taken not to proceed with the project at all.

15.10.2 Advance payment guarantees

These guarantees secure the repayment of an advance payment received by a contractor from an employer. The guarantor undertakes to pay a sum or sums of money not exceeding a certain amount upon receipt of a demand stating either that the contractor has failed to make the repayments or that a certain event specified in the guarantee has taken place.

15.10.3 Retention money bonds

Retention of part of the monthly payments due to the contractor is made by the employer under the provisions of sub-clause 60.2 in amounts calculated by applying the percentage of retention, stated in the Appendix to Tender, to the amount to which the contractor is entitled under the provisions of sub-clause 60.1. Such retention accumulates until the amount so retained reaches the limit of retention money also stated in the Appendix to Tender. It is then released in accordance with the provision of sub-clause 60.3, basically in two stages: half upon the issue of the taking-over certificate for the whole of the works and the other half upon the expiration of the defects liability period for the works.

In certain circumstances, upon the issue of the taking-over certificate, the employer may agree to release the whole of the retention fund against an on-demand or unconditional guarantee, to be issued on behalf of the contractor for the value of the funds released earlier than prescribed, that is half of the limit of retention money. Similar arrangements may be made between the contractor and his sub-contractors. The guarantor in this type of guarantee undertakes to pay a sum or sums of money not exceeding a certain amount upon receipt of a demand stating either that the contractor has failed to carry out his obligations under the contract or that a certain event specified in the guarantee has taken place.

15.10.4 Maintenance or defects liability bonds

These are bonds issued to guarantee that as soon as the construction or installation is complete, the contractor will fulfil his obligations throughout the defects liability or the commissioning and testing periods, respectively. Therefore, essentially, these securities provide a protection against defects in materials, workmanship and other aspects of the contract forming part of the contractor's obligations.

15.10.5 Company suretyship

This type of suretyship is transacted when a party, not necessarily a bank or an insurance company, acts as a guarantor for another in the lease purchase of equipment or plant. This transaction is extremely perilous for the surety since the beneficiary in this case is normally a finance company which has perfected a standard form of agreement in which the guarantee is made effective in respect of past and future transactions, irrespective of whether or not the surety has been made aware of them. In a recent case in Ireland, it was held that such a transaction is a legally valid one but it was at the same time described as three blank cheques, one for the past, one for the present, and another for the future. Those who are asked to act as surety in such transactions should be extremely careful of the undertakings to which they will become obligated.

15.11 Concluding remarks

The entire area of securities is marked with uncertainty, inconsistency and confusion, both conceptually and in respect of the terminology used to describe these securities in various parts of the world. This chapter is an attempt to provide a proper understanding of the topic and a clear distinction between the various securities required in construction. It also provides a check list of the pitfalls which must be avoided in such bonds and guarantees.

Chapter 16

Claims and Counterclaims

16.1 Introduction

The construction contract is unique in that it seeks to provide for a specific remedy in the event of any breach of the terms and conditions within its framework and/or for a contractual entitlement in respect of specified events. Therefore, it is essential that the parties and those who represent them fully comprehend the terms of the contract and the remedies available to them under it.

The genesis and development of the standard form of construction contract was and remains based on the need to redefine and reapportion the risk ascribed to the respective parties by the applicable law. By including a mechanism to give one party a certain remedy if a specified event arises, the risk of that event, which would otherwise remain with that party, is transferred to the other party. However, whether the remedy sought is in respect of a breach of the contract terms and conditions or for the occurrence of a specified event, all construction contracts place an obligation on the party who wishes to avail themselves of that remedy to follow a set procedure, which is referred to as 'the claims procedure'.

The first step of the claims procedure of any construction contract is the making of a claim. Consequently, the definition and legal basis of claims and counterclaims, and the process by which they can be made and through which the parties' rights can be exercised or are protected, also need to be understood.

16.2 Definition and legal basis of claims and counterclaims

In all construction contracts, claims and the right to claim play a significant role in the contractual relationship between the employer and the contractor. Curiously, for such a fundamental aspect of the contract, no express definition appears in the typical standard form of construction contract and it is rare to find a definition of 'a claim' in reference texts or authorities on construction contracts. A claim is defined in *The Oxford Companion to Law* as a general term for the assertion of a right to money, property, or to a remedy.[16.1] Strictly speaking then, whenever for example the contractor applies for his monthly interim payment for the original scope of the works, or whenever for example the employer writes to the contractor requiring him to remedy defective work, it would be a claim under this definition. In construction contracts, a claim is generally taken in practice to be an assertion for *additional* monies due to a party or for *extension* of *the Time for*

Completion. This interpretation of 'a claim' is borne out by the wording of the contractual provisions relating to claims. For example, clause 44 of the Fourth Edition of the Red Book provides for an entitlement to 'an *extension* of the Time for Completion of the Works or any Section or part thereof' (emphasis added), albeit that the word 'claim' appears nowhere in this clause; and sub-clause 53.1 includes the words '... if the Contractor intends to claim any *additional* payment...' (emphasis added). However, nowhere is there a reference to an assertion for a declaration, although such a claim is frequently made. Similarly, in the 1999 suite of the FIDIC Conditions of Contract, clause 2.5 states:

> 'If the Employer considers himself to be entitled to any payment under any Clause of these Conditions or otherwise in connection with the Contract, and/or to any *extension* of the Defects Notification Period, the Employer or the Engineer shall give notice and... The notice shall be given as soon as practicable after the Employer became aware of the event or circumstance giving rise to the claim.' (emphasis added)

Clause 20.1 of the 1999 suite states:

> 'If the Contractor considers himself to be entitled to any *extension* of the Time for Completion and/or any *additional* payment... the Contractor shall give notice to the Engineer, describing the event or circumstance giving rise to the claim.' (emphasis added)

It is clear then that these forms of contract seem to ignore claims arising from anything other than the assertion for an entitlement to payment or time.

A counterclaim is defined as an assertion made by a respondent party which can conveniently be examined and disposed of in an action originally initiated by the claimant party. It is not necessarily a defence, but a substantive claim against the claimant which could have grounded an independent action. The concept of convenience referred to here signifies that the background of the counterclaim is similar to that of the claim and results from the same set of facts and events. For instance, a party might be precluded from bringing a counterclaim in arbitration either because the claim is not subject to arbitration (i.e. arising from another contract or a different event) or by virtue of the conditions prescribed by the contract initiating an entitlement to claim. Furthermore, in the sense that the word is used, it would not include a defence where that defence does not itself give rise to an actionable claim against the claimant.

For the purposes of this chapter, both claims and counterclaims will be referred to as claims. Despite the complexity of attempting to categorise claims, it is useful to have an overview of the subject of claims in construction contracts.

Essentially, other than claims under statutory law, claims in construction contracts may be based on any one of four legal concepts and one non-legal concept. Therefore, if a claim is required to be categorised, and it is suggested that it should be, the categorisation could be done in accordance with the following five categories:

(a) *A claim under the contract*: The first category relates to a claim under the contract between the parties based on the grounds that should a certain event occur, then a claimant would be entitled to a remedy that is specified under a particular provision of the contract, subject to the effect of the applicable law. Such an event may be one of two types.

(i) First, it may be a specified event under the contract, which might or might not occur, where in certain defined circumstances the employer or the contractor is entitled to claim a designated remedy. For example, the contractor is entitled to claim an additional payment under the Red Book for tests in accordance with the provisions of sub-clause 36.4 (see Section 16.3).

(ii) Secondly, the specified event may be a breach of a particular provision in the contract entitling a claimant to a designated remedy if the terms of such provision are not, or are only partially, complied with. If the claim is successful, the particular provision in the contract would apply and the remedy could be in the form of a payment of a sum of money, or an extension of time, or some other benefit, or a combination of all three.

For example, it is stipulated in the Red Book that failure by the contractor to complete the works on a specified date would entitle the employer to deduct liquidated damages at a specified rate per day or week (see Section 16.3.8). In this connection, the provisions of the applicable law must be taken into consideration, for instance whether such damages are in effect a penalty, and if so, whether or not this can be treated as a valid claim.

(b) *A claim arising out of or in connection with the contract*: The second category relates to a claim arising not under, but out of or in connection with, the contract, where the remedy is not designated in the contract and the claimant needs to invoke a provision of the applicable law to obtain a remedy. Therefore, if the claim is valid, the remedy lies under the provisions of the applicable law of the contract, for example a claim for a breach of contract. Under English law, the remedy for a breach of contract would be under the principles governing damages, including those laid down in the *Hadley* v. *Baxendale* case.[16.2] (See also Chapter 3, Section 3.9(c)). The remedy in this case may extend to consequential damages, if foreseeable at the time of contracting, unless excluded by the contract. Furthermore, if the breach is of a serious nature, the aggrieved party might be entitled to cancel the contract and, if the breach amounts to the communication of an intention not to abide by the terms of the contract, such conduct could amount to a repudiation which would be open to acceptance by the innocent party.

Furthermore, if the employer terminates the contract, the contractor might have a lien over the works, depending on the terms of the contract, which would act as security for the payment of any money owed to it arising from the work performed pursuant to the contract.

Another example of a claim arising in connection with the contract, but not under it, is where one of the parties has misrepresented certain important facts in negotiations leading to the formation of the contract. In such case, the other party may use this as a basis for cancelling the contract or for claiming damages. However, the terms of the contract may exclude the right to cancel for a serious breach or misrepresentation. If, however, such misrepresentation constitutes fraud, the party making the representation would not be allowed to rely upon such an exemption clause.

(c) *A claim under the principles of the applicable law*: The third category relates to a claim arising under the application of the principles of the applicable law, either by the parties to the contract or against third parties. This could lead to a claim under the law of tort, or delict as it is referred to in some jurisdictions. The law applicable to a claim in tort/delict is not necessarily the same as the governing law of the contract. If the claim is successful, the remedy would typically be an award of general damages, the amount being dependent upon the particular circumstances of the case.

Depending on the applicable law, the parties may have concurrent claims arising from the contract and from tort/delict. Although the result will often be the same, the measurement of the loss is, in principle, different. Complicated questions arise between contracting parties as to the implementation of terms limiting the aggrieved party's entitlement to a claim brought in tort/delict. This might even apply to rights of action to persons who are not privy to the contract. For example, a party might wish to bring a claim in tort/delict rather than in contract because of time limitations; where there is an arbitration clause imposed by a contract; where there is an insurance policy covering particular claims; or where notice provisions under the contract affect the aggrieved party's entitlement to claim.

(d) *A claim arising out of the principle of* quantum meruit: The fourth category comprises claims where no contract exists between the parties,[16.3] or if one existed, it is deemed to be void.[16.4] It is based on the principle that an individual has the right to be paid a reasonable remuneration for work done. This is referred to in some legal systems as *quantum meruit* or 'as much as one has earned' and has been often equated to a claim for undue enrichment.[16.5] The principles of *quantum meruit* have also been applied to cases where there is a contract in existence but the price is not stipulated; instead the contract expressly provides that the amount to be paid will be based on a reasonable sum or the price will be agreed from time to time. In *Hudson's Building and Engineering Contracts*, it is stated:

> '... quantum meruit is frequently employed ... where a true contractual situation exists, in the sense of a request to do work accompanied by an intention to pay for it, and so supported by consideration, but where the price may not have been fixed at all, or with sufficient precision, by the contract, so that a promise to pay a reasonable price requires to be implied to give practical effect to the parties' intentions. ...'[16.6]

The case of *Constable Hart & Co. Ltd* v. *Peter Lind & Co. Ltd*[16.7] is one example where the court applied this principle. If the claim is successful, payment is assessed on the basis of a reasonable recompense of the cost of the work carried out by the contractor and may, although not necessarily, depending on the principles of the applicable law, include an element of overhead and profit. (See also Section 3.9(c) above).

Generally, the remedy for all the four categories of claims set out above would be sought through an action in arbitration or litigation unless the claim is settled amicably. Such action would usually necessitate the employment of lawyers and the outlay of large expenditure (see Section 16.3 below).[16.8]

(e) *A claim for* ex gratia *payment*: Finally there is the claim for an *ex gratia* payment (meaning out of kindness). Although claims for *ex gratia* payments are not claims which arise by virtue of a contractual entitlement, they are sometimes entertained by employers and engineers as a matter of expedience to avoid arbitration or litigation and, indeed, to maintain the goodwill necessary to complete the project successfully. There is no applicable legal basis for such payments, but rather some commercial sense or benefit in reaching a settlement between the parties without acceptance of liability.

In the remaining part of this chapter, only the first two categories of claim are considered, i.e. those which are made under the contract or which arise out of the contract, as the other categories of claim are beyond the provisions of the Red Book. Figure 16.1

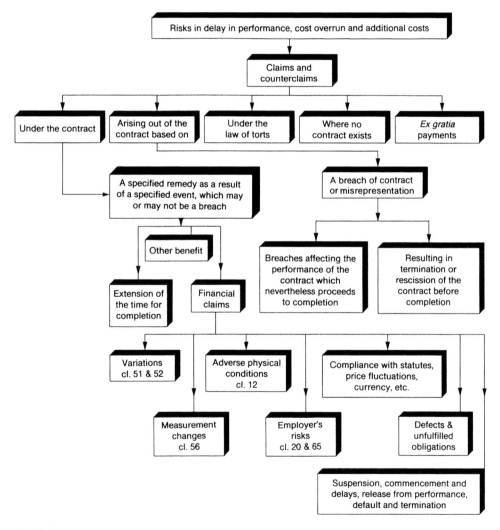

Fig. 16.1 Claims and counterclaims.

shows an analysis of these claims based on the provisions of the Red Book. Claims under the contract are presented, in the first place, to the engineer who should determine, impartially, the entitlement of the claimant in principle and in quantum. Such determination, if unacceptable to either the employer or the contractor, may be 'opened up, reviewed or revised as provided in Clause 67' (Settlement of Disputes). (See subclause 2.6 of the Red Book and Chapter 19). Claims arising out of or in connection with the contract were, until the 1992 reprint of the Red Book, not required to be put to the engineer and it was the employer who had to deal with such claims. However, the addition of the words 'or otherwise' after the words 'under the contract', as discussed in Section 16.4, means that all claims, whether under or arising out of or in connection with the contract, must first be presented to the engineer in any contract based on an edition of the Red Book issued since 1992.

The first step of the claims procedure of any construction contract is the making of a claim. Consequently, the definition and legal basis of claims and counterclaims, and the process by which they can be made and the parties' rights can be exercised or are protected, also need to be understood.

16.3 A claim under the contract and based on its provisions

As previously stated, the Red Book regulates the rights and obligations of the parties to the contract. Its provisions specify what should be done by the two parties or their agents and servants and the consequence if what ought to be done is either not done, or if done, is not done within the time specified.

Claims will very often arise in a traditional construction contract because it is perhaps the only contract where the price of the end result is defined before the process of 'production' even starts. Accordingly, in the competitive atmosphere of tendering which accompanies such a contract, little or no margins are left for future unknowns in a long and complex period of construction. Furthermore, as the rates and prices have to be based on certain assumptions which are, in turn, based on the provisions of the Conditions of Contract, any change between what was assumed and what actually happens may form a seed for a claim.

As already stated, claims under this category may be divided into:

(a) claims as a result of certain anticipated and specified events and for which a remedy is designated in the contract; and
(b) claims as a result of an event where a certain term of the contract is breached and for which a remedy is designated in the contract.

The remedy in both of these types of claim is designated in the contract and the claim may include the following:

(a) An assertion for financial compensation in respect of:
   (i) variations which include alterations, additions and omissions, as well as a change in any specified sequence or timing of construction of any part of the works;

(ii) measurement changes;
(iii) adverse physical conditions;
(iv) the employer's risks;
(v) compliance with statutes, regulations, price fluctuations, currency and other economic causes;
(vi) defects and unfulfilled obligations;
(vii) failure to commence, critical or non-critical delays, suspension of work, release from performance, default and termination; and
(viii) other miscellaneous specified events.
(b) An assertion for an extension of the time for completion of the contract works (see Chapter 17).
(c) An assertion for other benefit.

The claims in items (i) to (viii) in point (a) above are discussed below in Sections 16.3.1 to 16.3.8.

16.3.1 Variations

The nature of variations, the responsibility for them and their valuation are discussed in this section.

Nature of variations

Although some claims can be avoided by proper planning and risk management, claims for variations, as indeed claims in general, are inevitable since it is practically impossible to foresee every event that might occur during the construction period and to plan in advance for the consequences of such events. If that concept is accepted, then it is necessary to incorporate into the contract a mechanism for implementing changes which are found to be necessary or desirable during the construction period. These changes are permitted to be made to the works but not to the contract.

Furthermore, it is necessary to provide the engineer with authority to make such changes, as an agent of the employer. Clauses 51 and 52 in the Red Book provide for such a requirement by authorising the engineer to vary the works or any part thereof. They also regulate the rights and obligations of the parties in the event of such variations. The wording of sub-clause 51.1 confines this authority to make variations to the engineer. The first sentence of sub-clause 51.2 is in fact more specific in this respect in that the contractor is prevented from making such a variation without an instruction from the engineer. The engineer's authority in this respect is extremely wide extending to the 'form, quality or quantity of the Works or any part thereof'. However, in the case of additional work, such authority is explicitly restricted in paragraph (e) of sub-clause 51.1 to 'any kind necessary for the completion of the Works'. Accordingly, the engineer's authority to add further work to the contract is restricted so that it does not include work alien to the original concept of the contract or work which is not necessary for its completion.

An instruction is required from the engineer to the contractor to initiate a variation and such an instruction is required to be in writing in accordance with sub-clause 2.5, or, if given orally, then it should be confirmed in writing as provided for in that sub-clause. Some commentators have suggested that a drawing may constitute an instruction in writing if it is issued showing a change from previous drawings and the change may be defined as a variation. Of course, it would always be preferable to issue the drawing accompanied by a specific communication as to the nature of any contemplated variation. What is and what is not a variation is a major source of dispute in many projects and a frequent issue in arbitral proceedings.

Sub-clause 51.1 also provides that the engineer may vary the works or any part thereof if for any reason, other than it being necessary, he finds it appropriate to:

(a) increase or decrease the quantity of any work included in the contract but an instruction for such variation is not required if the actual quantities of the work envisaged at the time of tendering prove, on remeasurement, to be different from those recorded in the bill of quantities, see sub-clause 51.2.[16.9] The effect of such a variation must then be taken into account in determining whether or not the provisions of sub-clause 52.3 apply;
(b) omit such work provided it is not to be carried out by the employer or by another contractor;
(c) change the character or quality or kind of any such work;
(d) change the levels, lines, position and dimensions of any part of the works;
(e) execute additional work of any kind necessary for the completion of the works;
(f) change any specified sequence or timing of construction of any part of the works.

Paragraph (f) of sub-clause 51.1 which is a new provision under the Fourth Edition assumes that some specified sequence or timing of construction had already been made in the contract documents. It adds a new dimension to the meaning of a variation from previous editions of the Red Book in that it encompasses the timing of construction and its programme. Any change from the specified sequence or timing stated in a programme submitted by the contractor under clause 14 and instructed by the engineer under clause 51 would therefore qualify as a variation. Such a change may be in the form of a requirement to accelerate the work or to vary its timing. Whilst this provision is new in the Fourth Edition it is worth noting that it stops short of the provision of the Fifth and Sixth Editions of the ICE Forms of Contract for Works of Civil Engineering Construction which empower the engineer to order a change in the contractor's specified *'method'* of construction in addition to its timing. Such an order may be necessary to avoid continuing disruption on site as a result of an event for which the employer is liable. It is notable that the provisions of sub-clause 50.1 of the ICE Sixth Edition, published in 1991, were extended to include variations ordered after substantial completion of the works.

Sub-clause 51.1 of the Red Book also provides that any variation instructed by the engineer should not vitiate or invalidate the contract. It has been suggested that as a matter of business efficacy, this provision must be subject to an implied limitation of reasonableness in so far as instructions cannot stray 'outside the Contract'.[16.10] Extras, therefore, must not be of a certain value and type and must not be instructed at a time which would render the contract inapplicable.

Responsibility for variations

The allocation of liability for the value of any variation instructed by the engineer is dependent upon whether or not the necessity for it is due to some default of or breach of contract by the contractor. Accordingly, any additional costs attributable to a variation necessitated by some default of or breach of contract by the contractor are to be borne by him; see sub-clause 51.1.

Valuation of variations

Valuation of the variations referred to above is provided for under clause 52 of the Red Book which provides the rules for such valuation. In essence, they are:

(a) Within 14 days of the date of an instruction to vary the works in accordance with clause 51, and before commencement of such work, a notice is required to be given either:
 (i) by the contractor to the engineer of his intention to claim extra payment or a varied rate or price; or
 (ii) by the engineer to the contractor of his intention to vary a rate or price.

 It is clear by including this notice requirement in the proviso of clause 52.2 that a valid notice of intention to claim is a condition precedent to any additional payment. The importance of this condition can only be evaluated in the light of the provisions of the applicable law of the contract. Furthermore, in general terms, the treatment of this topic under civil law jurisdictions differs significantly from that under the common law.[16.11] In England, the relevance of such notice was highlighted by the Court of Appeal in 1965 in the *Tersons* case when it was held that a notice need only indicate the intention to make a claim and identify in general terms the additional work to which the claim will relate.[16.12] It is noteworthy that the wording of the clause in question in that case is different from that in the Fourth Edition of the Red Book from the point of view of the period within which it had to be served. The words of the clause in the *Tersons* case were: 'as soon after the date of the order as is practicable', as against the period of 14 days specified in the Red Book.

(b) Such notice is not required to be given where the variation entails an instruction to omit work.

(c) Varied work is valued in one of four different ways. The first is where the variation is valued at the rates and prices set out in the contract if, in the opinion of the engineer, these rates and prices are applicable to the items of varied work. In considering the applicability of the rates and prices in the contract to the varied work, the engineer would have to take into account the nature and amount of the varied work in addition to, presumably, the preliminary items which may be affected, the time when the variation is ordered, the method of its construction and its physical location compared with the other work under the contract (see Section 6.3.1 above).

(d) The second way in which a variation may be valued is where there are no applicable rates and prices in the contract, then the contract rates and prices are to be used as the basis for valuation so far as may be reasonable.

(e) The third way in which a variation may be valued applies if the contract rates and prices cannot be used as a basis for valuation; then the engineer is required to agree new suitable rates and prices through the procedure of 'due consultation' with the employer and the contractor.

Where no agreement is reached between the engineer and the contractor, then the engineer is required to fix such rates and prices as are, in his opinion, appropriate. Having done so, the engineer is required to notify the contractor accordingly with a copy to the employer.

Whilst no time limit is imposed on such agreement on suitable rates or prices, or any subsequent requirement to fix such rates or prices, in case of disagreement, the engineer is required under sub-clause 52.1 to determine provisional rates or prices for the purposes of on-account payments in any certificate issued under clause 60.

Sub-clause 60.2 provides that 'the Engineer shall ... certify ... the amount of payment to the Contractor which he considers due ...'. It may, therefore, be taken that whatever is certified under the monthly payment will have to incorporate properly assessed rates and prices for the various items of work executed and incorporated in the monthly statements of the contractor pursuant to sub-clause 60.1 'to which the Contractor considers himself to be entitled'.

Under the heading of applying new suitable rates or prices, sub-clause 52.2 provides for the possibility of changing the rates or prices of items of the works, other than varied work. It provides that if, in the opinion of the engineer, the nature or amount of any varied work relative to that of the whole of the works, or to any part thereof, is such that the rate or price contained in the contract for any item of the works, is rendered inappropriate or inapplicable, then a suitable new rate or price is to be agreed upon between the engineer and the contractor. This provision extends, therefore, to the rates or prices for 'any item of the works' and affects the rates or prices which are influenced by variations due to time delay, out of sequence working, changes in the method of execution and to a large extent the preliminary items of the bill of quantities, if any are included. For a proper adjustment of the rates and prices of these items, if such adjustment is found to be necessary, reliable knowledge and understanding of the make-up of these rates and prices is essential.

Once again, agreement on new suitable rates or prices is to be attempted under the provisions of sub-clause 52.2 after 'due consultation' by the engineer with the employer and the contractor. In the event of disagreement, the engineer is required to fix such other rate or price as is, in his opinion, appropriate in the particular case. The engineer is then required to notify the contractor accordingly with a copy to the employer.

As previously required under sub-clause 52.1, until such time as rates or prices are agreed or fixed, the engineer should determine provisional rates or prices to enable on-account payment to be included in any certificate issued under clause 60.

(f) The fourth way in which a variation may be valued applies where it is to be found, on the issue of the taking-over certificate for the whole of the works, that the value of varied work together with all adjustments upon measurement of the estimated quantities is in excess of 15 per cent of the 'Effective Contract Price'. The 'Effective Contract Price' is defined in sub-clause 52.3 of the Red Book as 'the Contract Price, excluding Provisional Sums and allowance for dayworks, if any'. Then, a sum is required to be added or deducted from the contract price as may be agreed between the engineer and the contractor. There is no guidance in the Red Book as to how this sum should be calculated. It is, however, apparent that the target of such adjustment is the bill of preliminary items which forms part of the contract price.

On page 117 of the Guide, referred to in Reference 9.1, there is reference to the purpose of this sub-clause where it is explained that 'in preparing a tender, a contractor may distribute his on-costs and profit in various ways,...', and an adjustment of these is necessary where the amount of work under the contract is varied beyond the specified percentage. For an accurate adjustment, however, it is imperative that the engineer should obtain details of the contractor's internal price make-up of these on-costs and profit items.

(g) Sub-clause 52.3, which deals with the situation where the value of variations exceeds 15 per cent of the effective contract price, defines the value of varied work and all adjustments upon measurements of the estimated quantities in the following manner:

   (i) varied work is defined in paragraph (a) of sub-clause 52.3 as all varied work valued under sub-clauses 52.1 and 52.2. This definition removes the confusion which existed in previous editions of the Red Book as to whether or not changes of quantity arising from measurements, in accordance with clause 56, should be included in the calculation of the value of the varied work. It is now clear that only varied work in accordance with sub-clause 52.1 and 52.2 should be included;

   (ii) all adjustments upon measurement of the estimated quantities is defined in paragraph (b) of sub-clause 52.3 as those adjustments which are 'set out in the Bill of Quantities, excluding Provisional Sums, dayworks and adjustments of price made under Clause 70'.

   This definition also clarifies in precise terms the value of additions to or deductions from the 'Effective Contract Price', or in other words from 'the Contract Price, excluding Provisional Sums and allowance for dayworks, if any';

   (iii) agreement between the engineer and contractor is once again based on the new concept of 'due consultation' by the engineer with the employer and the contractor. Where no such agreement can be reached, the sum to be added to or deducted from the contract price is required to be determined by the engineer having regard to the contractor's site and general overhead costs of the contract. The engineer, having determined such sum, is required to notify the contractor accordingly with a copy to the employer;

   (iv) finally, sub-clause 52.3 provides that the sum to be added to or deducted from the contract price is to be based only on the amount by which such additions or deductions are in excess of 15 per cent as fixed in sub-clause 52.3. This provision answers the question which was left unanswered in the previous editions of the Red Book. It is now clear that the adjustment to the contract price should be related only to the margin below or above the 15 per cent. Therefore, if for example, there is a reduction of 20 per cent in the contract price due to the valuation of variations under sub-clauses 52.1 and 52.2, the 'losses' sustained by the contractor due to this reduction should only be related to the margin between 20 per cent and 15 per cent, i.e., 5 per cent and not the whole of the 20 per cent.

(h) Sub-clause 52.4 deals with the situation where the engineer requires certain varied work to be carried out on a daywork basis. Daywork is usually covered by a provisional sum in the bill of quantities which may then be used for additional items for which no bill item is applicable. Where such a provisional sum is included in the bill of quantities, a daywork schedule of rates and prices is appended to the bill for pricing by the contractor. The procedure for payment in respect of daywork is set out in sub-clause 52.4.

Figure 16.2 shows a flow chart of the procedure to be followed where variations are found to be necessary and of the manner in which they are to be valued.

Therefore, in setting out comprehensively the manner in which variations are to be valued, the Red Book provides a remedy in the form of financial compensation in the anticipated event of a change to the works being instructed by the engineer, on behalf of the employer.

16.3.2 *Measurement changes*

Clauses 55 to 57 inclusive deal with 'Measurement' under the Fourth Edition of the Red Book. Clause 55, a clear and concise clause, provides the principles applicable to measurement of quantities. These are:

(a) That the contract is a measure and value contract and that the quantities of the various elements of the works are set out in a bill of quantities.
(b) The quantities of the bill are only estimated and are not to be taken as the actual or the correct quantities of the works to be executed by the contractor in fulfilment of his obligations under the contract. This means that there ought to be a recalculation of the contract price by establishing the actual quantities carried out at completion.
(c) It follows, therefore, that irrespective of whether or not the works are varied in accordance with clause 51, the contract price is adjustable. Such adjustment is dependent upon whether or not it is found necessary to change the quantities in the bill of quantities in order to complete the works.

Clause 56 sets out some procedural matters in connection with measurement of any part of the works. The clause begins with an express requirement that the engineer is to ascertain and determine by measurement the value of the works, such value being used as the basis of payment to the contractor under clause 60. The words 'ascertain and determine' mean a recalculation of the quantities actually carried out.

Claims and Counterclaims

In order to carry out such recalculation, clause 56 provides for two steps to be taken which are:

(a) When any part of the works is to be measured, the engineer is to give reasonable notice to the contractor's authorised agent to:
  (i) attend and assist the engineer in making such measurement; and
  (ii) supply any particulars required by the engineer.
(b) That the engineer should 'prepare records and drawings as the work proceeds and the Contractor, as and when called upon to do so in writing, shall, within 14 days, attend to examine and agree such records and drawings with the Engineer and shall sign the same when so agreed.'

Non-attendance by the contractor or an authorised representative on either of these two occasions would place a heavy burden of proof on him should he later wish to challenge any of the calculations of the engineer which are deemed to be correct under the provisions of this clause.

Where the contractor examines the records and drawings and wishes to disagree, he must, within 14 days of such examination, lodge with the engineer notice stating the elements which he alleges to be incorrect. On receipt of such notice, the engineer is required under clause 56 to review the records and drawings and either confirm or vary them.

Therefore, in providing for the engineer to measure and value the works as they proceed and to certify payment upon such valuation, the Red Book provides a remedy in the form of financial compensation in the anticipated event of a change to the quantities of the various elements of the works as set out in the contract bill of quantities.

16.3.3 Adverse physical obstructions or conditions

As a large proportion of work in civil engineering construction is usually carried out underground, the risk of encountering unforeseeable physical obstructions or physical conditions is quite high. Whilst it may be possible to investigate and establish the properties of the sub-soil at certain locations and extrapolate from that an overview of the whole site, the actual ground conditions for the whole of the works can only be established when the contractor excavates. If the unexpected happens and unforeseeable physical obstructions or physical conditions are encountered, the consequences can be enormous in terms of financial costs and time delays.

Clause 12 provides for this situation by allocating the risk of encountering such conditions to the employer. The reasons behind such allocation are many, including that if the risk is allocated to the contractor, the employer would have to pay for risks that might not eventuate through contingency sums that the contractor would be forced to include in his tender to take account of the unforeseeable. In any case, these reasons are essentially based on the criteria of control over the risk, in that it is the employer who:

(a) selects the site and the precise location of the works;
(b) has control over the design of the works and the timing of commencement of construction; and

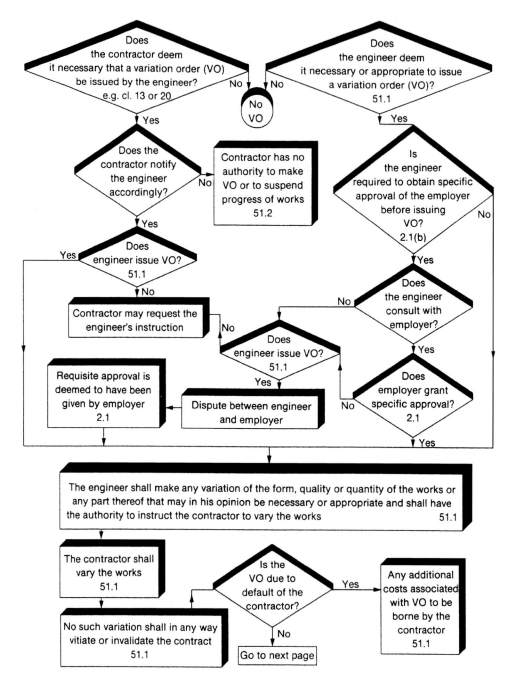

Fig. 16.2 Variation orders.

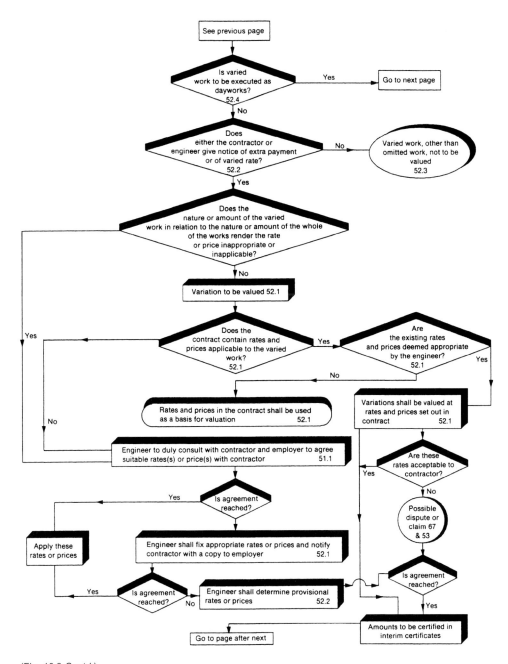

(Fig. 16.2 Contd.)

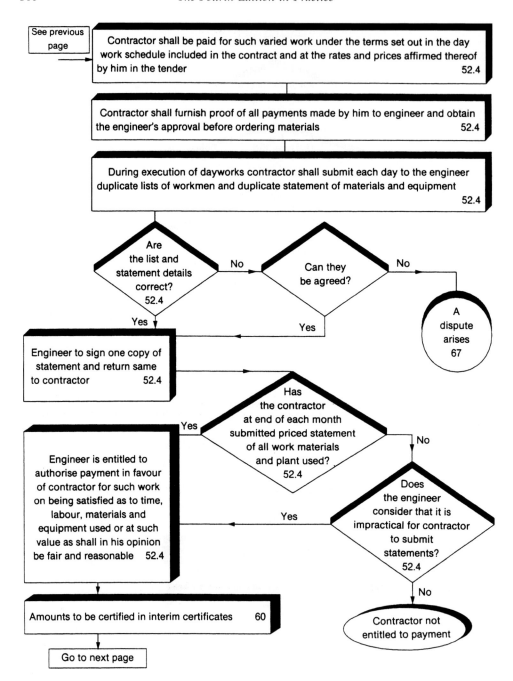

(Fig. 16.2 Contd.)

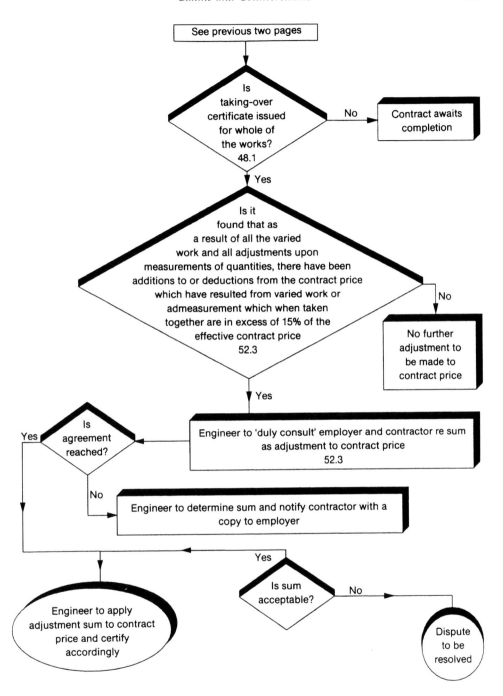

(Fig. 16.2 Contd.)

(c) can and has the opportunity to carry out whatever investigations he thinks necessary to safeguard himself against the unknown. Identification of the hazards and risks inherent in the ground conditions of a particular site is to a large extent under the control of the employer. Once identified, the hazards and risks, if unacceptable, can be mitigated or eliminated by the employer either directly through selection of an alternative site or indirectly by changing the design of the project.

It is generally recognised that contractors tendering for a particular project have neither the time nor the certainty of benefit to initiate extensive sub-soil investigation for a project for which they are participating in a tendering exercise. The contract for the construction of such a project might or might not be awarded to any such tenderer. Accordingly, the concept of clause 12 is based on the criterion that should the unexpected happen, then the cost of, and the time required in dealing with it would be paid to the contractor by the employer. The contractor should not therefore include in his tender for unforeseeable physical obstructions or physical conditions. This is perhaps best illustrated by the following quotation from the judgment of the Privy Council in the case of *Mitsui Construction Co. Ltd* v. *Attorney-General of Hong Kong* (1986):[16.13]

'Against this background of facts, if the contract documents were understood in the sense contended for by the Government, engineering contractors tendering for the work would have two options. They could either gamble on encountering more or less favourable ground conditions or they could anticipate the worst case and price their tenders accordingly. It is clear from what happened here that the worst case might double or more than double the time required to do the work with a consequent increase in time related costs. On this basis, tenderers gambling on favourable ground conditions would risk a large loss, while conversely, if all tenderers anticipated the worst case, but in the event reasonable conditions were encountered, the Government would be the losers. It follows that, if the Government are right, there is a large element of wagering inherent in this contract. It seems to their Lordships somewhat improbable that a responsible public authority on the one hand and responsible engineering contractors on the other, contracting for the execution of public works worth many millions of dollars, should deliberately embark on a substantial gamble.'

A construction contract should not be a gamble if disputes are to be avoided. Thus, clause 12 sets out the principles of risk allocation in connection with ground conditions and outlines the procedure when the risk of unforeseeable physical obstructions or physical conditions eventuates. Sub-clause 12.1 places on the contractor the responsibility of satisfying himself 'as to the correctness and sufficiency of the Tender and of the rates and prices stated (by him) in the Bill of Quantities . . .'. These rates and prices are deemed to cover all his obligations under the contract; see Chapter 13 for 'the proper execution and completion of the Works and the remedying of any defects therein'.

To satisfy himself as to the correctness and sufficiency of his tender where ground conditions are concerned, the contractor is deemed, amongst other things, to have complied with the provisions of clause 11 which include a requirement to inspect and examine 'the Site and its surroundings and information available in connection therewith', before submitting his tender. It is recognised in clause 11 that the contractor in making such inspection and examination is constrained by what 'is practicable, having regard to considerations of cost and time'. However, despite that restriction, the contractor is expected to consider the following specific matters in connection with the site, its surroundings and the information available on them:

'(a) the form and nature thereof, including the sub-surface conditions,
(b) the hydrological and climatic conditions,
(c) the extent and nature of work and materials necessary for the execution and completion of the Works and the remedying of any defects therein, and
(d) the means of access to the Site and the accommodation he may require and, in general, shall be deemed to have obtained all necessary information, subject as above mentioned, as to risks, contingencies and all other circumstances which may influence or affect his Tender.'

The employer, on the other hand, is expected under clause 11 to have made available to the contractor, before submitting his tender, all the available data on hydrological and sub-surface conditions which had been obtained from investigations relevant to the works. By express wording of sub-clause 11.1, the interpretation of such data is, however, the responsibility of the contractor and neither the employer nor his advisers should venture into this area.

The concept of clause 12 is based on the above requirements being fulfilled by the employer and the contractor. Accordingly, the criterion adopted in sub-clause 12.2 for a compensation event to the contractor is based on the discovery of physical obstructions or physical conditions which, in the contractor's opinion, were 'not foreseeable by an experienced contractor'. Attention should be given to the limitation imposed by the word 'physical' on any unforeseeable obstruction or condition. Thus administrative or other obstructions or conditions are not within the scope of sub-clause 12.2 provisions. It is also worth noting that the compensation event arises on the basis of the contractor's opinion, not the engineer's opinion, that the obstruction or condition in question was not one that was foreseeable by an experienced contractor.

Where such an event occurs, the contractor is required under sub-clause 12.2 to give notice to the engineer, with a copy to the employer, and this notice triggers the claim mechanism for compensation in money or time or both. The engineer is required, on receipt of such notice, to consider and conclude whether, in his opinion, such obstructions or conditions 'could not have been reasonably foreseen by an experienced contractor'.

If the engineer's opinion is that such obstructions or conditions could not have been reasonably foreseen by an experienced contractor, he is then required to determine, after 'due consultation' with the employer and the contractor:

'(a) any extension of time to which the Contractor is entitled under Clause 44, and
(b) the amount of any costs which may have been incurred by the Contractor by reason of such obstructions or conditions having been encountered, which shall be added to the Contract Price.'

In this way the Red Book provides a remedy in the form of financial compensation in the event of encountering unforeseeable physical obstructions or physical conditions.

After such determination is made, the engineer is therefore required to 'notify the Contractor accordingly, with a copy to the Employer' but such determination 'shall take account of any instruction which the Engineer may issue to the Contractor ... and any proper and reasonable measures acceptable to the Engineer which the Contractor may take in the absence of specific instructions from the Engineer'.

Accordingly, there is wide flexibility and risk sharing between the employer and the contractor in respect of dealing with the unexpected event. There is also flexibility in the type of action or position that the engineer may wish to adopt should an event occur which the contractor contends falls under the provision of sub-clause 12.2. This is quite different from the situation under clause 12 of the ICE Form, on which the Red Book was originally modelled, where specific requirements are stipulated in response to a notice under this clause.

An interesting change has been made in the wording of the Third Edition of the Red Book in the description of the obstructions or conditions. The words 'could ... not have been reasonably foreseen by an experienced contractor' as contained in the following sentence of the Third Edition:

> 'If, however, ... the Contractor shall encounter physical conditions ... or artificial obstruction, which ... could, in his opinion, not have been reasonably foreseen by an experienced contractor, ... ',

have been altered to 'were ... not foreseeable by an experienced contractor' in the Fourth Edition, most notably by the deletion of the word 'reasonably'. The significance of this change may have to be especially considered when these descriptive words remain unaltered later in the same paragraph in dealing with the reference to the opinion of the engineer.

The change in wording is particularly important in view of the fact that many authoritative commentators have criticised the indefinite meaning of the former phrase: 'could not have been reasonably foreseen by an experienced contractor'.

The comments include the following:

(a) The above phrase has been described as one which 'probably gives rise to the most frequent disputes of fact which come before engineering arbitrators in the United Kingdom' and that 'their application is likely to vary in accordance with the personal views of individual engineering arbitrators.'[16.14]

(b) The phrase was also questioned in terms of whether a claim is 'excluded only if an experienced contractor could have foreseen that the conditions or obstructions must occur, or is it sufficient that he could have foreseen that there was a possibility, however remote, that the conditions might occur?'[16.15]

(c) A further criticism was made in that the difficulty in ground conditions may be the result of inadequate pre-contract investigations or design which are the responsibility of the engineer. In having to determine the foreseeability of the obstructions or conditions, the engineer will find himself in a situation of a conflict of interest.[16.16]

(d) More recently, the phrase was questioned in respect of the 'hypothetical' nature of the experienced contractor, and the total dependence of the risk of unexpected adverse conditions on what this 'hypothetical person . . . ought to have (considered) at the time of the contract', and thus it is hardly surprising that 'this test gives rise to some very peculiar consequences'.

From this analysis, it was then concluded that 'this is why most substantial clause 12 claims must go to arbitration and why the clause itself is and always has been a contractual disaster.'.[16.17]

(e) It has also been said that the foreseeability concept prescribed by this phrase is 'subject to fair criticism on the ground that uncertainty is introduced. What is the baseline of deemed knowledge? What is the base date? What is the definition of "reasonable contractor" (at least we know the novice will be judged by the rules applicable to old hands)? What is the dividing line when the risk is foreseeable in general but not to the actual degree of severity? There are no part answers. These are areas that require judgement applied to unique circumstances. Consequently there will be room for disputation.'[16.18]

(f) On the other hand, it is argued that the policy behind a contract provision should be 'to allocate a risk to the party who is best able to take responsibility for it, either because it is fair to do so, or because that party is the better able to control the outcome', and that 'it is not the wording of the clause (clause 12) which causes the problems, but ignorance and the engineer's understanding of it.'.[16.19]

(g) Because of either the nature of the risk, or the wording of clause 12, or the type of disputes which may be triggered by events such as adverse ground conditions, the parties are driven to some illogical positions:

  (i) Experienced contractors who foresee such a risk as probable, and price their tenders taking into account that probability, lose the contract in a competitive tendering situation to the contractor with the lowest margin for such a risk. The contract may be lost, but for the employer, 'worse still, (it may be lost) to a class of contractor whose main skill lies in the successful advancement and prosecution of this type of claim.'.[16.20]

  (ii) As soon as the risk eventuates, the contractor is driven to concentrate on why it had not been foreseen instead of how to tackle its effect.

  (iii) The employer and the engineer, 'for whom the unforeseen conditions represent respectively a financial problem and an embarrassment, will be trying to show that some information existed at the tender stage (which the engineer might have been quite unaware of) to demonstrate that the conditions could have been foreseen'. (See Reference 16.17.)

  (iv) The contractor will attempt to show that if the conditions could have been foreseen by an experienced contractor, then they should have been foreseen by the engineer and that the engineer's design should have catered for the possibility of this risk eventuating. Accordingly, the contractor would attempt to argue that if an experienced engineer could not have reasonably foreseen the hidden risk, how could an experienced contractor have done so?

  If the engineer did, why then did he not expose it more positively to the employer at first and subsequently to the contractor?

In defence of the concept of clause 12, some commentators have warned against its deletion since:

— Encountering unforeseeable conditions and obstructions may invalidate the contract should there be no provision similar to clause 12 and should such conditions and obstructions present a situation far beyond the original contemplation of the parties.[16.21] In this connection, reference should also be made to the theory of exceptional circumstances in some Romano-Germanic jurisdictions, as discussed earlier in Chapter 8.
— Passing part of the responsibility of site investigations to the contractor might result in a reduced incentive for the employer to undertake the extensive investigations necessary for a final design. If this happens, it would certainly be to the detriment of the designer and his design as well as to the employer, in terms of cost and prompt completion of the project. Furthermore, if all contractors tendering for a project 'were to carry out an extensive survey, the cost, not only in money but in disruption and possibly damage to the site, would be enormous'. (See Reference 16.19.)
— The additional contingent sum which would be built into the contract price by prudent contractors in terms of the unknown factor would mean a likely overall price disadvantage to the employer.

Attempts have been made to resolve this problematic area by advocating the inclusion in the tender documents of a precise method of allocating the cost of the consequences of the risk of unexpected ground conditions, if and when it eventuates, thus eliminating the two extreme possibilities of the contractor either making excessive monetary gains or facing financial hardship. This can be done, it has been suggested, by incorporating into the tender documents a referenced list of adverse physical conditions where additional remuneration, either on the basis of cost or price, would be paid to the contractor should any of the listed adverse conditions be encountered.[16.22] Additional remuneration on the basis of price would be paid in accordance with tendered unit rates or prices for provisional quantities of work items included in a special part of the bill of quantities. Where the basis of such additional remuneration is cost, then the list of adverse conditions would serve as an itemised bill of compensation events under the terms of the contract and a provisional sum is placed against each such event. In both of these methods, the total of this special part of the bill of quantities is carried forward to the total of the contractor's priced offer.

The following advantages were envisaged in the above approach:

(a) A better comparison and analysis of tenders can be achieved. The tenders would include the tenderer's rates and prices for, or the cost of, dealing with the specified unexpected reference events. This would be of benefit to the employer in identifying the most cost-effective tender, inclusive of the cost of dealing with the specified events, as well as being an advantage to prudent contractors who would include such costs in their tender, irrespective of whether or not a special part is included in the bill of quantities.
(b) The contractor's priced offer would form a more realistic estimate of the cost of the project which is of benefit to the employer and the contractor alike.

(c) In compiling the list of unexpected events, the employer and his consulting engineer could be compelled to give more thought and time to identifying those unexpected events and to plan for dealing with the difficulties which may arise as a consequence of encountering them. Better planning of the design process and of the supervision of construction could thus be achieved.
(d) The enquiry into these unexpected events would lead to a more comprehensive site investigation which might eliminate part of the risk of the unknown.

Two serious objections to these proposals may be put forward:

— Deleting sub-clause 12.2 may be against the principles of the applicable law of the contract; see Chapter 8 in this book.
— The contractor in putting forward rates or prices for provisional events, must make certain assumptions. These assumptions extend to the quantity and complexity of the envisaged work, the plant required, whether or not it would be convenient to do such work simultaneously with other work, whether or not such work is of similar nature, whether or not a specialist type of labour and plant would be necessary, and so on. The risk to the employer and contractor in making the wrong assumptions is high and it is questionable as to whether or not the effect of such decision-making should be taken into account in the award of the contract.

There may, however, be some advantage in certain contracts adopting a compromise by leaving clause 12 unaltered but introducing the concept of a referenced list of unexpected events.

16.3.4 Employer's risks

Sub-clause 20.3 of the Red Book stipulates the procedure to be followed when loss or damage arises from any of the employer's risks as defined in sub-clause 20.4, or from a combination with other risks. In this connection, the provisions of sub-clause 22.2 and clause 65 should also be considered when assessing the effect of the employer's risks. In the event of any such loss or damage, the engineer is expected to inform the contractor whether or not he is required to rectify the loss or damage and if so, to what extent.

The engineer is required to determine an addition to the contract price for work done by the contractor in rectifying such loss or damage, such addition to be calculated under clause 52 for varied work. The engineer is then required to notify the contractor accordingly with a copy to the employer.

Where the loss or damage results from a combination of employer's and contractor's risks, the engineer must take into account their proportional responsibility.

Claims under this heading can therefore be easily formulated and made if the employer's risks are precisely defined in the Red Book. Problems will arise, however, under paragraph (h) of sub-clause 20.4 in which an attempt has been made to allocate some of the risks of the forces of nature to the employer. Once again, as in clause 12, the 'experienced contractor' who features prominently, is assisted by an ambiguous phrase which in this case is 'could not reasonably have been expected to take precautions'. In order to determine whether or not the contractor has a valid claim under this provision of the contract, it has first to be determined whether or not an experienced contractor could not reasonably have been expected to take precautions against the risk of the operation of the particular force of nature.

Unlike the ambiguity of clause 12, the impact of the problem in this case is slightly diminished by the fact that this risk is, in any case, required to be insured against under the provisions of clause 21. Accordingly, the problem will lie in the difference between the ability of the contractor to claim for the cost of rectifying any loss or damage resulting from an employer's risk, under clause 52 or a contractor's risk through the insurance policy provided under clause 21. Under clause 52, the contractor can claim for all his costs plus profit, whereas under the insurance policy he would forfeit an amount equivalent to the deductible under the provisions of the insurance policy together with, and depending upon the terms of the insurance policy, any profit he is expected to make.

Reference should also be made in this connection to Chapter 14 where the employer's risks are considered in greater detail.

16.3.5 Compliance with statutes, regulations, price fluctuations, currency and other economic causes

The provisions of clauses 26, 70, 71 and 72 of the Red Book regulate the relationship between the parties in connection with matters closely associated with the economic policy and the laws of the place where the project is constructed. Accordingly, supplementary clauses are drafted in Part II of the Red Book which provide for specific requirements of individual projects depending on their status, nature and size.

Clause 26 provides that the contractor should make all payments properly imposed under the relevant legal rules as mentioned in the clause. Two exceptions apply to these payments. The first is in respect of any compensation which may be payable for occupation of any land by the works, or any part thereof, as provided in sub-clause 22.2. The second is in respect of obtaining any planning, zoning or other similar permission required for the works to proceed. Both of these exceptions are the responsibility of the employer. Thus, claims may arise as a result of non-compliance with the provisions of clause 26. The Red Book does not prescribe a designated remedy in the form of financial compensation as a result of such non-compliance and so such claims for non-compliance come under the second category of claims as discussed at the beginning of this chapter, i.e. claims based on the grounds that a term of the contract had been breached but where the remedy is not designated. These claims are discussed in Section 16.4 below.

Sub-clause 70.1 deals with fluctuations in the costs of labour and materials or any other matters affecting the cost of construction. It is particularly important in projects of long duration and at times of inflation where the risk of increase in such costs is high. Unless wage rates are controlled by official bodies within the country where labour is employed, it is difficult to ascertain the fluctuations which occur. Similarly, unless the cost of materials is controlled by official indices showing the cost of various basic materials, it is extremely difficult to envisage a formula by which such costs could be calculated without the risk of serious abuse. Accordingly, once again, the opportunity is given in

Part II to draft specific clauses for the particular project under consideration and in many cases consideration should be given to allocating the risk of such fluctuation to the contractor on the basis of his ability to plan for minimising the effect of this risk.

Sub-clause 70.2 provides for cost fluctuation as a result of changes in legislation in the country where the project is to be constructed, which occur after the date 28 days prior to the latest date for submission of tenders for the contract. Such fluctuation is easily calculated, and the engineer is required to determine any additional or reduced cost after 'due consultation' with the employer and the contractor. Once determined, the additional or reduced costs should be notified by the engineer to the contractor with a copy to the employer. In this connection, reference should be made to Chapter 8 of this book and in particular to Reference 8.13 quoted therein.

Where clauses 71 and 72 are concerned, claims may be made in respect of currency exchange rate, variations and foreign currency proportions in accordance with the procedures provided and the particulars of the contract in question.

16.3.6 *Defects and unfulfilled obligations*

A defect has been defined as 'a failure of the completed project to satisfy the express or implied quality or quantity obligations of the construction contract.'.[16.23]

From the contractual point of view this is a succinct definition linking together the requirements of a number of sub-clauses in the Red Book dealing with defects. It is of benefit to consider these sub-clauses together but as they are scattered throughout the document, it is sometimes difficult to visualise their effect fully. However, before considering these sub-clauses, a controversial decision of the House of Lords in this connection is worthy of note. It is the decision in *Ruxley* v. *Forsyth*.[16.24]

A contract was entered into between Ruxley and Mr Forsyth for the construction of a swimming pool at Mr Forsyth's home. The pool was required to have a maximum depth of 7ft 6in (2.29m), but on completion it was discovered that the maximum depth was only 6ft (1.83m). Mr Forsyth sought to recover the sum of £21,560 in respect of the cost of demolishing the pool and reconstructing it to the depth originally specified. At first instance, the judge decided that as constructed, the pool was perfectly safe to dive into and that therefore the disadvantage of the reduced depth was totally disproportionate to the cost of demolition and reconstruction. He only awarded Mr Forsyth the sum of £2500 for loss of amenity. Mr Forsyth appealed. The Court of Appeal disagreed with the decision of the lower court and allowed the appeal on the basis that the only way in which the contractual objective of a depth of 7ft 6in could be achieved was by demolition of the pool and reconstructing it to the specified depth. Accordingly, Mr Forsyth was awarded the full amount of £21,560.

Ruxley appealed to the House of Lords which was faced with the question as to whether to award damages for the cost of providing what was promised or to award damages for the loss of value and amenity. Finding in favour of the contractor, the House of Lords concluded that the trial judge was justified in holding that it would be unreasonable to incur the cost of demolition of the existing pool and reconstruction of a new one. Whilst there are many repercussions to the detailed judgment of this case, its immediate impact should be considered with respect to clause 39 which requires the contractor to remove defective work and the fact that the Conditions do not permit the engineer to accept work not carried out in accordance with the contract. Another notable point in this case is the cost of litigation which for Ruxley alone reached the figure of £160,000.

Returning to the various sub-clauses under the Red Book which deal with defective work, these are set out below in a logical sequence:

(a) Sub-clause 36.1 defines the required quality of materials, plant and workmanship and provides that they are to be subjected, from time to time, to such tests as the engineer may require.
(b) Sub-clause 55.1 describes the quantities of the works to be executed by the contractor in fulfilment of his obligations under the contract.
(c) Sub-clauses 8.1 and 13.1 set out the contractor's general responsibilities within the contract, which include the obligation to remedy any defect in the works unless it is legally or physically impossible to do so.
(d) Sub-clause 37.4 incorporates the provision that as a result of an inspection or testing referred to in clause 37, the engineer can reject the materials or plant inspected, if he determines that they are defective or otherwise not in accordance with the contract. The engineer is then required to notify the contractor immediately stating his objections with reasons. The contractor is required to make good the defect promptly or ensure that the rejected items comply with the contract. All costs incurred by the employer due to any repetition of the tests performed must, after 'due consultation' with the employer and the contractor, be determined by the engineer. Such costs are recoverable from the contractor by the employer and may be deducted from any monies due or to become due to the contractor. As in other engineers' determinations, it is required that the contractor be notified accordingly, with a copy to the employer.

It is important to note that failure to comply with the engineer's notice referred to above, exposes the contractor to the sanctions under sub-clause 63.1. Upon certification by the engineer, the provisions of sub-clause 63.1 entitle the employer, after giving 14 days' notice to the contractor, to enter upon the site and the works and terminate the employment of the contractor (see also Section 16.3.7 below).
(e) Sub-clause 39.1 provides that the engineer is authorised to instruct the contractor to remove from the site any materials or plant which are not in accordance with the contract. As in sub-clause 37.4, failure by the contractor to comply with the engineer's instruction in this sub-clause, will expose the contractor to the provisions of clause 63.
(f) Sub-clauses 48.1 and 49.2 provide that once notified by the engineer, the contractor must remedy any defect in the works before the taking-over certificate or as soon as is practicable after the date of that certificate.

Sub-clause 49.2 also provides that the contractor is to remedy any defect, shrinkages or other faults as may be instructed by the engineer during the defects liability period or within 14 days after its expiration. Furthermore, it provides that besides the remedying of defects, the contractor is required to execute any work of amendment and reconstruction instructed by the engineer during the defects liability period or within 14 days after its expiration.

(g) Sub-clause 49.3 provides that the cost of remedying defects, shrinkages or other faults as well as that of any amendment and reconstruction work is to be borne by the contractor where the cause is that of a breach of contract by him. Otherwise, it is to be borne by the employer as an addition to the contract price in accordance with clause 52.

(h) Clause 50 gives authority to the engineer to instruct the contractor to search, under his direction, for the cause of any defect, shrinkage or other fault which appears at any time prior to the end of the defects liability period. It is obvious that this clause applies where the cause of the apparent defect is not known. Knowing the cause of a defect has two important purposes. The first is to determine the appropriate method of remedy and the second is to determine the liability.

The clause provides that if the defect is one for which the contractor is liable, then the cost of the search and of the remedy is to be borne by the contractor. Otherwise, the engineer, after 'due consultation' with the employer and the contractor, is required to determine the cost of such search which is to be added to the contract price. The provisions of sub-clause 49.3 apply to the cost of remedying the defect itself as referred to in item (f) above. The engineer is again required to notify the contractor accordingly, with a copy to the employer.

(i) Sub-clause 26.1 provides in its paragraph (a) that the contractor is required to conform in all respects with statutes, regulations, bylaws, etc., including the giving of notices and payment of all fees when remedying any defects in the works.

(j) Sub-clauses 39.2 and 49.4 provide for the situation where the contractor defaults in carrying out an instruction of the engineer under sub-clauses 39.1 and 49.2 respectively within the time specified or if not specified, then within a reasonable time. In such a case, the employer is entitled to employ and pay others to carry out the work instructed by the engineer. The employer is also entitled under sub-clause 64.1 to employ and pay others to carry out any remedial or other work which is, in the opinion of the engineer, urgently necessary for the safety of the works and which the contractor is unable or unwilling to do at once, whether during the execution of the works or during the defects liability period.

(k) Finally, sub-clause 60.3 provides, under its paragraph (b), that upon the expiration of the defects liability period, should there remain to be executed any work ordered pursuant to clauses 49 and 50 (defects liability), the engineer is entitled to withhold certification of as much of the balance of the retention money as shall, in his opinion, represent the cost of such work. The withholding of the certificate continues until such work is executed. If such work remains undone, then the provisions of sub-clause 49.4 apply, as described above.

Subject to the terms of the defects liability certificate, none of the above provisions supplants the rights of the parties to seek redress in respect of defects discovered during the execution of the works, or in the case of latent defects, after their completion. Sub-clause 62.2, in fact, specifically provides that the contractor and the employer is to remain liable for the fulfilment of their respective obligations under the provisions of the contract prior to the issue of the defects liability certificate.

Subject to the provisions of the applicable law of the contract, this express provision, combined with the obligation to complete, has in the case of claims for defective work or for defects, the valuable effect of causing time to start to run against the employer, for the purposes of the statutes of limitation, from the date of the defects liability certificate.

16.3.7 Failure to commence, delays, suspension of work, release from performance, default and termination

There are few designated remedies in the Red Book in the form of financial compensation as a result of the failure of either party to perform his obligations under the above heading. In most circumstances, the financial compensation is not designated and comes under the second category of claims as discussed at the beginning of this chapter, i.e., claims based on the grounds that a term of the contract had been breached but where the remedy is not designated. These claims are discussed in Section 16.4 below.

However, it is important to list the provisions of the Red Book which include designated remedies and these are as follows:

(a) Sub-clause 6.4 incorporates provisions which protect the contractor if he is delayed or incurs additional costs as a result of late issue of drawings or instructions in respect of which he has given notice in accordance with sub-clause 6.3. Sub-clause 6.5 provides for the situation where the contractor contributes to the delay in respect of which notice had been given under sub-clause 6.3. The engineer is required, after 'due consultation' with the employer and the contractor, to determine any extension of time and any amount of costs which should be added to the contract price.

In determining these elements, the engineer is required to take into account the contractor's contribution if any, to such delay. The engineer is then required to notify the contractor accordingly, with a copy to the employer.

(b) Clause 40 regulates the consequences of a suspension of the progress of the works or any part thereof when instructed by the engineer for a number of reasons specified in sub-clause 40.1. Figure 16.3 shows a flow chart of the procedure stipulated under clause 40 (and clause 69 in the event that the suspension lasts for more than 84 days). This chart shows that the contractor is entitled to financial reimbursement in respect of costs incurred and extension of time, determined by the engineer in consultation with the parties pursuant to sub-clause 40.2, unless the reason for the suspension is as prescribed in sub-paragraphs (a) to (d) of sub-clause 40.1. For a clear understanding of the procedure set out in the flowchart of Figure 16.3, it must be borne in mind that whilst the provisions of this clause only apply to a suspension ordered by the engineer, there is no obligation on him to give such an instruction. The authority to suspend the work is purely discretionary. Accordingly, no matter what the difficulties are, the engineer may decide that suspension of the progress of the works is not appropriate, in which case it is the contractor's duty to continue the work or as the case may be, for example, remove defective work and reconstruct without delay.

On the other hand, the contractor has the right to and may suspend work or reduce the rate of work in accordance with the provisions of sub-clause 69.4, if the employer fails to make a payment within the stipulated period under clause 60. If as a result of such an action by the contractor, he suffers delay or incurs costs, the engineer is required to determine, after 'due consultation' with the employer and the contractor the amount of such costs which are to be added to the contract price.

Where the progress of the works is suspended, it is essential that attention should be given to:

(i) partially completed work which is particularly vulnerable to hazards occurring during a temporary incomplete state for which it is not designed. The contractor's obligation to protect the works therefore assumes a different dimension and may require special action such as storage of certain items or costly and significant, though temporary, protective measures;

(ii) elements of the works in the process of being manufactured;

(iii) the equipment on the site; its detailed characteristics, such as, type, number, age, origin and value; the method of calculating the effect of suspension on its value; and the steps which must be taken to safeguard it against damage; and

(iv) any part of the works found to be defectively designed or executed.

Suspension of the works or of part of them may have to be ordered for a number of reasons which include: significant changes in design; additions of major elements to the works; unavailability of a part or parts of the site; the occurrence of one or more of the employer's special risks as specified under the contract, such as for example civil war; and temporary economic restraint experienced by the employer.

Whilst it may be financially prudent to suspend the works in any of the above circumstances, rather than to continue the work and face the inevitable claims afterwards, the financial consequences of a suspension order could be extremely costly to either or both the employer and the contractor. This is particularly so where heavy and expensive equipment is used on the project. The cost of placing such equipment on standby for a long period of time may end in costs approaching the value of the work intended to be executed using such equipment. Standby costs of equipment are usually calculated by reference to appropriate depreciation figures based on either a linear or non-linear relationship between the residual value of the equipment and the number of years of use.

(c) Sub-clause 42.1 provides for the requirement of the possession of the site from time to time, in a certain order and in such portions as may be necessary and subject to the requirements of the contract.

Failure by the employer to give possession of the site or portions of it which causes the contractor to suffer delay and to incur costs is governed by the provisions of sub-clause 42.2. The financial compensation, if any, is determined by the engineer, after 'due consultation' with the employer and the contractor. Once determined, the engineer is required to notify the contractor accordingly, with a copy to the employer.

(d) Clause 63 deals with the provisions which apply in the case of default by the contractor. Before dealing with the financial compensation arrangements provided for in this clause, it is important to note that the subject of default and its consequences is legally and technically a far-reaching and very complex subject. It is especially so where the language of the applicable law of the contract is different from that used in the Red Book. For example, when legal rules are translated from another language into English, they may contain terms such as 'rescission', 'repudiation',

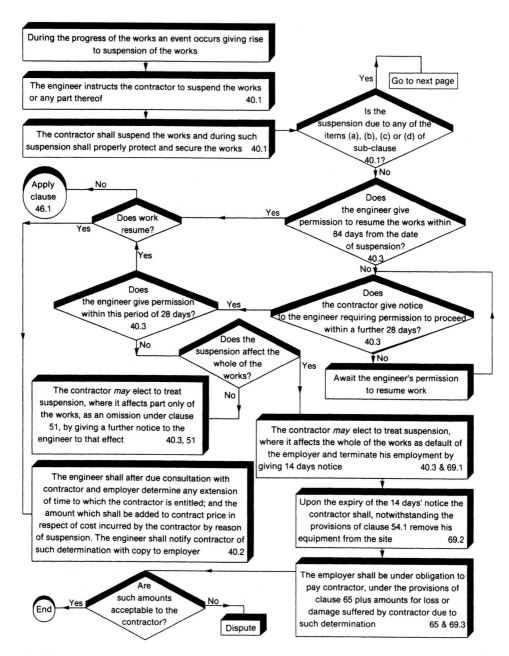

Fig. 16.3 Suspension.

Claims and Counterclaims

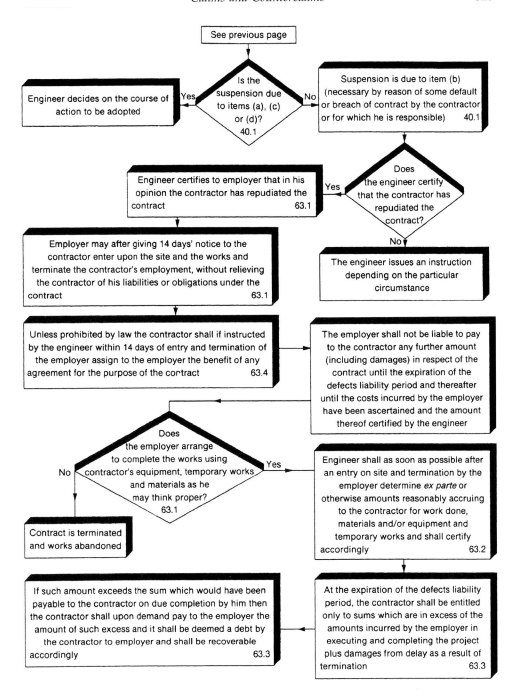

(Fig. 16.3 Contd.)

'frustration' or 'termination'. Each of these legal terms may have a specific meaning under the applicable law of the contract which in turn may have a certain impact on the provisions of clause 63. It is therefore essential that the terms used in the Red Book are carefully examined to establish their compatibility with the applicable law of the contract and with what is intended by the parties to be the consequence of default should it happen.

Clause 63 as worded in the Red Book is divided into four sub-clauses, the first of which defines when a default of the contractor occurs and establishes that the employer may, after giving 14 days' notice to the contractor, enter upon the site and the works and terminate the employment of the contractor. Sub-clause 63.1 also provides that such termination does not release the contractor from any of his obligations or liabilities under the contract and neither does it affect the rights nor authority conferred on the employer or the engineer. Following such termination, the employer may himself complete the works or may employ any other contractor to complete the works. It is also provided that he is entitled to certain rights against the contractor which include the use of the contractor's equipment, temporary works and materials.

The language of sub-clause 63.1 assumes that the employer would wish to complete the works. However, this may not be the case if, for example, the nature of the default is such that changes in design or execution must be implemented.

This brings into focus the question of whether or not the employer is able to apply the provisions of the applicable law of the contract to rescind the contract rather than terminate it under the provision of clause 63. In order to answer this question, the provisions of the applicable law of the contract must be identified and considered. For example, in certain jurisdictions, the court may grant time to the defaulting party to attend to the default or it may reject the application to rescind the contract if the obligation which had not been performed is insignificant in comparison with the obligation under the contract as a whole. In other jurisdictions, the fact that the provision of clause 63 permits the employer to terminate the employment of the contractor without affecting the rights and obligations of the parties is inconsistent with the more stringent definition of rescission under which the contract is treated as being at an end.

Another area of extreme importance is whether or not the contractor can be removed from the site once his employment is terminated by the employer. The contractor may object to such termination and the matter may result in a dispute which may remain unresolved until an arbitrator *or* a court holds that the termination is justified. The dispute resolution process, however, may take a long time. Accordingly, the provisions of clause 63 may have to be critically assessed.

The second and third sub-clauses of clause 63 must also be examined, for the purposes of the discussion under the heading of financial compensation as a result of default by the contractor. Sub-clause 63.2 provides that 'the Engineer shall, as soon as may be practicable after any such entry and termination by the Employer, fix and determine ex parte, ... and shall certify:' at the date of termination a valuation of work carried out under the contract, of unused or partially-used materials, of any contractor's equipment and of any temporary works.

Sub-clause 63.3 provides that the employer shall not be liable to pay to the contractor any amount

> 'until the expiration of the Defects Liability Period and thereafter until the costs of execution, completion and remedying of any defects, damages for delay in completion, if any, and all other expenses incurred by the Employer have been ascertained and the amount thereof certified by the Engineer'.

A claim for damages under the heading of breach of contract may also apply in the case of default by a contractor (see Section 16.4 below).

(e) Sub-clause 65.8 provides for payment if the contract is terminated under the terms of clause 65 (special risks). Details of the payment are contained in paragraphs (a) to (f) of this sub-clause. They also apply in the case of the provisions of sub-clause 66 where it is impossible or unlawful for either or both of the parties to meet their obligations under the contract. Such a situation arises when circumstances beyond the control of the parties occur after the issue of the letter of acceptance or where the parties are released from performance under the applicable law of the contract.

(f) Clause 69 deals with the provisions which apply in the case of default by the employer. Sub-clause 69.1 defines the events which render the employer in default and entitle the contractor to terminate his employment under the contract. The contractor may do so by giving notice of termination to the employer, with a copy to the engineer, 14 days after which the termination automatically takes effect.

As in the case of default of the contractor, the applicable law of the contract is also a critical element of the termination process where the employer is in default. See point (d) above. Furthermore, some provisions of clause 69 may not be appropriate. Paragraph (c) of sub-clause 69.1, for example, would have to be deleted where the employer is a government.

Sub-clause 69.3 provides the terms of payment on termination which are the same as those under clause 65, but, in addition, the employer is required to pay 'to the Contractor the amount of any loss or damage to the Contractor arising out of or in connection with or by consequence of such termination'.

16.3.8 *Other miscellaneous specified events*

The Red Book contains a number of provisions where financial compensation is specified should certain events occur other than those already discussed above. These include the following events arranged in the order in which they appear in the document:

(a) *Clause 17*: Where an error in setting out of the works is based on incorrect data supplied in writing by the engineer, the contractor is entitled to be paid an amount determined by the engineer in accordance with clause 52.

(b) *Clause 18*: Where the engineer requires the contractor to make boreholes or to carry out exploratory excavation, such requirement shall be instructed in accordance with clause 51 and accordingly valued and paid for in accordance with clause 52, unless an item or a provisional sum in respect of such work is included in the bill of quantities.

(c) *Clause 26*: The contractor is required under this clause to pay all fees in conformity with the provisions of national statutes, regulations, etc. Non-payment by the contractor will constitute a breach of contract, the remedy for which is stated to be an indemnity in favour of the employer against all penalties and liability of every kind for breach of these provisions.

Furthermore, it is stipulated in this clause that the employer is responsible for obtaining any planning, zoning or other similar permission required for the works to proceed. Failure to obtain these permissions will constitute a breach of contract by the employer, the remedy for which is designated as an indemnity in favour of the contractor in accordance with sub-clause 22.3, against all claims, proceedings, damages, costs, charges and expenses in connection therewith.

(d) *Clause 27*: Where fossils, coins, articles of value or antiquity and structures and other remains or things of geological or archaeological interest are discovered on the site, the engineer may issue an instruction for dealing with such articles. Should the contractor incur any costs (or delays) as a result of such an instruction, the engineer is required to determine, after 'due consultation' with the employer and the contractor, the amount of such costs incurred which are then added to the contract price (and an extension of time, if appropriate).

(e) *Sub-clause 28.1*: The contractor is responsible for dealing with patent and other rights and royalties in respect of plant, materials or contractor's equipment. Breach of this condition will invoke the designated remedy of indemnity in favour of the employer from and against all claims, proceedings for or on account of such infringement.

(f) *Sub-clause 29.1*: This sub-clause places an obligation on the contractor to limit the interference which his operations may cause to public or private thoroughfares or to properties adjoining the site. Failure to do so is a breach of contract which would invoke the designated remedy of indemnity to the employer in respect of all claims, proceedings, damages, costs, charges and expenses whatsoever arising out of, or in relation to any such matters for which the contractor is responsible.

(g) *Sub-clause 30.2*: The contractor is responsible for the cost of strengthening any bridges and altering or improving any road communicating with or on the routes to and from the site. Breach of this condition has the designated remedy of an indemnity to the employer against all claims for damage to any such roads or bridges.

(h) *Sub-clause 30.3*: Should the contractor fail to observe and perform his obligations under sub-clause 30.1 (avoidance of damage to roads) and any damage occurs, an amount should be determined by the engineer, after 'due consultation' with the employer and the contractor, to be recoverable from the contractor by the employer.

(i) *Sub-clause 31.2*: If the engineer requests in writing and the contractor provides facilities for other contractors pursuant to sub-clause 31.1 (opportunities for other contractors), the engineer is required to determine an addition to the contract price in accordance with clause 52.

(j) *Clause 36*: Where the cost of samples and tests are clearly intended by or provided for in the contract, the cost is to be borne by the contractor. However, where tests are neither intended by nor provided for under the contract, then sub-clause 36.5 requires the engineer to determine, after 'due consultation' with the employer and the contractor, an amount to be added to the contract price.

(k) *Sub-clause 38.2*: The engineer may require, as a consequence of some subsequent discovery, that work already covered up should be uncovered, inspected and tested. If such work is found to be in accordance with the contract, the engineer is required to determine, after 'due consultation' with the employer and the contractor, the amount of the contractor's costs which should then be added to the contract price.

(l) *Clause 43*: This clause provides that the works and, if applicable, any section thereof

> 'shall be completed in accordance with Clause 48, and within the time stated in the Appendix to Tender for the whole of the Works or the Section (as the case may be), calculated from the Commencement Date, or such extended date as may be allowed under Clause 44'.

Accordingly, it links time for completion with the taking-over certificate in clause 48, the Appendix to Tender and any extensions of time granted under clause 44. It should also be read in conjunction with the liquidated damages provision in clause 47. Clause 47 provides a designated remedy in the case of breach of the provisions of clause 43 by the contractor.

The amount of liquidated damages is determined by the employer before tenders are invited. Such determination should be based on a reasonable assessment of the actual damages which he would suffer should there be a delay. It is usual to have a limit to liquidated damages specified in the Appendix to Tender.

16.4 A claim arising out of or in connection with the contract

As previously discussed, in Section 16.1, this category of claim is based on grounds, for example, of a breach of a particular term of the contract or misrepresentation, the remedy for which is not specified in the Red Book. Accordingly, the remedy would have to be found under the provisions of the applicable law of the contract through arbitration or litigation, unless the claim is settled amicably. In addition to this basic difference, there are other differences between claims under this category and those that arise based on the provisions of the contract.

In specific terms, these differences depend on the applicable law of the contract but in general they would include the following differences where contracts are based on the Red Book:

(a) Until the 1992 Reprint of the Red Book, the engineer had no authority to deal with and determine the liability for claims on grounds, for example, of a breach of a particular term of the contract or misrepresentation where the remedy was not specified, unless the standard form of contract was amended to include a specifically drafted clause empowering him to do so under the contract. It was the employer who had to deal with such a claim as he is the other party to the contract. Of course, the employer may have appointed the engineer to deal with such claims on his behalf. If the claim was rejected by the employer, or by the engineer if appointed on his behalf, then any dispute arising in that connection would have had to be submitted to the engineer for a decision under clause 67.

However, in the 1992 reprint of the Red Book, the words 'or otherwise' were added at the end of sub-clauses 60.1, 60.6(b) and 60.8(a) after the words 'under the contract', thus enabling the engineer to deal with any sums to which the contractor may be entitled 'under the Contract or otherwise'. These sub-clauses relate to: monthly statements, final statement and final certificate, respectively.

(b) In the Third Edition of the Red Book, the requirement to give notice to the engineer of intention to make a claim does not generally apply to a claim for breach of contract or misrepresentation, unless the standard form of contract had been specifically amended to reflect an intention to that effect. If such an intention is not specified, then it is up to the parties to deal with the claim themselves, without the engineer's input, and failing agreement, the matter would have to be resolved through the dispute settlement mechanism in the contract. However, in the Fourth Edition of the Red Book, it is specifically required in sub-clause 53.1 that notice of intention to claim is required whether such claim is made 'pursuant to any Clause of these Conditions or otherwise'. The word 'otherwise' is used in this phrase to signify the inclusion of claims arising out of or in connection with the contract, for example for breach of contract or misrepresentation.

Reference in this connection should be made to Section 3.5 and, where the common law system is concerned, to Figure 3.1.

Examples of breach of contract claims where the remedy is not specified under the Red Book include all the provisions where obligations are placed on either or both of the parties. An example is clause 41 where the contractor is required to commence the works 'as soon as is reasonably possible after receipt of a notice to this effect from the Engineer . . .'. Failure to do so forms one ground of termination of the contract under sub-clause 63.1 and could result in a claim for breach of contract as well as for the remedies specified in sub-clause 63.2.

(c) Whereas the amount payable in respect of a claim under category (a) above should be easily calculable, the amount of damages payable under a category (b) claim is much more difficult to assess. Furthermore, the risk of estimating in advance an accurate assessment of such damages is too great.

If the claim is not resolved amicably, arbitrating or litigating over the amount of damages remains a significant and possibly costly gamble.

16.5 Procedure for claims for additional payment – clause 53

Clause 53 of the Fourth Edition of the Red Book is a new clause with greatly expanded provisions compared with those of sub-clause 52.5 in the Third Edition. It introduces an improved discipline into the area of claims and ensures that prompt attention is given to the events which may give rise to such claims. The clause sets out in detail a procedure for submitting and dealing with claims which should reduce, if not eliminate, a number of the contentious problems which have existed in the past.

16.5.1 Procedural steps

Figure 16.4 illustrates the discipline referred to in clause 53 in which it is required to:

(1) give a notice of intention to claim within 28 days of the event giving rise to the claim (note it is not after the consequences of the event). The notice is required whether the claim is pursuant to a clause of the Red Book or otherwise. It does not have to include any details of the claim itself;
(2) keep contemporary records (by the contractor);
(3) inspect the records (by the engineer);
(4) provide authority to instruct the contractor to keep further contemporary records (by the engineer);
(5) within 28 days of the notice or an agreed period, submit particulars of the claim in respect of amount and grounds upon which it is based (by the contractor);
(6) interim and accumulated accounts to be submitted for continuing effects (by the contractor);
(7) final accounts to be submitted at end (by the contractor);
(8) a copy of accounts to be sent to the employer, by the contractor, if so required by the engineer.

In respect of the contractor's obligation to give notice of his intention to claim, the discussion above in section 16.3.1 in respect of the *Tersons* case is relevant (see Reference 16.12). It is worth noting also that this notice is of the contractor's *intention* to claim for additional payment, the emphasis added to highlight that this notice is merely the contractor's signal to the engineer/employer that an event has arisen prompting the contractor to consider making an assertion of a right to additional payment in the future. Many contractors and engineers take this notice to constitute the claim itself. This would appear, at first glance, not to be the case, as borne out by the wording in sub-clause 53.2:

> 'Upon the happening of the event . . . the Contractor shall keep such contemporary records as may reasonably be necessary to support any claim he may *subsequently* wish to make' (emphasis added).

When does the claim manifest itself then? On the face of the above wording of sub-clause 53.2 it might be said that the claim comes into being only when the contractor submits the substantiation to support his claim. In the sentence after this wording, the engineer's obligation arising out of the notice is set out:

> '. . . the Engineer shall, on receipt of a notice . . . inspect such contemporary records and may instruct the Contractor to keep any further contemporary records as are reasonable and may be material to the claim of which notice has been given.'

This sentence does not oblige the engineer to assess the potential claim but does require him to direct his attention to the sort of records he will need in order to deal with it effectively. But what about the phraseology of the last part of this sentence – does the

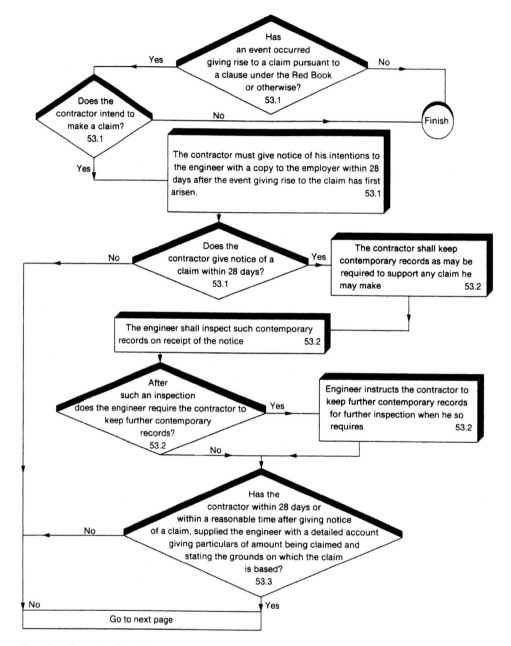

Fig. 16.4 Procedure for claims.

use of the wording *'the claim of which notice has been given'* mean that the notice is to be taken by the engineer to be a notice of claim and not merely of intention to claim? Sub-clause 53.4 gives the answer: the engineer (or an arbitrator) has the unilateral power and, indeed, a duty to assess the contractor's entitlement to additional payment

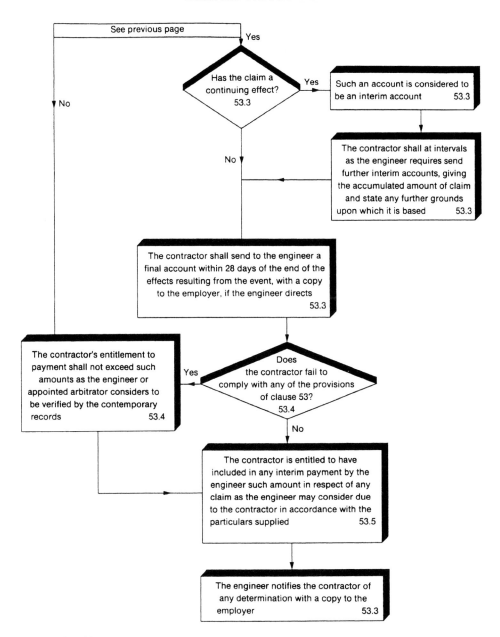

(Fig. 16.4 Contd.)

regardless of whether the contractor has failed to comply with any part of the claims procedure set out in clause 53, tempered only insofar as the claim can be verified by contemporary records. Therefore, this power/duty means that the engineer must treat the contractor's notice of intention to claim as a claim in its own right.

Some commentators have questioned the wisdom of sub-clause 53.4 which limits the contractor's entitlement to be paid to the amounts he can verify in the case of a failure to or a delay in notifying his claim or submitting accounts and particulars. The criticism is based on the fact that this limitation ought to apply in any case, even when there is no failure on the part of the contractor to comply with any of the provisions of clause 53. Sub-clause 53.4 is, however, of benefit in establishing that failure to comply with the provisions of sub-clauses 53.1, 53.2 or 53.3 does not bar the contractor's entitlement to additional payment to the extent that is verified by contemporary records (whether or not such contemporary records were brought to the engineer's notice).

Sub-clause 53.5 sets out the basis of payment of claims. It provides that the contractor is entitled to payments in respect of claims including interim payments for those claims having a continuing effect and should be dealt with in conjunction with sub-clauses 60.5 to 60.9. Accordingly, payment to the contractor for claims having a continuing effect does not have to await the settlement of the whole claim.

In this connection, however, it is important to note that under the provisions of sub-clause 60.9, the employer is not liable to the contractor for any matters arising out of the contract, unless the contractor has included a claim in respect thereof in his final statement and in the statement at completion, except in respect of matters arising out of the taking-over certificate.

16.5.2 Records

Actual and accurate records and information are necessary to establish the costs incurred as a result of any of the events leading to a claim and the provisions of sub-clause 53.2 of the Red Book must be complied with meticulously. It is recommended that such records include:

— Clause 14 programme setting out what the contractor had intended for the order, sequence and timing of the various activities at the time of tender.
— An estimate of resources and anticipated expenditure in units of time, which are required to achieve the clause 14 programme.
— Any update and revision programmes in accordance with events which may occur during the progress of the works, as required in sub-clause 14.2.
— Progress programmes setting out the progress of the various activities against the clause 14 programme.
— Records of actual resources and actual expenditure based on progress.
— Records of any resources which were standing or uneconomically employed.
— Records of overtime worked, and the cost thereof.
— Progress photographs and/or videos.
— Drawings register (with details of amendments and updates, if any).
— Site diaries.
— Approved minutes of meetings.
— Labour allocation sheets.
— Plant allocation sheets.

16.6 The presentation of claims

Whether a claim is submitted to the engineer or to an arbitrator as part of a statement of claim, it should be coherent but not wordy presented in a clear and logical manner. A well prepared and documented claim will save time in dealing with it and will have a better chance of success. It should have the following sections:

— Introduction providing details of the parties, the contract, the project, relevant dates and other pertinent information.
— Brief description of the works as required and specified in the relevant documents.
— Description of the claim events as they occurred and the nature of the resultant problems encountered.
— A section containing a statement of the facts, so that facts can be established.
— An analysis of the facts showing the legal basis upon which the claim is founded, citing the relevant provision and clause of the conditions of contract under which relief is claimed.
— Description of the procedural steps already taken in dealing with the claim event, including notices given, if any.
— Calculation of cost impact and the methods used.
— Calculation of the claimed extensions of time, if any, showing critical and non-critical delays encountered.
— Calculation of disruption experienced, if any, and the method used to calculate its impact.
— Appendices which should include all contemporaneous records and relevant correspondence relied upon.

16.7 Quantum

Although quantification of the type and extent of the remedy, whether designated under the contract or not, is dependent to some degree on what is permitted under the provisions of the applicable law of the contract, it is nevertheless accepted that the basic principle for the quantification of claims is that of damages under contract law, namely that the claimant should be put in the position he would have been in had the claim event not occurred. The quantification of claims where the remedy is specified under the conditions of contract depends on the precise wording of the provisions of the relevant clause or clauses forming the legal basis of the claim.

However, there could generally be additional costs which might be recoverable by the contractor as a result of a successful claim, depending on the nature of the claim event.

16.7.1 Heads of claim

In addition to the financial claims in respect of direct costs, the various heads of claim which may be encountered in practice include:

(a) On-site overhead costs: These are the costs incurred by the contractor where he is delayed and flow from the fact that he must maintain a continued site operation or presence and retain resources on site for longer than originally anticipated. They are referred to as site overheads and consist of the costs of administrative and supervisory staff on the site including but not limited to:
— Site management staff
— Health and safety officers and equipment
— Project-specific insurance, if any
— Trades foremen
— Plant and tools
— Welfare including cleaning etc.
— Lighting and power
— Storage areas/buildings and workshops
— Temporary works
— All contractor's site offices, mess cabins, etc. including equipment, consumables and communication charges
— Accommodation for the employer's representatives
— Sanitary accommodation
— Scaffolding
— Transport
— Sub-contractor prolongation cost claims

(b) Off-site overheads: Clause 1.1 (g)(i) of the 4th Edition of the Red Book defines 'cost' as 'all expenditure properly incurred or to be incurred, whether on or off the site *including overheads* and other charges properly allocable thereto but does not include any allowance for profit' (emphasis added). As stated above, it is generally accepted that where a contractor is delayed, he must maintain a continued site presence and retain resources on site for longer than originally anticipated. If these resources are not in a position to produce a corresponding increase in the company earnings, then there will be a shortfall in the contribution made by this site to defray head office overhead costs. Therefore, the off-site overheads cover contributions by individual contracts to the cost of maintaining the contractor's head office and non project-specific insurances. They are difficult to establish, especially in respect of a period of delay and disruption or prolongation of a particular contract, where a specific allocation of time to the various contracts is difficult to assess. Accordingly:

> 'What has to be calculated here is the contribution to off-site overheads... which the Contractor might reasonably have expected to earn with these resources if not deprived of them. The percentage to be taken for overheads... for this purpose is not therefore the percentage allowed by the Contractor in compiling the price for this particular contract, which may have been larger or smaller than his usual percentage and may or may not have been realised. It is not that percentage (i.e. the tendered percentage) that one has to take for this purpose but the average percentage earned by the Contractor on his turnover as shown by the Contractor's accounts.'[16.25]

This problem had led to the use of formulae, the Hudson, Emden and Eichleay formulae, which were developed for calculating off-site overheads where approximate figures can be obtained.[16.26] They have been the subject of scrutiny by the courts in various cases, amongst which is *Property and Land Contractors Ltd* v. *Alfred McAlpine Homes Ltd* case, where it was stated that the correct approach was to ascertain the actual loss or expense incurred and not any hypothetical loss or expense that might have been incurred.[16.27] The use of a formula should be approached with care as it does not establish the proof of such cost and is no substitute for accurate records which enable actual costs to be determined.

However, in the more recent case of *Norwest Holst Construction* v. *Cooperative Wholesale*,[16.28] Judge Thornton QC dealt with the use of formulae when evaluating overheads. The contractor had calculated their claim for off-site overheads by using the Emden formula. The arbitrator had satisfied himself that the certain underlying assumptions of this formula had been demonstrated, namely:

— the loss must be proved;
— the contractor must show that other work was available and was turned down because of the delay, and that that work would have contributed to overheads;
— the delay must not have been caused by variations to the contract that have increased the contract sum and so increased the contribution towards overheads;
— the overheads must not have been ones which would have been incurred anyway, regardless of contribution from work undertaken; and
— there must have been no change in market conditions that would have affected the contractor's ability to obtain other work and there must have been no alternative work to which the contractor could have diverted resources.

In this case the arbitrator had found that a delay to the contract period had involved the head office employees in some additional costs and that it was reasonable to assume that the extra time spent on the contract would otherwise have been spent productively on other contracts. A loss had therefore been proved. Once the fact of such loss had been proved, the arbitrator had considered the proof of the extent or quantification of that loss. The arbitrator found that the contractor had been trading reasonably profitably during a period of intense competition in the construction industry and that the loss of productive management time created a significant likelihood that overhead recovery on other contracts was diminished. He assessed the contractor's loss as being one-fifth of that allowable under the Emden formula.

It was argued before the court that actual loss must be proved and reliance was placed on the decision in *McAlpine*. It was held that the decision in *McAlpine* did not rule out the use of a formula to ascertain the loss and expense arising from the expenditure of additional overheads. The question therefore was how it could be applied in this case and whether a percentage of the formula could be used. The judge found that the arbitrator was satisfied as to all the requirements underlying the use of the Emden formula and so the use of the formula was appropriate. Reference was made to the decision in *Allied Maples Group Ltd* v. *Simmons & Simmons*.[16.29] It was held in that case that where the plaintiff claims that he himself would have acted in a particular way, he must prove it on a balance of probabilities. However, where the claim is that an independent third party would have acted in a particular way, he only has to prove that there was a 'real' or 'substantial' chance of the third party's action. The contractor had proved that third parties would have acted in a different way if more of its management time had been spent on these other contracts, so that the element of 'substantial chance' had been proved. It was therefore open to the arbitrator to take the Emden formula and discount it substantially.

(c) Adverse weather conditions: A claim may arise if as a result of a change in the timing of the execution of the works attributable to the employer, adverse weather conditions are encountered causing delay.
(d) Increased costs of labour, materials or equipment due to inflation or influences external to the contract, sometimes known as 'contract price fluctuation'.
(e) Finance charges and interest on delayed payment, if any, of sums due under the contract.
(f) Profit on direct costs.
(g) Loss of profits.
(h) Interest on late payments.

16.7.2 The global approach

Claims must be substantiated and the procedure for making them is detailed in clause 53 as described earlier in Section 16.5. In general, it is also necessary for the claimant to establish each and every head of claim by means of separate supporting evidence. Furthermore, it is generally necessary to link, in a logical manner, the individual breaches or grounds of a claim to the particular sums claimed.

However, where no other way to compute damages is found to be feasible, it may be permissible under the applicable law of the contract in some jurisdictions to submit one 'global claim' for a number of claim headings without showing the link between the breaches or grounds of claim to the particular sums claimed. A global claim is defined in *Hudson's Building and Engineering Contracts* as follows:[16.30]

> 'Global claims may be defined as those where a global or composite sum, however computed, is put forward as the measure of damage or of contractual compensation where there are two or more separate matters of claim or complaint, and where it is said to be impractical or impossible to provide a breakdown or subdivision of the sum claimed between those matters.'

These claims are sometimes referred to as total cost claims, especially in the United States of America. The major objections to these claims are highlighted in *Hudson*, referred to above, and include *inter alia* the following:

(a) Cost computation carried out in a global manner is neither evidence of a breach of contract nor of a ground for a claim entitlement. An example of this is the case of *Wharf Properties Ltd* v. *Eric Cumine Associates* where the court did not accept the presentation of a claim which obscured any examination of issues of causation despite its 'immense length and complication'.[16.31] There was no evidence presented of causation between the general breaches complained of and the damage suffered. The court found that '... The failure even to attempt to specify any discernible nexus between the wrong alleged and the consequential delay provides ... "no agenda for the trial".'[16.32]

(b) Even where the global claim is supported by evidence of breaches of contract or other valid ground of compensation, there may be many other possible explanations of the cost overrun, such as under pricing in the original tender, poor site organisation, poor cost control, inefficiency in carrying out the works, external matters such as labour or material shortages, strikes, weather, or other factors for which the claimant is contractually responsible and the defendant is not. An example of such a situation occurred in *Cervidone Construction Corp. v. US* where it was said that:

> 'A trial court must use the total cost method with caution and as a last resort. Under this method, bidding inaccuracies can unjustifiably reduce the contractor's estimated costs. Moreover, performance inefficiencies can inflate a contractor's costs. These inaccuracies and inefficiencies can thus skew accurate computation of damages.'[16.33]

(c) Global claims invariably avoid indicating the precise case to be met and enable the claimants to shift the practical onus of proving the extent of their damage, or lack of it, to the defendant or to the tribunal, if the tribunal insists on a proper particularisation and detailed and critical analysis of quantum. An example of this problem facing a defendant occurred in *McAlpine Humberoak Ltd* v. *McDermott International Inc* where the defendants had to carry out a retrospective and dissectional reconstruction by expert evidence of events almost day by day, drawing by drawing, technical query by technical query, and weld procedure by weld procedure to show that they were not responsible for the claimants' alleged losses.[16.34]

With the above objections in mind, an example of a successful global claim under English law goes back to 1967 in the case of *J. Crosby & Sons Ltd* v. *Portland Urban District Council*.[16.35] The conditions of contract used was the Fourth Edition of the ICE Form and one of the matters in dispute which was referred to arbitration by the contractor was a claim for delay and 'disorganisation'. The completion of the works had been delayed by 46 weeks due to a number of reasons, some of which entitled the contractor to additional time and/or money and some of which did not. The arbitrator stated:

> 'The result in terms of delay and disorganisation, of each of the matters referred to above was a continuing one. As each matter occurred its consequences were added to the cumulative consequences of the matters which had preceded it. The delay and disorganisation which ultimately resulted was cumulative and attributable to the combined effect of all these matters. It is therefore impracticable, if not impossible, to assess the additional expense caused by delay and disorganisation due to any one of these matters in isolation from other matters ...'.

The arbitrator held that the contractor was entitled to be paid in respect of 31 weeks of the overall delay and awarded the contractor a lump sum by way of compensation. This decision was upheld by the court and the employer's argument that the arbitrator must necessarily build up the sum by finding amounts due under each of the individual heads of claim upon which the contractor relied in support of his overall claim for delay and disorganisation (disruption) was rejected.

The boundaries of this decision were described as depending 'on an extremely complex interaction in the consequences of various denials, suspensions and variations' and where 'it may well be difficult or even impossible to make an accurate apportionment of the total extra cost between the several causative events'. Mr Justice Donaldson (as he then was) stated that in those limited circumstances there is no reason why an engineer or arbitrator:

'should not recognise the realities of the situation and make individual awards in respect of those parts of individual items of the claim which can be dealt with in isolation and a supplementary award in respect of the remainder of those claims as a composite whole'.

The principles stated in the *Crosby* case were approved in 1985 in *Leach v. London Borough of Merton*, albeit with strong reservations.[16.36] The court in that case held that a combined claim could be awarded, but it could only be made, first in those cases where the loss could not in reality be separated; and secondly, where apart from that practical impossibility, all the required contractual conditions for a claim had been satisfied in both cases.

These restrictions were considered and applied in an ICC arbitration case relating to disputes arising out of a contract under the Third Edition of the Red Book.[16.37] In that case it was concluded by the tribunal that the contractor:

'has made no attempt to distinguish between the additional costs it has incurred due to the late instruction [Clause 6(4) claims], and what might be suitable rates and prices under Clause 52(1) or what rates and prices might be unreasonable or inapplicable under Clause 52(2). The decision under *Crosby* clearly requires arbitrators to first make individual awards under the appropriate Clauses of the Contract and then, and only then, to contemplate making a composite award in respect of the remainder of the claims. The Arbitrators are accordingly unable to accept the submission that they should fix a suitable rate or price to reimburse the Claimant its losses and expenses under the provisions of Clause 52.'

In the more recent case of *Wharf Properties*, referred to above, the facts were unusual in the sense that there were serious causation issues to be tried between the parties where super-structure contractors sued the developers, who subsequently compromised the claim, and then brought proceedings for breach of contract alleging negligence against their professional advisers, claiming as damages the sum paid to the contractors and also to recover loss of rent suffered as a result of delay in the completion of the project. Thus, the causation issues between the developers and their professional advisers were quite complicated, but the damages claimed were in themselves precise and were recoverable in principle subject to establishing causation. However, the case lacked any evidence of causation between the general breaches alleged and the damages suffered.

The Hong Kong Court of Appeal struck out the statement of claim as not disclosing a reasonable cause of action. In their petition to the Privy Council, the developers relied on the *Crosby* and *Leach* cases as justifying their contention that particulars need not be given. It was held, by the Privy Council, that while the pleadings did disclose a cause of action, they failed to explain the nexus between the individual breaches and the sums claimed, and they should be struck out as embarrassing and prejudicial to a fair trial. The observations of the editors of Building Law Reports, on this case are worthy of note:

'It must therefore follow from the decision of the Privy Council in *Wharf Properties* v. *Eric Cumine Associates* that *Crosby* and *Merton* are to be confined to matters of quantum and then only where it is impossible and impracticable to trace the loss back to the event. The two cases are not authority for the proposition that a claimant can avoid providing a proper factual description of the consequences of the various events upon which reliance is placed before attempting to quantify what those consequences were to him.'[16.38]

A number of similar cases followed where the courts have stepped back somewhat to emphasise the principles of natural justice in such situations and to spell out the principles and the purpose of pleadings on which many of these cases were based and challenged. In *British Airways Pension Trustees* v. *Sir Robert McAlpine & Sons Ltd*, two defendants sought to strike out the statement of claim on the basis that, *inter alia* it did not set out the remedial cost of each alleged defect and did not ascribe to each defect the amount by which it contributed to the alleged diminution in value. The application failed, the English Court of Appeal found that the basic purpose of pleadings was to allow the opposing party to know the case it had to meet, but that the practice of seeking particulars where they were not necessary was to be discouraged. The Court of Appeal indicated in that case that the argument in *Wharf Properties* centred more on the adequacy of pleadings than on any principle that global claims were embarrassing *per se*.[16.39]

The case of *GMTC Tools & Equipment Ltd* v. *Yuasa Warwick Machinery Ltd* is another example of the same point. In that case, the English Court of Appeal decided that it was not open to a judge to require a party to prove causation by a particular method and that parties could prove their claim for damages as they wished. The point for decision is whether the case is put sufficiently for the opposing party to know the case it has to meet and not whether, once it is in an understandable form, it has good prospects of success.[16.40]

In the case of *John Doyle Construction Limited* v. *Laing Management (Scotland) Limited*,[16.41] the manner in which global claims for loss and expense may be presented was further considered. In his decision at first instance in April 2002, Lord MacFadyen confirmed that the logic of a global claim demanded that all the events which contribute to causing the global loss must be events for which the defendant is liable. He concluded that a global claim as such must fail if any material contribution to the causation of the global loss is made by a factor for which the defendant bears no legal liability. However, he went on to say that it did not follow that where a global claim had failed, no claim would succeed. There might be, within the evidence, a sufficient basis to find causal connections between individual losses and individual events. Alternatively, it might be possible to make a rational apportionment of part of the global loss to the causative events for which the defendant had been held responsible. Since those questions could not be answered until the trial of the issues between the parties, Lord MacFadyen refused to strike out the claim simply on the basis that it was pleaded as a global loss. He allowed the matter to proceed to trial. In the Inner House, Lord Drummond Young agreed with that conclusion and added further clarification. The starting point was that if a global claim were to succeed, the contractor must eliminate from the causes of his loss and expense, all matters that are not the responsibility of the employer. That position was however mitigated by three key considerations.

First, as noted by Lord MacFadyen, it may be possible to identify a causal link between particular events for which the employer is responsible and individual items of loss. In this way, in effect, parts of the claim are extracted from the global calculation of loss and separately allocated to individual events.

Secondly, Lord Drummond Young considered that if an event or events for which the employer is responsible could be described as the dominant cause of an item of loss, that would be sufficient to establish liability, notwithstanding the existence of other causes that are to some degree at least concurrent.

Thirdly, Lord Drummond Young noted that even if it cannot be said that events for which the employer is responsible are the dominant cause of the loss, it may be possible to apportion the loss between the causes for which the employer is responsible and other causes. Lord Drummond Young recognised that such an apportionment would be more readily obtained where the loss was being calculated by reference to delay in the works. Either the loss would be apportioned on the basis of the time during which each of the causes was operative, or responsibility could be divided on an equal basis. Lord Drummond Young also commented that in carrying out such an apportionment, where a concurrent cause of delay is the contractor's responsibility, it may be appropriate to deny him any recovery for the period of delay during which the contractor is in default. Finally, the court recognised that matters become more complex when considering disruption to the contractor's work. Nevertheless, apportionment will frequently also be possible in such cases and although that might result in a somewhat rough and ready result, Lord Drummond Young noted that the procedure did not seem to be fundamentally different in nature from that used in assessing contributory negligence. The alternative to such an approach was a strict view that if a contractor sustains a loss caused partly by events for which the employer is responsible and partly by other events, the contractor cannot recover anything because it cannot demonstrate that the whole of the loss is the responsibility of the employer. Lord Drummond Young plainly thought that that was an unacceptable conclusion. The practical difficulties of carrying out an exercise of apportionment should not prevent the contractor from recovering part of his loss and expense in such cases.[16.42]

16.8 Failure to follow the claims procedure

If the contractor does not follow the claims procedure, in particular he gives no notice of claim; does that preclude his right to that claim? This question is addressed in the Fourth Edition of the Red Book, as follows.

First, in claims for extension of time, clause 44.1 read together with clause 44.2 provides that the engineer may, but is not bound to, make a determination in respect of an extension of time if the contractor does not comply with the claims procedure set out in clause 44.2;

Secondly, in claims for additional payment, clause 53.4 expressly provides that the contractor retains his entitlement to additional payment regardless of non-compliance with the claims procedure of preceding sub-clauses, but such entitlement is dependent on the extent to which the engineer considers that the claim is verified by contemporary records.

In view of these contractual provisions, and in order to give unequivocal weight to the claims procedure in the contract, many employers amend the conditions of their construction contracts such that if the claims procedure is not followed strictly, the contractor loses his entitlement to additional payment or extension of time (whichever is the subject of his claim). In unamended construction contracts, therefore, compliance with the claims procedure is not a pre-condition to an entitlement to additional payment/ extension of time. How then can a claim arise? If the contractor communicates his assertion to additional payment or to an extension of time in a format other than by written notice, is such assertion a claim?

If it can be assumed correctly that it is the parties' intention in any construction contract that any communication which is a claim should place an obligation on the engineer to deal with it in his quasi-arbitral role, then could any communication that inherently gives rise to this obligation be taken as a claim? For example, if the contractor asserts additional payment, not by giving notice of claim but by including the claimed amount in an interim payment application, is the engineer not then required to treat this as a claim and examine it accordingly? Sub-clause 60.1(e) of the Fourth Edition of the Red Book places an obligation on the contractor to include in each monthly statement 'any other sum to which the Contractor may be entitled under the Contract or otherwise'. Upon receipt of the contractor's interim statement, the engineer is obliged to certify payment within a set time period. His duty in this certifying role is to 'exercise such discretion impartially within the terms of the contract and having regard to all the circumstances', as in sub-clause 2.6(c), namely the same duty imposed on him in giving a decision in respect of a claim or a dispute.

In light of the contractor's obligation to include asserted additional sums in his interim statement, and the engineer's duty to act impartially in certifying payment, there could be a strong argument in favour of a positive answer to the above question. (If the contractor asserts additional payment, not by giving notice of claim, but by including the claimed amount in an interim payment application, is the engineer not then required to treat this as a claim and examine it accordingly?) This argument is supported by the fact that, as discussed above, the engineer remains under obligation to assess the contractor's claims notwithstanding that the contractor has not followed the claims procedure.

Does this premise extend to a contractor's assertion for extension of time, not by giving notice of claim but by submitting a revised programme of the works including in it the period of extension to which he believes he is entitled? In the Fourth Edition of the Red Book, when the engineer gives his consent to a clause 14 programme, he is under a duty to exercise his discretion of consent impartially, as in sub-clause 2.6(a). Under sub-clause 14.2, the contractor is required to submit a revised programme, upon the engineer's request, showing modifications to the previously consented programme which are necessary to ensure completion of the works within the time for completion as it stands at that time. There is no entitlement for the contractor to extend the time for completion in this revised programme. Therefore, the engineer has no discretion to accept this programme, if it shows an extended time for the completion of the works, by giving his consent, and the contractor's only avenue to assert an entitlement to an extension of time is by complying with clause 44.2, namely by giving notice of his claim. In the event that the engineer inadvertently gives his consent to a revised programme, sub-clause 14.4 dictates that such consent does not relieve the contractor of his duties and responsibilities under the contract, including his duty to complete the works within the time for completion as it stands at that time. The time for completion can only be extended under clause 44 and by the engineer after due consultation with both the contractor and the employer.

Notwithstanding the above discussion, claims under the Red Book are assessed by the engineer who is required to be independent, and impartial in making his determination – a role often described as quasi-arbitral. By following the agreed claims procedure the contractor and the employer can be said to create the optimum structured conditions under the contract for the engineer to make a fair assessment. In the absence of such structured conditions there is the possibility that the effectiveness of the engineer's quasi-arbitral role is compromised and, perhaps then, the possibility that a mutually acceptable settlement of the claim cannot be reached.

16.9 Concluding remarks

In this chapter, the principles and the legal basis relating to claims and counterclaims were discussed. The categories of claim were considered in detail in the context of the provisions of the Red Book. The relevant clauses of the Red Book relating to claims for financial compensation were discussed together with the procedure for submitting a claim under the provisions of clause 53. The appropriate method for presenting a claim with all the necessary components was outlined and discussed. The necessity for linking cause and effect was explained with reference to decided legal cases and with particular reference to quantum and total cost or global claims. In the next chapter, claims for delay and extension of time are discussed.

Chapter 17

Delay in Completion and Claims for Extension of Time

17.1 Time is of fundamental importance

If time is not 'of the essence' in construction contracts,[17.1] it is certainly of fundamental importance. In practice, projects are required to be completed by a certain date and in the case of commercial projects, this usually means as soon as possible. In some cases, the design is conceived with a certain date for completion in mind as budgets, interest rates, rents, leases and saleability are worked into a formula by the promoter's or employer's financial advisers. Time is also of fundamental importance to the contractor in that he must assess his performance capabilities and resources to carry out and complete the works within a given time.

Whilst both the employer and the contractor would prefer the certainty of a fixed completion date, the reality of construction and the sensitivity of construction activities to delay events beyond the control of the contractor or of the parties mean that flexibility in respect of the completion date is typically provided for in all standard forms of construction contracts. This flexibility takes the form of a right to grant and be granted an extension of time. The right to an extension of time obviously benefits the contractor, since he will not be liable to pay liquidated damages for delay during the period for which time is validly extended. Nonetheless, the right to extend the time for completion is also of benefit to the employer. At common law, where there is a delay caused by the employer but no mechanism for extension of time under the contract, the contractor's obligation to complete the works by the contractual completion date is nullified, and the previously fixed time for completion no longer applies. This principle, sometimes known as the 'prevention principle', was first enunciated in the case of *Peak Construction (Liverpool) Ltd* v. *McKinney Foundations Ltd*.[17.2] In such circumstances the contractual completion date ceases to apply and is replaced by an obligation to complete the works within a reasonable time, a result often described as 'time at large'. It has been said that: 'The phrase "time at large" is much loved by contractors. It has about it ring of plenty; the suggestion that the Contractor has as much time as he wants to complete the works'.[17.3]

In order to reflect the fundamental importance of time in the contract, to maintain the certainty of a definite completion date and to retain right to levy liquidated damages in the event of the contractor's default in achieving such date, the employer should ensure that a proper mechanism for extension of time is included in the contract. Employers should also note that in the judgment in the *Peak* case it was stated as follows:

'The liquidated damages and extension of time clauses and printed forms of contract must be construed strictly *contra preferentum*. If the employer wishes to recover liquidated damages for failure by the contractors to complete on time in spite of the fact that some of the delay is due to the employer's own fault or breach of contract, then the extension of time clause should provide, expressly or by necessary inference, for an extension on account of such a fault or breach on the part of the employer'

which was confirmed in the recent Australian case of *Gaymark Investments* v. *Walter Construction Group Ltd*[17.4] (see further Section 17.3 below).

However, in practice, courts and arbitrators approach the subject of 'time at large' with caution. Experience shows that they will not readily allow a contractor to escape liquidated damages where there has been minor delay by the employer. An example of this is *Balfour Beatty Building Ltd* v. *Chestermount Properties Ltd*.[17.5] The contractor was in delay when the employer instructed additional work after the date for completion. The contractor argued that this exonerated him from all delay up to that point. The court refused to apply the prevention principle to its logical extreme in this situation and said there should be an apportioning of the delay.

If time is said to be at the heart of every construction contract, in order to identify whether or not the contractor is in compliance with the provisions of the contract in relation to time, all extensions of time entitlements in respect of events for which the employer is responsible have to be identified.

Whether or not completion is achieved by the designated date is a question which is accompanied by a matrix of risks, which under the provisions of the Red Book are shared by both the employer and the contractor. These provisions relating to questions of time are fundamental to the whole concept of risk sharing (see Chapter 7 and in particular Figure 7.2).

The sharing of time-related risks is based on the liability for the delay. If the delay is caused by an event for which the employer is liable, the risk is allocated to the employer. It is worth noting here that, if there is no express provision to extend the time for completion for a delay caused by an event which is not the fault of the employer, the contractor takes the risk of that delay.[17.6] The relevant clauses of the Red Book which deal with the sharing of the time-related risks and their consequences are clauses 43, 44, 46, 47 and 48.

17.2 Clauses 43, 44, 46, 47 and 48 of the Red Book

Clause 43 of the Red Book establishes the time by which the works must be completed. It requires the contractor to complete 'the whole of the Works and, if applicable, any Section required to be completed within a particular time . . . in accordance with Clause 48, within the time stated in the Appendix to Tender . . . , or such extended time as may be allowed under Clause 44'. 'Time for completion' is a defined term under sub-clause 1.1(c)(ii) of the Red Book and is calculated from the commencement date, which is the date of receipt of the notice to commence the works pursuant to clause 41. Failure to complete within the defined time limit is a breach of contract, the remedy for which is provided in clause 47.

Clause 44 deals with the events that permit an extension of the time for completion. Sub-clause 44.1 provides the following:

'In the event of
(a) the amount or nature of extra or additional work, or
(b) any cause of delay referred to in these Conditions, or
(c) exceptionally adverse climatic conditions, or
(d) any delay, impediment or prevention by the Employer, or
(e) other special circumstances which may occur, other than through a default of or breach of contract by the Contractor or for which he is responsible,
being such as fairly to entitle the Contractor to an extension of the Time for Completion of the Works, or any Section or part thereof, the Engineer shall, after due consultation with the Employer and the Contractor, determine the amount of such extension and shall notify the Contractor accordingly, with a copy to the Employer.'

Therefore, for example, a delay by the engineer in issuing a further drawing or instruction under the provisions of sub-clause 6.4 is an employer's risk. For such an event or any other event occurring as a result of a time-related risk allocated to the employer, the engineer is required to determine after 'due consultation' with the employer and the contractor the amount of the extension of time.

The extension that sub-clause 44.1 provides for is an extension of the time for completion of:

(i) the works, or
(ii) any section, or
(iii) any part of any section.

There are three important points of note arising here: first, each time the engineer grants an extension of time pursuant to sub-clause 44.1, the contractual time for completion increases cumulatively.

Secondly, the word 'fairly' in the first line after paragraph (e) of the quoted text of sub-clause 44.1 above, is not intended to be an open invitation to be construed as entitling the contractor to an extension of time without the application of a wealth of supporting scientific and legal principles that should be applied before such an extension is granted, see section 17.6 below. This second point is of great importance if the word 'fairly' is to be interpreted by arbitrators, since those who are not skilled in construction law and the principles referred to above might err and decide a dispute incorrectly. This need for appropriate knowledge is highlighted, for example, by the ICC Final Report on Construction Industry Arbitrations;[17.7] and the Society of Construction Law Protocol.[17.8]

Thirdly, and as a consequence to the principles referred to in the above second point, a contractor is entitled to an extension of time for non-culpable delays only to those activities which lie on the critical path which is associated with the stipulated completion dates. As such, where the contractor in his programme, submitted in accordance with clause 14, shows float in respect of an activity or sequence of activities, and a delay occurs for which he is not liable, he will have no entitlement to an extension of time until all such float has been eroded, at which point the affected activity or sequence of activities are said to lie on the critical path. This has led to the float sometimes being described as 'employer owned float'. The 'ownership' of float has caused many disputes in practice, the contractor arguing that he has included float in the programme as a contingency to give himself flexibility in planning and executing the works, and therefore it is he who 'owns' the float. If the contractor at the start, or in the early stages of the project, submits a clause 14 programme showing completion at a date earlier than that required by the contract, the engineer gives his consent to such a programme, and thereafter such early completion is delayed by the employer/engineer, then has the contractor a valid claim against the employer for preventing completion by the programmed date?

This question was answered in the English case of *Glenlion Construction* v. *The Guinness Trust*.[17.9] It was held that the contractor was entitled to plan completion of, and to complete, the works before the contractual date for completion but there was no term to be implied under the contract that the employer (and the engineer as the employer's agent) was obliged to plan and perform his duties so as to enable the contractor to finish by the earlier date. As a consequence, a contractor may choose to include only minimal float in his clause 14 programme, thereby ensuring that he would optimise the benefit to be gained by the entitlement provided under sub-clause 44.1. This, of course, means that the employer and the engineer are deprived of the benefit of a realistic programme for the project. It also means that there is no incentive (unless a bonus/incentive arrangement is expressly provided for under the contract, see Section 17.7.1 below) for the contractor to plan for, or to show in his clause 14 programme, completion at a date earlier than that required by the contract. Since, as already discussed above, time is of fundamental importance to both the contractor and the employer, this 'disincentive' has in theory the effect of compromising what might otherwise be a more efficient and therefore more economic project. On the other hand, experience has shown that employers often impose on contractors short times for completion and, therefore, perhaps in reality the effect of such 'disincentive' is not so significant as it would first appear.

If the engineer fails to follow the procedural requirements of clause 44.1, in particular as a result of due consultation with the parties, then it could be argued that the employer may lose his entitlement to liquidated damages under the contract. From Singapore, the case of *Assoland Construction Pte Ltd* v. *Malayan Credit Properties Pte Ltd* is of relevance here.[17.10] The contract required the architect to respond to the contractor's notice of claim for an extension of time within a specified period. In fact the architect failed to respond within the specified time and the court held that, as such, the architect's purported exercise of his power to grant an extension of time was invalid. There was no date from which liquidated damages could be computed and no liquidated damages were therefore recoverable.

As noted in Section 5.4 above, the engineer is considered as acting in the role of an agent of the employer in ensuring that the works are executed by the contractor in accordance with the contract and that the contractor is in possession of all required documentation for such execution. Therefore, sub-clause 44.1(d) also includes any delay, impediment or prevention by the engineer in performing this role as agent of the employer.

It can be said that sub-clause 44.1(e) is a 'catch-all' provision which in effect places the risk of delays caused by the occurrence of any special circumstance, other than that caused by the contractor, firmly in the hands of the employer. Being somewhat of an ambiguous provision it has, in practice, caused problems in interpretation and application. What type of circumstance should be deemed to be classed as 'special'? When might an otherwise ordinary circumstance become 'special'? It has been said that such a provision '... is an invitation to stretch ingenuity, and the results are unpredictable'.[17.11] It is worth noting here that, other than this 'catch all' provision, the fourth edition of the Red Book makes express provision for extension of time only for a limited number of specific events, see Section 17.3 below. The following events are not expressly catered for, as they are in some other standard forms of construction contracts:

(i) discrepancies in or divergence between contract documents;[17.12]
(ii) errors in drawings, technical specifications, items of reference for setting out of the works provided by the employer;[17.13]
(iii) changes in law/legislation;[17.14]
(iv) delays caused by other contractors employed by the employer, or nominated sub-contractors/suppliers;[17.15]
(v) delays caused by public bodies/authorities.[17.16]

Whether an occurrence of any such events would be classed as a 'special circumstance' or not is a matter for the engineer's consideration and at his discretion. However, such discretion must be exercised having regard to that which is 'such as fairly to entitle the Contractor to an extension of the Time for Completion of the Works', as stated in sub-clause 44.1, and his duty to act impartially pursuant to sub-clause 2.6.

In sub-clause 44.2, the contractor is required, within 28 days after an event described in sub-clause 44.1 had first arisen, to give notice of such event to the engineer with a copy to the employer. Within a further period of 28 days, the contractor is required to give detailed particulars of any extension of time to which he may consider himself entitled. This second period of 28 days may be extended by the engineer to such other reasonable time as may be agreed by him.

It is worth noting at this juncture that, unlike the case in respect of an assertion of entitlement to additional payment under clause 53, the word 'claim' appears nowhere in the text of clause 44 in respect of an assertion of entitlement to an extension of time. The reason for such an omission is not clear, most likely arising from the genesis of the Fourth Edition of the Red Book as described in Chapter 2. The 1999 Red Book has addressed this omission, in that clause 20.1, 'Contractor's Claims', covers both claims for additional payment and extension of the time for completion.

Whilst a notice is required under the provisions of sub-clause 44.2, thus appearing at first glance to make it a condition precedent to the contractor's entitlement to an extension of time under this clause, the actual wording used in the opening sentence of this sub-clause, namely '... the Engineer is not bound to make any determination unless the Contractor has ...', gives the engineer the discretion to allow such a claim in the event that the notice was served outside the period of 28 days or, indeed, in the absence of a notice. Should the engineer choose not to allow the claim, it might be argued that the contractor's claim would be time-barred or, by failing to give notice, that he had waived his right to an extension of time.

Such a scenario was considered in the recent Australian case of *Abigroup Contractors Pty Ltd* v. *Peninsula Balmain Pty Ltd*.[17.17] Peninsula (the employer) was a developer who appointed Abigroup, a building contractor, under a construction agreement to undertake reconstruction and refurbishment of two factory buildings by converting them to residential flats. Eighteen months into the contract the contractor initiated litigation proceedings seeking payment of a progress claim and shortly thereafter the employer filed a counterclaim for liquidated damages. The court ordered that the proceedings be referred to a referee for inquiry and report. The referee in his report granted extensions of time to the contractor for delays that were found to be excusable under the contract. The employer objected to this grant and applied to the court to have this part of the referee's report nullified on the basis that the contractual mechanism for claiming extensions of time was not followed by the contractor. It was accepted by the parties that said contractual mechanism had not been followed and the employer's argument was that, as this was a pre-condition to entitlement, no entitlement to extension of time arose. The Referee's view on this argument was that, apart altogether from this contractual mechanism, the clause pertaining to extensions of time gave the superintendent (engineer) a unilateral power of extension. The wording in the contract upon which the Referee based his view was as follows:

> '*Notwithstanding that the Contractor is not entitled to an extension of time* the Superintendent may at any time and from time to time before the issue of the Final Certificate by notice in writing to the Contractor extend the time for Practical Completion for any reason.' (emphasis added)

The court re-examined this wording and concluded that the superintendent did have a unilateral power of extension, the exercise of which was plainly (see emphasis above) not barred by the contractor's lack of entitlement to extension. The court stated: 'That power is additional to and separate from the regime under which the contractor may establish an entitlement to extension'. The referee's award of an extension of time to the contractor was upheld.

An observation made in the UK case of *Sindall Limited* v. *Solland* is also pertinent here:

> '... I remind myself that, not just as a matter of law, as set out by Vinelott J in *London Borough of Merton* v. *Leach* (1985) 32 BLR 51 at pages 89–90, but as a matter of established good practice, that a person in the position of a Contract Administrator has always to consider whether there are any factors known to him which might justify an extension of time, even though the contractor may not have given written notice of them.'[17.18]

Therefore, it would appear that the giving of notice (and subsequent provision of details) as required by sub-clause 44.2 may not be a condition precedent to the contractor's entitlement to an extension of time. Could then a contractor make an assertion for entitlement to an extension of time, not by giving notice but by submitting a revised programme of the works, including in it the period of extension to which he believes he is entitled? Under sub-clause 14.1 the programme of works is subject to the consent of the engineer. In granting his consent, the engineer is under a duty to exercise his discretion of consent impartially (sub-clause 2.6(a)). Under clause 14.2 the contractor is required to submit a revised programme (upon the engineer's request) showing modifications to the previously consented programme which are necessary to ensure completion of the works within the time for completion as it stands at that time. There is no entitlement for the contractor to extend the time for completion in this revised programme. Therefore, the engineer has no discretion to accept this programme by giving his consent and the contractor's only avenue to assert an entitlement to an extension of time is by complying with clause 44.2, namely by giving notice of his claim. In the event that the engineer inadvertently gives his consent to a revised programme, clause 14.4 dictates that such consent does not relieve the contractor of his duties and responsibilities under the contract, including his duty to complete the works within the time for completion as it stands at that time. The time for completion can only be extended under clause 44 and by the engineer after due consultation with both the contractor and the employer.

In view of the uncertainty arising from contractual provisions such as clause 44.1 and 44.2, and in order to give unequivocal weight to the claims procedure in the contract, many employers make express amendments to the conditions of contract such that if the claims procedure is not followed properly the contractor loses his entitlement to extension of time and the engineer has no discretion to consider the claim further. In the recent Scottish case of *City Inn* v. *Shepherd Construction*[17.19] the contract was based on a specially amended form of JCT Private With Quantities 1980. One of the special amendments provided that the contractor would not carry out any instruction issued by the architect which might cause delay or an increase in the contract sum unless he had first given notice within ten days giving various details. If he failed to do so, it was expressly stated that he would have no entitlement to an extension of time (and would be liable for liquidated damages for delay). The contractor, upon receipt of an architect's instruction, failed to give the necessary notice, applied for an extension of time for the additional works entailed by the instruction and was refused. The contractor argued that, in precluding his entitlement to an extension of time, the contractual provision was unenforceable as a penalty clause. It was held that the contractor was not obliged to give this notice; it was only necessary to do so if he wanted to seek an extension of time or additional payment. The provision was therefore not unenforceable as a penalty clause and the contractor was in breach of contract because he had failed to finish by the contractual completion date in circumstances in which he had taken no proper steps to obtain an extension of time.

However, the validity of time-barring a claim or removing the contractor's right to an extension of time where no notice is given at all, is not straightforward and is dependent on whether or not it is the employer who is liable for the event giving rise to the claim. This was considered in the recent case of *Gaymark Investments* v. *Walter Construction Group Ltd*.[17.20] Walter Construction Group was a building contractor and Gaymark Investments (the employer) a developer who appointed the contractor under a construction agreement to build a hotel, retail and office complex in Darwin, Australia. The contractor made a number of claims mainly based on variations, prolongation and disruption/acceleration; the employer counterclaimed for liquidated damages. The dispute was referred to arbitration and the arbitrator made an award in favour of the contractor and rejected the employer's counterclaim for liquidated damages. The arbitrator had concluded that:

(i) the employer was not entitled to anything by way of liquidated damages, basing his conclusion on his finding of fact that the contractor was delayed by causes for which the employer was responsible; these delays actually prevented the contractor from achieving practical completion by the date for practical completion; and the aforesaid delays constituted 'acts of prevention' by the employer with the result that there was no date for practical completion, time was at large and the contractor's obligation was then to complete within a reasonable time, and
(ii) the contractor's claim for extension of time was barred by its failure to meet the notification requirements of the extension of time clause in the contract.

The employer applied for leave to appeal from the arbitrator's decision. The court examined the concept of prevention by first referring to a statement of the principle given in the Australian case of *Turner Corporation Ltd* v. *Co-ordinated Industries Pty Ltd* (1994):[17.21]

> 'it is that a party to the contract has been prevented from fulfilling its contractual obligations by virtue of conduct of the other party. The consequence is that the "preventing party" cannot rely upon the failure by the other party to comply with its contractual obligations, even if the other party is otherwise in breach so that it could not have complied with its contractual obligations in any event. It is said this flows from a generally stated principle that a party cannot benefit from its own wrong. Whilst the so-called principle may be stated in general terms it seems to me it can only have that application, usually, in circumstances where the contract does not provide for the effect of breach causing prevention.'

The court examined the conditions of contract which related to extensions of time and noted that the standard-form conditions had been amended such that it afforded the superintendent (engineer) no general discretion to extend time in the absence of strict compliance with the notification requirements, notwithstanding that the contractor had been actually delayed by an act, omission or breach of contract for which the employer was responsible. The court considered the arbitrator's observation that where the employer was responsible for the delay and the contractor failed to apply for an extension, not only would the employer have been absolved from the need to pay the contractor's costs of the delay but the employer would also be able to recover liquidated damages for a delay of his own causing. The arbitrator had concluded that this was an absurd construction of the contractual provisions and the court concurred, stating that an award of liquidated damages to the employer would 'result in an entirely unmeritorious award . . . for delays of its own making'. However, the court also agreed with the arbitrator's finding that the contractor, in failing to comply with the contractual terms regarding the giving of notice of his claim, lost his entitlement to an extension of time. In this regard, the court followed the decision in *Turner Corporation Limited* v. *Austotel Pty Limited*,[17.22] namely:

> 'If the Builder having a right to claim an extension of time fails to do so, it cannot claim that the act of prevention which would have entitled it to an extension of the time for Practical Completion resulted in its inability to complete by that time. A party to a contract cannot rely upon preventing conduct of the other party where it failed to exercise a contractual right which would have negated the effect of that preventing conduct.'

Therefore, where the contract gives the engineer power to determine extension of time, but such power is qualified by the contractor's compliance with the contractual notification requirements, and where delays are caused by the employer but the contractor fails to give notice of his claim or is time-barred under the contract to claim entitlement of an extension for such delays, the employer shall have no entitlement to liquidated damages for such delay and the contractor shall have no entitlement to an extension (and so will bear his own costs arising from the employer's delay).

As noted above, clause 44.2 is in fact a proviso to the obligation on the engineer expressed in clause 44.1 where he is obliged to determine an extension of time in the certain listed circumstances. A failure on the part of the contractor to comply with sub-clause 44.2, therefore, has the effect of changing the engineer's function from an unequivocal obligation to determine and grant an extension of time to one of a discretion to so determine and grant. Apart from such dilution of the engineer's function in respect of the extension of time, there is no other sanction on the contractor in the event that he fails to give notice and/or to provide detailed particulars of his asserted entitlement.

Other standard forms of contract provide that if the contractor fails to comply with the procedure required by the contractor, then the engineer is entitled to take account of such failure in his determination, for example sub-clause 20.1 of the 1999 Red Book. The use of 'and' between sub-clause 44.2(a) and 44.2(b) is important and means that unless the contractor fulfils both the notice requirement and the detailed particular requirement, the engineer is not bound to make any determination. Accordingly, the notice together with the submission of detailed particulars constitutes a claim, as defined in Section 16.2 above, but could the notice on its own be strictly speaking a claim? That said, in practice, it would be unusual for the engineer to take the view that a notice is not a claim. Typically, in light of the discretion granted to him by virtue of the corollary of the words 'the Engineer is not bound to make any determination', a cautious engineer will take the contractor's notice as a claim in itself and, in the absence of detailed particulars from the contractor, he would determine the extension of time due based on the information available to him at the time.

Sub-clause 44.3 provides for the situation where the event giving rise to the delay has a continuing effect which makes it impracticable for the contractor to submit detailed particulars within the 28-day period referred to in sub-clause 44.2. The contractor having notified the engineer, with a copy to the employer, is required to submit interim particulars at intervals of not more than 28 days and to submit final particulars within 28 days following the end of the effects resulting from the event.

The engineer is required to make an interim determination of any extension of time to which the contractor is entitled, 'without undue delay'. On receipt of the final particulars, the engineer is required to review all circumstances and determine an overall extension of time in regard to the event. It is important to note that in undertaking the final review, the engineer is not permitted to decrease any extension of time already granted. This places the engineer in the difficult position of having to give his initial determination without undue delay and at the same time being restrained from adjusting it downward should it prove to be excessive. The necessity to impose such restriction on the engineer's discretion is understandable since having been given an extension of time, the contractor can plan his resources accordingly and should not be in a position of later discovering that any extension of time which he had already been granted has now been reduced. It does, however, mean that in practice many engineers are reluctant to grant significant interim extensions of time, preferring instead to wait until the full extent of the effect of the relevant delay is evident and then determining the final extension of time with the benefit of hindsight. This has the effect of undermining the ideology behind sub-clause 44.3.

It is clear that, apart from the restriction in sub-clause 44.3 and his duty to act impartially pursuant to sub-clause 2.6, the engineer is given a very wide discretion in determining what does or does not 'fairly entitle' the contractor to an extension of the time for completion.

Clause 46 of the Fourth Edition of the Red Book deals with delays where the contractor is not entitled to an extension of time and which result in a rate of progress of construction which is too slow to achieve the specified date for completion of the contract. This clause permits the engineer to notify the contractor requiring him to take such steps as are necessary to expedite progress of the works so as to comply with the time for completion.

Clause 47 provides that:

'if the Contractor fails to comply with the Time for Completion in accordance with Clause 48 . . . within the relevant time prescribed in Clause 43, then the Contractor shall pay to the Employer the relevant sum stated in the Appendix to Tender as liquidated damages for such default and not as a penalty . . .'.

The matter of liquidated damages is discussed in detail below at Section 17.6.

Clause 48 defines what completion means and specifies the procedure to be followed to reach that stage.

17.3 Relevant clauses of the Fourth Edition of the Red Book to an extension of time under Clause 44

Clauses of the Fourth Edition of the Red Book involving possible events which may entitle the contractor to an extension of time under clause 44 can be briefly identified as follows:

(a) sub-clauses 6.3 and 6.4: delay in supply of documents;
(b) sub-clause 12.2: adverse physical obstructions or physical conditions;
(c) sub-clause 27.1: fossils and articles of value or antiquity;
(d) sub-clause 36.5: tests required but not provided for;
(e) sub-clause 40.2: suspension of the progress of the works;
(f) sub-clause 42.2: failure to give possession of site;
(g) sub-clause 41.1: extension of time for completion;
(h) sub-clause 69.4: contractor's entitlement to suspend work or reduce rate of work.

The engineer in granting such an extension of time does not automatically provide an entitlement to the contractor to claim additional payment. There is no express link between clause 44 and any of the clauses which entitle the contractor to additional payment. Nevertheless, should he determine that such additional payment is due under any of the provisions of the contract, then the contractor may be entitled to such a payment even if he completes the works within the time for completion originally specified.

Any analysis of the question of delay and determination of responsibility in respect of an event leading to delay must bring into focus the programme required under clause 14 of the Red Book.

Sub-clause 14.1 provides that the contractor is required to submit to the engineer within a specified period, for his consent, a programme for the execution of the works, see Table 13.1 in Chapter 13. Sub-clause 14.1 does not, however, provide details of the form and contents of this programme but leaves these to be prescribed within reason by the engineer. In simple contracts, the traditional method of presenting a programme in the form of a bar chart may suffice. However, as the contract becomes more complex with a large number of interrelated activities involving intricate and laborious operations and extensive resources, the necessity for more sophisticated methods becomes obvious. In these days of the personal computer and with the advent of an array of advanced programming software in recent decades, there are numerous options available to be prescribed by the engineer. In addition, it would appear that each software programme requires an array of assumptions to be inputted. It has been said:

> 'in construction ... the value of a complete and well thought-out programme of the way the work is expected or planned to be carried out cannot be over-emphasised. At any one time, the as-planned programme for the period in question is the basis upon which many decisions will be based and upon which will be gauged the reasonableness of the period and the practicability of the contract's performance. It is also the standard by which progress will be measured, the source of identification of problems relating to integration of successive trades and activities, the tool for

analysing the impacts on time and cost of change, the tool for re-planning the project in the event of a change, and the principal method of demonstrating excusable delay and compensable disruption. Without an APP [as-planned programme] to act as a base-line, proper management of a project is difficult, if not impossible.'[17.23]

Therefore, the form and detail of the programme to be adopted should be given all due attention and consideration before being prescribed by the engineer pursuant to sub-clause 14.1. Unfortunately, given the fundamentally important role that the programme plays in construction contracts, there is little or no guidance available in any of the FIDIC forms of contract as to which type of form, presentation and content might be suitable for a particular project. Indeed, while the Fourth Edition of the Red Book allows the engineer to prescribe the form and detail of the programme, this is not provided for in the 1999 Red Book whereby the contractor is merely required to show the order in which he intends to carry out the works, to identify the timing of activities by nominated subcontractors, to show the sequence and timing of inspections and tests, and a report giving the methodology and labour/plant resources in support of the intended sequence. There is neither mention of network analysis nor of the critical path(s).

Therefore, it is left to the employer/engineer to specify exactly what he requires in respect of the programme. Whichever particular programming software is to be used, the network analysis involved and the critical path(s) which result should be clear and unambiguous. See further at Section 17.4. An observation by Judge Richard Seymour QC in *Royal Brompton Hospital NHS Trust* v. *Frederick A. Hammond and Others*[17.24] is worth noting here:

> '... there are a number of established ways in which a person who wishes to assess whether a particular event has or has not affected the progress of construction work can seek to do that. Because the construction of a modern building, other than one of the most basic type, involves the carrying out of a series of operations, some of which, possibly, can be undertaken at the same time as some of the others, but many of which can only be carried out in a sequence, it may well not be immediately obvious which operations impact upon which other operations. In order to make an assessment of whether a particular occurrence has affected the ultimate completion of the work, rather than just a particular operation, it is desirable to consider what operations, at the time the event with which one is concerned happens, are critical to the forward progress of the work as a whole ... the various different methods of making an assessment of the impact of unforeseen occurrences upon the progress of construction works are likely to produce different results, perhaps dramatically different results ... the accuracy of any of the methods in common use critically depends upon the quality of the information upon which the assessment exercise was based.'

The Fourth Edition of the Red Book does not provide any sanction for a failure on the part of the contractor to submit the programme within the timescale required by the contract. This leaves the engineer in a dilemma when the programme is not produced on time and may prejudice any future extension of time claim the contractor may have. To have credibility the contractor must usually be able to show that programme and progress times and/or sequence have been affected by the matter giving rise to his claimed entitlement to an extension of time. In practice, the absence of a proper programme leads to difficulties experienced by the contractor in establishing his claimed entitlement and the engineer is likely to question the reliability of a programme that is provided late or a programme produced retrospectively.

Under the Red Book the clause 14 programme is not a contract document. As noted previously, in Section 9.11, in the Fourth Edition of the Red Book the clause 14 programme is to be submitted for the engineer's 'consent' rather than his 'approval' (as appeared in the Third Edition of the Red Book). In giving his consent, the engineer shows that he has no objection to the contractor's programme but assumes no responsibility for it. From the judgment in the case of *GLC v. Cleveland Bridge & Eng. Co. Ltd*[17.25] it would appear that a term will not be implied that a contractor is required to work to the accepted programme.

The second part of sub-clause 14.1 gives authority to the engineer to require the contractor to submit in writing for his information a general description of the arrangements and methods which the contractor proposes to adopt for the execution of the works.

Sub-clause 14.2 provides that, when required by the engineer, the contractor is to provide updated and revised programmes showing modifications necessary to ensure completion of the works within the time for completion. In times past, there was a tendency for a lot of management resources and attention to be given to the initial clause 14 programme but after that updates of the programme were undertaken on a more cursory basis. This was understandable, if not excusable, before the advent of personal computers and the ease with which programming software enables updates to be produced on a regular basis and by personnel actually on site. Although not expressly required under this sub-clause, in order that each such programme can be properly used by both the contractor and the employer, and properly assessed by the engineer, each revised programme should also clearly show the revised network analysis and revision(s) to the critical path(s), if any. In the absence of such a programme evidencing the state of affairs for the time being, the contractor may find that his ability to prove an entitlement to extension of time is jeopardised. The consented programme is one of the main tools available to the engineer in understanding the contractor's current and planned construction logic and, accordingly, the impact of delay on such logic. The accuracy of his assessment of entitlement is likely to be compromised without a properly revised programme.

Figure 17.1 illustrates the link between the provisions of clauses 14, 44 and 46 of the Red Book. However, it is worth noting here that such a link is not expressly made in the wording of these clauses, albeit that clause 46 does refer to 'extension of time', the subject of clause 44. Again given the fundamentally important role that time and the programme play in construction contracts, it is regrettable that the 1996 Supplement to the Fourth Edition of the Red Book did not expressly provide the link between these three clauses. This is particularly so in respect of the production and submission of revised programmes pursuant to sub-clause 14.2 during the course of the works when events,

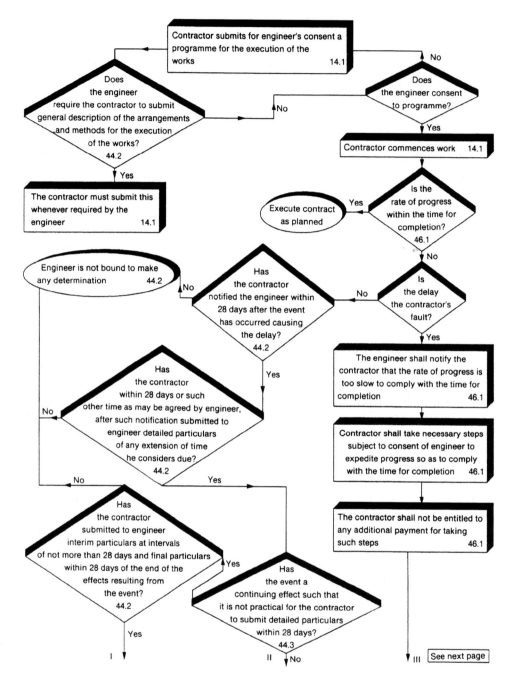

Fig. 17.1 Programme – time – delay – rate of progress.

Delay in Completion and Claims for Extension of Time 357

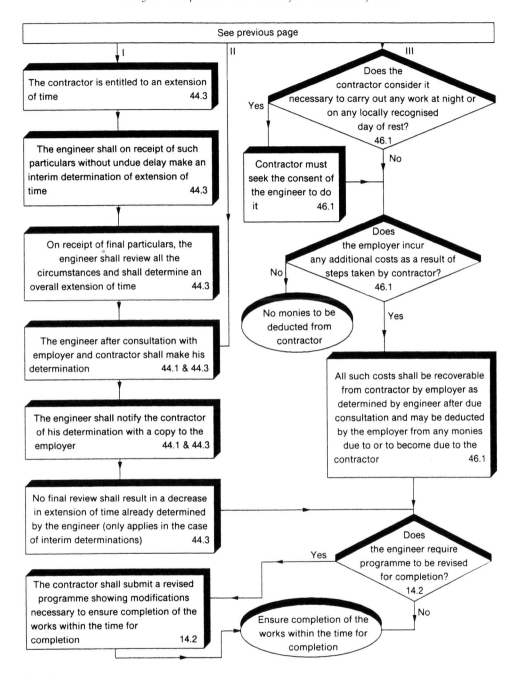

(Fig. 17.1 Contd.)

which may or may not lead to extension(s) of time, and the rate of progress have become evident to all the parties. If the revised programme at any given time is a *true* reflection of the state of affairs of the progression of the works to completion, then an analysis of the question of delay and determination of responsibility in respect of an event leading to delay may be properly undertaken.

17.4 Programming

Programming involves introducing the parameter of time into the work activities and ultimately into the project itself. As stated above, the traditional method of presenting a programme for the construction contract has been through a bar chart. This gives an outline plan of the timescale of a project broken down into a relatively small number of components, each made up of a collection of many activities. Each component may have its own bar chart.

The chart also provides the start date and completion date for each of the components or activities shown.

For the purpose of project control, the bar chart can show the progress actually achieved at any particular time but it is only useful at the lower level of management. It does not show how the various activities are connected except through sequence of listing and therefore the constraint imposed by the completion of one activity on the start of another cannot be accurately indicated or assessed.

As delay is costly to both the employer and the contractor, a more sophisticated method which can handle various details of each activity and the interrelationships between them should be used in all but the simplest of civil engineering projects. Network planning and control is such a method.

The details which can be highlighted in a network programme are:

(a) Independence of one activity from another requiring that neither the start nor the completion time of each is affected by the other.
(b) Sequence of activities showing when one activity cannot start before the completion of another.
(c) 'Burst' describing restrictive activities which when completed allow two or more activities to start.
(d) 'Merge' describing an activity which cannot start unless and until two or more other activities have ended.
(e) Combined burst and merge of activities describing the position when several activities cannot start until two or more immediately preceding activities have ended.
(f) Timing of activities in a unit of time which usually depends on the duration of the whole project, and as a general rule expressed as one per cent to two per cent of the timescale of the whole project. This operation also includes an identification of the start and end of each activity.

A project network analysis should be carried out in at least eight phases, as described below, but each may be sub-divided and expanded to give a more detailed picture:

(a) Planning phase where a network of all activities necessary for the completion of a project is planned and drawn up. An activity is an operation where time and resources are consumed and it is represented by an arrow.
(b) Project timing where estimates of duration of activities are calculated to determine as accurately as possible the project duration and to identify the activities which may prove to be critical. Duration estimates should be calculated without bias.
(c) Resource allocation where information is added to each activity duration to show the resources required to complete that activity within the projected duration.
(d) Allocation of work to sub-contractors and a programme of appointment to be followed with a schedule for production and approval of subcontractors' design, if any, and drawings.
(e) Pricing of the various elements of the works.
(f) Procurement of materials.
(g) Commencement on site.
(h) Project control where the actual progress on site is periodically measured against the network plan and where any corrective action can be identified and then taken. The network must then be updated in accordance with the actions taken, and a report can be periodically compiled to highlight the status of each activity at the particular time. The report may show:
   (i) delay in an activity and its effect on other activities and on the time for completion;
   (ii) new activities due to a variation and the effect on others and on the time for completion;
   (iii) resources which must be drafted to redress any new situation;
   (iv) any other change in the critical path network.

The network is divided into events which represent the end of all preceding activities and the start of a succeeding activity. Figure 17.2 shows an example of a network analysis with events labelled from 1 to 15 and activities of durations shown assigned to letters A to O. Figure 17.3 shows the critical path diagram associated with that network analysis.

The event is represented usually by a circle which carries a unique label shown in the left half of the circle. The right half of the circle is divided into quadrants; the upper quadrant carries the number of units of time representing the earliest event time and the lower quadrant carries the latest event time. These figures are calculated through a process called time analysis. Two sets of calculations are necessary for such analysis. The first set of calculations is carried out to find the figures which represent the earliest event time. These figures are calculated by adding from left to right through the network the durations of activities leading to each event. This will give the earliest possible time within which each event can be achieved. Where there is more than one path leading to an event, then the longest path in terms of time duration would provide the figure for the top right-hand quadrant of the event circle.

The second set of calculations is carried out in a reverse order from the end of the project back towards the start. By deducting the duration of events from the earliest time of the last event, one can obtain the latest possible time for each event to be completed

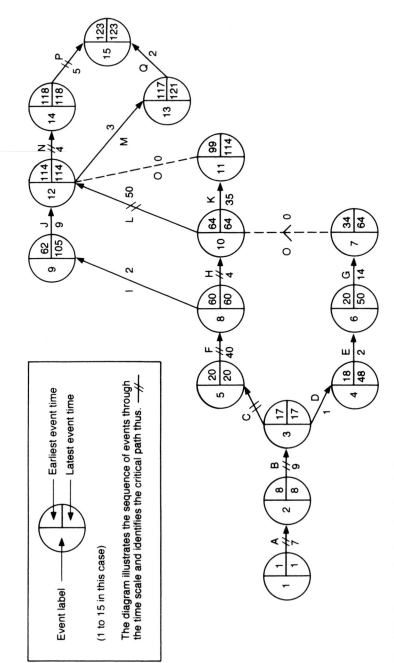

Fig. 17.2 An example of a network analysis.

Delay in Completion and Claims for Extension of Time 361

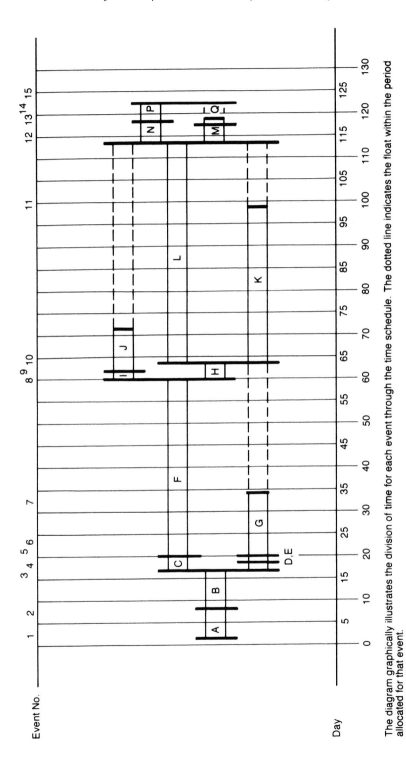

The diagram graphically illustrates the division of time for each event through the time schedule. The dotted line indicates the float within the period allocated for that event.

Fig. 17.3 Critical path diagram associated with the network analysis in Figure 17.2.

if the earliest time of the last event is not to be exceeded. If two or more paths lead to an event, then once again the longest path in terms of time duration is used.

When the time analysis is completed, the critical events and critical activities can be identified. A critical event is identified by the fact that its earliest and latest time are the same. A critical activity is identified when it joins two critical events and has a duration equal to the difference between the times of the critical events it joins.

A critical path joins critical events from the start of a project to its completion. Every project has at least one critical path and where there are more than one they all have the same duration.

A non-critical event has 'slack', which is the calculated time span within which the event must occur, that is, the difference between its latest and earliest times. A non-critical activity has 'float', which is the time available for an activity in addition to its duration. Since an activity has a start event and an end event, each of which has an earliest start and a latest start, there are four types of float. These are:

(a) Total float: the time by which an activity can be delayed or extended without affecting the total duration of the project. It is calculated as follows:

   Total float = latest end event time − earliest start event time − duration.

(b) Free float: the time by which an activity can be delayed or extended without delaying the start of any succeeding activity. It is calculated as follows:

   Free float = earliest end event time − earliest start event time − duration.

(c) Late free float: the time equal to latest end event time less latest start event time less duration and is of no practical significance. This is expressed as:

   Late free float = latest end event time − latest start event time − duration.

(d) Independent float: the time by which an activity can be delayed or extended without affecting the preceding or succeeding activities. It is calculated as follows:

   Independent float = earliest end event time − latest start event time − duration.

The identification of float in its different forms is of extreme importance when decisions have to be made on matters of time and resources to be employed whether during the planning stage or later during the execution of the works.

A critical path diagram showing critical events, critical activities, various forms of float and the resources planned for the execution of the works is a powerful source of project control, if properly devised and used. Such control can be exercised throughout planning, commencement of works and their execution, by periodically providing updated diagrams. Delay in an activity with an independent float longer than the period of delay would not result in delay to the time for completion. It may, however, cause disruption.

Events which may lead to delay can be avoided or their effects mitigated if a critical path diagram is available to place them in context. If they cannot be avoided or if their effect cannot be mitigated, they can perhaps be correctly allocated from the point of view of responsibility.

Major projects, nationally and internationally, have been completed on time when network analysis programmes were utilised by the contractor from the planning stage of the works and updated periodically until completion.[17.26]

Without the benefit of a critical path diagram before and after a delay event, it would be extremely difficult to precisely designate delay as critical or non-critical delay. This is because the question which is normally asked: 'Did a specific event cause a delay to the contractor in the execution of the works?' must be combined with: 'Should it have and if so to what extent?'

The term 'critical delay' is used here in the context of that which would delay the time for completion of the whole of the works or of a section required to be completed by a specific date.

In answering the last question posed above, the accuracy of the time analysis of the original critical path diagram drawn at the planning phase is of particular importance. This is so in view of the practice adopted by some contractors of submitting programmes with optimistic activity durations. There is some English case law supporting this practice, see *Walter Lawrence & Son Ltd* v. *Commercial Union Properties Ltd* and *Glenlion Construction Ltd* v. *Guinness Trust* as can be seen from the following quotation.[17.27]

> 'However Mr Ramsey for the contractors, Glenlion, readily conceded that a relevant fact is that both parties at the time of entering into the contract would have been well aware that Contractors frequently produce programmes that were over-optimistic.'

Research carried out in Australia has also shown this tendency.[17.28] A sample of 329 construction projects with a total value of more than $270 million was studied. The following comments were made:

> 'The main reason why so much excess time appears to be required in some classes of work is because these contract completion times tend to be optimistic, rather than because of fundamental differences in time requirements. The results show that writing in construction times known to be inadequate in hopes of spurring the contractor to greater endeavours has little influence on the times that are actually taken in practice. On the other hand, there is as yet no clear evidence that weighting in a long construction time will result in great improvement over one already conforming to reasonable standards unless at the same time control procedures are improved . . . One of the basic difficulties appears to be inadequate communication between the many people concerned in a project. Until more effective communication can be established, that is until the timing of work becomes mentally accepted as a real agreement on a common basis of understanding between the parties concerned and transfer of information is improved to permit adequate control against undue disruption from unessential changes there will be continued irritation, frustration and financial losses.'

It was also observed from the above-mentioned study that:

> 'Execution of the nature and extent of factors which affect the time of construction revealed that genuinely uncontrollable factors such as inclement weather account for

only 9% of the 47% overrun experienced overall in these projects. Faulty programming and organisation contributed 12%, much of which was caused by setting unrealistic targets in the first place. Tardiness in making decisions and granting approvals and faulty documentation are major sources of delay both in construction and during the design phase contributing a further 8% to the time overrun.

Variations caused extra construction or administrative work which led to 11% of the time overrun. Further investigation had revealed that clients generated a total of 41% of the gross value of variations (mainly additions) and appeared not to realise the extent to which they are disrupting the construction process.'

Accordingly, the employer or the engineer on his behalf must be fully acquainted with the contractor's programme and with its details which must be provided in such a manner to enable a proper assessment of its achievability or otherwise.

This is a difficult task in view of the fact that, as noted above, the programme is not a contractual document but only a valuable guide, and that it is only required to be submitted after the acceptance of the tender and the formation of the contract. Furthermore, as also noted above, there is no sanction specified in the Red Book should the contractor fail to submit a programme, or submit one without the full details that may be prescribed by the engineer. Of course, if the contractor chooses not to comply with the requirements of sub-clause 14.1, he would have a difficult time ahead should he wish to prove an entitlement to an extension of time and money in respect of any of the provisions in the contract. Irrespective of this difficulty, the engineer may wish to press for a programme to be submitted as it is considered the most efficient way of controlling many of the temporal aspects of a contract.

Two possible courses of action are available to him but both have extreme consequences. The first is under paragraph (b) of sub-clause 40.1, where he may suspend the progress of the works due to default of or breach of contract by the contractor. The second is in accordance with paragraph (d) of sub-clause 63.1 where, in his opinion, the contractor, despite previous warning from the engineer, in writing, is otherwise persistently or flagrantly neglecting to comply with any of his obligations under the contract.

Both of these courses of action would lead to an extremely serious situation should the contractor persistently refuse to submit a programme or the required details.

Whilst failure to comply with the programme is not in itself a breach of contract, it could show that the contractor has failed 'to proceed with the Works with due expedition and without delay', as required by sub-clause 41.1. It may also provide evidence for the application of sub-clause 46.1 which deals with the rate of progress.

17.5 Concurrent delays

The engineer, in assessing entitlement for an extension of time, must establish the responsibility as between the contractor and the employer for the delay in question. Such responsibility is seldom obvious, particularly in situations of concurrent delays. In simple terms, concurrent delays arise where the contractor is delayed by two critical events which occur contemporaneously or overlap, one for which he would be entitled to an extension of time under one of the headings listed in sub-clause 44.1 (a 'clause

44 event'), the other one being his own responsibility. There have been a number of recent UK court cases involving concurrent delays, whereby the phrase 'concurrent delay' has been applied to three scenarios.

In the first scenario the two events, one a clause 44 event and the other an event for which the contractor is liable, occur before either has had an effect on progress and either of which, had it happened on its own, would have caused delay. In such case, the contractor is entitled to an extension of time for the period of delay caused by the clause 44 event notwithstanding that the other event has a concurrent effect, as per the reasoning in part of the decision of *Henry Boot Construction (UK) Ltd* v. *Malmaison Hotel (Manchester) Ltd*[17.29] and confirmed in *Royal Brompton Hospital NHS Trust* v. *Frederick A. Hammond and Others*.[17.30]

In the second scenario, the event which is at the contractor's risk occurs first and has already caused delay when the clause 44 event occurs. The clause 44 event would have caused the contractor to be delayed, but in fact, because of the existing delay, it makes no difference. Therefore, there is no entitlement to an extension of time as per the reasoning in another part of the decision of *Henry Boot Construction* and also confirmed in the *Royal Brompton Hospital* case.

In the third scenario, the event which is at the contractor's risk occurs first and has already caused delay when the clause 44 event occurs, similar to the second scenario above. However, in this third scenario, the effect of the first event finishes before that of the clause 44 event so that the clause 44 event does make a difference and further delay occurs. This scenario was considered in *Balfour Beatty Building Ltd* v. *Chestermount Properties Ltd*.[17.31] It was held that the objective in assessing an extension of time was to assess whether the relevant event (equivalent to the clause 44 event) had caused further delay to the progress of the works, and if so how much. If there was in fact further delay to the overall completion date, the contractor was then entitled to an extension of time equal to the amount of such further delay.

The difficulty in practice is for the engineer to distinguish the situation where the clause 44 event occurs during a period of contractor's culpable delay, from the situation where the clause 44 event occurs before the other event has yet had time to have an effect. The engineer must consider all the circumstances surrounding the clause 44 event and, inevitably, network analysis will be necessary in all but the simplest cases.

17.6 Claims for both extension of time and money

As discussed earlier in Section 17.2, clause 44 of the Red Book sets out the events which may entitle the contractor to claim an extension of the time for completion of the works. A claim for an extension of time may or may not be associated with a claim for financial compensation or for that matter with a delay in a particular part of the works that, because it does not lie on the critical path, does not adversely affect the time for completion. Thus, in general terms, a contractor may have a claim for:

(a) time alone
(b) for money alone, or
(c) for both time and money.

A claim for time alone could be made in respect of, for example, experiencing bad weather on the site. A claim for money alone could be, for example, for being instructed to carry out an item of work using a more expensive material than that originally required. In a claim for both time and money, the contractor asks for an extension of time for the time for completion and for monetary compensation for the time lost.

There are three types of claim which may be made under the Red Book linking delay and money. The first type is associated with any critical delay which may occur in the completion of the works beyond the time for completion, and this is often referred to as a claim for prolongation.

The second type of claim is associated with a non-critical delay of an activity that has a positive float. There can be no claim for an extension of time for the time for completion, but there could be a claim for monetary compensation in this case, which will be limited to the effect of delay, if any, on the completion of that activity. For want of a better term, it is sometimes referred to as a claim for elongation.

The third type of claim is associated with the possible effect of a claim event on the efficiency of the execution of some part or parts of the works. This type of claim is referred to as a claim for disruption, which includes any reduction in the efficiency of the disrupted party's resources. The disrupted activity could be either a critical or a non-critical activity, both of which might be delayed as a result of the disruption. Disruption may also occur when the rate of progress of the execution of the works has to be accelerated, either to comply with the time for completion of the works or to advance it.

The term or concept of prolongation appears nowhere in either the Fourth Edition of the Red Book or in the 1999 suite of FIDIC contracts. Whilst disruption is mentioned at sub-clause 6.3 of the Fourth Edition, *'Disruption of progress'*, it is only in respect of the timing of the issue by the engineer of drawings and instructions, but it is not defined. Similarly, disruption is mentioned only once in the 1999 Red Book, in the text of sub-clause 8.5 in respect of delays caused by authorities, but again it is not defined. It is unfortunate that prolongation and disruption, and the concepts upon which they are based, are not dealt with adequately in the FIDIC forms of contract, especially in view of the fact that they have been a major source of disputes, arbitration and litigation in recent time. On the other hand, there are many clauses which deal with delay, for example, sub-clauses 6.3, 6.4, 12.2, 27.2, 36.5, 40.2 and 69.4 of the Fourth Edition of the Red Book, referred to earlier in Section 17.3.

The financial compensation, where there is prolongation or disruption, is not dealt with in the Red Book in a unified manner, as can be seen from the analysis made in Tables 9.1, 9.2 and 9.3 provided earlier in Chapter 9.

17.6.1 Prolongation

Prolongation may be defined as a critical delay which results when the time necessary to complete a critical activity is prolonged, as defined above in Section 17.4, thus extending the time for completion of the whole of the works. A prolongation claim is, therefore, a claim that a contractor would make when the time necessary to complete a critical activity becomes longer than that originally planned owing to the event giving rise to the critical delay. The delayed activity must be critical, i.e. it must start and finish on a pre-determined scheduled date with no float time or flexibility such that, if its start date or its end date were delayed without taking any subsequent corrective measures, such an activity would delay all the subsequent activities and, therefore, would ultimately delay the time for completion.

In the situation where the contractor programmes the works to achieve completion on a date before the contract completion date, he will not be entitled to prolongation costs for delays to this early completion date:

> 'It should be noted that ordinarily prolongation costs will only begin to be recoverable from the Completion Date. A provision either in the Bills or in the Contractor's Programme which provides for an earlier Completion Date but which does not form part of the Contract does not give rise to any time related contractual obligation. The Contractor cannot therefore claim prolongation costs commencing on his programmed Completion date if it is earlier than the Date for Completion stated in the Contract'.[17.32]

Delays in completion of the works might result in a number of added costs to the contractor for the additional time during which he remains on the site and also in respect of certain direct and indirect costs that he may have suffered. If such delay is determined by the engineer to be the responsibility of the employer, then the contractor would be entitled to an extension of time and a number of heads of claim for financial compensation can be pursued by the contractor. The extent of compensable costs will depend on the nature of the project and details of the circumstances surrounding the claim.

Sub-clause 44.1 of the Fourth Edition of the Red Book does not make clear what would constitute 'being fairly to entitle the Contractor to an extension of time', and this is left to the engineer's expected knowledge in the technical field and in construction law, discussed above; in the interpretation of the contract; his assessment of the circumstances leading to the claim; and his determination after 'due consultation' with the employer and the contractor. However, a strict interpretation would require that the delay be such that it extends the longest critical path to completion of the works, as discussed in Section 17.4 above. In effect this means that the contractor has to remain on site and has to expend his resources over a longer period than he had anticipated in his tender. Parts of these resources are spent in managing and carrying out the resourcing for a prolonged contract. The costs relating to prolongation may therefore include: off-site overheads, on-site overheads, financing of any retention fund, the cost of the extended periods for insurance and bonds required under the contract, any other financing charges and interest. Some of these heads of claim were referred to in Section 16.7.1 above.

There are limited circumstances where a non-culpable delay may affect a critical activity, thereby giving rise to an entitlement to an extension of the time for completion, but no entitlement to prolongation costs. In particular, occurrences of adverse weather and Acts of God events, see Section 17.2 above, are classed as 'neutral events' where the employer bears the risk of the delay incurred by virtue of the contractor's entitlement to an extension of time; and the contractor bears the costs of the prolongation of the project arising from such extension of time. In addition, the case of *Gaymark Investments* v. *Walter Construction Group Ltd*, referred to in Section 17.2 above, demonstrates that where it is a condition precedent of entitlement to an extension of time that the contractor gives notice of his claim, and where the relevant delay is caused by the employer but the contractor fails to give such notice or is time-barred to claim entitlement of an extension for such delay, the employer shall have no entitlement to liquidated damages for such delay and the contractor shall have no entitlement to an extension and so will bear his own costs arising from the employer's delay.

Finally, it may therefore be concluded that in the Fourth Edition of the Red Book, it is expected that those involved in dealing with extension of time claims, including the arbitrators that would be appointed if and when disputes are referred to arbitration, are sufficiently knowledgeable in many areas, including: the technical field relating to the project and in construction law; in the interpretation of the conditions of contract; and the assessment of the circumstances leading to the claim; so that they may be able to assess the meaning of the phrase 'being such as fairly to entitle the Contractor to an extension of the Time for Completion of the Works' in the first line after paragraph (e) of sub-clause 44.1.

17.6.2 Disruption

Disruption may be defined as the effect of an event or a number of events on the efficiency and loss of productivity of the execution of the works, irrespective of whether or not there had been a delay to a critical activity. Events that cause disruption during a construction project are very similar to those which are associated with delay and might include, for example, the employer stopping and starting the project, late design information, differing site conditions, delayed or unco-ordinated instructions for variations, employer instructed out-of-sequence work, overzealous inspections or testing, the presence of other contractors, damage caused to the works by other contractors, delayed or hindered access and late issue of or inaccurate drawings. Disruption claims are in themselves divided into two types of claim. In the first, the disruption caused is to a non-critical activity and hence the result is that the activity itself takes longer, but using the float time either before or after the date that the activity was to be performed. It will cost more because of the inefficiency in carrying out the activity but, because time is not critical, there is no delay in the time for completion and it will not entitle the contractor to an extension of time.

In the second, where continuous, extensive and cumulative disruption is such that the flexibility in time allowed to the relevant activity is exceeded, then this non-critical activity becomes a critical one and further disruption ends in critical delay and prolongation of the time for completion. In such a case the contractor is entitled not only to the monetary cost of that disruption but also to an extension of time, and the prolongation costs, if any, associated with such an extension.

Since disruption may not necessarily be accompanied by an overall delay of the time for completion, its assessment requires more than a consideration of the elongation of the particular event that had been disrupted.

Inefficiency, loss of productivity of labour and uneconomic use of equipment are aspects that come under the heading of disruption when they are caused by an event which is not the responsibility of the contractor. They are extremely difficult to assess because comprehensive records are essential, not only during the execution of the works but also records used for the compilation and the submission of the tender. Furthermore, as disruption may not necessarily be accompanied by overall delay, no reference can be used for its assessment. Calculations of the anticipated output figures of labour and equipment at the time of tendering are essential for the calculation of the actual disruption suffered, as this can only be proven by comparing these anticipated figures with those actually achieved.

A proper evaluation of a claim for disruption requires the following prerequisites:

(a) An identification and an analysis of each of the operations claimed to have been disrupted. It is not sufficient to simply state that the execution of the works has been disrupted.
(b) The cause and the manner in which disruption has occurred should be established.
(c) The figures for the anticipated output, the resources planned and the time required to achieve the completion of the disrupted operations as calculated in the tender have to be shown to be achievable.
(d) The effect of any inefficiency on the part of the disrupted party in carrying out the works should be properly calculated and its effect included in the calculations of disruption suffered.
(e) The number of hours actually logged in the time sheets for the disrupted operation has to be shown to be accurate.

Where records are available and are correct, then the cost of disruption can be simply calculated as the number of hours actually worked less that originally anticipated in the tender, with the result being multiplied by the cost of the particular resources disrupted per hour. Nevertheless, the contractor should also be able to demonstrate clearly that the number of hours originally anticipated were realistic and that he has adopted reasonable measures in order to mitigate the effects of the disrupting event(s).

Alternatively, the contractor might support his claim for disruption by comparing work performed in one period not adversely affected with the same or similar work performed in another period that was adversely affected by events causing loss of productivity and inefficiency. Records of a similar level and accuracy as those referred to above would have to be kept to credibly substantiate such an approach.

When acceleration of the progress of the works is required, the cost may include the expense of:

(a) working additional hours;
(b) providing additional labour;
(c) providing additional or different equipment;
(d) advancing the date of delivery of manufactured elements;

and the disruption caused by acceleration might also include the loss of productivity due to the dilution of supervision, more men per supervisor, and the lack of sufficient detailed design information to support the accelerated work.

In practice, many prolongation and disruption claims are presented as global claims. The nature and constraints imposed on such claims are discussed in Section 16.7.2 above.

17.7 Liquidated damages

Clause 47 of the Fourth Edition of the Red Book regulates the relationship between the employer and the contractor should there be a failure by the contractor to comply with the requirement to complete in accordance with clause 48 within the time for completion as specified in clause 43 and, if applicable, any extended time in accordance with clause 44.

Clause 47 also provides that should the contractor fail in his obligations as stated above:

> 'then the Contractor shall pay to the Employer the relevant sum stated in the Appendix to Tender as liquidated damages ... for every day or part of a day ... between the relevant Time for Completion and the date stated in a Taking-Over Certificate ... subject to the applicable limit stated in the Appendix to Tender'.

Liquidated damages may be defined as the genuine pre-estimate of all the losses which are likely to be incurred by the employer as a result of late completion of the works, calculated at the time of making the contract. The purpose of providing for a monetary payment in the event of late completion is essentially to compensate the employer for the damages incurred by the contractor's breach of contract in not completing the works on the specified date. The advantage of having a liquidated damages provision can be said to be that the damages payable by the contractor in culpable delay are limited and the employer who receives late completion does not have to prove his losses due to such delay. Additionally, the liquidated damages provision acts as an exhaustive remedy for damages for late completion. Therefore, the provision provides certainty for both parties, enabling them to assess and price the risk. As stated by Lord Woolf of the UK Privy Council in *Philips Hong Kong Ltd* v. *Attorney General of Hong Kong*:[17.33]

> 'Whatever the degree of care exercised by the draftsman it will still be almost inevitable that an ingenious argument can be developed for saying that in a particular hypothetical situation a substantially higher sum will be recovered than would be recoverable if the plaintiff was required to prove his actual loss in that situation. Such a result would undermine the whole purpose of parties to a contract being able to agree beforehand what damages are to be recoverable in the event of a breach of contract. This would not be in the interest of either of the parties to the contract since it is to their advantage that they should be able to know with a reasonable degree of certainty the extent of their liability and the risks which they run as a result of entering into the contract. This is particularly true in the case of building and engineering contracts. In the case of those contracts provision for liquidated damages should enable the employer to know the extent to which he is protected in the event of the contractor failing to perform his obligations ... As for the contractor, by agreeing to a provision for liquidated damages, he is seeking to remove the uncertainty as to the extent of his liability under the contract if he is unable to comply with his contractual obligations. ...'

The employer's ability to rely on the liquidated damages provision is lost if the employer prevents the contractor from completing by the completion date without any effective mechanism for extending time for completion. In such a case time is said to be at large, which was discussed earlier in Section 17.1 above.

What of the employer who decides to deduct liquidated damages on or just after the date corresponding to the time for completion for the time being, when the works are still under completion by the contractor?

Strictly speaking, the wording at the beginning of sub-clause 47.1, namely:

'If the Contractor fails to comply with the Time for Completion in accordance with Clause 48, for the whole of the Works or, if applicable, any Section within the relevant time prescribed by Clause 43, then the Contractor will pay to the Employer the relevant sum stated in the Appendix to Tender as liquidated damages for such default...',

would appear at first glance to entitle the employer to make such a deduction. Also, unlike the case in respect of other exchanges of money between the contractor and the employer, no certification is required by the engineer in respect of liquidated damages. However, in practice it is more often than not the case that, at the time in question, the contractor's full entitlement to extension(s) of time may not have been finally determined by the engineer and/or still awaits final review by the engineer pursuant to sub-clause 44.3. The reluctance of a cautious engineer to determine a significant interim extension of time where it cannot later be reduced in the final review, as noted above at Section 17.2, means that it is likely that the employer cannot be certain that the time for completion for the time being includes the contractor's full entitlement for extension of time. Should an employer act impulsively in deducting liquidated damages before this full entitlement has been determined by the engineer, then under the general law the contractor might be entitled to a refund of the amount of liquidated damages levied plus interest thereon.

The principle of liquidated damages is derived from the common law system and originally appeared in the First Edition of the Red Book through the ACE Form on which it was modelled. It is distinguished from a penalty, which is usually used in jurisdictions outside the common law system, in that a penalty is a sum which a party agrees to pay or forfeit in the event of a breach and is a fixed sum as a punishment, the threat of which is designed to prevent the breach, and not a pre-estimate of the probable loss.

The use of either of these two deterrents, i.e. 'liquidated damages' and 'penalties', is therefore linked to the applicable law of the contract.

17.7.1 Liquidated damages and penalties

Under the common law system, it is important to establish whether the sum entered is liquidated damages or a penalty because it has long been established that a penalty is subject to equitable jurisdiction. The courts of equity have taken the view that:

'... the promisee is sufficiently compensated by being indemnified for his actual loss, and that he acts unconscionably if he demands a sum which, though certainly fixed by agreement, may well be disproportionate to the injury.'[17.34]

The decision as to whether a sum represents liquidated damages or a penalty depends on the terms of the contract and the intention of the parties at the time of making the contract and not at the time of the breach. Therefore, the distinction between penalties and liquidated damages depends on:

> 'the intention of the parties to be gathered from the whole of the contract. If the intention is to secure performance of the contract by the imposition of a fine or penalty, then the sum specified is a penalty; but if, on the other hand, the intention is to assess the damages for breach of the contract, it is liquidated damages.'[17.35]

It must also be borne in mind that the expression used in the contract for the sum entered, i.e. whether liquidated damages or penalty, should not be taken as conclusive evidence of the parties' real intentions at the time of making the contract. The decision as to what it is in reality should be based on whether or not it is a genuine pre-estimate of the probable loss to result from the breach. Such a loss may be extremely difficult to quantify when dealing with civil engineering projects such as roads, water supply and sewerage works. Nevertheless, this remains a problem and the onus of showing that the specified sum is a penalty lies upon the party sued for its recovery.[17.36] In Reference 17.3 certain rules for guidance when distinguishing between liquidated damages and a penalty were summarised from the judgment in *Dunlop Pneumatic Tyre Co Ltd* v. *New Garage and Motor Co Ltd*.[17.37] This summary is very useful and hence it is quoted here:

> '(a) The conventional sum is a penalty if it is extravagant and unconscionable in amount in comparison with the greatest loss that could possibly follow from the breach.
> (b) If the obligation of the promisor under the contract is to pay a certain sum of money, and it is agreed that if he fails to do so he shall pay a larger sum, this larger sum is a penalty. The reason is that, since the damage arising from breach is capable of exact definition, the fixing of a larger sum cannot be a pre-estimate of the probable damage.
> (c) Subject to the preceding rules, it is a canon of construction that, if there is only one event upon which the conventional sum is to be paid, the sum is liquidated damages. This was held to be the case, for instance, where it was provided in a contract for the construction of sewerage works that, if the operations were not complete by 30 April, the contractor should pay £100 and £5 for every seven days during which the work was unfinished after that date.
> (d) If a single lump sum is made payable upon the occurrence of one or more or all of several events, some of which may occasion serious and others mere trifling damage, there is a presumption (but no more) that it is a penalty. This presumption, however, is weakened if it is practically impossible to prove the exact monetary loss that will accrue from a breach of the various stipulations. The sum fixed by the parties in such a case, if reasonable in amount, will be allowed as liquidated damages.'

In the recent Scottish case of *City Inn* v. *Shepherd Construction*[17.38] it was stated:

> 'The rule as to the unenforceability of penalty clauses ordinarily applied where a provision failed to qualify as liquidated damages because it did not constitute a genuine pre-estimate of the relevant loss. It might also apply where the provision was penal in the sense that it was unconscionable, not a true contractual provision but an attempt to punish; where the provision was incorporated in the contract *in terrorem*. In order for a provision to be classed as a penalty, it required to involve the concurrence of two events, namely (i) a breach of contract and (ii) a result or consequence which was regarded by the court as unconscionable in that it amounted to oppression or the imposition of a punishment. It was to be borne in mind that the rule against penalties was an exception to freedom of contract, and ought on that account to be kept within strict parameters.'

The very recent case of *Murray* v. *Leisureplay Plc*[17.39] usefully demonstrates the courts' approach in dealing with the question of whether a liquidated damages provision is deemed to be a penalty or not. The court first confirmed what previous authorities had shown, that the courts are slow to hold that a liquidated damages provision is a penalty, by saying:

> 'The rule against penalties is an exception to the general rule that the Court will enforce the terms of a lawfully made contract: *pacta sunt servanda*. Where a contract has been made between equal parties negotiating at arm's length, the Court may be reluctant to strike down a term freely entered into.'

The court then went on to refer to the guidance given in the *Dunlop* case and concluded:

> 'In general, a contractual provision that requires one party in the event of his breach of the contract to pay the other party a sum of money is unlawful as being a penalty unless the provision can be justified as a payment of liquidated damages, being a genuine pre-estimate of the loss that the innocent party will incur by reason of the breach. A liquidated damages clause will be held to be penal if it provides for payment of a sum greater than can be justified as a genuine pre-estimate of loss. Although the classic distinction is between payments stipulated *in terrorem* and genuine pre-estimates of damages (*Dunlop Pneumatic Tyre Company* v. *New Garage* [1915] AC 79), the phrase '*in terrorem*' adds nothing to the idea conveyed by the word 'penalty' and may obscure the fact that penalties may be undertaken by parties who are not 'terrorised' by the prospect of having to pay them. Whether a provision is penal is a matter of construction to be resolved by considering whether at the time the contract was made, the predominant function of the provision was to deter a party from breaking the contract or to compensate the innocent party for breach. The answer to this may be deduced by comparing the amount stipulated as payable on breach with the loss that might be sustained if a breach occurred. Although it is not sufficient merely to identify some situations in which the application of the provision could result in a party recovering more than his actual loss, if it is obvious that in

relation to some of the possible outcomes the liquidated damages are totally out of proportion to the losses that might be incurred, the failure to make special provision for these cases may result in the clause being held to be penal.'

Therefore, the latest thinking of the courts would appear to be that, to distinguish between liquidated damages and a penalty, the amount of liquidated damages stipulated in the contract should be compared with the actual loss that might have been incurred as a result of the breach. If the former is 'totally out of proportion' to the latter, then the liquidated damages provision acts as a penalty clause. Unfortunately, the court did not expand further into what, in practical terms, this phrase might mean.

Under common law, if a liquidated damages clause fails, the employer is not prevented from claiming general damages of his proven loss to an amount which would put him in the same position, so far as money can do, had the contractor completed within the specified time. It is worth sounding a cautionary note, however, as a result of a recent case where a 'nil' rate was entered against the liquidated damages clause in the appendix to a building contract. The issue that arose was whether this meant that the liquidated damages were simply nil pounds or, as the employer contended, that the effect of writing 'nil' was to exclude the whole of clause 24 (the liquidated damages clause in the particular form of contract used).

The employer's argument was rejected by the Court of Appeal. Lord Justice Nourse said:

'I think it clear . . . that if (1) Clause 24 is incorporated as part of the contract, and (2) the parties complete the relevant part of the Appendix, either by stating a rate at which the sum is to be calculated or, as here, by stating the sum is to be nil, then that constitutes an exhaustive agreement as to the damages which are or are not to be payable by the Contractor in the event of his failure to complete the works on time.'[17.40]

In jurisdictions outside the common law system, where penalty clauses for delay are recognised, the general principles of law which apply are those of placing the injured party economically in the same situation as he would have been had the contractual performance taken place. However, the question could arise as to whether it is better to include a clause in the contract between the parties, which would provide for a fixed amount per day or per week, thus evading the often complicated questions of establishing the loss suffered and its quantum, and thus reducing the tendency to end in dispute.

The penultimate sentence of sub-clause 47.1 makes it clear that the employer may deduct the amount of the liquidated damages from any monies due or to become due to the contractor. This, however, is 'without prejudice to any other method of recovery'. In practical terms, this means that the employer may take legal action to recover the amount of liquidated damages as a debt due. This sentence also makes clear that it is the employer who makes the deduction, if he so wishes. The deduction of liquidated damages is at his discretion and it has been known, albeit rarely, for an employer to waive his right to deduct liquidated damages where, for example, there is a continuing working relationship between the parties on other projects and in the interest of maintaining the financial viability of the contractor.

The engineer is required under sub-clause 48.1 to indicate the date on which, in his opinion, the works were substantially completed in accordance with the contract. This date establishes the extent of the contractor's culpable delay, enabling the amount of liquidated damages, if any, to be calculated. The engineer himself is not authorised to make deductions in respect of liquidated damages, as can be seen from paragraph (b) of sub-clause 60.2, thereby reserving the right to deduct liquidated damages exclusively to the employer.

The last sentence of sub-clause 47.1 provides that payment or deduction of the liquidated damages shall not relieve the contractor from his obligation to complete the works, or from any of his other obligations. This is included in the contract in order to avoid a situation where, because the right to deduct liquidated damages is the employer's sole remedy for the contractor's breach of contract in not achieving completion within the time for completion, it might otherwise be argued that in making a payment of liquidated damages the contractor has compensated the employer for such breach and, thereby, absolved himself from any remaining work/obligations under the contract (in particular, the remedying of defects during the defects liability period).

Sub-clause 47.2 deals with the situation where a taking-over certificate is issued for any part of the works or of a section thereof. The liquidated damages for delay, if any, in completion of the remainder of the works or of that section are to be reduced on a pro rata basis. If it is considered by the employer that a pro rata reduction would not be equitable, then an alternative provision should be included in Part II at the tender stage. This should be done in a very clear and precise manner as otherwise it may have a legal effect on the applicability of the whole clause.

Part II of the Red Book includes proposals for the inclusion of a bonus should this prove to be attractive from the employer's point of view. The advantage of providing the contractor with an incentive to complete the works on time should not be underestimated. In practice, the inclusion of a not insignificant bonus has been proven to be the greater motivator to contractors (on a day-to-day basis) to complete the works on time than the possibility that liquidated damages may be levied some time in the future.

17.8 The Society of Construction Law *Delay and Disruption Protocol*

In October 2002, following a significant consultation process with individuals and organisations involved in the construction industry, the Society of Construction Law (SCL) published its *Delay and Disruption Protocol*.[17.41] In doing so, SCL addressed and sought to bridge those considerable gaps in respect of delay, disruption, prolongation and programming in the common standard forms of contract, including the FIDIC forms as noted above, which had long been identified by practitioners and commentators alike, namely:[17.42]

(a) proper preparation and updating of a programme of the works, in a format and based on software suitable to the project in hand, so that it can be used effectively as a tool first by the contractor and secondly by the engineer in his determination of entitlement to extension of time, discussed in Section 17.4 above;

(b) the relationship between float shown in the programme of the works and entitlement to extension of time, discussed in Section 17.2 above;
(c) the requirement of proper provision in the contract for extension of time to avoid time becoming 'at large', discussed in Section 17.1 above;
(d) concurrent delays and entitlement to extension of time, discussed in Section 17.5 above;
(e) provision for entitlement to recover prolongation costs and a mechanism to evaluate such costs, discussed in Section 17.6 above;
(f) distinguishing between delay and disruption, discussed in Section 17.6 above;
(g) provision for disruption and for the recovery of costs arising from disruption, discussed in Section 17.6 above.

The purpose of the protocol is stated as providing a means by which the parties to a construction contract can resolve matters concerning the recovery by the contractor of an extension of time and/or compensation for the extra time taken to complete the project, thereby avoiding unnecessary disputes.

The protocol is not intended to be adopted as a contract document. Unless the employer and the contractor agree at pre-contract stage that the principles, or some of them, set out in the protocol will be adopted in a given contract, the protocol has no contractual standing. Nevertheless, it is a valuable document and it would be worthwhile for those drafting contracts based on the FIDIC standard forms to take note of the core principles and guidance advocated and, if appropriate, to make provision accordingly.

There may, at first glance, seem to be a lesser benefit to employers to adopt the recommendations of the protocol than that to contractors and to the engineer, in that it would appear to promote the increased use of resources in preparing claims for extension of time and compensation which might otherwise be engaged in executing the works. However, it is almost inevitable that some claims for extensions of time and additional payment will arise in every construction contract and there can be little doubt that the protocol provides very useful clarity in respect of such claims. Disputes are expensive and, therefore, any measure which can be taken to minimise disputes or to enable them to be resolved speedily must be of benefit to all the parties involved. The topic of delay and disruption was also touched on in the final report on construction industry arbitrations, see reference 19.87.

Chapter 18

Certificates and Payments

18.1 Introduction

Traditionally, it has been accepted, and embodied in the provisions of the standard forms of contract, that in admeasurement contracts payment to the contractor should be made periodically, on account, in accordance with a proper evaluation of the work done and materials supplied.[18.1] Clause 60 of the Red Book is the relevant clause in this connection and is worded on the basis of monthly payments against certificates issued by the engineer, referred to as an interim payment certificate.

Such monthly accounting and certification system is very helpful in providing a record and detailed information of the progress made on site and simulates the procedure in any properly run business organisation.

As provided in sub-clause 60.4 of the Red Book, each time an interim payment certificate is issued, a new evaluation of the work done should be carried out providing an opportunity to adjust or modify previous evaluations of earlier interim payment certificates. The adjustment can be made retrospectively taking into consideration, for example, any materials supplied, and paid for under previous certificates, becoming part of the completed works or any work carried out not to the satisfaction of the engineer.

As a rule, it is the intention of the Red Book that such certificates are binding on the employer in the absence of any cross claim from him against the certified sums. Furthermore, the certificate is a condition precedent to payment and is a necessary step in the prescribed procedure, subject to the arbitration clause and the arbitrator's power to open up, review and revise.

In this connection, it is notable that the engineer seems to have a continuing potential function under clause 67 of the contract in connection with deciding disputes between the parties until the expiry of the limitation period under the applicable law of the contract.[18.2]

There are five distinct types of certificate under the Red Book. They are as follows: interim payment certificate; taking-over certificate; defects liability certificate; final payment certificate; and certificate of valuation at date of termination. These certificates are discussed in this chapter.

18.2 Interim payment certificates

An interim payment certificate is defined under sub-clause 1.1(e)(iii) of the Red Book as 'any certificate of payment issued by the Engineer other than the Final Payment

Certificate'. It is required to be issued under the provisions of sub-clause 60.2 of the Red Book, by the engineer, within 28 days of him receiving the monthly statement from the contractor to certify to the employer the amount of payment which he considers due and payable to the contractor in respect of:

(a) the value of the permanent works executed;
(b) any other items in the bill of quantities including those for contractor's equipment, temporary works, dayworks and the like;
(c) the percentage of the invoice value of listed materials, as stated in the Appendix to Tender, and plant delivered by the contractor on the site for incorporation in the permanent works but not incorporated in such works;
(d) adjustments under clause 70 for cost fluctuations and subsequent legislation; and
(e) any other sum to which the contractor may be entitled under the contract.

The above amount is subject to the following deductions:

(a) a deduction in respect of a retention to be calculated by applying the percentage of retention, stated in the Appendix to Tender, to the amount to which the contractor is entitled (except for adjustments under clause 70), subject to the limit of retention money stated in the Appendix to Tender;
(b) a deduction in respect of any sums which may be due and payable by the contractor to the employer, other than pursuant to clause 47.

Before an interim payment certificate is issued, the engineer is required to give attention to the following matters:

(a) The amount certified, after the appropriate deductions are made, must be greater than the minimum amount for interim payment certificates as stated in the Appendix to Tender. Otherwise the engineer is not bound to certify under the provisions of sub-clause 60.2.
(b) Where a performance security is required, no amount is to be certified unless such a security is submitted by the contractor and approved by the employer as provided in sub-clause 60.2.
(c) Under sub-clause 60.4, the engineer is authorised to make any correction or modification in any previous certificate and if any work is not carried out to his satisfaction, to omit or reduce the value of such work in any interim certificate.

Once delivered to the employer, an interim payment certificate entitles the contractor to be paid within 28 days as provided under sub-clause 60.10, and within 56 days in the case of the final payment certificate.

The amount to be paid by the employer is equal to the amount shown on the certificate minus any deductions for liquidated damages under clause 47 and others required by law.

Should the employer fail to make payment within the time stated in sub-clause 60.10, the employer will be liable to pay interest on the amount overdue. Such interest is payable at the rate stated in the Appendix to Tender from the date by which such payment should have been made. The entitlement to interest is without prejudice to the contractor's right to suspend work or terminate his employment under the contract in accordance with clause 69.

18.3 Taking-over certificate

'Taking-Over Certificate' is defined under sub-clause 1.1(d)(ii) of the Red Book, but this definition is by necessity a very brief one as it simply refers to clause 48. Any elaboration on this definition would have to incorporate the effect of the issue of the taking-over certificate which includes a long list of provisions within the Red Book as described later in this section.

Clause 48 is divided into four parts: the first of which, sub-clause 48.1, deals with the certificate for 'the whole of the Works'; and sub-clauses 48.2, 48.3 and 48.4 deal with sectional or partial completion of the works.

Within 21 days of the date of delivery of a notice given by the contractor, the engineer is required, under the provisions of sub-clause 48.1, either to issue a taking-over certificate or give instructions in writing to the contractor specifying all the work which, in his opinion, is required to be done before the issue of such a certificate. The notice referred to above which initiates this process is given by the contractor to the engineer, with a copy to the employer, when the whole of the works have been substantially completed and have satisfactorily passed any tests on completion as prescribed by the contract. The notice must be accompanied by a written undertaking to finish, with due expedition, any outstanding work during the succeeding period of defects liability.

It is important to draw attention to the change of wording of sub-clause 48.1 in the Fourth Edition of the Red Book (see Part IV of this book). In particular, the words 'any final test that may be prescribed by the Contract' in the first sentence have been replaced by the words 'any Tests on Completion prescribed by the Contract'. 'Tests on Completion' is a defined term in paragraph (i) of sub-clause 1.1(d) as either those tests which are specified in the contract or otherwise agreed by the engineer and the contractor. As it is possible that the engineer and the contractor may fail to reach agreement on such tests and the matter would then have to be dealt with under clause 36, it is always better to specify such tests, if possible, in the contract documents. The purpose of these tests is three-fold:

(a) To consider the effect of time on the performance of any part of the works and to establish that a properly completed part of the works remains to be so without any damage or deterioration.
(b) To consider the effect of collective performance of various parts of the works. This would establish that not only each part of the works is properly constructed but also that all parts work together.
(c) To consider any special feature of the project and also any event which may occur during the construction period which may require a special test.

If and when the taking-over certificate is issued by the engineer, it must be copied to the employer, and it must state the date on which, in the opinion of the engineer, the works were substantially completed in accordance with the contract.

Sub-clauses 48.2 and 48.3 deal with the situation where the taking-over certificate is to be issued for:

(a) any section in respect of which a separate time for completion is provided in the Appendix to Tender; or
(b) any substantial part of the permanent works completed and occupied or used by the employer; or
(c) any part of the permanent works which the employer has elected to occupy or use prior to completion.

The certificate under sub-clause 48.2 is mandatory as demonstrated by the words 'the Engineer shall issue', whereas it is discretionary under sub-clause 48.3, since the engineer *may* issue such a certificate.

When a taking-over certificate is issued, it has the following effect on a number of the provisions of the Red Book:

(a) The responsibility for the care of the works passes from the contractor to the employer on the date of issue of the taking-over certificate, as provided in sub-clause 20.1.
(b) The scope of the insurance cover as provided under sub-clause 21.2 changes on the date of issue of the taking-over certificate.
(c) Upon the issue of the taking-over certificate, the contractor is required to clear away and remove all contractor's equipment, surplus material, etc. and leave the site and works clean as provided in sub-clause 33.1.
(d) Liquidated damages as provided for in sub-clause 47 apply in respect of the period between the relevant time for completion and the date stated in the taking-over certificate of the whole of the works or the relevant section.
(e) As provided in paragraph (a) of sub-clause 49.1, the defects liability period is calculated from the date of completion certified in accordance with clause 48.
(f) In accordance with sub-clause 49.2, the contractor is required to complete the work, if any, outstanding on the date stated in the taking-over certificate.
(g) Variations exceeding 15 per cent as provided in sub-clause 52.3 are calculated on the date of issue of the taking-over certificate for the whole of the works.
(h) Upon the issue of the taking-over certificate for the whole of the works, one half of the retention money shall be certified as provided under sub-clause 60.3.
(i) Not later than 84 days after the issue of the taking-over certificate in respect of the whole of the works, the contractor is required, under sub-clause 60.5, to submit to the engineer a statement at completion with supporting documents. Reference should also be made to sub-clause 60.9.

Attention should be drawn to the possibility that when the whole of the works has been substantially completed, the contractor may be prevented by reasons outside his control from carrying out the tests on completion. If such an event is envisaged, a fifth sub-clause should be added to clause 48 to deal with the consequences of such prevention. Part II of the Red Book provides a possible text in this connection.

18.4 Defects liability certificate

The defects liability certificate is required to be issued by the engineer, within 28 days after the expiration of the defects liability period, (or if different defects liability periods become applicable to different sections or parts of the permanent works, the expiration of the latest such period), or as soon thereafter as any works instructed, pursuant to clauses 49 and 50, have been completed to the satisfaction of the engineer. This requirement is in accordance with the provisions of sub-clause 62.1.

The issue of the defects liability certificate has an effect on the following provisions of the Red Book:

(a) In accordance with sub-clause 6.1, upon the issue of the defects liability certificate, the contractor is required to return to the engineer all drawings, specification and other documents provided under the contract.
(b) Under the provisions of sub-clause 10.2, no claim shall be made against the performance security provided by the contractor under sub-clause 10.1, after the issue of the defects liability certificate. Furthermore, such security is to be returned to the contractor within 14 days of the issue of the said defects liability certificate.
(c) In accordance with clause 61, only the defects liability certificate is deemed to constitute approval of the works.
(d) Not later than 56 days after the issue of the defects liability certificate, the contractor is required to submit to the engineer for his consideration a draft final statement with supporting documents as required in sub-clause 60.6.

18.5 Final payment certificate

Within 28 days after receipt of the final statement as provided in sub-clause 60.6 and the written discharge in accordance with sub-clause 60.7, the engineer is required to issue to the employer, with a copy to the contractor, a final payment certificate in accordance with sub-clause 60.8.

In the final payment certificate, the engineer is required to state:

(a) the amount which, in his opinion, is finally due under the contract; and
(b) after giving credit to the employer for all amounts previously paid by the employer and for all sums to which the employer is entitled under the contract, other than clause 47, the balance, if any, due from the employer to the contractor or from the contractor to the employer as the case may be.

Sub-clause 60.10 provides that within 56 days after the final payment certificate has been delivered to the employer, the amount due to the contractor, subject to clause 47,

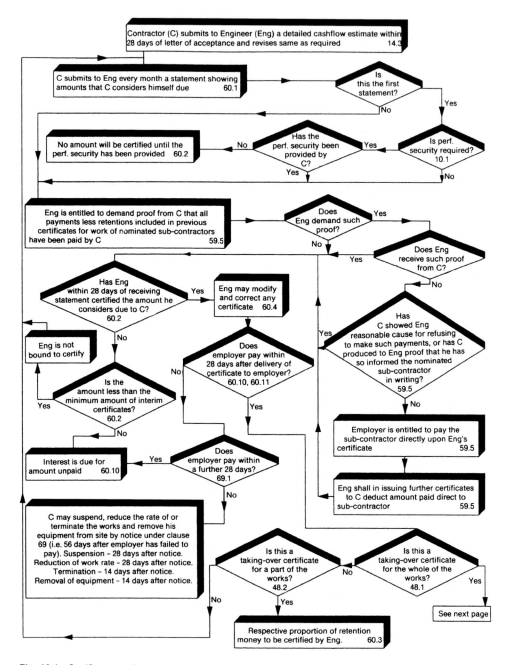

Fig. 18.1 Certificates and payments.

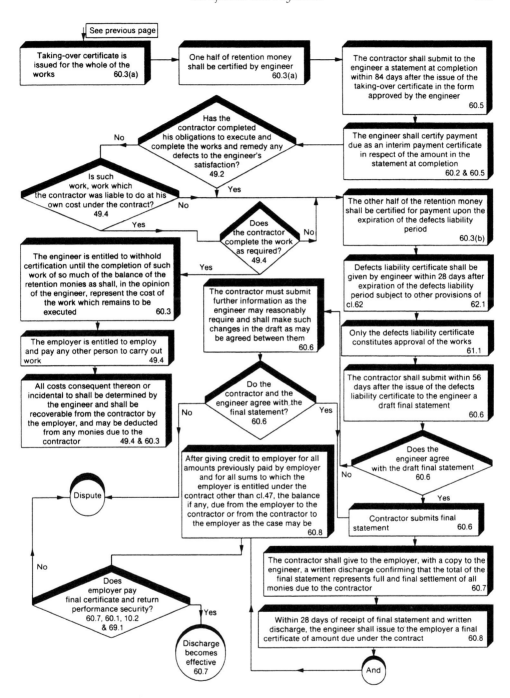

(Fig. 18.1 Contd.)

should be paid by the employer to the contractor. In the event of failure of the employer to make payment within the above period, the contractor is entitled to be paid interest at the rate stated in the Appendix to Tender.

18.6 The engineer is to certify a valuation at date of termination

Should the employer enter upon the site and the works and terminate the employment of the contractor under sub-clause 63.1 of the Red Book, the engineer is required as soon as practicable after any such termination to certify a valuation at date of termination. In certifying under sub-clause 63.2, the engineer is required to fix and determine *ex parte*, or by or after reference to the parties or after carrying out any investigation he may think fit, the following:

'(a) what amount (if any) had at the time of such entry and termination, been reasonably earned by or would reasonably accrue to the Contractor in respect of work then actually done by him under the Contract, and
(b) the value of any of the said unused or partially used materials, any Contractor's Equipment and any Temporary Works.'

18.7 Common requirements

All certificates are required to be in writing as provided for in sub-clause 1.5 of the Red Book which also provides that any certificate must not unreasonably be withheld or delayed. Where payments are concerned, Figure 18.1 shows a flow chart of the procedure to be followed under clauses 48 and 60.

18.8 Late certification

Sub-clause 60.10 of the Red Book provides that in the event of the employer's failure to make payments to the contractor within the stated time after an interim payment certificate, the employer is required to pay interest at a rate specified in the Appendix to Tender. The interest is applied to all sums unpaid from the date by which they should have been paid.

Whilst this provision deals with the employer's failure to pay against an interim payment certificate, it does not deal with the situation where the engineer fails to certify within the specified time after receiving the contractor's statement. To provide some safeguard for the contractor in such an event, FIDIC published in November 1996, as part of a supplement to the Red Book, an alternative wording to sub-clause 60.10. This alternative wording appears in Section C of the Supplement and varies the wording of sub-clause 60.10 by deleting in lines 3 and 4 of the sub-clause the words 'within 28 days after such Interim Payment Certificate has been delivered to the Employer' and substituting the words 'within 56 days after the Contractor's statement has been received by the Engineer'.

Chapter 19

Disputes Settlement by Arbitration

19.1 Introduction and glimpses from the past

One of the most important provisions in any commercial contract, but most especially where the parties are of different nationalities, is the provision of a fair method of dispute resolution.[19.1] Such a provision instils in the parties confidence that the contract as a whole will be justly operated, a fact which is all the more important due to the almost inevitable outcome that disputes will arise in complex, unique and long-term commercial contracts such as those in construction.

Many methods of resolving disputes have evolved over the centuries. The most familiar in civilised societies are:

(a) negotiation;
(b) litigation;
(c) arbitration; and
(d) alternative dispute resolution methods. The term 'alternative' is used in this book in the sense of alternative to litigation and arbitration, both of which are judicial processes of an adversarial character. This alternative is more properly conveyed by the term amicable.

Traditionally, alternative and amicable dispute resolution methods have been more popular than arbitration in the East, whilst arbitration has been used more extensively in the West. Despite that tradition, however, it can be said that for many centuries, and for many reasons, arbitration has been a successful alternative to litigation, as explained below. In the East, an old Buddhist advice to arbitrators contained five notable steps to making a decision: first, you must establish the truthfulness of the facts presented to you; secondly, you must ascertain that they fall within your jurisdiction; thirdly, you must enter the minds of the parties to the dispute so that the judgment to be rendered be a just one; fourthly, you must pronounce the verdict with kindness not harshness; and fifthly, you must judge with sympathy.[19.2] This advice can be said to apply to this day, but perhaps with the addition that you must analyse the matters with simplicity and logic.

Also in the East, Emperor Kang-hsi, one of the greatest Manchu emperors who ruled China from 1661 to 1722, was a proponent of arbitration as opposed to litigation and decreed:

'The Emperor, considering the immense population of the empire, and the great division of territorial property and the notoriously litigious character of the Chinese, is of the opinion that lawsuits would tend to increase to a frightful extent if people were not afraid of the tribunals and if they felt confident of always finding in them ready and perfect justice . . . I desire, therefore, that those who have recourse to the courts should be treated without any pity and in such a manner that they shall be disgusted with law and tremble to appear before a magistrate . . . In this manner the evil will be cut up by the roots; the good citizens who may have difficulties among themselves will settle them like brothers by referring to the arbitration of some old man or the mayor of the commune. As for those who are troublesome, obstinate and quarrelsome, let them be ruined in the law courts; that is the justice that is due to them.'[19.3]

In the West, in Ireland at least, the recorded history of arbitration dates back to the Brehon laws around the fifth century, if not before.[19.4] It continued to flourish and in the early eighteenth century, a society was formed in Dublin called the Ouzel Galley Society which was composed of Dublin merchants who had the responsibility for settling commercial disputes, under specific rules. The name of the society stems from a romantic adventure of a merchant vessel by that name. The vessel was dispatched in the year 1695 to the land now known as Syria and Lebanon by the Dublin firm of Ferris, Twig and Cash under Captain Massey.[19.5]

The year 1695 was a year of war, made doubly dangerous in the case of vessels bound for the Middle East by the activities of Algerian pirates who preyed on commercial traffic passing into and out of the Mediterranean.

The *Ouzel Galley* did not return by the time it was expected and nothing was heard of her. The vessel was given up for lost with all hands. The insurance underwriters paid up on its hull and cargo. All was well until five years later in 1700 the vessel reappeared and cast anchor in Dublin Port.

Captain Massey had a tale of high adventure to tell describing how the vessel had been attacked and captured in the Mediterranean by Algerian pirates who then used her as a pirate ship and how in time she amassed spoils on the high seas. Later the Dublin crew managed to repossess the vessel and sail home with the spoils.

But to whom did the return cargo belong? It did not belong to the owners because they had been fully compensated for their loss by the underwriters after the owners had made a formal act of abandoning their interest to them; but at the same time its return cargo did not belong to the underwriters because the abandonment of interest in the vessel to them covered the vessel and its outward cargo, not the return cargo with which the vessel was laden on its unexpected reappearance. While existing practices covered normal situations and made possible the amicable settlement of most contracts, all the uncertainties in marine insurance were to the fore in this unusual situation.

An action in law started in 1700. Litigation was notoriously slow and expensive, and the prospects of a speedy resolution seemed even less likely in a case involving issues as complex and as unusual as the *Ouzel Galley* presented. Wisely, the issue was settled in 1705 by the submission of the matter to the arbitration of a number of merchants. From this instance of arbitration in an isolated and complex case arose the Ouzel Galley Society as a permanent arbitration body of merchants. Eighteenth-century merchants everywhere in Western Europe quite frequently submitted complex matters to the arbitration of fellow merchants by private arrangement. The significance of the Ouzel Galley Society lay in the systemisation of arbitration by prominent merchants, who were widely respected, into a permanent private tribunal to which both parties to a commercial dispute could have recourse if they agreed. The attraction of such arbitration proceedings for merchants was that many disputes hinged less on fine points of law than on issues which could be more readily decided by practical knowledge for commodities and accepted practices. The law courts themselves would have had to resort to merchants for an opinion on matters such as the condition of goods or the practices in trade, and it was therefore only common sense to seek the arbitration of merchants whose standing would have made them the decisive witnesses in court proceedings.

Whilst it is accepted that the simplest and quickest way of resolving disputes is through negotiation, it is not in fact an easy method, especially if there is a clash of personalities behind the dispute, or if in the parties' opinion there are matters of principle at stake. Furthermore, until any of the other methods of dispute resolution have been invoked, the costs involved are rarely appreciated. In some cases, parties embark on litigation simply because they want their day in court; in other cases some are badly advised.

Internationally, the parties face a number of additional uncertainties, problems, risks and fears. These range from having to deal with people of different cultures, language, customs, laws and business practices to having to select foreign lawyers to deal with a foreign judicial process about whose neutrality and independence they may have serious doubts.[19.6] Furthermore, if the chosen forum for dispute resolution is litigation, then the courts of at least two different jurisdictions may compete for supremacy and would often have overlapping roles with consequential difficulties in the enforcement of foreign judgments.

It can be said that international commercial arbitration provides an answer to many of these problems, risks and fears and particularly in respect of the recognition and enforcement of foreign awards through the 1958 New York Convention and other similar international arbitral treaties. For that reason, it has been chosen as the preferred forum for dispute resolution in the international field and has been influential in facilitating international trade, investment and economic development around the world. The most important question, therefore, revolves around the features and the wording of a clause describing the manner in which disputes can be resolved with fairness, justice, speed and relative economy.

19.2 Advantages of arbitration

Where arbitration is chosen as the appropriate and most effective method of dispute resolution, such choice is made because of the many inherent advantages in arbitration, especially where there is a technical dispute to be resolved. These advantages are:

(a) The parties in dispute are in control of the identity of the person to whom they entrust the resolution of the dispute.

(b) The parties in dispute are able to choose as an arbitrator someone who is an expert in the field of the dispute, be it technical or legal. This has the potential to save significant time and costs since it does not become necessary to explain technical matters in graphic detail, which is otherwise the case where a judge seldom has expert knowledge.

(c) Unlike the procedure in litigation, pleadings are submitted at or close to the appointment of the arbitrator. This enables him to start with the case and to continue to be in close touch with the steps taken by both sides, thus allowing him to know the case as it develops and to become acquainted with the parties and their legal representatives who may be of different cultures, language, customs and business practices. The benefit of such knowledge should not be underestimated since, in practice, many arbitrations concern 'habitual disputes bred of strained relationships and mutual suspicion'.[19.7]

(d) The process can be expeditious and cost-effective, if the parties wish it to be so.

(e) The procedure is private, thus avoiding the disclosure of commercially sensitive information and any undesirable publicity or loss of reputation.

(f) Arbitration procedure is flexible and adaptable to the particular dispute or disputes.

(g) The conduct throughout is less formal than court procedure without abandoning courtesy.

(h) It is possible to arrange for an inspection of the project and the matters relating to the dispute.

(i) Depending upon the applicable law of the contract, the award rendered by the arbitrator in most jurisdictions is final and binding unless it can be shown that the arbitrator has erred in law or has misconducted the proceedings. Because arbitral awards are not set aside, except in the limited circumstances of the arbitrator's error in law or misconduct, they are much more likely to be final than the judgments of courts of first instance.[19.8] It is to be noted that the modern term for 'misconduct' is 'serious irregularity' as misconduct may give the impression of an act beyond what is intended.[19.9]

(j) It can be conducive to a negotiated settlement even at a late stage of the proceedings.

(k) In some jurisdictions, the arbitrator has greater power than the courts to open up, review and revise the decision of the engineer.[19.10]

(l) In international commercial contracts, the recognition and enforcement of a foreign arbitral award is much easier than a judgment of court. This is made possible by a number of international multilateral arbitral treaties and conventions. The most significant of these is the 1958 New York Convention on the Recognition and Enforcement of Foreign Arbitral Awards as it has been acceded to by over 100 different jurisdictions worldwide. It has been described as the single most important pillar on which the international arbitration edifice rests and that it 'perhaps could lay claim to be the most effective instance of international legislation in the entire history of commercial law'.[19.11] It is reported that an estimated 98 per cent of awards in international arbitration are honoured or successfully enforced and that enforcement by national courts has only been refused in less than 5 per cent of cases, justifying the claim that it is far easier to enforce arbitration awards than court judgments.[19.12]

(m) A developed and highly respected system of arbitration law, relating to national and international arbitration, supported by the judicial establishment and its policy-making organs is highly favourable on the international scene from the point of view of encouraging foreign investment into a developing country. It is seen as a safeguard to that investment against unjust and haphazard decision making.

(n) The flexibility of the arbitration procedure means that the arbitrator may make an interim or a partial award, or a number of such awards. If, which has become common practice in complex construction disputes, there is first an interim award in which liability is determined, with the decision regarding quantum to follow, it is often the case that the parties will make real attempts to, and frequently do, settle the case between them, thereby avoiding expending further time and costs in the arbitration. This flexibility is not available to the parties in litigation.

(o) Arbitration may take place in any country, in any language and with arbitrators of any nationality. With this flexibility, it is generally possible to structure a neutral procedure offering no undue advantage to any party.[19.13]

19.3 What is a dispute?

Most if not all commercial contracts, including construction contracts, contain provision for the resolution through arbitration of disputes which may arise between the parties during the course of the transaction. For example, once a dispute can be said to be in existence, clause 67 of the Fourth Edition of the Red Book governs the path from dispute to arbitration. Its opening words 'If a dispute... arises between the Employer and the Contractor...' means that the existence of a dispute is a precondition to the application of the provisions in the remainder of the clause. However, no express definition of 'dispute' appears in the Fourth Edition of the Red Book, or indeed in most typical standard forms of construction contract. For many years the meaning of dispute was taken from a comment made by Lord Denning in the important decision in *Monmouthshire County Council* v. *Costello & Kemple Ltd* where it was interpreted as follows:[19.14]

> 'The submission of a claim does not necessarily (although it may) cause a dispute to arise; the rejection of a claim will probably do so, but not always, for the Contractor might accept either wholly or in part the views expressed by the Engineer. If the claim is not one which is met with a clear rejection, but with a request for further information or even with a stalling reply, then no dispute may arise.'

So a dispute was said to come into being when the contractor did not accept the engineer's rejection of his claim.

However, a wider definition of 'dispute' in the context of arbitration emerged in the case of *Halki Shipping Corporation* v. *Sopex Oils Ltd*[19.15] where the Court of Appeal affirmed the judgment of Clarke J at first instance who held that:

> 'dispute in respect of a matter which under an arbitration agreement had to be referred to arbitration was to be given its ordinary meaning and included any claim which the other party refused to admit or did not pay whether or not there was an answer to the claim in fact or in law'.

Swinton Thomas LJ of the Court of Appeal said: '... there is a dispute once money is claimed unless and until the defendants admit that the sum is due and payable' citing a passage from the judgment of Templeman LJ in *Ellerine Bros Pty Ltd* v. *Klinger*:[19.16]

> 'It is not necessary, for a dispute to arise, that the defendant should write back and say "I don't agree". If, on analysis, what the plaintiff is asking or demanding involves a matter on which agreement has not been reached and which falls fairly and squarely within the terms of the arbitration agreement, then the applicant is entitled to insist on arbitration instead of litigation.'

Therefore, a dispute was deemed to come into being once it became evident that a claim was not to be settled. There was no need for its express rejection by the engineer, or a non-acceptance/rejection of such rejection by the claimant contractor.

This decision caused much confusion and since 1998 there have been numerous cases in the English courts where the meaning of dispute has been examined in detail (many of which, although in respect of challenges to statutory adjudications, are of equal relevance to arbitrations), the more notable being as follows:

(a) In *Fastrack Contractors Ltd* v. *Morrison Construction Ltd*[19.17] the judgment in *Halki* was considered but it was held that a claim of itself was not a dispute; it had to be rejected before a dispute came into being but the rejection could take the form of a refusal by the respondent party to answer the claim or the bare rejection of the claim 'to which there was no discernible answer in fact or in law'.
(b) In *Sindall Ltd* v. *Solland*[19.18] the judgment in *Fastrack* was referred to and it was concluded:

> 'This and other decisions concerning what may constitute a dispute for the purposes of statutory adjudication show that the absence of a reply (for example by a person in the position of Contract Administrator) may give rise to the inference that there was a dispute, e.g. where there was prevarication ... For there to be a dispute for the purposes of exercising the statutory right to adjudication it must be clear that a point has emerged from the process of discussion or negotiation has ended and that there is something which needs to be decided ... A person in the position of the Contract Administrator must be given sufficient time to make up its mind before one can fairly draw the inference that the absence of a useful reply means that there is a dispute.'

(c) In *Edmund Nuttall Ltd* v. *R.G. Carter Ltd*[19.19] the court considered the decisions in *Monmouthshire*, *Halki*, *Fastrack* and *Sindall* and held:

> 'a "dispute" is something different from a "claim" . . . it is instructive to consider the meaning of the word "dispute" as defined in *Shorter Oxford Dictionary*, 4th edition . . . : – "An oral or written discussion of a subject in which arguments for and against are put forward and examined. An instance of disputing or arguing against something or someone, an argument; a controversy; esp. a heated contention, a disagreement in which opposing views are strongly held. The act of disputing or arguing against something or someone; controversy, debate" . . . both the definitions in *Shorter Oxford Dictionary* and the decisions to which I have been referred in which the question of what constitutes a "dispute" has been considered have the common feature that for there to be a "dispute" there must have been an opportunity for the protagonists each to consider the position adopted by the other and to formulate arguments of a reasoned kind. It may be that it can be said that there is a "dispute" in a case in which a party which has been afforded an opportunity to evaluate rationally the position of an opposite party has either chosen not to avail himself of that opportunity or has refused to communicate the results of his evaluation. However, where a party has had an opportunity to consider the position of the opposite party and to formulate arguments in relation to that position, what constitutes a "dispute" between the parties is not only a "claim" which has been rejected, if that is what the dispute is about, but the whole package of arguments advanced and facts relied upon by each side.'

(d) In *Hitec Power Protection BV* v. *MCI World Com Limited*[19.20] Seymour J QC reiterated the wider concept of 'dispute':

> '. . . a dispute is something more than simply a rejection broadly of the claim, or a failure to respond to a claim. What a dispute, in the context of adjudication, amounts to, in my judgment, is a situation in which a claim has been made, there has been an opportunity for the protagonist each to consider the position adopted by the other and to formulate arguments of a recent kind.'

(e) In *Cowlin Construction Ltd* v. *CFW Architects*[19.21] what was taken to constitute a dispute in all the above cases was examined but it was concluded that the approach in *Halki* was preferred.

(f) In *Beck Peppiatt Ltd* v. *Norwest Holst Construction Ltd*[19.22] the court referred to a submission made by the claimant in that case, that the law with regard to what constitutes a dispute was in an 'unsatisfactory state' and that there was conflict between judges who adopted the *Halki* approach and those who followed the more traditional restricted approach. Forbes J held that:

> 'In my view, the law is satisfactorily stated by Judge Lloyd QC in his unreported decision of *Sindall* v. *Solland* dated June 2001,[19.23] in which he said this: "For there to be a dispute for the purposes of exercising the statutory right to adjudication it must be clear that a point has emerged from the process of discussion or

negotiation that has ended and that there is something which needs to be decided." As it seems to me, that is a statement of principle which is easily understood and is not in conflict with the approach of the Court of Appeal in Halki.'

(g) In *Costain Ltd* v. *Wescol Steel Ltd*[19.24] the court relied on the judgement in *Fastrack* as follows:

'The word "dispute" bears a range of meanings. I was referred to the decision of Judge Thornton QC in *Fastrack Contractors* v. *Morrison* [2000] BLR 168, which refers at page 178 to *Halki Shipping Corporation* v. *Sopex Oils Ltd* [1998] 1 WLR, CA. The latter case itself supports different meanings, or nuances of meaning, of the word as used in an arbitration clause. First, as including any claim which the other party refused to admit or did not pay (headnote, page 727). Second, a dispute exists where there is a claim which the defendant refuses to admit and refuses to pay (Page 746, letter A). Third, there is a dispute once money is claimed unless and until the defendants admit that the sum is due and payable (page 761, letter G). Fourth, there is a dispute where a party has refused to pay a sum which is claimed or has denied that it is owing. (page 761, letter H). Judge Thornton concluded (paragraph 28 of his judgment, at page 178) that a claim and its submission do not necessarily constitute a dispute, that a dispute only arises when a claim has been notified and rejected, and that a rejection can occur when an opposing party refuses to answer the claim.'

(h) In *Orange EBS Ltd* v. *ABB Ltd*[19.25] the court relied on the decision in *Halki* and *Beck Peppiatt* saying:

'The meaning of dispute adopted in *Halki* is that "there is a dispute once money is claimed unless and until the defendants admit that the sum is due and payable". I consider myself, therefore, bound by the decision in *Halki*. Further, given the weight which a decision of a puisne judge carries (and particularly given the position which Forbes J occupies in relation to the TCC) I consider I should give careful attention to the guidance in *Beck Peppiatt*.'

(i) In *Dean and Dyball Construction Ltd* v. *Kenneth Grubb Associates Ltd*[19.26] Judge Richard Seymour QC reiterated his own approach to the meaning of 'dispute' taken in *Edmund Nuttall*.

And finally,

(j) In *London & Amsterdam Properties Ltd* v. *Waterman Partnership Ltd*,[19.27] which concerned a dispute which had been adjudicated, the court undertook a comprehensive (and very useful) review of all the authorities encompassing both the narrow and the wide concepts of 'dispute' and confirmed that the *Halki* approach was the correct one in both arbitration and adjudication by saying:

'In my judgment the reasoning in *Halki* as to what constitutes a dispute in arbitration proceedings applies with equal effect in adjudication proceedings.'

The outcome, therefore, of this series of cases would appear to be that it is not necessary for a claim to be expressly rejected in order for a dispute to be said to have arisen. Once the engineer, as the employer representative, has been given sufficient opportunity to reply to the claim and it becomes clear to the contractor that the claim is not accepted, then the parties are deemed to be in dispute.

In view of the above, a dispute comprises either:

(a) a claim which is expressly rejected by the receiving party, such rejection not being accepted by the claiming party; or
(b) a claim which is met with protracted silence and the matter of the claim is not met by the receiving party, in which case the receiving party is deemed to have rejected the claim, such rejection not being accepted by the claiming party.

Therefore, whether the rejection of the claim is made expressly or by implication, for a dispute to be in existence there has to be a claim, a rejection and a non-acceptance of the rejection. A dispute can never be said to arise on the basis of a claim alone.

19.4 What is arbitration?

Arbitration is a process whereby parties in dispute agree to submit the matter in dispute to the decision of a person or persons in whom they have confidence and trust and undertake to abide by that decision. The prerequisites to a valid arbitration are:

(a) *The existence of a dispute*: A dispute must arise before the parties can enter into an arbitration. It is not, for example, an arbitration if two parties seek a decision in respect of a problem for which neither have an answer, as in the case of the evaluation of the cost of a certain item. If the cost of the item is not known a dispute has not yet arisen. A dispute will not arise until a decision is made by someone (for example, the engineer under the contract or an expert appointed by one of the parties) as to the cost, and that decision is challenged by one or both of the parties.[19.28] That there should be a dispute in existence between the parties before proceeding down the path to arbitration appears to be stating the obvious, but in fact the great number of litigation cases where the arbitrator's jurisdiction is challenged on this very point shows that it is not as straightforward as it would first appear. The *London & Amsterdam Properties* case discussed above is a prime example of a situation where the court had first to examine what was meant by 'dispute' and then to decide on the facts before it whether the matter in hand was actually in dispute.
(b) *Agreement to refer the dispute to arbitration when the dispute arises*: The parties in dispute must either agree or already have agreed to refer the dispute to a third party individual or tribunal for determination. Such arbitration agreement typically incorporates by reference a set of procedural rules which the parties and the third party individual/tribunal are required to follow in respect of their conduct during the arbitration and the manner by which the determination is reached.
(c) *Agreement to be bound by the award*: The parties in dispute must also agree to be bound by the award of the arbitrator who has been chosen or appointed.

(d) *Initiation of the arbitration*: A notice by one party to the other that he is commencing arbitration proceedings, and intends to refer the dispute that is in existence to an arbitrator, is required to be given to initiate the arbitration proceedings. In most, if not all jurisdictions, the date of such notice serves as the date of the commencement of arbitration. The date of commencement of arbitration is important from the point of view of the limitation period (see Chapter 3, Section 3.5.2) since the notice is equivalent to the issue of a writ. At the same time, or very shortly after serving this notice, very often the notifying party will propose an individual to act as arbitrator and so serves on the other party a notice to concur on the appointment of this individual as arbitrator.

The term 'arbitrator', wherever it appears in this paragraph and in the remainder of this chapter refers to a single arbitrator or an arbitral committee (or tribunal) made up of any number of individuals.[19.29]

19.5 The arbitration agreement

As can be seen in Section 19.4, the arbitration agreement is an essential part of the procedure of arbitration. Agreements to arbitrate are generally of two types and these are:

(a) *An ad hoc agreement*: An agreement where an arbitration agreement does not exist when a dispute arises between the parties, but later the parties in dispute agree to refer already existing disputes to arbitration is referred to as *ad hoc*. The parties in an ad hoc arbitration need not only agree to arbitrate but also, if such agreement is possible, agree on a detailed procedure to be followed suitable for the particular dispute within the legal framework set out in the relevant legislation controlling the conduct of arbitration (for example, in the UK, the Arbitration Act 1996). If such agreement is not reached on the procedure, the arbitrator will have to rule on the procedure as part of the matter in dispute.

(b) *The arbitration clause in an existing agreement*: This type of agreement is one where the parties to a certain contract, usually called the substantive contract, have already agreed to refer any and all of their future disputes to arbitration. In this type of agreement, a clause must be inserted in the underlying substantive contract to describe the steps to be followed when a dispute arises in order to reach and to complete the process of appointment of an arbitrator or a tribunal. In such a case, it would be advisable, if not imperative, to refer to the law to be applied, the place of arbitration and to a set of specific procedural rules to be followed once the arbitrator is appointed. Such procedural rules, if correctly formulated, must refer to the legislation which confers on the arbitrator the jurisdiction and powers necessary to ensure just, expeditious, economical and final enforceable determination of the dispute referred to him. Such a clause in a contract is called the arbitration clause.

When a dispute falls within the scope of the arbitration clause and if this clause is formulated correctly so that an arbitrator is appointed, then the underlying substantive contract, which includes the arbitration clause, together with a valid notice of appointment, is sufficient to ensure that the parties in dispute are bound by the award rendered by the arbitrator.

In some jurisdictions, in the absence of agreement to the contrary, one type of dispute which cannot be arbitrated under the arbitration clause is that which reaches into the essential validity of the substantive contract, i.e. as to whether or not there is a contract. If the contract is null and void, for instance due to fraud, it follows that all of its provisions, including the arbitration clause, are void and the arbitrator has no jurisdiction to adjudicate on any matter relating to that contract. A different situation arises, however, where a contract is terminated, as it has been accepted by most jurisdictions that the arbitration clause can survive the termination of the contract, thus enabling the arbitrator to hold that the contract between the parties has been frustrated or terminated without casting a shadow on his own status or jurisdiction as arbitrator.[19.30]

A large number of disputes arise after termination of a contract and it would therefore be illogical if the arbitration clause in such a contract was held to be terminated thus placing the whole dispute resolution process in a vacuum. The following quotation represents the proper analysis of the situation in the context of a contract where one party claims breach by the other:

'The contract is not put out of existence, though all further performance of the obligations undertaken by each party in favour of the other may cease. It survives for the purpose of measuring the claims arising out of the breach, and the arbitration clause survives for determining the mode of their settlement. The purposes of the contract have failed, but the arbitration clause is not one of the purposes of the contract.'[19.31]

It is now clearly provided in the Fourth Edition of the Red Book that any dispute including one which occurs 'before or after repudiation or other termination of the Contract' is included under clause 67 for the purpose of its settlement.

Furthermore, it should be noted that Article 6(2) of the Arbitration Rules of the ICC, named under clause 67 as the applicable Rules unless otherwise specified, provides:

'If the Respondent does not file an Answer, as provided by Article 5, or if any party raises one or more pleas concerning the existence, validity or scope of the arbitration agreement, the Court may decide, without prejudice to the admissibility or merits of the plea or pleas, that the arbitration shall proceed if it is prima facie satisfied that an arbitration agreement under the Rules may exist. In such a case, any decision as to the jurisdiction of the Arbitral Tribunal shall be taken by the Arbitral Tribunal itself. If the Court is not satisfied, the parties shall be notified that the arbitration cannot proceed. In such a case, any party retains the right to ask any court having jurisdiction whether or not there is a binding arbitration agreement.'

19.6 Sources of law in arbitration

In general, the rules which govern the conduct of and the procedure in arbitration in most jurisdictions are derived from the following separate sources.

19.6.1 General

Sources of law, as defined in Chapter 2, can be briefly stated as follows (not in hierarchical order):

(a) the constitution, where it exists;
(b) legislation and statute law;
(c) regulations and delegated or subordinate legislation;
(d) in the common law system judicial decisions recognised as precedents, and equity;
(e) in the Romano-Germanic system the judges' power of interpretation, and in some jurisdictions the binding effect of decisions of some higher courts;
(f) in Islamic law the Qur'an, the Sunna, the Ijma', the Qiyas, and in some jurisdictions the binding effect of decisions of some higher courts;
(g) practice and custom;
(h) international treaties and conventions;
(i) international law in the case of international arbitration;
(j) the arbitration agreement executed by the parties.

19.6.2 The arbitration agreement as a source of law

The arbitration agreement and any rules and procedures it incorporates, whether made before or after the dispute has arisen, governs the conduct of and the procedure in arbitration.

In the case of the Fourth Edition of the Red Book, the arbitration agreement is embodied in clause 67 and the procedural rules incorporated in the text of that clause are the Rules of Arbitration of the International Chamber of Commerce. These rules are discussed in more detail in Section 19.9. In the present context, suffice it to say that these rules import into the arbitration agreement a number of important provisions and where these are silent on a particular aspect, the parties themselves, or the arbitrator, may settle the procedural rule required. Article 15(1) of these rules states:

> 'The proceedings before the Arbitral Tribunal shall be governed by these Rules and, where these Rules are silent, by any rules which the parties or, failing them, the Arbitral Tribunal may settle on, whether or not reference is thereby made to the rules of procedure of a national law to be applied in the arbitration.'[19.32]

19.6.3 Practice and custom

The practice and custom of the commercial private sector and its arbitrators also governs the conduct of and the procedure in arbitration. In this context, it is relevant to quote the following passages, first from Sir M.J. Mustill and S.C. Boyd:[19.33]

> 'Many trades have developed their own, often idiosyncratic, ways of conducting arbitrations. The courts have shown themselves consistently willing, subject always to the dictates of natural justice, to recognise and sanction these individual procedures;'

and secondly from Mr Justice Megaw in *Orion Compania Espanola de Seguros* v. *Belfort Maatschappij voor Algemene Verzekgringeen*:

> 'the two arbitrators and the umpire here were gentlemen who were very experienced in the business of insurance, and it might well be thought that their view on a matter of this sort, even if it be strictly a matter of law, assuming they had directed themselves correctly in law, would be – and I see no reason why I should not say it – preferable to the view of the court.'[19.34]

In each and every arbitration, the sources of law listed in Section 19.5.1 would have to be examined in meticulous detail since the arbitrator cannot disregard the law or the contractual arrangements between the parties unless he is empowered to act as an *'amiable compositeur'* or to proceed *'ex aequo et bono'*.[19.35] Even then, this power should be exercised carefully ensuring that the rules of natural justice are upheld at all times, and within specific limits depending upon the relevant jurisdiction but always within the limits imposed by mandatory provisions and public policy rules. Furthermore, in certain jurisdictions such powers are not recognised and the arbitrator must, in general, apply a fixed and recognisable system of law.[19.36]

As an example, it is reported that in England the following policy applies:

> '... it is the policy of the law in this country that, in the conduct of arbitrations, arbitrators must in general apply a fixed and recognisable system of law ... and that they cannot be allowed to apply some different criterion such as the view of the individual arbitrator or umpire on abstract justice or equitable principles ....'[19.37]

In any event, the arbitrator must obtain a clear understanding of the interaction between the arbitration and the applicable law of the contract since recourse to the legal process may be needed at any of the various stages in a particular arbitration from the appointment of the arbitrator to rendering of the award.

Of course, there can be no doubt that courts and judges play an integral part in any effective arbitration system and where international arbitration is concerned, it is accepted that it cannot function without the assistance of national courts. Only they possess the coercive powers to enforce agreements to arbitrate, as well as the resultant awards.[19.38] Therefore, in order to be effective, written agreements to arbitrate must be enforceable as must arbitration awards. They are enforced, when necessary, by resort to the courts. There can also be no doubt that lawyers are an integral part of the arbitration process since they are, or should be, the experts in the application of the law, and in the procedural requirements necessary to attain fairness and justice.

19.7 The arbitrator

'Traditionally an appointment as arbitrator was regarded as having an honorary character: not so much that the arbitrator was expected to fulfil his duties without remuneration (although that was once the case) but that his appointment and agreement to act were at the same time a recognition of his standing and probity, and a recognition of the responsibilities owed by such persons to others engaged in the same trade . . .'.[19.39]

If the parties in their arbitration agreement prefer to choose the arbitrator rather than leave that task to the president of a named institution, the criteria behind that choice may be revealed by focusing on the duties and what is expected of such a person. In general terms, it has been said that an arbitrator must be fair, firm, formal, friendly, flexible and fast. It has also been said that the

> 'standards of behaviour expected of arbitrators are no less stringent than those demanded of judges; in fact arbitrators are expected to behave a shade better since judges are institutionally insulated by the established court-system, their judgments being also subjected to the corrective scrutiny of an appeal'.[19.40]

Furthermore, to quote from the experience of an eminent jurist and arbitrator on a comparison between the role of judges and arbitrators:

> 'Unlike judges, arbitrators must inevitably treat the parties and their lawyers as having something of the aura of a *clientele*, whose goodwill, understanding and respect for the tribunal's authority must be cultivated and preserved. One cannot overlook that, directly or indirectly, one has been chosen by the parties to decide the particular dispute in question. "Rent-a-judge" in the private forensic sector is necessarily different from sitting as a professional judge in the public sector. Similarly, the parties' lawyers will generally treat arbitral tribunals with far greater circumspection and consideration than they would display towards a judge.'[19.41]

A trained and experienced arbitrator will know that as well as the above requisites, the duties which must be taken care of are:

(a) to act fairly, independently and impartially, ensuring that each party is given sufficient and equal opportunity to present its case, and this, in international arbitration, includes a duty to act in a manner free from national, political and cultural prejudice;
(b) to weigh the evidence and reach a logical and reasoned decision and deal with technical and legal issues either personally or through advice;
(c) to initiate and plan a management structure for the effective and speedy resolution of the dispute; and
(d) to proceed diligently taking control of the arbitration proceedings as soon as the appointment has been made.

On the other hand, an arbitrator, to be validly appointed, must possess the qualifications required by the law and by the arbitration agreement. These are:

(a) he must have the legal capacity required by law of every person who assumes the office of arbitrator, although in some jurisdictions no such capacity is required;

(b) he must possess all the qualifications and none of the disqualifications prescribed by the arbitration agreement;
(c) he must be independent of the parties and thus free from any connection with them or the subject matter of the dispute;
(d) he must be impartial and thus free from bias either in favour of or against any of the parties or any issue in dispute.

In construction arbitrations, there are, however, two firmly held views as to who should be appointed as arbitrator, a lawyer or a construction professional. The first is held by those who believe that, as awards should be ultimately enforceable, importance and weight should be placed on the arbitrator being correct in law, as distinct from, and in comparison with, being correct in matters of fact, technical or otherwise. Furthermore, it is said that the judicial decision maker must be bold, imaginative and decisive and it is argued that a lawyer has the confidence and training to meet these requirements.

The second view is held by those who believe that the arbitrator should be selected from the technical field and are of the opinion that a technical arbitrator would be totally familiar with the subject matter of the dispute and the technical terminology employed. They believe that coupled with this an 'engineer' arbitrator possesses all the managerial skills necessary to take control of the proceedings from their inception, and to plan and devise an appropriate management structure for the particular dispute. In adopting this second view there is no departure from the original concept of arbitration which is based on the desire of commercial people to have their disputes resolved by experts in the field of their particular dispute.

Nevertheless, whichever background the arbitrator has, there are certain personal attributes which are conducive to an effective and efficient arbitration. These include:

(a) the ability and the patience to listen carefully and to have an open mind on all matters;
(b) to be authoritative and possess leadership qualities;
(c) to be friendly but not casual;
(d) to be firm but not overbearing;
(e) to be articulate, but not verbose;
(f) having general knowledge of the contractual and technical issues that may arise;
(g) to be flexible to consider different options that may be put to him by the parties;
(h) the capability of spontaneously handling procedural problems that usually arise during a hearing, which are in practice generally of a difficult nature;
(i) an aptitude for writing clear, concise and articulate awards.

19.8 The arbitration agreement under Clause 67 of the Red Book

The arbitration agreement in the Fourth Edition of the Red Book is embodied in clause 67. This clause, however, incorporates more than the arbitration agreement in that it provides in the first place a mechanism for the resolution of disputes through a two-tier system of reference to the engineer. The first tier is the submission of an assertion (claim) to the engineer for his determination. The second tier is the referral of the dispute arising from such assertion (to the extent that the engineer's determination is not agreed) back to the engineer for his decision. If the dispute is not resolved by the engineer's decision, the Fourth Edition has introduced a provision under sub-clause 67.2 which includes a mechanism for amicable settlement which must precede the reference to arbitration. In this way, the procedure of the dispute resolution mechanism of clause 67 incorporates a number of steps, each of which must be taken before proceeding to the next step, with arbitration as the ultimate step. These are discussed in detail in Section 19.8.1 below.

Clause 67 in the Fourth Edition has been revised in order to clarify a number of its provisions thereby closing some of the gaps which were discovered through the use of previous editions of the Red Book and also through some published awards,[19.42] and a number of articles.[19.43] These will be discussed in Section 19.8.1 of this chapter.

As mentioned earlier, the two-tier system which existed in previous editions of the Red Book has been maintained in the Fourth Edition, in its printed forms of 1987, 1988 and 1992. However, a supplement to the Fourth Edition of the Red Book (as reprinted in 1992), published by FIDIC in November 1996, contained in its Section A an alternative version to clause 67 of the Red Book. This alternative version provides for the establishment of a Dispute Adjudication Board, which may comprise either one or three members, to replace the engineer's traditional role of a decision maker in the settlement of disputes. As explained earlier in Chapter 10, and as will be discussed later in this section, there has been strong criticism in recent years of the role of the engineer as adjudicator or quasi-arbitrator and in order to address this, a serious alternative method for dispute resolution had to be developed. The particular alternative which is finding increasing favour in respect of the FIDIC type of contract for civil engineering construction is the appointment of an independent and impartial expert or experts at the beginning of a contract who keep in touch with work in progress by means of site visits at regular intervals and who are then available to act in resolving disputes should they arise. FIDIC, in its supplement, gives its full approval to this alternative to the engineer's former role in dispute adjudication. The supplement also includes a guide to the new alternative version of clause 67 and model terms of appointment and procedural rules for the Dispute Adjudication Board. Furthermore, it includes some necessary amendments to the Appendix to Tender corresponding to the new alternative version of clause 67. The new version of the clause, together with a discussion of Dispute Adjudication Boards is given in Chapter 26.

The multi-tier system is a feature which originates from the ICE Form and thus has not been appreciated by some professionals whose training is based on systems other than that used in England. Sometimes it is not even appreciated by those from within that system. This is perhaps mainly for two reasons. The first reason is the lack of precision in the language used, in that the wording of the clause is misunderstood, particularly: (a) in respect of the role of the engineer as a decision maker; and (b) in the definition of the word 'dispute'.

With regard to (a), reference should be made to Chapter 10 and in particular to the incorrect description given to the engineer's role as a decision maker when it is referred to as a quasi-arbitral role (see Section 10.9). With respect to (b), reference should be made to Section 19.3 above.

The second reason is that the engineer is expected to decide all disputes including those which call into question his own conduct. For instance, the contractor may claim that he has incurred additional costs due to late supply of drawings and/or information, or due to the engineer's alleged failure to approve workshop drawings promptly, or due to faulty design and specification. In such situations, should the engineer decide the dispute in favour of the contractor, he might expose himself to liability towards the employer. If he does not so decide and the contractor is in fact entitled to the claimed costs, the engineer would be in breach of his duty to act fairly and impartially. Accordingly, the system is criticised for this apparent conflict of interest in the decision-making process between the engineer's duty to the parties to act fairly and impartially and his own interest in avoiding liability.

Some commentators explain the origin of the inclusion of such dispute resolution within the engineer's function by reference to the ICE Form, upon which the Red Book had been modelled, and the English doctrine that the contractor cannot object to such situations if and when they arise since he:

'is taken to have known at the time of tendering and to have accepted, that the Engineer may be called upon to decide certain matters that will place him in a position of conflict between his duty to be fair to the contractor and his own self interest.'[19.44]

The following quotation is also relevant:

'Known interests do not disqualify the engineer from deciding between the contractor and employer. The contractor is in particular taken to know that the engineer will generally have prepared the contract; will have estimated the cost of the work and so will want to avoid extras; may have made mistakes in the plans involving extra cost which again he will want to keep down; and may wish to minimise the extension of time for any delay which he causes the contractor, for which he may be liable to the employer.'[19.45]

This situation is contrasted with the position of an American court which was faced with a similar question as to whether or not an architect/engineer is authorised to decide matters relating to his own alleged misconduct. The court stated:

'While paragraph 35 of the General Conditions . . . does give the architect the power . . . to adjust and determine disputes between the contractor and other contractors, it does not in our considered opinion, give the architect the power to pass judgment upon his own errors and omissions . . . To permit this would be an outrageous result not contemplated by the parties, and one not compelled by the language of the contract.'[19.46]

On the international scene, the view of the American court may be preferred to that under the English doctrine described above but then the matter can be referred to arbitration. As stated in the Guide (referred to in Reference 9.1) on page 125, 'the Engineer's decision . . .' is not final and binding, 'if either party challenges it in accordance with the procedure laid down in Clause 67'.

As a result of the above, it is important to deal with clause 67 in detail. It must be said at the outset, however, that most of those who criticise the effectiveness of the clause do so only when it has failed. Many disputes are resolved satisfactorily by the engineer on a day-to-day basis, often without even the need to draw them to the attention of the employer, and when they are that is the end of the matter, but these successes do not become part of the statistics.

The procedure set out in clause 67 only becomes operative when a dispute occurs. It begins with the words:

'If a dispute of any kind whatsoever arises between the Employer and the Contractor in connection with, or arising out of, the Contract or the execution of the Works . . .'

Before one can analyse the provisions of clause 67, it is necessary to go back in time and analyse the events which must have occurred to cause the dispute.

The evolvement of a dispute or disagreement is explained in detail in the penultimate paragraph of Section 19.3 above. The claim is an assertion for money or time or for both or it may be an assertion that responsibility or liability for a certain action or inaction is denied. It may also be an assertion that a certain view is incorrect or imprecise. It is only when a different point of view is firmly held by the other party in connection with the assertion, as a whole or in part, that a disagreement evolves and a dispute is generated. It is only then that the procedure of clause 67 becomes operative.

If this is accepted, as it should be,[19.47] there is one type of situation which remains outside the scope of this definition unless it is somehow developed into a dispute. This is where one party's claim or assertion remains unchallenged and no attempt to meet the other party's demand is made. In these circumstances, it has been held in most jurisdictions, for instance in England (see, for example, the case of *Halki Shipping*[19.48]) that a dispute has arisen which can be submitted to the engineer under clause 67. This is of course in line with the intention of the FIDIC Fourth Edition.

Another possible situation is where one party submits his claim or assertion to the engineer for determination under the first tier of the decision-making process and is then met with protracted silence. Once the engineer has been given adequate opportunity to reply and it becomes evident that the claim is being ignored and no response is likely to be forthcoming, either in terms of the validity of the claim or the determination of whatever is the subject of the claim, it would be reasonable for the contractor to suppose that his claim is not accepted, and so the engineer's silence is taken to be a form of rejection of the claim. A dispute is deemed to have arisen[19.49] and can therefore be submitted to the engineer under the second tier of clause 67.

It is interesting to note that the words 'any dispute or difference' in the previous editions of the Red Book have now been restricted to simply the word 'dispute'. This is perhaps an attempt by the draftsmen to emphasise the necessity of having an assertion and a rejection before the procedure in clause 67 can come into operation.

19.8.1 Procedure under clause 67

The procedure laid down in clause 67 in the event of a dispute can be summarised under the following headings.

Assertion or claim made

An assertion or a claim is made (by the claimant party) in connection with or arising out of the contract or the execution of the works including matters under the provision of the contract or in breach thereof, and it is rejected by the engineer (either expressly or by silence).

A dispute arises

To the extent that the claimant party does not accept the engineer's rejection of his claim, a dispute arises between the employer and the contractor. The Fourth Edition removed the reference in the previous edition to a dispute 'between the Engineer and the Contractor' in response to the criticism made that the engineer is not a party to the contract and therefore cannot be included in this agreement to arbitrate.

The dispute can be:

(a) of any kind whatsoever arising in connection with, or out of, the contract or the execution of the works; and
(b) including any opinion, instruction, determination, certificate or valuation of the engineer;
(c) arising during the execution of the works or after their completion; or
(d) arising before or after repudiation or other termination of the contract.

By using the word 'including' before the list in point (b) above, the dispute is not limited to any particular communication issued by the engineer, thus confirming that the scope of the dispute can encompass any aspect of the contract. Also, it should be noted that by providing such an extensive list, virtually every aspect of the engineer's administration of the contract and his communicative interaction with the contractor can be challenged; it is not limited to the engineer's determination of a claim asserted by one party under a particular clause of the contract.

Dispute referred in writing for a decision by the engineer

The dispute must, in the first place, be referred in writing to the engineer for his decision.[19.50] The words 'in the first place' in sub-clause 67.1 show that the contract envisages a number of steps towards resolution of the dispute but dictates that a referral to the engineer is the first step. The inference that can be drawn from this express obligation is that, since the engineer has no obligation in relation to deciding the dispute until and unless it is referred to him, the referral is a pre-condition to the engineer's decision and to proceeding any further in the dispute resolution process. Sub-clause 67.1 also requires that such referral is copied to the other party, stating that it is made in accordance with clause 67. This allows the other party to become aware of the case against him and gives him the opportunity to prepare his case in contemplation of resolution of the dispute, in compliance with the rules of natural justice. The term 'decision' in the Fourth Edition of the Red Book is reserved for use in clause 67 and no other clause contains a reference to a decision of the engineer.

The requirement that the reference to the engineer be in writing is a new provision, which eliminates the possible confusion as to what is and what is not a proper reference to the engineer.

The engineer's decision on the dispute

Following a valid reference, the engineer must, within a period of 84 days, give notice of his decision to the employer and the contractor; such notice must clearly state that the decision is given pursuant to the provisions of clause 67. This specific new reference to clause 67, in both the reference by the parties to a dispute to the engineer and in the notice of the engineer's decision, is aimed at eliminating a major source of confusion in practice. Many arbitrations under previous editions of the Red Book include allegations that such communication between either of the parties and the engineer or vice versa was not in effect a clause 67 communication.

The discretion that the engineer has in reaching his decision is governed by sub-clause 2.6(a), with three constraints as to how it is exercised: 'impartially', 'within the terms of the Contract', and 'having regard to all the circumstances'. In this regard he is often said to act as a quasi-arbitrator. His arbitral role is 'quasi' since, unlike an arbitrator, he is not empowered by the parties under the contract to look outside the contract, hence the wording 'within the terms of the Contract', and he does not draw support for any authority in this role from the applicable arbitration statute.

It is noteworthy here that, in reaching his decision, there is no obligation on the engineer to consult with the parties in his decision-making process, nor is he required to give either party the opportunity to present its case. Bearing in mind that the engineer's decision is final and binding on the parties unless and until it is revised or reversed by amicable settlement or arbitral award, perhaps it could be said that the rules of natural justice are not adequately provided for and therefore may be prejudiced. For that reason, a responsible engineer should take it upon himself to consult with the parties and give both parties the opportunity to present their case, although he has no contractual duty to do so.

If the engineer fails to give his decision within the specified time, either party may proceed to the next step in the dispute resolution process, namely the serving of notice of intention to commence arbitration.

The engineer's decision is a condition precedent to arbitration, as confirmed by the English court in the case of *J. T. Mackley & Company Ltd* v. *Gosport Marina Ltd*[19.51] which concerned a dispute under the ICE conditions of contract upon which the Red Book conditions are based. It was held:

> '... a decision of the Engineer is a condition precedent to the entitlement of a party to a contract which incorporates the ICE Conditions to refer a dispute to arbitration.'

Engineer's decision final unless referred to arbitration

Disputes involve claims or assertions by either party against the other party, either under particular provisions of the contract (for example, a claim for entitlement to an extension of time under sub-clause 44.1) or in respect of a breach by the other party of other provisions of the contract. It is important to recognise that sub-clause 67.1 does not impose restrictions on the type of dispute which can be submitted to the engineer for his decision. The Guide (referred to in Reference 9.1) on page 125 provides the following advice:

> 'It has been argued that, with respect to disputes having mainly a legal character, the Engineer could always refrain from giving a decision. More in line with the intention of the Clause would be that the Engineer gives his decision after having taken legal advice on the matter in dispute.'

This appears to be a change in policy or at least a clarification of a statement made in the Notes to the Third Edition of the Red Book, where it was stated with reference to the engineer's duties that:[19.52]

> 'The Engineer's task is to interpret the Contract as written and to determine the legal rights of either party.'

That statement in the Notes was taken by some commentators to mean that the engineer must refrain from giving a decision in respect of a purely legal dispute.[19.53] However, if that was a correct interpretation of the statement, one is faced with the problem that the opening words of clause 67 oblige the parties to submit all disputes, without restriction, to the engineer.

The above interpretation was in fact criticised by distinguished writers, one of whom stated:

> 'Some engineers suffer from the misconception that disputes other than those involving a claim under some specific clause in the contract do not fall within their right and duty to give a decision under this clause (Clause 67), for example a claim by the contractor for breach of contract.'[19.54]

Fortunately, this problem has now been resolved in the Guide.

The last paragraph of sub-clause 67.1 gives the decision of the engineer unequivocal weight by providing that, once given, it is final and binding on the parties unless either or both require that the matter in dispute be referred to arbitration. Therefore, even if one or both parties do not accept the engineer's decision, they are bound by it in its entirety right up until the point when they agree a settlement between them, or when an arbitrator issues his award, which has the effect of revising the engineer's decision. Neither party may derogate from what has been decided by the engineer in respect of the dispute, albeit that they are in the middle of negotiation or arbitration proceedings.

There is a strict timescale laid down in sub-clause 67.1 within which the dissatisfied party must notify the other party (copied to the engineer) of his intention to refer the dispute to arbitration. If this deadline is missed, then the engineer's decision becomes final and binding. This means that neither party has recourse to arbitration or to litigation in respect of the matter decided (only in respect of enforcement of the engineer's decision, see further below).

Notice of intention to commence arbitration

If dissatisfied with the decision of the engineer in respect of the dispute, the employer and the contractor have 70 days following the day on which notice of the engineer's decision was received, to give notice to the other party, with a copy to the engineer, of their intention to commence arbitration proceedings in respect of the dispute.

If the engineer fails to give notice of his decision within 84 days after the day on which he received the reference of a dispute, then either party may proceed to arbitration within 70 days after the day on which the 84 days expired, by giving notice to the other party (no longer to the engineer as provided in previous editions of the Red Book) of his decision to commence arbitration proceedings, with a copy to the engineer.

In order to be entitled to serve notice of intention to commence arbitration, the serving party must be dissatisfied with the engineer's decision in whole or in part. Accordingly, the contract provides that the notice shall be 'as to the matter in dispute', thereby requiring the notifying party to state what remains in dispute after the engineer's decision.

The notice of intention to commence arbitration proceedings has three important functions and these are:

(a) The notice establishes the entitlement of the party giving such notice to commence arbitration. It might be assumed from inclusion of the word 'intention' in the description of this notice that it is merely the mechanism whereby one party informs the other that he does not consider the matter in the dispute to be satisfactorily closed following the engineer's decision, and that he intends to proceed further down the path towards arbitration, rather than of itself being deemed to commence the arbitration. This assumption might also be said to be supported by Article 4(2) of the Rules of Arbitration of the International Chamber of Commerce (ICC), as incorporated into the arbitration clause by reference in sub-clause 67.3, which states that the date of commencement of the arbitration is deemed to be the date that a request is received by the ICC secretariat – see further at section 19.10 below. In fact, this assumption is incorrect, as borne out by two arbitration cases in 1986 (see paragraph (c) below). The notice of intention to commence arbitration is deemed to mark the commencement of the arbitration. Whilst the timescale of the procedure under the ICC Rules is measured from the date of receipt of the request, service of the notice marks the date from which time starts to run for the purposes of the limitation period under the applicable law of the contract. For this reason, perhaps the nomenclature of this notice, if not changed should be read as: notice to commence arbitration.

(b) If notice is not given, no arbitration proceedings may be instituted in respect of the matters in dispute, and the engineer's decision is final and binding on both parties. The notice is, therefore, a condition precedent to arbitration of the dispute in hand.

(c) The notice forms the mechanism by which arbitration is properly instituted. By expressly providing that the notice gives the serving party the entitlement to commence arbitration, the confusion arising from the wording of previous editions has been eliminated and the assumption referred to in paragraph (a) above is no longer valid. This confusion occurred when conflicting arbitral awards were issued in 1986 in two cases determining the question of 'whether a party had taken the required measures under Clause 67 ... to prevent the Engineer's decision from becoming final and binding'. The arbitrators acting under the ICC Rules of Arbitration came to opposite conclusions. In one, it was considered that 'a request for arbitration must be filed with the ICC Court of Arbitration' within the required period whereas in the other 'a notification to the Engineer requiring the dispute to be referred to arbitration' within the required period was considered sufficient.[19.55]

The time limits stated in sub-clause 67.1 are measured in days and are chosen to be divisible by seven to prevent the possibility of having a weekend at the end of the period. In previous editions the time limits were expressed as 90 days.

Introduction of procedure for amicable settlement

Where a notice of intention to commence arbitration in respect of a certain dispute, or a number of disputes, has been served, a procedure for an amicable settlement has to be followed before commencement of arbitration. Sub-clause 67.2 does not provide, however, for the type of amicable settlement to be followed nor by whom such procedures should be initiated.

The introduction in the Fourth Edition of the Red Book of this new step in the dispute resolution mechanism governed by clause 67 reflects FIDIC's view that, if at all possible, disputes should be resolved by the parties themselves. Of course, even if arbitration is commenced, the parties may settle the matter themselves at any time during the course of the proceedings. Nevertheless, by obliging the parties to follow a procedure with the aim of reaching amicable settlement, the provisions of clause 67.2 give the parties one last chance to settle the dispute between them before proceeding to arbitration. Although a contractual provision cannot oblige the parties to reach agreement, this sub-clause does put both parties under obligation to *attempt* to reach amicable settlement.

Should amicable settlement be initiated and the process appear to be moving towards a successful conclusion, but outside the stipulated period of 56 days, then the parties may agree to defer the commencement of arbitration pending the outcome of the negotiations. Otherwise, arbitration may be commenced on or after the 56th day after the day on which notice of intention to commence arbitration of such dispute was given, whether or not any attempt at amicable settlement had been made. The use of this phrase 'whether or not any attempt at amicable ...' at the end of sub-clause 67.2 eliminates any possibility of confusion as to what these attempts should be or whether or not they were made at all with obvious consequent problems for the parties and any appointed arbitrator pursuant to sub-clause 67.3. In reality, therefore, this sub-clause has little contractual weight and could be said to merely act as a reminder to the parties that perhaps direct communication between them, without the engineer, may lead to settlement of the dispute without the time and financial burden of entering into arbitration.

This provision in sub-clause 67.2 marks a new departure from arbitration towards alternative dispute resolution methods and since the phrase 'amicable settlement' is left undefined, it may include negotiation, mediation or conciliation, if the parties agree. Tripartite discussions involving the engineer may also take place if all the parties agree, especially if they are all involved in the matters in dispute. Furthermore, it serves as a provision expressly stated in the contract entitling some employers, who would otherwise be restricted, to follow that route without reproach.

Applicable rules

Should amicable settlement attempts fail to resolve the matters in dispute, then the procedure set out in sub-clause 67.3 becomes operative provided the notice of intention to commence arbitration has been served within the time specified.

This procedure provides that 'the dispute or disputes should be settled under the Rules of Conciliation and Arbitration of the International Chamber of Commerce in Paris, unless otherwise specified in the contract'. This last phrase marks another departure from previous editions of the Red Book in that should the parties now wish to use different rules from the ICC Rules they can do so by simply deleting the words 'unless otherwise specified . . .' following paragraph (b) of sub-clause 67.3, and inserting instead the title of the favoured rules. The ICC Rules and the procedure that is described in them are discussed in detail in Section 19.10 below.

Number of arbitrators to be appointed

The number of arbitrators is not restricted under sub-clause 67.3 allowing the parties to choose either a sole arbitrator or more than one depending on the circumstances of the case. However, if arbitration is to proceed under the Rules of the International Chamber of Commerce, then Article 8(2) of the 1998 Edition provides that:

> 'Where the parties have not agreed upon the number of arbitrators, the Court shall appoint a sole arbitrator, save where it appears to the Court that the dispute is such as to warrant the appointment of three arbitrators. In such case, the Claimant shall nominate an arbitrator within a period of 15 days from the receipt of the notification of the decision of the Court, and the Respondent shall nominate an arbitrator within a period of 15 days from the receipt of the notification of the nomination made by the Claimant.'

If a different set of rules is specified, then care must be taken to ensure that no dispute arises between the parties as to the number of arbitrators to be appointed and the procedure to be followed in such an appointment.

Rules under sub-clause 67.3

A number of rules are incorporated under sub-clause 67.3 and these are:

(a) The arbitrator has full power to open up, review and revise any decision, opinion, instruction, determination, certificate or valuation of the engineer when related to the dispute. This power is very wide and in some jurisdictions wider than the power available to courts of law.[19.56] However, it has now been explicitly restricted to the matters relating to the dispute, thus eliminating another confusion which existed in previous editions of the Red Book.

(b) Arbitration, once initiated, may take place before or after the completion of the works provided that the obligations of the employer, the engineer and the contractor are not altered by reason of the arbitration being conducted during progress of the works. This provision is a change from previous editions where arbitration could not be commenced until completion of the works was achieved.

(c) The parties are not limited in the arbitration proceedings to the evidence or arguments put before the engineer for the purpose of obtaining his decision under sub-clause 67.1.

(d) The decision of the engineer is not to disqualify him from being called as a witness and from giving evidence before the arbitrator on any matter relating to the dispute.

(e) Sub-clause 67.4 contains another new provision. It provides that failure of a party to comply with an engineer's decision which has become final and binding under clause 67 may itself be referred to arbitration. Such reference to arbitration is made directly under the provisions of sub-clause 67.3 and accordingly, a party's failure to comply with a final and binding decision of the engineer need not first be referred either to the decision of the engineer, or to an amicable settlement, under the clause.

In this way, the engineer's decision can be confirmed through an arbitration which, if international, could then be enforced by virtue of any international conventions on enforcement of foreign arbitral awards which might be applicable.

The above procedure is perhaps best illustrated by the use of a flow chart. Figure 19.1 provides such a chart showing the various steps leading from an assertion made to arbitration, and thus identifying the two-tier system discussed above.[19.57]

19.9 The 1996 supplement to the Fourth Edition of the Red Book

As explained above, the November 1996 supplement to the Fourth Edition of the Red Book, as reprinted in 1992, contained in its Section A an alternative version to clause 67 of the Red Book, providing for the establishment of a Dispute Adjudication Board. This is discussed in detail in Chapter 26.

Figure 19.2 shows a flow chart of this new procedure under the alternative version of clause 67 concerning the resolution of disputes using the Dispute Adjudication Board (the Board) under the 1996 Supplement to the Fourth Edition of the Red Book.

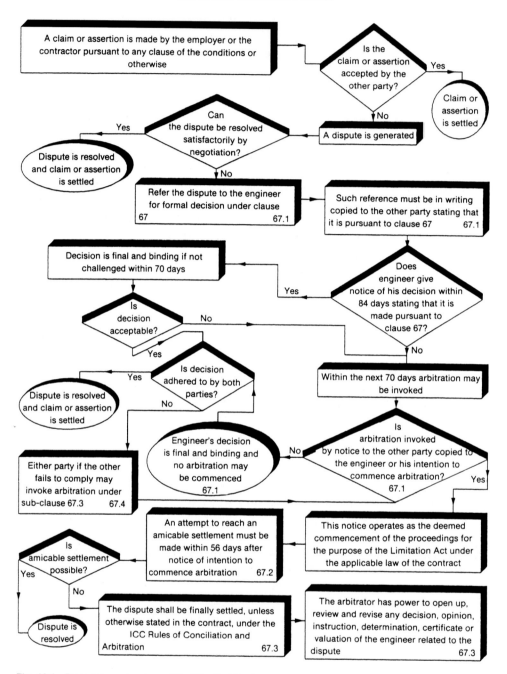

Fig. 19.1 Procedure under clause 67 of the Red Book.

There are a number of points in the rules and procedures in respect of this alternative which require a mention and to which attention must be given.

(a) It is extremely important that the parties agree the identity of the members of the Dispute Board prior to the commencement date, and their appointment should be treated as a priority since the requirement under sub-clause 67.1 of ensuring the appointment of the Board within 28 days of the commencement date is not supported by a specific mechanism in the contract. Furthermore, if the parties fail to agree the appointment of the Board within the stipulated 28 days, then first, there is no time limit within which the appointing authority must act and appoint; and secondly, a Dispute Board is imposed upon the parties rather than being chosen by them, which is contrary to the whole concept of a trusted, mutually agreed Dispute Board. Finally, in this regard, as recommended in the Guide to amended clause 67, if the appointing authority named in the Appendix to Tender is not the President of FIDIC, then the consent of such authority to act should have been previously obtained in writing.

(b) Sub-clause 67.1 requires Dispute Board members to be suitably qualified and the terms of appointment of the members demand a warranty that the member is so qualified. It follows from these requirements that potential candidates for Dispute Board membership will come from high calibre, experienced professionals who hold positions of authority in their particular field. However, the terms of appointment prohibit, and rightly so, a Dispute Board member, during his term of appointment, from entering into discussions with either of the parties or the engineer regarding employment by them after ceasing to be a Dispute Board member. Even though the appointment is a personal one, as provided in item 1(a) of the terms of appointment, the above requirements in practice will effectively exclude many senior members of professional firms and contracting companies from acting as Dispute Board members as the above prohibition must extend to their firms. This prohibition could last for a considerable period of time if the project is large and complex.

(c) The period of time allocated to the initial study of the project is, in practice, likely to be too short, especially in the case of complex and large projects, not only from the point of view of the technical documentation which must be studied but also from the Dispute Board member's remuneration point of view.

(d) Sub-clause 67.1 provides that the appointment of a Dispute Board member may be terminated only by mutual agreement of the parties. The question which does not seem to have been addressed is: what happens in the event that one of the parties wishes to terminate the appointment of a Dispute Board member whilst the other party does not?

(e) The Dispute Board is required under the provisions of sub-clause 67.2 to render its decision in 84 days after receipt of the reference. Are 84 days sufficient in the case of a complex dispute for a comprehensive procedure, adequate hearing, Dispute Board deliberation and reaching a fair and just decision? Or is it the intention that for such complicated and lengthy disputes, a summary decision would be rendered and if unacceptable, then the dispute must be referred to arbitration? It is notable that before a dispute can be referred to arbitration, a period of 168 days would have elapsed. This period is made up of 84 days for the Dispute Board to

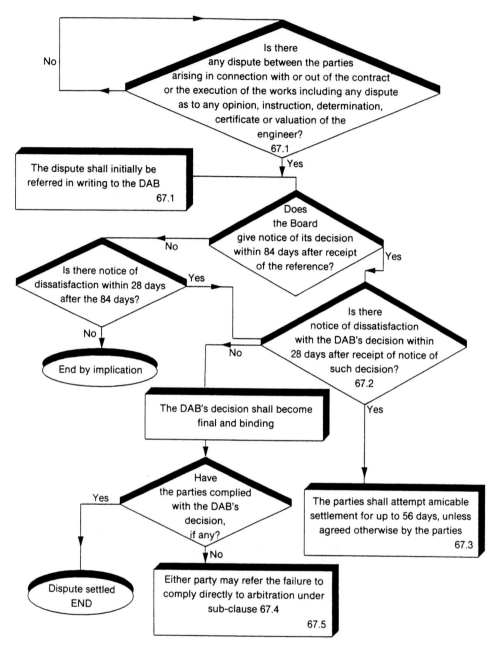

Fig. 19.2 Procedure under the provisions of the new alternative version of clause 67 contained in the 1996 Supplement.

Disputes Settlement by Arbitration 413

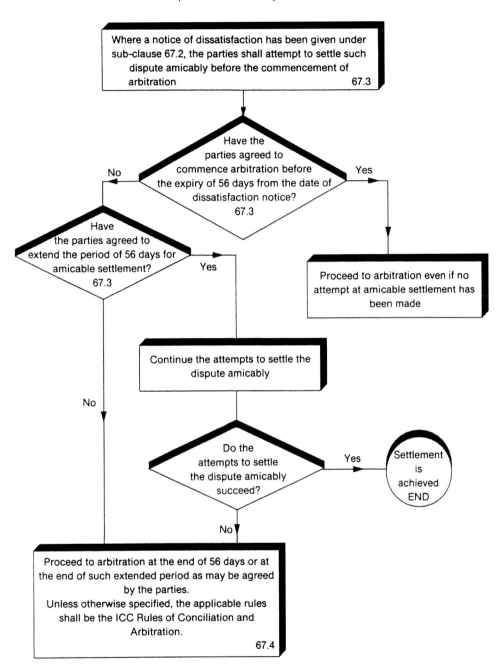

(Fig. 19.2 Contd.)

give its decision under sub-clause 67.2, plus 28 days for the parties to give notice of dissatisfaction with that decision, plus 56 days for the amicable settlement procedure under sub-clause 67.3. Of course, each of these periods could be shorter if the relevant procedural steps are completed earlier than allowed for.

(f) Item 3(e) of the terms of appointment provides that the Dispute Board member shall not give advice to either of the parties or to the engineer concerning the conduct of the project other than in accordance with the procedural rules. Item 5 of these terms provides that neither the employer, the contractor nor the engineer shall seek advice from or consultation with the Dispute Board member regarding the project otherwise than in the normal course of the Dispute Board's activities under the contract and the rules. Moreover, under item 5 of the procedural rules, the Dispute Board must not express any opinions during any hearing concerning the merits of any arguments advanced by the parties. These interrelated provisions lead to the following question: is a Dispute Board member prevented from voicing an opinion, if for example on a site visit, or indeed on examining a drawing, he encounters what he knows from his own knowledge and expertise to be incorrect?

(g) Under sub-clause 67.1 and in item 6 of the terms of appointment, the remuneration of all Dispute Board members is equal. In view of the fact that the chairman of the Board would have more onerous administrative duties and obligations than his co-members, a higher fee for the chairman should have been provided for under the rules.

(h) Under item 6 of the procedural rules, the contractor is required to pay the Dispute Board members' fees and expenses within 56 calendar days after receipt of their invoices. It is suggested that a preferred procedure for payment would be to establish a bank account in the form of security for costs from which payments are made against invoices submitted in advance to the employer and the contractor and released by authority of the chairman.

(i) Attention should be given to the fourth paragraph of sub-clause 67.2 which provides that the contractor, the employer and the engineer are required to give effect to the Board's decision unless and until such decision is revised through an amicable settlement or an arbitral award.

19.10 The ICC Rules of Arbitration

The revision of the ICC Rules of Arbitration and Conciliation in 1998 brought about a separation of both sets of Rules, and as such the Rules of Arbitration will be dealt with separate to the Rules of Conciliation, which are now called 'the ADR Rules' and are dealt with briefly in Chapter 20.

Unless otherwise expressly specified in the contract, the Rules of Arbitration of the International Chamber of Commerce (the ICC Rules) will apply to any arbitration proceeding under sub-clause 67.3 of the Fourth Edition of the Red Book, or sub-clause 67.4 of the alternative version of clause 67 in the 1996 Supplement.[19.58] The use of the ICC Rules has been maintained due to the numerous advantages it offers to the parties in dispute, and these will be discussed below in section 19.10.2.

The International Chamber of Commerce (ICC) is a non-governmental organisation founded in 1919 to serve world business through promoting international trade, services and investment; promoting the free market economy system based on the principle of competitiveness of business enterprise; and fostering the economic growth of developed and developing countries. Its members come from over 7,500 companies and business associations in more than 110 countries. The ICC's headquarters are in Paris and national committees from over 60 countries throughout the world represent its view to governments and business concerns.[19.59] Each year, ICC arbitrations are held in some 40 countries, in most major languages and with parties and arbitrators from over 100 countries.[19.60] As such, the ICC can truly be said to be a global organisation and, therefore, well suited to dealing with both domestic and international business disputes.

Attached to the ICC is the International Court of Arbitration of the ICC whose members, currently composed of 114 members from 78 countries, are appointed by the World Council of the ICC. For the day-to-day management of cases in many languages, the ICC Court is supported by a secretariat based at the headquarters of the ICC, in Paris. The International Court, which was founded in 1923, has built up unique experience and enjoys an outstanding position in the field of the settlement of international business disputes. Since its birth in 1923, the ICC has dealt with over 13,500 cases, of which about a third of them were dealt with in the past 10 years. Recent statistics show that approximately 600 new arbitration requests are filed with the secretariat every year, including parties from Africa, Asia, Europe, Latin America and North America.[19.61]

The Court has mandated itself to remain constantly alert to changes in the law and the practice of arbitration in all parts of the world and to adapt its working methods to the evolving needs of parties and arbitrators.[19.62]

The Court does not itself settle the disputes brought before it but it does organise, supervise and monitor the arbitration proceedings carried out by the arbitrator. Therefore, in that respect, the term 'Court' is misleading as it has neither judicial power nor does it render an award. It is the arbitrator or the tribunal appointed by the Court for each specific case, that exercises the judicial power throughout the arbitral proceedings. Neither the parties in dispute nor the arbitrator ever appear before the Court. It conducts its administrative functions through documentary material submitted to it by the parties and the arbitrator. The Court's function is stated in the forward to the ICC Rules as: 'to provide for the settlement by arbitration of business disputes of an international character in accordance with the Rules of Arbitration of the International Chamber of Commerce' but its functions also include, amongst others, the following.[19.63]

(a) deciding whether a prima facie agreement to arbitrate under the ICC Rules exists between the parties. However, any disagreement between the parties as to the validity of such a prima facie agreement is referred to the arbitrator(s) for his/their decision;

(b) taking all measures that may be required with regard to appointing, replacing or challenging the arbitrator(s). In that regard, the Court may refuse to confirm the appointment of the arbitrator(s) nominated by the parties if there is an apparent lack of independence or a question as to his/their competence or suitability for the particular arbitration case;

(c) determining the place of arbitration and the number of arbitrators where the parties have made no provision to that effect;
(d) ensuring that the arbitrators' terms of reference (such terms are discussed in detail in sub-section 19.10.1(g) below) are drafted promptly and in conformity with the ICC Rules;
(e) in the event of refusal by one of the parties to sign the terms of reference, setting a time limit on the expiry of which the arbitration proceeds regardless;
(f) supervising the proceedings conducted by the arbitrator(s) and ensuring that his/their award is made within the time limit originally set or, where necessary, as extended by the Court;
(g) fixing the amount of the advance payment to be requested and determining the arbitrators' fees, thus providing a buffer between the parties and the arbitrator(s) in respect of a matter which might well inhibit the independence and neutrality of the arbitrator(s); and
(h) scrutinising the draft award and, if necessary, laying down modifications as to the form of the award and, without affecting the arbitrators' liberty of decision, the Court may also draw their attention to points of substance.

19.10.1 *The procedure under the ICC Rules of Arbitration*

Since its inception, arbitration under the ICC Rules has been specifically tailored to handle international disputes, although the Rules are also technically acceptable in the case of domestic disputes.[19.64]

The ICC Rules essentially set out a framework for the arbitration procedure. To summarise briefly, the procedure in practice would follow the steps outlined below:

(a) The party wishing to refer a dispute to arbitration (the claimant) submits a request for arbitration to the secretariat in Paris. Under Article 4(3) of the ICC Rules the request must include certain information such as: the names, addresses and descriptions of the parties; the nature and circumstances of the dispute giving rise to the claim; the remedy and quantum, if known, being sought by the claimant; a copy of the arbitration agreement; details of the existing agreement, if any, between the parties or, if there is no agreement, a proposal as to the number of arbitrators; and proposals concerning the place of arbitration, the language to be used and the applicable law. Where there is agreement between the parties that there will be three arbitrators, the claimant must also nominate an arbitrator in the request. The claimant is required to submit to the secretariat, with the request, a non-refundable advance payment for administration costs, as required under Appendix III to the ICC Rules (currently in the amount of US$2,500).

Although Article 4(2) of the Rules stipulates that the date on which the request is received by the secretariat is deemed to be the date of commencement of the arbitral proceedings, a contradiction in terms exists with sub-clause 67.3 of the Conditions of Contract for the Red Book, as explained on page 406 above.[19.65]

(b) Upon receipt of the request the secretariat forwards a copy of the request (and all attachments) to the other party (the respondent) and, at this stage, the Secretary General of the secretariat may, under Article 30(1), request that the claimant make a provisional advance payment to cover the costs of arbitration up to the drawing up of the terms of reference (which are discussed at sub-section (g) below).

(c) The respondent is required to file with the secretariat an answer to the request, within 30 days of his receipt of the request (unless extended by the secretariat). Under Article 5(1) of the ICC Rules the answer must include certain information such as: the name, address and description of the respondent; comments on the claimant's statement in respect of the nature and circumstances of the dispute giving rise to the claim; a response to the remedy and quantum being sought by the claimant; comments on the agreement or the claimant's proposal, as the case may be, concerning the number of arbitrators; and comments on the claimant's proposals regarding the place of arbitration, the language to be used and the applicable law. Where there is agreement between the parties that there will be three arbitrators, the respondent is also required to nominate an arbitrator. If the respondent wishes to file any counterclaim(s) he must do so with his answer and, under Article 5(5), is required to give the nature of and circumstances of the dispute giving rise to the counterclaim(s) and the remedy and quantum, if known, being sought by the respondent under the counterclaim(s).

If the respondent refuses or fails to provide an answer the Court may decide that the arbitration will proceed regardless, but only after it has satisfied itself that *prima facie* a valid arbitration agreement is in place between the parties.

(d) Where the respondent has filed counterclaim(s) with the answer, the secretariat will forward a copy of such counterclaim(s) to the claimant who then has 30 days (unless extended by the secretariat), from the date he receives such copy, within which he must file his reply.

(e) The ICC Rules provide only for either one arbitrator or three arbitrators. Where the parties have agreed that the disputes shall be settled by a sole arbitrator, they may, by agreement, nominate the arbitrator for confirmation by the Court. Failing such agreed nomination within 30 days from the date that the request for arbitration has been received by the respondent, or within such additional time as may be allowed by the secretariat, the sole arbitrator will be appointed by the Court under Article 8(3) of the Rules. In the case of a tribunal, the Court will either appoint the arbitrators or confirm the appointment of the two arbitrators proposed by the parties. The Court appoints the chairman of the tribunal unless the parties have agreed upon another procedure for such appointment, in which case, such nomination will be subject to confirmation by the Court in pursuance of Article 9 of the Rules. As provided by Article 9(1) of the Rules:

> 'In confirming or appointing arbitrators, the Court shall consider the prospective arbitrator's nationality, residence and other relationships with the countries of which the parties or the other arbitrators are nationals and the prospective arbitrator's availability and ability to conduct the arbitration in accordance with these Rules...'

Having regard to the claimant's proposal and the respondent's comments on such proposal, the Court will also determine the place of the arbitration unless this issue has already been agreed on by the parties. The Court will also fix a monetary advance to cover the costs of arbitration and, upon receipt of the advance which is payable in equal amounts by the claimant and the respondent, the secretariat submits the file to the appointed arbitrator or tribunal.

(f) The Court requires a high standard of independence and impartiality in its arbitrators. Every prospective arbitrator is required to sign either of the following statements before being appointed:

> 'I am independent of each of the parties and intend to remain so: to the best of my knowledge, there are no facts or circumstances, past and present, that need be disclosed because they might be of such nature as to call into question my independence in the eyes of any of the parties.'

or

> 'I am independent of each of the parties and intend to remain so: however, in consideration of Article 7, paragraphs 2 and 3 of the ICC Rules of Arbitration, I wish to call your attention to the following facts or circumstances which I hereafter disclose because they might be of such a nature as to call into question my independence in the eyes of any of the parties. . . .'

Although only independence is mentioned in these declarations, the ICC requires impartiality as part of independence and refers to it in the Rules, see for example, Article 15(2).

Therefore, not only must each arbitrator be independent and impartial, he must also have regard and draw the Court and the parties' attention to any background involvement or circumstance which might be perceived by either party to compromise his independence, even if it does not in reality do so.

(g) The arbitrator's, or the tribunal's, first task once he/it receives the file from the secretariat is to draw up a document known as the terms of reference. He/it is required to draft this document on the basis of documents provided by the parties or in the presence of the parties, having regard to their most recent submissions. The terms of reference are a very useful tool, both for the arbitrator/tribunal and the parties, in that they summarise the state of the arbitration as it stands at that time. This is especially important in the case of large, complex disputes where the parties may already have been involved in protracted discussion and the issues in dispute may have lost clarity.

By virtue of Article 18(1) the terms of reference are required to contain certain information as a minimum: the names and descriptions of the parties; the addresses of the parties to which all formal notifications and communications are to be served; a summary of the claimant's claims and the respondent's counterclaims (if any), and the remedy, and quantum if known, sought for each; a list of the issues to be determined (unless deemed inappropriate by the arbitrator/tribunal); the name(s), address(es) and description(s) of the arbitrator(s); the place of arbitration; and details of the procedural rules to be adopted. As such, the terms of reference define the arbitrator's/tribunal's tasks and the procedural means for achieving them.

This document is then circulated as soon as possible to the parties for their approval and/or comments. The time available to the arbitrator/tribunal to finalise this document is limited, the approved and signed (by the arbitrator or each member of the tribunal, as the case may be, and by both of the parties) terms of reference are required to be transmitted to the Court within two months (unless extended by the Court) of the tribunal/arbitrator first receiving the file from the secretariat. Therefore, it is at this stage that the arbitrator/tribunal must show and exercise his/its managerial and leadership capabilities in order to fulfill this requirement of the Rules. By doing so, he/it is taking the opportunity to demonstrate to the parties that he/it has taken control of the dispute in hand and intends to proceed with the arbitration accordingly.

Once the signed terms of reference have been received by the Court, the Court informs the arbitrator/tribunal and the parties accordingly. If one of the parties refuses to take part in drafting the terms of reference or to sign it, then, nevertheless, the arbitrator/tribunal may submit the terms of reference to the Court for its approval. It is only when the Court has responded in respect of or approved the terms of reference that the arbitration may proceed and the arbitrator/tribunal may deal with the merits of the case. This document is a special feature of the ICC Rules and an extremely useful instrument in the conduct of international arbitration, see Section 19.10.3 below.

Although criticised by some, the terms of reference have four main advantages:

(i) technical: where the parties, their representatives and the arbitrators focus on the issues to be resolved and plan a management structure and framework for dealing with these issues, separating them into various categories to facilitate the effective conduct of the proceedings, with particular reference to the case in hand;

(ii) psychological: where the parties and their legal representatives, often from different legal backgrounds, meet and become aware of each other's points of view;

(iii) legal: where the parties, by their participation in the preparation of and the signing of the terms of reference, confirm their agreement to have their disputes resolved through the arbitration process; and

(iv) control: where the arbitrators in drafting their award, and subsequently the ICC Court in scrutinising the award, can easily establish not only that all the issues required to be resolved through the arbitral process have in fact been dealt with in the award, but also that the award has not departed from the scope of these issues.[19.66]

These advantages have sometimes induced experienced arbitrators to propose at the commencement of an arbitration under rules other than the ICC Rules an effective procedure that virtually includes a document similar to the terms of reference document.[19.67] It should be noted that some jurisdictions require there to be a document similar to the terms of reference document referred to above.

(h) The arbitrator/tribunal proceeds with the case in accordance with the ICC Rules and any other procedural rules agreed upon by the parties, either within the terms of reference or subsequently. If no procedural rules are agreed upon by the parties then the arbitrator/tribunal is empowered to decide which should be adopted for the arbitration in hand but, by virtue of Article 15(2) of the ICC Rules, in doing so must act impartially and fairly and ensure that each party has reasonable opportunity to present its case. In the absence of an agreement between the parties, the arbitrator/tribunal also has the power to determine the language(s) of the arbitration and the appropriate rules of law to be applied. If the parties so grant, the arbitrator/tribunal has the authority to decide disputes *ex aequo et bono* (according to what is right and good) or as *amiable compositeur* (according to legal principles that it believes to be just, without being limited to any national law), as in Article 17(3) of the Rules.

(j) The arbitrator/tribunal make a draft award, which must be submitted to the ICC Court for scrutiny and approval.

(k) The Court scrutinises the award, and as stated above may draw the attention of the arbitrator/tribunal to points of substance, and approves the final draft if it is satisfied with its form.

This scrutiny by the Court is another special feature of the ICC Rules in the form of a quality control process in which any potential weaknesses which may affect the enforcement of the award are brought to the attention of the arbitrator/tribunal.

(l) The award, having been signed by the arbitrator/tribunal, is notified to the parties by the secretariat of the Court.

The ICC Rules are kept up to date through periodic revisions. The latest Rules of Arbitration came into effect in 1998.

The present ICC Rules include 35 articles, which contain the provisions in accordance with which ICC arbitrations are conducted. They are arranged in a chronological manner following the sequence of arbitral proceedings. These articles deal with the following topics:

Introductory provisions
— Article 1   International Court of Arbitration;
— Article 2   Definitions;
— Article 3   Written notifications or communications; time limits;

Commencing the arbitration
— Article 4   Request for arbitration;
— Article 5   Answer to the request; counterclaims;
— Article 6   Effect of the arbitration agreement;

The arbitral tribunal
— Article 7   General provisions;
— Article 8   Number of arbitrators;
— Article 9   Appointment and confirmation of the arbitrators;
— Article 10 Multiple parties;
— Article 11 Challenge of arbitrators;
— Article 12 Replacement of arbitrators;

The arbitral proceedings
— Article 13   Transmission of the file to the arbitral tribunal;
— Article 14   Place of arbitration;
— Article 15   Rules governing the proceedings;
— Article 16   Language of the arbitration;
— Article 17   Applicable rules of law;
— Article 18   Terms of reference; procedural timetable;
— Article 19   New claims;
— Article 20   Establishing the facts of the case;
— Article 21   Hearings;
— Article 22   Closing of the proceedings;
— Article 23   Conservatory interim measures;

Awards
— Article 24   Time limit for the award;
— Article 25   Making of the award;
— Article 26   Award by consent;
— Article 27   Scrutiny of the award by the Court;
— Article 28   Notification, deposit and enforceability of the award;
— Article 29   Correction and interpretation of the award;

Costs
— Article 30   Advance to cover the costs of the arbitration;
— Article 31   Decision as to the costs of arbitration;

Miscellaneous
— Article 32   Modifed time limits;
— Article 33   Waiver;
— Article 34   Exclusion of liability;
— Article 35   General rule.

There are also three appendices to the ICC Rules, which deal with important matters such as the Court itself, its internal rules and the costs of the arbitration as follows:

Appendix I: Statutes of the International Court of Arbitration of the ICC
Article 1   Function
Article 2   Composition of the Court
Article 3   Appointment
Article 4   Plenary session of the Court
Article 5   Committees
Article 6   Confidentiality
Article 7   Modification of the rules of arbitration

Appendix II: Internal Rules of the International Court of Arbitration of the ICC
Article 1   Confidential character of the work of the International Court of Arbitration

Article 2	Participation of members of the International Court of Arbitration in ICC arbitration
Article 3	Relations between the members of the Court and the ICC National Committees
Article 4	Committee of the Court
Article 5	Court secretariat
Article 6	Scrutiny of arbitral awards

Appendix III: Arbitration Costs and Fees

Article 1	Advance on costs
Article 2	Costs and fees
Article 3	ICC as appointing authority
Article 4	Scales of administrative expenses and of arbitrator's fees

The text of these provisions is self-explanatory and easy to follow.

19.10.2 The advantages of the ICC Rules

The continued use of the ICC Rules of Arbitration in the successive editions of the Red Book, including the 1996 Supplement and the 1999 suite of contracts, is an obvious indication of the numerous advantages of these Rules in construction disputes. This observation is supported by the steady annual increase in the number of ICC arbitration cases.[19.68]

These advantages which have been commented on by many experienced users include the following:[19.69]

(a) *Institutional Rules administered and supervised by an internationally recognised and respected organisation*: Institutional arbitration has a number of benefits when compared with *ad hoc* arbitration, particularly on the international scene where there may be different systems of laws, jurisdictions, languages, geographic locations, cultures and other characteristics. In this regard, the following five can be recognised:
   (i) first, it is extremely useful to incorporate a set of established rules into the procedure at the commencement of an arbitration. This often becomes indispensable to the whole process, such as, for example, the effect of article 21(2) of the ICC Rules if one of the parties, although duly summoned, fails to appear; or the effect of article 6(4) of the Rules should there be a question of the arbitrator's jurisdiction;
   (ii) secondly, in institutional arbitration there is the necessary administrative support provided, for example in ICC arbitrations by the secretariat to the court, from the commencement of the arbitral proceedings until the tribunal is established;
   (iii) thirdly, institutions provide a procedure for the challenge to arbitrators, especially where the independence of the arbitrators is concerned;
   (iv) fourthly, institutions provide a buffer between the parties and the tribunal in respect of the financial arrangements obviating any possible embarrassment;
   (v) fifthly, the unique function of the ICC Court in its scrutiny of awards could contribute to their enforceability.[19.70]

(b) *Neutrality, impartiality and expertise of the International Court of the ICC and its secretariat*: The ICC Court and its secretariat are neutral and impartial. Both the members of the Court and members of its secretariat come from different nationalities, from the East and the West; and from the North and the South. The secretariat has a high ratio of multilingual staff to cases in hand who are responsible for administering these cases with computerised systems for case management and retrieval of 'ICC jurisprudence'. They offer objective advice to the parties, their lawyers and to the arbitrators. They may even assist in securing compliance with an award by a losing party.[19.71]

(c) *Flexibility and frequent updating of the Rules*: Article 15 of the ICC Rules provides great freedom to the arbitrators in the conduct of the proceedings. Such flexibility is essential for the variable nature of the potential arbitration cases which might come from any part of the world under some of the many possible other characteristics of arbitration. Furthermore, the ICC Rules are periodically examined and brought up to date should the need arise.

(d) *Access to internationally recognised and highly experienced arbitrators*: The ICC Court will only appoint as a chairman or sole arbitrator someone who has experience in ICC arbitrations. Through its system of seeking a nomination from national committees of the ICC, it has access to a large pool of qualified arbitrators from developing and developed countries all over the world. In deciding on the nationality of the committee to be invited to make such a nomination, the Court takes into account many factors, including the nationality of the parties, seat of arbitration, law of the contract, language of the arbitration, history of nominations made by the national committees, and the nature of the dispute.[19.72] For this reason, the Court is best suited to make the most fitting selection.

(e) *Party autonomy*: The ICC Rules and practice permit the parties a large amount of freedom in choosing various aspects of their arbitration case, such as the seat of arbitration, language, procedural law and arbitrators. Under the French legal system, the importance of the freedom of the will of the parties is emphasised, thus providing support in this regard to the ICC with its headquarters being located in France.

(f) *Quality of the awards*: As mentioned above in Section in 19.10.1, the scrutiny by the Court of the arbitrators' draft awards is a special feature of the ICC Rules in the form of a supervisory process in which any potential weakness which might affect the enforcement of the award is brought to the attention of the arbitrators. It also ensures a level of uniformity and consistency of the structure of the awards, that brings an element of certainty to the proceedings. This feature encourages compliance by the unsuccessful party with the award and if not complied with voluntarily, it contributes to the enforceability of the award.

(g) *Supplemental publications and reports*: The ICC Rules are periodically supported by high quality publications and reports drafted by working parties drawn from experts in the field of arbitration, appointed by the ICC Commission on International Arbitration. Recent titles include: 'Multi-party Arbitration', 'Interim and Partial Awards' and 'Terms of Reference'. These reports seek to clarify and elaborate on some of the special features unique to the ICC Rules, especially where they are misunderstood and/or misinterpreted. For example, article 18(1) of the ICC Rules provides that the terms of reference are to be drawn up on the basis of the documents and in the light of the parties' most recent submissions. Article 18(1) of the ICC Rules further provides that the terms of reference must contain a summary of the parties' respective claims. In connection with these two provisions, the parties are entitled to set down their respective claims as they define them, up to time of signature, and such signature of the document defining the terms of reference does not signify the parties' acceptance of the claims being made against them.[19.73] These provisions of article 18(1) are frequently misunderstood and in order to address this misunderstanding a report on the Terms of Reference was published by the ICC.[19.74]

Another misconception is the status of dissenting opinions. In ICC arbitration, there is only one award, whether unanimous, majority or a chairman's award, as can be seen from the provisions of article 25 of the ICC Rules. A dissenting arbitrator may submit a dissenting opinion, but it is just that: an opinion, not an award. This fact escapes many including some very experienced lawyers.[19.75] Hence, a report was published by the ICC on dissenting and separate opinions.[19.76] These are but two examples of such misconceptions.

19.10.3 *Some constructive criticism*

Against the advantages of the ICC Rules set out above, there is some constructive criticism levelled against the ICC system. Most worthy of note is: first, the administrative fees charged by the ICC Court add to the total costs of an arbitration and as they are collected in advance, the interest accrued is not credited to the parties. In a long-drawn-out arbitration with large sums in dispute and consequently a high advance on fees, these arrangements are considered to be disadvantageous to the parties. However, in reality, when compared with the total legal costs of the parties, these fees form only an insignificant proportion of the total. Furthermore, these fees are charged against a well worth service and the parties are given the opportunity to provide a bank guarantee against some of the advance on fees required to be paid.

The second and more important criticism relates to the policy of usually appointing lawyers rather than professionals of technical disciplines relating to the topic of the disputes. This is done in the tradition of the civil law of appointing lawyers in the belief that a lawyer is better qualified in international arbitration to deal with difficult problems of procedure and conflict of laws which may arise. Regrettably, this policy would appear to lose sight of the fact that, in construction disputes, the 'engineer' arbitrator, apart from his technical expertise, is more likely to be better qualified than a lawyer to deal with managerial aspects of an international arbitration. The managerial difficulties which arise are in themselves just as important, perhaps even more so, as those problems of procedure and conflict of laws. In any event, even if the arbitration required either technical or legal expertise, then the arbitrator is empowered to appoint an expert in the relevant field, whereas it would be inconceivable for an arbitrator to seek expert advice on managerial matters.

Furthermore, this policy departs from the original concept of arbitration based on the desire of parties in dispute to be judged by their peers. In doing so, it appears that the emphasis on industry or professionally orientated expertise among arbitrators is being eroded, if not lost.

The third criticism is related to the length of time usually taken by some of the administrative steps at the commencement of an arbitration leading to some delay in receipt of the arbitration file by the arbitrators. This problem has to a large extent been addressed in the 1998 revision of the ICC Rules by the inclusion of expedient timescales.

19.11 Why does arbitration in construction disputes continue to lose favour?

Users of international commercial arbitration place speed and economy high on the list of the benefits of arbitration as a method of dispute resolution. Regrettably, neither speed nor economy seems to be a feature of international construction arbitration today. As such, arbitration in fact is departing from its original goal and its procedures are increasingly evolving in the direction of court procedures.[19.77] This is mainly due to the diminution of some or all of the advantages of arbitration, referred to earlier in this chapter, to such an extent that arbitration has now been described as no more than a 'mimic' of litigation, but with the added disadvantage of having to pay the arbitrator and also pay for the hire of the arbitration rooms.

The reasons for this decline are many and are interrelated to a number of characteristic peculiarities which can be traced back to the nature of the dispute and/or to all the parties who play a part in the arbitral process.

Starting with the dispute, with the introduction of pre-arbitral methods of dispute resolution, such as the use of a Dispute Adjudication Board or Dispute Review Board, construction arbitrations deal largely with disputes which cannot be resolved except by arbitral award, either because they raise issues that go to the heart of the parties' relationships or important questions of principle or, in the case of a main or principal contract, are too complex to be resolved satisfactorily by an alternative method of dispute resolution. Construction disputes can generate difficult points of law relating to specialised forms of contract that are not known to those who are not involved in construction; and they still seem to need the examination of many more documents than other types of dispute. Many of the disputes that cannot be resolved by pre-arbitral methods will be those that encompass a multitude of issues of fact and opinion (leaving aside any questions of law), each of which merits consideration and a decision as if it were a separate arbitration. Moreover, the disputes themselves can involve difficult technical questions and occasionally difficult and novel questions of law. The arbitrator or tribunal has to decide how best such disputes are to be handled and such decision is crucial to securing a cost-effective arbitration and to retaining the confidence and co-operation of the parties.

Turning now to the arbitrator or tribunal, failure by him/it to take control of the arbitral proceedings at the outset, to maintain this control throughout and to devise a management framework for an effective procedure to be adopted throughout the arbitration is one of the major sources of the problems encountered in arbitration. This is particularly so in the case of complex arbitrations which, since the widespread adoption of alternative pre-arbitral methods of dispute resolution, are becoming the norm in the construction field. However, it must be mentioned that such failure is not always due to the inability of the arbitrator to perform his duties, but it is attributable, to some extent, to the parties themselves or to their legal representatives.

In this regard, as discussed above in Section 19.10.3, it should be reiterated that the arbitrator's managerial skills are of paramount importance in complex and lengthy construction arbitrations.

Moving onto the input of the legal representatives, the following quotation by an experienced construction lawyer, although referring specifically to arbitration in the United Kingdom, is worthy of note:

> 'Although enormous progress has been made recently, UK arbitrations, particularly those in the construction field, still leave much to be desired. All too often they are allowed to assume a life of their own and what is already financially and emotionally taxing for a client, becomes a positive nightmare. Arbitration grew out of the need of persons of commerce to have a commercially viable dispute resolution procedure which could cope with disputes which the Courts were singularly unsuited for. Regrettably, over the years, the arbitration process has been hijacked by lawyers. Now all too often it has become a pale imitation of High Court procedure, to which it is supposed to offer an alternative. What is more, this pale imitation does not possess the sanction available to the High Court. As a consequence most lay persons with practical experience of an arbitration are usually badly affected by that experience and claimants and defendants alike who have embarked upon arbitration emerge sadder, wiser and poorer people.'[19.78]

This view is echoed by another experienced lawyer and arbitrator in international construction disputes, identifying problems which play a major part in causing delay and consequent increase in cost and which must be addressed by the parties' legal representatives, such as:

— failing to meet deadlines;
— leaving no stone unturned in the interest of their clients irrespective of cost;
— agreeing to time extension by their opposite number knowing that they might want similar facility in the future; and
— availability of counsel, when briefed.[19.79]

> 'Now the law has come to be recognised as a vehicle for winning cases. Lawyers use legal argument before an arbitrator to overcome the technical shortcomings of their client's case. Expert witnesses write long reports, frequently based on disputed facts. The full disclosure of documents, relevant and irrelevant, is commonplace. Lawyers indulge in protracted openings, and arbitration has become a mirror of court procedure, with the consequent increase in costs. It is not surprising that there is now an increasing advocacy for mediation and conciliation as means of settling disputes.'[19.80]

Finally, considering the part played by the parties in delay and expense, the claimant contributes to these problems if and when he vigorously pursues all issues, large and small, some of a very trivial nature, irrespective of the cost and time involved compared with the sum claimed and that which might or might not be ultimately awarded. Moreover, should he grossly exaggerate the sums claimed, he diminishes any possibility of settlement and takes the debate from a matter of principle into avenues of quantum. As to the respondent, the question may well be asked: why should he speed up the arbitration process? Delaying tactics and obstructive behaviour are major factors in prolonging an arbitration and making it more costly.

It is important that the parties remind themselves that they have freely and voluntarily agreed under a commercial contract to have their disputes resolved in an arbitration. It is their money and time that are being expended when an arbitration becomes needlessly protracted. Each party should realise that it has a vested interest in co-operating with the arbitrator and in purposefully complying with the arbitration procedure. For that reason, each party should maintain control over how its case is being presented and should manage its representatives, so that each party's dealings with the other party and the arbitrator are kept within reasonable bounds and so that the conduct of the arbitration proceeds with as little hindrance as possible. Arbitration is by its nature an adversarial process, even under civil law and especially under common law, but nevertheless there is little to be gained in the long run by adopting the sort of practices described above which inevitably make the work of the arbitrator more difficult and, as such, the arbitration itself more time-consuming and more expensive.

It is more difficult to conceive a remedy than to define the shortcomings and to criticise, even more so where such remedy is to apply in general terms to arbitration with some of its objectives being in conflict with each other. However, in international arbitration there has been a recent evolution of 'an internationally accepted harmonised procedural jurisprudence ... and a procedure for the resolution of disputes which cuts right across past and present barriers between different philosophies and legal systems'.[19.81] If arbitration is to regain its full advantages, then some or all of such steps ought to be at least considered, for example:[19.82]

(a) The arbitrator should be selected on the basis of his qualifications which should include: appropriate training; ability to be proactive, manage an arbitration and devise an effective management framework; expertise in the area of the dispute; knowledge of the law of arbitration; time availability; and ability to take an active role in the arbitral process.
(b) In devising his management framework, the arbitrator should make use of as many as appropriate of the following procedures in order to give effect to the desired outcome of speed and economy:[19.83]
   (i) holding case management conferences at the commencement of an arbitration and at other specified stages of the case in order to plan and to monitor time and cost;
   (ii) separating the issues into categories and dealing with predominant issues before others in the hope of promoting settlement prior to the main hearing;
   (iii) arranging for full written submissions covering both fact and law from both parties, supported by all documents relied upon, properly referenced;
   (iv) the submission in writing and exchange of all witness statements and experts' reports in advance of any hearing, supported by documents relied upon;

(v) controlling the number of experts on the issues concerned, the time available for cross-examination and giving consideration as to whether experts should meet prior to writing their reports but after preparing a detailed agenda for the technical issues in question. Furthermore, consideration should be given as to whether experts ought to be examined by the experts of the other party or by the tribunal;

(vi) ensuring that experts know that in giving evidence to the tribunal, they owe their primary duty to the tribunal and not the parties;

(vii) applying limited timetable to oral hearings with equal sharing of time between the parties provided that such limitation does not offend the rules of natural justice which require each party to be given a reasonable opportunity to present evidence and argument and to test the case against him.[19.84] This 'reasonable opportunity' is considered by an eminent lawyer arbitrator in the following terms:

> 'But there is no right to conduct endless and exhaustive examination of witnesses, and there is considerable scope for an arbitrator to adopt a firm approach in determining how far the investigation of particular issues should be taken. In this regard, the arbitrator may need to expressly address the question of what is "reasonable opportunity", and if necessary hear the parties on the question. It is most unlikely that the court would seek to impose any different view'.[19.85]

(viii) in preparation for the hearings, the production of a joint and agreed bundle of documents, chronologically paginated, and in complex and large disputes, a much smaller core bundle;

(ix) at the hearing, taking written statements as evidence in chief and directly proceeding to cross-examination, in between short opening and closing statements from both parties which would then be supplemented by written post-hearing submissions;

(x) contemplating what, if any, other specific requirements peculiar to the particular project or dispute exist as these may vary widely.

(c) In devising his management framework, the arbitrator should be aware of the underlying forces which exist and consider such questions or matters as the following:
  (i) in only very few cases is the virtue of speed of dispute resolution an aspiration of both parties?
  (ii) as to cost, do the parties place a cost restraint on their lawyers?
  (iii) would party autonomy affect the influence that the parties' legal representatives could have on the arbitration procedure to be adopted?

Furthermore, the applicability of the following criteria to the dispute should be considered:[19.86]

(a) *Cost*: The procedure must have regard to cost and ensure that it is not disproportionate to the importance or monetary value of the case as a whole or of individual issues.

(b) *Speed*: The overall period from commencement to rendering an award should be reasonable in relation to the issues to be investigated and the need to bring the dispute to a result.

(c) *Hearing*: The time to be allowed for any hearing should be as short as is consistent with dealing properly with the issues. The rapidity which can be achieved on a hearing will be dependent on the degree of pre-preparation.

(d) *Use of expertise*: It is of the essence of the arbitration process that the expertise of the arbitrator should be used to the maximum possible extent.

(e) *Form and formalities*: No procedures ought to be adopted unless they are positively useful. Forms of procedure should not be adopted merely because they are familiar and/or available.

(f) *Interruptions*: These should be avoided wherever possible; it follows that the proceedings should be steered in a direction which minimises the possibility of unprogrammed delays.

And finally, in recognition of the perceived shortcomings of arbitration in construction disputes in recent years and based on the widely held view that some arbitrators and others may not be fully aware of how best to use the powers conferred by the 1998 ICC Rules to secure cost-effective arbitrations, in 2000 the Construction Arbitration Section of the ICC Commission on International Arbitration's Forum on 'Arbitration and New Fields' was commissioned to examine what techniques had been used successfully to control construction arbitrations. The Construction Arbitration Section canvassed nearly 40 arbitrators with established experience of construction arbitrations and other practitioners in this regard and prepared a first draft report in March 2000. A revised draft report was issued in November 2000 and made available on the ICC website for public comment.[19.87] The final report was published in 2001. The following summary of the main recommendations and suggestions as guidance for arbitrators is given for ease of reference (cross-references are given to the principal paragraphs of the report).

Composition of tribunal

(1) The tribunal should consist of people with proven experience in seeing how an international arbitration about a construction dispute is carried through from start to finish.   [Para 18]

(2) Sole arbitrators or chairmen should know how to write awards and should be able to construct an effective management framework for the arbitration.   [Paras 17–18]

(3) Some familiarity with computers is a distinct advantage, if not a necessity, and basic word-processing skills are now virtually indispensable.   [Paras 15–16]

(4) At the tender stage of projects whose value is not more than, say, US$20 million, the parties should consider whether their interests would be best served if a sole arbitrator were appointed. They should also consider appointing a sole arbitrator if the value of the claim is not large.   [Para 19]

The steps available prior to terms of reference

(5) The tribunal should obtain a chronology of events from each party, especially if there are claims for delay or disruption. It should itself prepare a composite chronology from the material provided by the parties, which it should then send to the parties and take up any discrepancies with them. The tribunal should thereafter maintain the chronology, amending it as the case develops, circulating the revisions and asking the parties to resolve any gaps in it.   [Para 21]

(6) The tribunal should not hesitate to seek information to enable it to create organisational charts, layouts and glossaries or to obtain other clarification where it is needed for the purposes of defining a claim or an issue.   [Para 22]

(7) Amplification of submissions may be needed where, for example, a party has not anticipated a point that has been taken by the other party or which the tribunal sees as likely to arise, e.g.

    (7.1) a point which might affect the jurisdiction of the tribunal, e.g. the identification of a contracting party;

    (7.2) where it is unclear whether notice of intention to claim has been given when it was required by the contract;

    (7.3) where a claim or defence is or may be barred in law, e.g. by prescription or limitation;

    (7.4) where it is not clear whether a claim has been referred to, considered or decided by the Engineer, DAB or DRB, or whether notice of dissatisfaction was given (e.g. under the FIDIC conditions);

    (7.5) where the amount of the claim is unclear.

However a tribunal is not obliged to seek clarification for the purposes of drawing up the terms of reference as in some cases such points should be left until later. In particular a tribunal should be careful before it asks a party to clarify the legal basis of a claim or defence as it may be for the tribunal to determine such a matter or for the other party to submit that there is no such basis.   [Paras 22–23]

The terms of reference

(8) A list of issues will be needed in all but the simplest case, not least because without such a list it will not be possible to decide on the future course of the arbitration. [Para 26]

(9) In construction arbitrations, to define the issues in broad terms may not help the parties (or the tribunal) where clear guidance is needed as to the issues for which proof or argument is required. Extracting those issues at an early stage is the primary task of the tribunal. For these reasons it is sensible for the tribunal to invite each party to set out its own list of issues before drafting the terms of reference. However, a list that is very long would not serve a useful purpose, in which case a working summary should be set out in the terms of reference but refined at the procedural or organisational meeting which must follow. That list should be revised and reissued by the tribunal as the case proceeds in consultation with the parties, e.g. at any further procedural meeting.   [Paras 26–27, 33]

(10) Unless the parties have already agreed on specific procedural rules no attempt should be made to do more than describe the rules in the usual general terms, and to leave them to be worked out at the procedural meeting.   [Paras 28 and 32]

Hearing date and timetable

(11) The tribunal should inform the parties of the likely hearing date (assuming a single hearing or the first of a series of hearings) at the time when the draft terms of reference are circulated so as to facilitate agreement on the date.   [Para 30]
(12) If a date cannot be agreed and has to be decided by the tribunal then it should be the earliest date practicable *for the parties*. Although in most typical construction arbitrations it may well be difficult or impossible to devise a timetable that will meet the six months contemplated by Article 24(1) of the ICC Rules, that period should not be ignored, especially where the dispute has already been processed by contractual dispute resolution machinery and has been the subject of settlement discussions when the points at issue may have been refined, and which will now only be resolved by an award (unless the dispute is of above-average complexity or requires more than one award). In arriving at a date (and also the procedure), the tribunal must of course take into account the financial position of each party (or those supporting a party) so far as it is known or to be inferred and the resources likely to be available to it.   [Paras 30–31]
(13) In arriving at any date for a hearing or any part of the provisional timetable, the tribunal should ensure that the parties have opportunities to take stock and to negotiate and some latitude must be allowed in case there is slippage.   [Paras 31, 46, 47]
(14) Time must also be set aside for the tribunal to be able to read all relevant material before the hearing (or any subsequent procedural meeting).   [Para 48]

Splitting the case

(15) Decisions about splitting a case into parts should be left until it is clear that it will be sensible and cost-effective to do so.   [Para 49]
(16) Before a decision is made about splitting a case, the claimant's case on both causation and quantification should be known so that it is clear how the costs and losses are said to have arisen. The tribunal should be sure that a decision favourable to the claimant on liability and causation will have significant financial consequences. If it is not sure, then the tribunal should not split a case, as one of the key reasons for a split is that a partial award is likely to lead to agreement on the remaining issues. Equally a tribunal must be satisfied that, if a decision were taken to examine the apparent basis in fact or law of a claim and if that basis were rejected, the claimant would not be able to present an alternative fall-back case.   [Paras 49, 50]

Procedure after the terms of reference

(17) The meeting at which the terms of reference are drawn up and signed should not be combined with the first procedural meeting, since discussions about procedure

and in particular the timetable can impede the settlement of the terms of reference, although the first procedural meeting should take place on the same occasion and should follow on immediately afterwards. [Paras 28–29]

(18) In cases of complexity it will be sensible to hold at least one further procedural meeting at which the timetable will be reviewed and difficulties discussed and the list of issues reconsidered. [Paras 33, 47]

(19) In cases where there have already been prior discussions, serious consideration should be given to proceeding directly to proof by requiring the parties to present submissions accompanied by the evidence that each considers necessary to establish its case (in the light of what is then known about the opposing case), both documentary and in the form of attested statements from witnesses. Unless the arbitration is 'fast-track' these submissions should not be submitted simultaneously but consecutively, with the claimant presenting its case first so that the defendant can reply to it. The timetable will therefore need to be fixed by the tribunal. The tribunal may then permit the parties to submit further submissions or evidence either of their volition or to meet requests or directions of the tribunal. Once this stage is complete the tribunal will be better able to draw up a list of the issues as they appear to it and to guide the parties as what is now required. [Paras 34 and 35]

Further working documents and schedules

(20) Some specialists favour the creation of a working document, by exchange between the parties, which records quite briefly the essential elements of each party's case. These 'schedules' are best used for typical claims for changes, for disputes about the value of work and for claims for work done improperly or not at all. Such schedules can be created and can travel on disk or via e-mail so they need not be cumbersome to handle. If fully and properly completed, schedules identify points that are not in dispute or which are irrelevant and thus expose the points that have to be decided. Schedules may also be used to extract the parties' cases on claims for delay (prolongation) and disruption but they require special care to be effective. Such schedule is of particular value where the claim is 'global'. [Paras 37–39]

(21) Even if a schedule is not used at this stage, it may be useful for one to be prepared (by the parties or the tribunal, or both) after the first submission of evidence or before the hearing takes place so as to find out what then needs investigation and decision. [Para 39]

Tests

(22) Where the complaint is about the unsuitability or malfunctioning of plant, equipment or work, the tribunal will need to ascertain what tests have already been carried out and whether the results are agreed or sufficient for the purposes of the arbitration. [Para 40]

(23) The tribunal should sanction tests that have not already been carried out (but it must be sure of the time needed for them). Although in the majority of cases a tribunal will seek to persuade a party of the value of a test, any test required by

it must be non-destructive if made without the consent of the party whose property is affected. The tribunal cannot and should not order any other tests of its own volition. Tests which the tribunal considers necessary and which are not permitted by the party that owns the property will have to be conducted by or for the tribunal elsewhere (if still practicable and likely to be of real value even if carried out off site), either as part of the tribunal's obligation to ascertain the facts (Article 20(1)) or by an expert appointed by it pursuant to Article 20(4). These recommendations apply whether the tests are carried out by an expert appointed by the tribunal or by a party (although once the arbitration has started any test carried out by an independent expert appointed by a party should be carried out jointly with any other expert and under the direction of the tribunal). Similar constraints apply to inspections of the site.   [Para 40]

Visits

(24) It can be very helpful to combine joint tests of plant with a visit by the tribunal, provided that there have been no material alterations since completion and the operating conditions are representative of those contemplated when the contract was made.   [Para 41]
(25) Visits can be expensive and difficult to arrange at a time convenient to the parties (who should be represented on it), especially if the tribunal comprises three people. All visits (and tests) must therefore be justifiable in terms of the real benefits and costs saved.   [Para 41]

Programmes and critical path networks

(26) Claims for delay and disruption require careful handling. It is important that the causative events are clearly identified and that events which occurred but which did not delay progress are isolated. The use of critical path network (CPN) techniques generally facilitates this process and should be required by the tribunal provided that they have already been used in the management of the project. Were CPN techniques not to have been used during the construction period of the project, the retrospective preparation of a CPN is almost always expensive and can produce misleading or unhelpful results. The processes used in preparing a retrospective CPN must be fully transparent. The parties and the tribunal must be informed about the logic at the basis of the CPN, the assumptions made and the data entered. This also requires that they all have access to the software used for the preparation of the CPN and its application.   [Paras 42–44]

Computation and quantification of claims

(27) Where the evidentiary justification of a claim has not already been provided in the statement of case (or prior to the proceedings) a claimant ought to be required to produce the primary documents that verify the amounts claimed, cross-referenced to the statement of case, and in a form that will readily enable the defendant to

know where the amounts come from and why they were incurred. The defendant will then have no excuse for not stating the reasons why liability does not exist or if liability exists, why the amounts claimed are nevertheless not due, e.g. because they were not caused by the events, because they were not incurred or because they were not reasonably incurred. In each case the reasons should be given. [Para 45]

Documents and document control

(28) The common law process of discovery (whether the English version or the United States version) in the form that it is practised in domestic fora has to be justified if it is to apply to an international arbitration. It does not otherwise have a place in ICC arbitrations. [Para 52]
(29) Documents produced by a party should be directly relevant to the issues as defined by the tribunal and should be confined to those which a party considers necessary to prove its case (or to dispose of the case of the other party) or which are referable to such principal documents in order to make them comprehensible. [Para 52]
(30) The tribunal should direct the parties to state at the time of each document's production (and if not then certainly in any pre-hearing submissions) what that document is intended to prove, as there was general agreement that the parties should first be required to ensure that they have produced all the documents which are needed for the proof of the points at issue. [Para 52]
(31) The tribunal may obviously call for further documents at any time in order to fulfil its duty to ascertain the facts. The procedural rules ought also to allow a party to request additional documents from another party and, if they are not provided, to seek an order from the tribunal, at which stage the legitimacy or reasonableness of the request or of the refusal will be decided. [Para 54]
(32) The tribunal should fix a cut-off date by which no further documents may be produced by any party, unless required by the tribunal or, exceptionally, permitted by it following a reasoned justification for the late introduction of the documents. [Para 55]

Document management

(33) Material such as pleadings, submissions, extracts from the key primary documentation, witness statements and reports from experts should be loaded on a CD-ROM. [Paras 16, 53–55, 64]
(34) The tribunal should require the parties in any event to organise the documents so as to avoid duplication and to enable them to be accessed easily. Such a procedural direction will need to be clear and precise since this useful practice is not yet widely recognised. For example, whether photocopied or on disk, inter-party correspondence (including instructions, requests for instructions and the like), the agreed records of meetings, programmes, agreed summaries of measurements, agreed summaries of valuations, drawings and other technical documents, ought to be contained in separate indexed files with the pages individually numbered so that additions can be made simply. [Paras 53–55]

Witnesses

(35) Subject to legal requirements and the wishes of the parties, evidence that is not contained in a document and which is necessary in order to prove or disprove a point in issue must be presented by means of a written statement from the witness, in that witness's own words (unless the witness is not able to do so), verified and signed by that witness. Where the evidence is not in the language of the arbitration an accredited translation must be provided.   [Para 56]

(36) It is usually sensible to provide for supplementary or additional statements of evidence of fact necessitated by the intended evidence of the other party (including expert evidence) to be exchanged within a short time of the principal statements so that all the evidence is in writing.   [Para 56]

(37) All witness statements should be exchanged in good time before the preparation of any pre-hearing submissions.   [Para 56]

Experts

(38) Where one or more members of the tribunal have been nominated or appointed for their expertise there should normally be no need for the tribunal to duplicate that expertise by appointing its own expert, unless the assessment of part of the case might take a considerable time. A decision of this kind has or may have important implications so it should normally be discussed with the parties.   [Paras 15, 58]

(39) It is always prudent to clarify whether or not expertise is required, why it is required, by whom it will be provided and when.   [Para 59]

(40) The tribunal should only appoint its expert if it is necessary since the costs of the tribunal's expert have to be borne by the parties. On the other hand in many cases it will be cost-effective to do so, for the opinion of that expert might render unnecessary any further expertise or it may identify the points upon which evidence or reports from witnesses or experts may be required.   [Para 59]

(41) The tribunal ought to decide whether it will appoint its own expert before it issues the provisional timetable under Article 18(4), since the timetable will be affected by the work of that expert.   [Para 59]

(42) The tribunal may need to differentiate between truly independent experts and consultants retained by the parties to assist in the preparation of the claim who may produce reports and give evidence in the arbitration and, for example, to be sure that any information obtained by such a consultant expert from a party and used in the formation of his evidence and opinions has been communicated to the other party and to the tribunal.   [Para 60]

(43) The tribunal ought either to draw up the terms of reference of the parties' experts (on the basis of the issues known to it) or to require the parties to agree a statement of the issues and of the facts (both agreed and assumed, e.g. as set out in the witness statements) upon which expert evidence is required. If the tribunal does not draw up the experts' terms of reference the experts should provide the tribunal with their terms of reference or instructions as received from their clients (subject to privilege) so as to ensure the expert is properly directed and motivated and that the opinions are reliable.   [Para 60]

(44) The experts, if independent of the parties, should discuss their views with each other preferably before preparing their reports, e.g. at a meeting (possibly chaired by the tribunal or, if the parties agree, a designated member) since most independent experts eventually agree about most things.   [Para 60]

(45) The tribunal must ensure that it is clear whether or not agreements between the experts bind the parties. If the tribunal were to chair the discussion it may be difficult for a party to question such an agreement. Reports must be confined to questions or issues where no agreement can be reached.   [Para 60]

General

(46) Whenever appropriate, all applications about procedural matters and which do not involve questions of substance should be made and decided by correspondence or telephone without a hearing.   [Para 46]

(47) Submissions should be numbered or arranged to match the submissions of the other party.   [Paras 34, 62]

(48) All submissions prior to a hearing should be in writing.   [Paras 61–62]

(49) All submissions should be full but concise and they should be delivered at the earliest possible occasion.   [Para 62]

Hearing

(50) The tribunal should either require the parties to decide how the time available within the period of the hearing should be allocated (in which case the parties will then be held to their agreement) or the tribunal should itself decide and adhere to a strict timetable, unless it is not just to do so. The tribunal must treat each party fairly, but that does not mean equality in terms of witness time, as opposed to the time for statements or submissions.   [Para 63]

(51) Prior to the hearing the parties should be required to agree which documents will be needed at the hearing, and which (if not already conveniently available) should be put on CD-ROM or assembled in the form of files (see above). Pre-hearing submissions, witness statements and any reports from experts should be cross-referenced to the documents.   [Para 64]

(52) Either only a nominal time should be allowed at the hearing for oral opening statements or there should be no opening statements.   [Para 62]

(53) Factual witnesses should be heard before the consideration of experts' reports since the questioning of a factual witness may require an expert for the modification or withdrawal of an opinion or provisional conclusion.   [Para 65]

(54) The time available at a hearing need not be used for closing submissions as these are frequently best presented in writing within a short period after the conclusion of the hearing. The period within which closing written submissions are to be delivered should be set by the tribunal well before the hearing (e.g. in the provisional timetable) and certainly in good time prior to its conclusion. No further submissions will be considered after that period.   [Para 66]

(55) The tribunal should make it clear that no new facts or opinions will be admitted once the hearing has taken place, unless specifically requested by it.   [Para 66]

This guidance, although drafted for arbitrators, would make useful reading also for parties who are either at the stage of contemplating an arbitration clause in a potential contract or the referral of a dispute to arbitration, whether or not the ICC Rules are to be adopted. The parties could then ensure that such recommendations are adopted by agreeing that they form part of the rules under which the arbitration is to be conducted.

19.12 Concluding remarks

This chapter provides an overview of the arbitral process and the significance of arbitration in international construction contracts, in general and also with particular reference to the FIDIC contract under the Red Book. It highlights the many advantages that arbitration has over other dispute resolution methods and in particular over litigation, the only other judicial process available to the parties. In Section 19.8, the specific features and some of the problems encountered in arbitration under the provisions of clause 67 of the Red Book were discussed. These features and problems combined with general dissatisfaction with the process of arbitration, in particular the excessive costs and delay, gave birth to a new alternative arbitration clause to that used under the Fourth Edition of the Red Book. This alternative was also needed to resolve at least some of the disputes which hitherto invariably ended in arbitration.

The new alternative arbitration clause is analysed and discussed in detail and a commentary is given on some of its aspects which raise questions to which there appears to be no answer given. In both versions of clause 67, the ICC Rules are stipulated if the dispute ends in arbitration. Thus, this chapter examines the ICC system for arbitration, the role of the International Court of Arbitration of the ICC, the ICC Rules themselves and their advantages and shortcomings.

The chapter then reverts to the existing essential problems facing arbitration in recent times and, drawing from successful experiences in procedures in international arbitration and the evolving jurisprudence, provides some guidance as to what ought to be done to reinstate the two most important advantages of the arbitral process, speed and economy.

In the following chapter, alternative dispute resolution (ADR) methods of dispute settlement, whether the letter A stands for alternative, appropriate, amicable, available or affordable, are discussed in detail with particular emphasis on conciliation and mediation. It is worth noting that the new alternative version of clause 67 retains the provision of the original sub-clause 67.2, providing for amicable settlement as an obligatory step between the Dispute Adjudication Board's decision and arbitration.

Chapter 20

Amicable Settlement Using Alternative Dispute Resolution

'Discourage litigation. Persuade your neighbours to compromise whenever you can. Point out to them how the nominal winner is often a real loser in fees, expenses and waste of time.'

Abraham Lincoln

20.1 Introduction

As discussed in Chapter 19, the main advantages of arbitration have traditionally been privacy, speed of resolution, cost effectiveness, convenience, finality, certainty and choice of tribunal. However, in recent years, some unhappy experiences in arbitration, especially in the construction field, have diminished the effect of these advantages, or at least some of them, in particular speed of resolution and cost effectiveness. The technical and legal journals contain examples of such experiences which have left some employers and contractors disenchanted with the arbitral process and led them to search for a more attractive method of dispute resolution.[20.1] This is particularly so in the United States where many large corporations and insurance companies have signed pledges to consider amicable methods of resolution when disputes arise. In the UK, the relatively new standard form of contract developed by the Institution of Civil Engineers, the Engineering and Construction Contract 'NEC', went into a second edition with no reference to arbitration as the ultimate method of dispute resolution. Instead, the authors of the second edition yielded to the pressure of protests on the erosion of the benefits of arbitration by referring any dispute that may arise to a 'tribunal', a term which could mean any dispute resolution forum.[20.2] This drastic measure was taken at the same time as judicial systems in some jurisdictions are moving towards a more efficient administration in the courts by borrowing from some of the successful, newly emerging jurisprudence of international arbitration.[20.3] Those involved in the arbitration field will have to take note of these developments if arbitration is to remain the leading method of dispute resolution in construction contracts.

One possible solution to the problem of the diminishing benefits of arbitration which was adopted by FIDIC in the fourth edition of the Red Book, in 1987, was the introduction of the idea of amicable settlement as a prerequisite step to arbitration. The authors of the Red Book, to their credit, adopted a more sensible approach than that of a total rejection of the arbitration process. Sub-clause 67.2 of the Fourth Edition of the Red

Book provides that arbitration should not be commenced unless an attempt has first been made by the parties to settle the dispute amicably. A period of 56 days is allowed for such an attempt to be made and this period is extendible by agreement of the parties. The 56-day period starts the day following that on which notice of intention to commence arbitration under sub-clause 67.1 is given. Arbitration may then be commenced, whether or not any attempt at amicable settlement had been made.

Since the publication of the Fourth Edition of the Red Book in 1987, many institutions followed the example of FIDIC. The Institution of Civil Engineers in London introduced conciliation procedures in its Minor Works Contract Form, first published in 1988. Its success prompted the introduction of such procedures in subsequent forms of contract: the sixth edition of the ICE Form and the ICE Design and Construct Contract. In Ireland, conciliation has now been introduced in the fourth edition of the Irish Conditions of Contract for works of civil engineering construction as a first mandatory step in resolving contractual disputes. The more recent forms of contract published by FIDIC in 1995 and 1999, and that for the sub-contract, also incorporate amicable dispute settlement methods in their conditions.

In addition, FIDIC introduced in November 1996 a Supplement to the Fourth Edition of the Red Book, as reprinted in 1992, which contains in its Section A an alternative version to clause 67 of the Red Book, providing for the establishment of a Dispute Adjudication Board which is in itself another form of an alternative method of dispute settlement. This new alternative version of clause 67 is discussed in detail in Chapters 19 and 26. The concept is also briefly discussed below in Section 20.7, but a whole chapter is provided on Dispute Boards in Chapter 26.

20.2 Methods of dispute settlement

Many methods of resolving commercial disputes have evolved over the centuries. The most familiar in civilised societies are:

(a) negotiation;
(b) mediation;
(c) conciliation;
(d) Dispute Boards, Dispute Review Boards or Experts, Claims Review Boards and Dispute Adjudication Boards or Experts;
(e) adjudication;
(f) mini trial;
(g) Pre Arbitral Referee Procedure (ICC);
(h) The ICC Expertise Procedure
(i) arbitration; and
(j) litigation.

The last two of the above methods lead to a solution that is imposed on the parties in dispute, through a court judgment or an arbitral award. The other methods are amicable and so the parties have a say in, and can control, the outcome of the dispute. They are more popularly known as 'alternative dispute resolution' or 'amicable dispute resolution' methods, and are usually referred to by the acronym ADR, the letter 'A' standing for alternative to litigation and arbitration. However, the letter 'A' could equally refer to appropriate, applicable, available, or affordable, depending on the context of its use. In this chapter, the first eight methods listed above are discussed in detail.

It should be pointed out that the best time to decide on the mandatory use of an amicable method of dispute resolution and the rules to be used for such method, is at the time of writing the contract agreement. The reason is that once a dispute has arisen, any proposal by one party towards amicable settlement may be perceived as a sign of weakness by the other party or parties and may lead to the opposite result: entrenchment. A further advantage of using amicable settlement as a mandatory step before reference to arbitration is the avoidance of any possible blame being attached to the decision maker who proposes amicable settlement of a dispute instead of a more adversarial method. These two reasons were influential in the decision by FIDIC to make amicable dispute settlement an obligatory step to be taken prior to arbitration.

Amicable dispute resolution methods are successful when the parties believe that the disputes in question are not 'black and white', and while good faith and trust still exist between them. On the other hand, these methods are generally less successful when emotions are running high and where the parties have no interest in a prompt settlement. In general, however, the ordinary person is not skilled in the art of negotiation and a third person is usually engaged to facilitate the dispute resolution process.

With the exception of direct negotiation, the above methods of dispute resolution differ from arbitration in that they involve a process whereby a third party is simply called upon to facilitate the process and to assist in reaching a settlement by issuing a non-binding evaluation of the dispute and a recommendation of how it could be resolved, or in the case of a dispute adjudication method a decision that is temporarily binding until and unless it is revised in a subsequent forum.

The advantages of these alternative methods can be summarised as follows:[20.4]

(a) Whilst the procedure may or may not affect the amount of settlement, it will more than likely affect the cost of achieving it.
(b) The parties are in greater control of their own destiny, thus avoiding any of the uncertain consequences of litigation or arbitration.
(c) The procedure tends to preserve business relationships and avoids the possibility of one party being viewed as the loser.
(d) The process is much faster than litigation or arbitration as it could be completed in a few days or a few weeks if the dispute is a complex one requiring the preparation of further documents or investigations.
(e) Arbitration or litigation may be pursued should the amicable method fail to produce the desired result. It is important, however, to remember that a written agreement should be signed by the parties to prevent information disclosed during the process from being used in subsequent litigation or arbitration. The parties must, however, be careful not to stipulate confidentiality in such broad terms that it becomes too restrictive to proceed to arbitration or litigation should settlement not be achieved or should it be found necessary to enforce the terms of any settlement agreement.

20.3 Direct negotiation

Wherever there is human endeavour, there is conflict. Conflict may be a conflict of interests, conflict of needs, conflict of opinion or simply a conflict of a desired outcome to a previous agreement. In such circumstances, the simplest and cheapest method to resolve the conflict is by negotiation. Direct negotiation between parties in dispute without the intervention of a third party is perhaps the most readily available method of dispute resolution and the most effective. It is effective because of the speed and economy of procedure with which a dispute may be resolved. In its simplest form, it consists of successively taking, and then giving up, a sequence of positions. Most, if not all, people negotiate on a daily basis without realising it, getting involved in decisions which influence them rather than accepting those which are dictated by others. Thus, the ordinary person constantly negotiates at home, at work and in other various daily transactions.

Negotiation may be defined as a process where two or more parties in conflict attempt to reach an agreement to settle their differences and where that agreement is such that all the parties involved are prepared to live with it and accept it.

Although the simplest and quickest method of solving disputes is through negotiation, it is not in fact easy, especially if there is a clash of personalities behind the dispute, or if in the parties' opinions there are matters of principle at stake. Furthermore, until any of the other methods of resolution have been invoked, the costs involved are rarely appreciated. In some cases, parties embark on litigation simply because they want to have their day in court; in others the parties are simply badly advised.

Negotiation may mean an element of trading or bargaining leading to a reduction in the parties' expectations to a level which is acceptable to all of them. The reduction in one party's expectation may however be greater than that of the other.

Under the FIDIC forms of contract, negotiation is conducted under the 'due consultation' requirements in the specific form of contract. Under the Fourth Edition of the Red Book, there are many instances where due consultation is required to be conducted by the engineer with the employer and the contractor, see section 9.3.1 above.

20.3.1 Negotiators

Negotiators who are involved in what is scientifically known as 'positional bargaining' are sometimes referred to as either soft or hard negotiators. To a soft negotiator, the goal is agreement; the other party is to be trusted; offers may be made and a bargaining position may be easily changed. To a hard negotiator, the goal is victory; others should be distrusted; no compromise should be offered; positions should be entrenched and if offers are rejected, threats may be used.

In its more sophisticated forms, direct negotiation may involve methods other than positional bargaining, and would certainly require the negotiator to be skilled and experienced.[20.5] As well as possessing expert knowledge of the matter in dispute, a skilled negotiator should have the ability to:

(a) listen to the other party and understand the point of view and the case being made;
(b) recognise the needs of the other party and identify his interests;

(c) express his thoughts clearly, both orally and in writing;
(d) think clearly and rapidly under pressure;
(e) persuade others;
(f) be patient;
(g) be flexible; and
(h) have the ability to control and to hide emotions.

20.3.2 *Distinguishing features of direct negotiation*

Direct negotiation which is successful in resolving disputes is distinguished by three main features:

(a) The difficulty of initiating direct negotiation has been overcome by one of the parties at an early stage after the dispute has arisen, despite the commonly held perception that to do so might be interpreted as a sign of weakness by the other party.
(b) The negotiators of both parties are skilled in the art of negotiation, knowledgeable in the subject matter of the dispute and experienced in the field.
(c) The agreement reached, if such agreement is at all possible, is efficiently produced with results which meet the legitimate interests of each disputant in resolving the conflicting interests fairly with an improved relationship.

Thus, it can be said that negotiation is a skill. Few of us are born with it and most of us become trained in its ways and means through day-to-day interaction with those around us. However, even if we were born with it, training is essential if we are to succeed in achieving a positive outcome in all the transactions we face.

20.3.3 *When should negotiation be used and what are the steps?*

Negotiation may be conducted by one person or a team of people from each of the parties in dispute whenever there is a desire to continue an existing relationship and whenever the cost is of concern. To resolve conflict through negotiation, the following steps or at least some of them should be used:

(a) *Find out the facts:*
    (i) What is the real issue?
    (ii) What is the history?
    (iii) What is really going on?
    (iv) Who are the people really involved?
(b) *Identify the needs of both parties:*
    (i) Know your own feelings (whether you are involved or not)
    (ii) What do the people involved really want?
    (iii) Why do they want it?
(c) *Make an assessment of the conflict:*
    (i) Is it a manageable size in the number of issues to be resolved? If not, break up the larger issues into a series of manageable elements.

(ii) What level has this conflict reached? If it has escalated beyond the level of maintaining trust, then more careful attention is needed.
(iii) Can you do anything? If you become involved, do you stand a reasonable chance of contributing to the resolution process?
(d) *Look for solutions:*
   (i) Help groups to clarify and test understanding.
   (ii) Facilitate stating of needs and desired outcomes.
   (iii) Negotiate towards agreement.
(e) *Agree action:*
   (i) Check that both parties have the same perception of the outcome.
   (ii) Follow up action agreed.

If one is asked why conflicts are not always resolved through negotiation, the answer may be traced to a lack of confidence in the success rate of winning through negotiation; difficulty in taking the first step; lack of confidence in one's ability to negotiate; and the fear that the other party is a better negotiator. Therefore, there is a need to increase confidence in the ability to negotiate in order to allay, or at least dampen, fears and to create an atmosphere of trust between the parties in conflict.

20.4 Mediation

Should negotiations fail between the parties, then a third party may be called upon to assist in finding common ground for compromise. This process can be either mediation or conciliation in that both mediation and conciliation are voluntary forms of dispute resolution where a 'neutral' party is appointed to facilitate negotiations between the parties in dispute and to act as a catalyst for them to reach a resolution. However, the difference between mediation and conciliation lies in the role played by the neutral party. In one, he simply performs the task of persuading the parties in dispute to change their respective positions in the hope of reaching a point where those positions coincide, a form of shuttle diplomacy without actively initiating any ideas as to how the dispute might be settled. In the other method, the neutral party takes a more active role probing the strengths and weaknesses of the parties' case, making suggestions, giving advice, finding persuasive arguments for and against each of the parties' positions, and creating new ideas which might induce them to settle their dispute. In this latter method, however, if the parties fail to reach agreement, the neutral party himself is then required to draw up and propose a solution which represents what, in his view, is a fair and reasonable compromise of the dispute. This is a fundamental difference between mediation and conciliation.[20.6]

Unfortunately, the two terms are used interchangeably and there is no universal agreement as to which of the two methods is mediation and which is conciliation. For the purpose of this book, the first method is referred to as mediation and the second is referred to as conciliation. This choice coincides with that made by various professional institutions in Europe where detailed and comprehensive rules of procedure have been published under the title of conciliation procedures.[20.7]

The remainder of this present section considers mediation, while the following section deals with conciliation.

Agreement on mediation may be made at the contract stage prior to any dispute taking place.[20.8] However, there should also be agreement on the forum of dispute resolution should the mediation process fail, otherwise the dispute would end in litigation with all its disadvantages.

Mediation is only marginally more expensive than direct negotiation, but it has the advantage that it exposes senior management to an independent view, which is extremely valuable whether it be adverse or favourable. In this regard, an adverse viewpoint may enable senior management to separate the people involved in the events which gave rise to the dispute from the dispute itself, and from their method of handling the problems which led to the dispute.

The advantages of mediation include informality, speed and economy, but more importantly perhaps, it often leads to an agreed settlement between the parties rather than an imposed award or judgment.

However, mediation has little chance of success unless the parties wish to mediate and have a considerable degree of mutual trust in each other's integrity and willingness to resolve the dispute. Apart from this, the main disadvantage of mediation is that the views of the mediator and any conclusion he may reach are not enforceable. Because of this, the success of the process of mediation depends to a large extent on the skills of the mediator. A definition of a mediator's role has been concisely given by the American Arbitration Association under six headings as follows:[20.9]

(a) The reconciliator, who brings parties together in order to engage in face-to-face discussions; opens channels of communication; and defuses hostility.
(b) The facilitator, who keeps discussions going by providing a neutral ground, arranging meetings, offering to chair them, helping to shape the agenda, simplifying procedures.
(c) The resource expander, who helps to gain access to necessary factual and legal information having an important bearing on the dispute; cuts through bureaucratic red tape.
(d) The interpreter/translator, who makes sure that each party understands what the other is saying; and increases perception and empathy between the parties.
(e) The trainer, who instructs the parties how to negotiate more effectively with each other through probing and questioning.
(f) The reality tester, who gets each party to look at how the other side sees the problem; makes each side think through and justify its facts, demands, positions and views; encourages the parties to assess the costs and benefits of either continuing or resolving the conflict; makes each party consider and deal with the other's arguments; raises doubts on rigid positions; and explores alternatives.

It is evident from the above that the mediator is to a large extent filling the position of the negotiator on both sides of the dispute and that a skilled mediator would have to possess the dexterity required in an expert negotiator, see Section 20.3 above.

Although mediation is not a regulated process, some mediation rules do exist and are published by a few organisations; for example, The Construction Industry Mediation

Rules are published by the National Construction Industry Arbitration Committee in the United States, and the Hong Kong Government Mediation Rules are administered by the Hong Kong International Arbitration Centre.[20.10] The Rules cover the whole process, from inception through the appointment of a mediator to the timetable and the agreement. The mediator is generally chosen by the parties as an impartial individual in whom they both place their trust and whom they believe will act in a completely independent manner. His fees are payable by both parties in equal shares prior to the handing out of his opinion. Although some rules are explicit in barring legal representatives from the mediation process,[20.11] one experienced point of view suggests that 'a party should proceed without reliance on an attorney (lawyer) only if an attorney thinks it a reasonable approach'.[20.12]

20.5 Conciliation

Unlike an arbitrator, a conciliator is not empowered to make a binding decision and this fact forms the main distinction between arbitration and conciliation. As with mediation, discussed in the previous section, this is also considered to be the main disadvantage of the conciliation procedure. However, it is notable that if a settlement agreement is reached as a result of mediation or conciliation, then that agreement is easier to enforce than an arbitrator's award because it would have been concluded through the parties' own choice.

Conciliation is a more formal process than mediation and it could generally involve the engagement of legal representatives, thus making it a more expensive process than mediation. There is, however, the added advantage that should no amicable solution be reached, the conciliator has the duty to attempt to persuade the differing parties to accept his own solution to the dispute. In fact, the description given in item (f) of the definition of the role of a mediator by the American Arbitration Association, referred to above, is more in line with the process of conciliation, as known in Europe.

There are a number of institutional rules for conducting conciliation. Amongst the more popular are:

(a) ICC Conciliation Rules, 1 January 1988. However, it should be noted that these Rules are now superseded by the ICC ADR Rules, in force as from 1 July 2001, see the Foreword to the 2001 set of Rules;
(b) The Chartered Institute of Arbitrators Conciliation Rules, 1 July 1981;
(c) The UNCITRAL Conciliation Rules, 4 December 1989;
(d) The ICE Conciliation Procedure, 1994;
(e) The Euro-Arab Chambers of Commerce Rules;
(f) The International Centre for the Settlement of Investment Disputes, ICSID Conciliation Rules; and
(g) The IEI Conciliation Procedure 2000.

It is worth noting that under most rules, neither a conciliator nor a mediator may later change his role to that of an arbitrator without the agreement of the parties. Such a possibility would inhibit the parties from confiding in the conciliator or mediator.

20.5.1 What is conciliation?

Similar to mediation, conciliation is a voluntary form of dispute resolution where a neutral party, the conciliator, is appointed to facilitate negotiation between the parties in dispute and to act as a catalyst for them to reach a resolution of their dispute.

Whilst it is generally accepted that conciliation is a non-binding form of dispute resolution, an agreement reached between the parties in dispute following a conciliation process becomes binding and has a better chance of being honoured than an arbitration award. Furthermore, conciliation allows the parties the freedom to explore ways of settling the dispute without commitment until they are ready and are prepared to commit themselves.

A party which is unhappy with the conciliation proceedings or with its outcome can opt out and proceed to arbitration or litigation, depending on the terms of the contract.

20.5.2 Why conciliation?

Like mediation, conciliation is considered to be one of the most informal dispute-resolution methods after direct negotiation, providing two important advantages over the more formal methods of dispute resolution. These are flexibility and choice. Flexibility does not, of course, mean lack of control since the process is always governed by rules of procedure and conduct, which are usually agreed upon by the parties in advance by inserting a reference to them in their contract. The rules may be institutional rules, or ad hoc rules chosen by the parties for a particular case.

As to the advantage of choice, conciliation, like mediation, is a process whereby the parties involved continue to be in control of the final outcome of the dispute resolution process. Thus, it is more likely that working relationships, whether business or otherwise, will survive this process, whereas they are unlikely to survive litigation or arbitration where the decision is imposed upon the parties by a judge or an arbitrator. Whether the outcome of a conciliation process is conceived by the parties or by the conciliator or by both, it ultimately has to be sanctioned by the parties themselves.

The ease with which multi-party disputes can be accommodated by conciliation is another important example of the flexibility and adaptability of the process. Such disputes cause immense problems in arbitration. For example, a neutral facilitator helped settle the 1986 Dupont Plaza Hotel fire case in San Juan, Puerto Rico, which involved more than 100 defendants and their insurers, thus saving an estimated $60 million in legal costs.

20.5.3 When should conciliation be used?

Conciliation should be used in, or is suited to, any dispute in which any or a combination of the following exist:

— a desire for a negotiated settlement which can be approved and sanctioned by those in charge of negotiations;
— a desire or a need to maintain an existing relationship;

— the need for privacy and confidentiality (this is one of the most attractive qualities of the conciliation process in that private conversations with the conciliator during private meetings are not divulged to the other side and are considered to be privileged and confidential so that the parties can feel free to confide fully in the conciliator);
— time is a matter for concern;
— the issues are complicated involving highly technical and interlinked problems;
— the costs are of concern; and
— there are more than two parties.

However, mediation or conciliation are generally not suitable where the following circumstances exist:

— there is a need for an authoritative interpretation of the law;
— there is a risk of harm to reputation, whether of an individual, a company or a product;
— there is a need to discourage similar future disputes; and
— the issues involved are of a criminal, constitutional or civil rights nature.

20.5.4 Who should be a conciliator?

The calibre of the conciliator is one of the most important factors contributing to the success of a conciliation process for a particular dispute. It is accepted that the ability, knowledge, experience and training of the conciliator contribute significantly to the success of the process. Bearing in mind that the instrument is only as good as the musician playing it, it is suggested that a skilled conciliator ought to possess as many as possible of the following attributes:

— honesty;
— impartiality, independence and neutrality;
— some knowledge of the law and detailed knowledge of the relevant conditions of contract and of the topic and the nature of the matters in dispute;
— ability to gain the trust and confidence of the parties;
— in international disputes, it is necessary to possess some expertise in cross-cultural communication and a sensitivity towards the customs and habits of various societies;
— the ability to analyse complex legal and technical issues quickly and logically;
— excellent communication and negotiating skills;
— poise;
— stamina and confidence;
— patience and tolerance;
— good listening skills; and
— a calm demeanour.

The above is not an exhaustive list. Specifically, the conciliator must ably perform at least three critical functions:

(a) a moderator for presentations of facts between the parties, keeping control of what will inevitably be an emotionally-charged atmosphere involving exchanges from those who were present during the execution of the contract;
(b) playing devil's advocate in private meetings with each party individually, pointing out weaknesses and strengths of the case as presented by them. This must however be done with a candour and objectivity; and
(c) a facilitator for actual negotiations, not only to carry back and forth the offers and counteroffers, but also to receive in confidence each party's bottom line.[20.13]

In this connection it must be emphasised that the role of the conciliator is very different from that of an arbitrator. Unlike an arbitrator, a conciliator has no power to impose a decision and he can only guide the parties to a settlement. The conciliator must inform himself of the case prior to meeting the parties to such an extent that he can engage in the above three functions from the start. The conciliator must also take the initiative in providing possible solutions and how to achieve them. He must be capable of educating the parties regarding the process of conciliation, ensuring that they can comply with the proposed procedures and requirements.

20.5.5 Who should attend the conciliation?

It is particularly important that the representatives from all parties at a conciliation should be of such calibre that they possess the decision-making authority to accept or reject proposals for settlement of the matters in dispute. They should be in possession of any authorisation required by their company or organisation to sign a document committing that company or organisation to the extent required. Such persons include external parties, for example insurers.

Of course, all those involved in the issues in the conciliation should be present to throw revealing light on the matters requiring resolution. No one should be excluded who can make a meaningful contribution or who has an interest in the outcome.

20.5.6 The conciliation process

In conciliation it is necessary for each party to carefully prepare a document containing the following material:

(a) *The facts:* A factual narrative of the events leading to the issues in the conciliation. One of the parties, usually the party initiating the process, should prepare a bundle containing documents which can be submitted jointly, such as the contract, the specification, any drawings necessary for understanding the issues involved, etc.
(b) *The issues:* It is necessary to identify the issues between the parties as clearly as possible.
(c) *The legal principles:* The legal basis supporting the case made by each of the parties should be set out in as clear a language as possible.
(d) *The remedy or remedies required and the time frame* within which the conciliation process may or should be conducted.

Like mediation, conciliation can take place in any location agreed between the parties, otherwise in a neutral location, which can accommodate both private and joint meetings with the conciliator.

Simple conciliation may take only a few hours. However, in construction the issues are generally complex and lengthy and may often involve short sessions over the course of a few weeks. For example, in multi-party mediation/conciliation the process could take several months with various parties meeting over consecutive periods of some weeks, each meeting lasting a few days.

There is no set format for the actual conciliation process. As a general rule, however, all conciliations involve a series of joint and separate meetings with the parties.[20.14]

The first step is a joint meeting in which all the parties in dispute are required to meet with the conciliator, who should describe the conciliation process and review the ground rules for participation, behaviour and confidentiality. The participants are expected to discuss matters such as: the role of the conciliator; identifying who will represent each party during the discussions; identifying who has authority to sign a final settlement; and the identification of documents to be exchanged. If litigation over the matter is pending, the parties and the conciliator may discuss what activities will be suspended and whether the court should be informed of such suspension and of the conciliation taking place.

At this point, if any side has any doubts about the conciliator's role or the procedure, that party can suggest modifications in the procedures or rules. The applicable conciliation rules are aimed at creating an atmosphere of co-operation and respect. The rules can include such matters as agreeing to have only one person speak at a time, the identity of that person from each party, setting an agenda, limiting the scope of the negotiation, defining the role of the conciliator, defining the use of private meetings, agreeing which documents are to be submitted to the conciliator, if any, establishing how confidentiality will be maintained, and stipulating how the sides will respond to media enquiries, if any.

This initial joint meeting also serves as an open forum for the parties to explain their positions and express how they believe the case should be resolved. At this stage, the conciliator begins to gather the facts, becomes more familiar with the case and develops his assessment of the interests and perceptions of the parties. Each party also has an opportunity to rebut the submissions of the other party under the supervision of the conciliator, who may ask questions. Throughout the process, participants should remember that patience is the key to a successful outcome. Detailed, and sometimes lengthy, presentations of the facts are crucial at the outset to inform the conciliator.

If the conciliator requests written statements from the parties, and/or documents, then another joint meeting may be scheduled to give each side an opportunity to prepare and gather such material as may be necessary. In this case, the parties may wait until the second joint meeting to discuss their positions regarding the dispute in detail.

After the initial airing of rules and views, the conciliation enters the problem-solving phase. During this phase, the conciliator may hold one or a series of private meetings or, in American terminology 'caucuses', with each party. The conciliator shuttles between the parties, probing each side's position, asking questions, establishing what each party is concerned about and its most desired and least desired outcome, assessing the merits of each argument, narrowing the issues by identifying what is important in an effort to understand the case from each side's perspective. At the same time, he works to defuse any hostility, in part by reframing the issues in objective language and in part by actively cooling down overheated discussions.

A major advantage of conciliation is that this process allows the parties to 'vent' their frustrations and engage in celebrating their 'day in court'. This therapeutic interaction often helps move parties towards settlement.

When the conciliator is satisfied that the private meetings phase is complete, the process is ready to be brought to an end. This means that the parties are ready to come together in a final joint session, or a number of sessions, to negotiate with each other the terms of an agreement. Prior to this final session, the conciliator usually helps each side develop several settlement options and a negotiation strategy which will move the dispute towards settlement. As stated earlier, the conciliator does not impose a settlement on the parties, even if he has been asked to render an opinion. The conciliator simply works with the parties and guides the process to remove barriers and smooth the way to settlement.

Any party may float a settlement proposal, which would usually have been discussed with the conciliator. If it has not been discussed in advance, the conciliator may interject an opinion if he feels the offer is inappropriate. If the conciliation involves many parties, it is more likely that the conciliator would initiate the settlement proposals.

Once a 'reasonable' offer is on the table, it becomes the basis for negotiation. If the parties are having difficulty achieving an agreement, the conciliator may request more private meetings with specific people from both sides to help bridge the gap or gaps.

If, at the end of a conciliation, a substantial difference persists over monetary issues, the conciliator may try to encourage the parties to make a final effort to close the gap, by allowing the conciliator to choose a monetary award between the last demand of the claimant and the last offer of the respondent.

If the case is settled in a way that is agreeable to all sides, the conciliator and/or the parties will draft a document spelling out the agreement and stipulating how it will be implemented. It may be circulated and edited as necessary.

As stated earlier, the outcome of a conciliation is non-binding. A proposed settlement may be rejected by any of the parties involved. If this occurs, the party objecting to the proposals may consent to work towards new settlement proposals, or it may give up and proceed to arbitration or litigation. However, it is important to note that conciliation more often than not leads to a durable settlement because the agreement has been forged by the parties themselves.

If all parties agree to, and sign, a settlement agreement at the conclusion of the conciliation, the parties are bound to uphold that agreement. The settlement agreement is a contract; an action alleging breach of contract may be brought if an agreement is not honoured. If litigation is pending, the settlement should be filed with the court so that it will be enforced without requiring a separate action for breach of contract.

If settlement is not achieved, then certain procedures require the conciliator to issue a recommendation which must state his solution to the dispute referred to him for conciliation. The recommendation must not disclose information which any party had provided in confidence. It should be based on the conciliator's opinion as to how the parties can best dispose of the dispute between them and need not necessarily be based on any principles of law. The conciliator is not usually required to give reasons for his recommendation, but he may choose to do so if he considers that such reasons would be helpful to the parties. If the recommendation is not rejected within a prescribed period, then it becomes final and binding on the parties, but if it is rejected, it becomes a nullity. In the latter case, the terms and provisions of the recommendation and any reasons given together with any disclosures specifically made for the purpose of the conciliation remain privileged and confidential and cannot be divulged or used in evidence in any subsequent forum.

20.6 Mini-trial procedure

The mini-trial procedure was developed by the Zurich Chamber of Commerce to meet the demand for alternative methods of dispute resolution to the traditional methods of litigation and arbitration. In the introductory remarks to its rules for mini-trial procedure, the salient features of the procedure are described as follows:[20.15]

(a) its aim is to settle disputes with the active co-operation of senior corporate officers of the parties as associate members of the mini-trial panel;
(b) it is quick, confidential and non-prejudicial;
(c) it is based on the consent of both parties;
(d) it concentrates on essentials;
(e) it maintains the dialogue between the parties.

At approximately the same time a similar procedure was developed in the United States when in 1984, the US Army Corps of Engineers developed a pilot programme designed to expedite the settlement of claims pending before the Board of Contract Appeals. The term coined to designate this pilot programme was 'mini-trial' since, although it is essentially an arbitration technique, it incorporated some characteristics of the judicial process.[20.16]

The mini-trial procedure was described in 1986 in a memorandum developed for the Assistant Secretary of the Army (Civil Works) as follows:

'Under the "mini-trial" procedure, top level management officials of each party voluntarily meet to present their best case and negotiate an expedited resolution to a pending Board of Contract Appeals case. The "mini-trial" is designed to resolve disputes arising from matters of fact rather than matters of law and to take no longer than three or four days. The procedure also provides for a neutral advisor who can assist the negotiators in understanding matters of law and assessing the merits of the claim. No transcript of "mini-trial" proceedings is maintained. Either party may withdraw from the "mini-trial" proceeding at anytime.'

As it developed, the mini-trial procedure has the following features:

(a) An independent and impartial adviser is appointed to take control of the proceedings, to act as adviser to the parties in dispute, to ask questions of witnesses, to provide comments if the parties so request, to enforce time limits and to act as chairman to two assistants who may be selected from among the senior corporate officers of both parties and who are expected to make an independent assessment of the issues in dispute.
(b) The mini-trial panel is expected to hear the parties and then to propose or to facilitate a settlement. If no settlement is reached or proposed within a reasonable time, then the panel should submit a recommendation either unanimously or by the chairman.
(c) The procedure is brief with only a few weeks allowed for the parties to prepare their case followed by a 'trial' of a few days' duration.
(d) Lawyers are permitted to represent the parties at the trial.
(e) A memorandum is exchanged between the parties and copied to the adviser two weeks prior to the trial, in which each party outlines its position on the dispute in question as well as all documentary evidence to be presented at the trial.
(f) The presentations are informal with rules of evidence not strictly adhered to. Cross-examination of witnesses is allowed but severely limited in duration.
(g) The proceedings are confidential and no transcript or recording is allowed. None of the material generated by the trial may be used as evidence in pending or future proceedings. The adviser is disqualified as a witness, consultant or expert for either party in later proceedings should there be any.

Parties intending to use mini-trial procedure to resolve disputes should either develop their own mini-trial procedure or adopt one of the following two framework procedures on offer:

(a) The Rules of the Center for Public Resources (CPR), New York;[20.17] or
(b) The Rules of the Zurich Chamber of Commerce, referred to in Reference 20.15.

20.7 Dispute Board, Dispute Review Board and Dispute Adjudication Board

Dispute Review Boards, Claims Review Boards and Dispute Adjudication Boards are titles given to the same concept of a three-member committee formed at the commencement of a construction project with one member appointed by each of the parties and the third selected by the other two appointed members or jointly by the employer and the contractor. These methods of Dispute Resolution are discussed in further detail in Chapter 26.

20.8 Adjudication

Adjudication can be defined as a process whereby an appointed neutral and impartial party is entrusted to take the initiative in ascertaining the facts and the law relating to a dispute and to reach a decision within a short period of time. A period of 28 days, extendable by a further 14 days has been named in recent legislation in the UK.[20.18] However, in the FIDIC procedure for the Dispute Adjudication Board, this period is set at 56 days or 84 days, depending on whether the Board is composed of one person or three, which in itself is a measure of the size and complexity of the project.[20.19]

The adjudicator should be suitably qualified in the topic of the dispute. He may meet and question the parties in dispute and their representatives either together or separately. He may visit the site and may request the production of documents or the attendance of certain individuals within specified times. Under certain procedural rules and in certain circumstances, the adjudicator is empowered to appoint experts, assessors or legal advisers to assist him in reaching a decision, but in such a case he must make such advice available to the parties.

The adjudication process is private and any information made available during the proceedings should not be released to third parties except insofar as it is necessary to implement the decision of the adjudicator, or as may be required in subsequent arbitral or legal proceedings.

The adjudicator's decision must be in writing and, subject to the rules adopted, may be accompanied by reasons. It is binding on the parties unless challenged within a specified period and then varied in an arbitration or litigation depending on the terms of the contract. If the decision is not challenged within the specified period, it then becomes final and binding.

20.9 The ICC Rules for Amicable Dispute Resolution

In July 2001, the International Chamber of Commerce published a set of new rules for the amicable resolution of international disputes. These new rules cover all dispute resolution techniques, other than arbitration, that are used by ICC, including mediation, neutral evaluation and mini-trials. The purpose of these new rules is to offer the commercial world a means of resolving disputes amicably, in the way best suited to their needs. A distinctive feature of the rules is the freedom that the parties are given to choose the technique they consider most conducive to settlement. Failing agreement by the parties on the dispute resolution method to be adopted, the fallback is mediation, unless the contract states otherwise.

Under the rules of the ICC, amicable dispute resolution (ADR) is treated as an entirely separate and distinct dispute resolution procedure from arbitration. They are two alternative methods of resolving disputes, although sometimes they may in fact be complementary to one another, for example, where the parties provide that when they do not reach an amicable settlement to a dispute, they should resort to arbitration. The parties can also use this in reverse order, for example, where they find that the dispute that exists between them could be settled in a more amicable way, the parties can switch from arbitration to ADR. The two services, nevertheless, are distinct and are provided by separate secretariats in the ICC headquarters in Paris.

The ADR rules replace the ICC Rules of Optional Conciliation that were in force from January 1988, and therefore, where the ICC receives a request for conciliation based upon the old Conciliation Rules, the ICC will ask the parties to reformulate their request in accordance with the new rules.

A Guide to these rules is attached to the ICC ADR Rules, explaining the characteristics of the rules within an institutional framework. It also sets out features of the administrative assistance of the ICC secretariat. The Guide sets out the characteristics of ICC ADR, which include the features of amicable dispute resolution in general, but it is worth referring in particular to those characteristics that are features of ICC ADR.

20.9.1 The Rules

The ICC ADR procedure is essentially a four-step programme for the parties, covering the whole process from the filing of the request for ADR to the ADR's termination. These four steps are as follows:

Step 1: Filing the request

ICC ADR proceedings can take place in two ways, as expressed in Article 2 of the Rules:

(i) The parties' intention to refer their dispute to the ICC ADR Rules can be contained in a prior agreement of the parties, either in the original contract agreement or in a later agreement between them; and
(ii) Where there is no agreement between the parties to refer their dispute to the ICC ADR Rules, the process can be initiated when one party submits a request for ADR to ICC, and the other party accepts such a request.

The request should contain specific information as detailed in Article 2(B) of the Rules, but in particular a description of the dispute including, if possible, an assessment of its value. This information should help ICC, where they have to appoint a neutral to the case, to determine the appropriate profile of the neutral to be appointed. Furthermore, it should also help the other party to the dispute to find out exactly what is being claimed against them.

Step 2: Selection of the neutral

Pursuant to Article 3 of the ICC ADR Rules, the selection of the neutral must be made. Ordinarily this is done by the parties; however, where they are unable to do so, or where the designated neutral does not agree to serve, ICC would appoint the neutral through an ICC National Committee or otherwise and notify the parties. Where ICC is the appointing authority, the parties still remain somewhat in control of the identity of the neutral, in that they would have a say on any desired qualifications and attributes that the neutral should have in their request for ADR. ICC should make reasonable efforts to take the parties' requirements into consideration when appointing a neutral. The success of the ADR process will largely depend on the abilities and skills of the neutral. As such, the parties should take care to ensure that the neutral has the professional skill to understand the dispute that exists between them and also the necessary human qualities needed to create a benevolent atmosphere between them so as to encourage an amicable resolution of the dispute.

Step 3: The ADR procedure

The third step in the ICC ADR proceedings is the conduct of the ADR procedure pursuant to Article 5 of the Rules. This is the operational or 'hands on' phase of the proceedings, and will involve first of all the neutral and the parties promptly discussing and seeking to agree upon and define[20.20] the ADR settlement technique to be used, i.e. mediation; neutral evaluation; mini-trial; any other settlement technique; or a combination of settlement techniques. Where the parties fail to make a final decision on a suitable method of dispute resolution, then, in accordance with the Rules, mediation will be the method that is to be used by the parties.

Step 4: Termination of the ADR proceedings

The fourth step in the ICC ADR Rules is the termination of the proceedings. The Rules set down, in Article 6, seven events that will determine that the ADR proceedings have been formally terminated upon the earlier of the following:

(a) When the parties sign a settlement agreement, their dispute will come to an automatic end. Such an agreement will be deemed to be binding upon the parties.
(b) When one or more of the parties formally writes to the neutral notifying him that they do not wish to pursue the ICC ADR procedure any further. Such notification can only be made after the first discussion between the parties and the neutral. The obligation upon the parties to attend the first meeting with the neutral before being able to terminate the ADR process, is to ensure that the ICC ADR process will have the maximum chance of success. In fact, it is near impossible for the parties to estimate the value of the ADR process without first attending the discussions with the neutral.
(c) Where the neutral writes to the parties, formally notifying them that the procedure that has been agreed between them after their first discussion, has been completed.
(d) Where the neutral writes to the parties, indicating that in his opinion, the ICC ADR proceedings will not result in an amicable settlement between the parties.
(e) When a time limit or period that has been set for the ICC ADR proceedings expires. This can occur where the parties, either in the original contract agreement or in a subsequent agreement between them, set a fixed time period within which the ADR proceedings must end. This is a very useful clause in that it gives both parties certainty in relation to when the proceedings will terminate. Of course, it is possible for the parties to extend such a time period if they wish to do so.
(f) Where the ICC formally notifies the parties that payments due by one or more parties in accordance with the Rules have not been paid. It should be noted that such a step should not take place until 15 days after the date on which the fees were due to be paid.
(g) Where the ICC formally notifies the parties that assignment of a suitable neutral was not possible or that it was not reasonably possible to appoint a neutral. This provision would be invoked where, for example, one party continuously objects to the appointment of a neutral.

20.9.2 Fees and costs

An overall matter that should be considered is that relating to the fees and costs of the ICC ADR process, which are dealt with in Article 4 of the rules and set out in an appendix to the rules in the form of a Schedule of ADR Costs. This schedule involves two independent sets of costs. The first is the administrative expenses of the ICC, which have an established maximum ceiling of US$10,000, as set out in paragraph B of the schedule. It is to be noted that under paragraph A of the schedule, the party or parties filing a request for ADR should include with the request a non-refundable registration fee of US$1,500 to cover the costs of processing the request.

Secondly, the second set of costs includes the fees of the neutral and his expenses. The fees are calculated on the basis of time reasonably spent by the neutral in carrying out his duties in the ADR proceedings. These fees are ordinarily established at the outset of the ADR process by ICC through consultation with the neutral and the parties on the basis of an hourly rate commensurate with the complexity of the dispute and any relevant circumstances that should be considered. By allowing the ICC to assume control of the costs, the parties and the neutral are assured of the expedient and smooth running of the ICC ADR procedure. It is important to note that the ADR proceedings will not begin until the first required payment is made by the parties.

Following receipt of a request for ADR, the ICC would request the parties to pay a deposit in an amount likely to cover the administrative expenses of the ICC and the fees and expenses of the neutral. This deposit is subject to subsequent readjustment(s). All deposits are borne in equal shares by the parties, unless they agree otherwise in writing. The parties' other expenditure shall remain the responsibility of that party.

20.9.3 General provisions

As in the arbitral process, the ICC ADR process, along with any relevant documentation, is entirely private and confidential, in accordance with Article 7 of the Rules. However, it should be noted that there are two exceptions to this general rule, namely: where the parties agree that some or all of the ADR proceedings will not be confidential; and where either party is compelled to disclose any given element of the ICC ADR proceedings under applicable law. This confidentiality requirement also covers the settlement agreement itself. There is an additional exception to the confidentiality of the settlement agreement, in that it may be disclosed where such disclosure is necessary for its implementation or enforcement. To ensure the confidentiality of the entire process, restrictions are placed upon the neutral from taking part in any future proceedings related to the dispute that is submitted to the ICC ADR proceedings. The neutral can only act as a judge, an arbitrator, an expert or a representative or an adviser of a party in such proceedings where both parties are agreeable to such a scenario and this agreement is expressed in writing. If there is no agreement, then the neutral cannot act in such a capacity. Furthermore, the neutral is forbidden from acting as a witness in any proceedings that are related to the dispute submitted to the ICC ADR procedure for resolution. The only exceptions to this rule are where the parties agree that the neutral can act as such, or where he is compelled to act in that capacity under the applicable law.

20.10 Pre-arbitral referee procedure

Compared with conciliation or mediation, this method is a relatively new procedure developed by the International Chamber of Commerce in Paris. The Rules, which were published in 1990, were designed to provide a procedure for recourse at very short notice to a third person, the 'referee', who is empowered to order provisional measures needed as a matter of urgency.[20.21]

The main features of these Rules can be summarised as follows:

(a) The Rules may be resorted to on the basis of a written agreement either before or after entering into the relevant contract from which the dispute may arise.
(b) The referee may be selected by the parties or be appointed by the chairman of the International Court of Arbitration of the ICC in the absence of such a selection.
(c) The referee is empowered to make widely varying orders of a binding nature unless and until a competent jurisdiction (court or arbitral tribunal) has decided otherwise.
(d) This procedure may be put into operation as a complementary role to that of the expert under the ICC Rules for Technical Expertise. The latter offers a method of quickly identifying, before the evidence is destroyed or changed, whether technical problems exist and, if so, their causes. The parties are thus provided with an objective statement of facts by an independent expert, which may serve either as an aid to reaching an amicable settlement or as reliable first-hand evidence in a subsequent forum, such as a court or an arbitral proceeding.[20.22]

20.11 The ICC Rules for Expertise

The ICC Centre for Expertise was originally established in 1976 and is governed by the ICC Rules for Expertise. It was established to facilitate international commercial relations and is part of the wide-ranging services that are offered by ICC. Since their publication, these rules have been amended and revised a number of times; the most recent were published in January 2003, replacing the 1993 version. The rules were established as an aid to the delicate task of selecting competent experts, thereby facilitating international commercial relations. It is commonly accepted that experts with specialised knowledge in technical, legal or financial fields may be needed in a variety of situations. Under the rules, the ICC International Centre for Expertise offers three distinct services in this regard. The Centre will propose experts; they will appoint experts; and they will administer the expertise proceedings.

The 2003 edition of the rules has introduced greater clarity and versatility in the conduct of expertise proceedings. Furthermore, all three of the Centre's services are clearly distinguished and methodically described. Expertise may be used in a variety of situations, for example, the removal of uncertainty in business operations; the assistance of an amicable settlement of disputes; and in connection with litigation or arbitration. In this connection, expertise is valuable as both a service in its own right and as a complementary service to other dispute resolution methods, including arbitration.

20.11.1 The Rules for Expertise

The Rules for Expertise are divided into five separate sections comprising the following provisions based on the functions to be carried out by the Centre, and two Appendices. These are as follows:

— Section I, *General Provisions*, comprising Article 1;
— Section II, *Proposal of Experts*, comprising Articles 2, 3, and 4;
— Section III, *Appointment of Experts*, comprising Articles 5, 6, 7 and 8;
— Section IV, *Administration of Expertise Proceedings*, comprising Articles 9, 10, 11, 12, 13 and 14;
— Section V, *Miscellaneous*, comprising Articles 15, 16 and 17;
— Appendix I, *Statutes of the Standing Committee of the ICC International Centre for Expertise*; and
— Appendix II, *Schedule of Expertise Costs*.

20.11.2 Functions of the ICC Centre for Expertise

As stated above, the International Centre for Expertise can perform the following functions.

The proposal of experts

This function is dealt with under Article 1(A) and also under all Articles of Section II of the Rules. ICC, upon the request from any person[20.23] to do so, may provide the name(s) of an expert(s) in a particular field of activity. Under Section I of the Rules, the Centre's role is limited to the proposal of an expert. Once the Centre has provided the name(s) of the expert(s), it is up to the person requesting a proposal to contact the expert directly and, as the case may be, agree with such expert(s) on the scope of the appropriate mission and fees. There is no obligation on the requesting person to use the services of the expert who is proposed by the Centre. This service may be used where a person needs an expert in connection with a contractual relation or where a party to an arbitration or even the arbitral tribunal needs the name of a potential expert witness. The costs for the proposal of an expert are set out in Article 4 of the Rules and Article 1 of Appendix II of the Rules at an amount of US$2,500, and this fee is non-refundable. Where the person requests the Centre to propose more than one expert, this fee will be multiplied by the amount of expert names that they require.

The appointment of experts

The appointment of an expert is dealt with under Article 1(B) and Section III of the Rules. In order for the Centre to accept this role, a 'Request for Appointment' must be submitted to the ICC secretariat in Paris; this has to be based upon an agreement between the parties to use the Centre as the appointing authority. The request for the appointment of an expert should include, amongst other factors, the following:

(a) the name, address, telephone and facsimile numbers and e-mail address of each person filing the request for appointment and of any other persons involved in the expertise;
(b) a statement that the requesting person is seeking the appointment of an expert by the Centre;
(c) a description of the field of activity of the expert to be appointed along with any desired qualifications of the expert, including but not limited to education, language skills and professional experience, and any undesired attributes of the expert;
(d) a description of any matters which would disqualify a potential expert;
(e) a description of the work to be carried out by the expert and the desired time frame for completing such work; and
(f) a copy of any agreement for the appointment of an expert by the Centre and/or of any other elements which form the basis for the request for appointment.

The costs for the appointment of an expert are set out in Article 8 of the Rules and in Article 2 of Appendix II of the Rules at an amount of US$2,500. This fee is non-refundable and should be paid by the party who makes the request for an expert.

The administration of expert proceedings

Article 1(C) and Section IV of the Rules make provision for the ICC to act as an administrative authority in expertise proceedings where the parties request them to do so; or where the Centre is otherwise satisfied that there is sufficient basis for administering expertise proceedings. The 'Request for Administration' of the expert proceedings should be submitted to the ICC secretariat in Paris and should contain the same details that the 'Request for Appointment' above contains. ICC, in administering expertise proceedings, will, under Article 9(5), offer the following services: the co-ordination between the parties and the expert; initiating the appropriate steps to encourage the expeditious completion of the expertise proceedings; supervising the financial aspects of the proceedings; appointment of an expert using the procedure referred to in Section III or confirmation of an expert agreed to by all of the parties; review of the form of the expert's report; notification of the expert's final report to the parties; and notification of the termination of the expertise proceedings. As well as the authority to appoint the expert, the Centre also has the power to replace or remove the expert where the expert has died, resigned, is unable to carry out his function as expert, or where all the parties request the Centre to do so. If a party objects to the expert's appointment on the basis that he does not have the necessary qualifications or that he is not fulfilling his functions in accordance with the Rules, or in a timely fashion, then the Centre, having considered the observations of the expert and the other party/parties, can replace the expert. In replacing the expert, it is within the discretion of the Centre to decide whether or not to adopt the original procedure of appointment.

The cost of the sole administration of the expertise proceedings is a non-refundable amount of US$2,500. The Centre has the discretion to set its own administrative expenses depending on the tasks that are carried out by them. There is a limit on these administrative expenses, however, in that they must not exceed 15% of the total expert's fee and they should not be less than US$2,500. The fees of the expert are calculated on the basis of the time the expert has reasonably spent on the expert procedure, at a daily rate that will be fixed by the Centre in consultation with the expert and the parties.

Under Article 12, the expert, having consulted with the parties, sets out his mission in a written document. This document cannot be inconsistent with anything in the Rules and will comprise the following:

(a) the names, addresses, telephone and facsimile numbers and e-mail address(es) of the parties;
(b) a list of issues to be treated in the expert's report;
(c) the name(s), address(es), telephone and facsimile numbers and e-mail address(es) of the expert or experts;
(d) the procedure to be followed by the expert and the place where the expertise should be conducted; and
(e) a statement indicating the language in which the proceedings will be conducted.

The expert should, as soon as possible after preparing his mission and consulting with the parties, prepare a provisional timetable for the conduct of the expertise proceedings. This should be communicated to both the parties and the Centre for Expertise. The ultimate task of the expert is to issue a written expert's report in which he denotes the findings that he made within the limits of his mission statement. This report can only be issued once the expert has heard the parties and/or allowed the parties to make written submissions. The expert's report will not be binding upon the parties unless the parties agree otherwise. Furthermore, the report will be admissible in any later judicial or arbitral proceedings in which all of the parties were parties to the expertise proceedings, again unless the parties agree otherwise. It goes without saying that any information that is given to the expert by the Centre or any party can only be used by him for the purposes of the expertise and will be treated by the expert as confidential. If a party to the expertise proceedings does not participate, this will not deprive the expert of the power to make findings and to issue his report, as long as the non-participating party has been given the opportunity to participate.

20.12 Concluding remarks

This chapter concludes Part III of this book, in which the Fourth Edition of the Red Book and its use in practice were discussed in detail. Part IV of the Book, which comprises a single chapter, deals with documents related to the Fourth Edition of the Red Book: other standard forms of contract published by FIDIC between 1993 and 1999, the third edition of the Yellow Book, the Orange Book and the Sub-contract Form.

Part IV

Other Documents Related to the Red Book

Chapter 21

FIDIC's Other Forms of Contract (1993–1999)

21.1 Introduction

Although the Contracts Committee of FIDIC was established in 1913, its activities were confined, until 1993, within the framework of the three Forms of Contract, the Red Book, the Yellow Book and the White Book. The Yellow Book, *Conditions of Contract for Electrical and Mechanical Works including Erection on Site*, was first published in 1963. It was first revised in 1980 and later in 1987 with the introduction of its third edition. The White Book, *Client/Consultant Model Services Agreement*, was first published in 1990 for general use in agreements of pre-investment and feasibility studies and in administration of construction and project management. It replaced the three documents previously used in formulating such agreements which were designated the International General Rules for Agreement.[21.1] The White Book was later revised with its second edition published in 1991 and its third edition published in 1998.

When the work on these three forms of contract was completed, the various committees involved in their preparation were disbanded and a single committee was reintroduced in 1993 with responsibility for producing new FIDIC contract documents and for updating those already produced. The new Contracts Committee through its various task groups focused on a number of areas which needed attention in the field of conditions of contracts. The main areas were: design-build; sub-contract work; and the use of the Red Book in building works as distinct from civil engineering. In the remaining sections of this chapter the following areas are considered: the main features of the third edition of the Yellow Book (an electrical and mechanical (E & M) contract), the first edition of the Orange Book (a design and build contract), and the first edition of the conditions of subcontract for works of civil engineering construction. The use of the Red Book in building works is discussed in Chapter 6 as part of the 1996 Supplement.

21.2 The Yellow Book, third edition

21.2.1 Background

It has always been recognised that the standard conditions of contract for civil engineering works do not properly provide for the requirements of all electrical and mechanical engineering projects, 'E & M Works'. The requirements for the two types of contract differ for many reasons inherent in their nature, as explained later in Section 21.2.2.

After the publication of the second edition of the Yellow Book in 1980, the comments received from various users around the world concentrated on the fact that the legal drafting of the conditions, which was based on the Anglo-Saxon system, made them difficult to comprehend, especially by people whose mother tongue is not English.[21.2] Many commentators considered it a higher priority to have a standard form of conditions of engineering contract understood by engineers rather than to have established legal meaning for the terms and expressions that were generally used in the first and second editions.[21.3] It was also thought to be more important that the employer and the contractor be able to understand the conditions and to have the project satisfactorily completed within the scheduled time and the approved budget, rather than to have a project facing difficulties but with conditions written in legal terms understood only by lawyers.

Accordingly, a major revision of the second edition of the Yellow Book was undertaken and the third edition of the E & M Form was published in 1987. Later, the Guide for its use was published in 1988 dealing with each individual clause of the Conditions.[21.4]

As in the Red Book, the third edition of the Yellow Book is issued in two parts, Part I – General Conditions and Part II – Special Conditions. However, Part II of the Yellow Book is much smaller than its equivalent in the Red Book and is divided into two sections.

Section A contains a number of sub-clauses where the provisions in the General Conditions refer to an alternative solution inserted in Part II. Thus, the provisions in Part I will prevail unless an alternative solution is given in Part II. Accordingly, if it is intended that there should be no changes to the provisions of Part I, then Section A in Part II should not be completed. But Section A of Part II must be completed wherever alternative solutions to the standard provisions of Part I are necessary. Section B simply provides the space to add any further special conditions which may be required for a particular project.

The Yellow Book also contains standard Forms of Tender and of Agreement.

The third edition of the Yellow Book was issued with a number of editorial amendments in May 1988. These amendments are listed at the end of the reprinted document.

The third edition of the E & M Form of Contract differs greatly from its previous edition in many important aspects, including the numbering and the sequence of the various clauses. Because of the numerous differences between civil engineering projects, on the one hand, and electrical and mechanical engineering projects, on the other, the third edition of the E & M Form of Contract also differs substantially from the Fourth Edition of the Red Book. However, it is notable that a considerable measure of compatibility and harmony between these two forms was maintained in order to permit their use jointly on a project comprising both civil engineering and electrical and mechanical works.

21.2.2 Differences in the nature of civil engineering and E & M engineering projects

Electrical and mechanical engineering projects differ from civil engineering projects and require a different set of standard conditions of contract for the following reasons:

(a) *The project itself*: Civil engineering projects generally involve the construction of works with bulk operations and underground excavations or underwater elements. Electrical and mechanical projects, on the other hand, involve the supply and erection of specialist work, including machinery and equipment.
(b) *Purpose and use*: Civil engineering projects are in general constructed for the benefit of society as a whole by a state or a semi-state organisation with no revenue generated at the end except in special circumstances. E & M projects on the other hand generally involve the operation of machinery and equipment for the purpose of generating a revenue.
(c) *Suitable tenderers*: In many cases, where electrical and mechanical works are involved, there would be relatively few organisations qualified to carry out the work successfully. Furthermore, in many instances the work would involve proprietary processes that are patented and cannot be reproduced except by those holding the patent rights or by their permission. On the other hand, a far greater number of construction companies would be eligible to tender for civil engineering projects.
(d) *Life span of the project*: The element of wear and tear has a much lesser effect on civil engineering projects than on electrical and mechanical projects. Furthermore, as technological developments are more frequent in the latter, this results in there being a greater probability of a project becoming obsolete. Accordingly, civil engineering projects tend to be designed and constructed with a much longer period of use in mind than E & M projects.
(e) *The design element*: Civil engineering projects have, traditionally, been designed by a consulting engineer or a firm of engineers who specialise in the design of the particular type of project, whereas electrical and mechanical projects contain to a large extent design elements by the supplier or the contractor or the manufacturer.
(f) *The execution of the project*: Civil engineering projects are constructed on-site. Electrical and mechanical works contain a major part of elements which must firstly be manufactured in a factory or a workshop off-site before erection can take place on-site.
(g) *Payment for the Works*: Civil engineering projects are generally paid for in accordance with a quantity of work executed which is estimated in a bill of quantities prepared prior to the commencement of work and finally established and ascertained after completion of the project. E & M projects are paid for in accordance with a schedule of rates for major items of machinery and equipment supplied and erected.

21.2.3 Essential features of the Yellow Book

The Yellow Book is drafted in such a way that it can be used successfully alongside the Red Book in projects comprising both civil engineering works and mechanical and electrical works. Thus for projects such as dams, water supply or sewerage works, the works could be separated into two separate contracts: one for civil engineering works and the other for electrical and mechanical works, with the contractor for either being allocated the overall responsibility for the whole project. This is possible since there are similar terms and functions used in both forms of contract, such as the presence of an engineer, the requirement for due consultation by the engineer prior to a number of similar circumstances, and the similarity of definitions of certain terms in both contracts. The format of the Yellow Book is also the same as that of the Red Book in that it is divided into two parts: Part I which contains the general conditions of contract and Part II which contains the conditions of particular application to the project in question. However, except for clause 1 of the Yellow Book which is divided into 37 sub-clauses, numbered from 1.1.1 to 1.1.37, all the other provisions of the Yellow Book are drafted as intact discrete units and not as subordinate clauses, although using the decimal point system, e.g. clauses 2.1 to 51.3. Each of these units has its own heading and other related units which are grouped under a group title.

Despite these similarities, however, the provisions of the Yellow Book differ greatly from those of the Red Book and can be distinguished in many respects. The essential distinguishing features of the Yellow Book are: the commencement date; the definition of the engineer; his impartiality; responsibility for design; the requirement of approval by the employer; the procedure for disputing the engineer's decisions and instructions; the procedure for his replacement; the contractor's general obligations; tests on completion; performance security; extension of the time for completion; due consultation; delay in completion and prolonged delay; risks and responsibility, damage to property and injury to persons, liability and insurance; *force majeure*; disputes and arbitration; and law and procedure. These features are discussed below.

Commencement date

Clause 1.1.1 of the conditions defines the commencement date as the latest of the following five dates:

(a) the date specified in the preamble as the date for commencement of the works; or
(b) the date when the contractor receives such payment in advance of the commencement of the works as may be specified in the terms of payment; or
(c) the date when the contractor receives notice of the issue of any import licence necessary for commencing performance of the contract; or
(d) the date when the contractor receives notice that any legal requirements necessary for the contract to enter into force have been fulfilled; or
(e) the date when the contractor receives notice that any necessary financial or administrative requirements specified in Part II as conditions precedent to commencement have been fulfilled.

It is necessary to provide the above set of alternative dates, since a major part of the works would normally be executed on the contractor's own premises or on premises other than the site.

In this connection and where international contracts are concerned, import permit licences and other formalities relating to the electrical and mechanical components of the project must be obtained prior to the time starting to run for the time for completion. Otherwise, delays would be inevitable.

Definition of the engineer

Under the Yellow Book, an engineer is appointed but in a different role from that under the Red Book. Clause 1.1.15 defines the engineer as 'the person appointed by the employer to act as Engineer for the purposes of the Contract and designated as such in the Preamble'. This is especially important with reference to clause 2 which sets out the engineer's duties, his power to delegate, the circumstances under which he is required to act impartially, his decisions and instructions, the procedure provided for disputing such decisions and instructions, and the contractor's consent as required should the employer decide to appoint any person to replace the engineer.

Impartiality of the engineer

Clause 2.4 provides that the engineer must act impartially whenever he is required to exercise his discretion by:

(a) giving his decision, opinion or consent; or
(b) expressing his satisfaction or approval; or
(c) determining value; or
(d) otherwise taking action which may affect the rights and obligations of the employer or the contractor.

In doing so, the engineer must have regard to all the circumstances surrounding the particular situation with which he is dealing. Accordingly, the engineer does not simply act as an agent of the employer but also as an impartial decision maker in these specific circumstances.

Design

It is to be noted that the engineer's duties do not include design as this is the contractor's responsibility, as provided in clause 8.1. The engineer is, however, required to comment on the contractor's detailed design of the plant and his proposed method of carrying out the works, both of which are set out in clauses 6 and 7 of the general conditions.

Specific approval of the employer

Clause 2.1 provides that the duties of the engineer are set out in the contract and that where the engineer is required, under the terms of his appointment by the employer, to obtain the specific approval of the employer before carrying out any of these duties, full particulars of such requirements should be set out in Part II of the Conditions.

Procedure for disputing the engineer's decisions and instructions

Clause 2.5 of the Form provides that the contractor must proceed with the works in accordance with the decision and instructions given by the engineer. The only remedy available to the contractor, should he wish to dispute any decision or instructions so given by the engineer, is to proceed under clauses 2.7 and 50.1. By the terms of clause 2.7, should the contractor wish to dispute or question any decision or instruction of the engineer under clause 2.5 or a written confirmation given under clause 2.6, he should give a reasoned notice to the engineer within 28 days of receipt of such decision, instruction or confirmation.

Clause 50.1 provides for the procedure to be followed in arbitration. This procedure is discussed in detail under the heading Disputes and Arbitration later in this section. However, it is notable in this regard that the two-tier procedure followed in clause 67 of the Red Book is not used in the Yellow Book. The contractor and the employer under the Yellow Book can proceed directly to arbitration, unless such reference is a result of dissatisfaction with a decision or instruction of the engineer. In that case, the notice of intention to refer the matter to arbitration must be issued within 56 days of such decision or instruction.

Replacement of engineer

Should the circumstances arise in the course of the contract where it is considered necessary by the employer to replace the engineer, the employer is not entitled to do so without the contractor's consent. This provision under clause 2.8 of the Yellow Book is included in order to prevent the appointment of someone who is not an impartial person to this sensitive position. However, it is not clear as to how the appointment of a replacement engineer is to be achieved if the contractor's consent is withheld. The only solution seems to be through the arbitration procedure in accordance with clause 50. However, this procedure requires time during which a vacuum would have already been created if there is no engineer in place or if the engineer is unable to act.

Contractor's general obligations

The contractor's general obligations are set out in clause 8.1 of the conditions and they are divided into seven headings which require him to:

— design;
— manufacture;
— deliver to site;
— set out the works;
— erect;
— test and commission the plant supplied;
— carry out the works within the time for completion.

It is relevant in this connection to compare these obligations with that simple but much wider overall obligation under the Red Book of having to complete the works irrespective of what might happen, with the exception of a few specified events, during the contract period.

Tests on completion

As it is usual for a project under the Yellow Book to contain a high element of plant for incorporation into the works, commissioning of such plant and ascertaining its performance is a major aspect of the contract. Clause 1.1.34 defines 'Tests on Completion' as those 'specified in the Contract or otherwise agreed by the Engineer and the Contractor to be performed before the Works are taken over by the Employer.' Therefore, it is essential that the necessary tests which must be successfully completed before the works are taken over are clearly specified in the contract documents. It would also be useful to set out the sanction to be imposed should these tests prove to be unsuccessful.

Performance security

Clause 10 of the conditions provides the details of the performance security required under the contract. Sub-clause 10.3 sets out the conditions which must be satisfied for a successful claim to be made by the employer under the terms of that security. Thus, the security provided is conditional on this procedure being followed.

Extension of the time for completion

Clause 26 sets out the conditions under which the contractor may become entitled to an extension of the 'Time for Completion' of the contract. It also sets out the procedure to be followed for such an extension to be granted. The engineer is authorised, after due consultation with the employer and the contractor, to grant from time to time such an extension. He may do so either in advance or retrospectively.

However, the clause is silent on whether such an extension, once given, can be reduced. Since such silence is not followed in the Red Book when the engineer is prevented from reducing any extension of time once granted, it could be interpreted that such reduction is permissible.

Due consultation

This term which is used throughout the Yellow Book marks a very important development of the conditions. As in the Red Book, its implications should be thoroughly understood not only by the employer and the contractor but also by the engineer.[21.5]

Clauses 27.1 and 27.2, delay in completion and prolonged delay

Clause 27.1 provides an alternative concept to the liquidated damages solution contained in the Red Book for the delay of the contractor in performing the contract. It provides that the employer is entitled to a reduction in the contract price for each day's delay and furthermore, after having reached a maximum reduction as stipulated in the preamble, clause 27.2 provides for the right of the employer to terminate the contract provided that the employer had notified the contractor requiring him to complete the works. However, the contractor can avoid such reduction under the provisions of clause 27.1 if 'it can be reasonably concluded from the circumstances that the Employer will suffer no loss'.

Clauses 37, 38 and 39, risks and responsibility; clauses 40 and 41, damage to property and injury to persons; clause 42, limitation of liability; and clause 43, insurance

For the first time, this section of a standard form of contract which embodies seven inter-related clauses has been written in clear and simple language and in a sequence that is logical.

However, unfortunately, there are a number of specific requirements which are embodied in the provisions of these clauses which are unacceptable to a number of employers as a matter of principle, and to a number of engineers as a matter of practicality.[21.6]

Force majeure

The term 'force majeure' covers a number of unforeseen circumstances which prevent, totally or partially, one or both parties from fulfilling their contractual obligations. Although this term is understood and accepted around the world, the definition and interpretation of the circumstances differ from one jurisdiction to another and the legal consequences of the events of *force majeure* differ accordingly. Thus, where the Yellow Book is expected to be used, the consequences of including the events set out in clause 44.1 as incidents of *force majeure* must be carefully considered under the applicable law of the contract as set out in clause 51 of the general conditions on law, procedure and language. Special attention must also be given to the consequences of a *force majeure* event and these are set out in sub-clauses 44.2 to 44.9 of the general conditions.

Disputes and arbitration:

Although quite different from clause 67 of the Red Book, clause 50 of the Yellow Book does also contain strict time limits which must be complied with if a dispute is to be resolved by arbitration.

As stated earlier, if a contractor wishes to dispute or question any decision or instruction given by the engineer, a notice must first be given under clause 2.7 of the conditions within 28 days of receipt of such decision or instruction. Once the process is initiated by that notice from the contractor, then the engineer must within a further 28 days confirm, reverse or vary such decision or instruction giving reasons for his action.

Should the employer or the contractor disagree with the action taken by the engineer or if the engineer fails to reply to the contractor's notice within the stipulated 28 days and the matter cannot be settled amicably, then the mechanism under clause 50 can be triggered.

It is worth noting that clause 2.7 does not have a provision for the employer to dispute or question an instruction or decision of the engineer in the first instance. The employer can participate in this process only if the contractor first disputes the engineer's decision or instruction. However, clause 50.1 refers to both the contractor and the employer which might create some confusion as to whether the employer can question an instruction of the engineer without referring the matter to arbitration. Furthermore, in circumstances where the employer does not become aware of the engineer's decision or instruction, then his remedy must lie in the provisions of his own contract with the engineer.

If the mechanism of clause 50 is to be initiated by the contractor, then this must be done within 56 days of the engineer's reply to the contractor's first notice under clause 2.7. As stated above, the employer can trigger the mechanism of clause 50.1 within 56 days of any decision or instruction of the engineer with which he is dissatisfied.

If no notice of intention to proceed to arbitration is issued within the prescribed 56 days, then the decision of the engineer becomes final.

Clause 50.2 provides that any reference to arbitration will be in accordance with the Rules of Conciliation and Arbitration of the ICC, unless other rules are specified in Part II of the Yellow Book. This clause also provides in similar terms to those under the Red Book that the arbitrator has full power to open up, review and revise any decision, instruction, or certificate of the engineer which has been referred to arbitration pursuant to clause 50.1.

Once the arbitration process is initiated, then clause 50.3 stipulates that performance of the contract should continue even where arbitration proceedings have commenced, unless suspension of the works had been or is ordered by the employer. Furthermore, the clause also provides that any payment due to be made by the employer should not be withheld on account of a pending reference to arbitration.

Clause 50.4 provides for a time limit in respect of the date of commencing arbitration proceedings stating that formal notice of arbitration must be given no later than 84 days after the issue of the final certificate of payment. In this connection, reference should be made to clause 33.11 which deals with the conclusiveness of the final certificate of payment.

Law and procedure

Clause 51 sets out the provision for the applicable law of the contract, the procedural law for any arbitration instituted under clause 50, the language and the place of any arbitration proceedings instituted. It is an extremely useful clause as it avoids many possible but unnecessary disputes and delays in the process of arbitration, if and when it is instituted.

21.3 The Orange Book, first edition

By the time the Orange Book was conceived, the construction industry had been flirting with design and build contracts for sometime in the form of package deals, but left them aside for use with special projects where the contractor possessed either specialist knowledge and expertise not available elsewhere or patent rights for a specific process. However, by the late 1980s, the design and build contract became attractive and was viewed as an alternative type of contract to the dispute-riddled, traditional form of contract. In response to demand, the Institution of Civil Engineers in London published its Standard Form of Contract for Design and Construct in 1992.[21.7] Notwithstanding its origin, this form of contract was not a version of the Sixth Edition of the ICE Conditions with some selected modifications, but a totally new form of contract.[21.8]

Subsequently, in response to international demand, FIDIC introduced its first edition of the Conditions of Contract for Design-Build and Turnkey in 1995. It is of interest to note that those who advocated the use of such contracts claimed that their benefits included the following:

(a) Lower costs of design and supervision due to a perceived close co-operation and involvement of designers and constructors during the whole process of design, construction and quality control.
(b) Anticipated shorter period of project implementation due to disposal of any need to separate the design phases from the tendering process and from the subsequent construction period. Work on the whole project could begin with the selection of the contractor and construction could be programmed so that it would not be affected by any delay in non-critical detailed design activities of some elements of the works.
(c) Merging the liabilities for design, material and workmanship into a single party, the contractor, thus eliminating any possible confusion or uncertainty as to who is responsible and liable for undesirable events.
(d) A fixed lump sum price which is less susceptible to cost over-run even when variations to the scope and extent of the works are permitted in the conditions of contract.

However, in contrast to the benefits claimed above, other experts foresaw that there were also some potential disadvantages which, for most projects, would have a more serious effect. These included the following:

(a) The loyalty and duty of the designer towards the employer is second to those afforded by him to the contractor.
(b) The cost of the preliminary designs which must be carried out by the unsuccessful tenderers in preparing their proposals would normally have to be paid for ultimately through the successful tender. It is usually included under an item, such as general overheads, as part of the head office costs. But, since it is clear that the proportion of unsuccessful bids would be much higher than that of successful ones, this cost element forms a significant burden on the employer.
(c) The design-build form of contract is inappropriate for some projects, if not all, as it stifles innovation both in design and in construction techniques. The contractor tends to use methods with which he is most familiar from past experience and knowledge.
(d) Using methods with which the contractor is most familiar from past experience and knowledge could result in ignoring recent advances in the science and art of design and construction, thus leading to more expensive construction costs. Furthermore, the temptation to short-cut design procedures would always be there

which would be implemented to refine and optimise the use of materials. If such short cuts are followed, materials would not be used to their optimum value resulting in more expensive construction.

(e) The employer's requirements could never be as explicit as in the traditional method of procurement. The more clearly the employer's requirements are specified, the more costly tendering becomes because these requirements must be taken into consideration in the preliminary design which must be submitted with the contractor's proposal as part of his tender.

There is a further problem for the employer if he issues a detailed and restrictive set of requirements in that this may result in weakening the contractor's design liability. Drafting the employer's requirements demands extreme care and a proper balance between what is explicitly stated and what is omitted. Such balance must be achieved fairly and sensitively, otherwise disputes might be generated or cultivated to a greater degree than that which already exists at the present time in the construction industry.

21.3.1 Background

The Orange Book has been drafted for use in construction contracts where the contractor is totally responsible and liable for design. It is essentially intended for international contracts, but with some minor modifications it is also suitable for use in domestic contracts. Its title refers to design-build contracts and turnkey contracts and its foreword sets out the difference between these two types of contract as envisaged by the draftsmen.

Design-build contracts include any combination of building work together with civil, mechanical and electrical engineering works. On the other hand, turnkey contracts include the provision of a fully-equipped facility, ready for operation at the turn of a key. Turnkey contracts typically include design, construction, fixtures, fittings and equipment to the extent defined in the contract documents. The foreword goes on to indicate that turnkey contracts are often financed by the contractor and may require him to operate the works for a few months' commissioning period, or for some years' operation on a build-operate-transfer basis. Part II of the Orange Book also includes an advice note on turnkey arrangements and sample wording for a contract financed by the contractor. As in the Red and Yellow Books, the Orange Book is divided into two Parts: Part I – General Conditions; and Part II – Conditions of Particular Application, which are usually prepared and completed for each individual contract.

The principles adopted in the preparation of these two parts of the Orange Book are set out in the introduction section to Part II of the document. It is usually envisaged that amendments to some provisions of Part I may have to be made, but not in the context of upsetting the balance of its inherent fairness. An important feature of the document is that when such amendments are implemented, they are not necessarily done through Part II of the Orange Book but simply through either activating or invalidating in the Appendix to Tender, certain provisions of Part I. For example, the provisions of: clause 11.4, 'Failure to Pass Tests after Completion'; clause 13.5, 'Plant and materials for the Permanent Works'; and others may be varied by insertion of the required effect in the Appendix to Tender.[21.9] Accordingly, special attention is required when completing or examining the provisions of the Appendix to Tender.

21.3.2 Differences in the nature of the Red and Yellow Books on the one hand and the Orange Book on the other

The concepts of these three books differ greatly in the allocation of the main functions discussed earlier in Chapter 6. First, the financial arrangements in the design-build and turnkey projects lend themselves to private finance; secondly, the entire design is entrusted to the contractor in the Orange Book; and thirdly, although the risks are shared between the employer and the contractor, the balance of risk-taking is generally shifted towards the contractor under the Orange Book. Accordingly, the fundamental differences between these three Forms of Contract may be summarised as follows:

(a) **The employer's representative**: Both the Red and the Yellow Books rely on the involvement of an impartial engineer who takes various specific roles during the project implementation cycle. In the case of the Red Book, the role of the engineer is explained in detail in Chapters 5 and 10 above. In the case of the Orange Book, the position of the engineer was replaced by an 'Employer's Representative' who has no duty to act impartially. Instead, he is the employer's agent with duties and authorities specified under clause 3 of the contract. When he is required to determine value, cost or extension of time, he is required to consult with the contractor in an endeavour to reach agreement. But, if agreement cannot be reached, the employer's representative is required, under clause 3.5, to determine the matter 'fairly, reasonably and in accordance with the Contract'.

(b) **Dispute resolution**: This shift in the status of the engineer meant that the principles of dispute resolution as contained in the Red and Yellow Books could not be maintained under the Orange Book. In particular, the mechanism of decision making as prescribed in clause 67 of the Red Book and its two-tier system based on the principle of the impartial engineer had to be abandoned in the Orange Book (see Chapter 19 above).

A completely different system of dispute resolution had to be adopted to fill this vacuum created in the Orange Book. This was done by incorporating the concept of dispute review boards as recommended in the Standard Bidding Documents of the World Bank for major contracts estimated to cost over US$50 million. For detailed analysis and discussion of this concept, its rules and procedures, reference should be made to Chapter 26.

In this regard, it is worthy of note that this concept was also adopted by FIDIC in the 1996 Supplement to the Red Book and the 1999 Red, Yellow and Silver Books, and is referred to as the Dispute Adjudication Board. For detailed analysis and discussion of the concept, its procedural rules and terms of appointment, reference should be made to Section 19.8 above and Chapter 26 below.

Accordingly, the Orange Book has a system of dispute resolution which may be divided into the following three steps:

(i) First, under clause 20.4, a dispute arising out of, or in connection with the contract, or with the execution of the works, must be referred in writing to an impartial Dispute Adjudication Board, 'DAB', for its decision. The Board, acting as a panel of experts, is required to render a reasoned decision within 56 days. The contractor and the employer are required to give effect forthwith to that decision unless and until the decision is revised in an amicable settlement or an arbitral award. The decision of the Board becomes final and binding on the parties within 28 days of the date of receipt of the Board's decision, unless a notice of dissatisfaction is given before that date.

(ii) The second tier of the dispute resolution procedure is activated when a notice of dissatisfaction is given by either party to the other, under clause 20.4. Following such notice being served, the provisions of clause 20.5 stipulate that 'the parties shall attempt to settle such dispute amicably before the commencement of arbitration'. Unless varied by the parties, 56 days are allowed for such amicable settlement to be completed. If unsuccessful in resolving their dispute amicably, the parties may proceed to arbitration as set out in item (iii) below.

(iii) Arbitration may be commenced, under sub-clause 20.6, only in the case of a dispute in respect of which:

'(a) the decision, if any, of the Dispute Adjudication Board has not become final and binding pursuant to clause 20.4; and
(b) amicable settlement has not been reached.'

The rules to be applied and other aspects of the arbitration procedure are referred to in the Appendix to Tender.

(c) *The employer's requirements*: The expression 'Employer's Requirements' is a defined term under sub-clause 1.1.1.2 of the Orange Book. It means 'the description of the scope, standard, design criteria (if any) and programme of work, as included in the Contract, and any alterations and modifications thereto in accordance with the Contract'. It is the most important document for the success or failure of the project and unless it is drafted with the utmost care, it could result in major disputes between the parties. It ranks third in priority of documents under clause 1.6 of the Orange Book and only after the contract agreement and the letter of acceptance. A check list of the main items to be included in this document appears on page 3 of Part II of the Orange Book. The employer's requirements as defined above should be drafted in a balanced manner as follows:

(i) they must be precise in their definition of what the employer requires, yet flexible enough to generate one of the main advantages of a design-build contract, namely the contractor's expert input into the design and building of the project;

(ii) the requirements should be demanding enough to enable the employer to choose successfully the most suitable contractor from amongst the tenderers, yet the tenderers should not be required, at the tender stage, to provide more than the necessary information for the correct decision to be made on the successful tenderer;

(iii) they must be sufficiently detailed to establish the purposes of the contract, yet concise enough not to limit the contractor's ability to design the works properly or to restrict his ingenuity in searching for the most appropriate solution;

(iv) they must be complete in prescribing all of the employer's requirements, including what the project would look like; how it is operated and maintained; the cost of operating it; the quality of the product, if any; the tests required to be run during and after construction and the rate of success of these tests; manuals to be prepared; and spare parts requirements and their costs. Yet, they must not relieve the contractor of his duties, obligations or responsibilities for the design, construction and completion of the whole project.

The Requirements are also the main source of information for the general obligations of the contractor as referred to in clause 4.1 of the Conditions. These obligations include fitness for the purposes for which the works are intended, which must be enunciated in the employer's requirements.

(d) *Fitness for purpose*: The standard of liability imposed on the contractor is specified under clause 4.1 of the Orange Book as that of fitness for purpose. This standard applies not only to the workmanship and materials used but also to the design as provided in clause 5 of the Conditions. The fitness for purpose with respect to design is a higher standard than that which usually applies to design professionals and than that covered by the professional indemnity insurance they must provide. In fact, it is extremely difficult, if not impossible, to obtain a professional indemnity cover to the standard of fitness for purpose.

(e) *Fixed lump sum price*: Clause 13.1 of the Orange Book provides that payment for the works shall be made on a fixed lump sum basis and that the contract price shall not be adjusted for changes in the cost of labour, materials or other matters. Clause 13.1 (d) provides that any quantities set out in the schedule are only estimated and are not to be taken as the actual and correct quantities of the works to be executed. Clause 13 also provides the principles for payment for the works and is divided into sixteen sub-clauses. Interim payments are intended to reflect the estimated value of the works executed up to the time of application for payment as indicated in clause 13.3 (a).

However, whilst the Orange Book is intended to be a fixed lump sum contract, the employer is empowered through his representation under clause 14 of the Conditions to initiate variations at any time during the contract period. The 'Contract Period' is a defined term meaning 'the period from Commencement Date to the date 365 days after the date on which the whole of the Works shall have been completed as certified by the Employer's Representative under Clause 10 (or as extended under Sub-Clause 12.3).' This is a new term quite different from its equivalent under both the Red and Yellow Books and it clearly removes the two illusions created in the minds of some people as to the intended meaning of 'maintenance' and 'defects liability'. The 365 days period is not the period of liability for defects and neither is it a period when maintenance of the project is provided. The period of liability is the limitation period as defined by reference to the applicable law of the contract. Maintenance and care of the project after taking over by the employer are provided by the employer.

(f) **Design**: Under the Orange Book, the design function is the contractor's responsibility as stipulated in clause 5 of the Conditions. It is contained by implication in what is referred to as the 'Construction Documents' which are defined in sub-clause 1.1.6.1 as 'all drawings, calculations, computer software (programmes), samples, patterns, models, operation and maintenance manuals, and other manuals and information of a similar nature, to be submitted by the Contractor.'

However, the design must first be initiated by the employer in a conceptual form and included in the employer's requirements as part of the tender documents. The conceptual design serves as the employer's expression of the project and might represent an input of no more than 10 per cent of the total necessary input.

This conceptual design should be checked by the tenderer, modified if necessary, but accepted if not, and developed into his proposal with the tender. If the tender is successful, the contractor is required, under clause 5.2 of the Conditions, to prepare the construction documents in sufficient detail to do all that is necessary to execute and complete the works. When each of the construction documents are considered ready for use, they are then required to be submitted to the engineer's representative for pre-construction review. This review which must be carried out within 21 days permits the verification of compliance with the contract at an early stage and also whether or not the design is proceeding in accordance with the employer's requirements. In this way, any corrective action could be taken at the earliest opportunity and before the commencement of any construction. However, it is notable that neither approval nor consent is specified as a condition precedent to commencement of construction. Although these words are mentioned in sub-clause 3.1 of the Orange Book, the contractor can proceed with the construction of the works in accordance with the construction documents once he has not received any comments by the date of expiry of the review period. In any case, clause 3.1 provides that

> 'Any proposal, inspection, examination, testing, consent, approval or similar act by the Employer's Representative (including absence of approval) shall not relieve the Contractor from any responsibility, including responsibility of his errors, omissions, discrepancies, and non-compliance with Sub-Clauses 5.3 and 5.4'.

Under the Orange Book, this complete responsibility for design includes the question of site data which is referred to under clause 4.9. Unlike the Red Book, no mention is made of the contractor's basis for his tender. Instead, the contractor is 'deemed to have obtained all necessary information as to risks, contingencies and all other circumstances which may influence or affect the Tender'.

(g) **Risk, responsibility, insurance and force majeure**: These topics are collected together and included logically in three consecutive clauses, 17, 18 and 19, of the Orange Book. Clause 17 which deals with risk and responsibility is logically constructed as it deals first with the risks of bodily injury, sickness, disease or death; and with injury to or destruction of physical property (other than the works), including consequential loss of use. These risks are allocated to the contractor but are limited to the extent that they are the result of a breach of a duty of care imposed by law.

Clause 17.1 provides for an indemnity by the contractor to the employer and his representative, their contractors, agents and employees in respect of these risks. It is equivalent in part to clause 22.1 of the Red Book, but more clearly set out and properly phrased in the context of the whole topic of risk, liability and indemnity for bodily injury and accidental loss or damage. In this connection, it must be remembered that clauses 17, 18 and 19 deal only with the accidental risks which might cause bodily injury or disease or death or damage or loss to physical property. Unfortunately, this principle is not made sufficiently clear in the Orange Book as there is no reference to risks of loss or damage in clause 17.5 'Contractor's Risks', leaving the unwary to assume that the contractor's risks are all risks inclusive of financial risks and the risks of loss or damage. Furthermore, there appears to be no explanation for this omission in the Guide to the use of FIDIC conditions of contract for design-build and turnkey.

The risks related to financial and/or time loss are not the subject of these clauses. Instead, they are spread in the other clauses of the Conditions. Such risks are allocated to the contractor and the employer in the same manner as the allocation of risks of accidental loss or damage: that is to the contractor unless specifically allocated to the employer. Accordingly, the events described in the following sub-clauses are allocated to the employer: clauses 4.24, 'Fossils'; 4.11, 'unforeseeable Sub-Surface Conditions'; 8.8, 'Consequences of Suspension'; 13.16, 'Changes in Legislation'; and 8.4, 'Delays caused by Authorities'. Clauses 17.2 to 17.5 are very similar to clause 20 of the Red Book and use the same language with slight improvement in the framework. The employer's risks are repeated verbatim from clause 20.4 of the Red Book except for the design risk which is naturally shifted in the Orange Book to the contractor's list.

Regarding risk and liability, three other important differences exist:

(a) Under sub-clause 17.4 of the Orange Book, a notice is required to be given by the contractor to the employer's representative when an employer's risk is either foreseen by or becomes known to the contractor. A further notice is required should the contractor suffer delay and/or incur cost as a result of an employer's risk eventuating. Under the Red Book only the second notice is required but only by reference to valuation under clause 52 thereof.
(b) The valuation of a claim made as a result of an employer's risk eventuating under the Orange Book is made under sub-clause 17.4 thereof which refers to cost, whereas clause 52 of the Red Book may include cost plus profit.
(c) Sub-clause 17.6 of the Orange Book is a new additional sub-clause which limits the liability of the contractor in respect of consequential and other specified losses except in certain circumstances stated therein.

It should be noted that the Orange Book has now been replaced with the 1999 Yellow Book, which is dealt with in Chapter 24 below.

21.4 The conditions of subcontract for works of civil engineering construction

As stated earlier, the conditions of subcontract for works of civil engineering construction were produced by FIDIC for the first time in 1994. Until then, contractors adapted either the Red Book itself or the Form of Contract issued by the English Federation of Civil Engineering Contractors known as the 'FCEC Form' and by colour as the 'Blue Form'.

The new FIDIC subcontract conditions were drafted in a compatible format and in language and terminology consistent with the Red Book. The document is red in colour to emphasise its harmony with the Red Book. Accordingly, the main contractor under the Red Book is the employer under the subcontract conditions with similar rights and obligations as those of the employer under the Red Book. The subcontractor, on the other hand, assumes the role of the contractor under the subcontract with similar rights and obligations as those of the contractor in the Red Book. However, it must be emphasised that some significant differences between the two forms of contract exist. The main features of the subcontract conditions are: format and interpretation; the relationship with the main contract; indemnities and insurance; payment; and the method of settlement of disputes. These features are briefly considered below.

21.4.1 Format

The subcontract conditions follow the same pattern as that of the other standard Forms of Contract published by FIDIC in that they are drafted in two parts. Part I contains the clauses of general application and is referred to as the General Conditions. Part II contains the clauses which must be specially drafted to suit each individual subcontract. When combined, Parts I and II form the conditions governing the rights and obligations of the main contractor and the subcontractor. Subject to minor modifications, this form of subcontract is also suitable for use when the subcontractor is nominated by the employer.

There are only 22 clauses in the subcontract form compared with the 72 contained in the Red Book. Accordingly, provisions of a number of clauses which appear in the Red Book are incorporated under a single clause in the subcontract form. For example, clause 2 of the subcontract combines the topics of the subcontractor's obligations, subcontracting and assignment of the subcontract, performance security and programming of the execution of the works. The insurance provisions are abbreviated and incorporated into clause 15, whereas settlement of disputes remains under a single clause, clause 19. Clauses 1, 4, 13, 15, 16 and 19 are discussed below.

21.4.2 Clause 1, definitions and interpretation

Clause 1 comprises six sub-clauses. Four of these, sub-clauses 1.2 to 1.5, are the same as their equivalent sub-clauses in the Red Book. Sub-clause 1.1 contains definitions of a number of standard terms used throughout the subcontract. In general, definitions of the terms used in the subcontract are similar to those in the Red Book and are grouped under similar sub-headings as those in the Red Book. However, by its very nature, sub-clause 1.1 of the subcontract differs in some respects from that in the Red Book and there are a number of new definitions which relate specifically to the subcontract.

Sub-clause 1.6 is the same as that numbered 2.5 in the Red Book.

21.4.3 Clause 4, 'main contract'

There are four sub-clauses under clause 4. Clause 4.1 provides that the subcontractor is deemed to have full knowledge of the provisions of the main contract. To that end, the contractor is required to make the main contract available to the subcontractor, but without its price details. He is also required to provide the subcontractor with copies of the Appendix to Tender and Part II of the Conditions and any conditions which differ from those in Part I of the main contract.

Sub-clause 4.2, 'Subcontractor's Responsibilities in Relation to Subcontract Works', provides that the subcontractor must carry out his contractual obligations so that no act or omission of his shall constitute cause or contribute towards any breach of the main contract by the contractor. Furthermore, it is stated that the subcontractor must assume and perform all the obligations and liabilities of the contractor under the main contract in relation to the subcontract works. This assumption of matching responsibility typifies the compatibility of the two documents.

Sub-clause 4.3, 'No Privity of Contract with Employer', confirms what might otherwise only be implied; that there is no privity of contract between the subcontractor and the employer under the main contract.

Sub-clause 4.4, 'Possible Effects of Subcontractor's Breaches of Subcontract', is one of several places in the subcontract form where the subcontractor must indemnify the contractor against any damages arising from a breach of the subcontract or misuse committed or liabilities incurred by the subcontractor (sub-clauses 5 and 13 respectively).

21.4.4 Clauses 13 and 15, 'indemnities; insurances'

The indemnity provisions relating to all losses and claims in respect of death of or injury to any person, or loss or damage to any property other than the subcontract works are contained in clause 13 of the subcontract. The insurance provisions are contained in sub-clause 15.1 and are obviously drafted in very broad terms leaving the details to be specified individually in Part II of the conditions, for each subcontract.

The details to be specified in Part II include:

— the risks to be covered by insurance by the subcontractor;
— the sums to be insured;
— the identity of the insured;
— the period of insurance.

However, the responsibility to insure against the employer's liability in respect of the subcontract works is placed on the subcontractor in Part I of the conditions.

The responsibility of the main contractor to provide insurance cover in respect of the whole of the works is outlined in sub-clause 15.2. Reference to this responsibility is also drafted in broad terms, but provision is made in Part II of the conditions for specific references particular to each subcontract. Sub-clause 15.2 also provides that the subcontract works are at the risk of the subcontractor until a taking-over certificate is issued in respect of the section of the works in which the subcontract works are contained. The responsibility for making good any loss or damage which may occur to the subcontract works during that period is allocated to the subcontractor. However, sub-clause 15.2 also provides that the subcontractor should be paid the amount of any claim which can be established under the main contractor's insurance cover in respect of certain specified events. These events are described in sub-clause 15.2 as those where any of the subcontract works, temporary works, materials or other belongings of the subcontractor are destroyed or damaged during the construction period of the main contract. Unfortunately, it is not clear what is intended by the word 'established' in that sub-clause and this must be deduced from the actual insurance policies provided under the main contract.

Accordingly, the whole area of insurance is left open and is totally dependent on the conditions negotiated between the main contractor and the subcontractor for each individual case. Therefore, in this connection, Part II of the subcontract conditions is of particular importance to those negotiating a subcontract, whether a main contractor or a subcontractor. The extent of the insurance cover to be provided by each of them must be clearly formulated and specified so that it can be obtained without creating any overlaps or gaps; the former could be costly and the latter ruinous.

21.4.5 Clause 16, 'payment'

This is another important clause in the subcontract form which is given prominence by its length and number of sub-clauses. There are six sub-clauses relating to six different topics as set out below.

Sub-clause 16.1, 'subcontractor's monthly payments'

In many respects, this sub-clause is similar to sub-clause 60.1 of the Red Book. The main difference is that the subcontractor's statement for payment is required to be submitted to the main contractor seven days after the end of each month, referred to as the 'Specified Day', compared with an unspecified period after the end of each month in clause 60 of the Red Book. The purpose of this time limitation is to ensure that the subcontractor's statement is received in time to enable the contractor to include its contents in his own submission to the engineer. The contractor may choose to alter this period of seven days to a shorter period by an entry in Part II of the conditions.[21.10] It is notable that this sub-clause establishes two new terms, 'Statement' and 'Specified Day', which are defined in sub-clause 16.1 of the main text of the conditions rather than in sub-clause 1.1, where the definitions are provided. Both of these terms are used several times in clause 16.

Sub-clause 16.2, 'contractor's monthly statements'

This sub-clause places an obligation on the contractor to include in 'his next statement for payment under the main contract', any appropriate amounts submitted by the sub-contractor in his statement to the main contractor.

Sub-clause 16.2 also provides that the contractor should include all sums certified but unpaid on account of the subcontract works in any proceedings he institutes against the employer to enforce payment. This provision is necessary from the sub-contractor's point of view since there is no privity of contract between the employer and the subcontractor. In this connection, reference should be made to sub-clause 4.3 of the subcontract conditions where this is explicitly stated.

Sub-clause 16.3, 'payment due; payment withheld or deferred; interest'

This sub-clause outlines the period within which the amounts included in the subcontractor's statement become due for payment. Subject to any specified deductions, the subcontractor is entitled to payment within 35 days of the specified day. There are a number of specified events which entitle the contractor to withhold or defer payment. In the event that he does so, the contractor is required to notify the subcontractor as soon as reasonably practicable, but not later than the date when such payment would otherwise have been payable.

It appears that since the conditions of subcontract were published, this period of 35 days has been found to be too short in relation to the events cited in the sub-clause. Until this is rectified in a subsequent revision, it is recommended that the period be increased to 70 days, by a suitable entry in Part II of the Conditions.[21.11]

The question of interest which may become due in certain specified circumstances is dealt with in the two penultimate paragraphs of sub-clause 16.3. In circumstances where interest is to be claimed, it is necessary for the subcontractor to give notice of such claim within seven days of the date when the sum claimed became payable, but was not actually paid.

Finally, this sub-clause incorporates the principle of sub-clause 60.2 of the Red Book in specifying that no amount becomes payable until the performance security, if required under the subcontract, has been provided by the subcontractor and approved by the main contractor. Therefore, it must be strongly recommended that the matter of acceptability of the performance security is determined at the earliest possible time and any question related to it immediately resolved if the subcontractor's cash flow is not to be affected.

Sub-clause 16.4, 'payment of retention money'

As suggested by its title, this sub-clause deals with the timing and method of payment to the subcontractor in respect of any retention money held under the contract.

Sub-clause 16.5, 'payment, of subcontract price and other sums due'

This sub-clause deals with the period of time by which the subcontractor is entitled to receive his final payment under the subcontract.

Sub-clause 16.6, 'cessation of contractor's liability'

As in sub-clause 60.9 of the Red Book, this sub-clause provides for the necessity for a notice of claim to be given to the main contractor as a prerequisite to establishing any liability in that respect. Such notice must be given before the issue of the defects liability certificate in respect of the main works.

21.4.6 Clause 19, 'settlement of disputes'

This clause follows the trend, set by FIDIC in its Fourth Edition of the Red Book, of requiring amicable settlement of disputes prior to arbitration. As in the Red Book, a period of 56 days is allowed for the parties to attempt settlement of their dispute in an amicable manner. The period of 56 days starts with a notice of dispute being given by one party to the other. In addition to the above, the first sub-clause of clause 19 provides that a dispute which has not been settled amicably 'shall be finally settled under the Rules of Conciliation and Arbitration of the International Chamber of Commerce by one or more arbitrators appointed under such Rules'.

Sub-clause 19.2 imposes the duty on the subcontractor to provide such information as may be required by the main contractor in relation to any dispute between the employer and the main contractor which touches or concerns the subcontract works. Sub-clause 19.2 also requires the subcontractor to attend any meetings in connection with such dispute as the contractor may reasonably request.

21.5 Other publications of FIDIC

In the last three sections, the main features of the Yellow Book, the Orange Book and the Subcontract Form as published by FIDIC have been set out and discussed. There are other publications of FIDIC which should be known to any person involved in construction, and particularly in international construction. However, they are outside the scope of this book, but it is appropriate that the more significant of these are mentioned here.

FIDIC's publications are divided into six sections, as follows:

1 Information about FIDIC.
2 Information for clients.
3 Agreements between clients and consulting engineers.
4 Information for consulting engineers.
5 Standard construction contract conditions.
6 Agreements between consulting firms.

Some of the above documents are available in languages other than English. Some are free of charge and others are for purchase at an affordable price, but all are a must to anyone who is involved in the construction industry.

Part V

The 1999 Green Book
The 1999 Red Book
The 1999 Yellow Book
The 1999 Silver Book
Dispute Boards

Chapter 22

The 1999 FIDIC Suite of Contracts

22.1 Introduction

As mentioned in Chapter 1, the FIDIC Conditions have been generally updated every ten years. Each update is usually preceded by about three years of redrafting and consultation. The task of updating the Fourth Edition of the Red Book and the Third Edition of the Yellow Book, both published in 1987, was therefore handed in 1994 to a Task Group within the FIDIC Contracts Committee. The Task Group was given the mandate to update the Red and the Yellow Books (Plant and Design-Build – for electrical and mechanical plant, and for building and engineering works, designed by the contractor) alongside a further book that was in the final stages of being drafted, the Orange Book (Design-Build and Turnkey). Another form of contract that was being contemplated at the time was one that could be used for privately financed projects and for the procurement of large projects throughout the world on a design-build/turnkey basis. It became ultimately known as the Silver Book.

The Orange Book was published in its first edition in 1995 under the title *Conditions of Contract for Design-Build and Turnkey*. However, the process of updating the Red and Yellow Books took place in a number of stages, which are described here. First, prior to the Task Group beginning the update, the group sought to establish exactly what matters the update should deal with. As a result, the Task Force nominated Reading University to conduct a survey of international construction projects and to find out how many of these were based on the FIDIC forms. This was done through a questionnaire that was sent to large consulting and construction firms and financial institutions in a number of countries throughout the world. The results of this survey were then examined by the Task Group, who took the most pertinent and relevant results and incorporated them in their ideas for the necessary changes that had to be made.

First, from the results of the Reading University survey, it was clear to the Task Force that one of the main problems with the FIDIC forms of contract was related to the role of the engineer. Half of the replies to the survey stated that they would prefer the engineer to be a more impartial figure, and the other half were of the opinion that it was contradictory for the engineer to be required to act impartially whilst he was employed by and/or on the payroll of the employer. This issue was something that the Task Group needed to address first if the update was to be a success.

Secondly, it is understood that the Task Group examined forms of contract conditions other than those published by FIDIC to see if the corresponding FIDIC documents were lacking in any way, or were leaving out important issues in a particular area.[22.1]

The third area that was looked into by the Task Group, in trying to decipher how they could best improve the FIDIC forms in use at the time, was to investigate any matters that had been brought to the attention of FIDIC by users of the various forms and any commentaries that had been written on such forms.

The immediate result of the work carried out in updating the Fourth Edition of the Red Book was the introduction of the 1996 Supplement, as discussed earlier in Chapter 19, which dealt with the problem of the role of the engineer. Subsequently, however, the group introduced a major change to the FIDIC contracts by the production of a totally new and different set of conditions of contract in September 1999, in the form of four new documents alongside those that have been in use at that time. The new suite of FIDIC contracts comprises the following forms:

(a) *The 1999 Red Book*: The Construction Contract (Conditions of Contract for Building and Engineering Works, Designed by the Employer) – General Conditions, Guidance for the Preparation of the Particular Conditions, Forms of Tender, Contract Agreement, and Dispute Adjudication Agreement;
(b) *The 1999 Yellow Book*: The Plant and Design-Build Contract (Conditions of Contract for Electrical and Mechanical Plant, and for Building and Engineering Works, Designed by the Contractor) – General Conditions, Guidance for the Preparation of the Particular Conditions, Forms of Tender, Contract Agreement and Dispute Adjudication Agreement; and
(c) *The Silver Book*: The EPC and Turnkey Contract (Conditions of Contract for EPC Turnkey Projects) – General Conditions, Guidance for the Preparation of the Particular Conditions, Forms of Tender, Contract Agreement and Dispute Adjudication Agreement.
(d) *The Green Book*: The Short Form of Contract – Agreement, General Conditions, Rules for Adjudication and Notes for Guidance.

Unlike the standard forms of contract published prior to September 1999, which were distinguished from each other on the basis of the type of project for which they were used, these new forms are distinguished on the basis of the allocation of the design function. As can be seen from the title of the 1999 Red and Yellow Books, the 1999 Red Book was drafted to be used for all types of building and engineering works designed by the employer (or designed on his behalf). The 1999 Yellow Book was designed to be used for all types of building and engineering works designed by the contractor. It is because of this new distinguishing characteristic that the 1999 Red and Yellow Books were not given a different colour, since FIDIC wished them to be identified by their respective allocation of the design function rather than by their colour. However, the experience gained since their publication in 1999, and despite FIDIC's wish, shows that these documents are found to be more easily identifiable in practice by their colour with the added tag of 'new' and/or the year of publication.[22.2]

Another intention of FIDIC or of the Task Group that was not realised was that the use of the Fourth Edition of the Red Book and the 1988 edition of the Yellow Book would diminish with time and that they would be phased out in favour of the newer 1999 forms of contract. The Task Group had intended that any new projects that were being contemplated at that time should immediately utilise the new forms of contract. However, the record of the sale of the documents shows that this did not happen; there appeared to be significant reservations on the part of those drafting new contracts about adopting a form that was untested and untried in the industry. Further, it is unlikely to happen to any significant extent in the future, particularly in the case of the Fourth Edition of the Red Book as amended by the 1996 Supplement, for many reasons. In the Middle East, for example, very few if any have ever considered abandoning the Fourth Edition of the Red Book in favour of the 1999 Red Book. The main reason is perhaps the fact that engineers and employers are by now very familiar and comfortable with the provisions of the Fourth Edition of the Red Book, and particularly their version of it if they had changed the standard form to suit their particular needs.

Furthermore, after 40 years of use, the meaning of most of the provisions of the Fourth Edition of the Red Book has become known through court decisions and some arbitral awards. Users were and are reluctant to start the process of adopting new conditions of contract as they believe, maybe rightly, that it is likely to cause further disputes and problems in an industry that is already overloaded with conflict. Perhaps it is for this reason that in the latest revision (May 2004) of its standard bidding document for the procurement of works of civil engineering works (published in January 1995), the World Bank maintained the basis of this document, the Fourth Edition of the Red Book, rather than revising it in reliance of the 1999 Red Book.

There are other reasons peculiar to the 1999 forms themselves; for example, it has transpired that the insurance clauses in the new forms (which are undoubtedly very important provisions in any construction contract) need considerable amendments. Many believe, rightly or wrongly, that the improvements that the FIDIC Task Force identified should have been incorporated in the wording of the Fourth Edition of the Red Book to produce a fifth edition rather than introducing a totally new form of contract, with a new numbering of clauses, sub-clauses and text. In any case it is important to remember that, although the new forms of contract have retained many of the principles and concepts of the old forms, the differences between them are too numerous and significant to consider the 1999 Red and Yellow Books a revision of the older forms.

These differences are both in format and in concept, as discussed in the following sections.

22.2 Differences in format

The format of the three major forms published in 1999, the Red, Yellow and Silver Books, followed that of the Orange Book. As stated in Chapter 21, the Orange Book is now replaced by the 1999 Yellow Book.

Furthermore, whereas the text of the Fourth Edition of the Red Book and the 1988 Yellow Book differed greatly, the 1999 Red, Yellow and Silver Books have been drafted with the same format and to a large extent their text is similar in its wording. However, perhaps the draftsman pursued this desire for similarity in wording too far in certain instances and it may be that problems will be encountered unless further wording is added or amendments made under the particular conditions. The need for such additional wording or amendment should become obvious on close scrutiny of a comparative analysis of the text of the three forms.[22.3]

There are vivid differences in the arrangement of the clauses and sub-clauses between the 1999 Red Book and the Fourth Edition of the Red Book. Each of the 1999 forms contains a total of 20 clauses, arranged on a more logical basis than that in the Fourth Edition of the Red Book, as they relate to discrete and distinct elements of a construction contract. This number of clauses can be compared to a total of 72 clauses that made up the Fourth Edition of the Red Book. Furthermore, there are 163 sub-clauses in the 1999 Red Book compared to 194 sub-clauses that were in the Fourth Edition of the Red Book.

It is clear from the above, therefore, that whereas in the Fourth Edition of the Red Book the clauses were shorter, under the 1999 forms the number of clauses has been drastically reduced, as has the number of sub-clauses. A reason for such a change could be the fact that clauses and sub-clauses in the 1999 Red Book contain a larger amount of text than the Fourth Edition, with a greater number of provisions. A graph highlighting the difference in the arrangement of clauses and sub-clauses between the Fourth Edition of the Red Book and the 1999 Red Book can be seen in Figure 23.2 of Chapter 23.

22.3 Differences in concept

The differences in concept between the Fourth Edition of the Red Book and the 1988 Yellow Book on the one hand, and the 1999 forms on the other, are more important than the differences in format and are more numerous. In the tradition of FIDIC, some of these conceptual differences are in favour of the employer and others are in favour of the contractor, to a large extent maintaining the balance of the risks allocated between the two parties. Twenty-one of these numerous differences are identified in a general approach here, and are then dealt with in more depth in Chapters 23 to 25 below. They are set out below not in the chronological sequence of the clause numbering, but in the order of importance:

(a) As stated above, the Fourth Edition of the Red Book and the 1988 Yellow Book were drafted for projects of different discipline of engineering; the former for civil engineering and the latter for electrical and mechanical works. The 1999 forms differ on the basis of the allocation of the design function and the existence of an engineer: the 1999 Red Book for a project designed by the employer (or designed on his behalf) with an engineer in place; and the 1999 Yellow Book for a project designed by the contractor with an engineer in place. The Silver Book was drafted for a project designed by the contractor but with no engineer in place.

(b) A fundamental change has been made to the role of the engineer in both the 1999 Red and Yellow Books. This change can be seen by comparing sub-clause 2.6 of the Fourth Edition of the Red Book with sub-clauses 3.1 and 3.5 of the 1999 Red Book. In the former the engineer is required to exercise the discretion granted to him under the contract impartially within the terms of the contract; in the latter the engineer is deemed to act for the employer unless expressly stated to the contrary but, if required to agree or determine any matter, is obliged to consult with the parties in an attempt to reach agreement and, failing agreement, to make a fair determination.

A similar difference exists between sub-clauses 2.1 and 2.4 of the 1988 Yellow Book and sub-clauses 3.1 and 3.5 in the 1999 Yellow Book. In the former, the engineer is to carry out his duties under the contract in an impartial and discretionary manner, having regard to all the relevant circumstances. Under the latter sub-clause, the engineer is deemed to act on the employer's behalf and if determining any issue in relation to the contract, he must consult with each party so as to achieve agreement between the parties, and if such an agreement is not reached, he should make a determination that is fair and reasonable in light of the relevant facts. This fundamental change forms a major departure by the 1999 forms of contract from the traditional impartial role of the engineer in the Fourth Edition of the Red Book and the Third Edition of the Yellow Book. The divergence between the 1999 forms and their predecessors grows later as the 1999 Yellow Book removes the design role from the engineer and entrusts it to the contractor; and further as the Silver Book eliminates the formal role of the engineer altogether and replaces it by that of the employer himself.

(c) Whilst all the forms of contract recognise the matrix of pure financial risks and make provision for the employer to require the contractor to provide a performance guarantee or a surety bond, the wording of the provision in the Fourth Edition of the Red Book differs greatly from that in the 1999 forms of contract. Under the 1999 Books the employer is required to return the performance security within 21 days of receiving a copy of the performance certificate (under sub-clause 11.9), whereas by virtue of clause 10.2 of the Fourth Edition of the Red Book the performance security is to be returned within 14 days of issuance of the defects liability certificate (under sub-clause 62.1).

(d) The performance security specified under the 1999 forms includes an on-demand bond,[22.4] which was not included in the examples attached to the Fourth Edition of the Red Book.

(e) Under sub-clause 2.4 of both the 1999 Red and Yellow Books, the risk of inadequate employer's financial arrangements has been recognised and, upon the contractor's request, the employer is required to provide evidence that, if the employer fails to comply with this requirement, then the contractor is entitled to suspend, or reduce the rate of, work and ultimately to terminate the contract.[22.5] There is no corresponding provision in respect of the employer's financial arrangements under either the Fourth Edition of the Red Book or the 1988 Yellow Book.

(f) One of the contractor's obligations in the 1999 forms of contract is to ensure that the works are fit for the purpose for which they were intended as stated in the contract conditions.[22.6] There is no such requirement in the either the Fourth Edition of the Red Book or the 1988 Yellow Book.[22.7]

(g) Whilst the concept of clause 12 of the Fourth Edition of the Red Book, in respect of the risks of unforeseeable physical obstructions or conditions, has been maintained in the 1999 Red and Yellow Books, both the negative and positive aspects of such risks are taken into consideration in the latter forms. Therefore, under the 1999 Books, in making a determination in respect of a contractor's claim for adverse unforeseen physical obstructions/conditions, the engineer is entitled to review whether other physical conditions encountered were more favourable than could reasonably have been expected at tender stage.[22.8]

(h) The term 'unforeseeable' is expressly defined in the 1999 forms of contract.[22.9] This definition is a useful attempt at bringing some certainty to the concept of unforeseeability. This concept relates to the allocation of risks between the parties under the FIDIC form of contract, namely that those risks that are foreseeable are borne by the contractor, those that are unforeseeable are borne by the employer. It would appear to derive from the similar concept which has been developed by the courts[22.10] under the law of negligence.

(i) The start of time limits in relation to interim payments to the contractor has been altered. Under sub-clause 14.7(b) of the 1999 Red Book, interim payment must be made within 56 days after the engineer receives the contractor's statement (and supporting documents), whereas in sub-clause 60.10 under the Fourth Edition of the Red Book such payment is required to be made within 28 days from the date of the employer's receipt of the interim payment certificate from the engineer.

(j) The employer under sub-clause 15.5 of the Red Book is entitled to terminate the contract at any time and for his convenience, by simply giving notice to the contractor of such termination. The only condition that applied to this provision is that the employer must not choose to terminate the contract in order to execute the works himself or to arrange for the works to be executed by another contractor.

(k) Some changes have been made in relation to the allocation of the risks as between the employer and the contractor. In particular, reference is made in the 1999 forms to sub-clause 17.3, which now includes terrorism as one of the employer's risks; and clause 19, which now expressly defines *force majeure* and the relief available to the parties in such an event.

(l) There is a new provision in the 1999 forms, entitled 'Limitation of liability', sub-clause 17.6, which limits the contractor's liability to the accepted contract amount, or to a sum stated in the particular conditions (if any).

(m) The insurance provisions of the 1999 forms of contract, clause 18, differ greatly from the provisions under the Fourth Edition of the Red Book and the 1988 Yellow Book, leaving the matter of insurance to be discussed and agreed at a meeting required to be held before the date of the letter of acceptance (second paragraph of sub-clause 18.1).

(n) Clause 19 of the 1999 forms entitled 'Force Majeure' replaces the 'special risks' concept of the Fourth Edition of the Red Book. Since *force majeure* is a legal concept which has slight but potentially significant application in different jurisdictions, intending employers should follow the recommendation given in the Guidance for the Preparation of Particular Conditions[22.11] to the 1999 forms of contract, namely that before inviting tenders they should verify that the wording of this clause is compatible with the law governing the contract.

(o) There is a new provision in the 1999 forms of contract requiring the employer to comply with a claims procedure for employer's claims, as can be seen from

(p) Strict time limits are imposed under the 1999 forms of contract if a claim is to be made by the contractor under sub-clause 20.1, in that notice of such a claim must be made within 28 days after the contractor became aware, or should have become aware, of the event or circumstance giving rise to the claim. Details of the claim with supporting particulars should be given within 42 days. If the contractor fails to give notice of a claim, the employer is discharged from all liability in connection with the claim.

(q) There is also a time limit under sub-clause 20.1 for the engineer under the 1999 Red and Yellow Books, and the employer under the Silver Book, to respond to the notification of a contractor's claim.

(r) Strict time limits are also imposed for other notices, for example the contractor must give 28 days' notice of the intended commencement of each sub-contractor's work,[22.12] and the employer is required to give not less than 7 days' notice of the commencement date.[22.13]

(s) The 1999 forms of contract have expressly introduced the concept of 'value engineering', which will if adopted accelerate completion of the works; reduce the cost to the employer of executing, maintaining or operating the works; improve the efficiency or value to the employer of the completed work or otherwise be of benefit to the employer.[22.14] The contractor may make a proposal incorporating value engineering at any time to the engineer; if approved it is valued under the variation procedure of sub-clause 13.3.

(t) There is a new provision in the 1999 Red and Yellow Books expressly entitling the contractor, in the event of late payment by the employer, to be reimbursed for financing charges calculated at an annual rate which is 3% above the discount rate of the central bank (in the country of the currency of payment) and compounded monthly.[22.15]

(u) A new step in the dispute settlement procedure, referral of the dispute to a Dispute Adjudication Board (DAB) similar to that introduced in the 1996 Supplement to the Fourth Edition of the Red Book, has been introduced under sub-clauses 20.2 to 20.4 of the 1999 forms of contract.[22.16] There are, however, a few important aspects that differentiate the DAB concept introduced in the 1996 Supplement and the 1999 forms. These are as follows:
  (i) The DAB in the 1999 Red Book is appointed by the date stated in the Appendix to Tender, which proposes 28 days after the commencement date;
  (ii) The Guidance for the Preparation of Particular Conditions to the 1999 Red Book includes an option of reverting to the traditional role of the engineer (and gives example sub-clauses to replace/amend sub-clauses 20.2 to 20.4); and
  (iii) The DAB in the 1999 Yellow and Silver Books is appointed by the date 28 days after a party gives notice of its intention to refer a dispute to a DAB, sub-clause 20.2.

It is worth noting here that the Silver Book is totally new and, to a large extent, forms a departure from FIDIC's established policy of providing standard forms of contract with balanced risk allocation. The risks in the Silver Book are mostly allocated to the contractor. This imbalance would obviously cause a large increase in the cost of the project, but perhaps a more stable final cost figure. The Silver Book can be distinguished from the other forms of contract by the absence of the function of the 'Engineer'.

The distinguishing characteristics of the Silver Book should not be taken as a criticism of its concept and application. In particular, the Silver Book was conceived in response to the need created by those who favoured the use of private finance for infrastructure projects, and grew as a result of the demands associated with BOT or BOOT projects and with the new ideas of mixing together design, construction and operation. This entailed demanding a fixed, lump-sum contract price with little or no risk of an increase if and when unexpected events took place. Of course, privately financed projects require financial viability with an assured return on the funds advanced. Therefore, although demanding a fixed, lump sum contract price means that the employer would be paying a higher price for the construction of the project, he would not normally object to having to do so if he were assured of an acceptable return on his total investment that would compensate for this additional cost. This form of contract is discussed further in Chapter 25.

Whilst the changes in format may not be sufficient to influence one's choice between the old and the new forms, the 21 significant changes in concept should. These changes in concept, and their implications, will be discussed further in Chapters 23 to 25 and the logic and rationale behind them will be reviewed.

The Green Book is also a new venture for FIDIC in that it is intended for smaller contracts of less than US$0.5 million. While this part of the book is essentially intended to deal with the three major forms published in 1999 (the Red, Yellow and Silver Books), it is appropriate to consider within this chapter the new Green Book as it has little in common with the other 1999 forms of contract.

22.4 The 1999 Green Book

The Green Book, also known as the Short Form of Contract, is intended to be used as a form of contract for engineering and building work of fairly simple or repetitive work of short duration with relatively small capital value,[22.17] but it may be suitable, subject to the type of work and circumstances, for contracts of greater value. The objective of the Green Book is for the contract to express in clear and simple terms traditional procurement concepts.[22.18]

Furthermore, the form is drafted in a flexible format that includes all essential commercial provisions and a variety of administrative arrangements, despite the fact that it is contained in only ten pages.

The flexibility of the Green Book is a significant feature of this form of contract, and reflects the fact that, although a simpler and shorter form of contract is appropriate for smaller value contracts, the circumstances and specifics of the intended project merit suitable amendments and provision. Such flexibility is centred on the Appendix, which appears at the beginning of the Green Book. The Appendix includes various default entries which, prior to inviting tenders, the employer is prompted to amend as he deems appropriate in order that the characteristics of his chosen contract form are incorporated. This is then signed by the selected contractor as part of the Agreement of the contract. One of the more important features of the Appendix is the fairly comprehensive range of options set out for valuation of the works: lump sum price, lump sum price with schedules of rates, lump sum price with bill of quantities, re-measurement with tender bill of quantities, and cost reimburseable. The Green Book does not envisage that the general conditions of contract will be supplemented by particular conditions to suit an individual contract[22.19] (unlike most of the other forms of FIDIC contracts). Nevertheless, in the 'Notes for Guidance' which appear at the end of the book but do not form part of the contract, alternative wording for certain clauses is given so that the employer may draft particular conditions if deemed necessary. In addition, various options are given in the Notes for Guidance so that, once the employer considers such options, he is guided to select what he needs and to delete what he does not, resulting in a contract form which crystallises his wishes.

It is envisaged in the Green Book that there will be no traditional 'Engineer' or 'Employer's Representative' in the formal sense that is used by FIDIC in most of its other forms of contract. Instead, a member of the employer's personnel is to be designated as the authorised person, who then takes over the functions usually performed by the engineer. However, the employer may also appoint an independent person to carry out certain duties, by naming such person in the Appendix or by notification to the contractor during the course of the contract, under sub-clause 3.2 of the general conditions 'Employer's Representative'.

The Notes for Guidance in respect of clause 3 are particularly worthy of note in this context. It would appear the intention behind sub-clause 3.2 is to enable employers to employ professional consultants to administer the contract, rather than to have an independent individual or firm appointed to act impartially between the parties. In fact the notes expressly state:

> 'There is no dual role or duty to be impartial. If an impartial Employer's Representative is required with a role similar to the traditional Engineer, then the following words could be used in the Particular Conditions . . .'.

The notes then give a suggestion of how sub-clause 3.2 should be amended, such that the employer's representative has a duty to act in a fair and impartial manner in connection with certain listed clauses, namely those dealing with instructions, giving consents, giving of notices, variations (instruction and evaluations), valuations (interim and final) of the works, and the determination and granting of extension of time.

The provisions of the Green Book are set out in 15 clauses following a similar format to that of the three 1999 major forms of contract (the Red, Yellow and Silver Books).

Clause 1 of the general conditions begins, like all the other forms of contract, with definitions of the words and expressions that are used in the Green Book and sets out the general provisions relating to interpretation; priority of documents; law; communications; and statutory obligations.

Clause 2 of the general conditions deals with the employer and his obligations in respect of the site and the obligation to provide assistance if the contractor requires any permits, licences or approvals in order to carry out the works. The contractor is required to comply with all instructions given by the employer and no approval by the employer (or his representative) will affect any of the contractor's obligations under the contract.

Clause 3 of the general conditions provides for the employer to appoint a member of his staff as an authorised person to act for him and, as already discussed above, also to appoint an employer's representative (who has no duty to the contractor in respect of fairness or impartiality, unless an amendment is included in the particular conditions, if any).

The contractor has general obligations under clause 4 of the general conditions to provide site supervision, all labour and materials, plant and machinery that are required for the construction of the works. If the contractor wishes to subcontract any of the works, he can only do so once he has obtained the consent of the employer under sub-clause 4.3 of the Green Book, but is forbidden from sub-contracting for the whole of the works.

It is also envisaged in the Green Book that, while the employer may provide the design himself or by others on his behalf, the contractor may undertake design to the extent specified in the Appendix and/or may have submitted a design with his tender. In either case the provisions of clause 5 of the general conditions, 'Design by contractor' apply. The contractor is thereby responsible for his tendered design and the design as specified in the Appendix which, he must ensure, is fit for the intended purposes defined within the contract. However, the employer retains responsibility for the specification and drawings.

Clause 6 of the general conditions deals with the employer's liabilities and is connected directly to, and has to be considered together with, clauses 13 and 14: 'Risk and Responsibility' and 'Insurance', respectively. If the logical sequence of the inter-relationship between the topics of 'Risk', 'Responsibility', 'Liability' and 'Insurance' is followed, one must start with risk and proceed to responsibility, then to liability and end with insurance. At this point, it is necessary to draw attention to the first of two serious problems with the Green Book, which relates to the somewhat mishandling of the three terms 'risks', 'responsibility' and 'liability'. In clause 13, which is the relevant clause to the first two of these topics, we find no mention of risk at all. In fact, other than in the title, the word 'risk' does not appear anywhere in the Green Book. But, on close scrutiny, it becomes apparent that the reference to 'an Employer's Liability' in the second paragraph of sub-clause 13.1 is intended to lead the reader to clause 6 of the form, where, somewhat inexplicably, it would appear that what are normally referred to as 'risks' in most forms of construction contracts are here referred to as 'liabilities'.[22.20]

It must be said in this context that it is rather peculiar that FIDIC, which pioneered the adoption of the natural flow of risk *to* responsibility *to* liability *to* insurance in its various forms of contract,[22.21] would now seem to be turning the clock back with its Green Book and confusing 'risk' with 'liability'. Even from a linguistic point of view, it is difficult to understand how risk could be confused with liability. They are two different concepts and two different terms, which are entirely different etymologically, scientifically and legally. As discussed in Chapter 7, 'risk' is technically defined as 'A combination of the probability, or frequency, of occurrence of a defined event and the magnitude of the consequences of the occurrence'. 'Liability', on the other hand, is defined and discussed in Chapter 11 as 'the legal concept of one party being subject to the power of another, or to a rule of law requiring something to be done or not done. This requirement to do something or not to do it can be compelled by legal process at the other party's instance. It is sometimes called subjection.'[22.22] Liability may arise either from a voluntary act or by force of some rule of law. Therefore, a person who enters into a contract becomes liable to perform what he has undertaken, or to pay for the counterpart performance, or otherwise to implement his part of the bargain.

Even if it were not wrong to use the term 'risk' and mean 'liability', and this is obviously wrong, the substitution of 'risks' with 'liabilities' is a negative step. The subject matters of risk and risk management are now part of a respected field of science and their principles should be strengthened and enhanced, rather than diluted in any form of contract.

In view of the above, it is submitted that 'Employer's Liabilities', as such words appear in the Green Book, should be read as 'Employer's Risks'. The following discussion is on that basis. It is generally accepted that the risks allocated to both the employer and the contractor in construction contracts are categorised into two types: those which result in physical loss, damage or personal injury; and those which result in pure economic loss and/or loss of time. However, clause 6 of the general conditions in the Green Book combines and lists both types of risks. Some of these risks are dealt with elsewhere in the document in more detail, such as the risk of *force majeure* in clause 13.2, whilst the majority remains with no further explanation.

It must be said that grouping in one clause all the employer's risks, including those that lead to pure economic and/or time loss and those that lead to physical loss/damage or personal injury, is a good idea provided they are properly referred to as 'risks'. At least then it is unlikely that the user will fall into the trap, set unwittingly in the 1999 Red, Yellow and Silver Books, of forgetting one of these types of risk, or confusing one with the other, see Chapter 23.[22.23] It would be very unfortunate for the contractor if, in a dispute situation, the arbitrator decided that a risk, which the contractor had during the course of the works assumed to be allocated to the employer, was in fact a contractor's risk. It would be even more regrettable for the employer to find that the arbitrator had decided the opposite.[22.24] For that reason perhaps, this idea of grouping all the employer's risks in one clause has now been adopted in the latest of the published FIDIC's forms, the Form of Contract for Dredging and Reclamation Works, published in 2001, which has also fortunately resolved the problem of the confusion between 'risk' and 'liability'. Thus, for anyone who is intent on using a form of FIDIC contract for smaller contracts, it would be advisable to base it on this latest form of FIDIC contract rather than the Green Book, or at least to replace clauses 6 and 13 of the Green Book with the corresponding relevant clauses in the Form of Contract for Dredging and Reclamation Works.

In both types of risk, namely that which results in physical loss/damage or personal injury, and that which results in pure financial loss and/or loss of time, neither clause 6 nor clause 14, dealing with insurance, answers the question as to whether the consequent liability of these risks is to be covered by an indemnity or by insurance. In connection with such consequent liability, reference should be made to sub-clause 10.4 of the Green Book which provides as follows:

'10.4 Right to Claim

If the Contractor incurs Cost as a result of any of the Employer's Liabilities, the Contractor shall be entitled to the amount of such Cost. If as a result of any of the Employer's Liabilities, it is necessary to change the Works, this shall be dealt with as a Variation.'

There is no explanation as to how the cost incurred by the contractor arising from 'the Employer's Liabilities' (which should, as submitted above, be read as 'the Employer's risks') is linked to the insurance provisions of the contract, other than in sub-clause 14.1 which merely provides that the contractor effect insurance for the liability of both parties for loss, damage or injury.

The second and equally serious problem with the Green Book is in respect of the gap created by the division of risks (referred to as 'liabilities' in this form of contract) between the employer and the contractor. The employer is allocated the risks as listed in clause 6 of the general conditions. By virtue of the second paragraph of sub-clause 13.1, the contractor is allocated the risks of (and is deemed to indemnify the employer against) all loss or damage happening to the works and of all claims or expense arising out of the works caused by 'a breach of the Contract, by negligence or by other default of the Contractor, his agents or employees'. To whom then are the other risks allocated, namely those that do not qualify within the meaning of an 'employer's risk' and those that cannot be described as 'a breach of the Contract' by the contractor? This problem is of a similar nature to that created in the 1999 Red, Yellow and Silver Books through their sub-clause 17.1(b)(ii), where the basis of indemnity is negligence rather than legal liability, see Chapter 23 below.[22.25] This gap in risk allocation ultimately creates a gap in the insurance cover for the project, unless it is specifically dealt with in the Appendix. There would appear to be, unfortunately, a widely held but misconceived view[22.26] in the construction industry that all the risk in the works lies automatically with the contractor unless expressly stated to be at the employer's risk in the contract. The Green Book, while it expressly lists some particular risks which are to be borne by the employer and the contractor, respectively, otherwise remains silent. In this regard it is no different from all of the other forms of FIDIC contracts, and indeed most standard forms of construction contract. The uncertainty that such absence of express allocation of risk brings has, in practice, led to much confusion and many claims and disputes when risks eventuated. Such confusion and conflict could and should be eliminated by expressly including a clear statement regarding the allocation of risks in the contract.

Sub-clause 13.2 of the general conditions deals with *force majeure*, which is a risk allocated to the employer in sub-clause 6.1(i), although it is referred to as 'a liability'. '*Force Majeure*' is defined in sub-clause 1.1.14 of the Green Book as 'an exceptional event or circumstance: which is beyond a Party's control, which such Party could not reasonably have provided against before entering into the Contract; which, having arisen, such Party could not reasonably have avoided or overcome; and, which is not substantially attributable to the other Party'.

Whatever the merit, desirability or necessity for such a clause and express definition in a major form of contract such as the new Red, Yellow or the Silver Books, it is suggested here that there is none for a 'simple contract of short duration with relatively small capital value' to which the Green Book is intended to be suitable.[22.27] The concept and principles of *force majeure* are well established and much used in the fields of law and insurance and, introducing such clause and definition may create undue complications from the legal and insurance points of view. For example, the FIDIC *force majeure* definition is wide and not limited to specified situations, but it has to be an 'exceptional' event beyond the control of a party; what constitutes 'exceptional' in any given circumstances is open to debate. It is further suggested that the appropriate method of dealing with the risks as captured by the intended meaning of sub-clause 13.2 of the Green Book is to designate them as exceptional risks leading to specific remedies under the contract, as is the case under clauses 65 and 66 of the Fourth Edition of the Red Book: 'Special Risks' and 'Release from Performance'.[22.28]

Whilst the development of the title of clause 66 of the old Red Book from 'Frustration' in its third edition to 'Release from Performance' in the fourth edition is outside the scope of this book, it is perhaps worth briefly exploring the difference between the two doctrines of frustration and *force majeure* with particular reference to construction and construction insurance.

With the exception of the White Book, the FIDIC construction contracts in their various forms have always been based on the premise that liability for non-performance of contractual obligations is a strict one. Failure to perform these required duties under the relevant contract gives the aggrieved party entitlement to a claim for damages. The rationale for the above rule in the FIDIC construction contract may lie in its common law origin, but in any case, except for specified events in the contract, the contractor is obliged to complete the contract.[22.29]

Where strict liability applies, why a party failed to fulfil its obligation is immaterial, and it is no defence for that party to plead that it has done its best.[22.30] As a party enters into contractual obligations freely, it accepts certain risks that are allocated to it and promises to bear these risks if and when they eventuate. In this way, the contracting parties are able to plan ahead with calculable certainty their schemes and arrange their business affairs. There are, however, specific risks that are beyond the capacity of a party to accept. In such circumstances, it would be better to name these risks and specify the method of dealing with and managing them. As construction contracts grow in size and complexity, such unacceptable risks become harder to identify and define in an explicit manner in the contract, hence the need for a doctrine of frustration or *force majeure* to excuse non-performance of promises. Frustration occurs whenever the law recognises that, without default of either party, a contractual obligation has become incapable of being performed because the circumstances in which performance is called for would render it a thing radically different from that which was undertaken by the contract: 'it was not this that I promised to do'.

As argued by those who advocate the use of a *force majeure* clause, the advantage of such a clause is that it offers to the parties, should they wish to avail themselves of it, the opportunity to escape from the narrowness of the doctrine of frustration by including within their clause an event which would not be sufficient to frustrate the contract. However, such a clause does give the court power to review each word of the whole of the clause.[22.31] Indeed, in certain jurisdictions it is argued that conflict as to the interpretation of a *force majeure* clause becomes a matter for litigation rather than arbitration.

It is said that the doctrine of frustration is much narrower than the doctrine of *force majeure* and that uncertainty is inherent in the former, but that such uncertainty might be eliminated to a large extent by the incorporation into a contract of a suitably drafted *force majeure* clause. Then, the enquiry of the court can be limited and focused on the terms of the clause rather than the whole general notion of what is reasonable and fair under the doctrine of frustration.[22.32]

It is suggested that it would be much more sensible and less likely to produce conflict in the first place, if neither is stated in the contract conditions, leaving the matter to the provisions of the applicable law in the relevant jurisdiction.

The flexibility of the Green Book is perhaps most apparent where the insurance provisions are concerned. This is due to the fact that the relevant clause, clause 14 of the general conditions, only specifies the general framework of the cover required, leaving the various details to be completed by the employer in the Appendix, with extensive freedom to include any insurance requirement and in any detail he deems fit.

When a risk eventuates, the consequences might be either insurable or not. Whether or not the risk is required to be insured, clause 6 is silent on how and when insurance, if available, is to be provided in respect of the 'Employer's Liabilities' or the risks specified therein. Thus, the link between risk, liability, indemnity and insurance is lost. This situation may be due to the idea expressed in sub-clause 14.2 that the employer should set out his precise requirements relating to the required insurance cover in the Appendix, but should not the Appendix then explain the relationship between sub-clauses 10.4 and 14.2?

It is, perhaps, smaller employers and contractors that are not usually fully versed in the complexities of construction insurance and, therefore, it is in smaller contracts for which the Green Book is intended that users require specific standard conditions to assist them in providing a balanced arrangement and one that would work without conflict when risk events lead to loss, damage and/or injury. Some of the points that may escape the attention of those who are not used to this topic include the following:

(a) Clause 14.1(c) requires insurance cover to be provided 'for liability of both parties and of any employer's representative ... except to the extent that liability arises from the negligence of the Employer, any Employer's representative or their Employees'. It is clear that there is no requirement to effect insurance against the negligence of the employer or the employer's representative. However, does this mean that an insurance cover is required for non-negligence of those named above?

(b) The term 'Works' is defined in sub-clause 1.1.19 of the conditions as meaning 'all the work and design (if any) to be performed by the Contractor including temporary work and any Variation'. As 'Works' would most probably include design carried out by the contractor, there is a standard requirement for professional indemnity insurance included in sub-clause 14.1(a). Details of such cover must be included in the Appendix and therefore space must be allocated in it for such details.

(c) What is the meaning of the phrase in clause 14.2 'evidence that any required policy is in force'? The uninitiated may feel that a letter from an insurance broker is sufficient. However, the wording of construction insurance policies differs greatly and it is meaningless to present or have presented as evidence anything other than the insurance policies themselves, including any endorsements issued and conditions attached.

Returning to clause 7 of the Green Book, the contractor is required under sub-clause 7.1 to begin carrying out the works on the commencement date and to proceed so as to comply with the time for completion stated in the Appendix. Sub-clause 7.2 requires the contractor to submit a programme in the form stated in the Appendix. This is the means by which the contractor guarantees that the works will be ready for taking over by the employer within the time for completion. Like the other 1999 FIDIC forms of contract, if the contractor's progress of the works is delayed because of the 'Employer's Liabilities' (employer's risks), the contractor will be entitled to an extension of time under the contract by virtue of sub-clause 7.3; and if the contractor fails to complete the works on time then he becomes liable to pay liquidated damages to the employer, though not referred to as such, for every day of delay by virtue of sub-clause 7.4.

The works are eligible to be taken over once the employer has notified the contractor that he considers the works eligible to be taken over under clause 8 of the Green Book. Alternatively, the employer may notify the contractor that, even though the works are not fully complete, they are nonetheless ready to be taken over.

Within the Appendix, the default entry for the period for notifying defects is 365 days but the employer is prompted to amend this period as appropriate. Under sub-clause 9.1 the employer can request the contractor to remedy any defects that he notices within this time period. These defects (provided they are the result of his own design or workmanship) are required to be rectified by the contractor at his own cost under sub-clause 9.1, while other defects are to be rectified by the contractor but then valued as a variation, also as per sub-clause 9.1. Sub-clause 9.2 permits the employer to instruct the uncovering and/or testing of work, the cost of which, if it transpires that the contractor's design or workmanship is not in accordance with the contract, lies with the contractor, but otherwise to be valued as a variation.

The employer can, of course, make any variation to the contract as he wishes under clause 10 of the Green Book. Once the employer has instructed the contractor about any variations he wishes to make to the contract, the contractor must value the variations in accordance with the procedure laid down in sub-clause 10.2 of the general conditions. He must then submit to the employer an itemisation of the cost of the variations that the employer has requested. The employer should then check the valuation and agree (if at all possible) upon the value. Importantly, if there is no agreement between the employer and the contractor in this regard, the employer will make the final decision on the value of the variation, in accordance with sub-clause 10.5.

In place of the strict claims procedure, notification requirements and corresponding time limits contained in clause 2.5 (employer's claims) and 20.1 (contractor's claims) of the 1999 Red, Yellow and Silver Books, sub-clause 10.3 of the Green Book merely requires each party to notify the other as soon as he becomes aware of anything which may delay or disrupt the works or give rise to additional payment. The contractor's entitlement to extension of time or additional payment is then said to be limited to that if he had given 'prompt notice'. What exactly 'prompt' means is open to question.

The general method of payment provided for under the Green Book is on a monthly basis, under clause 11 of the general conditions. The works are to be valued on the basis of the particular option chosen in the Appendix. The contractor is entitled to be paid monthly for the following:

(a) value of the works executed to the date of the invoice; and
(b) percentage stated in the Appendix of the value of materials and plant delivered to the site at a reasonable time;

but the employer is entitled to withhold retention at the rate stated in the Appendix. If payment to the contractor is delayed in any way, the contractor will be entitled to interest on any amount that remains overdue to him.

Clause 12 of the Green Book deals with three situations:

(a) where the contractor defaults from the contract;
(b) where the employer defaults from the contract; or
(c) where one of the parties to the contract is declared insolvent.

Clauses 13 'Risk and Responsibility' and 14 'Insurance' have been discussed above.

Clause 15 deals with resolution of disputes. It provides that if a dispute arises between the parties under the contract, then unless it is settled amicably, either party may refer the dispute to an adjudicator for a decision. The Notes for Guidance make some useful suggestions regarding the timing of appointment of the adjudicator, namely from the outset of the contract, and the country of origin of the person appointed. If either party is dissatisfied with the decision of the adjudicator, it can issue a notice of dissatisfaction under sub-clause 15.2 and refer the matter to arbitration by a single arbitrator under the rules as stated in the Appendix, which has the UNCITRAL Rules as a default entry.

Chapter 23

The 1999 Red Book

23.1 Introduction

As mentioned in Chapter 22, FIDIC's 1999 Red Yellow and Silver Books are very similar in format. They each contain 20 clauses, 17 of which have common titles and all of which have similar wording where the concepts match. The three clauses that carry different titles are: Clause 3 'The Engineer' in the 1999 Red and Yellow Books is changed to 'The Employer's Administration' in the Silver Book; Clause 5 'Nominated Subcontractors' in the 1999 Red Book is changed to 'Design' in the 1999 Yellow and Silver Books; and Clause 12 'Measurement and Evaluation' in the 1999 Red Book is changed to 'Tests after Completion' in the 1999 Yellow and Silver Books. Although the number of clauses in each of these three forms of contract has been drastically reduced from 72 in the Fourth Edition of the Red Book, and 51 in the Third Edition of the Yellow Book, the number of words in each of the new forms has largely increased. Figure 23.1 shows the number of words in Part I of the various forms of contract. However, the wording is simpler and clearer than in the older Books. The sentences are shorter and express fewer ideas, making them easier to understand. Figure 23.2 shows the number of sub-clauses in the 1999 Red Book.

In the words of Peter Booen, FIDIC's principal drafter of the 1999 forms of contract, the 1999 Red Book has been 'developed in the direction of "user-friendliness", where the "users" are the individuals who write and administer the contracts.'[23.1]

There are sufficiently valuable innovations and new concepts in the 1999 Red Book, some of which have been overdue for some time, to entice the construction industry to adopt them for use in a standard form. No other form of contract has managed to

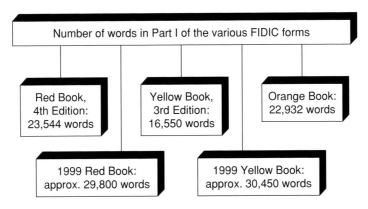

Fig. 23.1 Number of words in the FIDIC forms.

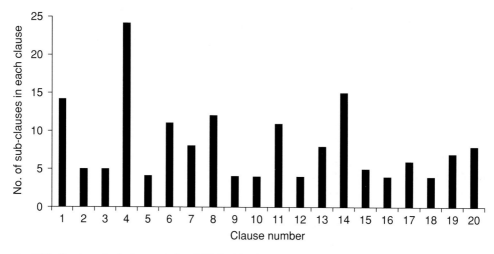

Fig. 23.2 Number of sub-clauses in the 1999 Red Book.

deal so successfully with some of the major issues dealt with under the 1999 Red Book. However, and perhaps inevitably in such work, in a number of provisions there are still some problems yet to be tackled and resolved by future amendments. In particular, amongst the amendments that are required are those to clauses 17 and 18, the risk and insurance clauses, as discussed in Sections 23.3.13 to 23.3.15 below.[23.2]

23.2 The 1999 Red Book: concepts and content

Although the 1999 Red Book is not a revision of the Fourth Edition of the Red Book, it embodies nearly all the concepts and principles of the Fourth Edition, but with different arrangements of text and some significant changes.

The most significant change is the departure of the 1999 Red Book from the traditional impartial role of the engineer, as explained in paragraph (b) of Section 23.3 and detailed further below. This departure, which had to incorporate with it a new mechanism for the impartial resolution of any dispute arising during the execution of the works, makes the 1999 Red Book a different type of contract from its predecessor. However, the function of design continues to remain with the engineer and no significant difference exists under that part of his role. It is only when we come to examine the 1999 Yellow Book that the difference from the Fourth Edition becomes even greater, as explained in Chapter 24.

Returning to the different arrangement of text, for those who are familiar with the Fourth Edition of the Red Book it is useful to identify the new location of its provisions, either in full or in part, under the 1999 Red Book and also to discuss the rationale behind the newer form. In the Fourth Edition of the Red Book, the provisions appear to follow no particular order or grouping, whereas the 1999 Red Book is rather different in this respect, with the clauses regrouped more logically. After definitions and general provisions, in clause 1, eight discrete subject areas can be identified, as follows:

(a) Clauses 2 to 4 incorporate and deal with the rights, duties and obligations of the triangle of employer, engineer and contractor;
(b) Clauses 5 to 7 incorporate and deal with resources, which includes:
 — Nominated sub-contractors
 — Staff and labour
 — Plant, materials and workmanship;
(c) Clauses 8 to 10 deal with time aspects:
 — Commencement, delays and suspension
 — Tests on completion
 — Employer's taking over;
(d) Clause 11 deals with defects liability;
(e) Clauses 12 to 14 deal with monetary terms:
 — Measurement and evaluation
 — Variations and adjustments
 — Contract price and payment;
(f) Clauses 15 and 16 deal with termination
 — By the employer
 — By the contractor (with suspension);
(g) Clauses 17 to 19 incorporate and deal with risk and responsibility, insurance and *force majeure*; and
(h) Clause 20 incorporates and deals with claims, disputes and their settlement.

Most of the concepts and principles in the Fourth Edition of the Red Book find their way into the 1999 Red Book, but under a different sub-clause and in a different location of the text. To locate corresponding sub-clauses between the two forms of contract, two tables are set out below. If the reader is familiar with the Fourth Edition of the Red Book, then Table 23.1 can be used to determine the location of a corresponding provision in the 1999 Red Book. If a provision in the 1999 Red Book is known, then Table 23.2 provides the equivalent sub-clause(s) of the Fourth Edition of the Red Book.

It must be noted however that:

(a) There are many new concepts and principles that have been injected into the 1999 Red Book (see Section 24.3 below).
(b) It is important to note that the text of some sub-clauses in the Fourth Edition of the Red Book has been deleted in the 1999 Red Book and thus a sub-clause in the older form of contract may not have a corresponding sub-clause in the newer form. Similarly, the 1999 Red Book includes additional provisions which do not have a corresponding sub-clause in the Fourth Edition of the Red Book. In either case, the word 'none' appears in the relevant table below.
(c) Although the text of some sub-clauses would appear to be similar, or sometimes very similar, between the Fourth Edition of the Red Book and the 1999 Red Book, the wording has in most cases been changed in the 1999 forms of contract and the new wording should be carefully examined to arrive at its proper intent. The differences could be of minor importance, but sometimes they could be very significant in their effect.

Table 23.1 The Fourth Edition of the Red Book and the corresponding conditions in the 1999 Red Book.

The Fourth Edition of the Red Book		Corresponding clauses in the 1999 Red Book	
Sub-clause	Title	Clause/sub-clause	Title
Definitions and Interpretation			
1.1	Definitions	1.1	Definitions
1.2	Headings and Marginal Notes	1.2	Interpretation
1.3	Interpretation	1.2	Interpretation
1.4	Singular and Plural	1.2	Interpretation
1.5	Notices, Consents, Approvals, Certificates and Determinations	1.3	Communications
Engineer and Engineer's Representative			
2.1	Engineer's Duties and Authority	3.1	Engineer's Duties and Authority
2.2	Engineer's Representative		None, but see sub-clause 3.2
2.3	Engineer's Authority to Delegate	3.2	Delegation by the Engineer
2.4	Appointment of Assistants	3.2	Delegation by the Engineer
2.5	Instructions in Writing	3.3	Instructions of the Engineer
2.6	Engineer to act Impartially		None, but see sub-clause 3.5
Assignment and Subcontracting			
3.1	Assignment of Contract	1.7	Assignment
4.1	Subcontracting	4.4	Subcontractors
4.2	Assignment of Subcontractor's Obligations	4.5	Assignment of Benefit of Subcontract
Contract Documents			
5.1	Language(s) and Law	1.4	Law and Language
5.2	Priority of Contract Documents	1.5	Priority of Documents
6.1	Custody and Supply of Drawings and Documents	1.8	Care and Supply of Documents
		1.11	Contractor's Use of Employer's Documents
6.2	One Copy of Drawings to be Kept on Site	1.8	Care and Supply of Documents
6.3	Disruption of Progress	1.9	Delayed Drawings or Instructions
6.4	Delays and Cost of Delay of Drawings	1.9	Delayed Drawings or Instructions
6.5	Failure by Contractor to Submit Drawings	1.9	Delayed Drawings or Instructions
7.1	Supplementary Drawings and Instructions		None, but see sub-clause 3.3
7.2	Permanent Works Designed by Contractor	4.1	Contractor's General Obligations
7.3	Responsibility Unaffected by Approval	4.9	Quality Assurance
		3.1(c)	Engineer's Duties and Authority
General Obligations			
8.1	Contractor's General Responsibilities	1.8	Care and Supply of Documents
		4.1	Contractor's General Obligations
8.2	Site Operations and Methods of Construction	4.1	Contractor's General Obligations
9.1	Contract Agreement	1.6	Contract Agreement
10.1	Performance Security	4.2	Performance Security
10.2	Period of Validity of Performance Security	4.2	Performance Security
10.3	Claims Under Performance Security	4.2	Performance Security
11.1	Inspection of Site	4.10	Site Data
		4.15	Access Route
12.1	Sufficiency of Tender	4.11	Sufficiency of the Accepted Contract Amount

(Table 23.1 Contd.)

The Fourth Edition of the Red Book		Corresponding clauses in the 1999 Red Book	
Sub-clause	Title	Clause/sub-clause	Title
12.2	Not Foreseeable Physical Obstructions or Conditions	4.12	Unforeseeable Physical Conditions
13.1	Work to be in accordance with the Contract	3.3	Instructions of the Engineer
		4.1	Contractor's General Obligations
		19.7	Release from Performance under the Law
14.1	Programme to be Submitted	4.1	Contractor's General Obligations
		8.3	Programme
14.2	Revised Programme	8.3	Programme
14.3	Cash Flow Estimate to be Submitted	14.4	Schedule of Payments (last paragraph)
14.4	Contractor not Relieved of Duties and Responsibilities	3.1(c)	Engineer's Duties and Authority
15.1	Contractor's Superintendence	4.3	Contractor's Representative
		6.8	Contractor's Superintendence
16.1	Contractor's Employees	6.8	Contractor's Superintendence
		6.9	Contractor's Personnel
16.2	Engineer at Liberty to Object	6.9	Contractor's Personnel
17.1	Setting Out	4.7	Setting Out
18.1	Boreholes and Exploratory Excavation	13.1(e)	Right to Vary
19.1	Safety, Security and Protection of the Environment	4.8	Safety Procedures
		4.18	Protection of the Environment
		4.22	Security of the Site
		6.7	Health and Safety
19.2	Employer's Responsibilities	2.3	Employer's Personnel
20.1	Care of Works	17.2	Contractor's Care of the Works
20.2	Responsibility to Rectify Loss or Damage	17.2	Contractor's Care of the Works (3rd paragraph)
20.3	Loss or Damage due to Employer's Risks	17.4	Consequences of Employer's Risks
20.4	Employer's Risks	17.3	Employer's Risks
21.1	Insurance of Works and Contractor's Equipment	18.2	Insurance for Works and Contractor's Equipment
21.2	Scope of Cover	18.1	General Requirements for Insurances
		18.2	Insurance for Works and Contractor's Equipment
21.3	Responsibility for amounts not recovered	18.1	General Requirements for Insurances (see last paragraph)
21.4	Exclusions	18.2	Insurance for Works and Contractor's Equipment (see paragraph (e))
22.1	Damage to Persons and Property	17.1	Indemnities
		18.3	Insurance against Injury to Persons and Damage to Property
22.2	Exceptions	18.3	Insurance against Injury to Persons and Damage to Property (see paragraph (d))
22.3	Indemnity by Employer	17.1	Indemnities
23.1	Third Party Insurance (including Employer's Property)	18.3	Insurance against Injury to Persons and Damage to Property
23.2	Minimum amount of Insurance	18.3	Insurance against Injury to Persons and Damage to Property (see 2nd paragraph)

(Table 23.1 Contd.)

The Fourth Edition of the Red Book		Corresponding clauses in the 1999 Red Book	
Sub-clause	Title	Clause/sub-clause	Title
23.3	Cross Liabilities	18.1	General Requirements for Insurances (see 4th paragraph)
24.1	Accident or Injury to Workmen	17.1	Indemnities
24.2	Insurance Against Accident to Workmen	18.4	Insurance for Contractor's Personnel
25.1	Evidence and Terms of Insurance	18.1	General Requirements for Insurances (see items (a) and (b) of 6th paragraph and also the 7th paragraph)
25.2	Adequacy of Insurances	18.1	General Requirements for Insurances (see paragraphs 8 and 9)
25.3	Remedy on Contractor's Failure to Insure	18.1	General Requirements for Insurances (see 3rd last paragraph)
25.4	Compliance with Policy Conditions	18.1	General Requirements for Insurances (see paragraphs 4, 8 and 9)
26.1	Compliance with Statutes, Regulations	1.13	Compliance with Laws
		6.4	Labour Laws
27.1	Fossils	4.24	Fossils
28.1	Patent Rights	17.5	Intellectual and Industrial Property Rights
28.2	Royalties	7.8	Royalties
29.1	Interference with Traffic Adjoining Properties	4.14	Avoidance of Interference
30.1	Avoidance of Damage to Roads	4.15	Access Route
		4.16(c)	Transport of Goods
30.2	Transport of Contractor's Equipment or Temporary Works	4.15	Access Route
		4.16	Transport of Goods
		4.17	Contractor's Equipment
30.3	Transport of Materials or Plant	4.15	Access Route
		4.16	Transport of Goods
30.4	Waterborne Traffic		None
31.1	Opportunities for Other Contractors	4.6	Co-operation
31.2	Facilities for Other Contractors	4.6	Co-operation
32.1	Contractor to Keep Site Clear	4.23	Contractor's Operations on Site
33.1	Clearance of Site on Completion	4.23	Contractor's Operations on Site
		11.11	Clearance of Site
Labour			
34.1	Engagement of Staff and Labour	6.1	Engagement of Staff and Labour
		6.11	Disorderly Conduct
35.1	Returns of Labour and Contractor's Equipment	6.10	Records of Contractor's Personnel and Equipment
Materials, Plant and Workmanship			
36.1	Quality of Materials, Plant and Workmanship	4.9	Quality Assurance
		4.19	Electricity, Water and Gas
		7.1	Manner of Execution
		7.2	Samples
		7.4	Testing
36.2	Cost of Samples	7.2	Samples
		7.4	Testing
36.3	Cost of Tests	7.4	Testing
36.4	Cost of Tests not Provided for	7.4	Testing

(Table 23.1 Contd.)

The Fourth Edition of the Red Book		Corresponding clauses in the 1999 Red Book	
Sub-clause	Title	Clause/ sub-clause	Title
36.5	Engineer's Determination where Tests not Provided for	7.4	Testing
37.1	Inspection of Operations	7.3	Inspection
37.2	Inspection and Testing	3.1	Engineer's Duties and Authority
		7.3	Inspection
		7.4	Testing
37.3	Dates for Inspection and Testing	7.4	Testing (see 4th paragraph)
37.4	Rejection	7.5	Rejection
37.5	Independent Inspection	3.2	Delegation by the Engineer (see 1st paragraph)
38.1	Examination of Work before Covering up	7.3	Inspection
38.2	Uncovering and Making Openings	7.3	Inspection (see last paragraph)
39.1	Removal of Improper Work, Materials or Plant	7.6	Remedial Work
		11.1	Completion of Outstanding Work and Remedying Defects
		11.2	Cost of Remedying Defects
		11.5	Removal of Defective Work
39.2	Default of Contractor in Compliance	7.6	Remedial Work
		11.4	Failure to Remedy Defects
Suspension			
40.1	Suspension of Work	8.8	Suspension of Work
40.2	Engineer's Determination following Suspension	8.9	Consequences of Suspension
		8.10	Payment for Plant and Materials in Event of Suspension
40.3	Suspension lasting more than 84 days	8.11	Prolonged Suspension
Commencement and Delays			
41.1	Commencement of Works	8.1	Commencement of Works
42.1	Possession of Site and Access Thereto	2.1	Right of Access to the Site
		4.15	Access Route
42.2	Failure to Give Possession	2.1	Right of Access to the Site
42.3	Rights of Way and Facilities	4.13	Rights of Way and Facilities
43.1	Time for Completion	8.2	Time for Completion
44.1	Extension of Time for Completion	8.4	Extension of Time for Completion
44.2	Contractor to Provide Notification and Detailed Particulars	20.1	Contractor's Claims
44.3	Interim Determination of Extension	8.4	Extension of Time for Completion (see 2nd paragraph)
45.1	Restriction on Working Hours	6.5	Working Hours
46.1	Rate of Progress	8.6	Rate of Progress
47.1	Liquidated Damages for Delay	8.7	Delay Damages
47.2	Reduction of Liquidated Damages	10.2	Taking Over of Part of the Works (see last paragraph)
48.1	Taking-Over Certificate	10.1	Taking Over of the Works and Sections
48.2	Taking Over of Sections or Parts	10.1	Taking Over of the Works
		10.2	Sections and Taking Over of Part of the Works
48.3	Substantial Completion of Parts	10.2	Taking Over of Part of the Works
48.4	Surfaces Requiring Reinstatement	10.4	Surfaces Requiring Reinstatement

(Table 23.1 Contd.)

The Fourth Edition of the Red Book		Corresponding clauses in the 1999 Red Book	
Sub-clause	Title	Clause/sub-clause	Title
Defects Liability			
49.1	Defects Liability Period	11	Defects Liability
49.2	Completion of Outstanding Work and Remedying Defects	11.1	Completion of Outstanding Work and Remedying Defects
49.3	Cost of Remedying Defects	11.2	Cost of Remedying Defects
49.4	Contractor's Failure to Carry Out Instructions	11.4	Failure to Remedy Defects
50.1	Contractor to Search	11.8	Contractor to Search
Alterations, Additions and Omissions			
51.1	Variations	12.4	Omissions
		13.1	Right to Vary
51.2	Instructions for Variations	12.4	Omissions
		13.1	Right to Vary (see last paragraph)
52.1	Valuation for Variations	12.3	Evaluation
		13.3	Variation Procedure
52.2	Power of Engineer to Fix Rates	12.3	Evaluation
		13.3	Variation Procedure
52.3	Variations Exceeding 15%	12.3	Evaluation
		13.3	Variation Procedure
52.4	Daywork	13.6	Daywork
Procedure for Claims			
53.1	Notice of Claims	20.1	Contractor's Claims
53.2	Contemporary Records	20.1	Contractor's Claims
53.3	Substantiation of Claims	20.1	Contractor's Claims
53.4	Failure to Comply	20.1	Contractor's Claims
53.5	Payment of Claims	14.3	Application for Interim Payment Certificates, see paragraph (f) and 20.1 (see 3rd last paragraph)
Contractor's Equipment, Temporary Works and Materials			
54.1	Contractor's Equipment, Temporary Works and Materials; Exclusive Use for the Works	4.17	Contractor's Equipment
54.2	Employer Not Liable for Damage	4.17	Contractor's Equipment
		17.2	Contractor's Care of the Works
54.3	Customs Clearance	2.2	Permits, Licences or Approvals (see paragraph (ii) of item (b))
54.4	Re-export of Contractor's Equipment	2.2	Permits, Licences or Approvals (see paragraph (iii) of item (b))
54.5	Conditions of Hire of Contractor's Equipment		None
54.6	Costs for the Purpose of Clause 63		None
54.7	Incorporation of Clause in Subcontracts		None
54.8	Approval of Materials not Implied		None
Measurement			
55.1	Quantities	12.1	Works to be Measured
		14.1	The Contractor's Price
56.1	Works to be Measured	12.1	Works to be Measured
		14.1	The Contractor's Price
57.1	Method of Measurement	12.2	Method of Measurement
		14.1	The Contractor's Price
57.2	Breakdown of Lump Sum Items	14.1	The Contract Price (see paragraph (d))

(Table 23.1 Contd.)

The Fourth Edition of the Red Book		Corresponding clauses in the 1999 Red Book	
Sub-clause	Title	Clause/sub-clause	Title
Provisional Sums			
58.1	Definition of 'Provisional Sum'	1.1	Definitions
58.2	Use of Provisional Sums	13.5	Provisional Sums
58.3	Production of Vouchers	13.5	Provisional Sums
Nominated Subcontractors			
59.1	Definition of 'Nominated Subcontractors'	5.1	Definition of 'Nominated Subcontractor'
59.2	Nominated Subcontractors; Objection to Nomination	5.2	Objection to Nomination
59.3	Design Requirements to be Expressly Stated	None	
59.4	Payments to Nominated Subcontractors	5.3	Payments to Nominated Subcontractors
		13.5	Provisional Sums
59.5	Certification of Payments to Nominated Subcontractors	5.4	Evidence of Payments
Certificates and Payment			
60.1	Monthly Statements	14.3	Application for Interim Payment Certificates
		14.5	Plant and Materials intended for the Works
60.2	Monthly Payments	14.3	Application for Interim Payment Certificates
		14.6	Issue of Interim Payment Certificates
60.3	Payment of Retention Money	14.9	Payment of Retention Money
60.4	Correction of Certificates	14.6	Issue of Interim Payment Certificates
60.5	Statement at Completion	14.10	Statement at Completion
60.6	Final Statement	14.11	Application for Final Payment Certificate
60.7	Discharge	14.12	Discharge
60.8	Final Payment Certificate	14.11	Application for Final Payment Certificate
		14.13	Issue of Final Payment Certificate
60.9	Cessation of Employer's Liability	14.14	Cessation of Employer's Liability
60.10	Time for Payment	14.7	Payment
		14.8	Delayed Payment
61.1	Approval only by Defects Liability Certificate	11.9	Performance Certificate
62.1	Defects Liability Certificate	11.9	Performance Certificate
62.2	Unfulfilled Obligations	11.10	Unfulfilled Obligations
Remedies			
63.1	Default of Contractor	15.1	Notice to Correct
		15.2	Termination by Employer
63.2	Valuation at Date of Termination	15.3	Valuation at Date of Termination
63.3	Payment after Termination	15.4	Payment after Termination
63.4	Assignment of Benefit of Agreement	15.2	Termination by the Employer (see the 3rd last paragraph)
64.1	Urgent Remedial Work	7.6	Remedial Work
Special Risks			
65.1	No Liability for Special Risks	19	*Force Majeure*
65.2	Special Risks	19	*Force Majeure*

(Table 23.1 Contd.)

The Fourth Edition of the Red Book		Corresponding clauses in the 1999 Red Book	
Sub-clause	Title	Clause/ sub-clause	Title
65.3	Damage to Works by Special Risks	19	*Force Majeure*
65.4	Projectile, Missile	19	*Force Majeure*
65.5	Increased Costs arising from Special Risks	17.4	Consequences of Employer's Risks
		19	*Force Majeure*
65.6	Outbreak of War	19	*Force Majeure*
65.7	Removal of Contractor's Equipment on Termination	19	*Force Majeure*
65.8	Payment if Contract Terminated	19	*Force Majeure*
Release from Performance			
66.1	Payment in Event of Release from Performance	19.7	Release from Performance under the Law
Settlement of Disputes			
67.1	Engineer's Decision		None, but see sub-clauses 20.2, 20.3 and 20.4 (dispute adjudication board)
67.2	Amicable Settlement	20.5	Amicable Settlement
67.3	Arbitration	20.6	Arbitration
67.4	Failure to Comply with Engineer's Decision		None, but see sub-clause 20.7
Notices			
68.1	Notice to Contractor	1.3	Communication
68.2	Notice to Employer and Engineer	1.3	Communication (see last paragraph)
68.3	Change of Address	1.3	Communication
Default of Employer			
69.1	Default of Employer	16.1	Contractor's Entitlement to Suspend Work
		16.2	Termination by Contractor
69.2	Removal of Contractor's Equipment	16.3	Cessation of Work and Removal of Contractor's Equipment
69.3	Payment on Termination	16.4	Payment on Termination
69.4	Contractor's Entitlement to Suspend Work	16.1	Contractor's Entitlement to Suspend Work
69.5	Resumption of Work	8.12	Resumption of Work
		16.1	Contractor's Entitlement to Suspend Work
Changes in Cost and Legislation			
70.1	Increase or Decrease of Cost	13.1	Right to Vary
		13.8	Adjustments for Changes in Cost
70.2	Subsequent Legislation	13.7	Adjustments for changes in Legislation
		13.8	Adjustments for Changes in Cost
Currency and Rates of Exchange			
71.1	Currency Restrictions	13.4	Payment in Applicable Currencies
		13.7	Adjustments for changes in Legislation
		14.15	Currencies of Payment
72.1	Rates of Exchange	13.4	Payment in Applicable Currencies
		14.15	Currencies of Payment
72.2	Currency Proportions	13.4	Payment in Applicable Currencies
		14.15	Currencies of Payment
72.3	Currencies of Payment for Provisional Sums	13.4	Payment in Applicable Currencies
		14.15	Currencies of Payment

Table 23.2 The 1999 Red Book and the corresponding conditions in the Fourth Edition of the Red Book.

The 1999 Red Book		Corresponding clauses in the Fourth Edition of the Red Book	
Sub-clause	Title	Clause/ sub-clause	Title
General Provisions			
1.1	Definitions	1.1	Definitions
1.2	Interpretation	1.2	Headings and Marginal Notes
		1.3	Interpretations
		1.4	Singular and Plural
1.3	Communications	1.5	Notices, Consents, Approvals, Certificates and Determinations
		68	Notices
1.4	Law and Language	5.1	Language(s) and Law
1.5	Priority of Documents	5.2	Priority of Contract Documents
1.6	Contract Agreement	9.1	Contract Agreement
1.7	Assignment	3.1	Assignment of Contract
1.8	Care and Supply of Documents	6.1	Custody and Supply of Drawings and Documents
		6.2	One Copy of Drawings to be Kept on Site
		8.1	Contractor's General Responsibilities
1.9	Delayed Drawings or Instructions	6.3	Disruption of Progress
		6.4	Delays and Cost of Delay of Drawings
		6.5	Failure by Contractor to Submit Drawings
1.10	Employer's Use of Contractor's Documents		None
1.11	Contractor's Use of Employer's Documents	6.1	Custody and Supply of Drawings and Documents
1.12	Confidential Details		None
1.13	Compliance with Laws	26.1	Compliance with Statutes and Regulations
1.14	Joint and Several Liability		None
The Employer			
2.1	Right of Access to the Site	42.1	Possession of Site Access thereto
		42.2	Failure to Give Possession
2.2	Permits, Licences or Approvals	54.3	Custom Clearance
		54.4	Re-export of Contractor's Equipment
2.3	Employer's Personnel	19.2	Employer's Responsibilities
2.4	Employer's Financial Arrangements		None
2.5	Employer's Claims		None
The Engineer			
3.1	Engineer's Duties and Authority	2.1	Engineer's Duties and Authority
		7.3	Responsibility Unaffected by Approval
		14.4	Contractor not relieved of Duties or Responsibilities
		37.2	Inspection and Testing
3.2	Delegation by the Engineer	2.2	Engineer's Representative
		2.3	Engineer's Authority to Delegate
		2.4	Appointment of Assistants
		37.5	Independent Inspection

(Table 23.2 Contd.)

The 1999 Red Book		Corresponding clauses in the Fourth Edition of the Red Book	
Sub-clause	Title	Clause/ sub-clause	Title
3.3	Instructions of the Engineer	2.5	Instructions in Writing
		7.1	Supplementary Drawings and Instructions
		13.1	Work to be in Accordance with Contract
3.4	Replacement of the Engineer		None, except by implication
3.5	Determinations	2.6	Engineer to Act Impartially
The Contractor			
4.1	Contractor's General Obligations	7.2	Permanent Works Designed by Contractor
		8.1	Contractor's General Responsibilities
		13.1	Work to be in accordance with Contract
		14.1	Programme to be Submitted
4.2	Performance Security	10.1	Performance Security
4.3	Contractor's Representative	15.1	Contractor's Superintendance
4.4	Subcontractors	4.1	Subcontracting, (some points)
4.5	Assignment of Benefit of Subcontract	4.2	Assignment of Subcontractor's Obligations
4.6	Co-operation	31.1	Opportunities for Other Contractors
4.7	Setting Out	17.1	Setting Out
4.8	Safety Procedures	19.1	Safety, Security and Protection of the Environment
4.9	Quality Assurance	36.1	Quality of Materials, Plant and Workmanship
4.10	Site Data	11.1	Inspection of Site
4.11	Sufficiency of the Accepted Contract Amount	12.1	Sufficiency of Tender
4.12	Unforeseeable Physical Conditions	12.2	Not Foreseeable Physical Obstructions or Conditions
4.13	Rights of Way and Facilities	42.3	Rights of Way and Facilities
4.14	Avoidance of Interference	29.1	Interference with Traffic and Adjoining Properties
4.15	Access Route	30.1	Avoidance of Damage to Roads
		30.2	Transport of Contractor's Equipment or Temporary Works
		30.3	Transport of Materials or Plant
		42.1	Possession of Site and Access thereto
4.16	Transport of Goods	30.2	Transport of Contractor's Equipment and Temporary Works
4.17	Contractor's Equipment	54.1	Contractor's Equipment
		54.2	Temporary Works and Materials; Exclusive Use for the Works and Employer not Liable for Damage
4.18	Protection of the Environment	19.1(c)	Safety, Security and Protection of the Environment
4.19	Electricity, Water and Gas		None
4.20	Employer's Equipment and Free-Issue Material		None

(Table 23.2 Contd.)

The 1999 Red Book		Corresponding clauses in the Fourth Edition of the Red Book	
Sub-clause	Title	Clause/ sub-clause	Title
4.21	Progress Reports		None
4.22	Security of the Site	19.1	Safety, Security and Protection of the Environment
4.23	Contractor's Operations on Site	32.1	Contractor to Keep Site Clear
		33.1	Clearance of Site on Completion
4.24	Fossils	27.1	Fossils
Nominated Subcontractors			
5.1	Definition of 'Nominated Subcontractor'	59.1	Definition of 'Nominated Subcontractors'.
5.2	Objection to Nomination	59.2	Nominated Subcontractors; Objection to Nomination
5.3	Payments to nominated Subcontractors	59.4	Payments to Nominated Subcontractors
5.4	Evidence of Payments	59.5	Certification of Payments to Nominated Subcontractors
Staff and Labour			
6.1	Engagement of Staff and Labour	34.1	Engagement of Staff and Labour
6.2	Rates of Wages and Conditions of Labour		None
6.3	Persons in Service of Employer		None
6.4	Labour Laws	26.1	Compliance with Statutes, Regulations
6.5	Working Hours	45.1	Restriction on Working Hours
6.6	Facilities for Staff and Labour		None
6.7	Health and Safety	19.1	Safety, Security and Protection of the Environment
6.8	Contractor's Superintendence	15.1	Contractor's Superintendence
		16.1(a)	Contractor's Employees
6.9	Contractor's Personnel	16.1	Contractor's Employees
		16.2	Engineer at Liberty to Object
6.10	Records of Contractor's Personnel and Equipment	35.1	Returns of Labour and Contractor's Equipment
6.11	Disorderly Conduct		None
Plant, Materials and Workmanship			
7.1	Manner of Execution	36.1	Quality of Materials, Plant and Workmanship
7.2	Samples	36.1	Quality of Materials, Plant and Workmanship
		36.2	Cost of Samples
7.3	Inspection	37.1	Inspection of Operations
		37.2	Inspection and Testing
		38	Examination of Work, etc.
7.4	Testing	36	Materials, Plant and Workmanship
		37.3	Dates for Inspection and Testing
7.5	Rejection	37.4	Rejection
7.6	Remedial Work	39	Removal of Improper Work, etc.
		64.1	Urgent Remedial Work
7.7	Ownership of Plant and Materials		None
7.8	Royalties	28.2	Royalties

(Table 23.2 Contd.)

The 1999 Red Book		Corresponding clauses in the Fourth Edition of the Red Book	
Sub-clause	Title	Clause/ sub-clause	Title
Commencement, Delays and Suspension			
8.1	Commencement of the Works	41.1	Commencement of Works
8.2	Time for Completion	43.1	Time for Completion
8.3	Programme	14.1	Programme to be Submitted
8.4	Extension of Time for Completion	44.1	Extension of Time for Completion
8.5	Delays Caused by Authorities		None
8.6	Rate of Progress	46.1	Rate of Progress
8.7	Delay Damages	47.1	Liquidated Damages for Delay
8.8	Suspension of Work	40.1	Suspension of Work
8.9	Consequences of Suspension	40.2	Engineer's Determination following Suspension
8.10	Payment for Plant and Materials in Event of Suspension	40.2	Engineer's Determination following Suspension
8.11	Prolonged Suspension	40.3	Suspension lasting more than 84 days
8.12	Resumption of Work	69.5	Resumption of Work
Tests on Completion			
9.1	Contractor's Obligations		See Clauses 36 and 37
9.2	Delayed Tests		See Clauses 36 and 37
9.3	Retesting		See Clauses 36 and 37
9.4	Failure to Pass Tests on Completion		See Clauses 36 and 37
Employer's Taking Over			
10.1	Taking Over of the Works and Sections	48.1	Taking-Over Certificate
		48.2	Taking Over of Sections or Parts
10.2	Taking Over of Parts of the Works	47.2	Reduction of Liquidated Damages
		48.2	Taking Over of Sections or Parts
		48.3	Substantial Completion of Parts
10.3	Interference with Tests on Completion		None
10.4	Surfaces Requiring Reinstatement	48.4	Surfaces Requiring Reinstatement
Defects Liability			
11.1	Completion of Outstanding Work and Remedying Defects	49.2	Completion of Outstanding Work and Remedying Defects
11.2	Cost of Remedying Defects	49.3	Cost of Remedying Defects
11.3	Extension of Defects Notification Period		None
11.4	Failure to Remedy Defects	49.4	Contractor's Failure to Carry Out Instructions
11.5	Removal of Defective Work	39.1	Removal of Improper Work, Materials or Plant
11.6	Further Tests		None
11.7	Right of Access		None
11.8	Contractor to Search	50.1	Contractor to Search
11.9	Performance Certificate	61.1	Approval only by Defects Liability Certificate
		62.1	Defects Liability Certificate
11.10	Unfulfilled Obligations	62.2	Unfulfilled Obligations
11.11	Clearance of Site	33.1	Clearance of Site on Completion

(Table 23.2 Contd.)

The 1999 Red Book		Corresponding clauses in the Fourth Edition of the Red Book	
Sub-clause	Title	Clause/ sub-clause	Title
Measurement and Evaluation			
12.1	Works to be Measured	56.1	Works to be Measured
12.2	Method of Measurement	57.1	Method of Measurement
12.3	Evaluation		None, but see sub-clause 52.1 in respect of valuation of variations
12.4	Omissions	51 and 52	Alterations, Additions and Omissions
Variations and Adjustments			
13.1	Right to Vary	18.1	Boreholes and Exploratory Excavation
		51.1	Variations
13.2	Value Engineering		None
13.3	Variation Procedure	52	Valuation of Variations, etc.
13.4	Payment in Applicable Currencies	71 and 72	Currency and Rates of Exchange
13.5	Provisional Sums	58	Provisional Sums
		59.4	Payments to Nominated Subcontractors
13.6	Daywork	52.4	Daywork
13.7	Adjustments for Changes in Legislation	70.2	Subsequent Legislation
		71.1	Currency Restrictions
13.8	Adjustments for Changes in Cost	70	Changes in Cost and Legislation
Contract Price and Payment			
14.1	The Contract Price	1.1	Definition
		55–57	Measurement
14.2	Advance Payment		None
14.3	Application for Interim Payment Certificates	60.1	Monthly Statements
		60.2	Monthly Payments
14.4	Schedule of Payments	14.3	Cash Flow Estimates to be Submitted
14.5	Plant and Materials intended for the Works	60.1(c)	Monthly Payments
14.6	Issue of Interim Payment Certificates	60.2	Monthly Payments
		60.4	Correction of Certificates
14.7	Payment	60.10	Time for Payment
14.8	Delayed Payment	60.10	Time for Payment
14.9	Payment of Retention Money	60.3	Payment of Retention Money
14.10	Statement at Completion	60.5	Statement at Completion
14.11	Application for Final Payment Certificate	60.6 or 60.8	Final Statement or Final Payment Certificate
14.12	Discharge	60.7	Discharge
14.13	Issue of Final Payment Certificate	60.8	Final Payment Certificate
14.14	Cessation of Employer's Liability	60.9	Cessation of Employer's Liability
14.15	Currencies of Payment	71–72	Currency and Rates of Exchange
Termination by Employer			
15.1	Notice to Correct	63.1	Default of Contractor
15.2	Termination by Employer	63.1(c)	Default of Contractor (see paragraph (c))
15.3	Valuation at Date of Termination	63.2	Valuation at Date of Termination
15.4	Payment after Termination	63.3	Payment after Termination
15.5	Employer's Entitlement to Termination		None

(Table 23.2 Contd.)

The 1999 Red Book		Corresponding clauses in the Fourth Edition of the Red Book	
Sub-clause	Title	Clause/ sub-clause	Title
Suspension and Termination by Contractor			
16.1	Contractor's Entitlement to Suspend Work	69.1	Default of Employer
		69.4	Contractor's Entitlement to Suspend Work
		69.5	Resumption of Work
16.2	Termination by Contractor	69.1	Default of Employer
16.3	Cessation of Work and Removal of Contractor's Equipment	69.2	Removal of Contractor's Equipment
16.4	Payment on Termination	69.3	Payment on Termination
Risk and Responsibility			
17.1	Indemnities	22.1	Damage to Persons and Property
		22.3	Indemnity by Employer
		24.1	Accident or Injury to Workmen
17.2	Contractor's Care of the Works	20.1	Care of Works
		20.2	Responsibility to Rectify Loss
		54.2	Employer not Liable for Damage
17.3	Employer's Risks	20.4	Employer's Risks
17.4	Consequences of Employer's Risks	20.3	Loss or Damage Due to Employer's Risks
		65.5	Increased Costs arising from Special Risks
17.5	Intellectual and Industrial Property Rights	28.1	Patent Rights
17.6	Limitation of Liability		None
Insurance			
18.1	General Requirements for Insurance	21.3	Responsibility for Amounts not Recovered
		25.1	Evidence and Terms of Insurances
		25.2	Adequacy of Insurances
		25.3	Remedy on Contractor's Failure to Insure
		25.4	Compliance with Policy Conditions
18.2	Insurance for Works and Contractor's Equipment	21.1	Insurance for Works and Contractor's Equipment
18.3	Insurance against Injury to Persons and Damage to Property	23	Third Party Insurance (including Employer Property), etc.
18.4	Insurance for Contractor's Personnel	24.2	Insurance Against Accident to Workmen
Force Majeure			
19.1	Definition of *Force Majeure*		None, but see clauses 65.2 and 65.4
19.2	Notice of *Force Majeure*		None, but see clause 65.1
19.3	Duty to Minimise Delay		None
19.4	Consequences of *Force Majeure*		None, but see clause 65.5
19.5	*Force Majeure* Affecting Subcontractor		None
19.6	Optional Termination		None, but see clauses 65.6–65.8
19.7	Release from Performance under the Law	13.1	Work to be in Accordance with the Contract
		66	Release from Performance

(Table 23.2 Contd.)

The 1999 Red Book		Corresponding clauses in the Fourth Edition of the Red Book	
Sub-clause	Title	Clause/sub-clause	Title
Claims, Disputes and Arbitration			
20.1	Contractor's Claims	44.2	Contractor to Provide Notification and Detailed Particulars
		53.1	Notice of Claims
		53.2	Contemporary Records
		53.3	Substantiation of Claims
		53.4	Failure to Comply
20.2	Appointment of the Dispute Adjudication Board		None
20.3	Failure to Agree Dispute Adjudication Board		None
20.4	Obtaining Dispute Adjudication Board's Decision		None, but see sub-clause 67.1 in respect of the engineer's decision
20.5	Amicable Settlement	67.2	Amicable Settlement
20.6	Arbitration	67.3	Arbitration
20.7	Failure to Comply with Dispute Adjudication Board's Decision		None (but it is similar to sub-clause 67.4)
20.8	Expiry of Dispute Adjudication Board's Appointment		None

As an example of wording that has been extensively changed in the 1999 Red Book, but to the same intent of the Fourth Edition of the Red Book, one may look at sub-clause 17.5 of the 1999 Red Book, which is intended to be a replacement of clause 28 of the Fourth Edition of the Red Book. Further examples may be considered, but there is no substitute for a comprehensive comparison between the two forms of contract. An excellent tool for the understanding of the 1999 Red, Yellow and Silver Books is *The FIDIC Contract Guide*, which was published by FIDIC in 2000.[23.3]

23.3 The 1999 Red Book: new concepts

It is clear from Tables 23.1 and 23.2 that there are a number of new sub-clauses representing new concepts that have been introduced into the provisions of the 1999 Red Book. These were discussed in general terms in Section 22.3 of Chapter 22 and the more pertinent new concepts are discussed in detail below.

23.3.1 Sub-clause 1.12: 'Confidential Details'

Sub-clause 1.12 requires the contractor to disclose all such confidential and other information as the engineer may reasonably require in order to verify the contractor's compliance with the contract. Contractors perceive this requirement as capable of presenting difficulties where the contractor has a duty of confidentiality to a third party in respect of the information required to be disclosed.²³·⁴ The solution to this problem seems to be explained in the *FIDIC Contracts Guide* where it is stated that such disclosure may be precluded under a part of the contract having priority over the contract conditions, which suggests that the parties need to discuss this matter some time prior to the contract signature and agree the extent of the privileged information.²³·⁵ Of course, any information made available by one party to the other should be kept confidential.

23.3.2 Sub-clause 2.4: 'Employer's Financial Arrangements'

A new provision has been added into the 1999 Red Book, which recognises the risk to the contractor of the employer's finances being such that he might not get paid on time or at all. Sub-clause 2.4 provides as follows:

> 'Employer's Financial Arrangements
> 2.4 The Employer shall submit, within 28 days after receiving any request from the Contractor, reasonable evidence that financial arrangements have been made and are being maintained which will enable the Employer to pay the Contract Price (as estimated at that time) in accordance with Clause 14 [Contract Price and Payment]. If the Employer intends to make any material change to his financial arrangements, the Employer shall give notice to the Contractor with detailed particulars.'

Inclusion of this new sub-clause seems entirely logical, for it is equitable that the employer should, if requested, reassure the contractor that he will be paid. This is a significant addition to the contract conditions from the contractor's point of view, especially where funding to the project is being provided by someone other than the employer. There is no definition, however, of what constitutes 'reasonable evidence' and this may become important in case of a dispute relating to its precise meaning due to the serious sanction that follows such failure. Sub-clause 2.4 provides that if the employer fails to furnish such evidence, the contractor is entitled to suspend the work or terminate the contract under sub-clause 16.1, 'Contractor's Entitlement to Suspend Work', and sub-clause 16.2, 'Termination by Contractor'. The seriousness of this sanction should also be considered by the parties in connection with the 28-day period available to the employer to provide the required evidence. Should the employer feel it necessary to restrict his obligation in this regard, then he may do so but such restriction must be notified to the contractor prior to execution of the contract as an amendment in the particular conditions of contract.²³·⁶

23.3.3 Sub-clause 2.5: 'Employer's Claims'

The 1999 Red Book differentiates between a claim made by the employer and a claim made by the contractor and sets up a procedure specifically for employer's claims. Sub-clause 2.5 expressly provides for employer's claims, whereas sub-clause 20.1 deals with contractor's claims. The provisions of this sub-clause apply to any claim:

'under any Clause of these Conditions or otherwise in connection with the Contract, and/or to any extension of the Defects Notification Period'.

Therefore, sub-clause 2.5 brings together claims for delay damages, other claims specifically provided for under the contract, see the list below; and claims for breach of contract by the contractor. Under sub-clause 2.5, the procedure for the employer's claims starts with a notice from the employer to the contractor, but the time within which this notice is required to be given is:

'as soon as practicable after the Employer became aware of the event or circumstances giving rise to the claim. A notice relating to any extension of the Defects Notification Period shall be given before the expiry of such period'.

Following notification by the employer, the engineer should then proceed in accordance with sub-clause 3.5 to agree or determine (i) the amount (if any) which the employer is entitled to be paid by the contractor, and/or (ii) the extension (if any) of the Defects Notification Period in accordance with sub-clause 11.3. Whilst this is quite different from the more onerous requirements imposed on a contractor submitting a claim, see sub-section 23.3.17 below, it is nevertheless considered to be offering a better protection to the contractor and thus preventing any unreasonable action by the employer.[23.7] In particular this protection is valuable where the last sentence of sub-clause 2.5 provides that:

'The Employer shall only be entitled to set off against or make any deduction from an amount certified in a Payment Certificate, or to otherwise claim against the Contractor, in accordance with the Sub-Clause.'

It is intended to exclude any right to make a deduction from sums otherwise payable to the contractor. The provisions of sub-clause 2.5 apply to a number of sub-clauses throughout the 1999 Red Book, as follows:

— Sub-clause 4.2   Performance Security;
— Sub-clause 4.19   Electricity, Water and Gas;
— Sub-clause 4.20   Employer's Equipment and Free Issue Material
— Sub-clause 7.5   Rejection
— Sub-clause 7.6   Remedial Work
— Sub-clause 8.6   Rate of Progress
— Sub-clause 8.7   Delay Damages
— Sub-clause 9.4   Failure to Pass Tests on Completion
— Sub-clause 11.3   Extension of Defects Notification Period
— Sub-clause 11.5   Failure to Remedy Defects
— Sub-clause 11.11   Clearance of Site
— Sub-clause 15.4   Payment after Termination
— Sub-clause 18.1   General Requirements for Insurances
— Sub-clause 18.2   Insurance for Works and Contractor's Equipment

There are four options that may apply after a determination. These are as follows:

(a) The employer could ask the engineer to deduct the amount in his calculation of the next payment certificate.
(b) The employer could ask the engineer to deduct the amount from a sum payable under a payment certificate.
(c) The employer could ask the engineer to make a separate claim against the contractor for payment.
(d) Either party may contest the determination.

In this connection, two matters should be noted: First, that sub-clause 14.3, 'Application for Interim Payment Certificates', provides that the application by the contractor for payment should include '(f) any other additions or deductions which may have become due under the Contract or otherwise'. Secondly, the employer is obliged to pay the contractor under sub-clause 14.7, 'Payment', and sanctions are provided for non-payment under sub-clauses 14.8, 'Delayed Payment', 16.1, 'Contractor's Entitlement to Suspend Work' and 16.2, 'Termination by Contractor'.

23.3.4 Clause 3: 'The Engineer'

Both the Fourth Edition of the Red Book and the 1999 Red Book feature an engineer in place, but differ in the allocation of duties, in particular with respect to his role. Clause 3 of the 1999 Red Book deals with the engineer. Sub-clause 3.1, 'The Engineer's Duties and Authority' makes it clear that:

> 'whenever carrying out duties or exercising authority, specified in or implied by the Contract, the Engineer shall be deemed to act for the Employer'.

This marks a fundamental change in the role of the engineer, as he acts as an agent of the employer at all times. The quasi-arbitral role assigned to him under the fourth edition of the Red Book is no longer applicable under the 1999 Red Book, as can be seen more fully from sub-clause 3.5 dealt with below.

If the employer wishes to impose restrictions on the engineer's authority, under sub-clause 3.1, such restrictions must be clearly enumerated in the Particular Conditions or agreed with the contractor, if they are to apply. Importantly, it should be noted that under sub-clause 3.1(c) any action of the engineer (listed as approval, check, certificate, consent, examination, inspection, instruction, notice, proposal, request, test, or similar act including absence of disapproval) will not relieve the contractor from his responsibilities under the contract, including responsibility for errors, omissions, discrepancies and non-compliances.

Sub-clause 3.2, 'Delegation by the Engineer', is similar in the 1999 Red Book to that in the Fourth Edition of the Red Book, except for the provision that:

> 'unless otherwise agreed by both Parties, the Engineer shall not delegate the authority to determine any matter in accordance with Sub-Clause 3.5 [Determinations]'.

Under sub-clause 3.2, the engineer is authorised to allocate duties and designate authority to assistants. Whenever such an assistant acts in accordance with the designated authority, the act has the same effect as though the engineer had performed it himself. As such, any such acts on the part of an assistant will not relieve the contractor from any responsibilities under the contract, but any failure to disapprove shall not constitute approval, and the contractor is entitled to refer any questionable determination or instruction of an assistant to the engineer himself for confirmation or revision.

Sub-clause 3.3, 'Instructions of the Engineer', is quite different from its predecessor, as it deals with and relates to the type of instructions that the engineer may give. It states that the engineer has the power to issue instructions to the contractor, which the contractor is obliged to carry out. Whilst it is expressly stated that such instructions should be given in writing, a procedure is laid out whereby if the engineer or a delegated assistant:

(i) gives an oral instruction;
(ii) receives written confirmation from the contractor of that oral instruction within two working days; and
(iii) does not reply by issuing a written instruction or rejection of the oral instruction within two working days of receiving the contractor's confirmation,

then the contractor's confirmation of the oral instruction shall be deemed to constitute the written instruction of the engineer/authorised assistant. This is a very practical provision, recognising as it does the necessity in the day-to-day activities on site of giving oral instructions. By providing a mechanism by which such oral instructions are converted to writing, it underlines the contractual importance of having written instructions and thereby avoids the confusion and potential for disputes that might otherwise arise.

Sub-clause 3.4, 'Replacement of the Engineer', is a new provision stating in express terms what could only be implied from sub-clause 1(1)(a)(iv) of the Fourth Edition of the Red Book.[23.8] However, as no explicit term existed in this connection in the Fourth Edition, the implication was that the engineer required the contractor's prior agreement to replace the engineer. The new words 'against whom the Contractor raises reasonable objection . . .' provide a different but more practical concept.

In the situation where the employer intends to replace the engineer, the contractor must receive notice of such intention no less than 42 days before the replacement is planned to take place under sub-clause 3.4. Such notice must include relevant details on the replacement engineer, for example, any experience he may have, and any duties and powers he would have in respect of the works. The contractor is then entitled to submit to the employer, a 'reasonable objection' to appointment of the replacement engineer, in such case the employer is not permitted to make the replacement. By implication, if no objection is received before the intended date of replacement the replacement engineer may be appointed by the employer and the contractor has no further right to object to the appointment for the remaining duration of the contract.

Sub-clause 3.5 embodies one of the most fundamental changes from the Fourth Edition of the Red Book to the 1999 Book and relates to the role of the engineer. Sub-clause 2.6 in the Fourth Edition of the Red Book requires the engineer to be impartial, as follows:

'Wherever, under the Contract, the Engineer is required to exercise his discretion by: . . . , he shall exercise such discretion impartially within the terms of the Contract and having regard to all the circumstances.'

However, sub-clause 3.5 in the 1999 Red Book provides as follows:

'3.5 Whenever these Conditions provide that the Engineer shall proceed in accordance with this Sub-Clause 3.5 to agree or determine any matter, the Engineer shall consult with each Party in an endeavour to reach agreement. If agreement is not achieved, the Engineer shall make a fair determination in accordance with the Contract, taking due regard of all relevant circumstances.'

The requirement of 'impartiality' has given way to a 'fair determination'. However, it is open to debate as to the precise meaning of 'fair determination'. The dictionary meaning of fair is 'just, unbiased, equitable in accordance with the rules'; and of fair mindedness is a 'sense of justice, impartiality'.[23.9] In the case of *Semco Salvage Marine Pte Ltd* v. *Lancer Navigation Ltd*[23.10] it was stated: 'I would hold that "fair" means "fair to both parties".'

There are 25 sub-clauses where the engineer is required to comply with sub-clause 3.5 in agreeing or determining the matter after consulting with each party in an endeavour to reach agreement. These sub-clauses are as follows:

— Sub-Clause 1.9   Delayed Drawings or Instructions
— Sub-Clause 2.1   Right of Access to the Site
— Sub-Clause 2.5   Employer's Claims
— Sub-Clause 4.7   Setting Out
— Sub-Clause 4.12   Unforeseeable Physical Conditions
— Sub-Clause 4.19   Electricity, Water and Gas
— Sub-Clause 4.20   Employer's Equipment and Free-Issue Material
— Sub-Clause 4.24   Fossils
— Sub-Clause 7.4   Testing
— Sub-Clause 8.9   Consequences of Suspension
— Sub-Clause 9.4   Failure to Pass Tests on Completion
— Sub-Clause 10.2   Taking Over of Part of the Works
— Sub-Clause 10.3   Interference with Tests on Completion
— Sub-Clause 11.4   Failure to Remedy Defects
— Sub-Clause 11.8   Contractor to Search
— Sub-Clause 12.3   Evaluation
— Sub-Clause 12.4   Omissions
— Sub-Clause 13.2   Value Engineering
— Sub-Clause 13.7   Adjustments for Changes in Legislation
— Sub-Clause 14.4   Schedule of Payments
— Sub-Clause 15.3   Valuation at Date of Termination
— Sub-Clause 16.1   Contractor's Entitlement to Suspend Work
— Sub-Clause 17.4   Consequences of Employer's Risks
— Sub-Clause 19.4   Consequences of *Force Majeure*
— Sub-Clause 20.1   Contractor's Claims

Under a few of the sub-clauses listed above, the procedure envisaged under sub-clause 3.5 is not initiated by a notice for a claim, for example under sub-clause 12.3 'Evaluation', the engineer is obliged to agree or determine the contract price of his own accord, with no mention of the existence of a claim in this regard. In most of the sub-clauses, however, the procedure is initiated by a notice of claim under that sub-clause.

Dealing first with the contractor's claims, sub-clause 1.9 relating to delayed drawings or instruction is a typical example. In all the sub-clauses which require the contractor to give notice of claim, such entitlement is subject to the contractor's compliance with sub-clause 20.1. In this way, the engineer's obligation in respect of determination of a contractor's entitlement under sub-clause 3.5 is closely merged with, if not dependent on, the contractor's obligations to give notice and provide detailed particulars under sub-clause 20.1. In such a case, the procedure is routed along the following steps:

(1) The contractor issues a notice under one of the sub-clauses describing the event or circumstance giving rise to the claim, for additional payment and/or extension of time;
(2) The engineer may after receiving such notice, monitor the record-keeping and/or instruct the contractor to keep further contemporary records (under sub-clause 20.1);
(3) Within 42 days of the event giving rise to the claim, the contractor shall send to the engineer a fully detailed claim with full supporting particulars of the basis of the claim and of the extension of time and/or the additional payment claimed (under sub-clause 20.1);
(4) Within 42 days of receiving the claim or any further particulars, the engineer is required to respond with approval, or with disapproval and detailed comments (under sub-clause 20.1). Such response should not be confused with the engineer's determination of the claim. This response is merely the engineer's opinion on the validity of the contractor's claim, and not the quantum of any additional payment and/or extension of time to which the contractor is entitled, as determined by the engineer;
(5) The engineer is required then to proceed in accordance with sub-clause 3.5, which provides for him to consult with each party in an endeavour to reach agreement on the claimed matter;
(6) If no agreement is achieved, the engineer shall make a fair determination in accordance with the contract, taking due regard of all relevant circumstances. The last paragraph of sub-clause 20.1 states that the requirements of this sub-clause are in addition to those of any other sub-clauses which apply to the claim, so that if the contractor does not comply with such aggregated requirements, the engineer's determination shall take account of the extent to which such failure to comply has prevented or prejudiced proper investigation of the claim;
(7) If either the employer or the contractor is dissatisfied with the engineer's determination, the provisions of sub-clause 20.4 apply.

It is most important to note here that the engineer is not under any time limit in performing the tasks in steps 5 and 6.

The procedure is different when dealing with the employer's claims, as it comprises the following steps:

(1) If the employer considers himself to be entitled to any payment under any clause of the conditions or otherwise in connection with the contract, and/or to any extension of the defects notification period, the employer or the engineer shall give to the contractor:
   (1.1) a notice; and
   (1.2) particulars.
(2) The notice should be given as soon as practicable after the employer becomes aware of the event or circumstances giving rise to the claim.
(3) The particulars should specify the clause or other basis of the claim, and should include substantiation of the amount and/or extension of time to which the employer considers himself to be entitled in connection with the contract.
(4) The engineer should then proceed in accordance with sub-clause 3.5, 'Determinations', to agree or determine:
   (4.1) the amount, if any, which the employer is entitled to be paid by the contractor, and/or
   (4.2) the extension, if any, of the defects notification period in accordance with sub-clause 11.3, 'Extension of Defects Notification Period'.
(5) The amount referred to in paragraph 4.1 above may be included as a deduction in the contract price and payment certificates.

In this connection, it is to be noted that:

— The employer shall only be entitled to set off against or make any deduction from an amount certified in a payment certificate, or to otherwise claim against the contractor, in accordance with this sub-clause.
— A notice relating to any extension of the defects notification period shall be given before the expiry of such period.
— A notice is not required for payments due under sub-clause 4.19, 'Electricity, Water and Gas', under sub-clause 4.20, 'Employer's Equipment and Free-Issue Material' or for other services requested by the contractor.

23.3.5 Sub-Clause 4.1(c): 'Fitness for Purpose'

The higher standard of liability of fitness for purpose applies whenever the contractor is required to design any part of the works under the 1999 Red Book.[23.11] Thus, it is specified in paragraph (c) of sub-Clause 4.1 of the 1999 Red Book that:

'the Contractor shall be responsible for such part and it shall, when the Works are completed, be fit for such purposes for which the part is intended are specified in the Contract'.

It should be noted that the 1999 Yellow and Silver Books require similar levels of design liability. Furthermore, reference should also be made to Clause 11.3 in all the 1999 Books, which entitles the employer to an extension of the defects notification period in the event that the works (or a section thereof) cannot be used for the purposes for which they were intended.

The purpose should be specified in detail in the employer's requirements, as these refer to purpose in their definition.[23.12] Of course it should be borne in mind that as the design function in the 1999 Red Book is allocated to the engineer, the contractor may only be allocated a design function relating to a specific part of the works. If the part designed by the contractor is not fit for the purpose, then limitation to his liability applies under sub-clause 17.6. It should be noted that when the contractor is required to design any part of the works, there should be a requirement to provide professional indemnity cover.

23.3.6 Sub-Clause 4.2: 'Performance Security'

All the 1999 Books recognise the matrix of pure economic risks between the parties (see Section 22.4 of Chapter 22 where economic risks are discussed in respect of the 1999 Green Book), and the wording of the provisions of sub-clause 4.2 relating to performance security differs greatly from its equivalent under the Fourth Edition of the Red Book. Briefly, these differences amount to:

(a) An on-demand guarantee or a surety bond is permitted under the 1999 Red Book, and model forms of each are annexed to the Red Book.
(b) The employer is not permitted to make a claim under the performance security except for amounts to which he is entitled in specified events.
(c) An indemnity is specified for the benefit of the contractor to the extent that the employer makes a claim under the Performance Security to which he is not entitled.
(d) The employer is required to return the Performance Security within 21 days after receiving a copy of the Performance Certificate, see sub-clause 11.9.

Due to the different type of security required, no such restriction on the employer's right to make a claim or on the contractor's indemnity is included in the Fourth Edition of the Red Book. Also, by virtue of clause 10.2 of the Fourth Edition of the Red Book, the performance security is to be returned within 14 days of the date of issue of the defects liability certificate (under sub-clause 62.1).

23.3.7 Sub-clause 4.12: 'Unforeseeable Physical Conditions'

The provisions of the famous clause 12 of the Fourth Edition of the Red Book have been maintained in the 1999 Red Book in the penultimate paragraph of sub-clause 4.12, but with a significant alteration, such that:

> '... before additional Cost is finally agreed or determined under subparagraph (ii), the Engineer may also review whether other physical conditions in similar parts of the Works (if any) were more favourable than could reasonably have been foreseen when the Contractor submitted the Tender. If and to the extent that these more favourable conditions were encountered, the Engineer may proceed in accordance with Sub-Clause

3.5 [Determinations] to agree or determine the reductions in Cost which were due to these conditions, which may be included (as deductions) in the Contract Price and Payment Certificates. However, the net effect of all adjustments under sub-paragraph (b) and all these reductions, for all the physical conditions encountered in similar parts of the Works, shall not result in a net reduction in the Contract Price.'

Therefore, the engineer may, in determining the issue of additional cost, under sub-clause 4.12, take into account any 'more favourable (conditions) than could reasonably have been foreseen when the Contractor submitted the Tender'. If the engineer decides that such favourable conditions were present, then he may take this information into account in making a determination on the issue of cost. The European International Contractors (EIC) argue that this provision has the potential to be extremely prejudicial to the contractor and that furthermore, the expressions 'similar parts of the Works' and 'more favourable' may be open to a different and wide-ranging version of interpretations.[23.13]

Sub-clause 4.12 deals with unforeseeable physical conditions and their effect on the contract and the contractor's obligations under the contract. Therefore, where unforeseeable physical conditions occur during the carrying out of the works, the contractor is entitled to an extension of time, subject to sub-clause 20.1, if the completion of the works will be or has been delayed as a result of such unforeseeable physical conditions. It is worthwhile to note that the 1999 Red Book follows on from the Fourth Edition of the Red Book and excludes climatic events from the definition of physical conditions that could entitle an award of extension of time and cost. The contractor may also be entitled to additional payment for any costs or losses that he incurs as a result of the unforeseeable physical conditions, again subject to sub-clause 20.1. However, once the contractor discovers the unforeseeable physical conditions he must give notice of such a situation, because if such notice is not issued he would be deprived of any entitlement to either an extension of time or any costs accrued. The vital component in this sub-clause is the statement that adverse physical conditions must not have been foreseeable. Under sub-clause 1.1.6.8, unforeseeable is defined as 'not reasonably foreseeable by an experienced contractor by the date for submission of the Tender'. It is clear that this sub-clause will protect the contractor if the actual conditions were not foreseeable, and it will also protect the employer, who, under sub-clause 4.10, is under an obligation to disclose to the contractor all information he has in relation to the site. Only where the extent of the adverse physical conditions is so serious so as to constitute *force majeure*, should the contractor not await instructions from the engineer. He will be expected to continue on with the works.

EIC also make the comment on the penultimate paragraph of sub-clause 4.12, where the contractor can provide evidence of the physical conditions foreseen in the tender calculation. They point out that a contractor should be aware that if he gives such evidence to the engineer, the engineer may take this information into account, but he is not bound by it in making a determination. They suggest that a better approach to this situation would be for the parties:

'to agree the foreseeable conditions beforehand. . . . this approach simplifies claims negotiations which is obviously to the benefit of both parties'.

23.3.8 Sub-Clause 4.21: 'Progress Reports'

Sub-clause 4.21 in the 1999 Red Book introduces the new requirement of the contractor having to produce 'progress reports' such that:

> 'Unless otherwise stated in the Particular Conditions, monthly progress reports shall be prepared by the Contractor and submitted to the Engineer in six copies.'

The sub-clause continues by prescribing that the first reports should cover the period up to the end of the calendar month following the commencement date and be submitted monthly thereafter, each within 7 days after the last day of the period to which it relates. This sub-clause also lists in detail what each report should contain.

23.3.9 Sub-Clause 13.2: 'Value Engineering'

The 1999 Red Book introduces the new concept of 'value engineering', which did not appear in the Fourth Edition of the Red Book. 'Value engineering' is described as being that which will, if adopted, result in any of the following: accelerate completion of the works; reduce the cost to the employer of executing, maintaining or operating the works; improve the efficiency or value to the employer of the completed work; or otherwise be of benefit to the employer. The contractor may make a proposal incorporating value engineering at any time to the engineer which, if approved, shall be valued under the variation procedure of sub-clause 13.3. As in connection with sub-clause 4.1, the matter of providing professional indemnity insurance should be considered.

23.3.10 Sub-Clause 14.7: 'Payment'

The time limits for 'Payment' are somewhat altered in sub-clause 14.7 of the 1999 Books such that the employer will be required to pay to the contractor:

> '... (b) the amount certified in each Interim Payment Certificate within 56 days after the Engineer receives the Statement and supporting documents; and (c) the amount certified in the Final Payment Certificate within 56 days after the Employer receives this Payment Certificate.'

The contractor is required to support each interim payment statement with the corresponding progress report. In the Fourth Edition of the Red Book, the employer was required to make an interim payment 28 days after receiving the engineer's interim payment certificate which, in turn, the engineer was required to issue 28 days after receiving the contractor's statement. Therefore, the 1999 Books eliminate the time taken for delivery of the engineer's interim payment certificate to the employer. In respect of final payment under the Fourth Edition of the Red Book, the employer was required to make payment 56 days after receiving the engineer's final payment certificate which, in turn, the engineer was required to issue 28 days after receiving the contractor's final statement. Therefore, in respect of final payment the 1999 Books do not differ greatly from the provisions in the Fourth Edition of the Red Book.

23.3.11 Sub-Clause 14.8: 'Delayed Payment'

Under the 1999 Books, if the employer fails to make payment within the timescales set down in sub-clause 14.7, then the contractor becomes entitled to recover his financing charges (those charges incurred by the contractor in not being able to repay a bank loan as soon as he would have if he had been paid on time) calculated at an annual rate which is 3% above the discount rate of the central bank (in the country of the currency of payment) and compounded monthly. This is a significant change to the Fourth Edition of the Red Book, where no express provision was made for recovery of finance charges, and the sanction on the employer for late payment was payment of interest at the rate stated in the Appendix to Tender from the date by which payment should have been made.

23.3.12 Sub-Clause 15.5: 'Employer's Entitlement to Termination'

There are significant changes in risk allocation in the 1999 Books, as can be seen for example in clauses 15.5 'Employer's Entitlement to Termination' and 19 'Force Majeure'. Sub-clause 15.5 introduces the concept that the employer can terminate the contract, at any time for no other reason or fault of any party, but just for his own convenience, by giving notice of such termination to the contractor. Clause 19 substitutes the concept of 'frustration' in the Fourth Edition of the Red Book with that of 'Force Majeure' (see further at Section 23.3.16 below).

23.3.13 Clause 17: 'Risk and Responsibility'

Although clause 17 of the 1999 Red, Yellow and Silver Books is entitled 'Risk and Responsibility', it encompasses other contractual provisions, including indemnities; limitation of liability; and the unrelated topic of intellectual and industrial property rights. In fact, it is suggested that clause 17 starts from the wrong end of the stick by dealing first with 'Indemnities', it back tracks to deal with 'Responsibility', then takes a further leap backwards returning to 'Risk', and finally marches on to 'Liability'. This illogical sequence hardly helps the non-lawyer professionals for whom these provisions are intended. The clause leaves even the expert in the field wondering about the purpose of this confused and baffling sequence.

The theory of risk has developed in the past 20 years or so to such an extent that it is now common knowledge that, for a contract to be performed in an effective manner, the inherent risks must be allocated to the contracting parties on some logical basis which should be made known to them. Thus, it has been said that the main purpose of a contract is to identify the principles of allocating the risks facing the contracting parties. Once these principles are identified, the consequences flow in the natural pattern of risk *to* responsibility *to* liability *to* indemnity *to* insurance. The format of clause 17 should, therefore, follow that same sequence, with the insurance provisions left to the next clause, i.e. clause 18, if it is desired that they should be presented separately.

Accordingly, it is suggested here that clause 17 ought to start with the provisions for 'Risk' and not with 'Indemnities' and sub-clause 17.3 should be 17.1. Furthermore, the wording of sub-clause 17.1 should then start by explaining that the risks included under clause 17 of the conditions of contract are only those risks of loss and damage and not the whole spectrum of the risks to which the project is exposed. The term 'employer's risks' in the context of this clause should therefore be replaced by 'employer's risks of loss and damage', since these risks are confined to those which lead to some form of accidental loss or damage to physical property or personal injury, which in turn may lead to financial and/or time loss risks, directly or through the other clauses of the contract.

If this explanation is not given and the mistake of referring to the risks under clause 17 as 'employer's risks' is not corrected, then there is serious danger that the reader, and of course the user, will conclude that having identified in clause 17 the employer's risks, all the other risks are the contractors' risks, including the contractual risks in the remaining provisions of the contract. This problem can be highlighted by reference to clause 17 of the Orange Book where the draftsman fell into that trap and stated expressly in sub-clause 17.5 that 'The Contractor's risks are all risks other than the Employer's Risks listed in Sub-Clause 17.3'. This mistake has led to many instances of misunderstanding, conflict and at least one serious arbitral proceeding, where the employer pointed out that by sub-clause 17.5 he bears no risks under the contract other than those specified in sub-clause 17.3.

Accordingly, it is essential to understand that the employer's risks, traditionally identified under sub-clause 20.3 of the Fourth Edition of the Red Book and those under sub-clause 17.3 of the 1999 Books, are only the amalgamation of risks which are beyond the control of either the contractor alone or both the contractor and the employer. Furthermore, these risks might have an implied resultant loss or damage to physical property or cause bodily injury, all of which are insurable. In contrast, very few of the other risks to which the project is exposed are insurable.

There are other problems in clause 17. The second problem lies in applying the general principle of allocation of the risks specified in sub-paragraph (h) of sub-clause 17.3 to the employer.[23.14] Whilst this does not form a departure from the Fourth Edition of the Red Book, it was hoped that the new suite of contracts would be up to date with developments in this field. The origin of this sub-paragraph goes back to the ACE Form of Contract which, as noted in Chapter 1 above, is the basis of the FIDIC Red Book. Whilst it is true that the contractor has no control over the events identified in this sub-paragraph, he is in control over their consequences and can instigate protection measures. The contractor can also mitigate any losses that might occur should any of these risks eventuate. Perhaps, more importantly, all the risks identified in sub-paragraph (h) represent events that are insurable and are generally required to be insured under the terms of the contract. The employer ultimately pays for such insurance through the contract provisions, leaving the contractor in charge of any necessary repair, its cost, and any claim negotiations with the insurers following the filing of such claims. These risks are not included as employer's risks in the ICE domestic contract or the other contracts rooted in it.[23.15]

The third problem in clause 17 of the 1999 Red, Yellow and Silver Books is the newly introduced restriction in sub-clause 17.1(b)(ii) of the contractor's indemnity to the employer for property damage. This indemnity is now based on the contractor's negligence rather than on his legal liability as was provided in clause 22.1 of the Fourth Edition of the Red Book.[23.16] This change is a retrograde step and copied from standard forms of contract for building works in the UK[23.17] without, seemingly, any benefit to either the contractor or the employer. The only beneficiary as a result of this change is the insurance market since to cover this gap a new policy is now needed, which is commonly referred to in the UK as the non-negligence insurance policy. As Mr Seppala explains in his article entitled 'FIDIC's New Standard Forms of Contract – *Force Majeure*, Claims, Disputes and Other Clauses',[23.18] it seems that in making this change, the draftsmen of clause 17 of the 1999 Books took comfort from a footnote in *Hudson's Building and Engineering Contracts* (1995), Vol. II, page 1437, where reference is made to both the Royal Institute of British Architects (RIBA) and the Institution of Civil Engineers (ICE) forms of contract. It is suggested here that the reference to the ICE form of contract in that footnote is incorrect, since civil engineering contracts do not distinguish between the indemnity required to be given by the contractor for property damage on one hand, and that for bodily injury, disease or death of any person on the other. In fact, the standard forms of contract for civil engineering construction in the UK or elsewhere do not impose the restriction now introduced.[23.19]

Finally, there are some minor problems of drafting in clause 17, which should be addressed for the proper understanding of what is intended by this clause. For example, sub-clause 17.2 is a 'Responsibility' clause, whereas sub-clause 17.5 is a 'Risk' clause. Therefore, there needs to be a statement as to proportional apportionment of indemnities when both employer and contractor have contributed to damage, loss or bodily injury. This would be particularly important where an indemnity clause is strictly interpreted under the applicable law of contract.

23.3.14 Sub-Clause 17.6: 'Limitation of Liability'

A new concept of 'Limitation of Liability' is now introduced in the three 1999 Books. Sub-clause 17.6 provides that the parties are not liable to each other

> 'for loss of use of any Works, loss of profit, loss of any contract or for any indirect or consequential loss or damage which may be suffered by the other Party in connection with the Contract, other than under Sub-Clause 16.4 [Payment on Termination] and Sub-Clause 17.1 [Indemnities] . . .'.

This introduction seeks to avoid the problems encountered by both employers and contractors in the construction industry, whereby they cannot fully insure their liabilities under the contract. As a matter of course, most if not all insurance policies these days have exclusions for indirect and consequential losses. The total liability of the contractor to the employer, under or in connection with the contract, other than in limited circumstances, is required to be stated in the particular conditions or (if a sum is not so stated) the accepted contract amount. However, this sub-clause is not intended to limit liability in any case of fraud, deliberate default or reckless misconduct by the defaulting party. It should be noted, however, that the contractor is only entitled to compensation for loss of profit in certain cases where the employer breaches the contract. EIC comment that this is an inequitable qualification and that the contractor should always be entitled to compensation for loss of profit and other damages, in the event of the employer breaching the contract.[23.20]

23.3.15 Clause 18: 'Insurance'

Clause 18 'Insurance' in the 1999 Books is significantly different from its equivalent in the Fourth Edition of the Red Book. Insurance is vital in order to protect both parties from the financial consequences of unexpected loss, damage, injury or liability. It is stated in the introduction to clause 18 in the *FIDIC Contracts Guide* that it would be a difficult task, and therefore the Red Book makes no attempt to:

> 'define the precise extent of insurance cover which will be reasonably available for all Works for which the publication may be used. Clause 18 specifies insurance requirements in terms of the cover which is typically available, but the Contractor may have to take account of the cover which is actually available.'

This means, therefore, that the employer in drafting the contract must take expert insurance advice in order to tailor the insurance policy/policies for the project in hand.

In general it is the contractor who is known as the 'insuring party', i.e. 'the party responsible for effecting and maintaining the insurance'.[23.21] The insuring party should insure the works in the names of both parties, and the cover shall apply separately to each insured as though a separate policy had been used for each of the parties. Neither party is entitled to alter the general terms of the insurance policy without first informing and getting the approval of the other party. If one of the parties fails to maintain an insurance policy, which under the terms of the contract it is obliged to keep, then the other party may effect insurance for the relevant coverage and pay the premiums due. The premium amount paid shall then be reimbursed to the party who paid it and the contract price will be adjusted accordingly.

It is important to include a commentary on clause 18 here, since, as explained earlier, insurance is the last of the contractual provisions in the chain of risk: responsibility; liability; indemnity; and insurance.

The first major problem in this clause is the fact that the 'Insuring Party', as defined in the contract, is not the same for all the insurance policies required under the contract and it may be either of the two parties, employer or contractor. This is a recipe for confusion, gaps and/or overlaps in the combined insurance package, which could cost the parties dearly. It could only be advantageous to those involved in the insurance market.

The second paragraph of clause 18 assumes that there would be a meeting between the parties prior to the date of the letter of acceptance at which the whole insurance package would be discussed and agreement would be reached on a policy or policies of insurance, which would 'take precedence over the provisions of (Clause 18)'. It remains to be seen how this provision would operate in practice and the effect it will have.

There are many drafting ambiguities in this clause which should be clarified if the contract is to be operated successfully. Examples are:

(a) Sub-clause 18.1 provides that 'Wherever the Employer is the insuring Party, each insurance shall be effected with insurers and in terms consistent with the details annexed to the Particular Conditions'.[23.22] What is intended by the term 'details'? If, as stated, these details are expected to furnish the terms of the insurances supplied by the employer, then surely this must mean that nothing less explicit than the policies of insurance themselves have to be annexed.
(b) Sub-clause 18.1 provides that 'When each premium is paid, ... the insuring Party shall submit evidence of payment to the other Party ...'. This wording does not provide the intended meaning. The wording 'When each premium is paid ...' is not sufficiently explicit, since there is usually no insurance cover unless the premium *has already been* paid. Payment of each insurance premium should be made to initiate or maintain the insurance cover and evidence should be provided whenever required and not when the premium is paid.
(c) Sub-clause 18.2(d) specifies the deductibles to be applied to the insurance cover for some of the employer's risks. Should the insurance cover for the contractor's risks be subject to no deductibles?
(d) The meaning of the phrase 'insurable at commercially reasonable terms' in sub-clause 18.2(d); in the last paragraph of sub-clause 18.2; and in sub-clause 18.3(d)(iii) is subject to many interpretations that would most likely lead to conflict.[23.23]

23.3.16 Clause 19: 'Force Majeure'

Force majeure is defined in sub-clause 19.1 as follows:

'an exceptional event or circumstance
(a) which is beyond a Party's control,
(b) which such Party could not reasonably have provided against before entering into the Contract,
(c) which, having arisen, such Party could not reasonably have avoided or overcome, and
(d) which is not substantially attributable to the other Party.'

In order for an event to be classified as *force majeure*, it must possess the following five qualities:

(a) the event must be an exceptional one;
(b) the event must be something that is outside the control of the party whom it affects;
(c) the affected party could not have reasonably provided against the event before the contract came into being;
(d) the affected party could not have reasonably avoided or overcome the event; and
(e) the event must not be substantially attributable to the other party.

Sub-clause 19.4 states that if the situation arises whereby the contractor is prevented from fulfilling his obligations as a result of *force majeure*, and he consequently suffers a delay, he may be entitled to an extension of time and any cost incurred. However, his entitlement to claim for such costs is limited to the events that are listed in sub-clause 19.1(ii) to (iv). The logic of this limitation is questionable bearing in mind that the events listed under paragraphs (ii) to (iv) of sub-clause 19.1 are not exhaustive.

Under sub-clause 19.7, where an event occurs during the duration of the contract which makes it impossible or unlawful for any contractual obligations under the contract to be carried out, or where the law of the contract allows the parties to be released from their obligations under the contract, then the parties may be discharged from further performance of the contract and the amount payable by the employer to the contractor should be the same as would be due under sub-clause 19.6 'Optional Termination, Payment and Release'.

A *force majeure* clause is an increasingly common feature of international contracts. It is the fashion, but is it necessary or even desirable? For FIDIC, I suspect that importing *force majeure* from the old Yellow and Orange Books into the new suite of contracts was a desire to show a closer position to the civil law concepts and a move away from the common law principles. Or, perhaps, it was the intention of FIDIC, by including an express definition of *force majeure* in the 1999 Books, that this somewhat complex concept might be clarified for engineering users of these forms of contract. If the truth be told, such a move is neither necessary nor desirable because:

(a) As noted in Chapter 22 in respect of the Green Book, the concept and principles of *force majeure* are well established and much used in the fields of law and insurance, and introducing such clause and definition may create undue complications from the legal and insurance points of view.
(b) The FIDIC *force majeure* definition is wide and not limited to specified situations, but it has to be an 'exceptional' event beyond the control of a party; what constitutes 'exceptional' in any given circumstances is open to debate.
(c) To duplicate what is usually provided for in the civil code of a civil law jurisdiction and then enlarge its scope does not prevent the parties from getting into a muddle and a contradictory situation.
(d) The original concept of the special risks in clause 65 of the Fourth Edition of the Red Book is all the protection that the contractor needs.
(e) Most of the risks, which now come under the FIDIC definition of *force majeure*, are insurable and required to be insured. Therefore, no real benefit accrues to the contractor from being protected by such a clause without having to slip into uncharted waters.

Therefore, whilst it must be agreed that the treatment of the risks specified in clause 19 should be a special one, it is erroneous to swing to the extreme end of the scale and designate them in the category of *force majeure*, particularly when that term has legal implications in certain jurisdictions. The answer for the purposes of these conditions of contract should be to designate as what they are, i.e. an exceptional set of risks with different treatment to that given to the normal set of risks to which the project is exposed.

23.3.17 Clause 20: 'Claims'

Sub-clause 20.1 provides the procedure to be followed whenever the contractor wishes to make a claim. Strict time limits apply: a notice must be served on the engineer within 28 days after the contractor became aware, or should have become aware, of the event giving rise to the claim. Subsequently, details of the claim must be submitted within 42 days. Due to the importance of the changes made in this connection, the text of the relevant part of sub-clause 20.1 is quoted below.

> 'Contractor's Claims
> 20.1 If the Contractor considers himself to be entitled to any extension of the Time for Completion and/or any additional payment, under any Clause of these Conditions or otherwise in connection with the Contract, the Contractor shall give notice to the Engineer, describing the event or circumstance giving rise to the claim. The notice shall be given as soon as practicable, and not later than 28 days after the Contractor became aware, or should have become aware, of the event or circumstance.
>
> If the Contractor fails to give notice of a claim within such period of 28 days, the Time for Completion shall not be extended, the Contractor shall not be entitled to additional payment, and the Employer shall be discharged from all liability in connection with the claim. Otherwise, the following provisions of this Sub-Clause shall apply.
>
> The Contractor shall also submit any other notices which are required by the Contract, and supporting particulars for the claim, all as relevant to such event or circumstance.'

Sub-clause 20.1 also allows the engineer to monitor the records being kept by the contractor to support his claim and provides for strict time limit on the engineer to respond to a contractor's claim. It provides inter alia the following:

> 'The Contractor shall keep such contemporary records as may be necessary to substantiate any claim, either on the Site or at another location acceptable to the Engineer. Without admitting the Employer's liability, the Engineer may, after receiving any notice under this Sub-Clause, monitor the record-keeping and/or instruct the Contractor to keep further contemporary records. The Contractor shall permit the Engineer to inspect all these records, and shall (if instructed) submit copies to the Engineer.
>
> Within 42 days after the Contractor became aware (or should have become aware) of the event or circumstance giving rise to the claim . . . the Contractor shall send to the Engineer a fully detailed claim which includes full supporting particulars of the basis of the claim and of the extension of time and/or additional payment claimed. . . .
>
> . . .
>
> Within 42 days after receiving a claim or any further particulars supporting a previous claim, or within such other period as may be proposed by the Engineer and approved by the Contractor, the Engineer shall respond with approval, or

with disapproval and detailed comments. He may also request any necessary further particulars, but shall nevertheless give his response on the principles of the claim within such time . . .'

The above-mentioned provisions are extremely important in seeking to reduce a number of usual conflicts that have persisted, and beleaguered the construction industry in the past.

23.3.18 Sub-clauses 20.2 to 20.4 'Appointment of the Dispute Adjudication Board'; 'Failure to Agree Dispute Adjudication Board'; 'Obtaining Dispute Adjudication Board's Decision'

Sub-clauses 20.2 to 20.4 in the 1999 Books introduce a new step, the Dispute Adjudication Board (DAB), in the dispute resolution mechanism prior to the amicable settlement procedure. This new step is, with minor variation, similar to that under the 1996 Supplement to the Fourth Edition of the Red Book. The 'quasi-arbitral' or adjudication role of the engineer is thus taken away from him and handed over to the DAB, which can be either a single person or a number of persons chosen with specific qualifications. The 1999 Books include details of the agreement that must be entered into between the parties and the members of the DAB.

However, the appointment of a DAB under sub-clause 20.2 differs from that under the 1996 Supplement to the Fourth Edition of the Red Book in a few important aspects:

(a) The DAB in the 1999 Red Book should be appointed by the date stated in the Appendix to Tender, which suggests 28 days after the commencement date.
(b) The Guidance to the 1999 Red Book includes the possibility of reverting to the traditional role of the engineer.
(c) The DAB in the 1999 Yellow Book is only appointed by a date 28 days after a party gives notice of its intention to refer a dispute to a DAB.

The whole subject of Dispute Adjudication Boards is dealt with in Chapter 26.

23.3.19 Sub-clauses 20.5 to 20.8: 'Amicable Settlement'; 'Arbitration'; 'Failure to Comply with Dispute Adjudication Board's Decision'; 'Expiry of Dispute Adjudication Board's Appointment'

There are three consequences to a properly given decision by the DAB under sub-clause 20.4 of the 1999 Red Book, and these are as follows:

(a) Any reasoned decision of the DAB given under sub-clause 20.4 of the 1999 Red Book is binding on the parties, who are then required to 'promptly give effect to it unless and until it shall be revised in an amicable settlement or an arbitral award' as described in sub-clauses 20.5 and 20.6 of the contract conditions.
(b) Sub-clause 20.4 also stipulates that 'If the DAB has given its decision and *no notice of dissatisfaction* has been given by either Party within 28 days after it received the DAB's decision, then the decision shall become final and binding upon both Parties'. (emphasis added.)

(c) However, if a notice of dissatisfaction is given by either party within 28 days after receiving the decision, setting out the matter in dispute and the reason(s) for dissatisfaction, then both parties are required to attempt to settle their dispute amicably, as stipulated under sub-clause 20.5, before commencement of arbitration. Furthermore, as set out in sub-clause 20.6, unless settled amicably, any dispute in respect of which the DAB's decision has not become final and binding may be finally settled by international arbitration.

As a result of the first consequence, sub-clause 20.7 deals with the possibility of non-compliance with the DAB's decision. However, although the clear intention of sub-clause 20.4 is that the DAB's decision should be complied with, unless and until it is revised in a subsequent forum, and irrespective of whether or not one of the parties is dissatisfied with it, sub-clause 20.7 is worded in such a way that it only deals with the event where the parties are satisfied with the decision. The draftsmen did not deal with circumstances where the parties are dissatisfied with the decision, leaving that situation without any prompt solution or elucidation. In this, the same error is made as that discussed at the end of section 19.8 above.

This is totally unsatisfactory as it leaves the matter open to the possibility of an erroneous conclusion that there is only a breach of contract, but no specified remedy under the contract for non-compliance with a decision that has carried a notice of dissatisfaction. This may even lead to the conclusion, by an arbitral tribunal appointed pursuant to sub-clause 20.7, that the provisions of the sub-clause do not apply to decisions where there is a notice of dissatisfaction and the decision of the DAB in that case cannot be enforced as a temporarily binding decision.

Unfortunately, the same problem exists in the 1999 Yellow and Silver Books, since the wording of sub-clause 20.7 is unchanged.

Accordingly, to resolve this conflict between sub-clauses 20.4 and 20.7, it is suggested that sub-clause 20.6 should start with the words 'Subject to the provisions of Sub-Clause 20.7' and the text of sub-clause 20.7 should be corrected to read as follows:

'Failure to Comply with Dispute Adjudication Board's Decision
In the event that a Party fails to comply with the Decision of the Dispute Adjudication Board, then the other Party may, without prejudice to any other rights it may have, refer the failure itself to arbitration under Sub-Clause 20.6 [Arbitration]. Sub-Clause 20.4 [Obtaining Dispute Adjudication Board's Decision] and Sub-Clause 20.5 [Amicable Settlement] shall not apply to this reference.'

In this way, the decision of the DAB is transformed into an arbitral award, with all the enforcement prerequisites attached to such a document, which must have been the intention of this sub-clause in the first place.

Without such a change in the wording of sub-clause 20.7, it is possible to think of a situation where a responding party to an arbitration pursuant to sub-clause 20.7 would counteract a claim made against it by submitting a counterclaim relating to the merits of the subject matter of the DAB's decision, thus forcing the whole case into arbitration without having to comply with the DAB's decision. This situation could not have been within the contemplation or the intention of the draftsman of clause 20.

In this connection, some have commented that a link should exist between the following two phrases in the fourth and fifth paragraphs of sub-clause 20.4: 'The decision shall be binding on both Parties, who shall promptly give effect to it unless and until it shall be revised . . .' and '. . . then either Party may, within 28 days after this period has expired, give notice to the other Party of its dissatisfaction'. That link should make it clear that irrespective of whether or not one of the parties is dissatisfied with the DAB's decision, they still have to 'promptly give effect to it'. Others have suggested that the problem of a contractor receiving a large award under a DAB decision, which is later reversed in arbitration should also be dealt with under clause 20 by requiring the contractor to provide an advance payment guarantee.

Sub-clause 20.8 deals with the situation where a dispute arises between the parties in connection with, or arising out of, the contract or the execution of the works, but there is no DAB in place, whether by reason of the expiry of the DAB's appointment or otherwise. In such a case, sub-clause 20.8 provides that sub-clauses 20.4 and 20.5 do not apply; and the dispute may then be referred directly to arbitration under sub-clause 20.6.

There are many such situations in practice, where the parties leave the matter of appointment of the DAB to drift away until it is too late to take advantage of the inbuilt intermediate mechanisms for dispute resolution, which are more cost and time effective than arbitration.

23.4 Some highlights of the 1999 Red Book

23.4.1 Sub-clause 1.1.4.3: 'Cost'

As in the Fourth Edition of the Red Book, the definition of 'Cost' excludes profit, but includes overhead charges and may include financing charges if the payment received is after expenditure.[23.24] Where profit is concerned, reference has to be made to the text of the clause itself where payment is mentioned. It is paid and added to the cost if the employer is responsible for the event but not in circumstances where neither party could be blamed for the event, for example, in sub-clauses 4.12, 4.24, 8.9, 13.7, 17.4 and 19.4. Whether the payment is for profit can be seen from Table 23.3 below. The amount of profit payable could be specified, if required to be included in the contract, in the particular conditions.

23.4.2 Clause 2: 'The Employer'

Under Clause 2 of the 1999 Books the obligations and responsibilities of the employer are set out in clear terms. While this clause is not exhaustive in listing the employer's obligations, for example sub-clause 1.13(a) requires that the employer shall have obtained all planning permits for the permanent works, it is a very useful grouping together and does include new provisions in respect of the employer's obligations and responsibilities.

Under sub-clause 2.1 of the 1999 Red Book, the employer must give the contractor right of access to and possession of all parts of the site where the works are to take place. Such access to and possession of the site is available to the contractor within a

Table 23.3 Cost, with or without profit.

Clause no.	Clause title	Event	Cost/profit
1.9	Delayed Drawings or Instructions	Late drawing or instruction	Cost and profit
2.1	Right of Access to the Site	Late access or possession	Cost and profit
4.7	Setting-Out	Error in specified reference points (not reasonably discoverable by experienced contractor)	Cost and profit
4.12	Unforeseeable Physical Conditions	Adverse unforeseeable physical condition	Cost
4.24	Fossils	Compliance with instructions	Cost
7.4	Testing	Instruction for additional passed test, or delay for which employer is responsible	Cost and profit
8.9	Consequences of Suspension	Suspension	Cost
10.2	Taking over of Parts of the Works	Employer's use of part of works without contractor's agreement	Cost and profit
10.3	Interference with Tests on Completion	Prevention of tests	Cost and profit
11.2	Cost of Remedying Defects	Defects not contractor's responsibility	Clause 13 applies
11.8	Contractor to Search	No contractor defect found	Cost and profit
12.4	Omissions	Omission of work	'Cost' incurred anyway
13.2	Value Engineering	Approved proposal changes design	50% of net saving made
13.7	Adjustments or Changes in Legislation	Changes in law	Cost
15.5	Employer's Entitlement to Termination	Employer terminates at will	Value of work and cost
16.1	Contractor's Entitlement to Suspend Work	Contractor suspends due to employer's default	Cost and profit
16.4	Payment on Termination	Contractor terminations due to employer's default	Cost, profit, loss, damage, release of bond and value
17.4	Consequences of Employer's Risk	Loss or damage to works etc. due to employer's risk	Cost and profit in two circumstances
19.4	Consequences of *Force Majeure*	Prevented from performing any obligation	Cost, except natural catastrophes
19.6	Optional Termination, Payment and Release	Prolonged prevention	Value and cost
19.7	Release from Performance under the Law	Impossible, unlawful or released by law	Value and cost

prescribed time period measured in days after the commencement date, which should be stated within the Appendix to Tender. Under the second paragraph of sub-clause 2.1, if no period of time is stated in the Appendix to Tender, then the employer must give the contractor access to the site within a reasonable time period so as to enable him to proceed with the programme for the works that is submitted under sub-clause 8.3. Failure by the employer to grant access of the site to the contractor in time may be said to constitute a substantial failure to perform his obligations under the contract, thereby entitling the contractor to terminate the contract under sub-clause 16.2. If the employer fails to give the contractor 'possession of any foundation, structure, plant or means of access' within the timescale stated in the specification, this would normally be deemed to constitute a less serious failure by the employer, entitling the contractor to compensation but not termination.

Under sub-clause 2.2, the employer is obliged upon request to provide assistance to the contractor in respect of obtaining permits, licences and approvals required under the applicable law. It expressly provides for such assistance where the contractor requires such permits, licences and approvals for the execution of the works (by reference to sub-clause 1.13), for the importation of goods and for the exportation of equipment. This is a new development; under the older FIDIC forms of contracts no such assistance was required of the employer and the obtaining of such permits, licences and approvals was the sole responsibility of the contractor.

Sub-clause 2.3 states expressly what was only to be implied in the older FIDIC forms of contract, namely the employer's responsibility to ensure that his personnel and other contractors co-operate with the contractor and comply with the safety and environmental precautions being taken on site.

The requirements of sub-clauses 2.4 'Employer's Financial Arrangements' and 2.5 'Employer's Claims' have been discussed in Sections 23.3.2 and 23.3.3, respectively.

23.4.3 Clause 4: 'The Contractor'

Under clause 4 of the 1999 Red Book, the contractor's responsibilities are set out. Sub-clause 4.1 lays down the contractor's general obligations to design execute and complete the works.

Under sub-clause 4.2, the contractor has an obligation to issue the employer with a performance security in the amount stated within the Appendix to Tender. If there is no amount stated within the Appendix to Tender, then sub-clause 4.2 does not apply to the contract. Security is a common requirement for the employer's protection.

Under sub-clause 4.3, the role of the contractor's representative is discussed. The contractor's representative is the person who is responsible for ensuring that all the contractor's obligations under the contract are carried out.

Sub-clause 4.4 sets out the criteria, which must be followed in relation to sub-contractors. The contractor is responsible for the acts or omissions of a sub-contractor or his agents. Consent by the engineer is required where the contractor proposes to employ a sub-contractor to carry out works in the contract, and that sub-contractor is not named in the contract.

Sub-clause 4.10 relates to site data. Here, the employer is required to have made available to the contractor at tender stage all relevant information/data in his possession in relation to sub-surface and hydrological conditions on the site, including any environmental aspects. Using this information and other data which the contractor is expected to have obtained before submitting his tender, the contractor is deemed to have satisfied himself as to the site and all conditions pertaining to the intended works.

Under sub-clause 4.19 and 4.20, the contractor is entitled to basic provisions, for example, water and power, that may already be available on the site, i.e. those provisions that belong to the employer, as long as the contract states that the contractor is entitled to use them. Sub-clause 4.20, in particular, sets out two categories of items that may be made available to the contractor by the employer. The first is 'Employer's

Equipment' and the second is 'Free-Issue Materials', i.e. the materials owned by the employer that are intended for use by the contractor free of charge.

Sub-clause 4.21 states that the contractor should issue monthly progress reports to the engineer from the beginning of the works up to and including the actual completion date for the works. The sub-clause details what the report should include; however, it is acknowledged that less-detailed reports may be more appropriate for smaller projects. This report should be submitted to the engineer alongside the contractor's statement for interim payment under sub-clause 14.3. It is important to note that, in accordance with sub-clause 14.7(b), the contractor may not be paid until all the above-mentioned documents have been submitted.

23.4.4 Clause 6: 'Staff and Labour'

Clause 6 of the 1999 Red Book enumerates the topic of employing staff and labour to carry out the works, which is stated under sub-clause 6.1 to be the responsibility of the contractor. The contractor also has the burden of:

(a) paying the staff;
(b) ensuring the conditions of labour meet with the required standards;
(c) complying with the labour laws that are applicable;
(d) ensuring that no work is carried out outside normal working hours and on days of rest;
(e) provision and maintenance for all accommodation and welfare facilities for all of his personnel;
(f) ensuring the maintenance of all health and safety standards of his staff;
(g) ensuring his personnel act in an orderly fashion and in no way act in an unlawful or riotous fashion;
(h) supervising the plans, arrangements, directions, management, inspection and testing of the works, for the duration of the works; and
(i) ensuring that his personnel are appropriately skilled, qualified and experienced for the job that they were hired to do.

23.4.5 Clause 7: 'Plant, Materials and Workmanship'

Sub-clause 7.1 expressly introduces an additional standard in respect of workmanship by requiring that the works not only must be carried out in accordance with the contract, but also:

'in a proper workmanlike and careful manner, in accordance with recognised good practice'

and

'with properly equipped facilities and non-hazardous Materials, except as otherwise specified in the Contract.'

The requirement that the works be undertaken as above is of a more objective standard, and significantly different, from that of the Fourth Edition of the Red Book, where workmanship was subject to the engineer's instructions.

Sub-clause 7.7 of the 1999 Red Book states that every item of plant and material will become the property of the employer when it is delivered to the site or when the contractor is entitled to payment for the plant and materials under sub-clause 8.10 of these conditions. It is important, from a legal point of view, to establish ownership of any plant or materials at the very outset of the works, or at the least upon delivery to the site. This means that, should the contractor run into financial difficulties during the course of the contract, such plant and materials cannot be repossessed by creditors. It also means, of course, that should the employer terminate the contract pursuant to sub-clause 15.2 of the contract, all plant and material is retained by him for a future contractor to use in completing the works.

23.4.6 Clause 8: 'Commencement, Delays and Suspension'

Under Clause 8 of the 1999 Red Book, the commencement delays and suspensions of the works under the contract are set out in full. Sub-clause 8.3 deals with the programme for the works which is to be submitted by the contractor to the engineer. It further states that whenever the programme is inconsistent with actual progress, a revised programme should be submitted that is in line with the actual progress of the works. Unless any other intention is expressed by the engineer, the contractor shall proceed with the works in accordance with the timetable set out in the programme. Of course, it is obvious that the programme should be in accordance with the terms of the contract.

Under sub-clause 8.4, the contractor becomes entitled to an extension of the time for completion of the works, subject to the claims procedure to be followed under sub-clause 20.1, if completion of the works is delayed as a result of:

'(a) a Variation (unless an adjustment to the Time for Completion has been agreed under Sub-Clause 13.3 [Variation Procedure]) or other substantial change in the quantity of any item of work included in the Contract,
(b) a cause of delay giving an entitlement to extension of time under a Sub-Clause of these Conditions,
(c) exceptionally adverse climatic conditions,
(d) Unforeseeable shortages in the availability of personnel or Goods caused by epidemic or governmental actions, or
(e) any delay, impediment or prevention caused by or attributable to the Employer, the Employer's Personnel, or the Employer's other contractors on the Site.'

It is important to note that the above list of ground for extension of time omits the reference to 'Other Special Circumstances' which featured in the Fourth Edition of the Red Book. Instead a new item is added, item (d), as set out above. The *FIDIC Contract Guide*, at page 174, explains the term 'Exceptionally Adverse' and how it could be established. In effect the above provisions refer to the sub-clauses of the 1999 Red Book in Table 23.4 below.

Table 23.4 Sub-clauses relating to extension of time.

Sub-clause no.	Title and event	Where costs are payable, is profit payable?
1.9	Delayed Drawings or Instructions: delay by the engineer in issuing drawing or instruction required by notice	Cost plus reasonable profit added to contract price
2.1	Right of Access to the Site: delay by the employer to give access to and possession of the site	Cost plus reasonable profit added to contract price
4.7	Setting Out: error in specified reference points, not reasonably discoverable by an experienced contractor	Cost plus reasonable profit added to contract price
4.12	Unforeseeable Physical Conditions: adverse unforeseeable physical conditions	Cost but no profit added to contract price
4.24	Fossils: compliance with discovery of fossils etc.	Cost but no profit added to contract price
7.4	Testing: delayed testing caused by employer. See also sub-clause 10.3	Cost plus reasonable profit added to contract price
8.5	Delays caused by Authorities: a delay caused by the authorities	Extension of time only
8.9	Consequences of suspension: a suspension initiated by employer. See also sub-clause 16.1	Cost but no profit added to contract price
10.3	Interference with Tests on Completion: interference by employer. See also sub-clause 7.4	Cost plus reasonable profit added to contract price
13	The time consequences of variations are dealt with in sub-clause 8.4(a)	See sub-clause 8.4(a)
13.7	Adjustments for Changes in Legislation: changes in Law	Cost but no profit added to contract price
16.1	Contractor's Entitlement to Suspend Work: a suspension initiated by the contractor. See also sub-clause 8.9	Cost plus reasonable profit added to contract price
17.4	Consequences of Employer's Risks: loss or damage to the works due to an employer's risk	Generally cost but no profit added to contract price
19.4	Consequences of *Force Majeure*: the contractor being prevented from performing an obligation	Cost but no profit added to contract price

The remedies open to contractors are based on the following three criteria depending on whether the employer is responsible for the event giving rise to the extension of time: time only; time and cost; and time, cost and profit. It is worthwhile noting that the procedure for claiming extension of time has been combined with that for additional payment under sub-clause 20.1.

Sub-clause 8.7 deals with the payment of delay damages by the contractor to the employer for late completion of the works (or a section of the works) by reference to the time for completion of the works, stated in the contract. The delay damages are calculated on a daily basis but an upper limit, measured as a percentage of the final contract price, is provided for in the Appendix to Tender. It should be noted that, even if delay damages are paid by the contractor, he is still under an obligation to complete the works or any other duties that he may have under the contract.

23.4.7 Clause 9: 'Tests on Completion'

Clause 9 of the 1999 Red Book deals with the contractor's duty to carry out tests on completion of the works. It also provides for a situation where the carrying out of such tests has been delayed by the employer and/or the contractor or where the works have to be retested if they fail the test the first time round.

Sub-clause 9.4 'Failure to Pass Tests on Completion' deals with the situation where the works, or a section thereof, fail to pass the tests on completion, even after having been repeated under sub-clause 9.3. The engineer is then entitled to:

'(a) order further repetition of Tests on Completion under Sub-Clause 9.3;
(b) if the failure deprives the Employer of substantially the whole benefit of the Works or Section, reject the Works or Section (as the case may be), in which event the Employer shall have the same remedies as are provided in sub-paragraph (c) of Sub-Clause 11.4 [Failure to Remedy Defects]; or
(c) issue a Taking-Over Certificate, if the Employer so requests.'

Under the last option, the employer may demand a reduced price by an amount that would reflect the reduced value to him as a result of this failure. The last paragraph of this sub-clause provides alternatives for calculating that reduction. However, the most drastic of the above three alternatives is (b) where the contractor is unable or unwilling to resolve the problems encountered in the testing programme and the parties are directed to paragraph (c) of sub-clause 11.4 'Failure to Remedy Defects'. Paragraph (c) of sub-clause 11.4 states that if the defect 'deprives the Employer of substantially the whole benefit of the Works or any major part of the Works', the employer can terminate the whole contract or the part 'which cannot be put to its intended use' and demand his money back together with interest and the cost of dismantling the offending work and returning it to the contractor.

This solution is extremely difficult to envisage in terms of a civil engineering or a building project. The *FIDIC Contracts Guide* does not throw much light on this provision, but it does describe this situation as 'most unlikely to occur'.

23.4.8 Clause 10: 'Employer's Taking Over'

Sub-clause 10.2 of the 1999 Red Book deals with the employer's taking over of any defined part or parts of the works. In this regard it mirrors all other forms of construction contracts. However, sub-clause 10.2 of the 1999 Red Book introduces a further right of takeover of the employer, namely a situation where the engineer, with the permission of the employer, may issue a taking-over certificate for any part of the permanent works. This means that the contractor's possession, and responsibilities for care, safety and insurance, of that particular part of the works come to an end. It would appear that a corresponding defects notification period does not start for this part upon taking over, since the definition of this term under sub-clause 1.1.3.7 refers only to taking over certified under sub-clause 10.1 and makes no reference to sub-clause 10.2. The contractor is entitled to recover any reasonable additional cost accrued to him as a result of such a takeover (subject to the claims procedure to be followed under sub-clause 20.1). The employer's right to take over any part of the works at his own discretion is qualified by the provision that he has no entitlement to use any part of the works until the engineer has issued a taking-over certificate for it. A sample of a taking-over certificate for parts of the works would read as follows:

'We hereby certify, in the terms of Sub-Clause 10.2 of the Conditions of Contract, that the following parts of the Works were completed in accordance with the Contract on the dates stated below. Except for minor outstanding work and defects [which include those listed in the attached Snagging List]:

[description of each part taken over; and state its completion date]'

23.4.9 Clause 11: 'Defects Liability'

Sub-clause 11.3 states that the employer is entitled, subject to sub-clause 2.5, to an extension of the defects notification period for the works as a whole or for a section of the works. However, such an extension will only be granted if the works, a section of the works or a major item of plant from the works, cannot be used for the purposes for which they were intended as a result of a defect or of damage. It should be noted that the maximum period of extension of time is 2 years measured from the date the defects notification period would otherwise have expired.

This clause is particularly designed for items of plant which should have a full period of 12 months of a defects notification period that is not diminished by interrupted failures. However, this extension may be abused in practice.

23.4.10 Clause 12: 'Measurement and Evaluation'

Clause 12 of the 1999 Red Book, is founded on the idea that the value of the works under the contract should be established by gauging the amount of each item under the method of measurement referred to in sub-clause 12.2 (which, in turn, refers to the contract bill of quantities or other applicable schedules of rates) and applying the appropriate rate per unit quantity or the appropriate lump sum price under sub-clause 12.3 'Evaluation'. In doing so, particular reference should be made to paragraph (a) of sub-clause 12.3 and also to paragraph (a) of sub-clause 13.1.

23.4.11 Clause 13: 'Variations and Adjustments'

Clause 13 of the 1999 Red Book deals with the possibility that variations and adjustments may be made to the original scope of the works under the contract. Under sub-clause 13.2, the contractor may submit a written proposal to the engineer in respect of value engineering of the works, detailing the ways in which he feels the contract may be completed sooner than originally planned, or the costs of the works may be reduced or the efficiency may be improved. It is worthwhile to note that the contractor is not under any kind of duty to submit such a proposal, and in fact it is likely that a contractor would be slow to do so, unless there was obvious remuneration to him. Furthermore, the fact that such a proposal must be done at his own expense may prove to be a hindrance rather than an encouragement for him to submit one. This sub-clause also covers the possibility that the contractor's proposal may involve a change in the employer's or the engineer's design of the works. Where the proposal includes a change in the design of part of the works and such a proposal is approved by the engineer, then unless the parties disagree, the contractor should design the proposed changed part of the works, sub-paragraphs (a) to (d) of sub-clause 4.11 will apply, and if the change results in a reduction in the contract value of the particular part, then the engineer will determine a fee that will be included in the contract price. Sub-clause 13.3 lays down the criteria that should be included in the contractor's proposal and the procedure that should be followed by both the contractor and the engineer in submitting and instructing on the value engineering proposal.

23.4.12 Clause 14: 'Contract Price and Payment'

Under the 1999 Red Book, sub-clause 14.1 deals with the monetary value of the contract. The value of the contract is established under sub-clause 12.3, as discussed above, but under sub-clause 14.1 the contractor is liable to pay all taxes and duties. Such payment will be included in the contract price, and therefore the only way that the agreed contract price may be varied in relation to such taxes and duties is if the legislation designating the amount of taxes or duties owed is altered after the contract price has been agreed. It is also worthwhile to note that, under the 1999 Red Book, the quantities set out in the bill of quantities are only deemed to be estimates of the actual quantities of work to be carried out. They are merely an attempt to define the extent of the contractor's obligations to carry out the works, and it is expressly stated that they should not be taken as correct for the purposes of final measurement to calculate the contract price.

Under sub-clause 14.2, the employer is obligated to make an advance payment of the overall contract price to the contractor. This advance payment is seen as an interest-free loan to enable the works to begin. However, payment of the advance does not have to take place until the employer receives an advance payment guarantee from the contractor as a form of security. Annex E to the 1999 Red Book gives an example form of this guarantee. It is important to note that in order for this sub-clause to apply, the advance payment amount must be stated in the Appendix to Tender. The contractor's advance payment guarantee must be valid up to the date on which the advance payment is due to be repaid. Furthermore, the guarantee must be extended if the advance payment has not been repaid 28 days before the guarantee is due to expire. If the contractor fails to extend the validity of the guarantee, and the advance has not been reimbursed, then under sub-clause 14.2 the employer is entitled to call in the guarantee. The advance payment is repaid by the contractor to the employer in the form of deductions in subsequent interim payments. The amount of the deduction in each interim payment is calculated at the percentage agreed by the parties, as stated in the Appendix to Tender, but if no percentage is so stated, sub-clause 14.2 provides a mechanism for such deductions.

Sub-clause 14.4 'Schedule of Payments' deals with a situation where the contract states that the contract price is to be paid in instalments. If the instalments in the contractual schedule of payments are not defined by reference to actual progress of the works, and if actual progress is slower than that upon which the schedule was based, the engineer becomes entitled to agree or determine (under sub-clause 3.5) revised instalments. If the contract does not provide for a schedule of payments, then the contractor must present non-binding estimates of the payments that he anticipates will become due for every three months, with revised estimates submitted at quarterly intervals. In this way, the employer has an advance indication of what amounts will become payable and can arrange his finances accordingly.

Sub-clause 14.6 deals with the issue of interim payment certificates. It is one of the roles of the engineer to issue to the employer an interim payment certificate, which details the amount that the engineer believes is due to the contractor every month, along with any documents supporting his belief. A copy of this certificate should also be issued to the contractor. The interim payment certificate is binding upon the employer and he must make payment in full, subject to sub-clause 14.7(b), regardless of any claim that he may have against the contractor. There are only two situations in which payment of the interim payment certificate may be withheld. These are as follows:

(a) where any item supplied or work done by the contractor does not conform with the terms of the contract, then the cost of correcting or replacing the work may be retained until correction or replacement has been completed by the contractor;
(b) if the contractor fails to carry out any work or commitment in line with the terms of the contract, and the engineer notifies him of this problem, then the monetary value of this work or commitment may be retained until the work or commitment has been carried out by the contractor.

This ensures that the contractor is not paid for works that are not in accordance with the contract, and that he has an incentive to properly complete such works.

Sub-clause 14.7 sets out the method and the timescale in which the employer is to pay the contractor for work done. If a situation arises where the contractor has not been paid during the construction of the works, sub-clause 14.8 states that he shall be entitled to claim interest on the amount that has not been paid from the time the date for payment was issued. The contractor is not obliged to give any notice in relation to this claim for interest.

23.4.13 Clause 15: 'Termination by Employer'

Under sub-clause 15.2, the employer can terminate the contract under the following instances:

(a) if the contractor does not act in accordance with sub-clause 4.2 'Performance Security', or with a notice to correct given under sub-clause 15.1;
(b) if the contractor abandons the works or clearly acts with the intention of not carrying out his commitments under the contract;
(c) if the contractor fails to proceed with the works in accordance with clause 8 'Commencement, Delays and Suspension', or
(d) if the contractor fails to comply with a rejection notice under sub-clause 7.5 or a remedial work notice under sub-clause 7.6 within 28 days of receiving either notice;
(e) if the contractor subcontracts the whole of the works or assigns the contract without the required agreement;

(f) if the contractor becomes insolvent, bankrupt or goes into liquidation or suchlike;

(g) if the contractor gives or offers to give a bribe, gift, gratuity, commission or other thing of value as an inducement or an award for doing or forbearing to do any action in relation to the contract, for showing or forbearing to show favour or disfavour *to any person* in relation to the contract, or if the contractor or any of his personnel give or offer to give any such inducement or reward as described above. The foregoing emphasis is added to highlight that the corruption governed by this provision is in respect of anybody whatsoever, and is not limited to the employer's or the engineer's personnel. However, it should be noted that lawful inducements and rewards to the contractor's personnel will not entitle the employer to terminate the contract.

In the first five situations above, the employer may terminate the contract once he has given the contractor 14 days' notice of such an intention. In the final two situations, the employer can by notice terminate the contract immediately. Furthermore, under sub-clause 15.5, the employer is entitled to terminate the contract at any time for his own convenience by giving notice of such termination to the contractor. However, there is a proviso in that the employer is not entitled to terminate the project under this particular sub-clause in order to carry out the works himself or to arrange for the works to be carried out by someone else.

23.4.14 *Clause 16: 'Suspension and Termination by Contractor'*

Under clause 16 of the 1999 Red Book, the contractor's entitlement to suspend and/or terminate the works is set out. Sub-clause 16.1 deals with the situation whereby the contractor becomes entitled to suspend the work if the engineer fails to issue interim payment certificates in accordance with sub-clause 14.6 or if the employer fails to comply with sub-clause 2.4 'Employer's Financial Arrangements' or sub-clause 14.7 'Payment'. In that case, the contractor may suspend the works until he has received the payment certificate or reasonable evidence of the employer's financial arrangements or payment, as the case may be. If the contractor suffers delay and/or incurs cost as a result of suspending the work, then upon giving notice to the engineer (and subject to complying with the claims procedure of sub-clause 20.1), he shall be entitled to an extension of time for any such delay and/or payment of any costs that he has incurred, as well as a 'reasonable profit'.

23.4.15 *Sub-clause 20.7: 'Failure to Comply with Dispute Adjudication Board's Decision'*

As explained in Section 23.3.19 above, it is worth noting that the problem in the wording of sub-clause 67.4 of the Fourth Edition of the Red Book, 'Failure to Comply with Engineer's Decision', also exists under sub-clause 20.7 of the 1999 Red Book. However, the decision in question under the 1999 Red Book is that of the Dispute Adjudication Board and not that of the engineer. The problem, which was explained in Chapter 19 under Section 19.8.1, relates to the situation where a notice of intention to proceed to arbitration had already been given, but the engineer's decision was not complied with. This problem in the wording of sub-clause 20.7 should be rectified by deleting the reference therein to the notice of dissatisfaction with the Board's decision.

23.4.16 *Guidance for the preparation of particular conditions*

Following the general conditions of contract (and their index) the 1999 Red Book includes a very useful section entitled 'Guidance for the Preparation of Particular Conditions'. This section is worth attention, by both employers and contractors, in order that the agreed provisions of an intended contract are best suited to the project in hand and the parties' proposed business transaction. In the unlikely event that no particular conditions are envisaged, these guidance notes provide an insight into the thinking behind the provisions of many of the general conditions and, in this way, an indirect interpretation of them. For example, sub-clause 1.14 'Joint and Several Liability' provides that if the contractor is a joint venture or consortium, then the members shall notify the employer of their leader who will then have authority to bind them during the course of the contract. The guidance notes explain this provision by noting that appointment of the leader (at an early stage in the contract) is necessary in order to have one point of contact and to avoid the employer's involvement in a dispute between the members of the joint venture/consortium. While this may be obvious to some users of this form of contract, it is a useful reminder of the implications of the provision.

The guidance notes are a particularly valuable reference in respect of the example sub-clauses that they provide. For example, under the heading of sub-clause 4.12 'Unforeseeable Physical Conditions' the guidance notes give example wording where the parties agree that the risk of sub-surface conditions is to be shared between them. Further, under the heading of clause 6 'Staff and Labour' the wording of additional sub-clauses is suggested where the parties wish to take account of the particular circumstances and culture of the locality of the site, namely: using foreign staff and labour; measures to be taken against insect and pest nuisance; alcoholic liquor or drugs; arms and ammunition; and festivals and religious customs. Furthermore, under the heading of clause 8 'Commencement, Delays and Suspension', FIDIC suggests the text of an additional sub-clause where the employer wishes to include incentives for the contractor to achieve early completion.

The above is not exhaustive and gives just a few examples of the many practical and constructive points raised in these guidance notes. It would be worth intending contractors, employers and engineers reviewing what is included in these guidance notes before entering into a contract under the 1999 Red Book.

Chapter 24

The 1999 Yellow Book

24.1 Introduction

Much of what is stated in Chapter 23 in relation to the 1999 Red Book also applies to the 1999 Yellow Book. However, there are variations between the two Books, which are dealt with below. The 1999 Yellow Book has been described as 'an appropriate model for the implementation of building projects in the private sector, and is being used increasingly for the delivery of infrastructure in the public sector.'[24.1] As the title of the 1999 Yellow Book suggests, FIDIC has put it forward as appropriate for use in contracts where the contractor is responsible for design, and relating to electrical and/or mechanical plant and/or other works which may include any combination of civil, mechanical, electrical and/or construction works. Perhaps FIDIC drafted this new form of contract in response to the difficulties experienced by employers and contractors in such contracts, as summarised by the following extract:

> 'In relation to mechanical and electrical works (and, to some extent, in relation to cladding works), it is not uncommon that a large element of the design be carried out by sub-contractors. The basis on which this is done (under traditional contracts) is not always well defined. One can have a contractor who believes that he is working to a straightforward traditional contract (with no design responsibility) who is, however, employing a sub-contractor who is [sic] obligations have been defined by reference to performance... Thus, it is not uncommon for a contractor under a traditional building contract in Ireland to find himself settled with an obligation in respect of some of the specialist works which is performance related. Needless to say, when this happens, it causes confusion and, sometimes an unwillingness to believe the legal advice offered.'[24.2]

FIDIC does not, however, promote the use of the 1999 Yellow Book over any other of its forms of contract, but insists that each form of contract has its own individual characteristics as well as its advantages and disadvantages. Therefore, each individual employer can decide which form he wants to use for a particular project. The European International Contractors (EIC), whilst acknowledging that the 1999 Yellow Book is a practical

> 'starting point for a design/build contract which includes both plant supply and construction works', also believe with some justification that 'such a broad application

[of the book] is inappropriate and [they] doubt the wisdom of trying to produce a standard form of contract for such a wide range of applications'.[24.3]

The range of application of this form of contract may be too wide if one considers that it might include contracts for just the supply and installation of plant and equipment, with no design or construction input required of the contractor. However, this may not be true if the range is expected to include projects where building and civil engineering works are required to be undertaken in conjunction with the supply and installation of plant and equipment. In their Guide to the 1999 Yellow Book, the EIC does acknowledge that there are improvements over previous forms of contract as well as retrogressions. The EIC welcomed a number of sub-clauses, including 2.4 'Employer's Financial Arrangements'; 2.5 'Employer's Claims'; 3.1 'Engineer's Duties and Authority'; 4.2 'Performance Security'; and 20.2 'Appointment of the Dispute Adjudication Board'. The latter category includes certain aspects of sub-clauses 1.12 'Confidential Details'; 4.1 'Contractor's General Obligations'; 4.5 'Nominated Subcontractors'; 4.7 'Setting Out'; 4.10 'Site Data'; 4.12 'Unforeseeable Physical Conditions'; 5.1 'General Design Obligations'; 15.5 'Employer's Entitlement to Termination'; and 20.1 'Contractor's Claims'.

24.2 Tendering under, and using, the 1999 Yellow Book

Tendering under, and using, the 1999 Yellow Book requires special skills on the part of the employer, the contractor and the engineer. In particular, unambiguously defining the scope of the design; the identity of who should be involved in the design; the extent of such design involvement; the purpose(s) for which the works are intended; and quality control, are all matters that normally do not feature in the more traditional forms of plant/equipment or construction contracts, so should be given special attention by the parties and agreed between them prior to signing the contract. The absence of such an agreement would most probably result in serious and costly misunderstandings and conflicts between the parties.

Further, the following are amongst the most controversial topics in design and build contracts that should be dealt with appropriately prior to entering into the contract. Indeed, in the tendering period they should be part of the document that defines the employer's requirements and later become part of the contract:

(a) a review of available site data and expert advice on what additional data is required;
(b) the employer's design concept of the proposed project and a review and development of such conceptual design;
(c) preliminary designs and drawings, which are sufficient to indicate sizes, arrangements, capacities and suchlike, for tender pricing;
(d) whether nominated sub-contractors, if any, are to be engaged on any design activities, and if so, who is in charge and responsible for their output;
(e) for plant and equipment manufactured off-site, and for structural steel or reinforced and pre-stressed concrete structures, workshop drawings and the approval process of these drawings;

(f) the approval process, in general, of all submissions by the contractor and the time limit for each particular step in this process;
(g) the type(s) of specification to be used and the standards to be adopted for each element of the project;
(h) whether it would be essential or helpful to have a limited bill of quantities or schedule of rates, with the more significant items priced;
(i) quality control and who is to provide that service, including the demarcation and the difference between supervision and superintendence;
(j) a more project-specific description of anticipated prolongation and disruption events than that provided in the general conditions of the 1999 Yellow Book and how to measure appropriately their different effects;[24.4]
(k) contractual arrangements, which are suitable for the parties, the circumstances of their transaction and the project, in order to provide for an improved working relationship and effective management of the project, such as partnering;
(l) the identity, and a plan for the appointment, of the members of the Dispute Board from the commencement date; and
(m) the coverage, conditions and obtaining of all necessary insurance policies, and responsibilities for these, must be given serious consideration by *both* parties, discussed in detail and agreed before the commencement date (particularly in light of the present wording of clause 18 of the 1999 Yellow Book).

24.3 The 1999 Yellow Book: The employer's requirements

The development of the 1999 Yellow Book marks a step away from the 1999 Red Book and a further step away from the Fourth Edition of the Red Book and the Third Edition of the Yellow Book. Not only does the 1999 Yellow Book take away the need for impartiality on the part of the engineer, it also removes the function of design from his role on the project. This is instead entrusted to the contractor. The engineer is left with the task of drawing up, on behalf of the employer and in skeleton form, the extent; scope; purpose; perhaps a preliminary or conceptual design; and other technical details, specifications, setting out details, required testing regimes and the design, construction, operational and maintenance criteria for the project. The skeleton form and technical details are expressed in a document referred to as the 'Employer's Requirements', which is a new contractual term incorporated into the provisions of the 1999 Yellow Book.

The employer's requirements are sometimes described as a type of 'output specification', in that the employer decides what he wants to achieve in respect of the project, setting out precisely the purpose(s) for which the permanent works are intended upon completion, and the contractor then designs the permanent works, taking full responsibility for the adequacy of these works in achieving the employer's purpose(s). This contrasts markedly with 'the input specification' of the traditional form of contracting where the employer/engineer has designed the works, taking full responsibility for its adequacy, and the contractor's responsibility is limited to construction in accordance with such specification.

The term 'Employer's Requirements' is defined in sub-clause 1.1.1.5 of the 1999 Yellow Book as:

'... the document entitled employer's requirements, as included in the Contract, and any additions and modifications to such document in accordance with the Contract. Such document specifies the purpose, scope, and/or design and/or other technical criteria, for the Works.'

The document specifying the employer's requirements is pivotal to the understanding of the concepts of the 1999 Yellow Book, as it forms part of the definition of the 'Contract' in sub-clause 1.1.1.1 and is expressly referred to in 24 other sub-clauses (of nine clauses) of the general conditions of contract, as follows: 1.1.6.9; 1.5; 1.8; 1.9; 1.11; 1.13; 2.1; 4.1; 4.6; 4.18; 4.19; 4.20; 5.1; 5.2; 5.4; 5.5; 5.6; 5.7; 6.1; 6.6; 7.8; 8.3; 9.1; and 17.5. Further, there are a number of additional sub-clauses dealing with technical provisions which refer to 'the Contract', but since it is the most important technical document of the contract, and takes priority over the other technical documents of the contract, effectively such reference is to the employer's requirements in the first instance, and the schedules or the contractor's proposal thereafter. These additional sub-clauses are as follows: 1.1.3.4; 1.1.3.6; 1.1.6.7; 4.7; 7.4; 8.2; 10.2 and 11.1. All these sub-clauses are discussed below.

The drafting of the employer's requirements document itself has proved, in practice, to be the main source of success or failure of the project and of the disputes that arise in projects under this form of contract.[24.5] Therefore, when the employer's requirements are drafted, and because they are to be presented as a contract document, it must be remembered that the document should take into account a number of contrasting characteristics, as set out below.

(a) The employer's requirements should be complete, including all parameters relating to the required shape, type, quality, tolerances, functional standards, safety criteria, and limits on the whole-life cost of the permanent works upon completion; the tests which must be successfully passed both during and after construction; the expected and required performance of the permanent works upon completion; the design-life and durability of the permanent works upon completion; how the permanent works are expected to be operated and maintained upon completion; the manuals to be supplied; and details of the spare parts required to be provided and their cost;
— *Yet* the engineer must not fall into the trap of specifying the parameters so stringently that such specification is overly biased towards a particular design, thereby inhibiting the contractor's capacity to undertake his own, perhaps more innovative, design. It must be remembered and care taken that by specifying all these parameters, the contractor's liability for the design must not be compromised, including the specified obligation of fitness for purpose.
(b) The employer's requirements must be drafted in a precise manner to define what the employer requires;
— *Yet* they should be flexible enough to generate one of the main advantages of a design and build contract, namely the contractor's expert and inventive input into the design and construction of the project.
(c) The employer's requirements should be demanding enough to enable the employer to successfully choose the most suitable contractor from amongst the tenderers invited to bid;

— *Yet* the tenderers should not be required, at the tender stage, to provide more than the necessary information for the correct decision on the successful tenderer.
(d) The employer's requirements must be sufficiently detailed to establish the purpose(s) of the project;
— *Yet* they must be concise enough so as not to limit the contractor's ability to properly design the works or to restrict his ingenuity in searching for the most appropriate solution.
(e) The employer's requirements must be drafted such that the tenderers have equal opportunities to present the most efficient and cost-effective design to achieve the employer's specified purpose(s) for the project;
— *Yet* precise enough so that the tendered designs can be evaluated by the employer on a comparative basis.
(f) The employer himself should give particular attention to the wording and contents of that part of the employer's requirements dealing with the purposes for which the works of the project are intended. These purposes need to be unambiguously described and explained in order that the contractor is fully aware of the extent of his obligations under the first paragraph of sub-clause 4.1 'Contractor's General Obligations' (fitness for purpose);
— *Yet* such defined purposes should consist of realistic expectations given the reluctance of many plant and equipment/construction contractors to agree to a fitness for purpose provision, and the inordinate time and cost terms that may appear in the tenders.
(g) The employer's requirements should be prepared, drafted and checked to ensure that they comprehensively and thoroughly include each of the matters which are expressly referred to in the general conditions (and particular conditions, where relevant) of contract;
— *Yet* since it is likely that the sections of this document which deal with different types of works may be drafted by professionals from different disciplinary backgrounds, they must be checked and double-checked to ensure that there is no direct and, more importantly, no *indirect* conflict or inconsistency between the technical requirements set out in them.

It is important now to refer to the text in the various sub-clauses mentioned above, where such text has a bearing on the employer's requirements, and to refer to the manner in which the term 'Employer's Requirements' plays a fundamental role in the provisions of the contract conditions. As mentioned above, the contents of the employer's requirements are referred to in no less than 20 sub-clauses of the general conditions of the 1999 Yellow Book, most of which involve the contractor's obligations under the contract. It is perhaps stating the obvious, but worth noting nonetheless, that if a certain matter is absent from the employer's requirements or, as is more likely, such matter has been incorrectly or inadequately set out in the employer's requirements, the contractor's obligation(s) in that regard are compromised, if not negated altogether. It is very important, therefore, that the employer/engineer thoroughly check the employer's requirements, prior to commencement of the tendering process, in order to satisfy himself that each of the matters referred to in the conditions of contract are properly contained in the employer's requirements.

The order that such sub-clauses are discussed below is the order in which they appear in the general conditions of the 1999 Yellow Book.

24.3.1 Sub-clause 1.1.3.4 'Tests on Completion'

The tests on completion are defined as those 'specified in the Contract', namely those specified in the employer's requirements in the first instance, and the schedules or the contractor's proposal thereafter, or agreed by the parties or instructed as a variation, which are to be undertaken by the contractor in accordance with clause 9 'Tests on Completion' prior to taking over of the works by the employer.

24.3.2 Sub-clause 1.1.3.6: 'Tests after Completion'

The tests after completion are defined as those 'specified in the Contract', namely those specified in the employer's requirements in the first instance, and the schedules or the contractor's proposal thereafter, which are to be undertaken by the contractor in accordance with clause 12 'Tests after Completion' following taking over of the works by the employer.

24.3.3 Sub-clause 1.1.6.7: 'Site'

The site is defined as those places where the works are to be executed and plant/materials are to be delivered, and any other places 'specified in the Contract', namely those specified in the employer's requirements in the first instance, and the schedules or the contractor's proposal thereafter.

24.3.4 Sub-clause 1.1.6.9: Definitions – 'Variation'

A variation under the 1999 Yellow Book is defined as any change to the employer's requirements or the works, which is instructed or approved as a variation under clause 13 'Variations and Adjustments'.

24.3.5 Sub-clause 1.5: General Provisions – 'Priority of Documents'

In this sub-clause, it is specified that the documents forming the contract are taken as mutually explanatory of one another. For the purposes of interpretation, the priority of the documents is then set out in a sequence, which places the employer's requirements in sixth position after (in order of the stated priority) the contract agreement (if any); the letter of acceptance; the letter of tender; the particular conditions; and the general conditions, but before the schedules; the contractor's proposal (submitted by the contractor as part of his tender) and any other documents forming part of the contract. In this position, the employer's requirements are the most important technical document of the contract. It replaces the specification and the drawings in the order of priority found in the 1999 Red Book.

24.3.6 Sub-clause 1.8: 'Care and Supply of Documents'

The contractor is required under the second paragraph of this sub-clause to keep, on the site, a copy of the contract, the publications named in the employer's requirements, the contractor's documents (the technical documents required to be prepared and submitted to the engineer/employer by the contractor), and variations and other communications given under the contract.

It is further provided in this sub-clause that if a party becomes aware of an error or defect of a technical nature in a document, which was prepared for use in executing the works, the party shall promptly give notice to the other party of such error or defect. An error in the employer's requirements themselves, given the crucial role they play in the contract, merits further provision in sub-clauses 1.9 and 5.1 of the contract conditions, as discussed below.

24.3.7 Sub-clause 1.9: General Provisions – 'Errors in the Employer's Requirements'

This sub-clause provides for the consequences of a contractor suffering delay and/or incurring cost as a result of an error in the employer's requirements, if an experienced contractor exercising due care would not have discovered the error when scrutinising the employer's requirements as the contractor is obliged to do shortly after the commencement date under sub-clause 5.1 'General Design Obligations'. Sub-clause 1.9 specifies that in such circumstance, the contractor should give notice to the engineer and should be entitled (subject to sub-clause 20.1 'Contractor's Claims' to:

(a) an extension of time for any such delay, if completion is or will be delayed, under sub-clause 8.4 'Extension of Time for Completion'; and/or
(b) payment of the costs incurred plus reasonable profit.

The entitlement to an extension of time is subject to completion being or expected to be delayed and any additional payment is required to be included in the contract price. Of course, in all circumstances under sub-clause 20.1, the engineer after receiving this notice is required to proceed in accordance with sub-clause 3.5 'Determinations', to agree or determine (i) whether and, if so, to what extent the error could not reasonably have been discovered by the contractor during his scrutiny of the employer's requirements; and (ii) the quantum of extension of time and/or additional payment arising from the error. It is important to recognise in this connection that the engineer is being asked under the provisions of this latter sub-clause to act fairly and to put himself in the position of an experienced contractor exercising due care when scrutinising the requirements under sub-clause 5.1 'General Design Obligations', which obviously belong to the contractor. The link between sub-clause 3.5 and sub-clause 20.1, and the procedure and steps to be followed by the contractor, the engineer and the employer in respect of such a claim for extension of time and/or additional payment under the 1999 Yellow Book, are the same as those under the 1999 Red Book, as explained under Section 23.3.4 of Chapter 23.

24.3.8 Sub-clause 1.11: General Provisions – 'Contractor's Use of Employer's Documents'

This sub-clause provides that, as between the parties, the employer shall retain the copyright and other intellectual property rights in the employer's requirements and other documents made by (or on behalf of) the employer. The text of this sub-clause continues to provide that the contractor may (at his own cost) copy, use and obtain communication of these documents for the purposes of the contract. However, the contractor is not permitted, without the employer's consent, to copy, use or communicate them to any third party, except as necessary for the purposes of the contract.

24.3.9 Sub-clause 1.13: General Provisions – 'Compliance with Laws'

This sub-clause requires the contractor, in performing the contract, to comply with the applicable laws. However, it is specified that, unless otherwise stated in the particular conditions, the employer shall have obtained, or shall obtain, the planning, zoning or similar permission for the permanent works, and any other permissions described in the employer's requirements as having been, or being, obtained by the employer. Further, it is provided that the employer is to indemnify and hold the contractor harmless against and from the consequences of any failure to obtain such permissions.

24.3.10 Sub-clause 2.1: The Employer – 'Right to Access to the Site'

As the title of this sub-clause suggests, it deals with the contractor's right of access to, and possession of, all parts of the site in the time(s) expressly stated in the Appendix to Tender or, if not, in time such as to enable the contractor to proceed with the works in accordance with his submitted programme. In certain circumstances there may be a need for the employer to give to the contractor possession of or access to particular structures or plant, in which case the employer is required to do so in the time and the manner described in the employer's requirements.

24.3.11 Sub-clause 4.1: The Contractor – 'Contractor's General Obligations'

This sub-clause deals with the contractor's obligation to design, execute and complete the 'Works' in accordance with 'the Contract', and to remedy any defects in the 'Works'. It also specifies that when completed, the 'Works' shall be fit for the purposes for which they are intended as defined in 'the Contract'. While the text of this sub-clause refers to 'the Contract', since the employer's requirements are the most important technical document of the contract, this provision effectively means that 'the Works' are to be in accordance with the employer's requirements in the first instance, and the schedules or the contractor's proposal thereafter and fit for the purposes defined therein.

The text of this sub-clause continues to explain that the

> 'Works... shall include any work which is necessary to satisfy the Employer's Requirements, Contractor's Proposal and Schedules, or is implied by the Contract, and all works which (although not mentioned in the Contract) are necessary for stability or for the completion, or safe and proper operation, of the Works'.

The inclusion in this definition of works which are to be *implied* by the express provisions of the contract should not be underestimated. For example, the contract may set out in detail the technical requirements for each of two connected elements but omit to give any detail as to the connection between them. The text quoted above means that the contractor will be expected to design and construct such a connection, and to have included for it in his tendered price, although no reference is made to it in the contract documents, of which the employer's requirements is the most relevant.

24.3.12 Sub-clause 4.6: The Contractor – 'Co-operation'

Under the first paragraph of this sub-clause, the contractor shall 'as specified in the Contract', in other words as specified in the employer's requirements in the first instance, and the schedules or the contractor's proposal thereafter, or as instructed by the engineer, allow:

(a) the employer's personnel;
(b) other contractors, if any; and
(c) public authority personnel to work on or near the site.

The contractor is made responsible for his construction activities on the site, but is required to co-ordinate his own activities with those of other contractors to the extent, if any, specified in the employer's requirements. Further, if the employer is required to give the contractor possession of or access to any particular foundation, structure or plant in accordance with the contractor's documents, the contractor must submit such documents to the engineer in the time and manner specified in the employer's requirements.

24.3.13 Sub-clause 4.7 'Setting Out'

The first sentence of this sub-clause requires the contractor to set out the works pursuant to the items of reference 'specified in the Contract', namely those specified in the employer's requirements in the first instance, and the schedules or the contractor's proposal thereafter, or notified to the contractor by the engineer.

24.3.14 Sub-clause 4.18: The Contractor – 'Protection of the Environment'

Under sub-clause 4.18, the contractor is required to ensure that emissions, surface discharges and effluent from its activities on site do not exceed the values indicated in the employer's requirements, and prescribed by the applicable laws.

24.3.15 Sub-clause 4.19: The Contractor – 'Electricity, Water and Gas'

The employer's requirements, by reference to this sub-clause, should give details of the availability, if any, and prices of supplies of electricity, water, gas and other services on the site which the contractor is entitled to use.

24.3.16 Sub-clause 4.20: The Contractor – 'Employer's Equipment and Free-Issue Material'

If the employer has undertaken to make equipment available for use by the contractor in the execution of the works, then the employer's requirements should set out the details, arrangements and prices in respect of such equipment. Unless otherwise stated in the employer's requirements, the employer remains responsible for his equipment, except that the contractor is responsible for each item of employer's equipment while any of his personnel is operating, driving or directing it, or in possession or control of it.

Further, if the employer has undertaken to supply, free of charge, the 'free-issue materials', the details of such should be stated in the employer's requirements.

24.3.17 Sub-clause 5.1: Design – 'General Design Obligations'

Together with sub-clause 4.1, the provisions of this sub-clause require the contractor to carry out, and be responsible for, the design of the works. It further provides that the design shall be prepared by qualified designers who are engineers or other professionals who shall comply with the criteria, if any, stated in the employer's requirements.

Upon receiving notice under sub-clause 8.1 'Commencement of Works', the contractor is required to scrutinise the employer's requirements, including design criteria and calculations, if any, and the items of reference mentioned in sub-clause 4.7 'Setting Out'. Within the period stated in the Appendix to Tender, calculated from the commencement date, the contractor shall give notice to the engineer of any error, fault or other defect found in the employer's requirements or these items of reference.

This sub-clause should be read in conjunction with sub-clauses 1.8 and 1.9, discussed above.

24.3.18 Sub-clause 5.2: Design – 'Contractor's Documents'

This sub-clause specifies that the contractor's documents shall comprise the technical documents specified in the employer's requirements, documents required to satisfy all regulatory approvals, and the documents described in sub-clauses 5.6 'As-Built Documents', and 5.7 'Operation and Maintenance Manuals'. Therefore, the employer's requirements should specify the following:

(a) which documents are to be supplied by the contractor as the 'Contractor's Documents', with particular reference to those required under sub-clauses 5.6 and 5.7; and
(b) which (if any) of such contractor's documents are to be submitted for review, or for approval under sub-clause 5.2, and other necessary information, for example, the extent of detail required for the submissions, and procedures and periods for the reviews.

It states further that, unless otherwise stated in the employer's requirements, that the contractor's documents shall be written in the language for communications defined in sub-clause 1.4 'Law and Language'.

The provisions of this sub-clause also provide that:

(a) If the employer's requirements specify that some contractor's documents are to be submitted to the engineer for review and/or for approval, they shall be submitted accordingly, together with a notice as described below.
(b) In the following provisions of this sub-clause,
　(i) 'review period' means the period required by the engineer for review and, if so specified, for approval, and
　(ii) the term 'Contractor's Documents' excludes any documents which are not specified as being required to be submitted for review and/or for approval.
(c) Unless otherwise stated in the employer's requirements, each review period shall not exceed 21 days, calculated from the date on which the engineer receives a contractor's document and the contractor's notice. This notice shall state that the contractor's document is considered ready, both for review and approval, if specified in the employer's requirements, and for use. The notice shall also state that the contractor's document complies with the contract, or the extent to which it does not comply.

24.3.19   Sub-clause 5.4: Design – 'Technical Standards and Regulations'

This sub-clause provides that the design, the contractor's documents, the execution and the completed works shall comply with the technical standards, building, construction and environmental laws of the country where the site is located; the laws applicable to the product being produced from the works; and other standards specified in the employer's requirements which are applicable to the works or defined by the applicable laws. In this way, the employer's requirements must be accurately comprehensive and unambiguous so as to form an extensive technical manual for use by the contractor in the design and construction of the works.

This sub-clause further provides that references 'in the Contract', namely those in the employer's requirements in the first instance, and the schedules or the contractor's proposal thereafter, to published standards shall be understood to be references to the edition applicable on the base date, unless stated otherwise.

24.3.20   Sub-clause 5.5: Design – 'Training'

Under this sub-clause, the employer's requirements should specify whether the contractor is required to carry out any training of the employer's personnel in the operation and maintenance of the permanent works following their completion. This sub-clause does not make provision for the type of details that should be included in the employer's requirements. It is suggested here that, if such training is required, the employer's requirements should set out clearly the extent of such training, including the qualifications, the technical knowledge and training experience necessary on the part of the trainer(s); the type/grade of personnel to be trained; their technical and/or professional qualifications, skills and experience; whether such personnel, despite being employer's personnel, would be paid a special allowance during the training period and if so, by whom and how much; whether the training is to take place on site or elsewhere; whether, and if so to what extent, such training is to be theoretical or 'hands-on'; the date by which such personnel should be available for training; and the effect of any delay in completion of the training or certain parts of the training on the taking over of the works by the employer. In the latter regard, sub-clause 5.5 provides that if 'the Contract' (namely, the employer's requirements) specifies that certain training is to be carried out before taking over, the works shall not be considered to be completed for the purposes of taking-over under sub-clause 10.1 'Taking Over of the Works and Sections' until such training has been completed.

24.3.21 Sub-clause 5.6: Design – 'As-Built Documents'

For the purposes of this sub-clause, the employer's requirements should specify the number of copies and types of the relevant as-built drawings showing all the works as executed, to be supplied to the engineer by the contractor, prior to the issue of any taking-over certificate under sub-clause 10.1 'Taking Over of the Works and Sections'.

24.3.22 Sub-clause 5.7: Design – 'Operation and Maintenance Manuals'

Under this sub-clause, the employer's requirements should specify any manuals, other than the operation and maintenance manuals in respect of the plant, if any, forming part of the permanent works, which are to be supplied to the engineer. While this sub-clause provides that the operation and maintenance manuals should be of sufficient detail for the employer to operate, maintain, dismantle, reassemble, adjust and repair the plant, the employer's requirements should also specify the detail to which the other manuals should be prepared. The works are not considered to be completed for the purposes of taking over under sub-clause 10.1 'Taking Over of the Works and Sections' until the engineer has received these documents.

24.3.23 Sub-clause 6.1: Staff and Labour – 'Engagement of Staff and Labour'

The provisions of this sub-clause refer to the employer's requirements, in the context of the contractor's obligation to make arrangements for the engagement of all staff and labour, local or otherwise, and for their payment, housing, feeding and transport, unless otherwise stated in the requirements.

24.3.24 Sub-clause 6.6: Staff and Labour – 'Facilities for Staff and Labour'

Once again, the provisions of this sub-clause refer to the employer's requirements, stating that, unless otherwise stated in the requirements, the contractor shall provide and maintain all necessary accommodation and welfare facilities for the contractor's personnel. The contractor shall also provide facilities for the employer's personnel if, and to the extent, stated in the employer's requirements.

24.3.25 Sub-clause 7.4: 'Testing'

This sub-clause applies to all the testing 'specified in the Contract', namely those specified in the employer's requirements in the first instance, and the schedules or the contractor's proposal thereafter, other than those to be undertaken after completion of the works. Therefore, if the employer requires tests to be carried out on the works, while under construction, all the necessary details and criteria for such testing must be accurately and sufficiently set out in the employer's requirements.

24.3.26 Sub-clause 7.8: Plant, Materials and Workmanship – 'Royalties'

If the contractor is not required to pay for all royalties, rents and other payments in respect of:

(a) natural materials obtained from outside the site; and
(b) the disposal of material from demolitions and excavations and of other surplus material, whether natural or man-made, except to the extent that disposal areas within the site are specified in the contract, then

the employer's requirements should deal with this matter and state so explicitly.

24.3.27 Sub-clause 8.2: 'Time for Completion'

This sub-clause states the contractor's general obligation to complete the works, or any section thereof, within the time for completion stated in the Appendix to Tender. By virtue of paragraph (b) of this sub-clause the contractor is also required to complete all the work which is stated in 'the Contract', namely that as specified in the employer's requirements in the first instance, and the schedules or the contractor's proposal thereafter, in order that the works (or any section thereof) are completed for the purposes of taking over under sub-clause 10.1 'Taking Over of the Works and Sections'.

24.3.28 Sub-clause 8.3: Commencement, Delays and Suspension – 'Programme'

This sub-clause requires the contractor to submit a detailed time programme to the engineer within 28 days after receiving the notice under sub-clause 8.1 'Commencement of Works'. The contractor shall also submit a revised programme whenever the previous programme is inconsistent with actual progress or with the contractor's obligations.

Following on from this general statement, the text of this sub-clause goes on in its paragraph (b) to require that each programme should include the periods for review of the contractor's documents under sub-clause 5.2 'Contractor's Documents', and for any other submissions, approvals and consents specified in the employer's requirements.

Further, the last paragraph of this sub-clause provides that if the engineer notifies the contractor at any time that a programme fails, inter alia, to comply with 'the Contract', namely the timing and sequence specified in the employer's requirements in the first instance, and the schedules or the contractor's proposal thereafter, the contractor shall submit a revised programme to the engineer in accordance with this sub-clause.

Accordingly, the employer's requirements should be sufficiently detailed to provide all required programming information clearly.

24.3.29 Sub-clause 9.1: Tests on Completion – 'Contractor's Obligations'

This sub-clause deals with the contractor's obligations in respect of the tests that are to be carried out on completion of the works in order that the employer can determine whether the works or any section thereof have reached the stage of taking over. These tests, and the criteria for passing them, must be specified in the employer's requirements. The fourth paragraph of this sub-clause states such requirement in the following terms:

> 'During trial operation, when the Works are operating under stable conditions, the Contractor shall give notice to the Engineer that the Works are ready for any other Tests on Completion, including performance tests to demonstrate whether the Works conform with criteria specified in the Employer's Requirements and with the Schedule of Guarantees.'

24.3.30 Sub-clause 10.2: 'Taking Over of Parts of the Works'

The second paragraph of this sub-clause provides that, unless and until the engineer has issued a taking-over certificate for it, the employer is not permitted to use any part of the works. However, if the employer wishes to use a part of the works as a temporary measure it must be specified in 'the Contract', namely in the employer's requirements, or agreed by both parties.

24.3.31 Sub-clause 11.1: 'Completion of Outstanding Works and Remedying Defects'

This sub-clause refers to the work necessarily to be completed by the contractor before the project and the 'Contractor's Documents' are in such condition as required and specified by the contract. It states that such condition shall be that as 'required by the Contract (fair wear and tear excepted)', namely that as stated in the employer's requirements in the first instance, and the schedules or the contractor's proposal thereafter.

24.3.32 Sub-clause 17.5: Risk and Responsibility – 'Intellectual and Industrial Property Rights'

In this sub-clause, as indeed under the same sub-clause of the 1999 Red Book, 'infringement' means an infringement or alleged infringement of any patent, registered design, copyright, trademark, trade name, trade secret or other intellectual or industrial property right relating to the works; and 'claim' means a claim or proceedings pursuing a claim alleging an infringement.

The third paragraph of this sub-clause provides that the employer shall indemnify and hold the contractor harmless against and from any claim alleging an infringement, as defined in this sub-clause, of any intellectual or industrial property right, which is or was inter alia an unavoidable result of the contractor's compliance with the employer's requirements.

Therefore, it is extremely important for the employer, and the engineer preparing the employer's requirements, to establish through legal advice from competent and qualified professionals in the area of patent, registered design, copyright, trade mark, trade name, trade secret or other intellectual or industrial property right, that the employer's requirements do not contain an infringement or will not lead to an infringement by the contractor in compliance with them.

It is perhaps obvious in light of the numerous provisions discussed above, but worth stating nonetheless, that the employer's requirements document is perhaps the document in the contract that requires the most input, care and attention by the employer and the engineer in a contract under the 1999 Yellow Book.

24.4 Comparison between the 1999 Yellow Book and the 1999 Red Book

As mentioned earlier, the concepts and content of the 1999 Yellow Book are similar to those of the 1999 Red Book, but differences do arise, mainly in respect of the function of design and the associated concepts of fitness for purpose and tests after completion in the 1999 Yellow Book. It is useful, therefore, to elaborate on these differences between the two forms, as discussed below.

24.4.1 Clause 1: General Provisions

Under clause 1 of the 1999 Yellow Book, the general provisions of the conditions of contract are set out.

Sub-clause 1.1.1.1 differs in the two forms of contract in that the 'Specification, the Drawings' under the 1999 Red Book have been replaced by the two terms 'Employer's Requirements' and 'the Contractor's Proposal' in the 1999 Yellow Book. Further, the 'Schedules' in the 1999 Red Book include a reference to 'the Bill of Quantities' whereas the reference is to a 'schedule of payment' in the 1999 Yellow Book. These changes necessitate a change not only in sub-clauses 1.1.1.5, 1.1.1.6, 1.1.1.7 and 1.1.1.10 where these terms are defined, but also throughout the whole of the 1999 Yellow Book.

Sub-clause 1.1.3.6, which defines 'Tests after Completion', refers to the particular conditions in the 1999 Red Book, whereas the same definition in the 1999 Yellow Book refers to clause 12 'Tests after Completion' where these tests are detailed.

Sub-clause 1.8 of both forms of contract makes provision for the care and supply of documents. In both, the contractor remains in care of the contractor's documents unless and until they are taken over by the employer. Under the 1999 Yellow Book, the contractor supplies a large number of documents to the employer and the engineer, particularly in respect of the design of the works for which he is responsible. In contrast, under the 1999 Red Book, the contractor is required to prepare and submit documents to a significantly lesser extent. Further, clause 1.8 of the 1999 Red Book provides that the specification and drawings of the works are to be kept in the care and custody of the employer, whereas the employer does not provide any specifications and drawings under the 1999 Yellow Book and so no mention of them is made in sub-clause 1.8.

Sub-clause 1.9 under the 1999 Yellow Book is quite different from that in the 1999 Red Book. The former deals with 'Errors in the Employer's Requirements', the latter with 'Delayed Drawings or Instructions'. Under both forms, the contractor becomes entitled to an extension of time and/or additional payment if the works are delayed as a result of information/data originating with the employer. Under the 1999 Yellow Book such contractor's entitlement arises from errors in the employer's requirements, whilst under the 1999 Red Book such entitlement arises as a result of late issuance of drawings or instructions. Corresponding to such difference, under sub-clause 1.9 of the 1999 Yellow Book, the engineer's determination of the contractor's entitlement must take into consideration whether, and to what extent, the errors in the employer's requirements should have been discovered during the contractor's scrutiny of this document (see Section 24.3.7 above). Under the 1999 Red Book the contractor's entitlement is limited to the extent that the engineer's failure to issue drawings or instructions on time arose from an error in, or delay in submission of, the contractor's documents.

24.4.2 Clause 3: The Engineer

The role of the engineer and his duties and authority in relation to the works are set out under Clause 3 of both the 1999 Yellow and Red Book. There are no differences in the provisions of this clause, except in the case of sub-clause 3.3 'Instructions of the Engineer'. The 1999 Yellow Book makes provision for instructions only, whereas the 1999 Red Book provides that the engineer may issue instructions *and* additional or modified drawings. Under the 1999 Yellow Book such instructions must be issued in writing. While it requires instructions to be, whenever practicable, in writing, the 1999 Red Book provides a mechanism in the event that any oral instruction is given (see Section 23.3.4 in Chapter 23).

24.4.3 Sub-clause 4.1: The Contractor – 'Contractor's General Obligations'

Sub-clause 4.1 under both forms of contract enumerates the contractor's general obligations, which are more extensive in the 1999 Yellow Book than they are in the 1999 Red Book. The first paragraph of sub-clause 4.1 of the 1999 Yellow Book requires the contractor to design, execute, complete and also remedy any defects in, the works. It further states that, upon completion, the works should be fit for the purposes for which, as stated under the contract, they were intended. This requirement forms one of the major characteristics of the 1999 Yellow Book and has had significant implications for the contractor. He cannot pass that liability for fitness for purpose on to any independent design professionals that he may appoint, whose standard of liability under most applicable laws is usually limited to one of skill, diligence and care. Furthermore, he cannot always insure against that liability, as the insurance market in certain jurisdictions does not usually provide a cover against the risks associated with that liability.[24.6] As a result, this particular provision of sub-clause 4.1 of the 1999 Yellow Book has received a negative reaction from contractor organisations, who have objected strongly to the high risk that such liability creates for contractors.[24.7] The premise of the 1999 Yellow Book, being a 'design-build' form of contract, is that the contractor is responsible for the design of, if not the whole of the works, a very significant proportion of it, so that when completed they are fit for the purposes for which they are intended, as defined in the contract. Therefore, the contract must define the employer's purposes for which the works are intended. However, although it is not entirely clear where such definition is expected to be found, the most appropriate place for it is in the employer's requirements document, which is itself defined as specifying 'the purpose, scope and/or design and/or other technical criteria, for the Works'.[24.8] It is worth noting that the words 'when completed' in sub-clause 4.1 mean what they say, namely that the fitness for purpose required is that which must be demonstrated when the works are completed and not during the design stage or many years later.

The 1999 Red Book also holds the contractor liable for fitness for purpose, unless provided otherwise in the particular conditions, but only to the extent that the contract requires him to design part of the works and only for that particular part of the works, paragraph (c) of sub-clause 4.1 refers. The contract must therefore clearly specify in the specification not only the extent of any design required to be carried out by the contractor, but also the purpose for which the part to be designed by him is intended.

Finally, the requirement of fitness for purpose under the 1999 Yellow Book should be considered together with the provisions of sub-clauses 5.1 'General Design Obligations' and 12.4 'Failure to Pass Tests after Completion', dealt with below. Whilst the intended purpose should be clearly and expressly specified in the contract, the provisions of these sub-clauses mean that there could be implied terms that might be added to the contract by the operation of the applicable law.

24.4.4 Sub-clause 4.4: 'Subcontractors', sub-clause 4.5: 'Nominated Subcontractors' and sub-clause 4.6: 'Co-operation'

As its heading suggests, sub-clause 4.4 deals with the topic of subcontractors. In this regard, the provisions in the 1999 Yellow Book are exactly the same as those in the 1999 Red Book, except that the final paragraph under this sub-clause in the 1999 Red Book, which entitles the employer to require the subcontract to be assigned to him under sub-clause 4.5, is omitted in the 1999 Yellow Book.

Sub-clause 4.5 of the 1999 Yellow Book deals with nominated subcontractors, a matter that is not usually compatible with or of particular relevance to design and build contracts. It is reduced from the full clause, clause 5, under the 1999 Red Book to a single paragraph to provide for the possibility that a need might arise for the contractor to follow an instruction from the engineer under clause 13 to enter into a subcontract with a specific person/company. Under the 1999 Yellow Book, the contractor does not have a duty to employ a nominated subcontractor as long as he raises a 'reasonable objection . . . with supporting particulars' to the nomination to the engineer as soon as he can.

Sub-clause 4.6 of the 1999 Yellow Book goes a little further than its equivalent sub-clause in the 1999 Red Book. Both require the contractor to co-operate with the employer's personnel and/or other contractors working on or near the site, but the 1999 Yellow Book further states that the contractor is responsible for his own construction activities on the site and must co-ordinate his own activities with those of other contractors working on the site. The 1999 Red Book is entirely silent on this point.

24.4.5 Clause 5: Design

Clause 5 of the 1999 Yellow Book is totally different from that under the 1999 Red Book, as the former deals with design whereas the latter deals with nominated subcontractors. In the 1999 Red Book, clause 5 provides a definition of nominated subcontractors, allows the contractor to object to a nominated subcontractor, obliges the contractor to pay the nominated subcontractors the amounts certified by the engineer, and provides that the contractor may be asked to provide evidence that previous payments have been made. Conversely, because it is intended that nominated subcontractors will be used only to a very limited extent under a design-build contract, there is minimal provision for them in the 1999 Yellow Book (see Section 24.4.4 above).

On the other hand, the 1999 Yellow Book is to be used for projects where most if not all of the design function is allocated to the contractor. As such, design forms a major and fundamental element of this form of contract and, as stated earlier, is an important departure from the traditional form of procurement where the employer is responsible for the design of the works. Clause 5 of the 1999 Yellow Book is divided into eight sub-clauses, which are now dealt with in turn.

The concept of the clause is explained in the *FIDIC Contracts Guide*[24.9] and it is useful to quote this here as it throws some light on various aspects of the 1999 Yellow Book and particularly those relating to design, such as the employer's requirements. The concept is based on dividing the design process into three stages, as follows:

'(a) Conceptual design by (or on behalf of) the Employer, for inclusion in the Employer's Requirements in order to define the Works. This might involve less than 10% of total design input, and it might be necessary to distinguish between early ideas and definite requirements.

(b) Preliminary design by (or on behalf of) each tenderer for inclusion in the Tender (including P&DB's [Plant and Design Build] Proposal). The Instructions to Tenderers should have indicated the extent of detail required, taking account of tenderers' understandable reluctance to incur excessive tendering costs if the likelihood of success seemed low.

(c) Final design for the working drawings (Contractor's Documents), which might involve two sub-stages: general arrangement drawings and detailed drawings. Note that the fourth paragraph of Sub-Clause 5.2 refers to each document being ready "for use", which could cover the completion of a general arrangement drawing which is to be used for the next design stage, but is not itself to be used for construction or manufacture.

If the Contractor is required to be responsible for a design provided by the Employer, tenderers must be allowed to check the design and to propose amendments.'

Sub-clause 5.1 in its opening paragraph, refers to the employer's requirements, which provide as follows:

> 'The Contractor shall carry out, and be responsible for, the design of the Works. Design shall be prepared by qualified designers who are engineers or other professionals who comply with the criteria (if any) stated in the Employer's Requirements. Unless otherwise stated in the Contract, the Contractor shall submit to the Engineer for consent the name and particulars of each proposed designer and design Subcontractor.'

The consent of the engineer must therefore be sought and obtained in respect of each proposed design professional as soon as possible so that the design activities are not delayed. The consent should not be unreasonably withheld or delayed, which is a requirement under sub-clause 1.3 'Communications'.

By virtue of the second paragraph of sub-clause 5.1, the contractor warrants that he, his designers and his design subcontractors have the necessary experience and capability to carry out the design of the works. He also undertakes that his designers will be available to attend discussions with the engineer in relation to the design of the works, until the end of the defects notification period.

The third paragraph of sub-clause 5.1 requires the contractor, when he receives notice under sub-clause 8.1 'Commencement of the Works' to scrutinise the employer's requirements and the items of reference referred to in sub-clause 4.7. If any error, fault or other defect is found in the employer's requirements or in the items of reference, the contractor should notify the engineer accordingly within a certain period of time (measured from the commencement date), which ought to be stated in the Appendix to Tender. This reference to the period stated in the Appendix to Tender appears to limit the right of the contractor to give notices of deficiencies in the employer's requirements only within that period. If so, what if errors, faults or defects are discovered after this period has expired. However, it is not FIDIC's intention that the contractor's right to give notice is so limited. In their *Contracts Guide*, FIDIC explain that:

> 'The third paragraph of Sub-Clause 5.1 therefore requires the Contractor to scrutinise the Employer's Requirements again, and notify the Engineer of any error, fault or other defect within the period stated in the Appendix to Tender. This period should be sufficient for the Contractor and his designer(s) to carry out a thorough check. Thereafter there remains the possibility that the Employer's Requirements are found to contain an error which could not previously have been discovered by an experienced Contractor exercising due care. In this event the Contractor may give the notice described in Sub-Clause 1.9.'

Accordingly, in the event that an error, fault or defect is discovered during the contractor's scrutiny of the employer's requirements the claim mechanism provided by sub-clause 1.9 'Errors in the Employer's Requirements' applies, see Section 24.3.7 above.

Sub-clause 5.2 sets out the procedural requirements for the contractor's documents, which include calculations, computer software (programs), drawings, manuals and models. Under the employer's requirements, the contractor's documents may be required to be submitted to the employer for review or approval. Although many might regard such provision as unnecessary interference in the design process, it is a valuable tool in ensuring that the contractor's design is proceeding in the right direction (rather than waiting until completion) and, therefore, beneficial to both the contractor and the employer.

The fifth paragraph of sub-clause 5.2 imposes time limits on the review process, if any, of the contractor's documents, namely (unless otherwise stated in the employer's requirements) each review period shall not exceed 21 days, calculated from the date on which the engineer receives the relevant contractor's document and the contractor's notice. Sub-clause 5.2 provides that:

> 'This notice shall state that the Contractor's Document is considered ready, both for review (and approval, if so specified) in accordance with this Sub-Clause and for use. The notice shall also state that the Contractor's Document complies with the Contract, or the extent to which it does not comply.'

The engineer may, within the review period, give notice to the contractor that a contractor's document fails to comply with the contract; then it is referred back to the contractor who should have it rectified, resubmitted for review or approval as originally required, in accordance with the fifth paragraph of sub-clause 5.2 at the cost of the contractor.

It is important to note that in the case of a contractor's document, which has been submitted for the engineer's approval, as specified, the execution of the relevant part of the works relating to that document should not commence until the engineer has approved that document.

It is worth noting also that the final paragraph of sub-clause 5.2 provides that:

> 'any such approval or consent, or any review (under this Sub-Clause or otherwise), shall not relieve the Contractor from any obligation or responsibility.'

Sub-clause 5.3 is a short sub-clause whereby the contractor undertakes that the design, the contractor's documents, the execution of and the completed works will be in accordance with the laws of the country where the works are taking place and with the contract documents, as altered or modified by variations.

Sub-clause 5.4 specifies in its opening sentence various criteria for the design, the contractor's documents and the works. These criteria include the country's technical standards, building, construction and environmental laws, and any other requirements stated in the employer's requirements. The laws and standards that apply, in respect of the works and each section, are those prevailing when the works or section are taken over by the employer under clause 10 'Employer's Taking Over'. References in the contract to published standards are to be read as references to the edition applicable on the base date, unless stated otherwise. However, should new standards/laws come into effect after the commencement of the works, the contractor should give notice to the engineer and, if appropriate, submit his proposals for compliance. If the engineer decides that such new standards/laws apply, and that they will constitute a variation to the contract, then the engineer should initiate a variation in accordance with clause 13 'Variations and Adjustments'.

Sub-clause 5.5 deals with the contractor's training of the employer's personnel, if such training is required as specified in the employer's requirements (see Section 24.3.20 above).

It is important to note that if the employer's requirements specify training that has to be carried out before taking over, the works would not be considered as completed for the purposes of sub-clause 10.1 'Taking Over of the Works and Sections' until the relevant training has been completed. The manner of performing such training can be gleaned from sub-clause 7.1 'Manner of Execution'.

'(a) in the manner (if any) specified in the Contract,
(b) in a proper workmanlike and careful manner, in accordance with recognised good practice, and
(c) with properly equipped facilities and non-hazardous Materials, except as otherwise specified in the Contract.'

It should be noted, however, that the performance of the obligation of training does not in itself have to achieve the standard of fitness for purpose, pursuant to sub-clause 4.1, since the result of training is to a large extent dependent on the quality of the persons chosen by the employer to be trained.

Sub-clause 5.6 requires the contractor to provide three record items, as follows:

(a) To prepare, and keep up-to-date, a complete set of 'as-built' records of the execution of the works, showing the exact as-built locations, sizes and details of the work as executed. These records should be kept on the site and should be used exclusively for the purposes of this sub-clause 5.6. Two copies should be supplied to the engineer prior to the commencement of the tests on completion. These records are factual and as such they do not have to be submitted for review under sub-clause 5.2, unless the employer's requirements specifically require these records to be submitted for that purpose.
(b) To prepare and provide to the engineer for review (in accordance with sub-clause 5.2 'Contractor's Documents') as-built drawings of the works, showing all the works as executed. The contractor shall obtain the consent of the engineer as to the size, the referencing system, and other relevant details of such drawings.
(c) Prior to the issue of any taking-over certificate, to supply to the engineer the specified numbers and types of copies of the relevant as-built drawings, in accordance with the employer's requirements. The third paragraph of sub-clause 5.6 specifies that the works are not to be considered as completed for the purposes of taking-over under sub-clause 10.1 'Taking Over of the Works and Sections' until the engineer has received these documents.

Sub-clause 5.7 requires the contractor to prepare operation and maintenance manuals for any plant in two phases. First, prior to commencement of the tests on completion, the contractor is required to supply to the engineer provisional operation and maintenance manuals in sufficient detail for the employer to operate, maintain, dismantle, reassemble, adjust and repair the plant. In the second phase, the contractor is required to supply final operation and maintenance manuals in such detail, and any other manuals specified in the employer's requirements for these purposes. In respect of these documents, the works will not be considered to be completed for the purposes of taking over under sub-clause 10.1 'Taking Over of the Works and Sections' until the engineer has received these finalised documents.

Sub-clause 5.8 provides that if any errors, omissions, ambiguities, inconsistencies, inadequacies or other defects are found in the contractor's documents, they and the works should be corrected at the contractor's cost, notwithstanding any consent or approval by the engineer under clause 5. The provisions of this sub-clause should be considered, together with the wording of sub-clauses 1.8 'Care and Supply of Documents'; 1.9 'Errors in the Employer's Requirements'; and 5.1 'General Design Obligations'.

24.4.6 *Sub-clause 7.5: Plant, Materials and Workmanship – 'Rejection'*

Clause 7 of both the 1999 Yellow and Red Books deals with plant, materials and workmanship in relation to the works. The difference between the provisions of the 1999 Red and Yellow Books in connection with this sub-clause arise in sub-clause 7.5 'Rejection'. Sub-clause 7.5, as its heading indicates, provides that if the engineer finds that any plant, materials or workmanship are defective or not in accordance with the contract, then the engineer may notify his rejection (with reasons) to the contractor, who is then required to rectify the problem. The difference between the 1999 Yellow and Red Books relates to the added function of design under the 1999 Yellow Book. Therefore, wherever the word 'workmanship' appears in this sub-clause of the 1999 Red Book, the words 'workmanship or design' appear in this sub-clause of the 1999 Yellow Book. Otherwise, the provisions of this sub-clause and clause 7 as a whole are identical. With regard to the provisions of this sub-clause, attention is drawn to the provisions of sub-clauses 15.5 'Notice to Correct' and 15.2 'Termination by Employer', where the employer may become entitled to terminate under paragraph (a) of the latter sub-clause if notice has been given under the former sub-clause that the contractor has failed to carry out any obligation under the contract.

24.4.7 *Sub-clause 8.3: Commencement, Delays and Suspension – 'Programme'*

As stated in Section 23.4.6 of Chapter 23 in respect of the 1999 Red Book, the contractor must, under sub-clause 8.3, submit a programme for the works to the engineer at the beginning of the works. Similar provision is made in sub-clause 8.3 of the 1999 Yellow Book but this programme differs in respect of additional items to be included in it, namely design, inspection, commissioning and trial operation. These functions precede the construction of the works on site in some respects, whilst they occur after completion of construction and erection in others. The programme under the 1999 Yellow Book assumes a much more high profile than that under the 1999 Red Book because of the inclusion of the design function into it, particularly with all the review/approval steps that have to be incorporated in it. However, the philosophy relating to the programme under the 1999 Yellow Book remains the same as that for the 1999 Red Book, namely:[24.10]

(a) details on the timing for reviews of the contractor's documents and for complying with any other deadlines for submissions, approvals and consents stipulated within the employer's requirements;[24.11]
(b) details regarding the order and timing of inspections and tests that are stated in the contract;
(c) an additional report enumerating the methods of construction that the contractor will adopt in carrying out the works, and also giving details about the contractor's personnel and the contractor's equipment that will be required on site during the various stages of the works.

In both forms of contract, the contractor is obliged to proceed with the works in accordance with the programme that he has submitted to the engineer unless the engineer notifies him that there is some error in it. If the engineer notifies the contractor that the programme is not consistent with the contract, the contractor must submit a revised programme.

24.4.8 Sub-clause 9.1: Tests on Completion – 'Contractor's Obligations'

This sub-clause sets out the contractor's obligations in respect of tests on completion. The term 'Tests on Completion' is a defined term under the 1999 Red, Yellow and Silver Books. It is defined in sub-clause 1.1.3.4 as:

> 'the tests which are specified in the Contract or agreed by both Parties or instructed as a Variation, and which are carried out under Clause 9 [Tests on Completion] before the Works or a Section (as the case may be) are taken over by the Employer'.

As under the 1999 Red Book, the contractor must under the 1999 Yellow Book carry out tests upon completion of the works and after providing the engineer with any documents that he is entitled to under sub-clause 5.6 'As-Built Documents' and sub-clause 5.7 'Operation and Maintenance Manuals'.

Sub-clause 9.1 under the 1999 Yellow Book is, however, far more extensive than its corresponding sub-clause under the 1999 Red Book, as it details the sequence of the tests on completion, as follows:

(a) pre-commissioning tests, which include the appropriate inspections and 'dry' or 'cold' functional tests to demonstrate that each item of plant can safely undertake the next stage, (b);
(b) commissioning tests, which include the specified operational tests to demonstrate that the works or a section can be operated safely and as specified, under all available operating conditions; and

(c) trial operation, which should demonstrate that the works or section perform reliably and in accordance with the contract.

The importance of the tests on completion in the 1999 Yellow Book is demonstrated by the fact that the term 'Tests on Completion' is referred to in 11 other sub-clauses of the 1999 Yellow Book: 5.6; 5.7; 8.2; 9.1; 9.2; 9.3; 9.4; 10.1; 10.2; 10.3 and 11.6. There are a number of points that should be remembered:

(a) The contractor should notify the engineer of the date after which he will be able to carry out the tests, no less than 21 days in advance, and the tests should be completed within 14 days from this notified date.
(b) During trial operation, when the works are operating under stable conditions, the contractor should notify the engineer that the works are ready for any other tests on completion, including any tests that may be required to be carried out to ensure that the works comply with the specifications listed within the employer's requirements.
(c) As soon as the works, or a section, have passed each of the tests on completion described in sub-paragraph (a), (b) or (c), referred to above, the contractor shall submit a certified report of the results of these tests to the engineer.

24.4.9 Sub-clause 11.2 and 11.6: Defects Liability – 'Cost of Remedying Defects and Further Tests' and 'Further Tests'

In respect of defects liability, the 1999 Yellow Book applies the same principles that apply in the 1999 Red Book. However, sub-clause 11.2 has the added requirement that any work done to remedy any defect in the works should be at the contractor's own risk and cost if such defect has arisen from improper operation or maintenance which is attributable to matters for which he is responsible (under sub-clauses 5.5 'Training'; 5.6 'As-built Documents'; and/or 5.7 'Operation and Maintenance Manuals', or otherwise).

If the contractor considers that none of the causes listed in sub-clause 11.2 applies, and that he is therefore entitled to be paid for remedying a defect or damage, he should respond promptly to the engineer's notice (under sub-clause 11.1 'Completion of Outstanding Work and Remedying Defects') requiring him to carry out any remedial work and in this response or under separate cover, whichever is more appropriate in the circumstances, give notice and detailed particulars of any claim he has in accordance with the procedure outlined in sub-clause 20.1 'Contractor's Claims'. However, it must be remembered that, whether or not the remedial works are at the contractor's cost or recoverable from the employer, the contractor is under a strict obligation to proceed with remedying the defects required of him, in order that they are completed before the expiry of the defects notification period, under sub-clause 11.1.

Under sub-clause 11.6 'Further Tests', it is stipulated that if the work of remedying any defect is to or may affect the performance of the works, the engineer may require any of the tests described in the contract to be repeated. Under the 1999 Yellow Book this includes tests on completion of the works, and/or tests carried out after completion of the works. Under the 1999 Red Book, there is no mention of 'Tests on

Completion' or 'Tests after Completion' in sub-clause 11.6; it merely refers to 'any of the tests described in the contract'.

24.4.10 Clause 12: Tests after Completion

The term 'Tests after Completion' is a defined term under the 1999 Red, Yellow and Silver Books, but the definition itself is slightly different in each. It is defined in sub-clause 1.1.3.6 of the 1999 Yellow Book as:

> 'the tests (if any) which are specified in the Contract and which are carried out under Clause 12 [Tests after Completion] after the Works or a Section (as the case may be) are taken over by the Employer.'

It is referred to in the 1999 Yellow Book in eight sub-clauses: 7.4; 11.6; 12.1; 12.2; 12.3; 12.4; 14.9 and 18.2. Under the 1999 Yellow Book, given its fundamental basis of the contractor's design to achieve fitness for purpose, a whole clause in the general conditions is devoted to tests after completion. In contrast, under the 1999 Red Book the only reference to tests after completion is in the definition of sub-clause 1.1.3.6 whereby the user is referred to the particular conditions. Clause 12 of the 1999 Yellow Book is entirely different to the corresponding clause under the 1999 Red Book entitled 'Measurement and Evaluation'. Clause 12 of the Yellow Book deals with the tests that are to be carried out after completion of the works, which are to be defined within the employer's requirements, and should be carried out as soon as possible after the employer takes over the works. Although not expressly stated in the general conditions of the 1999 Yellow Book, the objective of these tests is to determine if the works are performing to the standard required by the contract and comply with the criteria and purposes laid down in the employer's requirements. It should be noted, however, that while the contractor is responsible for carrying out the tests *on* completion, the employer is responsible for the tests *after* completion. The employer also provides all electricity, equipment, fuel, instruments, labour, materials and suitably qualified and experienced staff.

Sub-clause 12.1 'Procedure for Tests after Completion' lays down the procedure that should be followed when carrying out the tests after completion. This procedure entails that:

(a) the employer gives the contractor 21 days' notice of the date after which the tests would be carried out;
(b) the tests would then be carried out within 14 days after that date, unless otherwise agreed between the parties;
(c) finally, 'the results of the Tests after Completion shall be compiled and evaluated by both Parties. Appropriate account shall be taken of the effect of the Employer's prior use of the Works.'

Sub-paragraph (b) of the first paragraph of sub-clause 12.1 provides for these tests to be carried out in the presence of 'such Contractor's Personnel as either Party may reasonably request'. However, if the contractor does not attend at the time and place agreed, the employer may proceed with the tests, which would then be deemed to have been made in the contractor's presence. Under these circumstances, it is stated that 'the Contractor shall accept the readings as accurate'. The employer must follow and comply with the manuals supplied by the contractor in carrying out these tests and such guidance as the contractor may be required to give during the course of these tests. The contractor does not have to be present for the tests to be carried out; however, he must be invited to attend. Where the contractor has been invited but does not attend the tests, they will nonetheless be regarded as having been carried out in his presence and he must accept the results as accurate. On that basis, needless to say, it would be an imprudent contractor who would not attend such tests when invited.

Sub-clause 12.2 deals with the possibility of a delay in the tests in which case, if the contractor incurs cost as a result of any unreasonable delay by the employer to the tests after completion, he should, subject to sub-clauses 20.1 and 3.5,

(i) give notice to the engineer; and
(ii) be entitled to payment of any such cost plus reasonable profit to be included in the contract price.

Sub-clause 12.2 also provides that if, for reasons not attributable to the contractor, a test after completion of the works cannot be completed during the defects notification period, then the works or the relevant section shall be deemed to have passed this test after completion.

By virtue of sub-clause 12.3 'Retesting', where any part of the works fails the tests after completion, the tests may be carried out again under the same terms and conditions as the failed test.

Sub-clause 12.4 'Failure to Pass Tests after Completion' provides in its first paragraph a mechanism to compensate the employer for a failure to pass any one of the tests after completion, unless the contract does not define how these non-performance damages are to be determined for the particular failure. If these non-performance damages are not ascertainable, the first paragraph is inapplicable. Accordingly, the employer should define in the tender documents the minimum acceptable criteria or in other words, the maximum permissible extent of the test failure, together with a corresponding sum payable by the contractor as non-performance damages for this failure.

The second paragraph of sub-clause 12.4 deals with the difficulty which the employer may have in giving the contractor access to the permanent works in order to overcome a failure. The contractor remains liable to carry out the remedial work, unless he is instructed by the employer that right of access cannot be given until a convenient time and he does not receive reasonable notice of such time during the defects notification period.

24.4.11 Sub-clauses 13.1, 13.2 and 13.3: Variations and Adjustments – 'Right to Vary', 'Value Engineering' and 'Variation Procedure'

Clause 13 of the 1999 Red and Yellow Books dictates what is to happen where a variation or an adjustment is to be made to the contract. Whilst under both these forms of contract a variation may be made, either by an instruction or by a request for the contractor to submit a proposal at any time before the taking-over certificate has been issued, clause 13 of the 1999 Yellow Book differs from that of the 1999 Red Book in relation to some of its other provisions. Thus, the 1999 Yellow Book does not provide the list of variations, given in sub-clause 13.1 'Right to Vary' of the 1999 Red Book, that the engineer may instruct. Instead, this is replaced by a statement in the first paragraph of sub-clause 13.1 of the 1999 Yellow Book (which has the same heading as that of the 1999 Red Book), that any variation should not involve the omission of any work that is to be carried out for the employer by others. This is entirely appropriate in the design-build form of contract where, due to the very wide scope of responsibility assumed by the contractor, it would be impractical and probably inadvisable to attempt to exhaustively list the circumstances in which the engineer may initiate a variation.

The substance of the second paragraph of sub-clause 13.1 in the 1999 Yellow and Red Books is similar: the contractor is bound by each variation made to the contract unless he notifies the engineer with supporting particulars to the effect that he cannot readily obtain the goods that are required to carry out the variation. However, in the 1999 Yellow Book additional qualifications are included in respect of the binding effect of a variation, namely:

(a) the variation would reduce the safety or suitability of the works; or
(b) the variation will have an adverse impact on the achievement of the schedule of guarantees.

The result of such qualifications is the same in both forms of contract: upon receiving the contractor's notice, the engineer should cancel, confirm or vary the instruction in respect of the variation in hand.

Under sub-clause 13.2 'Value Engineering' of both the 1999 Red and Yellow Books, the contractor may submit a report to the engineer, in which the contractor sets out ways that he feels the contract may be completed sooner or at lesser cost than originally planned. However, as it is the contractor who is in charge of the design function under the 1999 Yellow Book, it is unlikely that such a proposal would be made in the way it is envisaged under this clause in the 1999 Red Book. Therefore, the 1999 Yellow Book does not contain the provisions included in the 1999 Red Book, which describe how to develop a proposal for value engineering, namely approval by the engineer, design by the contractor of the proposed part of the works to be changed, and the additional 'fee' payable for such change. The 1999 Yellow Book is silent in this regard.

Under sub-clause 13.3 'Variation Procedure' the procedure that should be followed where a variation is to be made to the contract is set out. As in the 1999 Red Book, the 1999 Yellow Book provides that the engineer shall request a proposal from the contractor before consenting to or instructing a variation. Following that request, once again, because of the design-build nature of the contract under the 1999 Yellow Book, the procedure is different to that under the 1999 Red Book. First, the contractor must include in his proposal a description of the proposed design and/or work to be performed. The 1999 Red Book merely refers to a description of the proposed work to be performed.

Both forms of contract require the contractor to include in his proposal a programme for the execution of the proposed change, and his proposal for any necessary modifications to the contract programme and the contractual time for completion.

Secondly, whereas the 1999 Red Book requires the contractor to further include in his proposal a proposition for evaluation of the variation, the 1999 Yellow Book requires that he include a proposition for adjustment to the contract price. Thirdly, the last paragraph of this sub-clause in the 1999 Yellow Book provides that, upon instructing or approving a variation, the engineer should proceed in accordance with sub-clause 3.5 'Determinations' to agree or determine adjustments to the contract price and the schedule of payments. These adjustments should include reasonable profit, and should take account of the contractor's submissions under sub-clause 13.2 'Value Engineering' if applicable. In contrast, the last paragraph of this sub-clause in the 1999 Red Book merely provides that each variation is to be evaluated in accordance with clause 12 'Measurement and Evaluation' unless the engineer instructs or proves otherwise in accordance with clause 13.

24.4.12 Sub-clause 14.1, 14.3 and 14.9: Contract Price and Payment – 'The Contract Price'; 'Application for Interim Payment Certificates' and 'Payment of Retention Money'

Clause 14 of both the 1999 Red and Yellow Books sets out the contractual terms in respect of the contract price and its payment. However, while sub-clause 14.1 'The Contract Price' has the same heading in both forms of contract; and provides that the contract price may be subject to any adjustments that are made under the contract and is deemed to include for all taxes, duties and fees for which the contractor is responsible; there are marked and fundamental differences. First, under the 1999 Yellow Book the contract price is the lump sum accepted by the employer at the close of the tender stage of the contract (unless stated otherwise in the particular conditions of the contract); under the 1999 Red Book the contract price is to be agreed or determined under sub-clause 12.3 'Evaluation'.

Secondly, in the 1999 Yellow Book it is stated that:

> 'any quantities or price data which may be set out in a Schedule shall be used for the purposes stated in the Schedule and may be inapplicable for other purposes'.

This should be compared with the provision of the 1999 Red Book which makes reference to the estimated quantities in the bill of quantities (or other schedule), which are not to be taken as the actual and measured quantities of the works and requires the contractor to provide a breakdown of each lump sum price in the schedules.

Thirdly, because clause 12 'Measurement and Evaluation' in the 1999 Red Book has been replaced by clause 12 'Tests after Completion' in the 1999 Yellow Book, there is an additional provision included under sub-clause 14.1 of the latter, namely if payment for part of the works is to be calculated in accordance with a quantity supplied or for work done, then the provisions for measurement and evaluation of such work done will be as stated in the particular conditions of the contract. The overall contract price will then be assessed accordingly, notwithstanding any variations or adjustments made under the contract.

Sub-clause 14.3 'Application for Interim Payment Certificates' under the 1999 Red Book is identical to that under the 1999 Yellow Book, except that the timing of the contractor's application is stated to be at the end of each month in the 1999 Red Book, whereas it is at the end of each payment period as stated in the contract for the Yellow Book, but if not stated then at the end of each month.

Under sub-clause 14.9 'Payment of Retention Money' both the 1999 Red and Yellow Books make provision for payment to the contractor of the retention money. The differences which exist between the two forms of contract under this sub-clause are as follows:

(a) Under the 1999 Yellow Book, no retention money will be released until not only a taking-over certificate has been issued, as is the case under the 1999 Red Book, but also until the works have passed all specified tests including the tests after completion.
(b) If a taking-over certificate is issued for a section of the works under the 1999 Yellow Book, the employer must pay a percentage of the first half of the retention money, such percentage as stated in the Appendix to Tender. There is no requirement whatsoever to pay any retention money if a taking-over certificate is issued for part of the works, to which the engineer is entitled under sub-clause 10.2 'Taking Over of Parts of the Works'. Under the 1999 Red Book, a proportion of the retention money must be released if a taking-over certificate is issued for a section *or part* of the works, the proportion to be calculated as 40% of the estimated contract value of the section/part divided by the estimated final contract price.
(c) Similarly, under the 1999 Yellow Book, after the expiry of the relevant defects notification period, if a taking-over certificate was issued for a section, the percentage as stated in the Appendix to Tender of the second half of the retention money shall be certified and paid promptly. Under the 1999 Red Book, after the expiry of the relevant defects notification period, if a taking-over certificate was issued for a section, the proportion, again calculated as 40% of the estimated contract value of the section divided by the estimated final contract price, of the second half of the retention money shall be certified and paid promptly.

24.4.13 Sub-clause 17.5: Risk and Responsibility – 'Intellectual and Industrial Property Rights'

Sub-clause 17.5 of both the 1999 Red and Yellow Books deals specifically with the provision of protection to all parties to the contract concerning their intellectual and industrial property rights. Under both forms of contract the employer is under an obligation to indemnify and protect the contractor against any claim alleging an infringement of such rights, where such an infringement is or was the result of the employer using any of the works for any purpose other than that stated within the contract or where the contractor uses the works in association with something that had not been provided by the contractor, unless the contractor knew of the use or the use was stated within the contract. The contractor is also protected where the infringement has arisen as an unavoidable result of the contractor's compliance with:

(a) the employer's requirements, under the 1999 Yellow Book; and
(b) the contract, under the 1999 Red Book.

The contractor has a corresponding obligation to indemnify the employer against any claim that occurs as a result of:

(a) the contractor's 'design, manufacture, construction or execution of the works'; use of the contractor's equipment; or proper use of the works, under the 1999 Yellow Book;
(b) the manufacture, use, sale or import of any goods, or the design for which the contractor is responsible, under the 1999 Red Book.

24.4.14 Clause 20.2: Claims, Disputes and Arbitration – 'Appointment of the Dispute Adjudication Board'

As in the 1999 Red Book, clause 20 'Claims, Disputes and Arbitration' of the 1999 Yellow Book introduces the relatively new concept of the Dispute Adjudication Board. However, the timing of the appointment of the Dispute Adjudication Board under sub-clause 20.2 of the 1999 Yellow Book, being within 28 days after a party gives notice to the other party of its intention to refer a dispute to a DAB under sub-clause 20.4 'Obtaining Dispute Adjudication Board's Decision', is significantly different from that under the 1999 Red Book, being by the date stated in the Appendix to Tender, the default time therein being 28 days after the contractual commencement date. Therefore, under the 1999 Yellow Book the DAB is referred to as an ad hoc DAB, whereas the DAB appointed under the 1999 Red Book is referred to as a standing DAB. It is suggested here that the appointment of an ad hoc DAB is unwise, as it would not achieve one of the main objectives of this mechanism which is the prevention of disputes during the course of the works and falls short of the second objective of effective disputes resolution. Detailed discussion on the advantages of appointing the DAB at the very early stages of a contract features in Chapter 26. However, it is important to note in this connection that the wording of sub-clause 20.8 in the Yellow Book is incorrect and required to be changed if an ad hoc DAB is to be appointed. This is because the occurrence of a dispute precedes the appointment of the DAB and therefore there would be no DAB in place when a dispute arises.

Chapter 25

The 1999 Silver Book

25.1 Introduction

As discussed in Chapter 22, the new Silver Book is a newcomer to the field. To understand the philosophy and reasons behind its conception, it is best to quote from the authoritative paper written by the chairman of FIDIC's Contracts Committee and leader of the Task Group that prepared the FIDIC 1999 Conditions of Contract.[25.1]

> 'Not only is it a fact of life that many employers have always demanded "fixed, lump sum contract prices", and that FIDIC did not have a suitable standard form to cater for such demand, but in recent years the trend has been towards private financing (not only of private investment and speculative projects, but also of public infrastructure projects). The prerequisites for obtaining private finance for a project are vastly different from those of obtaining government or other public money. Private financing requires that the project is independently viable in financial terms, and that there will be, so far as possible, an assured return on the finance provided. The lenders on a BOT or similar project will do their calculations showing the outlay over the construction period and the income over the succeeding operation period. For the return to be reasonably assured, the bases for their calculations will have to be as firm as possible. If the construction work costs more than reckoned (inclusive of any contingency allowance), then the calculations will not hold. If the construction time is longer than planned, then the income will not begin to come in on time, and the calculations will not hold. Therefore, such lenders have to ensure that the risks of cost and time overruns of the construction contract are limited as far as humanly possible. Such lenders are aware that contractors will have to charge a premium for carrying the additional risks necessary to provide the required greater security of construction cost and time. The premium in certain cases may reasonably be large. However, they would rather accept such premium and include it in their calculations before embarking on the project, than discover later on that the project is no longer viable and that they are incurring an overall loss.'

Thus, the Silver Book has been and is intended to be used for special projects. It is worthwhile to note that although the format is the same in all three books, the 1999 Red, Yellow and Silver Books, the Silver Book can be distinguished by the absence of the 'Engineer' and by its imbalanced allocation of the risks, a shift of a wide range of risks to the contractor. These distinguishing characteristics of the Silver Book should not be taken as a criticism of its concept and application. In particular, the Silver Book was conceived in response to the need created by those in favour of using private finance for infrastructure projects, and grew as a result of the demands associated with Build, Own and Transfer (BOT), or Build, Own, Operate and Transfer (BOOT) projects, and with the idea of mixing together design, construction and operation. This response entailed demanding a fixed, lump-sum contract price with little or no risk of an increase in cost if and when unexpected events took place. Of course, private finance requires a financially viable project with an assured return on the funds advanced. Although demanding a fixed, lump-sum contract with most of the risks being allocated to the contractor means that the employer would be paying a higher price for the construction of the project, he would not normally object to having to do so if he were assured of an acceptable return on his total investment outlay. In essence, under the Silver Book, the contractor has to take on board not only strict liability for design and the fitness for purpose standard of performance and liability, but he is also allocated the risk of 'any error, inaccuracy or omission of any kind' in the employer's requirements. As in the Yellow Book, the employer's requirements are set out in a document by that name specifying the purpose, scope, and/or design, and/or other technical criteria, for the works.

The Silver Book, if and when used, ought to be entered into with the utmost care, with all eyes open, focusing on the risks that have been shifted, from a balanced contract between the parties to one where the risks are mostly allocated to the contractor. It is intended for projects where extensive negotiations are entered into prior to the award of contract. The method of procurement has to be adopted to suit that concept. The criteria for risk allocation of control over the risk and/or its consequences is not used in the Silver Book. These risks are spread over a number of the sub-clauses of the Silver Book: 3.1, 3.5, 4.7, 4.12, 5.1, 5.8, 8.4, 17.3 and 20.1, which are dealt with below.

25.2 The 1999 Silver Book: the shifted risks

The Silver Book is closer in content to the 1999 Yellow Book than to the 1999 Red Book, mainly because of the allocation of the design function to the contractor in both the Silver and Yellow Books. By a comparison with the 1999 Yellow Book, the following sections identify the risks that have been traditionally allocated to the employer and now allocated to the contractor.

25.2.1 Sub-clause 3.1: The Employer's Administration – 'The Employer's Representative'

As mentioned above, under the Silver Book, the engineer has been replaced with the employer's representative. This is clear from sub-clause 3.1 where it states that:

> 'the Employer may appoint an Employer's Representative to act on his behalf under the Contract. . . . The Employer's Representative shall carry out the duties assigned to him, and shall exercise the authority delegated to him, by the Employer. Unless and until the Employer notifies the Contractor otherwise, the Employer's

Representative shall be deemed to have the full authority of the Employer under the Contract, except in respect of Clause 15 [Termination by Employer].'

Under the Yellow Book, the first paragraph of sub-clause 3.1 reads as follows:

'The Employer shall appoint the Engineer who shall carry out the duties assigned to him in the Contract. The Engineer's staff shall include suitably qualified engineers and other professionals who are competent to carry out these duties.'

Although, like the engineer under the Yellow Book, the employer's representative under the Silver Book is required to carry out the duties assigned to him, and shall exercise the authority delegated to him by the employer, his authority is potentially more comprehensive. The employer's representative has none of the restrictions in paragraphs (b) and (c) of sub-clause 3.1 of the Yellow Book. The employer's representative could be any person, including an independent consulting engineer or one of the employer's personnel. Furthermore, in that capacity, he acts for the employer and only looks after his interests.

Sub-clause 3.1 of the Silver Book also deals with the replacement of the employer's representative by the employer, who should give the contractor not less than 14 days' notice of the replacement's name, address, duties and authority, and of the date of appointment, but unlike sub-clause 3.4 of the Yellow Book, the contractor has no right in that situation to raise any objection, reasonable or otherwise.

25.2.2 Sub-clause 3.5: The Employer's Administration – 'Determinations'

As stated above, there is no engineer in place to deal with the usual tasks allocated to him, including the administration of the contract; the functions related to design in any of its three stages – conceptual design, preliminary design and final design; and the task of consultation or negotiation with both parties in cases of potential conflict. Instead, it is the employer who is involved in administration. Therefore, it is the employer who makes the determination when no agreement is reached between him and the contractor. Sub-clause 3.5 states in part the following:

'The Employer shall proceed in accordance with this Sub-Clause 3.5 to agree or determine any matter, the Employer shall consult with the Contractor in an endeavour to reach agreement. If agreement is not achieved, the Employer shall make a fair determination in accordance with the Contract, taking due regard of all relevant circumstances.'

It is unusual to find one of the parties in a contract being in charge of determining the respective rights of both parties, particularly when in dispute. However, it is to be noted that the second paragraph of sub-clause 3.5 reads as follows:

'The Employer shall give notice to the Contractor of each agreement or determination, with supporting particulars. Each Party shall give effect to each agreement or determination, unless the Contractor gives notice, to the Employer, of his dissatisfaction with a determination within 14 days of receiving it. Either Party may then refer the dispute to the DAB in accordance with Sub-Clause 20.4 [Obtaining Dispute Adjudication Board's Decision].'

The word 'fair' in the first paragraph should be noted, which provides some comfort to the contractor. However, there should be a further sense of comfort, since if the contractor is dissatisfied with the employer's determination, sub-clause 3.5 entitles him to give a notice of dissatisfaction to the employer within the period of 14 days of receiving the determination; and the matter is then referred to the Dispute Adjudication Board. As in the case of the 1999 Red and Yellow Books, the Board is impartial and independent of the employer, the contractor and the employer's representative, as provided for in Article 3 of the 'General Conditions of Dispute Adjudication Agreement', appended to the contract conditions.

It should also be noted, however, that there is no time limit imposed in sub-clause 3.5 of the Silver Book regarding the period within which agreement should be reached between the employer and the contractor on any matter that is the subject of consultation between them and that there is no format specified for the determination itself.

25.2.3 Sub-clause 4.7: The Contractor – 'Setting Out'

By comparison with the 1999 Red and Yellow Books, this sub-clause allocates the risk of errors in the 'Items of reference' to the contractor. It follows that any delay or cost suffered as a result will therefore be the contractor's responsibility under the Silver Book.

25.2.4 Sub-clause 4.12: The Contractor – 'Unforeseeable Difficulties'

This sub-clause provides a significant shift of a number of risks to the contractor, as clear from its text, which provides as follows:

'4.12 Unforeseeable Difficulties
Except as otherwise stated in the Contract:
(a) the Contractor shall be deemed to have obtained all necessary information as to risks, contingencies and other circumstances which may influence or affect the Works;
(b) by signing the Contract, the Contractor accepts total responsibility for having foreseen all difficulties and costs of successfully completing the Works; and
(c) the Contract Price shall not be adjusted to take account of any unforeseen difficulties or costs.'

The provisions of this sub-clause should be read in conjunction with those of sub-clause 4.10, which provides as follows:

'The Employer shall have made available to the Contractor for his information, prior to the Base Date, all relevant data in the Employer's possession on sub-surface and hydrological conditions at the Site, including environmental aspects. The Employer shall similarly make available to the Contractor all such data which come into the Employer's possession after the Base Date.

The Contractor shall be responsible for verifying and interpreting all such data. The Employer shall have no responsibility for the accuracy, sufficiency or completeness of such data, except as stated in Sub-Clause 5.1 [General Design Responsibilities].'

The reference to sub-clause 5.1 should be noted in that any data or information received by the contractor, from the employer or otherwise, shall not relieve the contractor from his responsibility for the design and execution of the works. However, sub-clause 5.1 continues to state that the employer shall be responsible for the correctness of certain specified portions of the employer's requirements and of specified data and information provided by, or on behalf of, the employer. These specified items appear in the last paragraph of sub-clause 5.1.

25.2.5 Sub-clause 5.1: Design – 'General Design Obligations'

It can be seen from the text of this sub-clause that the consequences of all unforeseen difficulties have been allocated to the contractor, irrespective of control of the risk or its consequences, if and when such risk eventuates.

As stated above, sub-clause 5.1 should also be read in conjunction with sub-clause 4.10 under which the contractor is responsible for verifying and interpreting data, including data or sub-surface and hydrological conditions at the site, including environmental aspects. The employer is not responsible for the accuracy, sufficiency or completeness of such data, except as stated at the end of sub-clause 5.1.

Furthermore, errors, inaccuracies or omission of any kind in the original employer's requirements, as originally included in the contract are deemed to be a contractor's risk, as can be seen from the text of the second paragraph of the sub-clause, which states, in part, the following:[25.2]

'The Employer shall not be responsible for any error, inaccuracy or omission of any kind in the Employer's Requirements as originally included in the Contract and shall not be deemed to have given any representation of accuracy or completeness of any data or information, except as stated below. Any data or information received by the Contractor, from the Employer or otherwise, shall not relieve the Contractor from his responsibility for the design and execution of the Works.'

25.2.6 Sub-clause 5.8: Design – 'Design Error'

The responsibility for design errors, etc. is shifted to the contractor by the provisions of this sub-clause, as its text provides the following:

'5.8 Design Error
If errors, omissions, ambiguities, inconsistencies, inadequacies or other defects are found in the Contractor's Documents, they and the Works shall be corrected at the Contractor's cost, notwithstanding any consent or approval under this Clause.'

Besides in sub-clause 5.8, 'error' is referred to in sub-clauses 1.8, 2.1, 3.3, 4.7 and 5.1 of the conditions and the responsibility attached to errors is generally that of the contractor, forming one of the most important concepts in the Silver Book. There is little said regarding the employer's requirements, including in sub-clause 5.1, and therefore the whole section in Chapter 24 on this subject applies to the Silver Book with the added risk of the responsibility for errors, inaccuracies or omission of any kind. Perhaps the highest risk is that of omission, rather than error, since it could affect the decision as to whether a variation is necessarily required. This part of the Silver Book, in particular, and the whole of clause 5, in general, marks a major shift in the risk allocation against the contractor and has received adverse comments from that end of the industry.[25.3]

25.2.7 Sub-clause 8.4: Commencement, Delay and Suspension – 'Extension of Time for Completion'

The entitlement of the contractor to an extension of the 'Time for Completion' under the Silver Book is limited by comparison with the 1999 Red and Yellow Books. This reduction in entitlement relates to the deletion of paragraphs (c) and (d) from the scope of sub-clause 8.4 of the 1999 Yellow Book. Paragraph (c) relates to 'exceptionally adverse climatic conditions' providing that these conditions do not constitute *force majeure*; and paragraph (d) relates to 'Unforeseeable shortages in the availability of personnel or Goods caused by epidemic or governmental actions'. The risks attached to these events under the Silver Book are thus allocated to the contractor.

Paragraph (b) of sub-clause 8.4 of the Silver Book is the same as that under the 1999 Red and Yellow Books and remains in text to include a cause of delay giving an entitlement to the contractor for an extension of time under any sub-clause of the contract conditions. However, the sub-clauses that entitle the contractor to claim for an extension of time in the Silver Book are fewer than those in the other two Books. Those that do not provide for an entitlement under the Silver Book, by comparison with the 1999 Yellow Book, are as follows:

— Sub-clause 1.9: Errors in the Employer's Requirements;
— Sub-clause 4.7: Setting Out; and
— Sub-clause 4.12: Unforeseeable Physical Conditions.

25.2.8 Sub-clause 17.3: Risk and Responsibility – 'Employer's Risks'

The consequence of any of the following risks, derived from the terms of the 1999 Red and Yellow Books, is borne by the contractor, since these risks have been omitted from the list of the Employer's Risks in the Silver Book:

'(f) use or occupation by the Employer of any part of the Permanent Works, except as may be specified in the Contract,
(g) design of any part of the Works by the Employer's Personnel or by others for whom the Employer is responsible, if any, and

(h) any operation of the forces of nature which is Unforeseeable or against which an experienced contractor could not reasonably have been expected to have taken adequate preventative precautions.'[25.4]

Whilst it may be logical that the risks attached to the employer's requirements under the Silver Book and the design risk would be allocated to the contractor, the shift of the other two risks quoted above is significant and should bring to the mind of the parties that high risk means high contingent price to cover it. However, such contingency could only be speculative since even an experienced contractor could not make a realistic risk assessment of the operation of the exceptional forces of nature.

25.2.9 Sub-clause 20.1: Claims, Disputes and Arbitration – 'Contractor's Claims'

With no engineer in place to deal with the contractor's claims, it is the employer who receives the claims when submitted under sub-clause 20.1 and it is he who responds with either approval or disapproval and with detailed comments. The effect of this is the same as that discussed under Section 26.2.2 above in respect of sub-clause 3.5. In this connection, it is worth noting that the comments made in Section 24.4.14 above regarding the wording of sub-clause 20.8 of the Yellow Book also apply to the Silver Book.

25.3 The 1999 Silver Book: concepts and content

A close scrutiny of the wording of the three forms of contract, the 1999 Red, Yellow and Silver Books, as presented in Chapter 27, will show the similarity of wording and format, but nonetheless the concepts of the Silver Book differ greatly in certain aspects from the Red and Yellow Books. Although it is closer to the Yellow Book, it is useful at this point to point out the differences when compared with the 1999 Red and Yellow Books.

25.3.1 Clause 1: General Provisions

Clause 1 of the 1999 Silver Book sets out the general provisions of the conditions of contract. Clause 1.1.1.1 is different in all three forms of contract, in many ways. First of all, under the Silver Book, the term 'Contract' is defined as 'the Contract Agreement, these Conditions, the Employer's Requirements, the Tender, and the further documents (if any) which are listed in the Contract Agreement'. The Red Book definition includes more documents including, for example, the specification and the drawings. Furthermore, there is no corresponding sub-paragraph in the Silver Book to paragraphs 1.1.1.3 'Letter of Acceptance' and sub-paragraph 1.1.1.4 'Letter of Tender' in the Yellow Book. Likewise, there is no corresponding sub-paragraph in the Silver Book, nor in the Yellow Book, to sub-paragraphs 1.1.1.6 'Drawings' and 1.1.1.7 'Schedules' in the Red Book.

Paragraph 1.1.1.3 in the Silver and Yellow Books defines the *'Employer's Requirements'*, a document that is not intended to form part of the terms of the Red Book. The term 'tender' is defined as 'the Contractor's signed offer for the Works and all other documents which the Contractor submitted therewith (other than these Conditions and the Employer's Requirements, if so submitted), as included in the Contract.' This is different to the Yellow Book where the term means the letter of tender and any other documents that the contractor submitted with the letter of tender. Sub-paragraph 1.1.1.5 defines the 'Performance Guarantees and the Schedules of Payments' as 'documents so named (if any), as included in the Contract'. There is no corresponding paragraph of this kind in the Red Book.

As mentioned above, a major difference between the Red and the Yellow Books on one hand and the Silver Book on the other is the removal of the role of the engineer in the latter. Therefore, where reference is made in the Red and Yellow Books to the 'Engineer' under the Silver Book, it appears as the 'Employer's Representative' or 'the Employer'. This is clear from sub-paragraph 1.1.2.6 where the term 'Employer's Personnel' is defined in the Silver Book as meaning 'the Employer's Representative, the assistants referred to in Sub-Clause 3.2 [Other Employer's Personnel] and all other staff, labour and other employees of the Employer and of the Employer's Representative; and any other personnel notified to the Contractor by the Employer or the Employer's Representative, as Employer's Personnel.'

The defects notification period is defined under paragraph 1.1.3.7. There are two differences between the definition given in the Silver Book and the one given in the Yellow Book. Under the Silver Book, the period for notification of defects within the works is that stated in the particular conditions. Under the Yellow Book, it is that stated in the Appendix to Tender. Secondly, under the Silver Book, where no such period is stated in the particular conditions, the period will be one year. This option does not exist under the Red Book.

It should be noted that there is no equivalent paragraph in the Silver Book which corresponds to paragraphs 1.1.4.1 'Accepted Contract Amount'; 1.1.4.4 'Final Payment Certificate'; 1.1.4.7 'Interim Payment Certificate'; and 1.1.4.9 'Payment Certificate' under the Yellow Book. Furthermore, under paragraph 1.1.4.1 of the Silver Book, the contract price is defined as being 'the agreed amount stated in the Contract Agreement for the design, execution and completion of the Works and the remedying of any defects, adjusted (if applicable) in accordance with the Contract.' The definition of contract price under the Yellow Book appears under sub-clause 14.1.

There is no equivalent text in the Silver Book which corresponds to paragraph 1.1.6.8 of the Red Book, which defines what the word 'Unforeseeable' means. This reflects the shift of that risk to the contractor.

Sub-clause 1.5 of the Silver Book is somewhat different to that same sub-clause in the Red Book. As mentioned above, the terms used in the Red Book, 'Letter of Acceptance', 'Letter of Tender', 'Specification', 'Drawing' and 'Schedules', are not used in the Silver Book and therefore these are not included in the list given in sub-clause 1.5 of the Silver Book which sets out the precedence of documents that are said to form the contract. Documents that form a contract under the Silver Book are, in order of precedence:

(i)   the contract agreement;
(ii) the particular conditions;
(iii) the general conditions;
(iv) the employer's requirements; and
(v)   the tender and any other documents forming part of the contract.

Under sub-clause 1.6 the contract comes into force and effect on the date stated in the contract agreement. This is in contrast to the Red Book, where the contract comes into being within 28 days after the contractor receives the letter of acceptance, unless otherwise agreed by the parties.

Sub-clause 1.9 under the Silver Book, confidentiality, is entirely different from that in the Red and Yellow Books, sub-clause 1.12 'Confidential Details'. Sub-clause 1.9 of the Silver Book deals with the confidentiality of the contract and states that both parties to the contract should treat all details of the contract between them as private and confidential, except to the extent necessary to carry out obligations under it or to comply with the applicable laws. Furthermore, the contractor is not entitled to publish, or to permit to be published, or to disclose any particulars of the works in any trade or technical paper or elsewhere without the previous agreement of the employer. Confidential details of the contract are dealt with in the Red and Yellow Books under sub-clause 1.12 where the contractor is required to disclose all such confidential and other information as the engineer may reasonably require in order to verify the contractor's compliance with the contract.

25.3.2   Clause 3: The Employer's Administration

Clause 3 of the 1999 Silver Book deals with the tasks and duties of the employer's representative, and other personnel. It goes on to deal with delegation, instructions and determinations. This clause is different from clause 3 under the Red and Yellow Books, which are entitled 'The Engineer'.

Sub-clause 3.2 is entitled 'Other Employer's Personnel' and states

> '3.2 The Employer or the Employer's Representative may from time to time assign duties and delegate authority to assistants, and may also revoke such assignment or delegation. These assistants may include a resident engineer, and/or independent inspectors appointed to inspect and/or test items of Plant and/or Materials. The assignment, delegation or revocation shall not take effect until a copy of it has been received by the Contractor.
>
> Assistants shall be suitably qualified persons, who are competent to carry out these duties and exercise this authority, and who are fluent in the language for communications defined in Sub-Clause 1.4 [Law and Language].'

Sub-clause 3.3 under the Silver Book states that all assistants appointed under sub-clause 3.2 above and the employer's representative can only issue instructions to the contractor to the extent that such instructions have been defined by the delegation. Furthermore, any approval, check, certificate, consent, examination, inspection, instruction, notice, proposal, request, test or similar act carried out by the employer's representative or an assistant in this connection, shall have the same effect as though the act had been an act of the employer. However, the following paragraphs apply:

'(a) unless otherwise stated in the delegated person's communication relating to such act, it shall not relieve the Contractor from any responsibility he has under the Contract, including responsibility for errors, omissions, discrepancies and non-compliances;
(b) any failure to disapprove any work, Plant or Materials shall not constitute approval, and shall therefore not prejudice the right of the Employer to reject the work, Plant or Materials; and
(c) if the Contractor questions any determination or instruction of a delegated person, the Contractor may refer the matter to the Employer, who shall promptly confirm, reverse or vary the determination or instruction.'

25.3.3 Clause 4: The Contractor

Clause 4 of the 1999 Silver Book deals with the obligations and duties of the contractor. The contractor is entitled under sub-clause 4.3 to appoint a 'Contractor's Representative' who would act on his behalf, issue instructions, receive instructions, delegate powers and functions or revoke any powers or functions on the contractor's behalf. If the contractor's representative is not named within the contract, then the contractor must submit to the employer the name and qualifications of the person he wishes to appoint as his representative. However, unlike the third paragraph of sub-clause 4.3 under the Yellow Book, sub-clause 4.3 under the Silver Book does not state that the whole time of the contractor's representative should be given to directing the contractor's performance of the contract.

Under sub-clause 4.4, it is forbidden for the contractor to subcontract the whole of the works. He is responsible for any acts or defaults that have been carried out by any subcontractor, his agents or his employees. The contractor must give the employer not less than 28 days' notice of any intended appointment of a subcontractor, including any relevant details in relation to the subcontractor's experience. He should also give the employer 28 days' notice of the intended commencement date of the subcontractor's work and the commencement of the subcontractor's work on site.

The employer is under an obligation under sub-clause 4.10 to make available to the contractor throughout the duration of the contract, all information that he may have on the hydrological and subsurface conditions that are present on the site. The contractor holds the sole responsibility for interpreting such documentation, and the employer will not be held liable for any inaccuracies within the above-mentioned documents. Under the 1999 Red and Yellow Books, the contractor is deemed to have obtained all other necessary information, to the extent which was practicable. This provision is not present in the Silver Book as a result of the greater extent of the contractor's responsibilities, particularly under sub-clause 4.12.[25.5]

Sub-clause 4.11 of the Silver Book states that the contractor shall be deemed to have satisfied himself that the contract price is correct and sufficient. There is an added provision in sub-clause 4.11 of the 1999 Red and Yellow Books that the accepted contract amount be based on any data, interpretations, necessary information, inspections, examinations, etc. This is not stated in the Silver Book.

It should be noted that under sub-clause 4.17 of the Silver Book, 'Contractor's Equipment', the contractor is responsible under the contract conditions for all of the contractor's equipment. The 1999 Red and Yellow Books continue on to state that the contractor is not entitled to remove any major items of the contractor's equipment without the consent of the engineer. This is not mentioned under the Silver Book.

25.3.4 Clause 5: Design

Clause 5 of the 1999 Silver Book deals with the procedure for the design process of the works. As quoted in section 24.4.5 above, which deals with clause 5 under the Yellow Book, the *FIDIC Contracts Guide* divides the design process under the Yellow and Silver Books into three stages identified in that section, although this is not stated within the contract conditions themselves. It is worthwhile to briefly repeat these three stages here:

(i) *The conceptual design stage:* this is the design that is done by the employer for inclusion in the employer's requirements in order to define the works. This stage does not involve a significant amount of design input and could be less than 10% of total design input.
(ii) *The preliminary design stage:* this is the design that is carried out by each tenderer for inclusion in the tender. The instructions to tenderers would ordinarily indicate the exact amount of detail required at this stage.
(iii) *The final design stage:* this stage is reserved solely for the working drawings, which may in itself involve two sub-stages: general arrangement drawings and detailed drawings.

Sub-clause 5.2 of the Silver Book defines what constitutes the contractor's documents. It is up to the contractor to prepare all of these documents and any other documents which may be necessary for the construction of the works. It is worthwhile to note that under the Yellow Book, the contractor's documents may be submitted to the engineer for his approval. Under the Silver Book there is no mention of the contractor's documents being submitted to the employer for his approval; they are merely submitted to him. Other than this difference, sub-clause 5.2 in both the Yellow and the Silver Books is the same.

25.3.5 Clause 6: Staff and Labour

The provisions of this clause are the same as in the 1999 Red and Yellow Books, other than the necessary changes required due to the nature of the three contract forms.

25.3.6 Clause 7: Plant, Materials and Workmanship

Clause 7 of the Silver Book deals with the general requirements in relation to plant for carrying out of the works, materials for the works and workmanship. Sub-clause 7.2 states that the contractor shall submit samples to the employer, for review in accordance with the procedures for Contractor's Documents described in sub-clause 5.2, 'Contractor's Documents', as specified in the contract and at the contractor's cost. Each sample shall be labelled as to origin and intended use in the works.

Under sub-clause 7.6 the employer has the power to instruct the contractor as follows:

> '7.6 Notwithstanding any previous test or certification, the Employer may instruct the Contractor to:
> (a) remove from the Site and replace any Plant or Materials which is not in accordance with the Contract,
> (b) remove and re-execute any other work which is not in accordance with the Contract, and
> (c) execute any work which is urgently required for the safety of the Works, whether because of an accident, unforeseeable event or otherwise.'

There is an important difference between the Silver Book and the 1999 Red and Yellow Books in this regard. Under the Red and Yellow Books, if the contractor fails to carry out any of the work that the employer has instructed him to do under this sub-clause, the employer has grounds for termination of the contract under sub-clause 15.2(c)(ii). However, under all three Books, if the contractor fails to carry out any of the above instructions issued by the employer, the employer has the power to employ and to pay other persons to carry out the work. Subject to sub-clause 2.5, the contractor shall pay to the employer all costs arising from this failure.

25.3.7 Clause 8: Commencement, Delays and Suspension

Clause 8 of the Silver Book deals with the commencement of work, any delays within the carrying out of the works, and suspension. Sub-clause 8.3 deals specifically with the programme for the works. Once the works have commenced, the contractor has to submit to the employer a time programme that he should revise whenever the previous programme is incorrect in relation to the progress of the works. Under the Silver Book, the programme should include the following:

> '8.3 The Contractor shall submit a time programme to the Employer within 28 days after the Commencement Date. The Contractor shall also submit a revised programme whenever the previous programme is inconsistent with actual progress or with the Contractor's obligations. Unless otherwise stated in the Contract, each programme shall include:
> (a) the order in which the Contractor intends to carry out the Works, including the anticipated timing of each major stage of the Works,

(b) the periods for reviews under Sub-Clause 5.2 [Contractor's Documents],
(c) the sequence and timing of inspections and tests specified in the Contract, and
(d) a supporting report which includes:
   (i) a general description of the methods which the Contractor intends to adopt for the execution of each major stage of the Works, and
   (ii) the approximate number of each class of Contractor's Personnel and of each type of Contractor's Equipment for each major stage.

Unless the Employer, within 21 days after receiving a programme, gives notice to the Contractor stating the extent to which it does not comply with the Contract, the Contractor shall proceed in accordance with the programme, subject to his other obligations under the Contract. The Employer's Personnel shall be entitled to rely upon the programme when planning their activities.

The Contractor shall promptly give notice to the Employer of specific probable future events or circumstances which may adversely affect or delay the execution of the Works. In this event, or if the Employer gives notice to the Contractor that a programme fails (to the extent stated) to comply with the Contract or to be consistent with actual progress and the Contractor's stated intentions, the Contractor shall submit a revised programme to the Employer in accordance with this Sub-Clause.'

The contractor should proceed in accordance with the programme subject to his other obligations under the contract. Furthermore, the contractor is under a duty to notify the employer of any future events or circumstances which may adversely affect the execution of the works. In this event, or if the employer gives notice to the contractor that the programme fails to comply with the contract, or that it is inconsistent with the actual progress of the works and the contractor's stated intentions, then the contractor must submit a revised programme to the employer that is in accordance with this sub-clause. The Red Book is different in this respect.

Sub-clause 8.4 of the Silver Book is dealt with in Section 25.2.7 above.

25.3.8 Clause 9: Tests on Completion

Clause 9 of the 1999 Silver Book deals with the tests that should be carried out once the works have been completed. These tests are carried out by the contractor in order to determine whether the works are ready for the employer to take them over. The specific tests that are to be carried out would be given in detail within the contract itself, so under the Silver Book, in the employer's requirements. The Silver and 1999 Yellow Books provide considerably more detail on the tests that are to be carried out than the Red Book, and it has been suggested that the reason for this is 'because of the importance of such tests in determining whether the works satisfy the Employer's Requirements. Under the Construction Book, tests on completion may be less important except in respect of any parts of the works designed by the contractor.'[25.6]

Sub-clause 9.4 in all three Books deals with the situation where the works or a section fails to pass the tests on completion. If this occurs under the Silver Book, the employer is first entitled to order further repetition of the tests on completion under sub-clause 9.3 'Retesting'. Secondly, if the failure deprives him of substantially the whole benefit of the works or a section, then he may reject the works or a section, whichever the case may be. In this case, the employer will have the same remedies that are provided for in sub-clause 11.4(c) 'Failure to Remedy Defects'. Thirdly, he can issue a taking-over certificate. Under the Red and Yellow Books, this taking-over certificate can only be issued if it has been requested by the employer.

25.3.9 Clause 10: Employer's Taking Over

Clause 10 of the 1999 Silver Book deals with the taking over of the works by the employer. There is a major difference between sub-clause 10.2 under the Silver Book and the corresponding sub-clause in the Red and Yellow Books. Under the Silver Book, parts of the works (other than sections) should not be taken over or used by the employer except as is stated in the contract or if it has been agreed by both parties. Under the Red and Yellow Books, a more in-depth procedure for the taking over of the works is set out.

Sub-clause 10.3 of all three Books sets out the procedure to be followed where there has been an interference with the tests to be carried out on completion. There is a different procedure under the three Books where the contractor is prevented from carrying out the tests on completion for more than 14 days by a cause for which the employer is responsible. Under the Silver Book, the contractor should carry out the tests on completion as soon as practicable. However, under the Red and Yellow Books the employer is deemed to have taken over the works or a section, whichever the case may be, on the date when the tests on completion would otherwise have been completed.

It is also worthwhile to note at this point that there is no corresponding clause under the Silver Book to sub-clause 10.4 'Surfaces Requiring Reinstatement' of the Red and Yellow Books.

25.3.10 Clause 11: Defects Liability

There is a significant difference between sub-clause 11.9 under the Silver Book and the corresponding sub-clause in the Red and Yellow Books, which deals with the performance certificate. Under all three Books, the performance of the contractor's obligations will not be considered to have been completed until the performance certificate has been issued to the contractor stating the date on which the contractor completed his obligations under the contract. Furthermore, the employer or the engineer (depending on which form is used) has to issue the performance certificate within 28 days after the latest date of the expiry dates of the defects notification periods or as soon after that as the contractor has supplied all the contractor's documents and completed and tested all the works, including remedying and defects. However, unlike the Red and Yellow Books, if the employer fails, under the Silver Book, to issue the performance certificate, then it is deemed to have been issued on the date 28 days after the date on which it should have been issued. Furthermore, sub-clauses 11.11 'Clearance of Site' and 14.14(a) 'Cessation of Employer's Liability' are inapplicable in this regard.

25.3.11 Clause 12: Tests after Completion

As clause 12 of the 1999 Red Book deals with 'Measurement and Evaluation', clause 12 of the Silver Book should be compared with that of the 1999 Yellow Book. The main difference between the two clauses is that whilst the contractor is required to carry out the tests after completion under the Silver Book, it is the employer who is required to carry out that task under the Yellow Book. Sub-clause 12.1 of the Silver Book states that it applies if the tests after completion are specified in the contract. Unless otherwise stated in the particular conditions, the employer is required to provide all electricity, fuel and materials for such tests and to also make available his personnel and plant. Furthermore, the contractor has to provide any other plant, equipment and suitably qualified and experienced staff that are necessary so as to carry out the tests after completion efficiently. It is a requirement under paragraph (c) of sub-clause 12.1 that the contractor should carry out the tests after completion in the presence of as many employer's and/or contractor's personnel as either party may reasonably request. Finally, once the results of the tests have been compiled and evaluated by the contractor he must prepare a detailed report.

Sub-clauses 12.2 'Delayed Tests' and 12.3 'Retesting' of the Silver Book are virtually the same as those under the Yellow Book.

25.3.12 Clause 13: Variations and Adjustments

Clause 13 of the 1999 Silver Book covers any variations and adjustments that may be made to the contract terms. Under sub-clause 13.5, the issue of provisional sums is dealt with and there is only slight difference between the Silver and the Yellow Books on this issue. First of all, there is no provision in the Yellow Book that the cost of plant, materials or services be deducted from the original provisional sums. Secondly, under the Yellow Book provision is made that if no percentage rate has been stated in the schedule, then the percentage rate that is stated in the Appendix to Tender will apply. This is not stated in the Silver Book because, as explained above, there is no Appendix to Tender in this form of contract.

Sub-clause 13.8 deals with the adjustment for changes in cost under all three Books. However, the provisions under the Yellow Book are considerably more extensive than those provided for in the Silver Book. Sub-clause 13.8 of the Silver Book rather simply states that if the contract price is to be adjusted as a result of rises and falls in the cost of labour, goods and other inputs to the works, then these adjustments are to be calculated in accordance with the particular conditions.

25.3.13 Clause 14: Contract Price and Payment

Sub-clause 14.1 of the Silver Book sets out the fundamental principles regarding payments that have to be made by the employer to the contractor. It differs significantly from that under the 1999 Red and Yellow Books. Sub-clause 1.1.4.1 of the Silver Book defines the term 'Contract Price' as the agreed amount stated in the contract, subject to adjustments in accordance with the contract, without any reference to sub-clause 14.1;

therefore, sub-clause 14.1 simply refers to the contract price as being the lump sum upon which payments are based unless otherwise stated in the particular conditions.

Paragraph (b) of sub-clause 14.1 provides for the contractor to pay any duties and taxes under the contract pursuant to sub-clause 1.13. The contract price is thus deemed to include these taxes and duties, together with associated administrative costs, based on the rates applicable at the base date. If these rates increase after the base date, then the provisions of sub-clause 13.7 would apply.

In this connection, it should be noted that the contract documents do not include 'Schedules' and therefore do not contain provisions similar to sub-clause 14.1 (c) and (d) of the 1999 Yellow Book. Furthermore, if part of the works is to be paid for in accordance with a quantity supply or work done, appropriate provisions must be included in the particular conditions.

Sub-clause 14.2 'Advance Payment' requires the employer to make an advance payment, as an interest-free loan for mobilisation and design, when the contractor submits a guarantee including the details stated in the particular conditions. The sub-clause then continues to explain a number of conditions relating to the contents of the particular conditions, should they fail to state the following:

(a) if they fail to state the amount of the advance payment, then this sub-clause shall not apply;
(b) if they fail to state the number and timing of instalments, then there shall be only one;
(c) if they fail to state the applicable currencies and proportions, then they shall be those in which the contract price is payable; and/or
(d) if they fail to state the amortisation rate for repayments, then it shall be calculated by dividing the total amount of the advance payment by the contract price stated in the contract agreement less provisional sums.

Sub-clause 14.2 also sets out how and when the advance payment should be repaid.

Sub-clause 14.3 deals with the applications for interim payments which should be submitted through a statement in six copies to the employer after the end of the period of payment stated in the contract, and if not stated, after the end of each month, in a form approved by the employer, showing in detail the amounts to which the contractor considers himself to be entitled, together with supporting documents which should include the relevant report on progress in accordance with sub-clause 4.21 'Progress Reports'. The sub-clause also sets out certain items that should be sent with the application.

Sub-clause 14.4 'Schedule of Payment' only becomes applicable if interim payments are to be based on a schedule of payments, except for its last paragraph.

Sub-clause 14.5 'Plant and Material intended for the Works' provides the conditions under which the contractor may be entitled to receive an interim payment for any plant and materials. The contractor will be entitled to such a payment even if the plant and materials are not even on the site. These conditions are: (a) the plant and materials are in the country and have been identified as the employer's property or (b) the contractor has given the employer evidence of insurance and a bank guarantee in an amount that is equal to the payment that the contractor is requesting.

Sub-clause 14.6 deals with interim payments. No money will be paid to the contractor until the employer has received and approved the Performance Security. It is worth noting that, according to the last paragraph of sub-clause 14.6, payment by the employer to the contractor cannot be taken as his acceptance, his approval, his consent or his satisfaction to the work done.

Sub-clause 14.7 sets out the procedure that the employer must follow in making payment to the contractor. Payment for the works can occur in three stages, as follows:

'(a) the first instalment of the advance payment within 42 days after the date on which the Contract came into full force and effect or within 21 days after the Employer receives the documents in accordance with Sub-Clause 4.2 [Performance Security] and Sub-Clause 14.2 [Advance Payment], whichever is later;
 (b) the amount which is due in respect of each Statement, other than the Final Statement, within 56 days after receiving the Statement and supporting documents; and
 (c) the final amount due, within 42 days after receiving the Final Statement and written discharge in accordance with Sub-Clause 14.11 [Application for Final Payment] and Sub-Clause 14.12 [Discharge].'

To all intents and purposes, clauses 14.8 to 14.15 in the Silver Book are similar or the same to those in the 1999 Yellow Book.

25.3.14 Clause 15: Termination by Employer

Clause 15 of the 1999 Silver Book deals with the procedure for termination of the contract by the employer. Under sub-clause 15.2, the employer is entitled to terminate the contract if the contractor:

(a) fails to comply with sub-clause 4.2 'Performance Security' or with a notice under sub-clause 15.1 'Notice to Correct';
(b) abandons the works or otherwise plainly demonstrates the intention not to continue performance of his obligations under the Contract;
(c) without reasonable excuse fails to proceed with the works in accordance with clause 8 'Commencement, Delays and Suspension'; or
(d) subcontracts the whole of the Works or assigns the Contract without the required agreement;
(e) becomes bankrupt or insolvent, goes into liquidation, has a receiving or administration order made against him, compounds with his creditors, or carries on business under a receiver, trustee or manager for the benefit of his creditors, or if any act is done or event occurs which (under applicable laws) has a similar effect to any of these acts or events; or
(f) gives or offers to give (directly or indirectly) to any person any bribe, gift, gratuity, commission or other thing of value, as an inducement or reward:
 (i) for doing or forbearing to do any action in relation to the contract, or
 (ii) for showing or forbearing to show favour or disfavour to any person in relation to the Contract,

or if any of the contractor's personnel, agents or sub-contractors gives or offers to give (directly or indirectly) to any person any such inducement or reward as is described in this sub-paragraph (f). However, lawful inducements and rewards to Contractor's Personnel shall not entitle termination.

Another example of where the employer can terminate the contract appears in sub-clause 15.2(c)(ii) of the Yellow Book: where the contractor without reasonable excuse fails to comply with a notice that is issued under sub-clauses 7.5 'Rejection' or 7.6 'Remedial Work', within 28 days of receiving it.

25.3.15 Clause 16: Suspension and Termination by Contractor

Sub-clause 16.1 of the 1999 Silver Book provides that the contractor may terminate the contract if the employer fails to comply with sub-clause 2.4 'Employer's Financial Arrangement' or sub-clause 14.7 'Timing of Payments', once he has given at least 21 days' notice to the employer. Sub-clause 16.2 of the Silver Book sets out the situations where the contractor can terminate the contract:

(a) the contractor does not receive the reasonable evidence within 42 days after giving notice under sub-clause 16.1 'Contractor's Entitlement to Suspend Work' in respect of a failure to comply with sub-clause 2.4 'Employer's Financial Arrangements';
(b) the contractor does not receive the amount due within 42 days after the expiry of the time stated in sub-clause 14.7 'Timing of Payments' within which payment is to be made (except for deductions in accordance with sub-clause 2.5 'Employer's Claims');
(c) the employer substantially fails to perform his obligations under the contract;
(d) the employer fails to comply with sub-clause 1.7 'Assignment';
(e) a prolonged suspension affects the whole of the works as described in sub-clause 8.11 'Prolonged Suspension'; or
(f) the employer becomes bankrupt or insolvent, goes into liquidation, has a receiving or administration order made against him, compounds with his creditors, or carries on business under a receiver, trustee or manager for the benefit of his creditors, or if any act is done or event occurs which (under applicable laws) has a similar effect to any of these acts or events.

There is no mention in the Silver Book of the other situation where the contractor can terminate the contract, which appears in the 1999 Yellow Book under sub-clause 16.2(b). It states that the contractor can terminate the contract where the engineer fails to issue the relevant payment certificate within 56 days after receiving a statement and supporting documents.

Chapter 26

Dispute Boards

With particular emphasis on FIDIC's Dispute Adjudication Board

26.1 Introduction

As stated in Chapter 19, one of the most important provisions in any commercial contract is the provision for a fair method of dispute resolution. Chapter 19 dealt specifically with two forms of dispute resolution that are alternatives to litigation and have existed for centuries: arbitration and conciliation. This chapter is dedicated to a relatively new concept of dispute resolution, Dispute Boards, which differ from all others in that they are intended to actively operate throughout the whole period of a contract, not only to resolve disputes but also, if at all possible, to prevent them from happening. It is appropriate to deal in depth with Dispute Boards, since all the standard FIDIC forms of contract between employer and contractor, other than the new Green Book, now incorporate this new mechanism as the first step in their dispute resolution framework.

In recent decades, there has been a trend towards implementing large construction projects with very significant capital costs, especially in the area of public sector infrastructure. The level of transactions involved has meant an inherent greater scope for conflict between the contracting parties and for an increase in the number and size of claims and eventually disputes. It became clear during the 1980s and early 1990s that society requires and needs not only appropriate dispute resolution mechanisms but also a method of avoiding disputes in the first place. It was against this background, explored more fully in Section 26.3 below, that FIDIC started with the introduction of the use of Dispute Boards in the Orange Book in 1995 and continued when this was incorporated into the 1992 Fourth Edition of the Red Book through its Supplement, published in 1996. The suite of FIDIC's three major contracts, the 1999 Red Book, the 1999 Yellow Book and the 1999 Silver Book, have all adopted this concept as the first step in the dispute resolution mechanism, albeit in different forms.

Three principal trends may be discerned from the introduction of Dispute Boards. First, one of the reasons why parties are willing to pay for a Dispute Board to act during the performance of a contract (by investigating and perhaps by recommendation or provisional decision to resolve a dispute or disputes) is that the cost is much cheaper than having an arbitration. The costs of any arbitration are, with reason, perceived to be high (even though they are likely to be no higher than any other form of binding dispute resolution that results in a readily enforceable decision). Obviously this is not a true comparison since a Dispute Board is not comparable to an arbitration and performs a different function. Secondly, the disputes that escape the filter of a Dispute Board (and any unsuccessful attempt to achieve a settlement) are likely to be relatively intractable, such as those which arise out of the termination of a contract. However, there may be other reasons why a Dispute Board does not achieve a settlement of the dispute. For example, the Board may have been unable to deal with the dispute as it would have wished and to have made a satisfactory recommendation or decision; or the parties may have been unwilling or unable to face up to the problem that gave rise to the dispute and to accept the financial or human consequences of a decision or recommendation from the Board; or, as happens, a party may simply be unable to meet its obligations or may be unreasonable and resort to dilatory tactics. Thirdly, the disputes will have been refined by that filtering process so that the points at issue should be clearer than they have been.[26.1]

To start with, it is interesting to explore the main advantages of the Dispute Board which gave rise to such prominence and success in a relatively short period of time. Before doing so, however, it is appropriate to define the term 'Dispute Board'.

In its present context, a Dispute Board is a board composed of one or more (usually three) independent and impartial professionals, who are qualified, experienced and knowledgeable in the technical field of the project, appointed at the commencement of a project to track its progress and to be available at short notice to prevent disagreements from escalating into disputes; and to resolve disputes should they arise. In order to do so effectively, ideally the Board should be appointed at the commencement of the project so as to become very quickly familiar with its technical and contractual characteristics, and then continue to monitor its progress until completion.

The three main types of Dispute Board, and the varieties of each type, are explained in detail at Section 26.4 below.

26.2 Main advantages of the Dispute Board

The mechanism of Dispute Boards achieved such prominence and success in a relatively short time because of the significant advantages they offer in comparison to more traditional forms of dispute resolution. The following main advantages are taken from personal experience and from the experience of practising Dispute Board members.[26.2]

(a) Dispute Boards bring objectivity and neutrality to a project by virtue of being appointed by the agreement of both parties and, if consisting of more than one member, by the balance of members' differing views.
(b) The Board members, being involved from the start, bring to the project focus, objectivity and an enthusiasm to preserve good working relations between the parties.
(c) Dispute Boards are established by consensus at a time when the parties are focused on agreement. The parties by jointly selecting the Board members accept that the Board members are impartial and independent and have the prerequisite qualification that the parties themselves set.

(d) The benefits of appointing the Dispute Board at the beginning of a project are:
   (i) its members have the opportunity to gain accumulated and intimate knowledge of the project and of its personnel;
   (ii) there is continued familiarisation of the site operations through periodic site visits;
   (iii) the ability to hold discussions regarding progress and any potential conflict; exercising dispute avoidance; enhanced opportunity for Board members to establish rapport and credibility with the parties, which greatly facilitates dispute resolution; and
   (iv) the opportunity for the Board members to get to know and work with each other before a dispute hearing is held.
(e) Trust and confidence in the integrity and ability of the Board members are in no doubt from the beginning of the project. There is, therefore, a greater likelihood that the employer–contractor relationship remains cordial throughout the construction of the project.
(f) Claims and responses are more carefully and realistically prepared than in arbitration, with a higher degree of credibility, as there is normally the desire not to appear frivolous in the eyes of the Board members. The result is that there are fewer spurious claims and fewer rejections, when the claims are meritorious.
(g) Procedural posturing, particularly by the representatives of the parties, sometimes to serve their own purposes, is rarely noticed in practice.
(h) The parties, knowing that sooner or later Board members would be copied with the correspondence relating to disputes, are less inclined to send acrimonious correspondence that can damage relationships. They are aware, possibly subconsciously, of the Board's reaction to such exchanges. The parties' approaches are thus tempered by their perception of the Board's view of their behaviour. Attitudes remain positive not adversarial.
(i) By the employer adopting the Board's mechanism in the bidding documents, tenderers are given a strong indication that fair play will prevail.
(j) The fact that a Dispute Board is in place promotes resolution of disputes by the parties themselves during the consultation process required under sub-clause 3.5 of the conditions of contract. The parties can and will anticipate the type of answer they might receive from the Board in many situations that they might encounter. An apt quotation was made in this connection, taken from John Milton in one of his sonnets of 1673: 'They also serve who only stand and wait'.[26.3] Board members serve even when they stand waiting and become a forceful instrument of dispute avoidance or at least minimising the outbreak of disputes.
(k) When a disagreement or dispute does arise, it is given early attention and addressed contemporaneously. Disagreements are settled as they arise rather than being left to fester and develop into intransigent disputes. This avoids the commonly encountered situation of the engineer being too busy to address a voluminous claim, where an inclination to reject in any event is not unknown, possibly in the hope that such action would make the claim go away. Delays occur which can result in aggravation, acrimony and the development of entrenched views. Opportunities to negotiate a settlement are lost. The Board members can prevent

this by their regular review of progress on claims so that parties' fantasies do not turn into expectations. Issues are isolated and contained, not being allowed to snowball into unmanageable proportions.

(l) Because of the Board members' familiarity with the project and the speed with which a hearing can be organised, facts are better understood by those presenting and adjudicating the dispute. Reconstruction of historical circumstances is greatly reduced. In most projects, senior construction personnel rarely remain after construction activities are complete; even if they remain with the same contractor, they are eager to move to their next interesting job, often depriving the arbitrator or judge of the benefit of their first-hand knowledge of events. With such individuals being on site, greater certainty prevails and the parties are usually content that the material germane to the issue has been revealed. Even unfavourable decisions are more readily accepted in good faith.

(m) Dispute Boards have both proactive and responsive duties: proactive in organising and managing their tasks in such a manner as to produce the most effective result; and responsive in responding to the parties' wishes when seeking either an informal opinion or a formal dispute resolution. Dispute Boards can proactively investigate the background to a dispute and take the initiative in ascertaining the facts and matters required for a decision, unlike arbitrators under common law who are entitled only to passively listen and make a decision based on the adversarial process of arguments between the parties.

(n) For meritorious claims, acceptance of the Board's decision results in earlier payments to the contractor, easing cash-flow difficulties. Furthermore, the employer will know where he stands in respect of his cash flow and budgets. With claims resolved as they arise, completion of the final account is usually quick and retention funds may be released earlier.

(o) Dispute Boards typically adopt a less formal procedure of dispute resolution than that of arbitration.

(p) The confidential and low-key procedures preserve good site relationships, which are vital for the remainder of the project. Furthermore, low-key procedure means minimal lawyer input and maximum technical reasoning. Face-saving settlement options are provided (with the Board being blamed for adverse decisions!) and neither party is seen as having to back down. The 'pay-up or we'll stop work' scenario, guaranteed to put an end to project partnering, is avoided.

(q) The use of Dispute Boards has been found to be a cost-effective, efficient and fast method of settling disputes.

26.3 Background and evolution

Historically, the Dispute Board concept originated in the late 1960s in the United States, in the form of what is now referred to as a Dispute Review Board. Its first recorded use was on Boundary Dam in Washington, where the process was a success.[26.4] At that time, escalating competition for construction work and a claim-conscious society led the US National Committee on Tunneling Technology in 1972 to sponsor a study to develop recommendations for improved contracting methods in the United States. It was a time of downturn in the construction industry, when more disputes began to arise. Employers and contractors began the search for a cost-effective, efficient and fast method of settling disputes. The study report, which was published in 1974, resulted in an increase in the awareness of the high cost of disputes and of their resolution, not only to the construction industry but also to the public in general.[26.5] Such high cost was in both monetary terms and time loss. The study report also provided many recommendations that were adopted by various owners and consulting engineers and affirmed that some mechanism was required to settle disagreements as they arose rather than leave them to fester and develop into intransigent disputes. Thus, the Dispute Review Board (DRB) concept became established.

With its success, the Dispute Review Board mechanism was used in other parts of the construction industry and in 1975 it was used on the construction of the second bore of the Eisenhower Tunnel in Colorado. It was reported that the mechanism was an overwhelming success and that the Board had heard three disputes, with an owner–contractor relationship remaining cordial throughout the construction of the project.[26.6] Other successful Dispute Review Boards followed and in time their use became more popular in the United States, not only in underground structures but also in highway, heavy civil, process and building construction. Statistics published in 2004 put the number of known projects with Dispute Review Boards since 1988 at about 1100, valued at over $75 billion, where approximately 1300 disputes have been heard and resolved.[26.7]

Internationally, the first project that incorporated the use of a Dispute Review Board was the El Cajon Hydro Project in Honduras, in 1980.[26.8] However, since then use has increased as a result of various successes in large construction projects, in various countries, including Australia, Bangladesh, Canada, Denmark, Egypt, Hong Kong, India, Ireland, Italy, People's Republic of China, Lesotho, New Zealand, Poland, Uganda, the UK and the United States of America. The huge capital investment worldwide in construction and engineering projects also added to the increase in the use of Dispute Boards. This investment reached incredible figures and in 2003 the global construction spending was US$3.663 trillion with an expected increase of 4.6% for 2004. The expected increase is much greater in some countries, such as India at 10.1%, China at 8.4% and Brazil also at 8.4%.[26.9] Similar figures were attained in areas other than construction; so for example, global transactions in international banking have increased by over 16 times in a period of 18 years, from US$265 billion to US$4.3 trillion. These levels of transactions could only mean an inherently greater scope for conflict between contracting parties and an increase in the number and size of claims and, eventually, disputes. As in the United States, it became clear that society needs not only appropriate dispute resolution mechanisms but also a method of avoiding disputes in the first place.

The use of Dispute Boards allowed parties to focus on how they might adequately and effectively prevent such disputes from developing in the first place, and from fusing into an intractable mass that is incapable of resolution in a sensible manner. Furthermore, if disputes could not be prevented, then the need was turned to how best to resolve them both quickly and effectively.

As the use of Dispute Boards increased, its numerous advantages, as set out in Section 26.2 above, became more recognised.

In 1995, the use of Dispute Review Boards in international construction received major encouragement when the World Bank, the largest financing agency in the international field, produced the first edition of its standard bidding documents for the procurement of works of civil engineering construction, 'SBDW', and included as one of its mandatory provisions the use of a Dispute Review Board for the resolution of disputes between the employer and the contractor.

The popularity of the Dispute Boards increased further when FIDIC adopted the Dispute Board mechanism in the dispute resolution clause, clause 20, of its Orange Book, issued in 1995 for design/build projects. It is appropriate, however, to mention that the type of Board chosen by FIDIC was not a Dispute Review Board, but a Dispute Adjudication Board. Then, almost immediately, FIDIC published its Supplement to the Fourth Edition of the Red Book in November 1996, by which it provided for the establishment of a Dispute Adjudication Board to replace the engineer's traditional role of a decision maker or quasi-arbitrator in the settlement of disputes. This significant change in FIDIC's policy towards the role of the engineer occurred as a result of the strong criticism in the preceding years of the role of the engineer as adjudicator or quasi-arbitrator. As explained earlier, much of the criticism related to the fact that the engineer was appointed by the employer with the contractor having no say in that appointment.[26.10]

In September 1999, the Dispute Board mechanism was included as the first step in the dispute settlement procedure for the new suite of FIDIC's standard forms of contract conditions, the new Red, Yellow and Silver Books.

More recently, the mechanism has also been endorsed by a number of financial and arbitral institutions, such as international lending agencies other than the World Bank, and more importantly by the International Chamber of Commerce (ICC). The ICC, after two years of study and discussion, adopted in June 2004 the Dispute Board mechanism, as discussed further in Section 26.4.3 below, and published a set of rules to complement its two other major rules: The Arbitration Rules and the ADR Rules.[26.11] The latter were published in 2001 as a further contribution by ICC to the international business community in the field of dispute resolution services.[26.12]

26.4 Types of Dispute Boards

There are three major types of Dispute Board in use at the present time:

- Dispute Review Board;
- Dispute Adjudication Board; and
- Combined Dispute Board.

26.4.1 Dispute Review Board

As explained above, the innovative concept of a Dispute Review Board was to continuously track the progress of a project and assist the parties in resolving any points of contention between them, should they arise. Many felt, and practice confirmed, that in reality the very existence of such a Board during the construction of a project would reduce to a minimum the number of disputes arising.[26.13] The basic principles, objectives and procedures of the concept of Dispute Review Boards in the international field were summarised as follows:[26.14]

(a) The Board members visit the site periodically, but at least three times every year, to keep abreast of construction activities and problems and of any developing potential claims. At the end of each visit, the Board provides a report to both parties.
(b) The terms of reference given to the Board provide, as a general rule, that a claim is referred to the Board only after the engineer's decision and when either of the two parties has expressed its non-acceptance of that decision.

  The first submission to the Board, therefore, should be a written statement of claim by the appellant party, complete with relevant correspondence and other documentation of the appellant's choosing. A copy is provided for each member of the Board and for the other party.

  The respondent is then afforded an opportunity to submit, in writing and copied to all concerned, his reason for rejection of the appellant's claim. A short period of time is allowed for this step. This is followed, within a similar period of time, by the appellant's rebuttal of the respondent's reasons for rejection. The respondent is then allowed a further period in which to file his written rebuttal. This step concludes the submission of written representations to the Board.
(c) The Board may hold hearings, review the project records and take testimony from the parties and their technical representatives. Upon completing its deliberations, the Board issues a non-binding recommendation to the parties. This review of the dispute and recommendation by the Board is a condition precedent to implementation of the contract's dispute clause. No party is obliged to accept the Board's recommendation, in which case the dispute may proceed to another method of dispute resolution, generally arbitration. Even if there is an intermediate step, if the parties (or one of them) do not accept the Board's recommendation, the ultimate solution is arbitration in accordance with the rules selected under the contract.
(d) If neither party sends a written notice to the other party (copied to the DRB) expressing its dissatisfaction with a recommendation within a certain number of days of receiving it, it is deemed to become a binding decision, the parties are required to comply with its terms, and they waive any right of recourse they may have against it.

Therefore, in essence, the process involves an amicable settlement of disputes leaving it to the parties to decide whether or not they want to accept the Board's non-binding recommendation. Thus, it can be said that the adopted concept of a Dispute Review Board was a consensual form of dispute resolution where a recommendation could be a step towards an assessment of the quantum, or towards providing the parties with a certain amount of time before they decided whether or not to take the dispute further in the dispute resolution ladder to an arbitral tribunal or a court.

26.4.2 Dispute Adjudication Board

The role of a Dispute Adjudication Board (DAB) in dispute settlement is neither consensual nor amicable in nature. It is a decision-making role, like that of the traditional engineer under clause 67 of the Fourth Edition of the Red Book. When FIDIC adopted the Dispute Board concept in its 1996 Supplement to the 1992 Fourth Edition of the Red Book, it was in fact reallocating the role of adjudication of disputes, which had belonged until then to the traditional engineer under clause 67.1, to an independent, impartial and neutral Dispute Board. That adjudication role required the traditional engineer to determine disputes of any kind between the employer and the contractor

> 'in connection with, or arising out of, the Contract or the execution of the Works, whether during the execution of the works or after their completion and whether before or after repudiation or other termination of the Contract, including any dispute as to any opinion, instruction, determination, certificate or valuation of the Engineer ...'.

The wording of clause 67.1 of the 1996 Supplement makes clear that the adjudication role is allocated to the Dispute Adjudication Board, which is neither amicable nor one leading to a consensual settlement of disputes, as is the role of a Dispute Review Board.[26.15]

In fact, it was not feasible for the Dispute Adjudication Board under the FIDIC contracts to render a non-binding recommendation. The Board's decision had to have the same effect as the decision of the traditional engineer, i.e. a binding effect on the employer and the contractor. This can be seen from the second paragraph of clause 67.1 of the 1992 Fourth Edition of the Red Book, which provided as follows:

> '... and the Contractor and the Employer shall give effect forthwith to every such decision of the Engineer unless and until the same shall be revised, as hereinafter provided, in an amicable settlement or an arbitral award'.

The same wording appeared in the fourth paragraph of clause 67.2 of the 1996 Supplement to the Fourth Edition of the Red Book. Thus, the board could not act as a Dispute Review Board and had to be named differently: Dispute Adjudication Board. However, the word 'adjudication' was an unfortunate inclusion in the name given to the board, because it confused the activities of the board under the FIDIC contracts with the process of adjudication as it was being developed in the UK.[26.16]

To the same effect and in a separate, and just as important, consideration, the dispute adjudication binding concept stems from FIDIC's traditional view that a binding determination is needed in order to continue the progress of the construction work.

There is, therefore, a very significant difference between the concept of a Dispute Adjudication Board and that of a Dispute Review Board. Whereas a Dispute Review Board delivers a non-binding recommendation, the Dispute Adjudication Board renders a decision that has to be implemented as indicated in the relevant wording of clause 67.2 of the Supplement, referred to above. Technically, it is referred to as a decision that is temporarily binding in that it remains binding unless and *until* it is revised, in an amicable settlement or an arbitral award. This type of Dispute Board is discussed in detail in Section 26.5 below.

26.4.3 Combined Dispute Board

As stated earlier, the International Chamber of Commerce in Paris, ICC, recently developed a third Dispute Board concept.[26.17] It is referred to as a Combined Dispute Board, followed by ICC arbitration, if required. It is intended to give the parties flexibility of choice between a recommendation of a Dispute Review Board and a temporarily binding decision of a Dispute Adjudication Board. It involves the following principles:

(a) All disputes arising out of or in connection with the contract are submitted, in the first instance, to the Combined Dispute Board in accordance with the ICC Dispute Board Rules.
(b) For any given dispute, the Combined Dispute Board delivers a recommendation unless the parties agree that it shall render a decision or it decides to do so upon the request of a party and in accordance with the Rules, see below at the end of this section.
(c) If any party fails to comply with a recommendation or a decision when required to do so, pursuant to the Rules, the other party may refer the failure itself to arbitration under the ICC Rules of Arbitration.
(d) If any party sends a written notice to the other party and to the Combined Dispute Board expressing its dissatisfaction with a recommendation or a decision as provided for in the Rules, or if the Combined Dispute Board does not issue the recommendation or decision within the time limit provided for in the Rules, or if the Combined Dispute Board is disbanded pursuant to the Rules, the dispute shall be finally settled under the ICC Rules of Arbitration.

Therefore, in general terms a referral to a Combined Dispute Board is treated as a request for a non-binding recommendation. In certain circumstances the parties may, by agreement, find it necessary to request the Combined Dispute Board to render a temporarily binding decision. However, where the parties differ in their approach to the type of determination they require, a party may request the Combined Dispute Board to render such a decision and the Board will have to decide on the appropriateness of rendering a decision rather than a recommendation, taking into consideration whether there is an important or urgent matter that is interfering or threatening to interfere with the progress of the works or the fulfilment of the parties' contractual obligations. This mechanism is an intermediate approach between that of the Dispute Review Board and the Dispute Adjudication Board.

26.5 Varieties of Dispute Boards

Within the three types of Dispute Boards mentioned in Section 26.4 above, there are varieties common between all of the three types, as in sub-sections 26.5.1 and 26.5.2

below. There are also different varieties that are peculiar to each type of the Dispute Boards mentioned above as in sub-section 26.5.3 below. For anyone who is involved in the setting up of a Dispute Board for a particular contract or project, it is imperative to realise their characteristics, advantages and disadvantages or shortcomings, since some boards that have been used with success in certain circumstances, may prove unsuccessful in others. The varieties that are common to all three types of Dispute Boards are as follows.

26.5.1   Varieties based on the number of Dispute Board members

A Dispute Board may be composed of one or more members depending on the size and complexity of the project. The World Bank provides certain guidelines which have to be followed on any World Bank funded project. The Bank requires a three-member Board for projects over US$50 million and either a single or a three-member Board for smaller projects. In general, however, there are other varieties, as follows:

(1) Single-member Dispute Boards

The Dispute Board mechanism can be operated successfully with only one member when both parties are comfortable with one trusted individual who is available and who possesses all the required qualifications, or when the project is too small in size and value to merit a three-member Dispute Board.[26.18] However, it might not be so easy to find such an individual and even if found, any determination rendered by him or her would be the result of a single viewpoint and individual experience.

(2) Two-member Dispute Boards

The parties sometimes prefer, from cost and effective management points of view, a two-member Dispute Board. In such a case, the two members would be of different backgrounds and qualifications, for example one technical member and the other legally qualified. Where the Board members fail to reach a unanimous decision, the casting vote is allocated in accordance with and on the basis of the type of dispute before the Board. In such a case the two members will have to be appointed jointly by the parties to prevent any feeling of allegiance to the appointing party.

(3) Three-member Dispute Boards

This is the most common variety of Dispute Boards in international construction, with the employer and the contractor each selecting one member of the Board. The chairperson of the Board, being the third member, is then selected either by the two board members or jointly by the parties. This variety provides the most balanced decision-making process and despite its higher cost is perhaps the most appropriate variety of Dispute Boards for medium and large-size projects.

(4) More than three-member Dispute Boards

A Dispute Board with more than three members may be used on very large, complex projects with long periods of construction and many construction disciplines, for example an airport or a hydroelectric power plant. The main reason for this variety is usually the large number of disciplines involved in the project, and it is particularly useful where specific areas of engineering expertise form a major role in the project. Two recent examples of this type of Board are Hong Kong Airport with a seven-member Board[26.19] and the Channel Tunnel Rail Link project which had two panels, a construction panel and a financial panel. The members of such Boards are selected for their relevant expertise and experience in the different construction disciplines of the project and, as in the smaller Board, the chairperson is either selected by the already appointed Board members or jointly by the parties. The rights and obligations of all Board members are usually the same, but it is the chairperson who selects, amongst the Board members, those who should be called upon, usually three, to determine any particular dispute, depending on their relevant disciplines. The obvious disadvantage of this variety of Dispute Board is the increased cost, but although there is an evident increase in the total cost of providing a larger Board, such increase in cost, which is directly related to the increase of Board members employed, i.e. 133% when the number increases from three to seven, would normally be much lower than the increase in the cost of the project that necessitates more than three members, from say US$50million to US$20billion.[26.20] The increase in the cost of the project in this example is 39,000%.

26.5.2 Varieties based on the date of appointment of the Dispute Board members

(1) Appointment linked to a specific date in the contract

If all the benefits of the Dispute Board mechanism are to be gained during the construction of a project, then a Dispute Board should be appointed at the commencement of the contract, referred to usually as a Standing Dispute Board. The later such an appointment is made, the lesser are the benefits, see Section 26.2 above.

In certain contracts, the date of appointment of the Dispute Board is stated other than in the conditions of contract, such as the case in the 1999 Red Book, where the parties are required to jointly appoint the Dispute Board members by a date stated in the Appendix to Tender. In this variety, the appointment of the Dispute Board is generally made for the whole period of the contract.

(2) Appointment linked to when a dispute has arisen, ad hoc Dispute Board

In certain circumstances, the Dispute Board mechanism is initiated only after a dispute has arisen, as provided in the 1999 Yellow Book:

> 'by the date 28 days after a Party gives notice to the other Party of its intention to refer a dispute to a DAB in accordance with Sub-Clause 20.4.'

In this variety, referred to as ad hoc Dispute Board, unless otherwise agreed by the parties, the term of appointment of the Dispute Board expires when it has given its decision on the dispute referred to it under the relevant clause of the contract. Most of those experienced in this type of work agree that this variety of Dispute Board dissipates the very essence of the advantages offered by the Dispute Board mechanism, as stated in variety 26.5.2(1) above. In particular, it limits the opportunity for Board members to establish rapport and credibility with the parties and for the Board members to get to know and work with each other before a dispute hearing is held. The only reason given for adopting this variety is a saving in costs of maintaining a Dispute Board throughout the period of construction. This saving is in reality very much false economy. In some cases, a Dispute Board is selected on a standby basis until called upon to determine a dispute. Thus, the Board would be named in the contract but not called upon to act until a dispute has arisen, saving the time taken to agree on the identity of the Board. Once again, in such circumstances most of the benefits of the concept of Dispute Boards are lost.

26.5.3 Varieties peculiar to the type of Dispute Board

For each type of Dispute Board, there are varieties peculiar to that type. For the Dispute Review Board, these varieties include the possibility that if the recommendation is not accepted, the role of the Board is transformed into a conciliator's role, either under an already established multi-tier dispute resolution clause or otherwise. For the Dispute Adjudication Board, the varieties include the possibility of appointing the engineer under the contract as the sole Dispute Board member, as provided in the Guidance Notes for the Preparation of Particular Conditions of Contract of the 1999 FIDIC Red Book.[26.21]

26.6 Dispute Adjudication Boards under the FIDIC contracts

The 1996 Supplement to the 1992 Fourth Edition of the Red Book introduced for the first time the concept of Dispute Adjudication Board into the Red Book. The Supplement was issued in three sections. Section A of the Supplement, entitled 'Dispute Adjudication Board', is intended, if used, to replace clause 67 of the 1992 Fourth Edition of the Red Book in its entirety by incorporating the use of a Dispute Adjudication Board. Section A also includes the following documents:

(1) Guide to Amended Clause 67;
(2) Terms of appointment for a Board of three members; and
(3) Procedural Rules of the Dispute Adjudication Board (of three members); and of one member.

Details of Section A of the Supplement are explained in Chapter 19, Section 19.8 above. As can be seen, the provisions of the 1999 FIDIC forms of contract for Dispute Adjudication Boards are totally different from those for Section A of the Supplement. Therefore, the remaining part of this chapter will be devoted to these provisions of the 1999 FIDIC forms of contract.

Each of the 1999 FIDIC forms of contract using the Dispute Adjudication Board mechanism, i.e. the 1999 Red Book, the 1999 Yellow Book and the 1999 Silver Book, refer to the Dispute Adjudication Boards in six parts, as follows:

(a) Clause 20 of the conditions of contract: sub-clauses 20.2, 20.3 and 20.8 deal with the appointment of the Dispute Adjudication Board and its duration; and sub-clause 20.7 deals with failure to comply with the Dispute Adjudication Board's decision;
(b) Appendix to the form of contract entitled 'General Conditions of Dispute Adjudication Agreement';[26.22]
(c) Annex to the Appendix to the form of contract, entitled 'Procedural Rules', which contains nine rules in the 1999 Red Book but only six in the 1999 Yellow and Silver Books;
(d) The Guidance Notes for the Preparation of Particular Conditions relating to clause 20;[26.23]
(e) Letter of tender, which includes the Appendix to Tender, except for the 1999 Silver Book where no appendix is included; and
(f) Dispute Adjudication Agreement Form, pages vi and vii.

The above documents differ slightly from one form of contract to the other. The differences in clause 20 are set out in Chapter 27. The differences in the other five parts referred to above emanate from two causes:

(i) The appointment of the members of a Dispute Adjudication Board in the 1999 Red Book is based on the standing variant, whereas it is based on the ad hoc Board variant in the 1999 Yellow and Silver Books. Thus, the following provisions apply in the 1999 Red Book:
— The date from which the dispute adjudication agreement takes effect is the latest of the following dates:
(a) the commencement date, as defined in the contract;
(b) when the employer, the contractor and the member have each signed the agreement; or
(c) when the employer, the contractor and each of the other members (if any) have respectively each signed the agreement.
— A member of the Dispute Adjudication Board who wishes to resign must give not less than 70 days' notice of resignation to the employer and to the contractor, and the dispute adjudication agreement shall terminate upon the expiry of such period.
— Each member of the Dispute Adjudication Board is obliged to become conversant with the contract and with the progress of the works, and of any other parts of the project of which the contract forms part, by studying all documents received, which shall be maintained in a current working file.
— Each member of the Dispute Adjudication Board is obliged to be available to give an opinion on any matter relevant to the contract when requested by both

the employer and the contractor, subject to the agreement of the other members, if any.
— The employer and the contractor are obliged to provide each member with appropriate security for a sum equivalent to the reasonable expenses to be incurred by that member whenever they refer a dispute to the Dispute Adjudication Board (under sub-clause 20.4 of the conditions of contract).
— There are also specific provisions for payment, termination and disputes.
(ii) References to 'the Engineer' in the 1999 Red and Yellow Books are deleted in the Silver Book and replaced, where appropriate by 'the Employer's Representative'.

In the following sections of this chapter, the various aspects of the Dispute Adjudication Board mechanisms under the various FIDIC forms are discussed.

26.7 The role of the Dispute Adjudication Board

The duties of the Dispute Adjudication Board members will normally begin once their appointment has been confirmed and their agreements signed. Their main role is two-fold:

(a) to prevent disagreements from becoming disputes by providing an opinion, if the parties jointly refer a matter for the Dispute Adjudication Board or if they consult it on any matter under the seventh paragraph of sub-clause 20.2 of the conditions of contract, see Section 26.12 below; and
(b) to resolve disputes (of any kind) that may arise at any stage throughout the duration of the project between the parties in connection with, or arising out of, the contract or the execution of the works, including any dispute as to any certificate, determination, instruction, opinion or valuation of the engineer, under sub-clause 20.4 of the conditions of contract, see Section 26.13 below.

Thus, the role of the Dispute Adjudication Board includes both proactive and responsive duties: proactive in organising and managing its tasks in such a manner as to produce the most effective result; and responsive in responding to the parties' wishes when seeking either an informal opinion or a formal dispute resolution.

In order to perform and fulfil its obligations and its role effectively within the short time prescribed in the contract, it is important that the Dispute Adjudication Board follows the progress of the project on a step-by-step basis through regular updates and reports from the parties, usually on a monthly basis, regular site visits and meetings. The main purpose of site visits is to enable the Dispute Adjudication Board to become and remain acquainted with the progress of the works and of any actual or potential problems or claims, but it could also serve as a predetermined time for holding special meetings with the parties or formal hearings, see Section 26.10.

The role and duties of the Dispute Adjudication Board normally end when the date of discharge, as referred to in sub-clause 14.12 of the conditions of contract, becomes effective.

26.8 Establishment of the Dispute Adjudication Board

As stated above, and it is worth repeating here, for a Dispute Adjudication Board to give its best service it should be appointed from the commencement of the contract, whether the works are being undertaken on site or off site. The difference between an appointment for a project where construction work commences first on site and that where most of the work is done initially off site, such as in a mechanical/electrical project, may be reflected in the level of the retainer fees, which are discussed in Section 26.14 below. Therefore, it is suggested that an ad hoc Dispute Adjudication Board should not be used in preference to a standing Dispute Adjudication Board if the only reason for doing so is the saving in costs of maintaining a Dispute Adjudication Board on the project. It is extremely useful to be able to view the events as they happen and to witness the technical and physical conditions on the site, instead of assembling factual evidence, which is said, in a hearing, to have occurred many years earlier. Therefore, the appropriate and logical time for a Dispute Adjudication Board to be set up is as stated in the Appendix to Tender of the 1999 Edition of the Red Book, and not as an ad hoc Dispute Adjudication Board, as in the 1999 edition of the Yellow and Silver Books. The former allows the Dispute Adjudication Board to follow the contract step by step and to be able to prevent disputes or resolve them as soon as they arise.

Establishing the Dispute Adjudication Board entails a process whereby the members are selected and agreed or approved by the parties. It is a necessary requirement in the General Conditions of Dispute Adjudication Agreement and in the Procedural Rules[26.24] that the Dispute Adjudication Board members should be independent and impartial within the full meaning of these characteristics. They should also be seen to be acting independently and impartially at all times.[26.25] A member of a Dispute Adjudication Board should of course serve both parties equally and impartially, irrespective of who appointed him, and should not be selected to, appointed for, or act as a party representative on the Dispute Adjudication Board. It is also important to remember that a member of a Dispute Adjudication Board is not appointed as a consultant and therefore should never attempt to redesign any part of the project or advise the engineer on design aspects or the contractor on methods of construction. Neither should he be approached to do so by either of the parties. Furthermore, by paragraph 2 of sub-clause 20.2 of the Conditions of Contract, a member should be a 'suitably qualified person'. By paragraph 5 of sub-clause 20.2 of the Conditions of Contract, the agreement between the parties and a Board member incorporates, by reference, the General Conditions of Dispute Adjudication Agreement appended to the Conditions of Contract. Condition 4 of these General Conditions places the following onerous obligations on Dispute Adjudication Board members:[26.26]

'The Member shall:
(a) have no interest financial or otherwise in the Employer, the Contractor or the Engineer, nor any financial interest in the Contract except for payment under the Dispute Adjudication Agreement;
(b) not previously have been employed as a consultant or otherwise by the Employer, the Contractor or the Engineer, except in such circumstances as were disclosed

in writing to the Employer and the Contractor before they signed the Dispute Adjudication Agreement;
(c) have disclosed in writing to the Employer, the Contractor and the Other Members (if any), before entering into the Dispute Adjudication Agreement and to his/her best knowledge and recollection, any professional or personal relationships with any director, officer or employee of the Employer, the Contractor or the Engineer, and any previous involvement in the overall project of which the Contract forms part;
(d) not, for the duration of the Dispute Adjudication Agreement, be employed as a consultant or otherwise by the Employer, the Contractor or the Engineer, except as may be agreed in writing by the Employer, the Contractor and the Other Members (if any);
(e) comply with the annexed procedural rules and with sub-clause 20.4 of the Conditions of Contract;
(f) not give advice to the Employer, the Contractor, the Employer's Personnel or the Contractor's Personnel concerning the conduct of the Contract, other than in accordance with the annexed procedural rules;
(g) not while a Member enter into discussions or make any agreement with the Employer, the Contractor or the Engineer regarding employment by any of them, whether as a consultant or otherwise, after ceasing to act under the Dispute Adjudication Agreement;
(h) ensure his/her availability for all site visits and hearings as are necessary;
(i) become conversant with the Contract and with the progress of the Works (and of any other parts of the project of which the Contract forms part) by studying all documents received which shall be maintained in a current working file;
(j) treat the details of the Contract and all the DAB's activities and hearings as private and confidential, and not publish or disclose them without the prior written consent of the Employer, the Contractor and the Other Members (if any); and
(k) be available to give advice and opinions, on any matter relevant to the Contract when requested by both the Employer and the Contractor, subject to the agreement of the Other Members (if any).'

The success of the Dispute Adjudication Board depends on the parties' confidence in the integrity and expertise of the members and in particular in the chairman of the Board. In particular, conditions (i) and (k) must be taken by implication as giving rise to necessary technical qualifications of a prospective member of the Dispute Adjudication Board. To become 'conversant with the Contract and progress of the Works' necessitates certain technical knowledge, experience and educational/professional qualifications in the member. The majority of disagreements brought to the attention of Dispute Adjudication Boards will have some significant technical input and so there is a strong possibility that a member who fails to fully appreciate the technical problems encountered will fail in his duties. To become conversant with the contract also requires each member to have a certain amount of knowledge in contract administration and the ability to comprehend and interpret contractual provisions. Without such qualifications, the parties might lose confidence in the member, the Dispute Adjudication Board or, in the worst case, the process as a whole.

Furthermore, condition (k) implies that each member should be capable of giving advice and opinions on 'any matter relevant to the Contract'. This involves both the technical and contractual aspects of the project, combined with knowledge and experience in the various methods of dispute resolution. In particular, it would be useful if the Board member had the experience of acting as an arbitrator.

The effectiveness of a member's participation in the Board's duties and in the dispute resolution process will be influenced by the personal qualities of conduct that he brings. Therefore, besides the technical and contractual qualifications, such personal qualities of conduct are just as important and necessary for a member to have. These may be implied generally from the procedural rules and the above conditions and in particular from conditions (e) and (k). Open mindedness; respect for others and their opinions no matter how different they are to his own; ability to listen; logical and coherent thinking; abstention from commenting on the design or methods of construction or making suggestions as to how these could have been done better; patience; flexibility; inventiveness in organisation and case management; and ability to work within a team, are all prerequisites to the success of a Dispute Adjudication Board. A successful Board will have a good relationship with all personnel on site and in particular with the engineer or the employer's representative, as the case may be.

As to the chairman of the Dispute Adjudication Board, it is important to emphasise the administrative and management roles he plays. He is responsible for organising and conducting not only hearings but also correspondence, site visits and meetings, and for making arrangements on behalf of the Board in general. The chairman must, therefore, possess managerial skills and aptitude for writing concise and articulate appraisals, reports and opinions. Whether technically or legally qualified, he should be knowledgeable in many of the contractual and technical issues that may arise. He should be prepared to lead discussions, particularly at meetings with the parties or at hearings, and should be articulate but not verbose. He should also be authoritative and possess leadership qualities. In particular he should have the ability and patience to listen carefully and to have an open mind on all matters; to be flexible to consider different options; to be friendly but not casual; to be firm but not overbearing; to be able to conduct his Board as a team striving for unanimity, but at the same time to be able to delegate the work as appropriate to the person best qualified to contribute; and finally he should be capable of spontaneously handling procedural problems that usually arise during a hearing, which are in practice generally of a difficult nature.

In international contracts, the chairman should be mindful and considerate of foreign culture and customs; of the fact that many languages may be spoken; and of the religious differences and beliefs of those involved in the project. Furthermore, it should be remembered that it is customary, but not necessary, for none of the Board members to be of the same nationality as the parties themselves in order to be seen to have impartiality.

Where the number of members of the Dispute Adjudication Board is concerned, sub-clause 20.2 of the 1999 Edition of all the FIDIC forms stipulates that the

'DAB shall comprise, as stated in the Appendix to Tender, either one or three suitably qualified persons ("the members").

If the number is not so stated and the parties do not agree otherwise, the DAB shall comprise three persons.' This sub-clause continues by specifying the method of selection of the Board members, stating:

'each Party shall nominate one member for the approval of the other Party. The Parties shall consult both these members and shall agree upon the third member, who shall be appointed to act as chairman.'

The sub-clause deals also with the possibility that a list of potential members might be included in the contract, in which case

'the members shall be selected from those on the list, other than anyone who is unable or unwilling to accept appointment to the DAB.'

Other than the two methods referred to in sub-clause 20.2, the standard form of contract may be amended so that the members of the Dispute Adjudication Board may be selected by other methods, including the following:

(a) In the first method outlined in sub-clause 20.2, as referred to above, the right to object, with reasons and within reasonable limits, to each other's selection should be included in the process in order to provide total freedom of choice. In some cases, where the proposal is for a single-member Board, a specific name is given in the tender documents, but the right to object in that case ought to be preserved on the basis of a reasoned statement to be made if and when the tender is accepted.
(b) In the second method referred to above in sub-clause 20.2, it may be added that the selection shall be carried out on the basis of a scoring system where each party gives a score to each member on the list and the person with the highest score is automatically appointed.
(c) The parties may each provide a list of a number of potential members from which the other party selects a member and, once the members so selected are appointed, the two appointed members select the chairman of the Board, either from the remaining names on the two lists, or otherwise by agreeing on a person not listed.
(d) The parties may seek the nomination of the Board members from the president or chairman of an appointing authority such as FIDIC or the ICC.
(e) The parties may agree to the selection of the chairman of the Board first and then once appointed, he selects the other members of the Board in consultation with both parties.

It is important to note that if the parties fail to agree on the identity of the Board members, then by sub-clause 20.3 of the conditions of contract 'the appointing entity or official named in the Appendix to Tender shall, upon the request of either or both parties, and after due consultation' make the necessary appointments. The term 'due consultation' means consultation with both parties to the extent required by the nature of the difficulty encountered. This element of the Appendix to Tender should be filled with the utmost care as the success of the Dispute Adjudication Board depends on the parties' confidence in the expertise of the members and in particular in the chairman of the Board who has the duty of conducting the process fairly and firmly including any meetings, visits to site, and any hearings to resolve disputes referred to the Dispute Adjudication Board.

Finally, in this connection, if for one reason or another the appointment of the Dispute Adjudication Board is delayed despite the provision in the contract requiring such appointment to take place within a certain period of time, such delay would be unfortunate for various reasons:

(a) the Board members would have missed part of the construction or erection period;
(b) a dispute or indeed many disputes may already have arisen at this stage and therefore a tense atmosphere might already exist between the parties;
(c) the parties would not have had sufficient knowledge of the procedure to be followed in a dispute adjudication mechanism; and
(d) the Board members would not have dealt with each other and established a teamwork relationship before having to work together formally in hearing a dispute.

26.9 Obligations of the parties and the members of the Dispute Adjudication Board

The obligations of the parties towards the members of the Dispute Adjudication Board, and vice versa, are spread throughout the six parts of each of the 1999 FIDIC forms of contract, as referred to in Section 26.6 above. Dealing with these obligations in the order in which they appear throughout these documents yields 22 obligations.[26.27] The first 11 are obligations of the parties and the remaining 11 are obligations of the members of the Dispute Adjudication Board, as set out below.

26.9.1 Obligations of the parties towards members of the Dispute Adjudication Board

(1) Remuneration

This obligation is dealt with in three of the documents referred to in Section 26.6 above and they should be read together. The sixth paragraph of sub-clause 20.2 of the conditions of contract sets out the obligation of the parties to remunerate Board members, including the remuneration of any expert whom they consult, in equal shares. Clause 6 of the General Conditions of Dispute Adjudication Agreement and the Dispute Adjudication Agreement also deal with the same obligation.

(2) Referral of a disagreement for an opinion

The seventh paragraph of sub-clause 20.2 of the conditions of contract requires each of the parties to refrain from referring a matter to the Dispute Adjudication Board for its opinion or consulting it on any matter without the agreement of the other party. Similarly, under article 5 of the General Conditions of Dispute Adjudication Agreement, the parties and their personnel undertake not to request advice from or consultation with a member of the Dispute Adjudication Board regarding the contract, unless agreement is reached between all of those mentioned to do so and then only in accordance with the documents relevant to the working of the Board.

(3) Replacement to a Board member

The eighth and ninth paragraphs of sub-clause 20.2 of the conditions of contract require the parties to appoint a replacement to a Board member whenever that replacement is required under the relevant documents.

(4) Obligation to maintain presence of Board members

Under the 1999 Red Book, but not the 1999 Yellow and Silver Books, unless agreed otherwise, the final paragraph of sub-clause 20.2 of the conditions of contract places the obligation on the parties to maintain the presence of the Board members until the date of discharge, as referred to in sub-clause 14.12 of the conditions of contract, becomes effective.

(5) Providing additional information, etc.

The parties are required by the third paragraph of sub-clause 20.4 of the conditions of contract to make available promptly to the Board members all such additional information, further access to the site and appropriate facilities as the Dispute Adjudication Board may require for the purposes of making a decision on such dispute. It must be remembered that there is no statutory instrument to regulate the activities of a Dispute Adjudication Board as there is in the case of arbitration. As members are deemed not to be acting as arbitrator(s), the effect is that none of the provisions of the relevant Arbitration Act applies to the work undertaken by a Dispute Adjudication Board.

(6) Binding effect of decisions of the Dispute Adjudication Board

As the decision of the Board is binding on the parties, they are required to give effect promptly to it unless and until it shall be revised in an amicable settlement or an arbitral award as described in the contract conditions. Furthermore, unless the contract has already been abandoned, repudiated or terminated, the contractor undertakes to continue to proceed with the works in accordance with the contract.

(7) Undertakings by the parties towards the Board

Under the second paragraph of condition 5 of the General Conditions of Dispute Adjudication Agreement, the parties undertake to each other and to each Board member that the member shall not, except as otherwise agreed in writing by the parties and the Board member and the other members (if any):

(a) be appointed as an arbitrator in any arbitration under the contract;
(b) be called as a witness to give evidence concerning any dispute before arbitrator(s) appointed for any arbitration under the contract; or
(c) be liable for any claims for anything done or omitted in the discharge or purported discharge of the member's functions, unless the act or omission is shown to have been in bad faith.

(8) Indemnity by the parties

Under the third paragraph of condition 5 of the General Conditions of Dispute Adjudication Agreement, the parties have the obligation to jointly and severally indemnify and hold each member harmless against and from claims from which he/she is relieved from liability under the second paragraph of that article.

(9) Advance on costs to be provided by the parties

Under the fourth paragraph of condition 5 of the General Conditions of Dispute Adjudication Agreement, whenever the parties refer a dispute to the Board under sub-clause 20.4 of the conditions of contract, which will require the member to make a site visit and attend a hearing, they have the obligation of providing appropriate financial security for a sum equivalent to the reasonable expenses to be incurred by the member.

(10) Attendance at and arrangements of site visits

Under the third rule of the Procedural Rules, site visits shall be attended by the employer, the contractor and the engineer and are co-ordinated by the employer in co-operation with the contractor. The employer is required to ensure the provision of appropriate conference facilities and secretarial and copying services.

(11) Furnishing of documents to Board members

Under the fourth rule of the Procedural Rules, the employer and the contractor are required to furnish the Board members with one copy of all documents which the Dispute Adjudication Board may request, including contract documents, progress reports, variation instructions, certificates and other documents pertinent to the performance of the contract. All communications between the Board members and either party are required to be copied to the other party. If the Dispute Adjudication Board comprises more than one member, the employer and the contractor shall send copies of the requested documents and any communications to each member of the Board. Furthermore, documents relating to programmes and revisions to them, together with the necessary data and assumptions used in generating these programmes, are of particular importance to the work of the Dispute Adjudication Board.

26.9.2 Obligations of the members of Dispute Adjudication Board towards the parties

(1) Decisions of the Dispute Adjudication Board

The Board members are required by the fourth paragraph of sub-clause 20.4 of the conditions of contract to give their reasoned decision in respect of any dispute referred to them within 84 days after receiving such reference, or within such other period as may be proposed by the Board and approved by both parties.

(2) Notice of resignation

By condition 2 of the General Conditions of Dispute Adjudication Agreement, a Board member wishing to resign is required to give not less than 70 days' notice to the parties.

(3) Duties are not to be assigned or sub-contracted

By condition 2 of the General Conditions of Dispute Adjudication Agreement, a Board member undertakes not to assign or subcontract any part of the Dispute Adjudication Agreement without the prior written agreement of all the parties to it and of the other members (if any).

(4) Impartiality and independence of Board members

By condition 3 of the General Conditions of Dispute Adjudication Agreement, a Board member warrants and agrees that he/she is and shall be impartial and independent of the parties and the engineer. The Board member also undertakes to promptly disclose, to each of the parties and to the other Board members (if any), any fact or circumstance which might appear inconsistent with this warranty and agreement of impartiality and independence.

(5) Information and undertakings by the Board members

A Board member is under the obligation to provide the information and undertakings specified in condition 4 of the General Conditions of Dispute Adjudication Agreement, quoted in Section 26.8 above.

(6) Site visits

Under the first rule of the Procedural Rules, unless otherwise agreed by the parties, the Board members are under the obligation to visit the site at intervals of not more than 140 days, including times of critical construction events, at the request of either of the parties. Further, unless otherwise agreed by them and the Dispute Adjudication Board, the period between consecutive visits shall not be less than 70 days, except as required to convene a hearing as described below. It is also important to note that the Dispute Adjudication Board should visit the site at significant events or when difficult problems occur. Availability for such visits is one of the reasons for payment of the retainer part of the fee structure.

(7) Agenda and timing of site visits

Under the second rule of the Procedural Rules, the timing of and agenda for each site visit shall be as agreed jointly by the Dispute Adjudication Board and the parties. However, in the absence of agreement, the agenda becomes the responsibility of the Board, an example of which is provided in Section 26.10 below.

(8) Report

Under the third rule of the Procedural Rules, the Dispute Adjudication Board is required to prepare at the conclusion of each site visit and before leaving the site, a report on its activities during the visit and should provide the employer and the contractor with copies of that report.

(9) Procedure at hearings

Under the fifth rule of the Procedural Rules, if any dispute is referred to the Dispute Adjudication Board in accordance with sub-clause 20.4 of the conditions of contract, the Dispute Adjudication Board is required to give notice of a decision, and to:

(a) act fairly and impartially as between the employer and the contractor, giving each of them a reasonable opportunity of putting his case and responding to the other's case, and
(b) adopt procedures suitable to the dispute, avoiding unnecessary delay or expense.

(10) Date and place of the hearings

Under the sixth rule of the Procedural Rules, if the Dispute Adjudication Board is to conduct a hearing, the Board is required to decide on the date and place of the hearing and may request the presentation of written documentation and arguments from the employer and the contractor to be made prior to or at the hearing.

(11) Procedures relating to decisions

Under the ninth rule of the Procedural Rules, the Board members are under an obligation not to express any opinion during any hearing concerning the merits of any arguments advanced by the parties. The Dispute Adjudication Board is required to make and give its decision in accordance with sub-clause 20.4 of the conditions of contract, or as otherwise agreed by the parties in writing. Furthermore, if the Dispute Adjudication Board comprises three members:

(a) it shall convene in private after a hearing, in order to have discussions and prepare its decision;
(b) it shall endeavour to reach a unanimous decision, and if this proves to be impossible, the applicable decision shall be made by a majority of the members. The majority members may require the minority member to prepare a written report for submission to the parties; and
(c) if a member fails to attend a meeting or hearing, or to fulfil any required function, the other two members may nevertheless proceed to make a decision, unless:
   (i) either the employer or the contractor does not agree that they do so, or
   (ii) the absent member is the chairman and he/she instructs the other members to not make a decision.

26.10 Powers of the Dispute Adjudication Board

Under the seventh rule of the Procedural Rules, except as otherwise agreed in writing by the parties, the Dispute Adjudication Board shall have power to adopt an inquisitorial procedure; to refuse admission to hearings or audience at hearings to any person other than representatives of the employer, the contractor and the engineer; and to proceed in the absence of any person whom the Board is satisfied had received notice of the hearing; but it shall have discretion to decide whether and to what extent this tripartite power may be exercised.

Furthermore, under the eighth rule of the Procedural Rules, the employer and the contractor empower the Dispute Adjudication Board, amongst other things, to:

(a) establish the procedure to be applied in deciding a dispute;
(b) decide upon the Board's own jurisdiction and the scope of any dispute referred to it;
(c) conduct any hearing as it thinks fit, not being bound by any rules or procedures other than those contained in the contract and the Procedural Rules;
(d) take the initiative in ascertaining the facts and matters required for a decision;
(e) make use of its own specialist knowledge, if any;
(f) decide upon the payment of financing charges in accordance with the contract;
(g) decide upon any provisional relief such as interim or conservatory measures; and
(h) open up, review and revise any certificate, decision, determination, instruction, opinion or valuation of the engineer, relevant to the dispute.

These powers are extremely wide and virtually give the Dispute Adjudication Board power to be the master of all procedural aspects that may arise during the time in which it is carrying out its role. In particular, the words 'amongst other things' in Rule 8 should be noted.

26.11 Procedures relating to site visits and meetings

Once the Dispute Adjudication Board becomes familiar with the contract and the technical aspects of the project, as required under article (i) of the General Conditions of Dispute Adjudication Agreement, it is appropriate to arrange the first site visit and meeting, which would serve as an opportunity for introductions and meeting the personnel on site. Rules 1, 2 and 3 of the Procedural Rules deal with site visits and related arrangements, timing, attendance, agenda, report, etc. In the interests of efficiency and convenience for all involved, typically a Dispute Adjudication Board meeting with the parties' representatives would be arranged to coincide with the date of a site visit.

A typical agenda for the first site visit and meeting may take the format shown in Figure 26.1 below. This typical agenda also serves as a guideline and a pattern for subsequent visits and meetings.

For subsequent site visits, the agenda should be adjusted to include special sections on matters relating to programming; on anticipated or potential problems; and on the status of current disputes and difficulties. On all site visits, the Board members would normally be given a brief progress update at the meeting followed by a site inspection, in particular for those areas where potential difficulties had arisen or are expected to arise. The Board members may also request and convene special sessions with the parties in order to ask questions and obtain additional information. Such sessions are of particular importance in technically complex projects and in practice usually result in clarification of matters that would otherwise become problem areas.

One important procedural aspect that should, if possible, be decided at the very beginning of the project or at the first site visit, is whether or not the Board members could convene such meetings with each of the parties separately, rather than jointly, in the presence of both parties. Experience has shown that separate meetings with each party are extremely useful in allowing the parties to open their minds to the Board members and alert them to possible problems that they may not wish to discuss, or are apprehensive about discussing, in the presence of the other party. The usual objection to this procedure is the fear that a party may not know what is said by the other party to the Board members, in which case they would not be able to provide their own point of view in response. However, this objection would not have much force if the Dispute Board members are experienced and knowledgeable in this field. They would normally check any sensitive statement made to them before treating it as accurate and before depending on it in any future action or inaction. Obviously, separate meetings with the Board members can only take place with the agreement of both parties, but if they do agree such a procedure, it could lead them to establishing a firm base of trust between the Board members and each of the parties. However, this procedure should not be used where a formal referral of a dispute is made under sub-clause 20.4 of the contract

First Meeting of the Dispute Adjudication Board

Project: ....
Date of meeting: ....   Place of meeting: ....   Venue:............
Time at commencement of meeting: ....
Time at end of meeting:........................

Names of attendees:	Position	Affiliation	Contact numbers	Addresses
1. ...	....	....	....	....
2. ...	....	....	....	....
3. ...	....	....	....	....

Item No.	Topic	Action by
1.	Attendance/Introductions	DB
2.	Approval of Draft Agenda	All
3.	Review of DB Role/Conditions of Contract, Guidelines, Procedure and Agreement: (a) Purpose of DB and its advantages (b) Relationship of DB to the contract dispute resolution mechanism (c) Periodic DB meetings and dates (d) Procedure for informal opinions (e) Procedure for formal decisions and hearings (f) Obligations of the parties and the DB	DB
4.	Administrative matters: (a) Details of contact with each party and person in charge (b) Ensure that the necessary formalities are in place (c) DB meeting minutes (d) Copies of periodic progress reports and key contract documents for DB members • List of the personnel involved and their job titles • Invoicing procedures relating to fees and expenses	DB
5.	Project familiarisation briefing: (a) Contractor organisation; key project and personnel and their roles (b) Employer organisation; key project personnel and their roles (c) Description/scope/contract price/major items of work (d) Unique features (e) Schedule: network analysis, contract completion date (f) Milestones and targets	The Parties

6.	Contractor's discussion of
	(a) Current status and problems, if any
	(b) Projected schedule for next three months
	    (i) Design
	    (ii) Construction
	(c) Pending issues, if any
	(d) Anticipated issues and potential problems, if any
		Contractor

7.	Employer's discussion of
	(a) Work progress
	    (i) Design
	    (ii) Construction
	(b) Pending issues (if any)
	(c) Anticipated issues (if any)
	(d) Perceived problem areas
		Employer

8.	Other items as advised beforehand	All
9.	Are meetings required? If so, set date/time of next meeting	All
10.	Site visits	All

Fig. 26.1 Agenda for the first meeting and site visit, to be adjusted for other visits.

conditions for the Board's decision. Both parties should be present throughout the whole process, from the referral of the dispute until the Dispute Adjudication Board's decision, whether in a site inspection, at meetings or at the hearing.

Any matters of concern which either of the parties require to be discussed at a future meeting or site visit should be listed in a letter, sent by the parties to the Dispute Adjudication Board, a reasonable period of time before the visit takes place, together with a list of matters to be discussed, which should be included in the agenda.

It should be noted that, as required under the third rule of the Procedural Rules, the Dispute Adjudication Board should prepare a report at the conclusion of each site visit and before leaving the site. This report should state what had occurred at the site visit and, if appropriate, include suggestions relating to the matters discussed and those to be followed in the future. These suggestions may be procedural, administrative or relate to disagreements on which an opinion had been jointly requested by the parties and rendered by the Dispute Adjudication Board.

In circumstances where a disagreement or a dispute had arisen, a meeting or a hearing as the case may demand could take place on site once the routine aspects of the visit are over. Such meetings and hearings are described below.

26.12 Procedures relating to referral of a matter to the Board for its opinion

Sub-clause 20.2 of the conditions of contract, in its seventh paragraph, provides for an informal joint reference by both parties to the Dispute Adjudication Board for its opinion. It should be noted that the provisions of this sub-clause are initiated by the *joint* reference of the parties, so that the Dispute Adjudication Board has no authority to consider any matter for its opinion which has been referred unilaterally by one of the parties, and indeed neither party is authorised to do so. A matter that has already matured into a dispute cannot be referred to the Board in this manner since a dispute can only be referred to the Board for its decision under sub-clause 20.4 of the Conditions of Contract. Furthermore, in view of condition 5 of the General Conditions of Dispute Adjudication Agreement, which prohibits the parties and their personnel from requesting 'advice from or consultation with' the Board members, 'otherwise than in the normal course of the DAB's activities under the contract . . .', such matter referred to the Board for its opinion would most likely be a disagreement or a difference of opinion, but not a dispute. Unfortunately, there is nothing more said in the FIDIC's documents regarding this important aspect of the role of the Dispute Adjudication Board in preventing matters from escalating into disputes. However, the intention is clear. When matters of contention are still being discussed between the parties, the parties may agree, preferably prior to the reference to the engineer for determination under sub-clause 3.5, to refer such matter to the Dispute Adjudication Board for its opinion. By providing its opinion on the matter in contention or on the disagreement, the Dispute Adjudication Board may throw a revealing light on the rights and obligations of the parties and thus prevent the matter from becoming a dispute. In doing so, the Board can either ask each party to provide a position paper, or request each party to present its case orally at an informal meeting or a hearing. The Board may provide its opinion either in writing or orally, as required by the parties. The Board's opinion is not binding and as it is specifically based on the presentations made, its boundaries are limited by the contents of such presentations and, therefore, the Board would not be precluded from subsequently deviating from that opinion should the matter develop into a dispute, which may then be referred to the Board under sub-clause 20.4 of the contract conditions.

A typical situation for such an opinion would be when there is a disagreement between the contractor and the engineer on the interpretation of a provision in the technical specification or in the contract conditions. The view of the Board members would be extremely helpful in order to clarify the situation and give an objective interpretation from the perspective of wide experience, which would normally prevent the matter from developing into a dispute. Such an opinion is persuasive in nature and not coercive. It would be wise to begin the opinion by saying 'based on the material placed in front of us on this point, the board is of the opinion that it means x, y and z'. In this way, the Board is not restricted from adopting a different view if the matter develops into a full blown dispute.

It is unfortunate that this process, as provided in sub-clause 20.2 of the 1999 Red Book, is not available under the 1999 Yellow and Silver Books, since as stated earlier, the Board appointed under those forms of contract is of the ad hoc variety and as such it is only appointed when a dispute has arisen and therefore the Board is not available to deal with informal referrals. However, the parties should seriously consider changing that ad hoc appointment under these forms to a standby appointment or at least adopt this informal approach whenever the need arises.

This informal process is to a certain extent similar to the process of conciliation/ mediation. Whilst it is of great benefit to the parties, the Board members must be sensitive and avoid usurping the role of the engineer or the professional advisers and must listen and carefully assess the parties' contentions.

26.13 Procedures relating to referral of a dispute to the Board for its decision

If a dispute develops and is clearly a dispute and not just a difference of opinion or a disagreement that has not yet matured into a dispute, then either party may refer it to the Board for resolution under sub-clause 20.4 of the contract conditions. Such referral must be in writing and must be copied to the other party and to the engineer stating that it is made under sub-clause 20.4 of the conditions of contract.

Although the requirements of sub-clause 20.4 of the contract conditions are clearly set out, it is of great concern that in many arbitration proceedings the respondent party claims that the matter referred is outside the jurisdiction of the tribunal as it is a matter that has not yet matured into a dispute. It is therefore worth repeating what was explained in Chapter 16, that a dispute is only generated when a claim by either party is rejected by the engineer and that rejection is rejected. This principle has been maintained and is accepted as recently as 2002.[26.28]

Under the 1999 edition of the Red Book, that process is carried out through sub-clause 3.5 of the contract conditions. A claim is formally accepted or rejected when the engineer makes a 'determination in accordance with the Contract, taking due regard of all relevant circumstances'. Therefore, the steps that should be followed in a formal referral to the Dispute Adjudication Board are as follows:

(1) A claim is submitted to the engineer under a relevant provision of the contract 'or otherwise in connection with the Contract' and in accordance with either sub-clause 20.1 (for contractor's claims) or sub-clause 2.5 (for employer's claims) of the conditions of contract. This submission necessitates a notice to be given either by the employer/engineer to the contractor under sub-clause 2.5 or by the contractor to the engineer under sub-clause 20.1. It should be remembered that a notice must satisfy the provisions of sub-clause 1.3 of the contract conditions, and therefore must be in writing and properly delivered.
(2) The engineer is required to proceed under sub-clause 3.5 to agree or determine the matters claimed, after consulting 'with each party in an endeavour to reach an agreement'.
(3) If no agreement is achieved, the engineer is required to 'make a fair determination in accordance with the Contract, taking due regard of all relevant circumstances'.
(4) The determination of the engineer is required under sub-clause 3.5 to be in the form of a 'notice to both Parties of each . . . determination, with supporting particulars'. Once again the notice should be given as indicated in sub-clause 1.3 of the contract conditions.
(5) It is only if and when the engineer's determination is rejected by either party that a dispute comes into existence between the parties. This is best done by a letter together with a reference to the Dispute Adjudication Board under sub-clause 20.4 of the contract conditions.
(6) The dispute is referred in writing to the Dispute Adjudication Board for its decision, with copies to the other party and the engineer, stating that it is made under sub-clause 20.4. Where the Dispute Adjudication Board is composed of more than one member, the date of receipt of the referral by the chairman of the Dispute Adjudication Board is deemed to be the date of receipt by the Board.
(7) The parties are required to 'promptly make available to the board all additional information, further access to the site, and appropriate facilities, as the board may require for the purposes of making a decision on such dispute'. Considering the limited time available to the Dispute Adjudication Board to make its decision, see below, it is important that the parties react without delay to the Board's requests and co-operate in a timely manner.
(8) The decision of the Board should be made 'within 84 days after receiving' the reference, 'or within such other period as may be proposed by the DAB and approved by both Parties', stating that it is rendered under sub-clause 20.4 of the contract conditions. The decision of the Board should be reasoned and is binding upon the parties, subject to step 9 below.

(9) The parties should 'promptly give effect to it unless and until it shall be revised in an amicable settlement or an arbitral award as described' in the remaining part of clause 20 of the contract conditions.
(10) Dissatisfaction by either party with the Board's decision propagates the other dispute resolution mechanisms in the conditions of contract, i.e. amicable settlement and arbitration. Such dissatisfaction must be notified to the other party within 28 days after receipt of the decision. If the Board 'fails to give its decision within the period of 84 days (or as otherwise approved) after receiving such reference, then either Party may, within 28 days after this period has expired, give notice to the other Party of its dissatisfaction'. This notice of dissatisfaction should state that it is given under sub-clause 20.4 of the conditions of contract, setting out the matter in dispute and the reason(s) for dissatisfaction. Further, neither party is entitled to commence arbitration of a dispute unless a notice of dissatisfaction is given, as provided above.
(11) If no notice of dissatisfaction has been given by either party 'within 28 days after it received the DAB's decision, then the decision shall become final and binding upon both Parties'.

Rules 5 to 9 of the Procedural Rules provide for the procedure when a dispute is referred to the Board. Under Rule 5, the Board must act fairly and impartially providing each party with a reasonable opportunity of putting its case and responding to the other party's case. In order for the Board to reach its decision, and as part of step 7 above, it is usual to conduct an oral hearing as referred to in Rule 6 of the Procedural Rules, either on the application of either party or at the direction of the Board. Under Rules 5(b), 7 and 8 of the Procedural Rules, the Board is empowered to establish the procedure that it believes is suitable to the particular dispute and is not bound by any rules or procedures other than those contained in the contract and in the Procedural Rules. Furthermore, by Rule 7 the Board is empowered to adopt an inquisitorial role and to that extent it lessens the adversarial atmosphere at the hearing. This is a very useful instrument, providing the Board with the power to ask questions and enquire into the causes of the dispute. To facilitate this role at the hearing, the Board might adopt a seating arrangement that places the parties side by side opposite the Board rather than the usual confrontational seating arrangement with the parties facing each other across the table.

The hearing is usually less formal than that in arbitration or in a court action. Each party would have presented a statement of its case with supporting documents a few days before the hearing date, preferably in sequential manner, first by the claimant followed by the respondent. The statement should be in the form of a setting out of the facts and the contractual provisions on which the party's case is based, with reasons as to why these provisions correctly apply to the facts. The Board should allocate time before the hearing to read and comprehend these statements so that any further documents it requires may be requested in time for the hearing. Further, the Board, with the help of these statements and the parties' input, should establish the issues that are to be determined and have them agreed by the parties. The Board may also compile a list of questions they may have to be asked of the parties at the hearing.

Hearings are usually divided between matters of principle and matters of quantum, because in general, matters of quantum are easier to negotiate and settle between reasonable parties once the matters of principle are settled.

At the hearing, the claimant party followed by the respondent party should set out their position clearly, concisely and comprehensively. Following these presentations, the Board members usually ask questions in clarification. Witnesses of fact in support of the claimant are then introduced, followed by any independent experts that may have prepared reports for the hearing. Cross-examination would normally be part of the procedure, but in some cases it could be carried out through the chairman of the Board. The same procedure is then followed by the respondent. Closing statements are usually essential in this brief process in order to ensure that each party has replied to the other party's contentions. An oral hearing is ordinarily held on site for the duration of approximately 1 to 4 days.

Before the hearing is formally closed, the Board would be well advised to establish that each party has had a reasonable opportunity to put its case forward and to respond to the other party's case, which is the requirement referred to under Rule 5(a) of the Procedural Rules.

After the formal closure of the hearing, the Board is required to comply with rule 9 of the Procedural Rules, which provides as follows:

'The DAB shall not express any opinions during any hearing concerning the merits of any arguments advanced by the Parties. Thereafter, the DAB shall make and give its decision in accordance with Sub-Clause 20.4, or as otherwise agreed by the Employer and the Contractor in writing. If the DAB comprises three persons:
(a) it shall convene in private after a hearing, in order to have discussions and prepare its decision;
(b) it shall endeavour to reach a unanimous decision: if this proves impossible the applicable decision shall be made by a majority of the Members, who may require the minority Member to prepare a written report for submission to the Employer and the Contractor; and
(c) if a Member fails to attend a meeting or hearing, or to fulfil any required function, the other two Members may nevertheless proceed to make a decision, unless:
    (i) either the Employer or the Contractor does not agree that they do so, or
    (ii) the absent Member is the chairman and he/she instructs the other Members to not make a decision.'

26.14 Remuneration of the members of the Dispute Adjudication Board

Under sub-clause 20.2 of the contract conditions, the terms of the remuneration of either the sole member or each of the three members of the Dispute Adjudication Board, including the remuneration of any expert whom the Board consults, is mutually agreed upon by the parties when agreeing the terms of appointment of the Board members. Each party is responsible for paying one-half of this remuneration.

Under Condition 6 of the General Conditions of Dispute Adjudication Agreement, payment to the Board members is divided into four categories, as follows:

(a) A retainer fee per calendar month, which shall be considered as payment in full for:
   (i) being available on 28 days' notice for all site visits and hearings;
   (ii) becoming and remaining conversant with all project developments and maintaining relevant files;
   (iii) all office and overhead expenses including secretarial services, photocopying and office supplies incurred in connection with his duties; and
   (iv) all services performed hereunder except those referred to in sub-paragraphs (b) and (c) of this clause.
(b) A daily fee which shall be considered as payment in full for:
   (i) each day or part of a day up to a maximum of two days' travel time in each direction for the journey between the member's home and the site, or another location of a meeting with the other members (if any);
   (ii) each working day on site visits, hearings or preparing decisions; and
   (iii) each day spent reading submissions in preparation for a hearing.
(c) All reasonable expenses incurred in connection with the member's duties, including the cost of telephone calls, courier charges, faxes and telexes, travel expenses, hotel and subsistence costs: a receipt shall be required for each item in excess of 5% of the daily fee referred to in sub-paragraph (b).
(d) Any taxes properly levied in the country on payments made to the member (unless a national or permanent resident of the country).

Furthermore, under the same Condition 6 of the General Conditions of Dispute Adjudication Agreement, the following provisions apply:

(i) The retainer fee is paid with effect from the last day of the calendar month in which the dispute adjudication agreement becomes effective, until the last day of the calendar month in which the taking-over certificate is issued for the whole of the works.
(ii) With effect from the first day of the calendar month following the month in which the taking-over certificate is issued for the whole of the works, the retainer fee shall be reduced by 50%. This reduced fee shall be paid until the first day of the calendar month in which the member resigns or the dispute adjudication agreement is otherwise terminated.
(iii) The retainer and daily fees are as specified in the dispute adjudication agreement. Unless it specifies otherwise, these fees remain fixed for the first 24 calendar months, and shall thereafter be adjusted by agreement between the employer, the contractor and the member, at each anniversary of the date on which the dispute adjudication agreement became effective.
(iv) The member is to submit invoices for payment of the monthly retainer and air fares quarterly in advance. Invoices for other expenses and for daily fees shall be submitted following the conclusion of a site visit or hearing. All invoices shall be accompanied by a brief description of activities performed during the relevant period and should be addressed to the contractor.
(v) The contractor is to pay each of the member's invoices in full within 56 calendar days after receiving each invoice and shall apply to the employer (in the statements

under the contract) for reimbursement of one-half of the amounts of these invoices. The employer shall then pay the contractor in accordance with the contract.
(vi) If the member does not receive payment of the amount due within 70 days after submitting a valid invoice, the member may (i) suspend his services (without notice) until the payment is received, and/or (ii) resign his appointment by giving notice under clause 7.

Unfortunately, neither the General Conditions of Dispute Adjudication Agreement nor the Dispute Adjudication Agreement specify an hourly rate. They only specify a daily rate. This is an omission that sometimes causes a problem, since in a number of situations, particularly in the case of the chairman of the Dispute Adjudication Board, the time devoted would be measurable in hours rather than days and arguments may then result as to how many working hours there are per day. Ideally, this omission should be rectified in the agreement.

In general, the hourly/daily rate used for calculating the fees are the same for all the Board members. However, the chairman, if he were more highly qualified than the other members of the Board, is sometimes paid a higher rate. In this connection, it should be noted that although the rates of pay in various jurisdictions differ greatly, the nationality of the members of the Dispute Adjudication Board should not be taken into account in deciding the applicable hourly/daily rate. However, the chairman would in any case receive a higher payment due to the greater number of hours he would have to devote to his assignment.

The retainer fee normally amounts to a multiple of the daily rates per month, for example the Schedule of Fees issued by the World Bank fixes the multiple at three. The fee is normally paid three to four months in advance. This multiple should vary with the complexity of the project and be related to the conditions and undertakings that a Board member accepts in taking the assignment. Under the 1999 FIDIC Red Book, these conditions and undertakings are onerous as can be seen from conditions (d), (g), (h) and (k) as quoted in Section 26.8 above, which bind the member:

(d) not to be employed as a consultant or otherwise by either party or the engineer, except as may be agreed in writing by them and the other members (if any); presumably this condition includes acting as an arbitrator;
(g) not to enter into discussions or make any agreement with either party or the engineer regarding employment by any of them, whether as a consultant or otherwise, after ceasing to act under the Dispute Adjudication Agreement;
(h) to ensure his availability for all site visits and hearings that are necessary; and
(k) to be available to give advice and opinions on any matter relevant to the contract when requested by both parties, subject to the agreement of the other members (if any).

26.15 Cost of maintaining the members of the Dispute Adjudication Board

The cost of maintaining a Dispute Adjudication Board depends on many factors, but in general terms it includes the direct costs of fees and expenses of the Board members and the indirect costs of the parties' personnel time devoted to dealing with the Board members. However, these costs are much less and therefore compare extremely well with the likely cost of an arbitration case in the international scene. It is reported that an estimate of the cost of Dispute Review Boards may range between 0.05% of the final construction contract cost, for relatively dispute-free projects, to a maximum of 0.25% for difficult projects with disputes. Furthermore, considering only projects that refer disputes to the Board or that had difficult problems, the cost ranges from 0.05% to 0.26% with an average of 0.15% of final construction contract cost, including an average of four dispute recommendations.[26.29] These figures would also apply in the case of Dispute Adjudication Boards.

26.16 The decision of the Dispute Adjudication Board

The sole Board member, or in the case of a more than one member Board, the chairman, is usually entrusted with the task of drafting the formal decision of the Board after deliberation with his co-members, and as required under Rule 9 of the Procedural Rules. In the case of a more than one member Board the first draft of the decision is then discussed at a Board meeting where the members should endeavour to reach a unanimous decision, in fulfilment of Rule 9(b) of the Procedural Rules. From personal experience, the draftsman of the decision should remember the following important points and recommendations:

(a) The Board must address all the issues raised in the parties' submissions and either accept or reject them, in part or totally.
(b) The party who makes an allegation has the burden of proof, on the basis of the balance of probability. If that party fails to discharge its burden, the allegation fails.
(c) In accepting or rejecting an issue raised, the decision must deal with the
   — facts when the evidence in the submissions made are contradictory;
   — interpretation of the contractual provisions of the contract argued in the parties' submissions;
   — any points of law that may be at issue; and
   — quantum, if applicable.
(d) The Board is entitled to draw adverse conclusions if its requests were not complied with.
(e) The Board should be mindful of the possibility that its decision could be rejected by one or both of the parties in dispute and, if that dispute is not resolved in the subsequent amicable dispute resolution mechanism under the contract, it could end up in arbitration. Therefore, the decision should be drafted with the intention of providing a future arbitral tribunal with the maximum assistance.
(f) One of the aims of the decision is to persuade the reader that the arguments made by the parties were understood by the Board, analysed and reasoned, and that a determination was reached for reasons properly explained. Omitting to deal with or not deciding an argument presented by the parties in their oral and written

submissions would suggest that the Board was evading the issue and this is totally unacceptable.

(g) Whilst there is no standard structure or framework for a Board's decision, there should be consistency and uniformity of the structure and framework in a project where many decisions may be rendered. However, it is suggested that such uniformity should use chronology as a criterion. Chronology would entail the structure following the pattern of a story. As Lord Justice Donaldson stated *'Much of the art of giving a judgment lies in telling a story logically, coherently and accurately. That is something that requires skill, but it is not a legal skill and it is not necessarily advanced by legal training . . .'*[26.30]

(h) As in telling a story in a clear presentation, the decision should start at the beginning with an introductory section providing details of the parties and the contract, and then give a section on the background setting forth the facts relevant to the issues dealt with in the decision. This background is usually followed by the contractual provisions that are relevant to those facts and the arguments presented by the parties, with each point argued noted and the response to it from the other party equally presented. Finally, the Board applies its mind to analysing the disputed facts and the argued legal provisions, giving its reasons for the conclusions reached. The dispositive section of the decision, i.e. what the decision orders and directs the parties to do, could be placed either at the beginning of the decision or at the end, noting that it is the section that the parties would want to read first.

(i) Finally, having completed their decision and had it word-processed, the Board members would be well advised to leave it aside overnight. Rarely should a Board decision be released without a review and/or some reflection to correct the last few mistakes. As Chaucer said 'In wikked haste is no profit'.[26.31]

Should it transpire that there will be a majority decision of the Board, and a minority report, the minority report should be first submitted to the majority Board members so that they can take it into consideration when they finalise their majority decision.

Part VI

Comparison between text of the three 1999 Major Books: Red, Yellow and Silver Books

Chapter 27

A Precise Record of the Alterations, Omissions and Additions in the 1999 Yellow and Silver Books as compared with the 1999 Red Book

The wording of the 1999 three major Books is very similar in that wherever possible the same wording was used in all three forms of contract. However, the similarity in wording conceals essential differences between these three forms necessary because of the very different concepts for which these forms were drafted.

In order to explain these differences and the meaning of the various clauses of the three forms, FIDIC published an excellent Guide in connection with the 1999 forms of contract.[27.1] It is a pity, however, that despite its considerable size, the Guide does not provide a comparative analysis of the text of these forms and does not denote the differences that exist between one form and another. The commentary was, in the main, confined to the fundamental changes implemented. A table is set out below using the 1999 Red Book as a basis to which any difference in the text of the 1999 Yellow and Silver Books is marked by adopting the following method:

(a) The text of each paragraph whose wording differs in the three major Books is shown in the table below, either in full or in part depending on the extent of the differences;
(b) The text of paragraphs of similar wording in the three books is not shown in the table;
(c) Wording altered from that in the 1999 Red Book in both the Yellow and Silver Books is shown in **bold** and *italics*, under all three Books;
(d) Any word in the 1999 Red Book that was omitted in either the 1999 Yellow or the 1999 Silver Books is shown <u>underlined</u> in the 1999 Red Book and the location of that omission is marked by superimposing the symbol ▲ in the location of the omitted text of the Yellow and Silver Books;
(e) New words added to the text of either of the 1999 Yellow or the 1999 Silver Books are shown in ***bold*** and ***italics*** and the symbol ▲ is inserted in the relevant location of the 1999 Red Book; and
(f) Where the text of either the 1999 Yellow or the 1999 Silver Books is the same as that in the 1999 Red Book, it is shown as 'No change'.

Note that as bold text is used in the original text of the three forms of contract, bold lettering on its own is not part of the identification markings used above.

No commentary is given in this part of the book. Instead, the reader should use the Guide published by FIDIC[27.1] and Chapter 9 of this book for complete understanding of what has been done.

A Precise Record of Alterations, Omissions and Additions

Clause/ sub-clause	1999 Red Book	1999 Yellow Book	The Silver Book
1.1.1	The Contract	No change	No change
1.1.1.1	'Contract' means the Contract Agreement, the Letter of Acceptance, the Letter of Tender, these Conditions, the Specification, the Drawings, the Schedules, and the further documents (if any) which are listed in the Contract Agreement or in the Letter of Acceptance.	1.1.1.1 'Contract' means the Contract Agreement, the Letter of Acceptance, the Letter of Tender, these Conditions, the Employer's Requirements, the Schedules, the Contractor's Proposal, and the further documents (if any) which are listed in the Contract Agreement or in the Letter of Acceptance.	1.1.1.1 'Contract' means the Contract Agreement, these Conditions, the Employer's Requirements, the Tender, and the further documents (if any) which are listed in the Contract Agreement.
1.1.1.2	'Contract Agreement' means the contract agreement (if any) referred to in Sub-Clause 1.6 [Contract Agreement]. ▲	1.1.1.2 'Contract Agreement' means the contract agreement (if any) referred to in Sub-Clause 1.6 [Contract Agreement]. ▲	1.1.1.2 'Contract Agreement' means the contract agreement ▲ referred to in Sub-Clause 1.6 [Contract Agreement], including any annexed memoranda.
1.1.1.3	'Letter of Acceptance' means the letter of formal acceptance, signed by the Employer, of the Letter of Tender, including any annexed memoranda comprising agreements between and signed by both Parties. If there is no such letter of acceptance, the expression 'Letter of Acceptance' means the Contract Agreement and the date of issuing or receiving the Letter of Acceptance means the date of signing the Contract Agreement.	1.1.1.3 'Letter of Acceptance' No change	No equivalent text
1.1.1.4	'Letter of Tender' means the document entitled letter of tender, which was completed by the Contractor and includes the signed offer to the Employer for the Works.	1.1.1.4 'Letter of Tender' No change	No equivalent text

Clause/sub-clause	1999 Red Book	1999 Yellow Book	The Silver Book
1.1.1.5	'**Specification**' means the document entitled specification, as included in the Contract, and any additions and modifications to the specification in accordance with the Contract. Such document specifies the Works.	**1.1.1.5** '**Employer's Requirements**' *means the document entitled employer's requirements, as included in the Contract, and any additions and modifications to such document in accordance with the Contract. Such document specifies the purpose, scope, and/or design and/or other technical criteria, for the Works.*	1.1.1.3 '**Employer's Requirements**' *means the document entitled employer's requirements, as included in the Contract, and any additions and modifications to such document in accordance with the Contract. Such document specifies the purpose, scope, and/or design and/or other technical criteria, for the Works.*
1.1.1.6	'**Drawings**' means the drawings of the Works, as included in the Contract, and any additional and modified drawings issued by (or on behalf of) the Employer in accordance with the Contract.	*No equivalent text*	*No equivalent text*
1.1.1.7	'**Schedules**' means the document(s) entitled schedules, completed by the Contractor and submitted with the Letter of Tender, as included in the Contract. Such document may include the Bill of Quantities data, lists, and schedules of rates and/or prices.	**1.1.1.6** '**Schedules**' *means the document(s) entitled schedules, completed by the Contractor and submitted with the Letter of Tender, as included in the Contract. Such document may include data, lists and schedules of **payments** and/or prices.*	*No equivalent text*
	No equivalent text	**1.1.1.7** '**Contractor's Proposal**' *means the document entitled proposal, which the Contractor submitted with the Letter of Tender, as included in the Contract. Such document may include the Contractor's preliminary design.*	*No equivalent text*
1.1.1.8	'**Tender**' means the Letter of Tender and all other documents which the Contractor submitted with the Letter of Tender, as included in the Contract.	**1.1.1.8** '**Tender**' No change	1.1.1.4 '**Tender**' *means the Contractor's signed offer for the Works and all other documents which the Contractor submitted therewith (other than these Conditions and the Employer's Requirements, if so submitted), as included in the Contract.*

A Precise Record of Alterations, Omissions and Additions 641

1.1.1.10	**'Bill of Quantities'** and **'Daywork Schedule'** mean the documents so named (if any) which are comprised in the Schedules.	**1.1.1.5 'Performance Guarantees'** and **'Schedule of Payments'** mean the documents so named (if any), as included in the Contract.
1.1.2	Parties And Persons	No change
1.1.2.3	**'Contractor'** means the person(s) named as contractor in the Letter of Tender accepted by the Employer and the legal successors in title to this person(s).	**1.1.2.3 'Contractor'** No change
1.1.2.4	**'Engineer'** means the person appointed by the Employer to act as the Engineer for the purposes of the Contract and named in the Appendix to Tender, or other person appointed from time to time by the Employer and notified to the Contractor under Sub-Clause 3.4 *[Replacement of the Engineer]*.	**1.1.2.4 'Employer's Representative'** means the person named by the Employer in the Contract or appointed from time to time by the Employer under Sub-Clause 3.1 *[The Employer's Representative]*, who acts on behalf of the Employer.
1.1.2.6	**'Employer's Personnel'** means the Engineer, the assistants referred to in Sub-Clause 3.2 *[Delegation by the Engineer]* and all other staff, labour and other employees of the Engineer and of the Employer; and any other personnel notified to the Contractor, by the Employer or the Engineer, as Employer's Personnel.	**1.1.2.5 'Employer's Personnel'** No change
		1.1.2.6 'Employer's Personnel' means the Employer's Representative, the assistants referred to in Sub-Clause 3.2 *[Other Employer's Personnel]* and all other staff, labour and other employees of the Employer and of the Employer's Representative; and any other personnel notified to the Contractor, by the Employer or the Employer's Representative, as Employer's Personnel.
1.1.3	Dates, Tests, Periods and Completion	No change
1.1.3.2	**'Commencement Date'** means the date notified under Sub-Clause 8.1 *[Commencement of Works]*.	**1.1.3.2 'Commencement Date'** means the date notified under Sub-Clause 8.1 *[Commencement of Works]*, unless otherwise defined in the Contract Agreement.

Clause/sub-clause	1999 Red Book	1999 Yellow Book	The Silver Book
1.1.3.6	'Tests after Completion' means the tests (if any) which are specified in the Contract and which are carried out *in accordance with the provisions of the Particular Conditions after the Works or a Section* (as the case may be) are taken over by the Employer.	1.1.3.6 'Tests after Completion' means the tests (if any) which are specified in the Contract and which are carried out *under Clause 12 [Tests after Completion]* after the Works or a Section (as the case may be) are taken over by the Employer.	1.1.3.6 'Tests after Completion' means the tests (if any) which are specified in the Contract and which are carried out *under Clause 12 [Tests after Completion]* after the Works or a Section (as the case may be) are taken over by the Employer.
1.1.3.7	'Defects Notification Period' means the period for notifying defects in the Works or a Section (as the case may be) under Sub-Clause 11.1 [Completion of Outstanding Work and Remedying Defects], as stated in the *Appendix to Tender* (with any extension under Sub-Clause 11.3 [Extension of Defects Notification Period]), calculated from the date on which the Works or Section is completed as certified under Sub-Clause 10.1 [Taking Over of the Works and Sections].	1.1.3.7 'Defects Notification Period' No change	1.1.3.7 'Defects Notification Period' means the period for notifying defects in the Works or a Section (as the case may be) under Sub-Clause 11.1 [Completion of Outstanding Work and Remedying Defects], as stated in *the Particular Conditions* (with any extension under Sub-Clause 11.3 [Extension of Defects Notification Period]), calculated from the date on which the Works or Section is completed as certified under Sub-Clause 10.1 [Taking Over of the Works and Sections]. *If no such period is stated in the Particular Conditions, the period shall be one year.*
1.1.4	Money and Payments	No change	No change
1.1.4.1	'Accepted Contract Amount' means the amount accepted in the Letter of Acceptance for the execution and completion of the Works and the remedying of any defects.	1.1.4.1 'Accepted Contract Amount' No change	No equivalent text
1.1.4.2	'Contract Price' means the price defined in Sub-Clause 14.1 [The Contract Price], and includes adjustments in accordance with the Contract.	1.1.4.2 'Contract Price' No change	1.1.4.1 '*Contract Price' means the agreed amount stated in the Contract Agreement for the design, execution and completion of the Works and the remedying of any defects, adjusted (if applicable) in accordance with the Contract.*

A Precise Record of Alterations, Omissions and Additions

1.1.4.3	'**Cost**' means all expenditure reasonably incurred (or to be incurred) by the Contractor, whether on or off the Site, including overhead and similar charges, but does not include profit.	1.1.4.3 '**Cost**' No change	*1.1.4.2 'Cost'* No change
1.1.4.4	'**Final Payment Certificate**' means the payment certificate issued under Sub-Clause 14.13 *[Issue of Final Payment Certificate]*.	1.1.4.4 '**Final Payment Certificate**' No change	*No equivalent text*
1.1.4.5	'**Final Statement**' means the agreed statement defined in Sub-Clause 14.11 *[Application for Final Payment Certificate]*.	1.1.4.5 '**Final Statement**' No change	*1.1.4.3 'Final Statement'* No change
1.1.4.6	'**Foreign Currency**' means a currency in which part (or all) of the Contract Price is payable, but not the Local Currency.	1.1.4.6 '**Foreign Currency**' No change	*1.1.4.4 'Foreign Currency'*
1.1.4.7	'**Interim Payment Certificate**' means a payment certificate issued under Clause 14 *[Contract Price and Payment]*, other than the Final Payment Certificate.	1.1.4.7 '**Interim Payment Certificate**' No change	*No equivalent text*
1.1.4.8	'**Local Currency**' means the currency of the Country.	1.1.4.8 '**Local Currency**' No change	*1.1.4.5 'Local Currency'* No change
1.1.4.9	'**Payment Certificate**' means a payment certificate issued under Clause 14 *[Contract Price and Payment]*.	1.1.4.9 '**Payment Certificate**' No change	*No equivalent text*
1.1.5	Works and Goods	No change	No change
1.1.5.4	'**Permanent Works**' means the permanent works to be ▲ executed by the Contractor under the Contract.	1.1.5.4 '**Permanent Works**' No change	1.1.5.4 '**Permanent Works**' means the permanent works to be *designed and* executed by the Contractor under the Contract.

Clause/sub-clause	1999 Red Book	1999 Yellow Book	The Silver Book
1.1.6	Other Definitions	No change	No change
1.1.6.1	'Contractor's Documents' means the calculations, computer programs and other software, drawings, manuals, models and other documents of a technical nature (if any) supplied by the Contractor under the Contract. ◄	1.1.6.1 'Contractor's Documents' means the calculations, computer programs and other software, drawings, manuals, models and other documents of a technical nature (if any) supplied by the Contractor under the Contract; *as described in Sub-Clause 5.2 [Contractor's Documents]*.	1.1.6.1 'Contractor's Documents' means the calculations, computer programs and other software, drawings, manuals, models and other documents of a technical nature ◄ supplied by the Contractor under the Contract; *as described in Sub-Clause 5.2 [Contractor's Documents]*.
1.1.6.8	'Unforeseeable' means not reasonably foreseeable by an experienced contractor by the date for submission of the Tender.	1.1.6.8 'Unforeseeable' No change	No equivalent text
1.1.6.9	'Variation' means any change to the ◄ Works, which is instructed or approved as a variation under Clause 13 *[Variations and Adjustments]*.	1.1.6.9 'Variation' means any change to the *Employer's Requirements or the* Works, which is instructed or approved as a variation under Clause 13 *[Variations and Adjustments]*.	*1.1.6.8 'Variation'* means any change to the *Employer's Requirements or the* Works, which is instructed or approved as a variation under Clause 13 *[Variations and Adjustments]*.
1.3	Communications	No change	No change
	Wherever these Conditions provide for the giving or issuing of approvals, certificates, consents, determinations, notices and requests, these communications shall be: (a) in writing and delivered by hand (against receipt), sent by mail or courier, or transmitted using any of the agreed systems of electronic transmission as stated in the *Appendix to Tender*; and (b) delivered . . . issued.	No change	Wherever these Conditions provide for the giving or issuing of approvals, certificates, consents, determinations, notices and requests, these communications shall be: (a) in writing and delivered by hand (against receipt), sent by mail or courier, or transmitted using any of the agreed systems of electronic transmission as stated in the *Particular Conditions*; and (b) delivered . . . issued

1.4	Law and Language	No change	No change
	The Contract shall be governed by the law of the country (or other jurisdiction) stated in the **Appendix to Tender**.		The Contract shall be governed by the law of the country (or other jurisdiction) stated in the **Particular Conditions**.
	If there are versions of any part of the Contract which are written in more than one language, the version which is in the ruling language stated in the **Appendix to Tender** shall prevail.		If there are versions of any part of the Contract which are written in more than one language, the version which is in the ruling language stated in the **Particular Conditions** shall prevail.
	The language for communications shall be that stated in the **Appendix to Tender**. If no language is stated there, the language for communications shall be the language in which the Contract (or most of it) is written.		The language for communications shall be that stated in the **Particular Conditions**. If no language is stated there, the language for communications shall be the language in which the Contract (or most of it) is written.
1.5	Priority of Documents	No change	No change
	The documents . . . sequence: (a) the Contract Agreement (if any), (b) the Letter of Acceptance, (c) the Letter of Tender, (d) the Particular Conditions, (e) these General Conditions, **(f) the Specification,** (g) the Drawings, and (h) the Schedules and any other documents forming part of the Contract.	The documents . . . sequence: (a) the Contract Agreement (if any), (b) the Letter of Acceptance, (c) the Letter of Tender, (d) the Particular Conditions, (e) these General Conditions, **(f) the Employer's Requirements,** ◄ **(g) the Schedules, and** **(h) the Contractor's Proposal** and any other documents forming part of the Contract.	The documents . . . sequence: (a) the Contract Agreement ◄ ◄ ◄ (b) the Particular Conditions, (c) these General Conditions, **(d) the Employer's Requirements,** ◄ **(e) the Tender** and any other documents forming part of the Contract.
	If an ambiguity or discrepancy is found in the documents, the Engineer shall issue any necessary clarification or instruction.	No change	No equivalent text

Clause/sub-clause	1999 Red Book	1999 Yellow Book	The Silver Book
1.6	Contract Agreement	No change	No change
	The Parties shall enter into a Contract Agreement within 28 days after the Contractor receives the Letter of Acceptance, unless they otherwise agree. The Contract Agreement shall be based upon the form annexed to the Particular Conditions. The costs of stamp duties and similar charges (if any) imposed by law in connection with entry into the Contract Agreement shall be borne by the Employer.	No change	*The Contract shall come into full force and effect on the date stated in the Contract Agreement. The costs of stamp duties and similar charges (if any) imposed by law in connection with entry into the Contract Agreement shall be borne by the Employer.*
1.8	Care and Supply of Documents	No change	No change
	The Specification and Drawings shall be in the custody and care of the Employer. Unless otherwise stated in the Contract, two copies of the Contract and of each subsequent Drawing shall be supplied to the Contractor, who may make or request further copies at the cost of the Contractor.	*No equivalent text*	*No equivalent text*
	Each of . . . Documents. The Contractor . . . times. If a Party . . . Defect.	No change No change No change	No change No change No change
1.9	***Delayed Drawings or Instructions***	***1.9 Errors in the Employer's Requirements***	***1.9 Confidentiality***
	[Different text in the three Books]	*[Different text in the three Books]*	*[Different text in the three Books]*

1.11	Contractor's Use of Employer's Documents	No change	No change
	As between the Parties, the Employer shall retain the copyright and other intellectual property rights in the Specification, the Drawings and other documents made by (or on behalf of) the Employer. The Contractor may . . . Contract.	As between the Parties, the Employer shall retain the copyright and other intellectual property rights in the *Employer's Requirements* and other documents made by (or on behalf of) the Employer. The Contractor may . . . Contract.	As between the Parties, the Employer shall retain the copyright and other intellectual property rights in the *Employer's Requirements* and other documents made by (or on behalf of) the Employer. The Contractor may . . . Contract.
1.12	Confidential Details	No change	No change
	The Contractor shall disclose all such confidential and other information as the Engineer may reasonably require in order to verify the Contractor's compliance with the Contract.	No change	**The Contractor shall not be required to disclose, to the Employer, any information which the Contractor described in the Tender as being confidential. The Contractor shall disclose any other information which the Employer may reasonably require in order to verify the Contractor's compliance with the Contract.**
1.13	Compliance with Laws	No change	No change
	The Contractor shall, in performing the Contract, comply with applicable Laws. Unless otherwise stated in the Particular Conditions:	The Contractor shall, in performing the Contract, comply with applicable Laws. Unless otherwise stated in the Particular Conditions:	The Contractor shall, in performing the Contract, comply with applicable Laws. Unless otherwise stated in the Particular Conditions:
	(a) the Employer shall have obtained (or shall obtain) the planning, zoning or similar permission for the Permanent Works, and any other permissions described in the Specification as having been (or being) obtained by the Employer; and the Employer shall indemnify and hold the Contractor harmless against and from the consequences of any failure to do so; and	(a) the Employer shall have obtained (or shall obtain) the planning, zoning or similar permission for the Permanent Works, and any other permissions described in the *Employer's Requirements* as having been (or being) obtained by the Employer; and the Employer shall indemnify and hold the Contractor harmless against and from the consequences of any failure to do so; and	(a) the Employer shall have obtained (or shall obtain) the planning, zoning or similar permission for the Permanent Works, and any other permissions described in the *Employer's Requirements* as having been (or being) obtained by the Employer; and the Employer shall indemnify and hold the Contractor harmless against and from the consequences of any failure to do so; and

Clause/sub-clause	1999 Red Book	1999 Yellow Book	The Silver Book
	(b) the Contractor shall give all notices, pay all taxes, duties and fees, and obtain all permits, licences and approvals, as required by the Laws in relation to the execution and completion of the Works and the remedying of any defects; and the Contractor shall indemnify and hold the Employer harmless against and from the consequences of any failure to do so.	(b) the Contractor shall give all notices, pay all taxes, duties and fees, and obtain all permits, licences and approvals, as required by the Laws in relation to the *design*, execution and completion of the Works and the remedying of any defects; and the Contractor shall indemnify and hold the Employer harmless against and from the consequences of any failure to do so.	(b) the Contractor shall give all notices, pay all taxes, duties and fees, and obtain all permits, licences and approvals, as required by the Laws in relation to the *design*, execution and completion of the Works and the remedying of any defects; and the Contractor shall indemnify and hold the Employer harmless against and from the consequences of any failure to do so.
Clause 2	The Employer	The Employer	The Employer
2.1 Right of Access to the Site	The Employer shall give the Contractor right of access to, and possession of, all parts of the Site within the time (or times) stated in the *Appendix to Tender*. The right and possession may not be exclusive to the Contractor. If, under the Contract, the Employer is required to give (to the Contractor) possession of any foundation, structure, plant or means of access, the Employer shall do so in the time and manner stated in the *Specification*. However, the Employer may withhold any such right or possession until the Performance Security has been received.	No change The Employer shall give the Contractor right of access to, and possession of, all parts of the Site within the time (or times) stated in the *Appendix to Tender*. The right and possession may not be exclusive to the Contractor. If, under the Contract, the Employer is required to give (to the Contractor) possession of any foundation, structure, plant or means of access, the Employer shall do so in the time and manner stated in the *Employer's Requirements*. However, the Employer may withhold any such right or possession until the Performance Security has been received.	No change The Employer shall give the Contractor right of access to, and possession of, all parts of the Site within the time (or times) stated in *the Particular Conditions*. The right and possession may not be exclusive to the Contractor. If, under the Contract, the Employer is required to give (to the Contractor) possession of any foundation, structure, plant or means of access, the Employer shall do so in the time and manner stated in the *Employer's Requirements*. However, the Employer may withhold any such right or possession until the Performance Security has been received.

A Precise Record of Alterations, Omissions and Additions 649

	If no such time is stated in the **Appendix to Tender**, the Employer shall give the Contractor right of access to, and possession of, the Site **within such times as may be required to enable the Contractor to proceed in accordance with the programme submitted under Sub-Clause 8.3 [Programme]**.	No change	If no such time is stated in the **Particular Conditions**, the Employer shall give the Contractor right of access to, and possession of, the Site with **effect from the Commencement Date**.
	If the Contractor suffers delay ... Cost or profit	No change	No change
2.5	Employer's Claims	No change	No change
	If the Employer considers himself to be entitled to any payment under any Clause of these Conditions or otherwise in connection with the Contract, and/or to any extension of the Defects Notification Period, **the Employer or the Engineer** shall give notice and particulars to the Contractor. However, notice is not required for payments due under Sub-Clause 4.19 [Electricity, Water and Gas], under Sub-Clause 4.20 [Employer's Equipment and Free-Issue Material], or for other services requested by the Contractor.		If the Employer considers himself to be entitled to any payment under any Clause of these Conditions or otherwise in connection with the Contract, and/or to any extension of the Defects Notification Period, **he** shall give notice and particulars to the Contractor. However, notice is not required for payments due under Sub-Clause 4.19 [Electricity, Water and Gas], under Sub-Clause 4.20 [Employer's Equipment and Free-Issue Material], or for other services requested by the Contractor.
	The Notice ... such period.		The Notice ... such period.
	The particulars ... [Extension of Defects Notification Period].		The particulars ... [Extension of Defects Notification Period].
	This amount may be included as a deduction in the Contract Price and Payment Certificates. The Employer shall only be entitled to set off against or make any deduction from an amount **certified in a Payment Certificate**, or to otherwise claim against the Contractor, in accordance with this Sub-Clause. ▲		**The Employer may deduct this amount from any moneys due, or to become due, to the Contractor.** The Employer shall only be entitled to set off against or make any deduction from an amount **due to the Contractor**, or to otherwise claim against the Contractor, in accordance with this Sub-Clause **or with sub-paragraph (a) and/or (b) of Sub-Clause 14.6 [Interim Payments]**.

Clause/sub-clause	1999 Red Book	1999 Yellow Book	The Silver Book
Clause 3	The Engineer	The Engineer	The Employer's Administration
3.1	Engineer's Duties and Authority	Engineer's Duties and Authority	The Employer's Representative
	The Employer shall appoint the Engineer who shall carry out the duties assigned to him in the Contract. The Engineer's staff shall include suitably qualified engineers and other professionals who are competent to carry out these duties. ▲ The Engineer shall have no authority to amend the Contract. The Engineer may exercise the authority attributable to the Engineer as specified in or necessarily to be implied from the Contract. If the Engineer is required to obtain the approval of the Employer before exercising a specified authority, the requirements shall be as stated in the Particular Conditions. The Employer undertakes not to impose further constraints on the Engineer's authority, except as agreed with the Contractor. However, whenever the Engineer exercises a specified authority for which the Employer's approval is required, then (for the purposes of the Contract) the Employer shall be deemed to have given approval.	No change	The Employer may appoint an Employer's Representative to act on his behalf under the Contract. In this event, he shall give notice to the Contractor of the name, address, duties and authority of the Employer's Representative. ▲ The Employer's Representative shall carry out the duties assigned to him, and shall exercise the authority delegated to him, by the Employer. Unless and until the Employer notifies the Contractor otherwise, the Employer's Representative shall be deemed to have the full authority of the Employer under the Contract, except in respect of Clause 15 [Termination by Employer]. If the Employer wishes to replace any person appointed as Employer's Representative, the Employer shall give the Contractor not less than 14 days' notice of the replacement's name, address, duties and authority, and of the date of appointment.

	Except as otherwise stated in these Conditions: (a) whenever carrying out duties or exercising authority, specified in or implied by the Contract, the Engineer shall be deemed to act for the Employer; (b) the Engineer has no authority to relieve either Party of any duties, obligations or responsibilities under the Contract; and (c) any approval, check, certificate, consent, examination, inspection, instruction, notice, proposal, request, test, or similar act by the Engineer (including absence of disapproval) shall not relieve the Contractor from any responsibility he has under the Contract, including responsibility for errors, omissions, discrepancies and non-compliances.		*[see Sub-Clause 3.3 (a)]*
3.2	*Delegation by the Engineer*	*Delegation by the Engineer*	*Other Employer's Personnel*
	The Engineer may from time to time assign duties and delegate authority to assistants, and may also revoke such assignment or delegation. These assistants may include a resident engineer, and/or independent inspectors appointed to inspect and/or test items of Plant and/or Materials. The assignment, delegation or revocation shall be in writing and shall not take effect until copies have been received by both Parties. However, unless otherwise agreed by both Parties, the Engineer shall not delegate the authority to determine any matter in accordance with Sub-Clause 3.5 *[Determinations]*.	No change	◄ *If the Employer considers himself to be entitled to any payment under any Clause of these Conditions or otherwise in connection with the Contract, and/or to any extension of the Defects Notification Period, the Employer or the Engineer shall give notice and particulars to the Contractor. However, notice is not required for payments due under Sub-Clause 4.19 [Electricity, Water and Gas], under Sub-Clause 4.20 [Employer's Equipment and Free-Issue Material], or for other services requested by the Contractor.*

Clause/ sub-clause	1999 Red Book	1999 Yellow Book	The Silver Book
	Assistants shall . . . [Law and Language]. Each assistant, to whom duties have been assigned or authority has been delegated, shall only be authorised to issue instructions to the Contractor to the extent defined by the delegation. Any approval, check, certificate, consent, examination, inspection, instruction, notice, proposal, request, test, or similar act by an assistant, in accordance with the delegation, shall have the same effect as though the act had been an act of the Engineer. However: (a) any failure to disapprove any work, Plant or Materials shall not constitute approval and shall therefore not prejudice the right of the Engineer to reject the work, Plant or Materials; (b) if the Contractor questions any determination or instruction of an assistant, the Contractor may refer the matter to the Engineer, who shall promptly confirm, reverse or vary the determination or instruction. [Sub-Clause 3.1(c) is comparable with 3.3(a) of the Silver Book]	▲	3.3 **Delegated Persons** *All these persons, including the Employer's Representative and assistants, to whom duties have been assigned or authority has been delegated, shall only be authorised to issue instructions to the Contractor to the extent defined by the delegation. Any approval, check, certificate, consent, examination, inspection, instruction, notice, proposal, request, test, or similar act by a delegated person, in accordance with the delegation, shall have the same effect as though the act had been an act of the Employer. However:* *(a) unless otherwise stated in the delegated person's communication relating to such act, it shall not relieve the Contractor from any responsibility he has under the Contract, including responsibility for errors, omissions, discrepancies and non-compliances;* *(b) any failure to disapprove any work, Plant or Materials shall not constitute approval, and shall therefore not prejudice the right of the Employer to reject the work, Plant or Materials; and* *(c) if the Contractor questions any determination or instruction of a delegated person, the Contractor may refer the matter to the Employer, who shall promptly confirm, reverse or vary the determination or instruction.*

A Precise Record of Alterations, Omissions and Additions

	Instructions of the Engineer	Instructions of the Engineer	3.4 Instructions
3.3	The Engineer may issue to the Contractor (at any time) instructions **and additional or modified Drawings** which may be necessary for the execution of the Works and the remedying of any defects, all in accordance with the Contract. The Contractor shall only take instructions from the Engineer, or from an assistant to whom the appropriate authority has been delegated under this Clause. If an instruction constitutes a Variation, Clause 13 [*Variations and Adjustments*] shall apply. The Contractor shall comply with the instructions given by the Engineer or delegated assistant, on any matter related to the Contract. Whenever practicable, *their* instructions shall be given in writing. If the Engineer or a delegated assistant: (a) gives an oral instruction, (b) receives a written confirmation of the instruction, from (or on behalf of) the Contractor, within two working days after giving the instruction, and (c) does not reply by issuing a written rejection and/or instruction within two working days after receiving the confirmation, then the confirmation shall constitute the written instruction of the Engineer or delegated assistant (as the case may be).	The Engineer may issue to the Contractor (at any time) instructions ◀ which may be necessary for the execution of the Works and the remedying of any defects, all in accordance with the Contract. The Contractor shall only take instructions from the Engineer, or from an assistant to whom the appropriate authority has been delegated under this Clause. If an instruction constitutes a Variation, Clause 13 [*Variations and Adjustments*] shall apply. The Contractor shall comply with the instructions given by the Engineer or delegated assistant, on any matter related to the Contract. ▲*These* instructions shall be given in writing. ◀	*The **Employer** may issue to the Contractor* ◀ *instructions* ◀ *which may be necessary for the **Contractor to perform his obligations under the Contract**. Each instruction shall **be given in writing and shall state the obligations to which it relates and the Sub-Clause (or other term of the Contract) in which the obligations are specified**. If any such instruction constitutes a Variation, Clause 13* [*Variations and Adjustments*] *shall apply.* *The Contractor shall **take instructions from the Employer, or from the Employer's Representative or an assistant to whom the appropriate authority has been delegated under this Clause.***

Clause/sub-clause	1999 Red Book	1999 Yellow Book	The Silver Book
3.4	Replacement of the Engineer	3.4 Replacement of the Engineer	*See last paragraph of Sub-Clause 3.1 above*
	If the Employer intends to replace the Engineer, the Employer shall, not less than 42 days before the intended date of replacement, give notice to the Contractor of the name, address and relevant experience of the intended replacement Engineer. The Employer shall not replace the Engineer with a person against whom the Contractor raises reasonable objection by notice to the Employer, with supporting particulars.	No change	◄
3.5	Determinations	Determinations	Determinations
	Whenever these Conditions provide that the *Engineer* shall proceed in accordance with this Sub-Clause 3.5 to agree or determine any matter, the *Engineer* shall consult with *each Party* in an endeavour to reach agreement. If agreement is not achieved, the *Engineer* shall make a fair determination in accordance with the Contract, taking due regard of all relevant circumstances. The *Engineer* shall give notice to **both Parties** of each agreement or determination, with supporting particulars. Each Party shall give effect to each agreement or determination **unless and until revised under Clause 20 [Claims, Disputes and Arbitration]**.	No change	Whenever these Conditions provide that the *Employer* shall proceed in accordance with this Sub-Clause 3.5 to agree or determine any matter, the *Employer* shall consult with *the Contractor* in an endeavour to reach agreement. If agreement is not achieved, the *Employer* shall make a fair determination in accordance with the Contract, taking due regard of all relevant circumstances. The *Employer* shall give notice to **the Contractor** of each agreement or determination, with supporting particulars. Each Party shall give effect to each agreement or determination, unless **the Contractor gives notice, to the Employer, of his dissatisfaction with a determination within 14 days of receiving it. Either Party may then refer the dispute to the DAB in accordance with Sub-Clause 20.4 [Obtaining Dispute Adjudication Board's Decision]**.

A Precise Record of Alterations, Omissions and Additions 655

Clause 4	The Contractor		
4.1	Contractor's General Obligations	No change	
	The Contractor shall design (to the extent specified in the Contract), execute and complete the Works in accordance with the Contract and with the Engineer's instructions, and shall remedy any defects in the Works. ▲	The Contractor shall design, ▲ execute and complete the Works in accordance with the Contract ▲, *and shall remedy any defects in the Works. When completed, the Works shall be fit for the purposes for which the Works are intended as defined in the Contract.*	The Contractor shall design, ▲ execute and complete the Works in accordance with the Contract ▲, *and shall remedy any defects in the Works. When completed, the Works shall be fit for the purposes for which the Works are intended as defined in the Contract.*
	The Contract shall . . . of defects.	The Contract shall . . . of defects.	The Contract shall . . . of defects.
	▲	*The Works shall include any work which is necessary to satisfy the Employer's Requirements, Contractor's Proposal and Schedules, or is implied by the Contract, and all works which (although not mentioned in the Contract) are necessary for stability or for the completion, or safe and proper operation, of the Works.*	*The Works shall include any work which is necessary to satisfy the Employer's Requirements, Contractor's Proposal and Schedules, or is implied by the Contract, and all works which (although not mentioned in the Contract) are necessary for stability or for the completion, or safe and proper operation, of the Works.*
	The Contractor shall be responsible for the adequacy, stability and safety of all Site operations and of all methods of construction. Except to the extent specified in the Contract, the Contractor (i) shall be responsible for all Contractor's Documents, Temporary Works, and such design of each item of Plant and Materials as is required for the item to be in accordance with the Contract, and (ii) shall not otherwise be responsible for the design or specification of the Permanent Works.	The Contractor shall be responsible for the adequacy, stability and safety of all Site operations, ▲ of all methods of construction *and of all the Works.* ▲	The Contractor shall be responsible for the adequacy, stability and safety of all Site operations, ▲ of all methods of construction *and of all the Works.* ▲

Clause/sub-clause	1999 Red Book	1999 Yellow Book	The Silver Book
	The Contractor shall, whenever required by the Engineer, submit details of the arrangements and methods which the Contractor proposes to adopt for the execution of the Works. No significant alteration to these arrangements and methods shall be made without this having previously been notified to the Engineer. If the Contract specifies that the Contractor shall design any part of the Permanent Works, then unless otherwise stated in the Particular Conditions: (a) the Contractor shall submit to the Engineer the Contractor's Documents for this part in accordance with the procedures specified in the Contract; (b) these Contractor's Documents shall be in accordance with the Specification and Drawings, shall be written in the language for communications defined in Sub-Clause 1.4 [*Law and Language*], and shall include additional information required by the Engineer to add to the Drawings for coordination of each Party's designs; (c) the Contractor shall be responsible for this part and it shall, when the Works are completed, be fit for such purposes for which the part is intended as are specified in the Contract; and	No change ◀	The Contractor shall, whenever required by the *Employer*, submit details of the arrangements and Methods which the Contractor proposes to adopt for the execution of the Works. No significant alteration to these arrangements and methods shall be made without this having previously been notified to the *Employer*. ◀

656 *Comparison of text in major 1999 Books*

A Precise Record of Alterations, Omissions and Additions 657

	(d) prior to the commencement of the Tests on Completion, the Contractor shall submit to the Engineer the 'as-built' documents and operation and maintenance manuals in accordance with the Specification and in sufficient detail for the Employer to operate, maintain, dismantle, reassemble, adjust and repair this part of the Works. Such part shall not be considered to be completed for the purposes of taking-over under Sub-Clause 10.1 *[Taking Over of the Works and Sections]* until these documents and manuals have been submitted to the Engineer.		
4.2	Performance Security	No change	No change
	If an amount is not stated in the *Appendix* to Tender, this Sub-Clause shall not apply.	No change	If an amount is not stated in the ***Particular Conditions*** to Tender, this Sub-Clause shall not apply.
	The Contractor shall deliver the Performance Security to the Employer within 28 days *after receiving **the Letter of Acceptance, and shall send a copy to the Engineer.***	No change	The Contractor shall deliver the Performance Security to the Employer within 28 days after ***both Parties have signed the Contract Agreement.***
	The Employer shall return the Performance Security to the Contractor within 21 days *after receiving **a copy of the Performance Certificate.***	No change	The Employer shall return the Performance Security to the Contractor within 21 days after ***the Contractor has become entitled to receive the Performance Certificate.***

Clause/sub-clause	1999 Red Book	1999 Yellow Book	The Silver Book
4.3	Contractor's Representative	No change	No change
	Unless the Contractor's Representative is named in the Contract, the Contractor shall, prior to the Commencement Date, submit to the *Engineer* for consent the name and particulars of the person the Contractor proposes to appoint as Contractor's Representative.	No change	Unless the Contractor's Representative is named in the Contract, the Contractor shall, prior to the Commencement Date, submit to the *Employer* for consent the name and particulars of the person the Contractor proposes to appoint as Contractor's Representative.
	The Contractor shall not, without the prior consent of the *Engineer*, revoke the appointment of the Contractor's Representative or appoint a replacement.	No change	The Contractor shall not, without the prior consent of the *Employer*, revoke the appointment of the Contractor's Representative or appoint a replacement. ◄
	<u>The whole time of the Contractor's Representative shall be given to directing the Contractor's performance of the Contract. If the Contractor's Representative is to be temporarily absent from the Site during the execution of the Works, a suitable replacement person shall be appointed, subject to the Engineer's prior consent, and the Engineer shall be notified accordingly.</u>	No change	
4.4	Subcontractors		
	The Contractor shall be responsible for the acts or defaults of any Subcontractor, his agents or employees, as if they were the acts or defaults of the Contractor. *Unless otherwise stated in the Particular Conditions:*	No change	The Contractor shall be responsible for the acts or defaults of any Subcontractor, his agents or employees, as if they were the acts or defaults of the Contractor. *Where specified in the Particular Conditions, the Contractor shall give the Employer not less than 28 days' notice of:*

A Precise Record of Alterations, Omissions and Additions 659

	(a) the Contractor shall not be required to obtain consent to suppliers of Materials, or to a subcontract for which the Subcontractor is named in the Contract; (b) *the prior consent of the Engineer shall be obtained to other proposed Subcontractors;* (c) *the Contractor shall give the Engineer not less than 28 days' notice of the intended date of the commencement of each Subcontractor's work, and of the commencement of such work on the Site; and* (d) <u>each subcontract shall include provisions which would entitle the Employer to require the subcontract to be assigned to the Employer under Sub-Clause 4.5 [Assignment of Benefit of Subcontract] (if or when applicable) or in the event of termination under Sub-Clause 15.2 [Termination by Employer].</u>	(a) the Contractor shall not be required to obtain consent to suppliers of Materials, or to a subcontract for which the Subcontractor is named in the Contract; (b) *the prior consent of the Engineer shall be obtained to other proposed Subcontractors; and* (c) *the Contractor shall give the Engineer not less than 28 days' notice of the intended date of the commencement of each Subcontractor's work, and of the commencement of such work on the Site.* ▲	(a) *the intended appointment of the Subcontractor, with detailed particulars which shall include his relevant experience,* (b) *the intended commencement of the Subcontractor's work, and* (c) *the intended commencement of the Subcontractor's work on the Site.* ▲
4.5	Assignment of Benefit of Subcontract	*Nominated Subcontractors*	*Nominated Subcontractors*
	If a Subcontractor's obligations extend beyond the expiry date of the relevant Defects Notification Period and the Engineer, prior to this date, instructs the Contractor to assign the benefit of such obligations to the Employer, then the Contractor shall do so. Unless otherwise stated in the assignment, the Contractor shall have no liability to the Employer for the work carried out by the Subcontractor after the Assignment takes effect.	In this Sub-Clause, 'nominated Subcontractor' means a Subcontractor whom the **Engineer**, under Clause 13 *[Variations and Adjustments]*, instructs the Contractor to employ as a Subcontractor. The Contractor shall not be under any obligation to employ a nominated Subcontractor against whom the Contractor raises reasonable objection by notice to the **Engineer** as soon as practicable, with supporting particulars.	In this Sub-Clause, 'nominated Subcontractor' means a Subcontractor whom the **Employer**, under Clause 13 *[Variations and Adjustments]*, instructs the Contractor to employ as a Subcontractor. The Contractor shall not be under any obligation to employ a nominated Subcontractor against whom the Contractor raises reasonable objection by notice to the **Employer** as soon as practicable, with supporting particulars.

Clause/sub-clause	1999 Red Book	1999 Yellow Book	The Silver Book
4.6	Co-operation	No change	No change
	The Contractor shall, as specified in the Contract or as instructed by the **Engineer**,	No change	The Contractor shall, as specified in the Contract or as instructed by the **Employer**,
	Any such instruction shall constitute a Variation if and to the extent that it causes the Contractor to incur Unforeseeable Cost. ◀	No change	Any such instruction shall constitute a Variation if and to the extent that it causes the Contractor to incur **Cost in an amount which was not reasonably foreseeable by an experienced contractor by the date for submission of the Tender**.
	◀	The Contractor shall be responsible for his Construction activities on the Site, and shall co-ordinate his own activities with those of other contractors to the extent (if any) specified in the Employer's Requirements.	The Contractor shall be responsible for his construction activities on the Site, and shall co-ordinate his own activities with those of other contractors to the extent (if any) specified in the Employer's Requirements.
	the Contractor shall submit such documents to the **Engineer** in the time and manner stated in the **Specification**.	the Contractor shall submit such documents to the **Engineer** in the time and manner stated in the **Employer's Requirements**.	the Contractor shall submit such documents to the **Employer** in the time and manner stated in the **Employer's Requirements**.
4.7	Setting Out	No change	No change
	The Contractor shall set out the Works in relation to original points, lines and levels of reference specified in the Contract or notified by the Engineer.	No change	The Contractor shall set out the Works in relation to original points, lines and levels of reference specified in the Contract. ◀
	The Employer shall be responsible for any errors in these specified or notified items of reference, but the Contractor shall use reasonable efforts to verify their accuracy before they are used.	No change	◀

A Precise Record of Alterations, Omissions and Additions 661

	If the Contractor suffers delay and/or incurs Cost from executing work which was necessitated by an error in these items of reference, and an experienced contractor could not reasonably have discovered such error and avoided this delay and/or Cost, the Contractor shall give notice to the Engineer and shall be entitled subject to Sub-Clause 20.1 *[Contractor's Claims]* to: (a) an extension of time for any such delay, if completion is or will be delayed, under Sub-Clause 8.4 *[Extension of Time for Completion]*, and (b) payment of any such Cost plus reasonable profit, which shall be included in the Contract Price. After receiving this notice, the Engineer shall proceed in accordance with Sub-Clause 3.5 *[Determinations]* to agree or determine (i) whether and (if so) to what extent the error could not reasonably have been discovered, and (ii) the matters described in sub-paragraphs (a) and (b) above related to this extent.		
4.9	Quality Assurance	No change	No change
	Details of all procedures and compliance documents shall be submitted to the *Engineer* for information before each design and execution stage is commenced. When any document of a technical nature is issued to the *Engineer*, evidence of the prior approval by the Contractor himself shall be apparent on the document itself.		Details of all procedures and compliance documents shall be submitted to the *Employer* for information before each design and execution stage is commenced. When any document of a technical nature is issued to the *Employer*, evidence of the prior approval by the Contractor himself shall be apparent on the document itself.

Clause/ sub-clause	1999 Red Book	1999 Yellow Book	The Silver Book
4.10	Site Data	No change	No change
	The Employer shall have made available to the Contractor for his information, prior to the Base Date, all relevant data in the Employer's possession on sub-surface and hydrological conditions at the Site, including environmental aspects. The Employer shall similarly make available to the Contractor all such data which come into the Employer's possession after the Base Date. <u>The Contractor shall be responsible for interpreting all such data.</u>	No change	The Employer shall have made available to the Contractor for his information, prior to the Base Date, all relevant data in the Employer's possession on sub-surface and hydrological conditions at the Site, including environmental aspects. The Employer shall similarly make available to the Contractor all such data which come into the Employer's possession after the Base Date. ▲
	To the extent which was practicable (taking account of cost and time), the Contractor shall be deemed to have obtained all necessary information as to risks, contingencies and other circumstances which may influence or affect the Tender or Works.	No change	***The Contractor shall be responsible for verifying and interpreting all such data. The Employer shall have no responsibility for the accuracy, sufficiency or completeness of such data, except as stated in Sub-Clause 5.1 [General Design Responsibilities].*** ▲
	To the same extent, the Contractor shall be deemed to have inspected and examined the Site, its surroundings, the above data and other available information, and to have been satisfied before submitting the Tender as to all relevant matters, including (without limitation):	No change	
	(a) the form and nature of the Site, including sub-surface conditions, (b) the hydrological and climatic conditions,		

A Precise Record of Alterations, Omissions and Additions

	(c) the extent and nature of the work and Goods necessary for the execution and completion of the Works and the remedying of any defects, (d) the Laws, procedures and labour practices of the Country, and (e) the Contractor's requirements for access, accommodation, facilities, personnel, power, transport, water and other services.	
4.11	The Contractor shall be deemed to: (a) have satisfied himself as to the correctness and sufficiency of the **Accepted Contract Amount**, and (b) <u>have based the Accepted Contract Amount on the data, interpretations, necessary information, inspections, examinations and satisfaction as to all relevant matters referred to in Sub-Clause 4.10 [Site Data]</u>. Unless otherwise stated in the Contract, the **Accepted Contract Amount** covers all the Contractor's obligations under the Contract (including those under Provisional Sums, if any) and all things necessary for the proper design, execution and completion of the Works and the remedying of any defects.	No change The Contractor shall be deemed to have satisfied himself as to the correctness and sufficiency of the **Contract Price**. ◀ Unless otherwise stated in the Contract, the **Contract Price** covers all the Contractor's obligations under the Contract (including those under Provisional Sums, if any) and all things necessary for the proper design, execution and completion of the Works and the remedying of any defects.
4.12	Unforeseeable Physical Conditions	No change Unforeseeable Difficulties
	In this Sub-Clause, 'physical conditions' means natural physical conditions and man-made and other physical obstructions and pollutants, which the Contractor encounters at the Site when executing the Works, including sub-surface and hydrological conditions but excluding climatic conditions.	No change ***Except as otherwise stated in the Contracts:*** ***(a) the Contractor shall be deemed to have obtained all necessary information as to risks, contingencies and other circumstances which may influence or affect the Works;***

Clause/ sub-clause	1999 Red Book	1999 Yellow Book	The Silver Book
	If the Contractor encounters adverse physical conditions which he considers to have been Unforeseeable, the Contractor shall give notice to the Engineer as soon as practicable. This notice shall describe the physical conditions, so that they can be inspected by the Engineer, and shall set out the reasons why the Contractor considers them to be Unforeseeable. The Contractor shall continue executing the Works, using such proper and reasonable measures as are appropriate for the physical conditions, and shall comply with any instructions which the Engineer may give. If an instruction constitutes a Variation, Clause 13 [*Variations and Adjustments*] shall apply. If and to the extent that the Contractor encounters physical conditions which are Unforeseeable, gives such a notice, and suffers delay and/or incurs Cost due to these conditions, the Contractor shall be entitled subject to Sub-Clause 20.1 [*Contractor's Claims*] to: (a) an extension of time for any such delay, if completion is or will be delayed, under Sub-Clause 8.4 [*Extension of Time for Completion*], and (b) payment of any such Cost, which shall be included in the Contract Price.		**(b) by signing the Contract, the Contractor accepts total responsibility for having foreseen all difficulties and costs of successfully completing the Works; and** **(c) the Contract Price shall not be adjusted to take account of any unforeseen difficulties or costs.**

A Precise Record of Alterations, Omissions and Additions 665

After receiving such notice and inspecting and /or Investigating these physical conditions, the Engineer shall proceed in accordance with Sub-Clause 3.5 [*Determinations*] to agree or determine (i) whether and (if so) to what extent these physical conditions were Unforeseeable, and (ii) the matters described in sub-paragraphs (a) and (b) above related to this extent.

However, before additional Cost is finally agreed or determined under sub-paragraph (ii), the Engineer may also review whether other physical conditions in similar parts of the Works (if any) were more favourable than could reasonably have been foreseen when the Contractor submitted the Tender. If and to the extent that these more favourable conditions were encountered, the Engineer may proceed in accordance with Sub-Clause 3.5 [Determinations] to agree or determine the reductions in Cost which were due to these conditions, which may be included (as deductions) in the Contract Price and Payment Certificates. However, the net effect of all adjustments under sub-paragraph (b) and all these reductions, for all the physical conditions encountered in similar parts of the Works, shall not result in a net reduction in the Contract Price.

The Engineer may take account of any evidence of the physical conditions foreseen by the Contractor when submitting the Tender, which may be made available by the Contractor, but shall not be bound by any such evidence.

Clause/sub-clause	1999 Red Book	1999 Yellow Book	The Silver Book
4.17	Contractor's Equipment		
	The Contractor shall not remove from the Site any major items of Contractor's Equipment without the consent of the Engineer. However, consent shall not be required for vehicles transporting Goods or Contractor's Personnel off Site.	No change	◄
4.18	Protection of the Environment		
	The Contractor shall ensure that emissions, surface discharges and effluent from the Contractor's activities shall not exceed the values indicated in the **Specification**, and shall not exceed the values prescribed by applicable Laws.	No change	The Contractor shall ensure that emissions, surface discharges and effluent from the Contractor's activities shall not exceed the values indicated in **the Employer's Requirements**, and shall not exceed the values prescribed by applicable Laws.
4.19	Electricity, Water and Gas		
	The Contractor shall be entitled to use for the purposes of the Works such supplies of electricity, water, gas and other services as may be available on the Site and of which details and prices are given in the **Specification**.	The Contractor shall be entitled to use for the purposes of the Works such supplies of electricity, water, gas and other services as may be available on the Site and of which details and prices are given in the **Employer's Requirements**.	The Contractor shall be entitled to use for the purposes of the Works such supplies of electricity, water, gas and other services as may be available on the Site and of which details and prices are given in the **Employer's Requirements**.
	The quantities consumed and the amounts due (at these prices) for such services shall be agreed or determined by the Engineer in accordance with Sub-Clause 2.5 [Employer's Claims] and Sub-Clause 3.5 [Determinations].	No change	The quantities consumed and the amounts due (at these prices) for such services shall be agreed or determined ◄ in accordance with Sub-Clause 2.5 [Employer's Claims] and Sub-Clause 3.5 [Determinations].

A Precise Record of Alterations, Omissions and Additions 667

4.20	Employer's Equipment and Free-Issue Material		
	The Employer shall make the Employer's Equipment (if any) available for the use of the Contractor in the execution of the Works in accordance with the details, arrangements and prices stated in the **Specification**. Unless otherwise stated in the **Specification**:	The Employer shall make the Employer's Equipment (if any) available for the use of the Contractor in the execution of the Works in accordance with the details, arrangements and prices stated in the *Employer's Requirements*. Unless otherwise stated in the *Employer's Requirements*:	The Employer shall make the Employer's Equipment (if any) available for the use of the Contractor in the execution of the Works in accordance with the details, arrangements and prices stated in the *Employer's Requirements*. Unless otherwise stated in the *Employer's Requirements*:
	The appropriate quantities and the amounts due (at such stated prices) for the use of Employer's Equipment shall be agreed or determined by the Engineer in accordance with Sub-Clause 2.5 *[Employer's Claims]* and Sub-Clause 3.5 *[Determinations]*.	No change	The appropriate quantities and the amounts due (at such stated prices) for the use of Employer's Equipment shall be agreed ▲ or determined in accordance with Sub-Clause 2.5 *[Employer's Claims]* and Sub-Clause 3.5 *[Determinations]*.
	The Employer shall supply, free of charge, the 'free-issue materials' (if any) in accordance with the details stated in the **Specification**.	The Employer shall supply, free of charge, the 'free-issue materials' (if any) in accordance with the details stated in the *Employer's Requirements*.	The Employer shall supply, free of charge, the 'free-issue materials' (if any) in accordance with the details stated in the *Employer's Requirements*.
4.21	Progress Reports		
	Each report shall include:	No change	Each report shall include:
	(a) charts and detailed descriptions of progress, including each stage of design (if any), Contractor's Documents, procurement, manufacture, delivery to Site, construction, erection <u>and</u> testing; *and including these stages for work by each nominated Subcontractor (as defined in Clause 5 [Nominated Subcontractors])*.	(a) charts and detailed descriptions of progress, including each stage of design, Contractor's Documents, procurement, manufacture, delivery to Site, construction, erection, ▲ testing, *commissioning and trial operation*;	(a) charts and detailed descriptions of progress, including each stage of design, Contractor's Documents, procurement, manufacture, delivery to Site, construction, erection, ▲ testing, *commissioning and trial operation*;

Clause/sub-clause	1999 Red Book	1999 Yellow Book	The Silver Book
4.23	Contractor's Operations on Site		
	The Contractor shall confine his operations to the Site, and to any additional areas which may be obtained by the Contractor and agreed by the *Engineer* as working areas.	No change	The Contractor shall confine his operations to the Site, and to any additional areas which may be obtained by the Contractor and agreed by the *Employer* as working areas.
	Upon the issue of a Taking-Over Certificate ▲ the Contractor shall clear away and remove, from that part of the Site and Works to which the Taking-Over Certificate refers, all Contractor's Equipment, surplus material, wreckage, rubbish and Temporary Works.	No change	Upon the issue of the Taking-Over Certificate for the Works, the Contractor shall clear away and remove ▲ all Contractor's Equipment, surplus material, wreckage, rubbish and Temporary Works.
4.24	Fossils		
	The Contractor shall, upon discovery of any such finding, promptly give notice to the *Engineer*, who shall issue instructions for dealing with it. If the Contractor suffers delay and/or incurs Cost from complying with the instructions, the Contractor shall give a further notice to the *Engineer* and shall be entitled subject to Sub-Clause 20.1 *[Contractors Claims]* to:	No change	The Contractor shall, upon discovery of any such finding, promptly give notice to the *Employer*, who shall issue instructions for dealing with it. If the Contractor suffers delay and/or incurs Cost from complying with the instructions, the Contractor shall give a further notice to the *Employer* and shall be entitled subject to Sub-Clause 20.1 *[Contractors Claims]* to:
	After receiving this further notice, the *Engineer* shall proceed in accordance with Sub-Clause 3.5 *[Determinations]* to agree or determine these matters.	No change	After receiving this further notice, the *Employer* shall proceed in accordance with Sub-Clause 3.5 *[Determinations]* to agree or determine these matters.

A Precise Record of Alterations, Omissions and Additions 669

Clause 5	Nominated Subcontractors	Design	Design
5.1	Definition of 'nominated Subcontractor'	General Design Obligations	General Design Obligations
	In the Contract, 'nominated Subcontractor' means a Subcontractor: (a) who is stated in the Contract as being a nominated Subcontractor, or (b) whom the Engineer, under Clause 13 [Variations and Adjustments], instructs the Contractor to employ as a Subcontractor.	The Contractor shall carry out, and be responsible for, the design of the Works. Design shall be prepared by qualified designers who are engineers or other professionals who comply with the criteria (if any) stated in the Employer's Requirements. Unless otherwise stated in the Contract, the Contractor shall submit to the Engineer for consent the name and particulars of each proposed designer and design Subcontractor. The Contractor warrants that he, his Designers and design Subcontractors have the Experience and capability necessary for the design. The Contractor undertakes that the designers shall be available to attend discussions with the Engineer at all reasonable times, until the expiry date of the relevant Defects Notification Period. Upon receiving notice under Sub-Clause 8.1 [Commencement of Works], the Contractor shall scrutinise the Employer's Requirements (including design criteria and calculations, if any) and the items of reference mentioned in Sub-Clause 4.7 [Setting Out]. Within the period stated in the Appendix to Tender, calculated from the Commencement Date, the Contractor shall give notice to the Engineer of any error, fault or other defect found in the Employer's Requirements or these items of reference.	The Contractor shall be deemed to have scrutinised, prior to the Base Date, the Employer's Requirements (including design criteria and calculations, if any). The Contractor shall be responsible for the design of the Works and for the accuracy of such Employer's Requirements (including design criteria and calculations), except as stated below. The Employer shall not be responsible for any error, inaccuracy or omission of any kind in the Employer's Requirements as originally included in the Contract and shall not be deemed to have given any representation of accuracy or completeness of any data or information, except as stated below. Any data or information received by the Contractor, from the Employer or otherwise, shall not relieve the Contractor from his responsibility for the design and execution of the Works. However, the Employer shall be responsible for the correctness of the following portions of the Employer's Requirements and of the following data and information provided by (or on behalf of) the Employer:

Clause/sub-clause	1999 Red Book	1999 Yellow Book	The Silver Book
		After receiving this notice, the Engineer shall determine whether Clause 13 [Variations and Adjustments] shall be applied, and shall give notice to the Contractor accordingly. If and to the extent that (taking account of cost and time) an experienced contractor exercising due care would have discovered the error, fault or other defect when examining the Site and the Employer's Requirements before submitting the Tender, the Time for Completion shall not be extended and the Contract Price shall not be adjusted.	
5.2	*Objection to Nomination*	*Contractor's Documents*	No change
	The Contractor shall not be under any obligation to employ a nominated Subcontractor against whom the Contractor raises reasonable objection by notice to the Engineer as soon as practicable, with supporting particulars. An objection shall be deemed reasonable if it arises from (among Other things) any of the following matters, unless the Employer agrees to indemnify the Contractor against and from the consequences of the matter:	*The Contractor's Documents shall comprise the technical documents specified in the Employer's Requirements, documents required to satisfy all regulatory approvals, and the documents described in Sub-Clause 5.6 [As-Built Documents] and Sub-Clause 5.7 [Operation and Maintenance Manuals]. Unless otherwise stated in the Employer's Requirements, the Contractor's Documents shall be written in the language for Communications defined in Sub-Clause 1.4 [Law and Language].*	*The Contractor's Documents shall comprise the technical documents specified in the Employer's Requirements, documents required to satisfy all regulatory approvals, and the documents described in Sub-Clause 5.6 [As-Built Documents] and Sub-Clause 5.7 [Operation and Maintenance Manuals]. Unless otherwise stated in the Employer's Requirements, the Contractor's Documents shall be written in the language for Communications defined in Sub-Clause 1.4 [Law and Language].*
	(a) there are reasons to believe that the Subcontractor does not have sufficient competence, resources or financial strength;	*The Contractor shall prepare all Contractor's documents, and shall also prepare any other documents necessary to instruct the Contractor's Personnel.*	*The Contractor shall prepare all Contractor's documents, and shall also prepare any other documents necessary to instruct the Contractor's Personnel.* ▲

(b) the subcontract does not specify that the nominated Subcontractor shall indemnify the Contractor against and from any negligence or misuse of Goods by the nominated Subcontractor, his agents and employees; or

(c) the subcontract does not specify that, for the subcontracted work (including design, if any), the nominated Subcontractor shall:

(i) undertake to the Contractor such obligations and liabilities as will enable the Contractor to discharge his obligations and liabilities under the Contract, and

(ii) indemnify the Contractor against and from all obligations and liability arising under or in connection with the Contract and from the consequences of any failure by the Subcontractor to perform these obligations or to fulfil these liabilities. | <u>The Employer's Personnel shall have the right to inspect the preparation of all these documents, wherever they are being prepared.</u>

If the Employer's Requirements specify that some Contractor's Documents are to be submitted to the Engineer for review and/or for approval, they shall be submitted accordingly, together with a notice as described below. In the following provisions of this Sub-Clause, (i) 'review period' means the period required by the Engineer for review and <u>(if so specified) for approval</u>, and (ii) 'Contractor's Documents' exclude any documents which are not specified as being required to be submitted for review <u>and/or for approval</u>.

Unless otherwise stated in the Employer's Requirements, each review period shall not exceed 21 days, calculated from the date on which the Engineer receives a Contractor's Document and the Contractor's notice. This notice shall state that the Contractor's Document is considered ready, both for review <u>(and approval, if so specified)</u> in accordance with this Sub-Clause and for use. The notice shall also state that the Contractor's Document complies with the Contract, or the extent to which it does not comply. | If the Employer's Requirements specify that some Contractor's Documents are to be submitted to the Employer for review, they shall be submitted accordingly, together with a notice as described below. In the following provisions of this Sub-Clause, (i) 'review period' means the period required by the Employer for review ▲ and (ii) 'Contractor's Documents' exclude any documents which are not specified as being required to be submitted for review ▲.

Unless otherwise stated in the Employer's Requirements, each review period shall not exceed 21 days, calculated from the date on which the Employer receives a Contractor's Document and the Contractor's notice. This notice shall state that the Contractor's Document is considered ready, both for review ▲ in accordance with this Sub-Clause and for use. The notice shall also state that the Contractor's Document complies with the Contract, or the extent to which it does not comply. |

Clause/sub-clause	1999 Red Book	1999 Yellow Book	The Silver Book
		The Engineer may, within the review period, give notice to the Contractor that the Contractor's Document fails (to the extent stated) to comply with the Contract. If a Contractor's Document so fails to comply, it shall be rectified, resubmitted and reviewed (and, if specified, approved) in accordance with this Sub-Clause, at the Contractor's cost.	*The Employer may, within the review period, give notice to the Contractor that the Contractor's Document fails (to the extent stated) to comply with the Contract. If a Contractor's Document so fails to comply, it shall be rectified, resubmitted and reviewed ▲ in accordance with this Sub-Clause, at the Contractor's cost.*
		For each part of the Works, and except to the extent that the prior approval or consent of the Engineer shall have been obtained:	*For each part of the Works, and except to the extent that the Parties otherwise agree:*
		(a) in the case of a Contractor's Document which has (as specified) been submitted for the Engineer's approval: *(i) the Engineer shall give notice to the Contractor that the Contractor's Document is approved, with or without comments, or that it fails (to the extent stated) to comply with the Contract;* *(ii) execution of such part of the Works shall not commence until the Engineer has approved the Contractor's Document; and*	▲

A Precise Record of Alterations, Omissions and Additions

(iii) the Engineer shall be deemed to have approved the Contractor's Document upon the expiry of the review periods for all the Contractor's Documents which are relevant to the design and execution of such part, unless the Engineer has previously notified otherwise in accordance with sub-paragraph (i);

(b) execution of such part of the Works shall not commence prior to the expiry of the review periods for all the Contractor's Documents which are relevant to its design and execution;

(c) execution of such part of the Works shall be in accordance with these reviewed (and, if specified, approved) Contractor's Documents; and

(d) if the Contractor wishes to modify any design or document which has previously been submitted for review (and, if specified, approved), the Contractor shall immediately give notice to the Engineer. Thereafter, the Contractor shall submit revised documents to the Engineer in accordance with the above procedure.

If the Engineer instructs that further Contractor's Documents are required, the Contractor shall prepare them promptly.

Any such approval or consent, or any review (under this Sub-Clause or otherwise) shall not relieve the Contractor from any obligation or responsibility.

(a) execution of such part of the Works shall not commence prior to the expiry of the review periods for all the Contractor's Documents which are relevant to its design and execution;

(b) execution of such part of the Works shall be in accordance with these Contractor's Documents, as submitted for review; and

(c) if the Contractor wishes to modify any design or document which has previously been submitted for review ◀ the Contractor shall immediately give notice to the Employer. Thereafter, the Contractor shall submit revised documents to the Employer in accordance with the above procedure.

◀

Any such agreement (under the preceding paragraph or any review (under this Sub-Clause or otherwise) shall not relieve the Contractor from any obligation or responsibility.

Clause/ sub-clause	1999 Red Book	1999 Yellow Book	The Silver Book
5.3	*Payments to nominated Subcontractors* The Contractor shall pay to the nominated Subcontractor the amounts which the Engineer certifies to be due in accordance with the subcontract. These amounts plus other charges shall be included in the Contract Price in accordance with subparagraph (b) of Sub-Clause 13.5 [Provisional Sums], except as stated in Sub-Clause 5.4 [Evidence of Payments].	*Contractor's Undertaking* The Contractor undertakes that the design, the Contractor's Documents, the execution and the completed Works will be in accordance with: (a) the Laws in the Country, and (b) the documents forming the Contract, as altered or modified by Variations.	*Contractors Undertaking* The Contractor undertakes that the design, the Contractor's Documents, the execution and the completed Works will be in accordance with: (a) the Laws in the Country, and (b) the documents forming the Contract, as altered or modified by Variations.
5.4	*Evidence of Payments* Before issuing a Payment Certificate which includes an amount payable to a nominated Subcontractor, the Engineer may request the Contractor to supply reasonable evidence that the nominated Subcontractor has received all amounts due in accordance with previous Payment Certificates, less applicable deductions for retention or otherwise. Unless the Contractor: (a) submits this reasonable evidence to the Engineer, or (b) (i) satisfies the Engineer in writing that the Contractor is reasonably entitled to withhold or refuse to pay these amounts, and (ii) submits to the Engineer reasonable evidence that the nominated Subcontractor has been notified of the Contractor's entitlement, then	*Technical Standards and Regulations* The design, the Contractor's Documents, the execution and the completed Works shall comply with the Country's technical standards, building, construction and environmental Laws, Laws applicable to the product being produced from the Works, and other standards specified in the Employer's Requirements, applicable to the Works, or defined by the applicable Laws. All these Laws shall, in respect of the Works and each Section, be those prevailing when the Works or Section are taken over by the Employer under Clause 10 [Employer's Taking Over]. References in the Contract to published standards shall be understood to be references to the edition applicable on the Base Date, unless stated otherwise.	*Technical Standards and Regulations* The design, the Contractor's Documents, the execution and the completed Works shall comply with the Country's technical standards, building, construction and environmental Laws, Laws applicable to the product being produced from the Works, and other standards specified in the Employer's Requirements, applicable to the Works, or defined by the applicable Laws. All these Laws shall, in respect of the Works and each Section, be those prevailing when the Works or Section are taken over by the Employer under Clause 10 [Employer's Taking Over]. References in the Contract to published standards shall be understood to be references to the edition applicable on the Base Date, unless stated otherwise.

5.5	the Employer may (at his sole discretion) pay, direct to the nominated Subcontractor, part or all of such amounts previously certified (less applicable deductions) as are due to the nominated Subcontractor and for which the Contractor has failed to submit the evidence described in sub-paragraphs (a) or (b) above. The Contractor shall then repay, to the Employer, the amount which the nominated Subcontractor was directly paid by the Employer.	If changed or new applicable standards come into force in the Country after the Base Date, the Contractor shall give notice to the Engineer and (if appropriate) submit proposals for compliance. In the event that: (a) the Engineer determines that compliance is required, and (b) the proposals for compliance constitute a variation, then the Engineer shall initiate a Variation in accordance with Clause 13 [Variations and Adjustments].	If changed or new applicable standards come into force in the Country after the Base Date, the Contractor shall give notice to the Employer and (if appropriate) submit proposals for compliance. In the event that: (a) the Employer determines that compliance is required, and (b) the proposals for compliance constitute a variation, then the Employer shall initiate a Variation in accordance with Clause 13 [Variations and Adjustments].
	◄	Training	Training
	◄	The Contractor shall carry out the training of Employer's Personnel in the operation and maintenance of the Works to the extent specified in the Employer's Requirements. If the Contract specifies training which is to be carried out before taking-over, the Works shall not be considered to be completed for the purposes of taking-over under Sub-Clause 10.1 [Taking Over of the Works and Sections] until this training has been completed.	The Contractor shall carry out the training of Employer's Personnel in the operation and maintenance of the Works to the extent specified in the Employer's Requirements. If the Contract specifies training which is to be carried out before taking-over, the Works shall not be considered to be completed for the purposes of taking-over under Sub-Clause 10.1 [Taking Over of the Works and Sections] until this training has been completed.
5.6	◄	As-Built Documents	As Built Documents
	◄	The Contractor shall prepare, and keep up-to-date, a complete set of 'as-built' records of the execution of the Works, showing the exact as-built locations, sizes and details of the work as executed. These records shall be kept on the Site and shall be used exclusively for the purposes of this Sub-Clause. Two copies shall be supplied to the Engineer prior to the commencement of the Tests on Completion.	The Contractor shall prepare, and keep up-to-date, a complete set of 'as-built' records of the execution of the Works, showing the exact as-built locations, sizes and details of the work as executed. These records shall be kept on the Site and shall be used exclusively for the purposes of this Sub-Clause. Two copies shall be supplied to the Employer prior to the commencement of the Tests on Completion.

Clause/sub-clause	1999 Red Book	1999 Yellow Book	The Silver Book
5.7	◀	In addition, the Contractor shall supply to the *Engineer* as-built drawings of the Works, showing all Works as executed, and submit them to the *Engineer* for review under Sub-Clause 5.2 [Contractor's Documents]. The Contractor shall obtain the consent of the Engineer as to their size, the referencing system, and other relevant details. Prior to the issue of any Taking-Over Certificate, the Contractor shall supply to the *Engineer* the specified numbers and types of copies of the relevant as-built drawings, in accordance with the Employer's Requirements. The Works shall not be considered to be completed for the purposes of taking-over under Sub-Clause 10.1 [Taking Over of the Works and Sections] until the Engineer has received these documents.	In addition, the Contractor shall supply to the *Employer* as-built drawings of the Works, showing all Works as executed, and submit them to the *Employer* for review under Sub-Clause 5.2 [Contractor's Documents]. The Contractor shall obtain the consent of the Employer as to their size, the referencing system, and other relevant details. Prior to the issue of any Taking-Over Certificate, the Contractor shall supply to the *Employer* the specified numbers and types of copies of the relevant as-built drawings, in accordance with the Employer's Requirements. The Works shall not be considered to be completed for the purposes of taking-over under Sub-Clause 10.1 [Taking Over of the Works and Sections] until the Employer has received these documents.
	◀	Operation and Maintenance Manuals	Operation and Maintenance Manuals
		Prior to commencement of the Tests on Completion, the Contractor shall supply to the *Engineer* provisional operation and maintenance manuals in sufficient detail for the Employer to operate, maintain, dismantle, reassemble, adjust and repair the Plant.	Prior to commencement of the Tests on Completion, the Contractor shall supply to the *Employer* provisional operation and maintenance manuals in sufficient detail for the Employer to operate, maintain, dismantle, reassemble, adjust and repair the Plant.

5.8		The Works shall not be considered to be completed for the purposes of taking-over under Sub-Clause 10.1 [Taking Over of the Works and Sections] until the *Engineer* has received final operation and maintenance manuals in such detail, and any other manuals specified in the Employer's Requirements for these purposes.	The Works shall not be considered to be completed for the purposes of taking-over under Sub-Clause 10.1 [Taking Over of the Works and Sections] until the *Employer* has received final operation and maintenance manuals in such detail, and any other manuals specified in the Employer's Requirements for these purposes.
	◄	Design Error	Design Error
	◄	If errors, omissions, ambiguities, inconsistencies, inadequacies or other defects are found in the Contractor's Documents, they and the Works shall be corrected at the Contractor's cost, notwithstanding any consent or approval under this Clause.	If errors, omissions, ambiguities, inconsistencies, inadequacies or other defects are found in the Contractor's Documents, they and the Works shall be corrected at the Contractor's cost, notwithstanding any consent or approval under this Clause.
Clause 6	Staff and Labour		
6.1	Engagement of Staff and Labour	No change	No change
	Except as otherwise stated in the **Specification**, the Contractor shall make arrangements for the engagement of all staff and labour, local or otherwise, and for their payment, housing, feeding and transport.	Except as otherwise stated in **the Employer's Requirements**, the Contractor shall make arrangements for the engagement of all staff and labour, local or otherwise, and for their payment, housing, feeding and transport.	Except as otherwise stated in **the Employer's Requirements**, the Contractor shall make arrangements for the engagement of all staff and labour, local or otherwise, and for their payment, housing, feeding and transport.

Clause/sub-clause	1999 Red Book	1999 Yellow Book	The Silver Book
6.5	Working Hours	No change	No change
	No work shall be carried out on the Site on locally recognised days of rest, or outside the normal working hours stated in the Appendix to Tender, unless: (a) otherwise stated in the Contract, (b) the **Engineer** gives consent, or (c) the work is unavoidable, or necessary for the protection of life or property or for the safety of the Works, in which case the Contractor shall immediately advise the **Engineer**.	No change	No work shall be carried out on the Site on locally recognised days of rest, or outside normal working hours ▲, unless: (a) otherwise stated in the Contract, (b) the **Employer** gives consent, or (c) the work is unavoidable, or necessary for the protection of life or property or for the safety of the Works, in which case the Contractor shall immediately advise the **Employer**.
6.6	Facilities for Staff and Labour		
	Except as otherwise stated in the **Specification**, the Contractor shall provide and maintain all necessary accommodation and welfare facilities for the Contractor's Personnel. The Contractor shall also provide facilities for the Employer's Personnel as stated in the **Specification**.	Except as otherwise stated in the **Employer's Requirements**, the Contractor shall provide and maintain all necessary accommodation and welfare facilities for the Contractor's Personnel. The Contractor shall also provide facilities for the Employer's Personnel as stated in the **Employer's** Requirements.	Except as otherwise stated in the **Employer's Requirements**, the Contractor shall provide and maintain all necessary accommodation and welfare facilities for the Contractor's Personnel. The Contractor shall also provide facilities for the Employer's Personnel as stated in the **Employer's** Requirements.
6.7	Health and Safety	No change	No change

A Precise Record of Alterations, Omissions and Additions 679

6.9	The Contractor shall send, to the *Engineer*, details of any accident as soon as practicable after its occurrence. The Contractor shall maintain records and make reports concerning health, safety and welfare of persons, and damage to property, as the *Engineer* may reasonably require.	No change	The Contractor shall send, to the *Employer*, details of any accident as soon as practicable after its occurrence. The Contractor shall maintain records and make reports concerning health, safety and welfare of persons, and damage to property, as the *Employer* may reasonably require.
	Contractors Personnel	No change	No change
	The Contractor's Personnel shall be appropriately qualified, skilled and experienced in their respective trades or occupations. The *Engineer* may require the Contractor to remove (or cause to be removed) any person employed on the Site or Works, including the Contractor's Representative if applicable, who: (a) . . .	No change	The Contractor's Personnel shall be appropriately qualified, skilled and experienced in their respective trades or occupations. The *Employer* may require the Contractor to remove (or cause to be removed) any person employed on the Site or Works or Works, including the Contractor's Representative if applicable, who: (a) . . .
6.10	Records of Contractor's Personnel and Equipment	No change	No change
	The Contractor shall submit, to the *Engineer*, details showing the number of each class of Contractor's Personnel and of each type of Contractor's Equipment on the Site. Details shall be submitted each calendar month, in a form approved by the *Engineer*, until the Contractor has completed all work which is known to be outstanding at the completion date stated in the Taking-Over Certificate for the Works.	No change	The Contractor shall submit, to the *Employer*, details showing the number of each class of Contractor's Personnel and of each type of Contractor's Equipment on the Site. Details shall be submitted each calendar month, in a form approved by the *Employer*, until the Contractor has completed all work which is known to be outstanding at the completion date stated in the Taking-Over Certificate for the Works.

Clause/ sub-clause	1999 Red Book	1999 Yellow Book	The Silver Book
Clause 7	Plant, Materials and Workmanship	No change	No change
7.2	Samples	No change	No change
	The Contractor shall submit the following samples of Materials, and relevant information, to the Engineer for *consent prior to using the Materials in or for the Works*: (a) manufacturer's standard samples of Materials and samples specified in the Contract, all at the Contractor's cost, and (b) additional samples instructed by the Engineer as a Variation. Each sample shall be labelled as to origin and intended use in the Works.	The Contractor shall submit the following samples of Materials, and relevant information, to the Engineer for *review in accordance with the procedures for Contractor's Documents described in Sub-Clause 5.2 [Contractor's Documents]*: (a) manufacturer's standard samples of Materials and samples specified in the Contract, all at the Contractor's cost, and (b) additional samples instructed by the Engineer as a Variation. Each sample shall be labelled as to origin and intended use in the Works.	The Contractor shall submit *samples to the Employer, for review in accordance with the procedures for Contractor's Documents described in Sub-Clause 5.2 [Contractor's Documents], as specified in the Contract and at the Contractor's cost. Each sample shall be labelled as to origin and intended use in the Works.* ▲
7.3	Inspection	No change	No change
	The Employer's Personnel shall at all reasonable times: (a) have full access to all parts of the Site and to all places from which natural Materials are being obtained, and (b) during production, manufacture and construction (at the Site and ▲ elsewhere), be entitled to examine, inspect, measure and test the materials and workmanship, and to check the progress of manufacture of Plant and manufacture of Materials.	No change	The Employer's Personnel shall at all reasonable times: (a) have full access to all parts of the Site and to all places from which natural Materials are being obtained, and (b) during production, manufacture and construction (at the Site and, to the extent specified in the Contract, elsewhere), be entitled to examine, inspect, measure and test the materials and workmanship, and to check the progress of manufacture of Plant and production and manufacture of Materials.

			The Contractor shall give the Employer's Personnel. . . .
		▲ The Contractor shall give notice to the *Engineer* whenever any work is ready and before it is covered up, put out of sight, or packaged for storage or transport. The *Engineer* shall then either carry out the examination, inspection, measurement or testing without unreasonable delay, or promptly give notice to the Contractor that the *Engineer* does not require to do so. If the Contractor fails to give the notice, he shall, if and when required by the *Engineer,* uncover the work and thereafter reinstate and make good, all at the Contractor's cost.	In respect of the work which Employer's Personnel are entitled to examine, inspect, measure and/or test, the Contractor shall give notice to the *Employer* whenever any work is ready and before it is covered up, put out of sight, or packaged for storage or transport. The *Employer* shall then either carry out the examination, inspection, measurement or testing without unreasonable delay, or promptly give notice to the Contractor that the *Employer* does not require to do so. If the Contractor fails to give the notice, he shall, if and when required by the *Employer,* uncover the work and thereafter reinstate and make good, all at the Contractor's cost.
7.4		Testing	No change
		The Contractor shall agree, with the *Engineer,* the time and place for the specified Testing of any Plant, Materials and other parts of the Works.	No change
		The *Engineer* may, under Clause 13 [*Variations and Adjustments*], vary the location or the details of specified tests, or instruct the Contractor to carry out additional tests. If these varied or additional tests show that the tested Plant, Materials or workmanship is not in accordance with the Contract, the cost of carrying out this Variation shall be borne by the Contractor, notwithstanding other provisions of the Contract.	The *Employer* may, under Clause 13 [*Variations and Adjustments*], vary the location or the details of specified tests, or instruct the Contractor to carry out additional tests. If these varied or additional tests show that the tested Plant, Materials or workmanship is not in accordance with the Contract, the cost of carrying out this Variation shall be borne by the Contractor, notwithstanding other provisions of the Contract.

Clause/ sub-clause	1999 Red Book	1999 Yellow Book	The Silver Book
	The *Engineer* shall give the Contractor not less than 24 hours' notice of the *Engineer's* intention to attend the tests. If the *Engineer* does not attend at the time and place agreed, the Contractor may proceed with the tests, unless otherwise instructed by the *Engineer*, and the tests shall then be deemed to have been made in the *Engineer's* presence.	No change	The *Employer* shall give the Contractor not less than 24 hours' notice of the *Employer's* intention to attend the tests. If the *Employer* does not attend at the time and place agreed, the Contractor may proceed with the tests, unless otherwise instructed by the *Employer*, and the tests shall then be deemed to have been made in the *Employer's* presence.
	If the Contractor suffers delay and/or incurs Cost from complying with these instructions or as a result of a delay for which the Employer is responsible, the Contractor shall give notice to the *Engineer* and shall be entitled subject to Sub-Clause 20.1 [*Contractor's Claims*] to:		If the Contractor suffers delay and/or incurs Cost from complying with these instructions or as a result of a delay for which the Employer is responsible, the Contractor shall give notice to the *Employer* and shall be entitled subject to Sub-Clause 20.1 [*Contractor's Claims*] to:
	After receiving this notice, the *Engineer* shall proceed in accordance with Sub-Clause 3.5 [*Determinations*] to agree or determine these matters.	No change	After receiving this notice, the *Employer* shall Proceed in accordance with Sub-Clause 3.5 [*Determinations*] to agree or determine these matters.
	The Contractor shall promptly forward to the *Engineer* duly certified reports of the tests. When the specified tests have been passed, the *Engineer* shall endorse the Contractor's test certificate, or issue a certificate to him, to that effect. If the *Engineer* has not attended the tests, he shall be deemed to have accepted the readings as accurate.	No change	The Contractor shall promptly forward to the *Employer* duly certified reports of the tests. When the specified tests have been passed, the *Employer* shall endorse the Contractor's test certificate, or issue a certificate to him, to that effect. If the *Employer* has not attended the tests, he shall be deemed to have accepted the readings as accurate.
7.5	Rejection	No change	No change

A Precise Record of Alterations, Omissions and Additions 683

	If, as a result of an examination, inspection, measurement or testing, any Plant, Materials ◄ or workmanship is found to be defective or otherwise not in accordance with the Contract, the ***Engineer*** may reject the Plant, Materials ◄ or workmanship by giving notice to the Contractor, with reasons. The Contractor shall then promptly make good the defect and ensure that the rejected item complies with the Contract. If the ***Engineer*** requires this Plant, Materials ◄ or workmanship to be retested, the tests shall be repeated under the same terms and conditions.	If, as a result of an examination, inspection, measurement or testing, any Plant, Materials, design or workmanship is found to be defective or otherwise not in accordance with the Contract, the ***Engineer*** may reject the Plant, Materials, design or workmanship by giving notice to the Contractor, with reasons. The Contractor shall then promptly make good the defect and ensure that the rejected item complies with the Contract If the ***Engineer*** requires this Plant, Materials, design or workmanship to be retested, the tests shall be repeated under the same terms and conditions.	If, as a result of an examination, inspection, measurement or testing, any Plant, Materials, design or workmanship is found to be defective or otherwise not in accordance with the Contract, the ***Employer*** may reject the Plant, Materials, design or workmanship by giving notice to the Contractor, with reasons. The Contractor shall then promptly make good the defect and ensure that the rejected item complies with the Contract. If the ***Employer*** requires this Plant, Materials, design or workmanship to be retested, the tests shall be repeated under the same terms and conditions.
7.6	Remedial Work	No change	No change
	Notwithstanding any previous test or certification, the ***Engineer*** may instruct the Contractor to: The Contractor shall comply with the instruction within a reasonable time, which shall be the time (if any) specified in the instruction, or immediately if urgency is specified under sub-paragraph (c).	No change No change	Notwithstanding any previous test or certification, the ***Employer*** may instruct the Contractor to: ◄
	If the Contractor fails to comply with ***the*** instruction, ◄ the Employer shall be entitled to employ and pay other persons to carry out the work.	No change	If the Contractor fails to comply with ***any such*** instruction, which complies with Sub-Clause 3.4 [Instructions], the Employer shall be entitled to employ and pay other persons to carry out the work.
7.8	Royalties	No change	No change

684   *Comparison of text in major 1999 Books*

Clause/sub-clause	1999 Red Book	1999 Yellow Book	The Silver Book
	Unless otherwise stated in the **Specification**, the Contractor shall pay all royalties, rents and other payments for:	Unless otherwise stated in the **Employer's Requirements**, the Contractor shall pay all royalties, rents and other payments for:	Unless otherwise stated in the **Employer's Requirements**, the Contractor shall pay all royalties, rents and other payments for:
Clause 8	Commencement, Delays and Suspension	No change	No change
8.1	Commencement of Work	No change	No change
	The Engineer shall give the Contractor not less than 7 days' notice of the Commencement Date. Unless otherwise stated in the Particular Conditions, the Commencement Date shall be within 42 days after the Contractor receives the Letter of Acceptance.	No change	**Unless otherwise stated in the Contract Agreement:** **(a) the Employer shall give the Contractor not less than 7 days' notice of the Commencement Date; and** **(c) the Commencement Date shall be within 42 days after the date on which the Contract comes into full force and effect under Sub-Clause 1.6 [Contract Agreement].**
	The Contractor shall commence the ▲ execution of the Works as soon as is reasonably practicable after the Commencement Date,	The Contractor shall commence the **design and** execution of the Works as soon as is reasonably practicable after the Commencement Date,	The Contractor shall commence the **design and** execution of the Works as soon as is reasonably practicable after the Commencement Date,
8.3	Programme	No change	No change
	The Contractor shall submit a detailed time programme to the **Engineer** within 28 days after *receiving the notice under Sub-Clause 8.1 [Commencement of Works]*. The Contractor shall also submit a revised programme whenever the previous programme is inconsistent with the actual progress or with the Contractor's obligations. ▲ Each programme shall include:	No change	The Contractor shall submit a ▲ time programme to the **Employer** within 28 days after **the Commencement Date**. The Contractor shall also submit a revised programme whenever the previous programme is inconsistent with actual progress or with the Contractor's obligations. Unless otherwise stated in the Contract, each programme shall include:

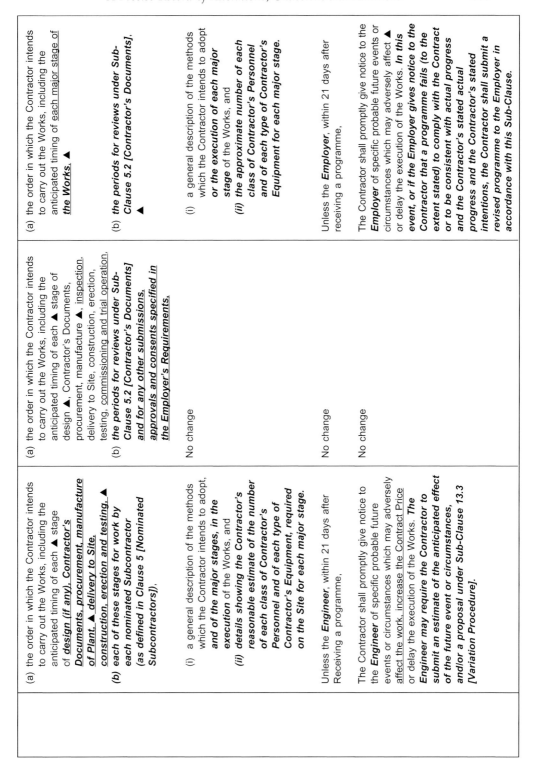

Clause/sub-clause	1999 Red Book	1999 Yellow Book	The Silver Book
	If, at any time, the Engineer gives notice to the Contractor that a programme fails (to the extent stated) to comply with the Contract or to be consistent with actual progress and the Contractor's stated intentions, the Contractor shall submit a revised programme to the Engineer in accordance with this Sub-Clause.		◄
8.4	Extension of Time for Completion	No change	No change
	(a) a Variation (unless an adjustment to the Time for Completion has been agreed under Sub-Clause 13.3 *[Variation Procedure]*) or other substantial change in the quantity of any item of work included in the Contract,	(a) a Variation (unless an adjustment to the Time for Completion has been agreed under Sub-Clause 13.3 *[Variation Procedure]* ◄,	(a) a Variation (unless an adjustment to the Time for Completion has been agreed under Sub-Clause 13.3 *[Variation Procedure]* ◄,
	(c) exceptionally adverse climatic conditions, (d) Unforeseeable shortages in the availability of personnel or Goods caused by epidemic or governmental actions, or	No change	◄
	(e) any delay, impediment or prevention caused by or attributable to the Employer, the Employer's Personnel, or the Employer's other contractors on the Site.	No change	(c) any delay, impediment or prevention caused by or attributable to the Employer, the Employer's Personnel, or the Employer's other Contractors on the Site.
	If the Contractor considers himself to be Entitled to an extension of the Time for Completion, the Contractor shall give notice to the ***Engineer*** in accordance with Sub-Clause 20.1 *[Contractor's Claims]*. When determining each extension of time under Sub-Clause 20.1, the ***Engineer*** shall review previous determinations and may increase, but shall not decrease, the total extension of time.	No change	If the Contractor considers himself to be entitled to an extension of the Time for Completion, the Contractor shall give notice to the ***Employer*** in accordance with Sub-Clause 20.1 *[Contractor's Claims]*. When determining each extension of time, the ***Employer*** shall review previous determinations and may increase, but shall not decrease, the total extension of time.

A Precise Record of Alterations, Omissions and Additions

8.5	Delay Caused by Authorities	No change	No change
	(c) the delay or disruption was *Unforeseeable*,	No change	(c) the delay or disruption was ***not reasonably foreseeable by an experienced contractor by the date for submission of the Tender***,
8.6	Rate of Progress	No change	No change
	If, at any time:	No change	If, at any time:
	(a) actual progress is too slow to complete within the Time for Completion, and/or (b) progress has fallen (or will fall) behind the current programme under Sub-Clause 8.3 *[Programme]*,		(a) actual progress is too slow to complete within the Time for Completion, and/or (b) progress has fallen (or will fall) behind the current programme under Sub-Clause 8.3 *[Programme]*,
	other than as a result of a cause listed in Sub-Clause 8.4 *[Extension of Time for Completion]*, then the **Engineer** may instruct the Contractor to submit, …		other than as a result of a cause listed in Sub-Clause 8.4 *[Extension of Time for Completion]*, then the **Employer** may instruct the Contractor to submit, …
	Unless the **Engineer** notifies otherwise, the Contractor shall adopt these revised methods, which may require increases in the working hours and/or in the numbers of Contractor's Personnel and/or Goods, at the risk and cost of the Contractor. If these revised methods cause the Employer to incur additional costs, the Contractor shall subject to Sub-Clause 2.5 *[Employer's Claims]* pay these costs to the Employer, in addition to delay damages (if any) under Sub-Clause 8.7 below.	No change	Unless the **Employer** notifies otherwise, the Contractor shall adopt these revised methods, which may require increases in the working hours and/or in the numbers of Contractor's Personnel and/or Goods, at the risk and cost of the Contractor. If these revised methods cause the Employer to incur additional costs, the Contractor shall subject to Sub-Clause 2.5 *[Employer's Claims]* pay these costs to the Employer, in addition to delay damages (if any) under Sub-Clause 8.7 below.

Clause/sub-clause	1999 Red Book	1999 Yellow Book	The Silver Book
8.7	Delay Damages	No change	No change
	If the Contractor fails to comply with Sub-Clause 8.2 *[Time for Completion]*, the Contractor shall subject to Sub-Clause 2.5 *[Employer's Claims]* pay delay damages to the Employer for this default. These delay damages shall be the sum stated in the **Appendix to Tender**, which shall be paid for every day which shall elapse between the relevant Time for Completion and the date stated in the Taking-over-Certificate. However, the total amount due under this Sub-Clause shall not exceed the maximum amount of delay damages (if any) stated in the **Appendix to Tender**.	No change	If the Contractor fails to comply with Sub-Clause 8.2 *[Time for Completion]*, the Contractor shall subject to Sub-Clause 2.5 *[Employer's Claims]* pay delay damages to the Employer for this default. These delay damages shall be the sum stated in the **Particular Conditions**, which shall be paid for every day which shall elapse between the relevant Time for Completion and the date stated in the Taking-over-Certificate. However, the total amount due under this Sub-Clause shall not exceed the maximum amount of delay damages (if any) stated in the **Particular Conditions**.
8.8	Suspension of Work	No change	No change
	The **Engineer** may at any time instruct the Contractor to suspend progress of part or all of the Works.	No change	The **Employer** may at any time instruct the Contractor to suspend progress of part or all of the Works.
	The **Engineer** may also notify the cause for the suspension.	No change	The **Employer** may also notify the cause for the suspension.
8.9	Consequences of Suspension	No change	No change
	If the Contractor suffers delay and/or incurs Cost from complying with the **Engineer's** instructions under Sub-Clause 8.8 *[Suspension of Work]* and/or from resuming the work, the Contractor shall give notice to the **Engineer** and shall be entitled subject to Sub-Clause 20.1 *[Contractor's Claims]* to:		If the Contractor suffers delay and/or incurs Cost from complying with the **Employer's** instructions under Sub-Clause 8.8 *[Suspension of Work]* and/or from resuming the work, the Contractor shall give notice to the **Employer** and shall be entitled subject to Sub-Clause 20.1 *[Contractor's Claims]* to:

A Precise Record of Alterations, Omissions and Additions

	After receiving this notice, the **Engineer** shall proceed in accordance with Sub-Clause 3.5 *[Determinations]* to agree or determine these matters.	No change	After receiving this notice, the **Employer** shall proceed in accordance with Sub-Clause 3.5 *[Determinations]* to agree or determine these matters.
8.10	Payment for Plant and Materials in Event of Suspension	No change	No change
	(b) the Contractor has marked the Plant and/or Materials as the Employer's property in accordance with the **Engineer's** instructions.	No change	(b) the Contractor has marked the Plant and/or Materials as the Employer's property in accordance with the **Employer's** instructions.
8.11	Prolonged Suspension	No change	No change
	If the suspension under Sub-Clause 8.8 *[Suspension of Work]* has continued for more than 84 days, the Contractor may request the **Engineer's** permission to proceed. If the **Engineer** does not give permission within 28 days after being requested to do so, the Contractor may, by giving notice to the **Engineer**, treat the suspension as an omission under Clause 13 *[Variations and Adjustments]* of the affected part of the Works.	No change	If the suspension under Sub-Clause 8.8 *[Suspension of Work]* has continued for more than 84 days, the Contractor may request the **Employer's** permission to proceed. If the **Employer** does not give permission within 28 days after being requested to do so, the Contractor may, by giving notice to the **Employer**, treat the suspension as an omission under Clause 13 *[Variations and Adjustments]* of the affected part of the Works.
8.12	Resumption of Work	No change	No change
	After the permission or instruction to proceed is given, **the Contractor and the Engineer** shall jointly examine the Works and the Plant and Materials affected by the suspension.	No change	After the permission or instruction to proceed is given, **the Parties** shall jointly examine the Works and the Plant and Materials affected by the suspension.

Clause/sub-clause	1999 Red Book	1999 Yellow Book	The Silver Book
Clause 9	Tests on Completion		
9.1	Contractor's Obligations	No change	No change
	The Contractor shall carry out the Tests on Completion in accordance with this Clause and Sub-Clause 7.4 *[Testing]*, after providing the documents in accordance with **sub-paragraph (d) of Sub-Clause 4.1 [Contractor's General Obligations]**.	The Contractor shall carry out the Tests on Completion in accordance with this Clause and Sub-Clause 7.4 *[Testing]*, after providing the documents in accordance with **Sub-Clause 5.6 [As-Built Documents] and Sub-Clause 5.7 [Operation and Maintenance Manuals]**.	The Contractor shall carry out the Tests on Completion in accordance with this Clause and Sub-Clause 7.4 *[Testing]* after providing the documents in accordance with **Sub-Clause 5.6 [As-Built Documents] and Sub-Clause 5.7 [Operation and Maintenance Manuals]**.
	The Contractor shall give to the *Engineer* not less than 21 days' notice of the date after which the Contractor will be ready to carry out each of the Tests on Completion. Unless otherwise agreed, Tests on Completion shall be carried out within 14 days after this date, on such day or days as the *Engineer* shall instruct.	No change	The Contractor shall give to the *Employer* not less than 21 days' notice of the date after which the Contractor will be ready to carry out each of the Tests on Completion. Unless otherwise agreed, Tests on Completion shall be carried out within 14 days after this date, on such day or days as the *Employer* shall instruct.
	◀	Unless otherwise stated in the Particular Conditions, the Tests on Completion shall be carried out in the following sequence:	Unless otherwise stated in the Particular Conditions, the Tests on Completion shall be carried out in the following sequence:
		(a) pre-commissioning tests, which shall include the appropriate inspections and ('dry' or 'cold') functional tests to demonstrate that each item of Plant can safely undertake the next stage. (b);	(a) pre-commissioning tests, which shall include the appropriate inspections and ('dry' or 'cold') functional tests to demonstrate that each item of Plant can safely undertake the next stage. (b);
		(b) commissioning tests, which shall include the specified operational tests to demonstrate that the Works or Section can be operated safely and as specified, under all available operating conditions; and	(b) commissioning tests, which shall include the specified operational tests to demonstrate that the Works or Section can be operated safely and as specified, under all available operating conditions; and

		(c) trial operation, which shall demonstrate that the Works or Section perform reliably and in accordance with the Contract.	(c) trial operation, which shall demonstrate that the Works or Section perform reliably and in accordance with the Contract.
	◄	During trial operation, when the Works are operating under stable conditions, the **Engineer** that the Works are ready for any other Tests on Completion, including performance tests to demonstrate whether the Works conform with criteria specified in the Employer's Requirements and with the Schedule of Guarantees.	During trial operation, when the Works are operating under stable conditions, the **Employer** that the Works are ready for any other Tests on Completion, including performance tests to demonstrate whether the Works conform with criteria specified in the Employer's Requirements and with the Schedule of Guarantees.
	◄	Trial operation shall not constitute a taking-over under Clause 10 [Employer's Taking Over]. Unless otherwise stated in the Particular Conditions, any product produced by the Works during trial operation shall be the property of the Employer.	Trial operation shall not constitute a taking-over under Clause 10 [Employer's Taking Over]. Unless otherwise stated in the Particular Conditions, any product produced by the Works during trial operation shall be the property of the Employer.
	In considering the results of the Tests on Completion, **the Engineer shall make allowances** for the effect of any use of the Works by the Employer on the performance or other characteristics of the Works. As soon as the Works, or a Section, have passed any Tests on Completion ▲, the Contractor shall submit a certified report of the results of these Tests to the **Engineer**.	In considering the results of the Tests on Completion, **the Engineer shall make allowances** for the effect of any use of the Works by the Employer on the performance or other characteristics of the Works. As soon as the Works, or a Section, have passed each of the Tests on Completion described in sub-paragraph (a), (b) or (c), the Contractor shall submit a certified report of the results of these Tests to the **Engineer**.	In considering the results of the Tests on Completion, **appropriate allowances shall be made** for the effect of any use of the Works by the Employer on the performance or other characteristics of the Works. As soon as the Works, or a Section, have passed each of the Tests on Completion described in sub-paragraph (a), (b) or (c), the Contractor shall submit a certified report of the results of these Tests to the **Employer**.
9.2	Delayed Tests	No change	No change

Clause/sub-clause	1999 Red Book	1999 Yellow Book	The Silver Book
	If the Tests on Completion are being unduly delayed by the Contractor, the **Engineer** may by notice require the Contractor to carry out the Tests within 21 days after receiving the notice. The Contractor shall carry out the Tests on such day or days within that period as the Contractor may fix and of which he shall give notice to the **Engineer**.	No change	If the Tests on Completion are being unduly delayed by the Contractor, the **Employer** may by notice require the Contractor to carry out the Tests within 21 days after receiving the notice. The Contractor shall carry out the Tests on such day or days within that period as the Contractor may fix and of which he shall give notice to the **Employer**.
9.3	Retesting		
	If the Works, or a Section, fail to pass the Tests on Completion, Sub-Clause 7.5 [Rejection] shall apply, and the **Employer** or the Contractor may require the failed Tests, and Tests on Completion on any related work, to be repeated under the same terms and conditions.	No change	If the Works, or a Section, fail to pass the Tests on Completion, Sub-Clause 7.5 [Rejection] shall apply, and the **Employer** or the Contractor may require the failed Tests, and Tests on Completion on any related work, to be repeated under the same terms and conditions.
9.4	Failure to Pass Tests on Completion		
	If the Works, or a Section, fail to pass the Tests on Completion repeated under Sub-Clause 9.3 [Retesting], the **Engineer** shall be entitled to:	No change	If the Works, or a Section, fail to pass the Tests on Completion repeated under Sub-Clause 9.3 [Retesting], the **Employer** shall be entitled to:
	(a) order further repetition of Tests on Completion under Sub-Clause 9.3; (b) if the failure deprives the Employer of substantially the whole benefit of the Works or Section, reject the Works or Section (as the case may be), in which event the Employer shall have the same remedies as are provided in sub-paragraph (c) of Sub-Clause 11.4 [Failure to Remedy Defects]; or		(a) order further repetition of Tests on Completion under Sub-Clause 9.3; (b) if the failure deprives the Employer of substantially the whole benefit of the Works or Section, reject the Works or Section (as the case may be), in which event the Employer shall have the same remedies as are provided in sub-paragraph (c) of Sub-Clause 11.4 [Failure to Remedy Defects]; or

			(c) issue a Taking-Over Certificate ▲.
	(c) issue a Taking-Over Certificate, if the Employer so requests.		
Clause 10	Employer's Taking Over	No change	
10.1	Taking Over of the Works and Sections	No change	No change
	The Contractor may apply by notice to the Engineer for a Taking-Over Certificate not earlier than 14 days before the Works will, in the Contractor's opinion, be complete and ready for taking over.	No change	The Contractor may apply by notice to the Employer for a Taking-Over Certificate not earlier than 14 days before the Works will, in the Contractor's opinion, be complete and ready for taking over.
	The Engineer shall, within 28 days after receiving the Contractor's application:	No change	The Employer shall, within 28 days after receiving the Contractor's application:
	If the Engineer fails either to issue the Taking-Over Certificate or to reject the Contractor's application within the period of 28 days,	No change	If the Employer fails either to issue the Taking-Over Certificate or to reject the Contractor's application within the period of 28 days,
10.2	Taking Over of Parts of the Works	No change	No change
	The Engineer may, at the sole discretion of the Employer, issue a Taking-Over Certificate for any part of the Works. The Employer shall not use any part of the Works (other than as a temporary measure which is either specified in the Contract or agreed by both Parties) unless and until the Engineer has issued a Taking-Over Certificate for this part. However, if the Employer does use any part of the Works before the Taking-Over Certificate is issued:	No change	Parts of the Works (other than Sections) shall not be taken over or used by the Employer, except as may be stated in the Contract or as may be agreed by both Parties. ◄

Clause/ sub-clause	1999 Red Book	1999 Yellow Book	The Silver Book
	(a) the part which is used shall be deemed to have been taken over as from the date on which it is used, (b) the Contractor shall cease to be liable for the care of such part as from this date, when responsibility shall pass to the Employer, and (c) if requested by the Contractor, the Engineer shall issue a Taking-Over Certificate for this part. After the Engineer has issued a Taking-Over Certificate for a part of the Works, the Contractor shall be given the earliest opportunity to take such steps as may be necessary to carry out any outstanding Tests on Completion. The Contractor shall carry out these Tests on Completion as soon as practicable before the expiry date of the relevant Defects Notification Period. If the Contractor incurs Cost as a result of the Employer taking over and/or using a part of the Works, other than such use as is specified in the Contract or agreed by the Contractor, the Contractor shall (i) give notice to the Engineer and (ii) be entitled subject to Sub-Clause 20.1 [Contractor's Claims] to payment of any such Cost plus reasonable profit, which shall be included in the Contract Price. After receiving this notice, the Engineer shall proceed in accordance with Sub-Clause 3.5 [Determinations] to agree or determine this Cost and profit.		

A Precise Record of Alterations, Omissions and Additions 695

If a Taking-Over Certificate has been issued for a part of the Works (other than a Section), the delay damages thereafter for completion of the remainder of the Works shall be reduced. Similarly, the delay damages for the remainder of the Section (if any) in which this part is included shall also be reduced. For any period of delay after the date stated in this Taking-Over Certificate, the proportional reduction in these delay damages shall be calculated as the proportion which the value of the part so certified bears to the value of the Works or Section (as the case may be) as a whole. The Engineer shall proceed in accordance with Sub-Clause 3.5 *[Determinations]* to agree or determine these proportions. The provisions of this paragraph shall only apply to the daily rate of delay damages under Sub-Clause 8.7 *[Delay Damages]*, and shall not affect the maximum amount of these damages.		
10.3	Interference with Tests on Completion	No change
	If the Contractor is prevented, for more than 14 days, from carrying out the Tests on Completion by a cause for which the Employer is responsible, **the Employer shall be deemed to have taken over the Works or Section (as the case may be) on the date when the Tests on Completion would otherwise have been completed.**	No change
		If the Contractor is prevented, for more than 14 days, from carrying out the Tests on Completion by a cause for which the Employer is responsible, ***the Contractor shall carry out the Tests on Completion as soon as practicable.***
	If the Contractor suffers delay and/or incurs Cost as a result of this delay in carrying out the Tests on Completion, the Contractor shall give notice to the **Engineer** and shall be entitled subject to Sub-Clause 20.1 *[Contractor's Claims]* to:	No change
		If the Contractor suffers delay and/or incurs Cost as a result of this delay in carrying out the Tests on Completion, the Contractor shall give notice to the **Employer** and shall be entitled subject to Sub-Clause 20.1 *[Contractor's Claims]* to:

Clause/ sub-clause	1999 Red Book	1999 Yellow Book	The Silver Book
	After receiving this notice, the **Engineer** shall proceed in accordance with Sub-Clause 3.5 [*Determinations*] to agree or determine these matters.	No change	After receiving this notice, the **Employer** shall proceed in accordance with Sub-Clause 3.5 [*Determinations*] to agree or determine these matters.
10.4	Surfaces Requiring Reinstatement	No change	◄
	Except as otherwise stated in a Taking-Over Certificate, a certificate for a Section or part of the Works shall not be deemed to certify completion of any ground or other surfaces requiring reinstatement.	No change	◄
Clause 11	Defects Liability	No change	No change
11.1	Completion of Outstanding Work and Remedying Defects	No change	No change
	(a) complete any work which is outstanding on the date stated in a Taking-Over Certificate, within such reasonable time as is instructed by the **Engineer**, and (b) execute all work required to remedy defects or damage, as may be notified by (or on behalf of) the Employer on or before the expiry date of the Defects Notification Period for the Works or Section (as the case may be).	No change	(a) complete any work which is outstanding on the date stated in a Taking-Over Certificate, within such reasonable time as is instructed by the **Employer**, and (b) execute all work required to remedy defects or damage, as may be notified by ▲ the Employer on or before the expiry date of the Defects Notification Period for the Works or Section (as the case may be).
	If a defect appears or damage occurs, **the Contractor shall be notified accordingly, by (or on behalf of) the Employer.**	No change	If a defect appears or damage occurs, the **Employer shall notify the Contractor accordingly.**

A Precise Record of Alterations, Omissions and Additions 697

11.2	Cost of Remedying Defects	No change	No change
	(a) **any design for which the Contractor is responsible,** (b) Plant, Materials or workmanship not being in accordance with the Contract, or ▲ (c) failure by the Contractor to comply with any other obligation. If and to the extent that the work is Attributable to any other cause, **the Contractor shall be notified promptly by (or on behalf of) the Employer**, and Sub-Clause 13.3 *[Variation Procedure]* shall apply.	(a) **the design of the Works, other than a part of the design for which the Employer is responsible (if any),** (b) Plant, Materials or workmanship not being in accordance with the Contract, (c) improper operation or maintenance which was attributable to matters for which the Contractor is responsible (under Sub-Clauses 5.5 to 5.7 or otherwise), or (d) failure by the Contractor to comply with any other obligation. No change	(a) **the design of the Works,** (b) Plant, Materials or workmanship not being in accordance with the Contract, (c) improper operation or maintenance which was attributable to matters for which the Contractor is responsible (under Sub-Clauses 5.5 to 5.7 or otherwise), or (d) failure by the Contractor to comply with any other obligation. If and to the extent that the work is attributable to any other cause, **the Employer shall give notice to the Contractor accordingly**, and Sub-Clause 13.3 *[Variation Procedure]* shall apply.
11.4	Failure to Remedy Defects	No change	No change
	(b) require the Engineer to agree or determine a reasonable reduction in the Contract Price in accordance with Sub-Clause 3.5 *[Determinations]*; or	No change	(b) ▲ agree or determine a reasonable reduction in the Contract Price in accordance with Sub-Clause 3.5 *[Determinations]*; or
11.6	Further Tests	No change	No change
	If the work of remedying of any defect or Damage may affect the performance of the Works, the **Engineer** may require the repetition of any of the tests described in the Contract ▲.	If the work of remedying of any defect or damage may affect the performance of the Works, the **Engineer** may require the repetition of any of the tests described in the Contract, including Tests on Completion and/or Tests after Completion.	If the work of remedying of any defect or damage may affect the performance of the Works, the **Employer** may require the repetition of any of the tests described in the Contract, including Tests on Completion and/or Tests after Completion.

Clause/sub-clause	1999 Red Book	1999 Yellow Book	The Silver Book
11.7	Until the Performance Certificate has been issued, the Contractor shall have *such right of access to the Works as is reasonably required in order to comply with this Clause*, except as may be inconsistent with the Employer's reasonable security restrictions.	Until the Performance Certificate has been issued, the Contractor shall have *the right of access to all parts of the Works and to records of the operation and performance of the Works*, except as may be inconsistent with the Employer's reasonable security restrictions.	Until the Performance Certificate has been issued, the Contractor shall have *the right of access to all parts of the Works and to records of the operation and performance of the Works*, except as may be inconsistent with the Employer's reasonable security restrictions.
11.8	Contractor to Search		
	The Contractor shall, if required by the *Engineer*, search for the cause of any defect, under the direction of the *Engineer*. Unless the defect is to be remedied at the cost of the Contractor under Sub-Clause 11.2 *[Cost of Remedying Defects]*, the Cost of the search plus reasonable profit shall be agreed or determined by the Engineer in accordance with Sub-Clause 3.5 *[Determinations]* and shall be included in the Contract Price.	No change	The Contractor shall, if required by the *Employer*, search for the cause of any defect, under the direction of the *Employer*. Unless the defect is to be remedied at the cost of the Contractor under Sub-Clause 11.2 *[Cost of Remedying Defects]* the Cost of the search plus reasonable profit shall be agreed or determined ▲ in accordance with Sub-Clause 3.5 *[Determinations]* and shall be added to the Contract Price.
11.9	Performance Certificate		
	Performance of the Contractor's obligations shall not be considered to have been completed until the *Engineer* has issued the Performance Certificate to the Contractor, stating the date on which the Contractor completed his obligations under the Contract.	No change	Performance of the Contractor's obligations shall not be considered to have been completed until the *Employer* has issued the Performance Certificate to the Contractor, stating the date on which the Contractor completed his obligations under the Contract.

		The *Engineer* shall issue the Performance Certificate within 28 days after the latest of the expiry dates of the Defects Notification Periods, or as soon thereafter as the Contractor has supplied all the Contractor's Documents and completed and tested all the Works, including remedying any defects. *A copy of the Performance Certificate shall be issued to the Employer.* ▲	No change	The *Employer* shall issue the Performance Certificate within 28 days after the latest of the expiry dates of the Defects Notification Periods, or as soon thereafter as the Contractor has supplied all the Contractor's Documents and completed and tested all the Works, including remedying any defects. *If the Employer fails to issue the Performance Certificate accordingly:* (a) the Performance Certificate shall be deemed to have been issued on the date 28 days after the date on which it should have been issued, as required by this Sub-Clause, and (b) Sub-Clause 11.11 [Clearance of Site] and sub-paragraph (a) of Sub-Clause 14.14 [Cessation of Employer's Liability] shall be inapplicable.
Clause 12	*Measurement and Evaluation*		*Tests After Completion*	*Tests After Completion*
12.1	*Works to be Measured*	The Works shall be measured, and valued for payment, in accordance with this Clause. Whenever the Engineer requires any part of the Works to be measured, reasonable notice shall be given to the Contractor's Representative, who shall: (a) *promptly either attend or send another qualified representative to assist the Engineer in making the measurement,* and (b) supply any particulars requested by the Engineer.	*Procedure for Tests after completion* If Tests after Completion are specified in the Contract, this Clause shall apply. Unless otherwise stated in the Particular Conditions, the Employer shall: (a) provide all electricity, equipment, fuel, instruments, labour, materials, and suitably qualified and experienced staff, as are necessary to carry out the Tests after Completion efficiently, and (b) carry out the Tests after Completion in accordance with the manuals supplied by the Contractor under Sub-Clause 5.7 [Operation and Maintenance Manuals]	*Procedure for Tests after completion* If Tests after Completion are specified in the Contract, this Clause shall apply. Unless otherwise stated in the Particular Conditions: (a) the Employer shall provide all electricity, fuel and materials, and make the Employer's Personnel and Plant available; (b) the Contractor shall provide any other plant, equipment and suitably qualified and experienced staff, as are necessary to carry out the Tests after Completion efficiently; and

Clause/sub-clause	1999 Red Book	1999 Yellow Book	The Silver Book
	If the Contractor fails to attend or send a representative, the measurement made by (or on behalf of) the Engineer shall be accepted as accurate.

Except as otherwise stated in the Contract, wherever any Permanent Works are to be measured from records, these shall be prepared by the Engineer. The Contractor shall, as and when requested, attend to examine and agree the records with the Engineer, and shall sign the same when agreed. If the Contractor does not attend, the records shall be accepted as accurate.

If the Contractor examines and disagrees the records, and/or does not sign them as agreed, then the Contractor shall give notice to the Engineer of the respects in which the records are asserted to be inaccurate. After receiving this notice, the Engineer shall review the records and either confirm or vary them. If the Contractor does not so give notice to the Engineer within 14 days after being requested to examine the records, they shall be accepted as accurate. | and such guidance as the Contractor may be required to give during the course of these Tests; and in the presence of such Contractor's Personnel as either Party may reasonably request.

The Tests after Completion shall be carried out as soon as is reasonably practicable after the Works or Section have been taken over by the Employer. The Employer shall give to the Contractor 21 days' notice of the date after which the Tests after Completion will be carried out. Unless otherwise agreed, these Tests shall be carried out within 14 days after this date, on the day or days determined by the Employer.

If the Contractor does not attend at the time and place agreed, the Employer may proceed with the Tests after Completion, which shall be deemed to have been made in the Contractor's presence, and the Contractor shall accept the readings as accurate.

The results of the Tests after Completion shall be compiled and evaluated by both Parties. Appropriate account shall be taken of the effect of the Employer's prior use of the Works. | (c) the Contractor shall carry out the Tests after Completion in the presence of such Employer's and/or Contractor's Personnel as either Party may reasonably request.

The Tests after Completion shall be carried out as soon as is reasonably practicable after the Works or Section have been taken over by the Employer. The Employer shall give to the Contractor 21 days' notice of the date after which the Tests after Completion will be carried out. Unless otherwise agreed, these Tests shall be carried out, within 14 Days after this date, on the day or days determined by the Employer.

The results of the Tests after Completion shall be compiled and evaluated by the Contractor, who shall prepare a detailed report. Appropriate account shall be taken of the effect of the Employer's prior use of the Works. |

12.2	Method of Measurement	Delayed Tests	Delayed Tests
	Except as otherwise stated in the Contract and notwithstanding local practice: (a) measurement shall be made of the net actual quantity of each item of the Permanent Works, and (b) the method of measurement shall be in accordance with the Bill of Quantities or other applicable Schedules.	If the Contractor incurs Cost as a result of any unreasonable delay by the Employer to the Tests after Completion, the Contractor shall (i) give notice to the Engineer and (ii) be entitled subject to Sub-Clause 20.1 [Contractor's Claims] to payment of any such Cost plus reasonable profit, which shall be included in the Contract Price. After receiving this notice, the Engineer shall proceed in accordance with Sub-Clause 3.5 [Determinations] to agree or determine this Cost and profit. If, for reasons not attributable to the Contractor, a Test after Completion on the Works or any Section cannot be completed during the Defects Notification Period (or any other period agreed upon by both Parties), then the Works or Section shall be deemed to have passed this Test after Completion.	If the Contractor incurs Cost as a result of any unreasonable delay by the Employer to the Tests after Completion, the Contractor shall (i) give notice to the Employer and (ii) be entitled subject to Sub-Clause 20.1 [Contractor's Claims] to payment of any such Cost plus reasonable profit, which shall be added to the Contract Price. After receiving this notice, the Employer shall proceed in accordance with Sub-Clause 3.5 [Determinations] to agree or determine this Cost and profit. If, for reasons not attributable to the Contractor, a Test after Completion on the Works or any Section cannot be completed during the Defects Notification Period (or any other period agreed upon by both Parties), then the Works or Section shall be deemed to have passed this Test after Completion.
12.3	Evaluation	Retesting	Retesting
	Except as otherwise stated in the Contract, the Engineer shall proceed in accordance with Sub-Clause 3.5 [Determinations] to agree or determine the Contract Price by evaluating each item of work, applying the measurement agreed or determined in accordance with the above Sub-Clauses 12.1 and 12.2 and the appropriate rate or price for the item.	If the Works, or a Section, fail to pass the Tests after Completion: (a) sub-paragraph (b) of Sub-Clause 11.1 [Completion of Outstanding Work and Remedying Defects] shall apply, and (b) either Party may then require the failed Tests, and the Tests after Completion on any related work, to be repeated under the same terms and conditions.	If the Works, or a Section, fail to pass the Tests after Completion: (a) sub-paragraph (b) of Sub-Clause 11.1 [Completion of Outstanding Work and Remedying of Defects] shall apply, and (b) either Party may then require the failed Tests, and the Tests after Completion on any related work, to be repeated under the same terms and conditions.

Clause/ sub-clause	1999 Red Book	1999 Yellow Book	The Silver Book
	For each item of work, the appropriate rate or price for the item shall be the rate or price specified for such item in the Contract or if there is no such item, specified for similar work. However, a new rate or price shall be appropriate for an item of work if: (a) (i) the measured quantity of the item is changed by more than 10% from the quantity of this item in the Bill of Quantities or other Schedule, (ii) this change in quantity multiplied by such specified rate for this item exceeds 0.01% of the Accepted Contract Amount, (iii) this change in quantity directly changes the Cost per unit quantity of this item by more than 1%, and (iv) this item is not specified in the Contract as a 'fixed rate item'; or (b) (i) the work is instructed under Clause 13 [Variations and Adjustments], (ii) no rate or price is specified in the Contract for this item, and (iii) no specified rate or price is appropriate because the item of work is not of similar character, or is not executed under similar conditions, as any item in the Contract.	If and to the extent that this failure and re-testing are attributable to any of the matters listed in sub-paragraphs (a) to (d) of Sub-Clause 11.2 [Cost of Remedying Defects] and cause the Employer to incur additional costs, the Contractor shall subject to Sub-Clause 2.5 [Employer's Claims] pay these costs to the Employer.	If and to the extent that this failure and retesting are attributable to any of the matters listed in sub-paragraphs (a) to (d) of Sub-Clause 11.2 [Cost of Remedying Defects] and cause the Employer to incur additional costs, the Contractor shall subject to Sub-Clause 2.5 [Employer's Claims] pay these costs to the Employer.

		Failure to Pass Tests after Completion	Failure to Pass Tests after Completion
	Each new rate or price shall be derived from any relevant rates or prices in the Contract, with reasonable adjustments to take account of the matters described in sub-paragraph (a) and/or (b), as applicable. If no rates or prices are relevant for the derivation of a new rate or price, it shall be derived from the reasonable Cost of executing the work, together with reasonable profit, taking account of any other relevant matters. Until such time as an appropriate rate or price is agreed or determined, the Engineer shall determine a provisional rate or price for the purposes of Interim Payment Certificates.		
12.4	Omissions	If the following conditions apply, namely: (a) the Works, or a Section, fail to pass any or all of the Tests after Completion, (b) the relevant sum payable as nonperformance damages for this failure is stated (or its method of calculation is defined) in the Contract, and (c) the Contractor pays this relevant sum to the Employer during the Defects Notification Period, then the Works or Section shall be deemed to have passed these Tests after Completion.	If the following conditions apply, namely: (a) the Works, or a Section, fail to pass any or all of the Tests after Completion, (b) the relevant sum payable as nonperformance damages for this failure is stated (or its method of calculation is defined) in the Contract, and (c) the Contractor pays this relevant sum to the Employer during the Defects Notification Period, then the Works or Section shall be deemed to have passed these Tests after Completion.
	Whenever the omission of any work forms part (or all) of a Variation, the value of which has not been agreed, if: (a) the Contractor will incur (or has incurred) cost which, if the work had not been omitted, would have been deemed to be covered by a sum forming part of the Accepted Contract Amount; (b) the omission of the work will result (or has resulted) in this sum not forming part of the Contract Price; and (c) this cost is not deemed to be included in the evaluation of any substituted work;		

Clause/sub-clause	1999 Red Book	1999 Yellow Book	The Silver Book
	then the Contractor shall give notice to the Engineer accordingly, with supporting particulars. Upon receiving this notice, the Engineer shall proceed in accordance with Sub-Clause 3.5 [Determinations] to agree or determine this cost, which shall be included in the Contract Price.	If the Works, or a Section, fail to pass a Test after Completion and the Contractor proposes to make adjustments or modifications to the Works or such Section, the Contractor may be instructed by (or on behalf of) the Employer that right of access to the Works or Section cannot be given until a time that is convenient to the Employer. The Contractor shall then remain liable to carry out the adjustments or modifications and to satisfy this Test, within a reasonable period of receiving notice by (or on behalf of) the Employer of the time that is convenient to the Employer. However, if the Contractor does not receive this notice during the relevant Defects Notification Period, the Contractor shall be relieved of this obligation and the Works or Section (as the case may be) shall be deemed to have passed this Test after Completion. If the Contractor incurs additional Cost as a result of any unreasonable delay by the Employer in permitting access to the Works or Plant by the Contractor, either to investigate the causes of a failure to pass a Test after Completion or to carry out any adjustments or modifications, the Contractor shall (i) give notice to the Engineer and (ii) be entitled subject to Sub-Clause 20.1 [Contractor's Claims] to payment of any such Cost plus reasonable profit, which shall be included in the Contract Price.	If the Works, or a Section, fail to pass a Test after Completion and the Contractor proposes adjustments or modifications to the Works or such Section, the Contractor may be instructed by (or on behalf of) the Employer that right of access to the Works or Section cannot be given until a time that is convenient to the Employer. The Contractor shall then remain liable to carry out the adjustments or modifications and to satisfy this Test, within a reasonable time of receiving notice by (or on behalf of) the Employer of the time that is convenient to the Employer. However, if the Contractor does not receive this notice during the relevant Defects Notification Period, the Contractor shall be relieved of this obligation and the Works or Section (as the case may be) shall be deemed to have passed this Test after Completion. If the Contractor incurs additional Cost as a result of any unreasonable delay by the Employer in permitting access to the Works or Plant by the Contractor, either to investigate the causes of a failure to pass a Test after Completion or to carry out any adjustments or modifications, the Contractor shall (i) give notice to the Employer and (ii) be entitled subject to Sub-Clause 20.1 [Contractor's Claims] to payment of any such Cost plus reasonable profit, which shall be added to the Contract Price.

A Precise Record of Alterations, Omissions and Additions

				After receiving this notice, the Engineer shall proceed in accordance with Sub-Clause 3.5 [Determinations] to agree or determine this Cost and profit.	After receiving this notice, the Employer shall proceed in accordance with Sub-Clause 3.5 [Determinations] to agree or determine this Cost and profit.
	Clause 13	Variations and Adjustments		No change	No change
	13.1	Right to Vary		No change	No change
			Variations may be initiated by the **Engineer** at any time prior to issuing the Taking-Over Certificate for the works, either by an instruction or by a request for the Contractor to submit a proposal. ◀	Variations may be initiated by the **Engineer** at any time prior to issuing the Taking-Over Certificate for the works, either by an instruction or by a request for the Contractor to submit a proposal. A variation shall not comprise for the omission of any work which is to be carried out by others.	Variations may be initiated by the **Employer** at any time prior to issuing the Taking-Over Certificate for the works, either by an instruction or by a request for the Contractor to submit a proposal. A variation shall not comprise for the omission of any work which is to be carried out by others.
			The Contractor shall execute and be bound by each Variation, unless the Contractor promptly gives notice to the **Engineer** stating (with supporting particulars) that ◀ the Contractor cannot readily obtain the Goods required for the Variation. ◀ Upon receiving this notice, the **Engineer** shall cancel, confirm or vary the instruction.	The Contractor shall execute and be bound by each Variation, unless the Contractor promptly gives notice to the **Engineer** stating (with supporting particulars) that (i) the Contractor cannot readily obtain the Goods required for the Variation, or (ii) it will reduce the safety or suitability of the Works, or (iii) it will have an adverse impact on the achievement of the Schedule of Guarantees. Upon receiving this notice, the **Engineer** shall cancel, confirm or vary the instruction.	The Contractor shall execute and be bound by each Variation, unless the Contractor promptly gives notice to the **Employer** stating (with supporting particulars) that (i) the Contractor cannot readily obtain the Goods required for the Variation, (ii) it will reduce the safety or suitability of the Works, or (iii) it will have an adverse impact on the achievement of the Performance Guarantees. Upon receiving this notice, the **Employer** shall cancel, confirm or vary the instruction. ◀
			Each Variation may include: (a) changes to the quantities of any item of work included in the Contract (however, such changes do not necessarily constitute a Variation), (b) changes to the quality and other characteristics of any item of work,		

Clause/ sub-clause	1999 Red Book	1999 Yellow Book	The Silver Book
	(c) changes to the levels, positions and/or dimensions of any part of the Works, (d) omission of any work unless it is to be carried out by others, (e) any additional work, Plant, Materials or services necessary for the Permanent Works, including any associated Tests on Completion, boreholes and other testing and exploratory work, or (f) changes to the sequence or timing of the execution of the Works. The Contractor shall not make any alteration and/or modification of the Permanent Works, unless and until the Engineer instructs or approves a Variation.		
13.2	Value Engineering	No change	No change
	The Contractor may, at any time, submit to the *Engineer* a written proposal which (in the Contractor's opinion) will, if adopted, (i) accelerate completion, (ii) reduce the cost to the Employer of executing, maintaining or operating the Works, (iii) improve the efficiency or value to the Employer of the completed Works, or (iv) otherwise be of benefit to the Employer. If a proposal, which is approved by the Engineer, includes a change in the design of part of the Permanent Works, then unless otherwise agreed by both Parties:	No change ◀	The Contractor may, at any time, submit to the *Employer* a written proposal which (in the Contractor's opinion) will, if adopted, (i) accelerate completion, (ii) reduce the cost to the Employer of executing, maintaining or operating the Works, (iii) improve the efficiency or value to the Employer of the completed Works, or (iv) otherwise be of benefit to the Employer. ◀

	(a) the Contractor shall design this part, (b) sub-paragraphs (a) to (d) of Sub-Clause 4.1 [*Contractor's General Obligations*] shall apply, and (c) if this change results in a reduction in the contract value of this part, the Engineer shall proceed in accordance with Sub-Clause 3.5 [*Determinations*] to agree or determine a fee, which shall be included in the Contract Price. This fee shall be half (50%) of the difference between the following amounts: (i) such reduction in contract value, resulting from the change, excluding adjustments under Sub-Clause 13.7 [*Adjustments for Changes in Legislation*] and Sub-Clause 13.8 [*Adjustments for Changes in Cost*], and (ii) the reduction (if any) in the value to the Employer of the varied works, taking account of any reductions in quality, anticipated life or operational efficiencies. However, if amount (i) is less than amount (ii), there shall not be a fee.	
13.3	Variation Procedure	No change
	If the ***Engineer*** requests a proposal, prior to instructing a Variation, the Contractor shall respond in writing as soon as practicable, either by giving reasons why he cannot comply (if this is the case) or by submitting:	No change If the ***Employer*** requests a proposal, prior to instructing a Variation, the Contractor shall respond in writing as soon as practicable, either by giving reasons why he cannot comply (if this is the case) or by submitting:

Clause/ sub-clause	1999 Red Book	1999 Yellow Book	The Silver Book
	(a) a description of the proposed ▲ work to be performed and a programme for its execution, (b) the Contractor's proposal for any necessary modifications to the programme according to Sub-Clause 8.3 *[Programme]* and to the Time for Completion, and (c) the Contractor's proposal for *evaluation of the Variation*.	(a) a description of the proposed design and/or work to be performed and a programme for its execution, (b) the Contractor's proposal for any necessary modifications to the programme according to Sub-Clause 8.3 *[Programme]* and to the Time for Completion, and (c) the Contractor's proposal for *adjustment to the Contract Price*.	(a) a description of the proposed design and/or work to be performed and a programme for its execution, (b) the Contractor's proposal for any necessary modifications to the programme according to Sub-Clause 8.3 *[Programme]* and to the Time for Completion, and (c) the Contractor's proposal for *adjustment to the Contract Price*.
	The *Engineer* shall, as soon as practicable after receiving such proposal (under Sub-Clause 13.2 *[Value Engineering]* or otherwise), respond with approval, disapproval or comments. The Contractor shall not delay any work whilst awaiting a response.	No change	The *Employer* shall, as soon as practicable after receiving such proposal (under Sub-Clause 13.2 *[Value Engineering]* or otherwise), respond with approval, disapproval or comments. The Contractor shall not delay any work whilst awaiting a response.
	Each instruction to execute a Variation, with any requirements for the recording of Costs, shall be issued by the *Engineer* to the Contractor, who shall acknowledge receipt.	No change	Each instruction to execute a Variation, with any Requirements for the recording of Costs, shall be issued by the *Employer* to the Contractor, who shall acknowledge receipt.
	Each Variation shall be evaluated in accordance with Clause 12 [Measurement and Evaluation], unless the Engineer instructs or approves otherwise in accordance with this Clause.	*Upon instructing or approving a Variation, the Engineer shall proceed in accordance with Sub-Clause 3.5 [Determinations] to agree or determine adjustments to the Contract Price and the Schedule of Payments. These adjustments shall include reasonable profit, and shall take account of the Contractor's submissions under Sub-Clause 13.2 [Value Engineering] if applicable.*	Upon instructing or approving a Variation, the *Employer* shall proceed in accordance with Sub-Clause 3.5 *[Determinations]* to agree or determine adjustments to the Contract Price and the Schedule of Payments. These adjustments shall include reasonable profit, and shall take account of the Contractor's submissions under Sub-Clause 13.2 *[Value Engineering]* if applicable.

A Precise Record of Alterations, Omissions and Additions

13.5	Provisional Sums	No change	No change
	Each Provisional Sum shall only be used, in whole or in part, in accordance with the **Engineer's** instructions, and the Contract Price shall be adjusted accordingly. The total sum paid to the Contractor shall include only such amounts, for the work, supplies or services to which the Provisional Sum relates, as the **Engineer** shall have instructed. For each Provisional Sum, the **Engineer** may instruct:	Each Provisional Sum shall only be used, in whole or in part, in accordance with the **Engineer's** instructions, and the Contract Price shall be adjusted accordingly. The total sum paid to the Contractor shall include only such amounts, for the work, supplies or services to which the Provisional Sum relates, as the Engineer shall have instructed. For each Provisional Sum, the **Engineer** may instruct:	Each Provisional Sum shall only be used, in whole or in part, in accordance with the **Employer's** instructions, and the Contract Price shall be adjusted accordingly. The total sum paid to the Contractor shall include only such amounts, for the work, supplies or services to which the Provisional Sum relates, as the **Employer** shall have instructed. For each Provisional Sum, the **Employer** may instruct:
	(b) Plant, Materials or services to be purchased by the Contractor, <u>from a nominated Subcontractor (as defined in Clause 5 [Nominated Subcontractors]) or otherwise;</u> and for which there shall be included in the Contract Price:	(b) Plant, Materials or services to be purchased by the Contractor, ▲ for which there shall be included in the Contract Price:	(b) Plant, Materials or services to be purchased by the Contractor, ▲ for which there shall be added to the Contract Price less the original Provisional Sums:
	(ii) a sum for overhead charges and profit, calculated as a percentage of these actual amounts by applying the relevant percentage rate (if any) stated in the **appropriate Schedule**. <u>If there is no such rate, the percentage rate stated in the Appendix to Tender shall be applied.</u>	No change	(ii) a sum for overhead charges and profit, calculated as a percentage of these actual amounts by applying the relevant percentage rate (if any) stated in the ***Contract***. ▲
	The Contractor shall, when required by the **Engineer**, produce quotations, invoices, vouchers and accounts or receipts in substantiation.	No change	The Contractor shall, when required by the **Employer**, produce quotations, invoices, vouchers and accounts or receipts in substantiation.

Clause/ sub-clause	1999 Red Book	1999 Yellow Book	The Silver Book
13.6	Daywork	No change	No change
	For work of a minor or incidental nature, the *Engineer* may instruct that a Variation shall be executed on a daywork basis.	No change	For work of a minor or incidental nature, the *Employer* may instruct that a Variation shall be executed on a daywork basis.
	Before ordering Goods for the work, the Contractor shall submit quotations to the *Engineer*.	No change	Before ordering Goods for the work, the Contractor shall submit quotations to the *Employer*.
	Except for any items for which the Daywork Schedule specifies that payment is not due, the Contractor shall deliver each day to the *Engineer* accurate statements in duplicate which shall include the following details of the resources used in executing the previous day's work:	No change	Except for any items for which the Daywork Schedule specifies that payment is not due, the Contractor shall deliver each day to the *Employer* accurate statements in duplicate which shall include the following details of the resources used in executing the previous day's work:
	One copy of each statement will, if correct, or when agreed, be signed by the *Engineer* and returned to the Contractor. The Contractor shall then submit priced statements of these resources to the *Engineer*, prior to their inclusion in the next Statement under Sub-Clause 14.3 *[Application for Interim Payment Certificates]*.	No change	One copy of each statement will, if correct, or when agreed, be signed by the *Employer* and returned to the Contractor. The Contractor shall then submit priced statements of these resources to the *Employer*, prior to their inclusion in the next Statement under Sub-Clause 14.3 *[Application for Interim Payment Certificates]*.
13.7	Adjustment for Changes in Legislation	No change	No change
	After receiving this notice, the *Engineer* shall proceed in accordance with Sub-Clause 3.5 *[Determinations]* to agree or determine these matters.	No change	After receiving this notice, the *Employer* shall proceed in accordance with Sub-Clause 3.5 *[Determinations]* to agree or determine these matters.

13.8	Adjustment for Changes in Cost	No change	No change
	In this Sub-Clause, 'table of adjustment data' means the completed table of adjustment data included in the Appendix to Tender. If there is no such table of adjustment data, this Sub-Clause shall not apply.		*If the Contract Price is to be adjusted for rises or falls in the cost of labour, Goods and other inputs to the Works, the adjustments shall be calculated in accordance with the provisions in the Particular Conditions.*
	If this Sub-Clause applies, the amounts payable to the Contractor shall be adjusted for rises or falls in the cost of labour, Goods and other inputs to the Works, by the addition or deduction of the amounts determined by the formulae prescribed in this Sub-Clause. To the extent that full compensation for any rise or fall in Costs is not covered by the provisions of this or other Clauses, the Accepted Contract Amount shall be deemed to have included amounts to cover the contingency of other rises and falls in costs.		◄
	The adjustment to be applied to the amount otherwise payable to the Contractor, as valued in accordance with the appropriate Schedule and certified in Payment Certificates, shall be determined from formulae for each of the currencies in which the Contract Price is payable. No adjustment is to be applied to work valued on the basis of Cost or current prices. The formulae shall be of the following general type:		◄
	$$P_n = a + b\frac{L_n}{L_o} + c\frac{E_n}{E_o} + d\frac{M_n}{M_o} + \ldots$$		◄

Clause/ sub-clause	1999 Red Book	1999 Yellow Book	The Silver Book
	where:		
	'**Pn**' is the adjustment multiplier to be applied to the estimated contract value in the relevant currency of the work carried out in period 'n', this period being a month unless otherwise stated in the Appendix to Tender;		◀
	'**a**' is a fixed coefficient, stated in the relevant table of adjustment data, representing the non-adjustable portion in contractual payments;		◀
	'**b**', '**c**', '**d**', . . . are coefficients representing the estimated proportion of each cost element related to the execution of the Works, as stated in the relevant table of adjustment data; such tabulated cost elements may be indicative of resources such as labour, equipment and materials;		◀
	'**Ln**', '**En**', '**Mn**', . . . are the current cost indices or reference prices for period 'n', expressed in the relevant currency of payment, each of which is applicable to the relevant tabulated cost element on the date 49 days prior to the last day of the period (to which the particular Payment Certificate relates); and		◀
	'**Lo**', '**Eo**', '**Mo**', . . . are the base cost indices or reference prices, expressed in the relevant currency of payment, each of which is applicable to the relevant tabulated cost element on the Base Date.		◀

The cost indices or reference prices stated in the table of adjustment data shall be used. If their source is in doubt, it shall be determined by the Engineer. For this purpose, reference shall be made to the values of the indices at stated dates (quoted in the fourth and fifth columns respectively of the table) for the purposes of clarification of the source: although these dates (and thus these values) may not correspond to the base cost indices.

In cases where the 'currency of index' (stated in the table) is not the relevant currency of payment, each index shall be converted into the relevant currency of payment at the selling rate, established by the central bank of the Country, of this relevant currency on the above date for which the index is required to be applicable.

Until such time as each current cost index is available, the Engineer shall determine a provisional index for the issue of Interim Payment Certificates. When a current cost index is available, the adjustment shall be recalculated accordingly.

If the Contractor fails to complete the Works within the Time for Completion, adjustment of prices thereafter shall be made using either each index or price applicable on the date (i) 49 days prior to the expiry of the Time for Completion of the Works, or (ii) the current index or price; whichever is more favourable to the Employer.

Clause/ sub-clause	1999 Red Book	1999 Yellow Book	The Silver Book
	The weightings (coefficients) for each of the factors of cost stated in the table(s) of adjustment data shall only be adjusted if they have been rendered unreasonable, unbalanced or inapplicable, as a result of Variations.		◄
Clause 14	Contract Price and Payment		
14.1	The Contract Price	No change	No change
	Unless otherwise stated in the Particular Conditions:	Unless otherwise stated in the Particular Conditions:	Unless otherwise stated in the Particular Conditions:
	(a) *the Contract Price shall be agreed or determined under Sub-Clause 12.3 [Evaluation] and be subject to adjustments in accordance with the Contract;*	(a) *the Contract Price shall be the lump sum Accepted Contract Amount and be subject to adjustments in accordance with the Contract;*	(a) *payment for the Works shall be made on the basis of the lump sum Contract Price, subject to adjustments in accordance with the Contract;*
	(c) any quantities which may be set out in the Bill of Quantities or other Schedule are estimated quantities and are not to be taken as the actual and correct quantities: ◄ (i) of the Works which the Contractor is required to execute, or (ii) for the purposes of Clause 12 [Measurement and Evaluation]; and	(c) *any quantities which may be set out in a Schedule are estimated quantities and are not to be taken as the actual and correct quantities of the Works which the Contractor is required to execute;* and ◄	◄
	(d) *the Contractor shall submit to the Engineer, within 28 days after the Commencement Date, a proposed breakdown of each lump sum price in the Schedules. The Engineer may take account of the breakdown when preparing Payment Certificates, but shall not be bound by it.*	(d) *any quantities or price data which may be set out in a Schedule shall be used for the purposes stated in the Schedule and may be inapplicable for other purposes.*	◄

A Precise Record of Alterations, Omissions and Additions

	▲	However, if any part of the Works is to be paid according to quantity supplied or work done, the provisions for measurement and evaluation shall be as stated in the Particular Conditions. The Contract Price shall be determined accordingly, subject to adjustments in accordance with the Contract.	▲
14.2	Advance Payment	No change	No change
	The Employer shall make an advance payment, as an interest-free loan for mobilization ▲, when the Contractor submits a guarantee in accordance with this Sub-Clause. *The total advance payment, the number and timing of instalments (if more than one), and the applicable currencies and proportions, shall be as stated in the Appendix to Tender.* *Unless and until the Employer receives this guarantee, or if the total advance payment is not stated in the Appendix to Tender, this Sub-Clause shall not apply.*	The Employer shall make an advance payment, as an interest-free loan for mobilization and design, when the Contractor submits a guarantee in accordance with this Sub-Clause. *The total advance payment, the number and timing of instalments (if more than one), and the applicable currencies and proportions, shall be as stated in the Appendix to Tender.* *Unless and until the Employer receives this guarantee, or if the total advance payment is not stated in the Appendix to Tender, this Sub-Clause shall not apply.*	The Employer shall make an advance payment, as an interest-free loan for mobilization and design, when the Contractor submits a guarantee in accordance with this Sub-Clause *including the details stated in the Particular Conditions. If the Particular Conditions do not state:* *(a) the amount of the advance payment, then this Sub-Clause shall not apply;* *(b) the number and timing of instalments, then there shall be only one;* *(c) the applicable currencies and proportions, then they shall be those in which the Contract Price is payable; and/or* *(d) the amortisation rate for repayments, then it shall be calculated by dividing the total amount of the advance payment by the Contract Price stated in the Contract Agreement less Provisional Sums.*

Clause/sub-clause	1999 Red Book	1999 Yellow Book	The Silver Book
	The Engineer shall *issue an Interim Payment Certificate for* the first instalment after receiving ▲ a Statement (under Sub-Clause 14.3 *[Application for Interim Payment Certificates]* and after the Employer receives *(i)* the Performance Security in accordance with Sub-Clause 4.2 [Performance Security] and *(ii)* a guarantee in amounts and currencies equal to the advance payment. This guarantee shall be issued by an entity and from within a country (or other jurisdiction) approved by the Employer, and shall be in the form annexed to the Particular Conditions or in another form approved by the Employer. ▲	No change	The Employer shall *pay the* first instalment after receiving (i) a Statement (under Sub-Clause 14.3 *[Application for Interim Payments]*), ▲ *(ii)* the Performance Security in accordance with Sub-Clause 4.2 [Performance Security], and *(iii)* a guarantee in amounts and currencies equal to the advance payment. This guarantee shall be issued by an entity and from within a country (or other jurisdiction) approved by the Employer, and shall be in the form annexed to the Particular Conditions or in another form approved by the Employer. <u>Unless and until the Employer receives this guarantee, this Sub-Clause shall not apply.</u>
	The Contractor shall ensure that the guarantee is valid and enforceable until the advance payment has been repaid, but its amount shall be progressively reduced by the amount repaid by the Contractor <u>as indicated in the Payment Certificates.</u>	No change	The Contractor shall ensure that the guarantee is valid and enforceable until the advance payment has been repaid, but its amount shall be progressively reduced by the amount repaid by the Contractor ▲.
	The advance payment shall be repaid through *percentage* deductions in *Payment Certificates. Unless other percentages are stated in the Appendix to Tender:* *(a) deductions shall commence in the Payment Certificate in which the total of all certified interim payments (excluding the advance payment and deductions and repayments of retention) exceeds ten per cent (10%) of the Accepted Contract Amount less Provisional Sums; and*	No change	The advance payment shall be repaid through *proportional* deductions in *interim payments. Deductions shall be made at the amortization rate stated in the Particular Conditions (or, if not so stated, as stated in sub-paragraph (d) above), which shall be applied to the amount otherwise due (excluding the advance payment and deductions and repayments of retention), until such time as the advance payment has been repaid.*

A Precise Record of Alterations, Omissions and Additions 717

	(b) deductions shall be made at the amortization rate of one quarter (25%) of the amount of each Payment Certificate (excluding the advance payment and deductions and repayments of retention) in the currencies and proportions of the advance payment, until such time as the advance payment has been repaid.	No change	No change
14.3	Application for Interim Payment Certificates	No change	
	The Contractor shall submit a Statement in six copies to the **Engineer** after the end of **each month**, in a form approved by the **Engineer**, showing the detail of the amounts to which the Contractor considers himself to be entitled, together with supporting documents which shall include the report on the progress during this month in accordance with Sub-Clause 4.21 *[Progress Reports]*.	The Contractor shall submit a Statement in six copies to the **Engineer** after the end of **the period of payment stated in the Contract (if not stated, after the end of each month)**, showing the detail of the amounts to which the Contractor considers himself to be entitled, in a form approved by the **Engineer**, together with supporting documents which shall include the report on the progress during this month in accordance with Sub-Clause 4.21 *[Progress Reports]*.	The Contractor shall submit a Statement in six copies to the **Employer** after the end of **the period of payment stated in the Contract (if not stated, after the end of each month)**, showing the detail of the amounts to which the Contractor considers himself to be entitled, in a form approved by the **Employer**, together with supporting documents which shall include the report on the progress during this month in accordance with Sub-Clause 4.21 *[Progress Reports]*.
	(a) the estimated contract value of the Works executed and the Contractor's Documents produced up to the end of the month (including Variations but excluding items described in sub-paragraphs *(b) to (g)* below);	No change	(a) the estimated contract value of the Works executed and the Contractor's Documents produced up to the end of the month (including Variations but excluding items described in sub-paragraphs *(b) to (f)* below);
	(c) any amount to be deducted for retention, calculated by applying the percentage of retention stated in the Appendix to Tender to the total of the above amounts, until the amount so retained by the Employer reaches the limit of Retention Money (if any) stated in the Appendix to Tender;	No change	(c) any amount to be deducted for retention, calculated by applying the percentage of retention stated in the **Particular Conditions** to the total of the above amounts, until the amount so retained by the Employer reaches the limit of Retention Money (if any) stated in the **Particular Conditions**;

Clause/sub-clause	1999 Red Book	1999 Yellow Book	The Silver Book
	(e) any amounts to be added and deducted for Plant and Materials in accordance with Sub-Clause 14.5 [Plant and Materials Intended for the Works];	No change	◄
	(f) any other additions or deductions which may have become due under the Contract or otherwise, including those under Clause 20 [Claims, Disputes and Arbitration]	No change	(e) any other additions or deductions which may have become due under the Contract or otherwise, including those under Clause 20 [Claims, Disputes and Arbitration]
	(g) the deduction of amounts certified in all previous Payment Certificates.	No change	(f) the deduction of amounts certified in all previous Payment Certificates.
14.4	Schedule of Payments	No change	No change
	(a) the instalments quoted in this schedule of payments shall be the estimated contract values for the purposes of sub-paragraph (a) of Sub-Clause 14.3 [Application for Interim Payment Certificates] ◄	No change	(a) the instalments quoted in the Schedule of Payments shall be the estimated contract values or the purposes of sub-paragraph (a) of Sub-Clause 14.3 [Application for Interim Payments ◄], subject to Sub-Clause 14.5 [Plant and Materials intended for the Works]; and
	(b) Sub-Clause 14.5 [Plant and Materials intended for the Works] shall not apply; and	No change	◄
	(c) if these instalments are not defined by reference to the actual progress achieved in executing the Works, and if actual progress is found to be less than that on which this schedule of payments was based, then the *Engineer* may proceed in accordance with Sub-Clause 3.5 [Determinations] to agree or determine revised instalments, which shall take account of the extent to which progress is less than that on which the instalments were previously based.	No change	(b) if these instalments are not defined by reference to the actual progress achieved in executing the Works, and if actual progress is found to be less than that on which this schedule of payments was based, then the *Employer* may proceed in accordance with Sub-Clause 3.5 [Determinations] to agree or determine revised instalments, which shall take account of the extent to which progress is less than that on which the instalments were previously based.

	If the Contract does not include a schedule of payments, the Contractor shall submit non-binding estimates of the payments which he expects to become due during each quarterly period.	If the Contract does not include a Schedule of Payments, the Contractor shall submit non-binding estimates of the payments which he expects to become due during each quarterly period.	If the Contract does not include a Schedule of Payments, the Contractor shall submit non-binding estimates of the payments which he expects to become due during each quarterly period.
14.5	Plant and Materials Intended for the Works	No change	No change

	If this Sub-Clause applies, Interim Payment Certificates shall include, under subparagraph (e) of Sub-Clause 14.3, (i) an amount for Plant and Materials which have been sent to the Site for incorporation in the Permanent Works, and (ii) a reduction when the contract value of such Plant and Materials is included as part of the Permanent Works under sub-paragraph (a) of Sub-Clause 14.3 [Application for Interim Payment Certificates].
	
	If the lists referred to in sub-paragraphs (b)(i) or (c)(i) below are not included in the Appendix to Tender, this Sub-Clause shall not apply.
	
	The Engineer shall determine and certify each addition if the following conditions are satisfied:
	
	(a) the Contractor has:
	    (i) kept satisfactory records (including the orders, receipts, Costs and use of Plant and Materials) which are available for inspection, and
	    (ii) submitted a statement of the Cost of acquiring and delivering the Plant and Materials to the Site, supported by satisfactory evidence;
		No change	*If the Contractor is entitled, under the Contract, to an interim payment for Plant and Materials which are not yet on the Site, the Contractor shall*
			
			Nevertheless not be entitled to such payment unless:
			
			(a) the relevant Plant and Materials are in the Country and have been marked as the Employer's property in accordance with the Employer's instructions; or
			(b) the Contractor has delivered, to the Employer, evidence of insurance and a bank guarantee in a form and issued by an entity approved by the Employer in amounts and currencies equal to such payment. This guarantee may be in a similar form to the form referred to in Sub-Clause 14.2 [Advance Payment] and shall be valid until the Plant and Materials are properly stored on Site and protected against loss, damage or deterioration.

Clause/sub-clause	1999 Red Book	1999 Yellow Book	The Silver Book
	and either: (b) the relevant Plant and Materials: (i) are those listed in the Appendix to Tender for payment when shipped, (ii) have been shipped to the Country, en route to the Site, in accordance with the Contract; and (iii) are described in a clean shipped bill of lading or other evidence of shipment, which has been submitted to the Engineer together with evidence of payment of freight and insurance, any other documents reasonably required, and a bank guarantee in a form and issued by an entity approved by the Employer in amounts and currencies equal to the amount due under this Sub-Clause: this guarantee may be in a similar form to the form referred to in Sub-Clause 14.2 [Advance Payment] and shall be valid until the Plant and Materials are properly stored on Site and protected against loss, damage or deterioration; or (c) the relevant Plant and Materials: (i) are those listed in the Appendix to Tender for payment when delivered to the Site, and		

A Precise Record of Alterations, Omissions and Additions 721

	(ii) have been delivered to and are properly stored on the Site, are protected against loss, damage or deterioration, and appear to be in accordance with the Contract. *The additional amount to be certified shall be the equivalent of eighty percent of the Engineer's determination of the cost of the Plant and Materials (including delivery to Site), taking account of the documents mentioned in this Sub-Clause and of the contract value of the Plant and Materials.* *The currencies for this additional amount shall be the same as those in which payment will become due when the contract value is included under subparagraph (a) of Sub-Clause 14.3 [Application for Interim Payment Certificates]. At that time, the Payment Certificate shall include the applicable reduction which shall be equivalent to, and in the same currencies and proportions as, this additional amount for the relevant Plant and Materials.*		
14.6	Issue of Interim Payment Certificates	No change	
	No amount will be <u>certified</u> or paid until the Employer has received and approved the Performance Security. Thereafter, the *Engineer* shall, within 28 days after receiving a Statement and supporting documents, issue to the Employer an Interim Payment Certificate which shall state the amount which the Engineer fairly determines to be due, with supporting particulars.	No change	Interim Payments
			No amount will be ▲ paid until the Employer has received and approved the Performance Security. Thereafter, the *Employer* shall within 28 days after receiving a Statement and supporting documents, ***give to the Contractor notice of any items in the Statement with which the Employer disagrees, with supporting particulars. Payment due shall not be withheld, except that:***

Clause/sub-clause	1999 Red Book	1999 Yellow Book	The Silver Book
14.7	(b) if the Contractor was or is failing to perform any work or obligation in accordance with the Contract, and had been so notified by the **Engineer**, the value of this work or obligation may be withheld until the work or obligation has been performed.	No change	(b) if the Contractor was or is failing to perform any work or obligation in accordance with the Contract, and had been so notified by the **Employer**, the value of this work or obligation may be withheld until the work or obligation has been performed.
	The **Engineer** may in any Payment Certificate make any correction or modification that should properly be made to any *previous Payment Certificate*. *A Payment Certificate* shall not be deemed to indicate the **Engineer's** acceptance, approval, consent or satisfaction.	No change	The **Employer** may, by any payment, make any correction or modification that should properly be made to any *amount previously considered due*. *Payment* shall not be deemed to indicate the **Employer's** acceptance, approval, consent or satisfaction.
	Payment	No change	Timing Of Payments
	▲ The Employer shall pay to the Contractor:	No change	Except as otherwise stated in Sub-Clause 2.5 [*Employer's Claims*], the Employer shall pay to the Contractor:
	(a) the first instalment of the advance payment within 42 days after *issuing the Letter of Acceptance or within 21 days after receiving* the documents in accordance with Sub-Clause 4.2 [*Performance Security*] and Sub-Clause 14.2 [*Advance Payment*], whichever is later;	No change	(a) the first instalment of the advance payment within 42 days after *the date on which the Contract came into full force and effect or within 21 days after the Employer receives* the documents in accordance with Sub-Clause 4.2 [*Performance Security*] and Sub-Clause 14.2 [*Advance Payment*], whichever is later;
	(b) the amount *certified in each Interim Payment Certificate* within 56 days after *the Engineer receives* the Statement and supporting documents; and	No change	(b) the amount *which is due in respect of each Statement, other than the Final Statement*, within 56 days after *receiving* the Statement and supporting documents; and

A Precise Record of Alterations, Omissions and Additions 723

		(c) the amount certified in the Final Payment Certificate within 56 days after the Employer receives this Payment Certificate.	No change	*(c) the final amount due, within 42 days after receiving the Final Statement and written discharge in accordance with Sub-Clause 14.11 [Application for Final Payment] and Sub-Clause 14.12 [Discharge].*
14.8	Delayed Payment	No change	No change
		If the Contractor does not receive payment in accordance with Sub-Clause 14.7 [Payment], the Contractor shall be entitled to receive financing charges compounded monthly on the amount unpaid during the period of delay. This period shall be deemed to commence on the date for payment specified in Sub-Clause 14.7 [Payment], irrespective (in the case of its sub-paragraph (b)) of the date on which any Interim Payment Certificate is issued.	No change	If the Contractor does not receive payment in accordance with Sub-Clause 14.7 [Timing of Payments], the Contractor shall be entitled to receive financing charges compounded monthly on the amount unpaid during the period of delay. ▲
14.9	Payment of Retention Money	No change	No change
		When the Taking-Over Certificate has been issued for the Works, ▲ the first half of the Retention Money shall be *certified by the Engineer for payment* to the Contractor. If a Taking-Over Certificate is issued for a Section or part of the Works, *a proportion* of the Retention Money shall be certified and paid ▲. This proportion shall be two-fifths (40%) of the proportion calculated by dividing the estimated contract value of the Section or part, by the estimated final Contract Price.	When the Taking-Over Certificate has been issued for the Works, and the Works have passed all specified tests (including the Tests after Completion, if any), the first half of the Retention Money shall be *certified by the Engineer for payment* to the Contractor. If a Taking-Over Certificate is issued for a Section ▲, *the relevant percentage of the first half* of the Retention Money shall be certified and paid when the Section passes all tests. ▲	When the Taking-Over Certificate has been issued for the Works, and the Works have passed all specified tests (including the Tests after Completion, if any), the first half of the Retention Money shall be *paid* to the Contractor. If a Taking-Over Certificate is issued for a Section ▲, *the relevant percentage of the first half* of the Retention Money shall be ▲ paid when the Section passes all tests. ▲

Clause/sub-clause	1999 Red Book	1999 Yellow Book	The Silver Book
	Promptly after the latest of the expiry dates of the Defects Notification Periods, the outstanding balance of the Retention Money shall be *certified by the Engineer for payment* to the Contractor. If a Taking-Over Certificate was issued for a Section, *a proportion* of the second half of the Retention Money shall be certified and paid promptly after the expiry date of the Defects Notification Period for the Section. This proportion shall be two-fifths (40%) of the proportion calculated by dividing the estimated contract value of the Section by the estimated final Contract Price.	Promptly after the latest of the expiry dates of the Defects Notification Periods, the outstanding balance of the Retention Money shall be *certified by the Engineer for payment* to the Contractor. If a Taking-Over Certificate was issued for a Section, *the relevant percentage* of the second half of the Retention Money shall be certified and paid promptly after the expiry date of a Defects Notification Period for the Section. ◀	Promptly after the latest of the expiry dates of the Defects Notification Periods, the outstanding Balance of the Retention Money shall be *paid* to the Contractor. If a Taking-Over Certificate was issued for a Section, *the relevant percentage* of the second half of the Retention Money shall be ▲ paid promptly after the expiry date of a Defects Notification Period for the Section. ◀
	However, if any work remains to be executed under Clause 11 *[Defects Liability]* ▲, the *Engineer* shall be entitled to withhold certification of the estimated cost of this work until it has been executed.	However, if any work remains to be executed under Clause 11 *[Defects Liability]* or Clause 12 *[Tests after Completion]*, the *Engineer* shall be entitled to withhold certification of the estimated cost of this work until it has been executed.	However, if any work remains to be executed under Clause 11 *[Defects Liability]* or Clause 12 *[Tests after Completion]*, the *Employer* shall be entitled to withhold the estimated cost of this work until it has been executed.
	When calculating these proportions, no account shall be taken of any adjustments under Sub-Clause 13.7 [Adjustments for Changes in Legislation] and Sub-Clause 13.8 [Adjustment for Changes in Cost].	*The relevant percentage for each Section shall be the percentage value of the Section as stated in the Appendix to Tender. If the percentage value of a Section is not stated in the Appendix to Tender, no percentage of either half of the Retention Money shall be released under this Sub-Clause in respect of such Section.*	*The relevant percentage for each Section shall be the percentage value of the Section as stated in the Contract. If the percentage value of a Section is not stated in the Contract, no percentage of either half of the Retention Money shall be released under this Sub-Clause in respect of such Section.*

14.10	Statement at Completion	No change	Within 84 days after receiving the Taking-Over Certificate for the Works, the Contractor shall submit to the **Employer** six copies of a Statement at completion with supporting documents, in Accordance with Sub-Clause 14.3 *[Application for Interim Payment* ▲*]*, showing:
	Within 84 days after receiving the Taking-Over Certificate for the Works, the Contractor shall submit to the **Engineer** six copies of a Statement at completion with supporting documents, in accordance with Sub-Clause 14.3 *[Application for Interim Payment Certificates]*, showing:	Within 84 days after receiving the Taking-Over Certificate for the Works, the Contractor shall submit to the **Engineer** six copies of a Statement at completion with supporting documents, in accordance with Sub-Clause 14.3 *[Application for Interim Payment Certificates]*, showing:	
	The **Engineer** shall then **certify** in accordance with Sub-Clause 14.6 *[Issue of Interim Payment Certificates]* ▲.	No change	The **Employer** shall then **give notice to the Contractor** in accordance with Sub-Clause 14.6 *[Interim Payments]* and make payment in accordance with Sub-Clause 14.7 *[Timing of Payments]*.
14.11	Application for Final Payment Certificates	No change	Application for Final Payment
	Within 56 days after receiving the Performance Certificate, the Contractor shall submit, to the **Engineer**, six copies of a draft final statement with supporting documents showing in detail in a form approved by the **Engineer**:	No change	Within 56 days after receiving the Performance Certificate, the Contractor shall submit, to the **Employer**, six copies of a draft final statement with supporting documents showing in detail in a form approved by the **Employer**:
	If the **Engineer** disagrees with or cannot verify any part of the draft final statement, the Contractor shall submit such further information as the **Engineer** may reasonably require and shall make such changes in the draft as may be agreed between them. The Contractor shall then prepare and submit to the **Engineer** the final statement as agreed.	No change	If the **Employer** disagrees with or cannot verify any part of the draft final statement, the Contractor shall submit such further information as the **Employer** may reasonably require and shall make such changes in the draft as may be agreed between them. The Contractor shall then prepare and submit to the **Employer** the final statement as agreed.

Clause/sub-clause	1999 Red Book	1999 Yellow Book	The Silver Book
	However if, following discussions between the **Engineer and the Contractor** and any changes to the final draft final statement which are agreed, it becomes evident that a dispute exists, **the Engineer shall deliver to the Employer (with a copy to the Contractor) an Interim Payment Certificate for the agreed parts of the draft final statement**. Thereafter, if the dispute is finally resolved under Sub-Clause 20.4 *[Obtaining Dispute Adjudication Board's Decision]* or Sub-Clause 20.5 *[Amicable Settlement]*, the Contractor shall then prepare and submit to the Employer (with a copy to the Engineer) a Final Statement.	No change	However, if, following discussions between **the Parties** and any changes to the final draft final statement which are agreed, it becomes evident that a dispute exists, **the Employer shall pay the agreed parts of the draft final statement in accordance with Sub-Clause 14.6 *[Interim Payments]* and Sub-Clause 14.7 *[Timing of Payments]*.** Thereafter, if the dispute is finally resolved under Sub-Clause 20.4 *[Obtaining Adjudication Board's Decision]* or Sub-Clause 20.5 *[Amicable Settlement]*, the Contractor shall then prepare and submit to the Employer ▲a Final Statement.
14.13	Issue of Final Payment Certificate	No change	Final Payment
	Within 28 days after receiving the Final Statement and written discharge in accordance with Sub-Clause 14.11 [Application for Final Payment Certificate] and Sub-Clause 14.12 [Discharge], the Engineer shall issue, to the Employer, the Final Payment Certificate which shall state: (a) the amount which is finally due, and (b) after giving credit to the Employer for all amounts previously paid by the Employer and for all sums to which the Employer is entitled, the balance (if any) due from the Employer to the Contractor or from the Contractor to the Employer, as the case may be.	No change	*In accordance with sub-paragraph (c) of Sub-Clause 14.7 [Timing of Payments], the Employer shall pay to the Contractor the amount which is finally due, less all amounts previously paid by the Employer and any deductions in accordance with Sub-Clause 2.5 [Employer's Claims].* ▲

A Precise Record of Alterations, Omissions and Additions 727

	If the Contractor has not applied for a Final Payment Certificate in accordance with Sub-Clause 14.11 [Application for Final Payment Certificate] and Sub-Clause 14.12 [Discharge], the Engineer shall request the Contractor to do so. If the Contractor fails to submit an application within a period of 28 days, the Engineer shall issue the Final Payment Certificate for such amount as he fairly determines to be due.	No change	◂
14.15	Currencies of Payment	No change	No change
	The Contract Price shall be paid in the currency or currencies named in the **Appendix to Tender**.	No change	The Contract Price shall be paid in the currency or currencies named in the **Contract Agreement**.
	(i) the proportions or amounts of the Local and Foreign Currencies, and the fixed rates of exchange to be used for calculating the payments, shall be as stated in the **Appendix to Tender**,	No change	(i) the proportions or amounts of the Local and Foreign Currencies, and the fixed rates of exchange to be used for calculating the payments, shall be as stated in the **Contract Agreement**,
	(b) payment of the damages specified in the **Appendix to Tender** shall be made in the currencies and proportions specified in the **Appendix to Tender**,	No change	(b) payment of the damages specified in the **Particular Conditions** shall be made in the currencies and proportions specified in the **Particular Conditions**;
	(e) if no rates of exchange are stated in the **Appendix to Tender**,	No change	(e) if no rates of exchange are stated in the **Contract**,
Clause 15	Termination by Employer	No change	No change
15.1	Notice to Correct	No change	No change
	If the Contractor fails to carry out any Obligation under the Contract, the **Engineer** may by notice require the Contractor to make good the failure and to remedy it within a specified reasonable time.	No change	If the Contractor fails to carry out any obligation under the Contract, the **Employer** may by notice require the Contractor to make good the failure and to remedy it within a specified reasonable time.

Clause/sub-clause	1999 Red Book	1999 Yellow Book	The Silver Book
15.2	Termination by the Employer	No change	No change
	(c) without reasonable excuse fails: (i) to proceed with the Works in accordance with Clause 8 *[Commencement, Delays and Suspension]*, or (ii) to comply with a notice issued under Sub-Clause 7.5 *[Rejection]* or Sub-Clause 7.6 *[Remedial Work]*, within 28 days after receiving it.	No change	(c) without reasonable excuse fails ◄ to proceed with the Works in accordance with Clause 8 *[Commencement, Delays and Suspension]*. ◄
	The Contractor shall then leave the Site and deliver any required Goods, all Contractor's Documents, and other design documents made by or for him, to the **Engineer**.	No change	The Contractor shall then leave the Site and deliver any required Goods, all Contractor's Documents, and other design documents made by or for him, to the **Employer**.
15.3	Valuation at Date of Termination	No change	No change
	As soon as practicable after a notice of termination under Sub-Clause 15.2 *[Termination by Employer]* has taken effect, the **Engineer** shall proceed in accordance with Sub-Clause 3.5 *[Determinations]* to agree or determine the value of the Works, Goods and Contractor's Documents, and any other sums due to the Contractor for work executed in accordance with the Contract.	No change	As soon as practicable after a notice of termination under Sub-Clause 15.2 *[Termination by Employer]* has taken effect, the **Employer** shall proceed in accordance with Sub-Clause 3.5 *[Determinations]* to agree or determine the value of the Works, Goods and Contractor's Documents, and any other sums due to the Contractor for work executed in accordance with the Contract.
15.4	Payment after Termination	No change	No change
	(b) withhold further payments to the Contractor until the costs of ◄ execution, completion and remedying of any defects, damages for delay in completion (if any), and all other costs incurred by the Employer, have been established, and/or	(b) withhold further payments to the Contractor until the costs of design, execution, completion and remedying of any defects, damages for delay in completion (if any), and all other costs incurred by the Employer, have been established, and/or	(b) withhold further payments to the Contractor until the costs of design, execution, completion and remedying of any defects, damages for delay in completion (if any), and all other costs incurred by the Employer, have been established, and/or

A Precise Record of Alterations, Omissions and Additions 729

Clause 16	Suspension and Termination by Contractor	No change	No change
16.1	Contractor's Entitlement to Suspend Work	No change	No change
	If the Engineer fails to certify in accordance with Sub-Clause 14.6 *[Issue of Interim Payment Certificates]* or the Employer fails to comply with Sub-Clause 2.4 *[Employer's Financial Arrangements]* or Sub-Clause 14.7 *[Payment]*, the Contractor may, after giving not less than 21 days' notice to the Employer, suspend work (or reduce the rate of work) unless and until the Contractor has received the Payment Certificate, reasonable evidence or payment, as the case may be and as described in the notice.		If the ▲ **Employer** fails to comply with Sub-Clause 2.4 *[Employer's Financial Arrangements]* or Sub-Clause 14.7 *[Timing of Payments]* the Contractor may, after giving not less than 21 days' notice to the Employer, suspend work (or reduce the rate of work) unless and until the Contractor has received the ▲ reasonable evidence or payment, as the case may be and as described in the notice.
	If the Contractor subsequently receives such Payment Certificate, evidence or payment (as described in the relevant Sub-Clause and in the above notice) before giving a notice of termination, the Contractor shall resume normal working as soon as is reasonably practicable.	No change	If the Contractor subsequently receives such ▲ evidence or payment (as described in the relevant Sub-Clause and in the above notice) before giving a notice of termination, the Contractor shall resume normal working as soon as is reasonably practicable.
	If the Contractor suffers delay and/or incurs Cost as a result of suspending work (or reducing the rate of work) in accordance with this Sub-Clause, the Contractor shall give notice to the **Employer** and shall be entitled subject to Sub-Clause 20.1 *[Contractor's Claims]* to:		If the Contractor suffers delay and/or incurs Cost as a result of suspending work (or reducing the rate of work) in accordance with this Sub-Clause, the Contractor shall give notice to the **Employer** and shall be entitled subject to Sub-Clause 20.1 *[Contractor's Claims]* to:
	After receiving this notice, the **Engineer** shall proceed in accordance with Sub-Clause 3.5 *[Determinations]* to agree or determine these matters.		After receiving this notice, the **Employer** shall proceed in accordance with Sub-Clause 3.5 *[Determinations]* to agree or determine these matters.

Clause/sub-clause	1999 Red Book	1999 Yellow Book	The Silver Book
16.2	Termination Contractor	No change	No change
	(b) the Engineer fails, within 56 days after receiving a Statement and supporting documents, to issue the relevant Payment Certificate.	No change	◄
	(c) the Contractor does not receive the amount due under an Interim Payment Certificate within 42 days after the expiry of the time stated in Sub-Clause 14.7 [Payment] within which payment is to be made (except for deductions in accordance with Sub-Clause 2.5 [Employer's Claims]),	No change	(b) the Contractor does not receive the amount due ◄ within 42 days after the expiry of the time stated in Sub-Clause 14.7 [Payment] within which payment is to be made (except for deductions in accordance with Sub-Clause 2.5 [Employer's Claims]),
	(e) the Employer fails to comply with Sub-Clause 1.6 [Contract Agreement] or Sub-Clause 1.7 [Assignment],	No change	(d) the Employer fails to comply with ◄ Sub-Clause 1.7 [Assignment],
	(f)	No change	(e)
	(g)	No change	(f)
	In any of these events or circumstances, the Contractor may, upon giving 14 days' notice to the Employer, terminate the Contract. However, in the case of sub-paragraph (f) or (g), the Contractor may by notice terminate the Contract immediately.	No change	In any of these events or circumstances, the Contractor may, upon giving 14 days' notice to the Employer, terminate the Contract. However, in the case of sub-paragraph (e) or (f), the Contractor may by notice terminate the Contract immediately.
16.3	Cessation of Work and Removal of Contractor's Equipment	No change	No change
	(a) cease all further work, except for such work as may have been instructed by the *Engineer* for the protection of life or property or for the safety of the Works.	No change	(a) cease all further work, except for such work as may have been instructed by the *Employer* for the protection of life or property or for the safety of the Works.

A Precise Record of Alterations, Omissions and Additions

Clause 17	Risk and Responsibility			
17.1	Indemnities	No change	No change	No change
	(a) bodily injury, sickness, disease or death of any person whatsoever arising out of or in the course of or by reason of the Contractor's design (if any). (i) arises out of or in the course of or by reason of the Contractor's design (if any). (ii) is attributable to any negligence, wilful act or breach of the Contract by the Contractor, the Contractor's Personnel, their respective agents, or anyone directly or indirectly employed by any of them..	No change	(a) bodily injury, sickness, disease or death of any person whatsoever arising out of or in the course of or by reason of the ▲ design ▲. (i) arises out of or in the course of or by reason of the ▲ design ▲.	(a) bodily injury, sickness, disease or death of any person whatsoever arising out of or in the course of or by reason of the ▲ design ▲. (i) arises out of or in the course of or by reason of the ▲ design ▲. (ii) is attributable to any negligence, wilful act or breach of the Contract by the Employer, the Employer's Personnel, their respective agents, or anyone directly or indirectly employed by any of them..
17.2	Contractor's Care of the Works	No change	No change	No change
	If a Taking-Over Certificate is issued (or is deemed to be issued) for any Section or part of the Works, responsibility for the care of the Section or part shall then pass to the Employer.	No change		If a Taking-Over Certificate is issued (or is deemed to be issued) for any Section ▲ of the Works, responsibility for the care of the Section ▲ shall then pass to the Employer.
17.3	Employer's Risks	No change	No change	No change
	(f) use or occupation by the Employer of any part of the Permanent Works, except as may be specified in the Contract, (g) design of any part of the Works by the Employer's Personnel or by others for whom the Employer is responsible, if any, and (h) any operation of the forces of nature which is Unforeseeable or against which an experienced contractor could not reasonably have been expected to have adequate preventative precautions.			▲

Clause/sub-clause	1999 Red Book	1999 Yellow Book	The Silver Book
17.4	Consequences of Employer's Risks	No change	No change
	If and to the extent that any of the risks listed in Sub-Clause 17.3 above results in loss or damage to the Works, Goods or Contractor's Documents, the Contractor shall promptly give notice to the **Engineer** and shall rectify this loss or damage to the extent required by the **Engineer**.	No change	If and to the extent that any of the risks listed in Sub-Clause 17.3 above results in loss or damage to the Works, Goods or Contractor's Documents, the Contractor shall promptly give notice to the **Employer** and shall rectify this loss or damage to the extent required by the **Employer**.
	If the Contractor suffers delay and/or incurs Cost from rectifying this loss or damage, the Contractor shall give a further notice to the **Engineer** and shall be entitled subject to Sub-Clause 20.1 *[Contractor's Claims]* to:	No change	If the Contractor suffers delay and/or incurs Cost from rectifying this loss or damage, the Contractor shall give a further notice to the **Employer** and shall be entitled subject to Sub-Clause 20.1 *[Contractor's Claims]* to:
	(b) payment of any such Cost, which shall be included in the Contract Price. <u>In the case of sub-paragraphs (f) and (g) of Sub-Clause 17.3 *[Employer's Risks]*, reasonable profit on the Cost shall also be included.</u>	No change	(b) payment of any such Cost, which shall be added to the Contract Price. ▲
	After receiving this further notice, the **Engineer** shall proceed in accordance with Sub-Clause 3.5 *[Determinations]* to agree or determine these matters.	No change	After receiving this further notice, the **Employer** shall proceed in accordance with Sub-Clause 3.5 *[Determinations]* to agree or determine these matters.
17.5	Intellectual and Industrial Property Rights	No change	No change
	(a) an unavoidable result of the Contractor's compliance with the **Contract**, or	(a) an unavoidable result of the Contractor's compliance with the **Employer's Requirements**, or	(a) an unavoidable result of the Contractor's compliance with the **Employer's Requirements**, or

A Precise Record of Alterations, Omissions and Additions 733

	The Contractor shall indemnify and hold the Employer harmless against and from any other claim which arises out of or in relation to (i) **the manufacture, use, sale or import of any Goods, or (ii) any design for which the Contractor is responsible.**	The Contractor shall indemnify and hold the Employer harmless against and from any other claim which arises out of or in relation to (i) **the Contractor's design, manufacture, construction or execution of the Works, (ii) the use of Contractor's Equipment, or (iii) the proper use of the Works.**	The Contractor shall indemnify and hold the Employer harmless against and from any other claim which arises out of or in relation to (i) **the Contractor's design, manufacture, construction or execution of the Works, (ii) the use of Contractor's Equipment, or (iii) the proper use of the Works.**
17.6	Limitation of Liability	No change	No change
	The total liability of the Contractor to the Employer, under or in connections with the Contract other than under Sub-Clause 4.19 [Electricity, Water and Gas], Sub-Clause 4.20 [Employer's Equipment and Free Issue Material], Sub-Clause 17.1 [Indemnities] and Sub-Clause 17.5 [Intellectual and Industrial Property Rights], shall not exceed the sum stated in the Particular Conditions (if a sum is not so stated) the **Accepted Contract Amount.**		The total liability of the Contractor to the Employer, under or in connections with the Contract other than under Sub-Clause 4.19 [Electricity, Water and Gas], Sub-Clause 4.20 [Employer's Equipment and Free Issue Material], Sub-Clause 17.1 [Indemnities] and Sub-Clause 17.5 [Intellectual and Industrial Property Rights], shall not exceed the sum stated in the Particular Conditions (if a sum is not so stated) the **Contract Price stated in the Contract Agreement.**
Clause 18	Insurance	No change	No change
18.1	These terms shall be consistent with any terms agreed by both Parties before **the date of the Letter of Acceptance.**	No change	These terms shall be consistent with any terms agreed by both Parties before **they signed the Contract Agreement.**
	The relevant insuring Party shall, within the respective periods stated in the **Appendix to Tender** (calculated from the Commencement Date), submit to the other Party:	No change	The relevant insuring Party shall, within the respective periods stated in the **Particular Conditions** (calculated from the Commencement Date), submit to the other Party:
	When each premium is paid, the insuring Party shall submit evidence of payment to the other Party. Whenever evidence or policies are submitted, the insuring Party shall also give notice to the Engineer.	No change	When each premium is paid, the insuring Party shall submit evidence of payment to the other Party. ▲

Clause/ sub-clause	1999 Red Book	1999 Yellow Book	The Silver Book
18.2	Insurance for Works and Contractor's Equipment	No change	No change
	The insuring Party shall maintain this insurance to provide cover until the date of issue of the Performance Certificate, for loss or damage for which the Contractor is liable arising from a cause occurring prior to the issue of the Taking-Over Certificate, and for loss or damage caused by the Contractor in the course of any other operations (including those under Clause 11 *[Defects Liability]* ▲).	The insuring Party shall maintain this insurance to provide cover until the date of issue of the Performance Certificate, for loss or damage for which the Contractor is liable arising from a cause occurring prior to the issue of the Taking-Over Certificate, and for loss or damage caused by the Contractor in the course of any other operations (including those under Clause 11 *[Defects Liability]* <u>and</u> Clause 12 *[Tests after Completion]*).	The insuring Party shall maintain this insurance to provide cover until the date of issue of the Performance Certificate, for loss or damage for which the Contractor is liable arising from a cause occurring prior to the issue of the Taking-Over Certificate, and for loss or damage caused by the Contractor in the course of any other operations (including those under Clause 11 *[Defects Liability]* <u>and</u> Clause 12 *[Tests after Completion]*).
	(d) shall also cover loss or damage <u>to a part of the Works which is attributable to the use or occupation by the Employer of another part of the Works, and loss</u> or damage from the risks listed in sub-paragraphs (c), <u>(g)</u> and <u>(h)</u> of Sub-Clause 17.3 *[Employer's Risks]*, <u>excluding (in each case) risks which are not insurable at commercially reasonable terms</u>, with deductibles per occurrence of not more than the amount stated in the ***Appendix to Tender*** (if an amount is not so stated, this sub-paragraph (d) shall not apply), and		(d) shall also cover loss or damage ▲ from the risks listed in sub-paragraph (c) ▲ of Sub-Clause 17.3 *[Employer's Risks]*, ▲ with deductibles per occurrence of not more than the amount stated in the ***Particular Conditions*** (if an amount is not so stated, this sub-paragraph (d) shall not apply), and
18.3	Insurance against Injury to Persons and Damage to Property	No change	No change

	This insurance shall be for a limit per Occurrence of not less than the amount stated in the **Appendix to Tender**, with no limit on the number of occurrences. If an amount is not stated in the **Appendix to Tender**, this Sub-Clause shall not apply.	No change	This insurance shall be for a limit per occurrence of not less than the amount stated in the **Particular Conditions**, with no limit on the number of occurrences. If an amount is not stated in the **Contract**, this Sub-Clause shall not apply.
18.4	Insurance for Contractor's Personnel	No change	No change
	The Employer and the Engineer shall also be indemnified under the policy of insurance, except that this insurance may exclude losses and claims to the extent that they arise from any act or neglect of the Employer or of the Employer's Personnel.		The Employer ▲ shall also be indemnified under the policy of insurance, except that this insurance may exclude losses and claims to the extent that they arise from any act or neglect of the Employer or of the Employer's Personnel.
Clause 19	Force Majeure	No change	No change
19.4	Consequences of Force Majeure	No change	No change
	After receiving this notice, the **Engineer** shall proceed in accordance with Sub-Clause 3.5 [Determinations] to agree or determine these matters.	No change	After receiving this notice, the **Employer** shall proceed in accordance with Sub-Clause 3.5 [Determinations] to agree or determine these matters.
19.6	Optional Termination, Payment and Release	No change	No change
	Upon such termination, the **Engineer** shall determine **the value of the work done and issue a Payment Certificate which shall include**:	No change	Upon such termination, the **Employer shall pay to the Contractor**:

Clause/ sub-clause	1999 Red Book	1999 Yellow Book	The Silver Book
Clause 20	Claims, Disputes and Arbitration	No change	No change
20.1	If the Contractor considers himself to be entitled to any extension of Time for Completion and/or any additional payment, under any Clause of these Conditions or otherwise in connection with the Contract, the Contractor shall give notice to the *Engineer* describing the event or circumstance giving rise to the claim.	No change	If the Contractor considers himself to be entitled to any extension of Time for Completion and/or any additional payment, under any Clause of these Conditions or otherwise in connection with the Contract, the Contractor shall give notice to the *Employer* describing the event or circumstance giving rise to the claim.
	The Contractor shall keep such contemporary records as may be necessary to substantiate any claim, either on the Site or at another location acceptable to the *Engineer*. Without admitting the Employer's liability, the *Engineer* may, after receiving any notice under this Sub-Clause, monitor the record-keeping and/or instruct the Contractor to keep further contemporary records. The Contractor shall permit the *Engineer* to inspect all these records, and shall (if instructed) submit copies to the *Engineer*.	No change	The Contractor shall keep such contemporary records as may be necessary to substantiate any claim, either on the Site or at another location acceptable to the *Employer*. Without admitting ▲ liability, the *Employer* may, after receiving any notice under this Sub-Clause, monitor the record-keeping and/or instruct the Contractor to keep further contemporary records. The Contractor shall permit the *Employer* to inspect all these records, and shall (if instructed) submit copies to the *Employer*.
	Within 42 days after the Contractor became aware (or should have become aware) of the event or circumstance giving rise to the claim, or within such other period as may be proposed by the Contractor and approved by the *Engineer*, the Contractor shall send to the *Engineer* a fully detailed claim which includes full supporting particulars of the basis of the claim and of the extension of time and/or additional payment claimed.	No change	Within 42 days after the Contractor became aware (or should have become aware) of the event or circumstance giving rise to the claim, or within such other period as may be proposed by the Contractor and approved by the *Employer*, the Contractor shall send to the *Employer* a fully detailed claim which includes full supporting particulars of the basis of the claim and of the extension of time and/or additional payment claimed.

A Precise Record of Alterations, Omissions and Additions 737

(b) the Contractor shall send further interim claims at monthly intervals, giving the accumulated delay and/or amount claimed, and such further particulars as the *Engineer* may reasonably require; (c) the Contractor shall send a final claim within 28 days after the end of the effects resulting from the event or circumstance, or within such other period as may be proposed by the Contractor and approved by the *Engineer*.	No change
Within 42 days after receiving a claim or any further particulars supporting a previous claim, or within such other period as may be proposed by the *Engineer* and approved by the Contractor, the *Engineer* shall respond with approval or disapproval and detailed comments.	No change
Each *Payment Certificate* shall include such amounts for any claim as have been reasonably substantiated as due under the relevant provision of the Contract.	No change
The *Engineer* shall proceed in accordance with Sub-Clause 3.5 *[Determinations]* to agree or determine (i) the extension (if any) of the Time for Completion (before or after its expiry) in accordance with Sub-Clause 8.4 *[Extension of Time for Completion]*, and/or (ii) the additional payment (if any) to which the Contractor is entitled under the Contract.	No change
	(b) the Contractor shall send further interim claims at monthly intervals, giving the accumulated delay and/or amount claimed, and such further particulars as the *Employer* may reasonably require; (c) the Contractor shall send a final claim within 28 days after the end of the effects resulting from the event or circumstance, or within such other period as may be proposed by the Contractor and approved by the *Employer*.
	Within 42 days after receiving a claim or any further particulars supporting a previous claim, or within such other period as may be proposed by the *Employer* and approved by the Contractor, the *Employer* shall respond with approval or disapproval and detailed comments.
	Each *interim payment* shall include such amounts for any claim as have been reasonably substantiated as due under the relevant provision of the Contract.
	The *Employer* shall proceed in accordance with Sub-Clause 3.5 *[Determinations]* to agree or determine (i) the extension (if any) of the Time for Completion (before or after its expiry) in accordance with Sub-Clause 8.4 *[Extension of Time for Completion]*, and/or (ii) the additional payment (if any) to which the Contractor is entitled under the Contract.

Clause/sub-clause	1999 Red Book	1999 Yellow Book	The Silver Book
20.2	Appointment of the Dispute Adjudication Board	No change	No change
	The Parties shall jointly appoint a DAB by the date **stated in the Appendix to Tender**.	The Parties shall jointly appoint a DAB by the date **28 days after a Party gives notice to the other Party of its intention to refer a dispute to a DAB in accordance with Sub-Clause 20.4.**	The Parties shall jointly appoint a DAB by the date **28 days after a Party gives notice to the other Party of its intention to refer a dispute to a DAB in accordance with Sub-Clause 20.4.**
	The terms of the remuneration of either the sole member or each of the three members, including the remuneration of any expert whom the DAB consults, shall be mutually agreed upon by the Parties when agreeing the terms of appointment. Each Party shall be responsible for paying one-half of this remuneration. If at any time the Parties so agree, they may jointly refer a matter to the DAB for it to give its opinion. Neither Party shall consult the DAB on any matter without the agreement of the other Party.	The terms of the remuneration of either the sole member or each of the three members, ◂ shall be mutually agreed upon by the Parties when agreeing the terms of appointment. Each Party shall be responsible for paying one-half of this remuneration. ◂	The terms of the remuneration of either the sole member or each of the three members, ◂ shall be mutually agreed upon by the Parties when agreeing the terms of appointment. Each Party shall be responsible for paying one-half of this remuneration. ◂
	If at any time the Parties so agree, they may appoint a suitably qualified person or persons to replace (or to be available to replace) any one or more members of the DAB.	If at any time the Parties so agree, they may appoint a suitably qualified person or persons to replace ◂ any one or more members of the DAB.	If at any time the Parties so agree, they may appoint a suitably qualified person or persons to replace ◂ any one or more members of the DAB.
	◂	The replacement shall be appointed in the same manner as the replaced person was required to have been nominated or agreed upon, as described in this Sub-Clause.	The replacement shall be appointed in the same manner as the replaced person was required to have been nominated or agreed upon, as described in this Sub-Clause.

A Precise Record of Alterations, Omissions and Additions 739

	If any of these circumstances occurs and no such replacement is available, a replacement shall be appointed in the same manner as the replaced person was required to have been nominated or agreed upon, as described in this Sub-Clause.		◄
	Unless otherwise agreed by both Parties, the appointment of the DAB (including each member) shall expire when the *discharge referred to in Sub-Clause 14.12 [Discharge] shall have become effective*.	Unless otherwise agreed by both Parties, the appointment of the DAB (including each member) shall expire when the *DAB has given its decision on the dispute referred to it under Sub-Clause 20.4 [Obtaining Dispute Adjudication Board's Decision], unless other disputes have been referred to the DAB by that time under Sub-Clause 20.4, in which event the relevant date shall be when the DAB has also given decisions on those disputes.*	Unless otherwise agreed by both Parties, the appointment of the DAB (including each member) shall expire when the *DAB has given its decision on the dispute referred to it under Sub-Clause 20.4 [Obtaining Dispute Adjudication Board's Decision], unless other disputes have been referred to the DAB by that time under Sub-Clause 20.4, in which event the relevant date shall be when the DAB has also given decisions on those disputes.*
20.3	Failure to Agree Dispute Adjudication board	No change	No change
	... then the appointing entity or official named in the *Appendix to Tender* shall, upon the request of either or both of the Parties and after due consultation with both Parties, appoint this member of the DAB.	No change	... then the appointing entity or official named in the *Particular Conditions* shall, upon the request of either or both of the Parties and after due consultation with both Parties, appoint this member of the DAB.

Clause/sub-clause	1999 Red Book	1999 Yellow Book	The Silver Book
20.4	Obtaining Dispute Adjudication Board's Decision	No change	No change
	If a dispute (of any kind whatsoever) arises between the Parties in connection with, or arising out of, the Contract or the execution of the Works, including any dispute as to any certificate, determination, instruction, opinion or valuation of the **Engineer**, ▲ either Party may refer the dispute in writing to the DAB for its decision, with copies to the other Party and the Engineer.	If a dispute (of any kind whatsoever) arises between the Parties in connection with, or arising out of, the Contract or the execution of the Works, including any dispute as to any certificate, determination, instruction, opinion or valuation of the Engineer, then after a Dab has been appointed pursuant to Sub-Clause 20.2 [Appointment of the DAB] and 20.3 [Failure to Agree DAB] either Party may refer the dispute in writing to the DAB for its decision, with copies to the other Party and the Engineer.	If a dispute (of any kind whatsoever) arises between the Parties in connection with, or arising out of, the Contract or the execution of the Works, including any dispute as to any certificate, determination, instruction, opinion or valuation of the Employer, then after a Dab has been appointed pursuant to Sub-Clause 20.2 [Appointment of the DAB] and 20.3 [Failure to Agree DAB] either Party may refer the dispute in writing to the DAB for its decision, with copies to the other Party ▲.
	Both Parties shall promptly make available to the DAB all such additional information, further access to the Site, and appropriate facilities, as the DAB may require for the purposes of making a decision on such dispute.	Both Parties shall promptly make available to the DAB all ▲ information, further access to the Site, and appropriate facilities, as the DAB may require for the purposes of making a decision on such dispute.	Both Parties shall promptly make available to the DAB all ▲ information, further access to the Site, and appropriate facilities, as the DAB may require for the purposes of making a decision on such dispute.

A Precise Record of Alterations, Omissions and Additions 741

	Within 84 days after receiving such reference, ◄ or within such other period as may be proposed by the DAB and approved by both Parties, the DAB shall give its decision, which shall be reasoned and shall state that it is given under this Sub-Clause. ◄	Within 84 days after receiving such reference, or the advanced payment referred to in Clause 6 of the Appendix – General Conditions of the Dispute Adjudication Agreement, whichever date is later, or within such period as may be proposed by the DAB and approved by both Parties, the DAB shall give its decision, which shall be reasoned and shall state that it is given under this Sub-Clause. However, if neither of the Parties has paid in full the invoices submitted by each member pursuant to Clause 6 of the Appendix, the DAB shall not be obliged to give its decision until such invoices have been paid in full.	Within 84 days after receiving such reference, or the advanced payment referred to in Clause 6 of the Appendix – General Conditions of the Dispute Adjudication Agreement, whichever date is later, or within such period as may be proposed by the DAB and approved by both Parties, the DAB shall give its decision, which shall be reasoned and shall state that it is given under this Sub-Clause. However, if neither of the Parties has paid in full the invoices submitted by each Member pursuant to Clause 6 of the Appendix, the DAB shall not be obliged to give its decision until such invoices have been paid in full.
20.6	Arbitration	No change	No change
	The arbitrator(s) shall have full power to open up, review and revise any certificate, determination, instruction, opinion or valuation of the *Engineer*, and also any decision of the DAB, relevant to the dispute. Nothing shall disqualify the Engineer from being called as a witness and giving evidence before the arbitrator(s) on any matter whatsoever relevant to the dispute.	No change	The arbitrator(s) shall have full power to open up, review and revise any certificate, determination, instruction, opinion or valuation of *(or on behalf of)* the *Employer*, and also any decision of the DAB, relevant to the dispute. ◄

Appendices

Appendix A

Editorial Amendments in the 1988 Reprint of the Fourth Edition of the Red Book

Following publication of the Fourth Edition in 1987 of the Conditions of Contract for Works of Civil Engineering Construction, a number of editorial amendments were agreed by FIDIC. The amendments were incorporated during a 1988 reprinting and the list below clarifies the differences between the 1988 reprint and the original document.

Foreword	The last sentence of the first paragraph previously read 'The Conditions are equally suitable for use; on domestic contracts.'
Page 6	Sub-Clause 10.1. A comma has been inserted after the word 'Contract' in the second line. The third sentence previously read 'Such security shall be in such form as may be agreed between the Employer and the Contractor.'
Page 11	Sub-Clause 22.1 (b) was previously one complete paragraph, i.e. there was no space between the words 'other than the Works),' and the remainder of the Sub-Clause.
Page 15	Sub-Clause 31.2 (c) was previously one complete paragraph, i.e. there was no space between the words '... nature for any such,' and the remainder of the Sub-Clause.
Page 20	Sub-Clause 44.3. The penultimate sentence was previously, 'In both such cases the Engineer shall notify the Contractor accordingly, with a copy to the Employer.'
Page 21	Sub-Clause 49.1 (a). The word 'substantial' has been deleted.
Page 29	Sub-Clause 60.3 (b) was previously two paragraphs, the second beginning with the words 'Provided also that if at such time ...'
Page 30	Sub-Clause 60.5. The word 'The' has been inserted at the beginning of the final paragraph.
Page 35	Sub-Clause 67.1. In the eighth line of the third paragraph, a comma has been inserted after the word 'provided'. In the second line of the fourth paragraph, the word 'notice' replaces the word 'notification'.

Page 38   Reference to Part II. In the third line, the words 'and (iv)' have been inserted after paragraph (a) (i).

[NB Although neither the Tender nor the Agreement is included in this book, the amendments are included below for completeness.]

Tender

Item 3   The word 'Works' has been capitalised.

Agreement

Line 4	Inverted commas have been inserted following the words 'the Employer.'
Line 6	Inverted commas have been inserted before the word 'the' instead of before the word 'Contractor.'
Line 8	The word 'Contractor' has been capitalised.
Line 9	The words 'Tender by the Contractor' were previously 'Tender by Contractor'.
Line 11	The word 'Agreement' has been capitalised.
Last lines	The Agreement previously ended with the words 'Binding Signature of Employer' and 'Binding Signature of Contractor'.

Appendix B

Further Amendments in the 1992 Reprint of the Fourth Edition of the Red Book

The following amendments have been made to the 1988 Reprint of the Fourth Edition of the Conditions of Contract for Works of Civil Engineering Construction. The amendments of the 1988 Reprint are shown in Appendix A. In addition, some minor changes in the use of punctuation marks (commas, semicolons, colons and stops), as well as the use of the words 'or' and 'and' have been introduced to attain uniformity in the style of all clauses. These minor changes which improve the style, but which have no effect on the meaning of Clauses, are listed at the end of this Appendix.

Foreword	The eighth paragraph previously referred to the anticipated publication of the 'Guide to the Use of FIDIC Conditions of Contract for Works of Civil Engineering Construction'.
Page 2	Sub-Clause 1.1, sub-para (e). Definitions (iii) 'Interim Payment Certificate' and (iv) 'Final Payment Certificate' have been added.
Page 6	Sub-Clause 8.1. Second paragraph has been added.
Page 7	Sub-Clause 12.2. Marginal note. The word 'Adverse' has been changed to read 'Not Foreseeable' (also amended in the Contents and the Index).
Page 8	Sub-Clause 13.1. Last sentence has been shortened by deleting the words 'or, subject to the provisions of Clause 2, from the Engineer's Representative.', and adding the words '(or his delegate).'.
	Sub-Clause 15.1, para 1. Last sentence has been shortened by placing a full stop after the word 'Engineer', deleting the words 'or subject to the provisions of Clause 2, the Engineer's Representative.'.
Page 10	Sub-Clause 21.1, sub-para (a). The words '(the term "cost" in this context shall include profit)' have been added.
Page 11	Sub-Clause 21.4, sub-para (a). The word 'where' has been corrected to read 'whether'.
Page 18	Sub-Clause 40.3. The word 'written' has been deleted at the end of the first line.

Page 19	Sub-Clause 42.3. The word 'wayleaves' has been changed to read 'rights of way' in the text and marginal note (also amended in the Contents and the Index).
Page 29	Sub-Clause 60.1, sub-para (e). The words 'or otherwise' have been added at the end.
	Sub-Clause 60.2. The words 'certify to the Employer' have been changed to read 'deliver to the Employer an Interim Payment Certificate stating', the word 'thereof' has been changed to read 'of such statement' and the word 'he' has been changed to read 'the Engineer'. Sub-para (b). The words 'Interim Certificates' have been changed to read 'Interim Payment Certificates'.
	Sub-Clause 60.3, sub-para (b). In the eighth line, the word 'ordered' has been changed to read 'instructed'.
	Sub-Clause 60.4. The words 'interim certificate' in the first and fourth lines, and the word 'certificate' in the second line, have been changed to read 'Interim Payment Certificate'.
Page 30	Sub-Clause 60.5. In the second line, after the word 'Engineer', the words 'six copies of' have been added.
	Sub-Clause 60.6. In the second line, after the word 'consideration', the words 'six copies of' have been added. Sub-para (b). The words 'or otherwise' have been added at the end. At the end of the sub-clause, the final paragraph has been added.
	Sub-Clause 60.7 and Sub-Clause 60.8 (text and marginal note). The words 'Final Certificate' have been changed to read 'Final Payment Certificate' (also amended in the Contents and the Index).
	Sub-Clause 60.8(a). The words 'or otherwise' have been added. Sub-Clause 60.8(b). The words 'under the Contract other than Clause 47' have been changed to read 'other than under Clause 47'.
Page 31	Sub-Clause 60.10. In the first and fourth lines, the words 'interim certificate' have been changed to read 'Interim Payment Certificate'. In the fifth and sixth lines, the words 'Final Certificate' have been changed to read 'Final Payment Certificate'. The words 'or otherwise' have been added at the end.
Page 33	Sub-Clause 65.6. In the ninth line, the words 'and to the operation of Clause 67' have been changed to read 'and Clause 67'.
Page 34	Sub-Clause 66.1. In the second line the word 'party' has been changed to read 'or both parties'. In the third line between the words 'his' and 'contractual' the words 'or their' have been added. In the fourth line after the word 'then', the words 'the parties shall be discharged from

	the Contract, except as to their rights under this Clause and Clause 67 and without prejudice to the rights of either party in respect of any antecedent breach of the Contract, and' have been added.
Page 35	Sub-Clause 67.2. The words 'arbitration of such dispute shall not be commenced unless an attempt has first been made by the parties to settle such dispute amicably' have been changed to read 'the parties shall attempt to settle such dispute amicably before the commencement of arbitration'. The words 'whether or not any attempt at amicable settlement thereof has been made' have been changed to read 'even if no attempt at amicable settlement thereof has been made'.
Page 37	Sub-Clause 69.1, sub-para (d). The words 'unforseen reasons, due to economic dislocation' have been changed to read 'unforeseen economic reasons'.
	Sub-Clause 69.4. In the second line of the second paragraph, the word 'cost' has been changed to read 'costs'.
Page 38	REFERENCE TO PART II. In the third line, the words '5.1 part' have been changed to read '5.1 (part)'.
Tender	Paragraph 1. In the last line, the word 'sums' has been changed to read 'sum'.
Appendix	In the ninth line, the words 'and Plant' have been added. In the twelfth line, the word 'Payment' has been added. In the thirteenth line, the words 'per annum' have been added.
Editorial amendments	For page 35, after the words 'Sub-Clause 67. 1 the first sentence has been inserted.

The minor changes made in the 1992 reprint of the Fourth Edition of the Red Book

The minor changes which were made in punctuation and in the use of the words 'and' and 'or' in the 1992 reprint of the Fourth Edition of the Red Book are set out below for the sake of completeness. They are also incorporated in the comparative text of the Third and Fourth Editions in Chapter 23.

Part I changes

Sub-clause 2.3	After the semicolon at the end of sub-paragrasph (a), insert 'and'.
Sub-clause 2.6	At the end of sub-paragraphs (a) and (b), delete the word 'or'.
Sub-clause 4.1	At the end of paragraph (a), delete the word 'or'.
Sub-clause 11.1	At the end of (b), after the word 'required', add a comma.

Sub-clause 12.2	In the margin, delete the penultimate word 'and' and substitute 'or'.
Sub-clause 16.1	After the word 'therein', add a colon.
Sub-clause 19.1	At the end of (a), delete the word 'and'.
Sub-clause 20.4	
sub-par. (c)	After the words 'toxic explosive', delete the comma.
sub-par. (g)	Add at the end: 'and'.
Sub-clause 21.1	At the end of (a) after 'include profit' add a comma; and at the end of (b) after the word 'nature', add a comma followed by 'and'.
Sub-clause 21.4	At the end of the second line, after the word 'by', add a colon; and at the end of (c), after 'thereof', add 'or'.
sub-par. (c)	After the words 'toxic explosive', delete the comma.
Sub-clause 22.2	At the end of (c), after 'Contract', add 'and'.
Sub-clause 31.2	At the end of (a), delete 'or'.
Sub-clause 36.1	At the end of the first line, after the words 'shall be', add a colon.
Sub-clause 36.3	At the end of the first line, after the words 'test is', add a colon.
Sub-clause 36.4	At the end of the first line, after the words 'which is', add a colon; and at the end of (a), delete 'or'.
Sub-clause 40.1	After the words 'suspension is', add a colon; and at the end of both (a) and (b), delete 'or'.
Sub-clause 40.2	After the word 'determine', add a colon.
Sub-clause 42.1	At the end of (a), delete 'and'; at the end of (b), after the words 'the contractor', add a comma; and in the next line between 'and' and 'subject' insert a comma.
Sub-clause 44.1	At the end of the first line, after the words 'event of', add a colon; and at the end of (a), (b) and (c), delete 'or'.
Sub-clause 48.2	At the end of (a), delete 'or'.
Sub-clause 49.1	At the end of (b), after the words 'so certified', add a comma.
Sub-clause 49.2	At the end of (a) between 'date' and 'and', insert a comma.
Sub-clause 49.3	At the end of (a), delete 'or'.
Sub-clause 51.1	At the end of (e), after the comma, add 'or'.
Sub-clause 58.2	At the end of (a), after the comma, add 'and'.
Sub-clause 59.4	At the end of (b), after the semicolon, add 'and'.
Sub-clause 59.5	After the words 'unless the Contractor', add a colon; and at the end of (a) between 'payments' and 'and' insert a comma.

Sub-clause 60.1 After the words 'respect of', add a colon; at the end of (a), (b) and (c), add a comma and at the end of (d), add a comma followed by 'and'.

Sub-clause 60.3 In (b) delete the comma following the word 'time' in the third sentence.

Sub-clause 60.5 After the words 'by the Engineer' delete the comma and substitute a colon. At the end of sub-paragraph (a), add a comma. After the word 'due' in sub-paragraph (b), insert a comma. After the word 'Contract' in sub-paragraph (c), insert a comma.

Sub-clause 60.6 After the words 'by the Engineer' delete the comma and substitute a colon.

Sub-clause 60.8 After the word 'stating', add a colon.

Sub-clause 63.1 At the end of (a), at the end of (b) (ii) and at the end of (c), delete 'or'.

Sub-clause 65.1 At the end of (a), delete 'or'.

Sub-clause 65.3 At the end of (b), after the words 'Contractor's Equipment', add a comma.

Sub-clause 65.8 The first letter of each of sub-paragraphs (a) to (f) should be altered from a capital to lower case; at the end of sub-paragraphs (a) to (d), add a semicolon; and at the end of sub-paragraph (e), add a semicolon followed by 'and'.

Sub-clause 67.3 At the end of (b), after the words 'Sub-Clause 67.2', add a comma.

Sub-clause 69.1 At the end of (a) and (b), delete 'or'.

Sub-clause 69.4 In the second paragraph after the word 'determine', add a colon. Replace 'cost' with 'costs'.

In the Reference to Part II, '5.1 part' has become '5.1 (part)'.

In the Tender, paragraph 1, 'sums' in the last line has become 'sum'.

Appendix entries have been changed as follows:

Percentage of invoice value listed materials *and plant*	60.1(c)	per cent
Minimum amount of Interim *Payment* Certificates	60.2	
Rates of interest upon unpaid sums	60.10	per cent *per annum*

Part II changes

Sub-clause 21.1 At the end of (a), after 'include profit', add a comma followed by 'and'.

Clause 60 (Payment to contractor)
    At the end of the first line, after the words 'shall be made' add a colon; and at the end of (b), add 'and'.

Clause 60 (Payment to employer)
    At the end of the second line, after the words 'shall be made' add a colon; and at the end of (c), add 'and'.

Sub-clause 70.1 (Second example)
    In (a)(i) at the end of the first sub-paragraph commencing following 'or quarry', alter the full stop to a semi-colon; and at the end of the sub-paragraph 'Basic Rate' after the words 'carried out', alter the full stop to a semi-colon and add 'and'. In (a)(ii) at the end of sub-paragraph (a), delete 'also'.

Sub-clause 70.1 (Second example)
    In (b)(i) at the end of the first paragraph, after the words 'the Works', change the full stop to a semicolon, and at the end of the sub-paragraph 'Basic Prices', after the words 'of tenders', change the full stop to a semicolon and add 'and'.

Sub-clause 70.1 (Third example)
    In (a) after 'publication', delete the dash which follows the colon; at the end of (i), add a comma; at the end of (ii), add a comma followed by 'or', and at the end of (iii), add a full stop.

Sub-clause 70.1 (Third example)
    In (b) at the end of (i), change the full stop to a comma and add 'and'.

Appendix C

Part II – Conditions of Particular Application

Clause 1

Sub-Clause 1.1 – Definitions

(a) (i) The Employer is (*insert name*)
(a) (iv) The Engineer is (*insert name*)

If further definitions are essential, additions should be made to the list.

Clause 2

Sub-Clause 2.1 – Engineer's Duties

EXAMPLE
The Engineer shall obtain the specific approval of the Employer before carrying out his duties in accordance with the following Clauses of Part 1:

(a) Clause (*insert applicable number*)
(b) Clause (*insert applicable number*)
(c) Clause (*insert applicable number*)

This list should be extended or reduced as necessary. In some cases the obligation to obtain the approval of the Employer may apply to only one Sub-Clause out of several in a Clause or approval may only be necessary beyond certain limits, monetary or otherwise. Where this is so, the example wording must be varied.

If the obligation to obtain the approval of the Employer could lead to the Engineer being unable to take action in an emergency, where matters of safety are involved, an additional paragraph may be necessary.

EXAMPLE
Notwithstanding the obligation, as set out above, to obtain approval, if, in the opinion of the Engineer, an emergency occurs affecting the safety of life or of the Works or of adjoining property, he may, without relieving the Contractor of any of his duties and responsibilities under the Contract, instruct the Contractor to execute all such work or to do all such things as may, in the opinion of the Engineer, be necessary

to abate or reduce the risk. The Contractor shall forthwith comply, despite the absence of approval of the Employer, with any such instruction of the Engineer. The Engineer shall determine an addition to the Contract Price, in respect of such instruction, in accordance with Clause 52 and shall notify the Contractor accordingly, with a copy to the Employer.

Clause 5

Sub-Clause 5.1 – Language/s and Law

(a) The language is (*insert as applicable*)
(b) The law is that in force in (*insert name of country*)

If necessary (a) above should be varied to read:

The languages are (*insert as applicable*)

and there should be added

The Ruling Language is (*insert as applicable*)

Sub-Clause 5.2 – Priority of Contract Documents

Where it is decided that an order of precedence of all documents should be included, this Sub-Clause may be varied as follows:

EXAMPLE
Delete the documents listed 1–6 and substitute:

(1) the Contract Agreement (if completed);
(2) the Letter of Acceptance;
(3) the Tender;
(4) the Conditions of Contract Part II;
(5) the Conditions of Contract Part I;
(6) the Specification;
(7) the Drawings; and
(8) the priced Bill of Quantities

or

Where it is decided that no order of precedence of documents should be included, this Sub-Clause may be varied as follows:

EXAMPLE
Delete the text of the Sub-Clause and substitute:

The several documents forming the Contract are to be taken as mutually explanatory of one another, but in the case of ambiguities or discrepancies the priority shall be that accorded by law. If, in the opinion of the Engineer, such ambiguities or discrepancies make it necessary to issue any instruction to the Contractor in explanation or adjustment, the Engineer shall have authority to issue such instruction.

Clause 9

Where it is decided that a contract Agreement should be entered into and executed the form must be annexed to these Conditions as stated in Sub-Clause 9.1 of Part I of these Conditions.

A suitable form is annexed to Part I – General Conditions.

Clause 10

Sub-Clause 10.1 – Performance Security

Where it is decided that a performance security should be obtained by the Contractor, the form must be annexed to these Conditions as stated in Sub-Clause 10.1 of Part I of these Conditions.

Two example forms of performance security are given on pages 7, 8 and 9. The Clause and wording of the example forms may have to be varied to comply with the law of the Contract which may require the forms to be executed under seal.

Where there is provision in the Contract for payments to the Contractor to be made in foreign currency, Sub-Clause 10.1 of Part I of these Conditions may be varied.

EXAMPLE
After the first sentence, insert the following sentence:

The security shall be denominated in the types and proportions of currencies stated in the Appendix to Tender.
Where the source of the performance security is to be restricted, an additional Sub-Clause may be added.

EXAMPLE SUB-CLAUSES

Source of Performance Security 10.4
The performance security, submitted by the Contractor in accordance with Sub-Clause 10.1, shall be furnished by an institution registered in (*insert the country where the Works are to be executed*) or licensed to do business in such country.

or

Source of Performance Security 10.4
Where the performance security is in the form of a bank guarantee, it shall be issued by:

(a) a bank located in the country of the Employer, or
(b) a foreign bank through a correspondent bank located in the country of the Employer.

Clause 11

Where the bulk or complexity of the data, or reasons of security enforced by the country where the Works are to be executed, makes it impracticable for the Employer to make all data available with the Tender Documents and inspection of some data by the Contractor at an office is therefore expected, it would be advisable to make the circumstances clear.

EXAMPLE SUB-CLAUSE

Access to Data 11.2
Data made available by the Employer in accordance with Sub-Clause 11.1 shall be deemed to include data listed elsewhere in the Contract as open for inspection at (insert particulars of the office or offices where such data is stored)

Sub-Clause 11.1 – Inspection of Site

For a Contract comprising dredging and reclamation work the Clause may be varied as follows:

EXAMPLE
In the first paragraph, delete the words 'hydrological and sub-surface' and substitute 'hydrographic and sub-seabed'.
   In the second paragraph, under (a) delete the word 'sub-surface' and substitute 'sub-seabed' and under (b) delete the word 'hydrological' and substitute 'hydrographic'.

Clause 12

Sub-Clause 12.2 – Not Foreseeable Obstructions or Conditions

For a Contract comprising dredging and some types of reclamation work the Sub-Clause may require to be varied.

EXAMPLE
Delete the word 'other than climatic conditions on the Site,'.

Clause 14

Sub-Clause 14.1 – Programme to be Submitted

   The time within which the programme shall be submitted shall be (*insert number*) days.

Sub-Clause 14.3 – Cash Flow Estimate to be Submitted

The time within which the detailed cash flow estimate shall be submitted shall be (*insert number*) days.

In both examples given above it is desirable for consistency with the rest of the Conditions that the number of days inserted should be a multiple of seven.

EXAMPLE PERFORMANCE GUARANTEE
By this guarantee We _____ whose registered office is at _____ (hereinafter called 'the Contractor') and _____ whose registered office is at _____ (hereinafter called 'the Guarantor') are held and firmly bound unto _____ (hereinafter called 'the Employer') in the sum of _____ for payment of which sum the Contractor and the Guarantor bind themselves, their successors and assigns jointly and severally by these presents.

Whereas the Contractor by an Agreement made between the Employer of the one part and the Contractor of the other part has entered into a Contract (hereinafter called 'the said Contract') to execute and complete certain Works and remedy any defects therein as therein mentioned in conformity with the provisions of the said Contract.

Now the Condition of the above-written Guarantee is such that if the Contractor shall duly perform and observe all the terms provisions conditions and stipulations of the said Contract on the Contractor's part to be performed and observed according to the true purport intent and meaning thereof or if on default by the Contractor the Guarantor shall satisfy and discharge the damages sustained by the Employer thereby up to the amount of the above-written Guarantee then this obligation shall be null and void but otherwise shall be and remain in full force and effect but no alteration in terms of the said Contract or in the extent or nature of the Works to be executed, completed and defects therein remedied thereunder and no allowance of time by the Employer or the Engineer under the said Contract nor any forbearance or forgiveness in or in respect of any matter or thing concerning the said Contract on the part of the Employer or the said Engineer shall in any way release the Guarantor from any liability under the above-written Guarantee. Provided always that the above obligation of Guarantor to satisfy and discharge the damages sustained by the Employer shall arise only

(a) on written notice from both the Employer and the Contractor that the Employer and the Contractor have mutually agreed that the amount of damages concerned is payable to the Employer or
(b) on receipt by the Guarantor of a legally certified copy of an award issued in arbitration proceeding carried out in conformity with the terms of the said Contract that the amount of the damages is payable to the Employer.

Signed on _____ Signed on _____
on behalf of _____ on behalf of _____
by _____ by _____
in the capacity of _____ in the capacity of _____
in the presence of _____ in the presence of _____

EXAMPLE SURETY BOND FOR PERFORMANCE

Know all Men by these Presents that (name and address of Contractor) _____
as Principal (hereinafter called 'the Contractor') and (name, legal title and address of Surety) _____ as Surety (hereinafter called 'the Surety'), are held and firmly bound unto (name and address of Employer) _____ as Obligee (hereinafter called 'the Employer') in the amount of _____ for the payment of which sum, well and truly to be made, the Contractor and the Surety bind themselves, their successors and assigns, jointly and severally, firmly by these presents.

Whereas the Contractor has entered into a written contract agreement with the Employer dated the _____ day of _____ 19____ for (name of Works) _____ in accordance with the plans and specifications and amendments thereto, to the extent herein provided for, are by reference made part hereof and are hereinafter referred to as the Contract.

Now, therefore, the Condition of this Obligation is such that, if the Contractor shall promptly and faithfully perform the said Contract (including any amendments thereto) then this obligation shall be null and void; otherwise it shall remain in full force and effect.

Whenever Contractor shall be, and declared by Employer to be, in default under the Contract, the Employer having performed the Employer's obligations thereunder, the Surety may promptly remedy the default, or shall promptly:

(1) Complete the Contract in accordance with its terms and conditions; or
(2) Obtain a bid or bids for submission to the Employer for completing the Contract in accordance with its terms and conditions, and upon determination by the Employer and Surety of the lowest responsive bidder, arrange for a contract between such bidder and Employer and make available as work progresses (even though there should be a default or succession of defaults under the contract or contracts of completion arranged under this paragraph) sufficient funds to pay the cost of completion less the balance of the Contract Value; but not exceeding, including other costs and damages for which the Surety may be liable hereunder, the amount set forth in the first paragraph hereof. The term 'balance of the Contract Value', as used in this paragraph, shall mean the total amount payable by Employer to Contractor under the Contract, less the amount properly paid by the Employer to Contractor; or
(3) Pay the Employer the amount required by Employer to complete the Contract in accordance with its terms and conditions any amount up to a total not exceeding the amount of this Bond.

The Surety shall not be liable for a greater sum than the specified penalty of this Bond.

Any suit under this Bond must be instituted before the issue of the Defects Liability Certificate.

No right of action shall accrue on this Bond to or for the use of any person or corporation other than the Employer named herein or the heirs, executors administrators or successors of the Employer.

Signed on _____	Signed on _____
on behalf of _____	on behalf of _____
by _____	by _____
in the capacity of _____	in the capacity of _____
in the presence of _____	in the presence of _____

Clause 15

Where the language in which the Contract documents have been drawn up is not the language of the country in which the Works are to be executed, or where for any other reason it is necessary to stipulate that the Contractor's authorised representative shall be fluent in a particular language, an additional Sub-Clause may be added.

EXAMPLE SUB-CLAUSES

Language Ability of Contractor's Representative 15.2
The Contractor's authorised representative shall be fluent in (insert name of language).

or

Interpreter to be Made Available 15.2
If the Contractor's authorised representative is not, in the opinion of the Engineer, fluent in (insert name of language), the Contractor shall have available on Site at all times a competent interpreter to ensure the proper transmission of instructions and information.

Clause 16

Where the language in which the Contract documents have been drawn up is not the language of the country in which the Works are to be executed, or where for any other reason it is necessary to stipulate that members of the Contractor's superintending staff shall be fluent in a particular language, an additional Sub-Clause may be added.

EXAMPLE SUB-CLAUSE

Language Ability of Superintending Staff 16.3
A reasonable proportion of the Contractor's superintending staff shall have a working knowledge of (*insert name of language*) or the Contractor shall have available on Site at all times a sufficient number of competent interpreters to ensure the proper transmission of instructions and information.

Where there is a desire, but not a legal requirement, that the Contractor makes reasonable use of materials from or persons resident in the country in which the Works are to be executed, an additional Sub-Clause may be added.

Employment of Local Personnel 16.4

The Contractor is encouraged, to the extent practicable and reasonable, to employ staff and labour from sources within (*insert name of country*).

Clause 18

Sub-Clause 18.1 – Boreholes and Exploratory Excavation

For a Contract comprising dredging and reclamation work the Sub-Clause may require to be varied.

EXAMPLE
Add second sentence as follows:

Such exploratory excavation shall be deemed to include dredging.

Clause 19

Sub-Clause 19.1 – Safety, Security and Protection of the Environment

Where a Contract includes dredging the possibility of pollution should be given particular attention and additional wording may be required. For example, where fishing and recreation areas might be influenced, the Contractor should be required to plan and execute the dredging so that the effect is kept to a minimum. Where there is a risk of chemical pollution from soluble sediments in the dredging area, for instance in a harbour, it is important that sufficient information is provided with the Tender documents. Responsibilities should be clearly defined.

Clause 21

Sub-Clause 21.1 – Insurance of Works and Contractor's Equipment

Where there is provision in the Contract for payments to the Contractor to be made in foreign currency, this Sub-Clause may be varied.

EXAMPLE
Add final sentence as follows:

The insurance in paragraphs (a) and (b) shall provide for compensation to be payable in the types and proportions of currencies required to rectify the loss or damage incurred.

Where it is decided to state the deductible limits for the Employer's Risks, this Sub-Clause may be varied.

EXAMPLE
Add to paragraph (a) as follows:

and with deductible limits for the Employer's Risks not exceeding (*insert amount*)

Clauses 21, 23 and 25, Insurances Arranged by Employer

In certain circumstances, such as where a number of separate contractors are employed on a single project, or phased take-over is involved, it may be preferable for the Employer to arrange insurance of the Works, and Third Party insurance. In such case, it must be clear in the Contract that the Contractor is not precluded from taking out any additional insurance, should he desire to do so, over and above that to be arranged by the Employer.

Tenderers must be provided at the Tender stage with details of the insurance to be arranged by the Employer, in order to assess what provision to make in their rates and prices for any additional insurance, and for the amount of policy deductibles which they will be required to bear. Such details shall form part of the Contract between the Employer and the Contractor.

Example wording to allow for the arrangement of insurance by the Employer is as follows:

EXAMPLE

Clause 21
Delete the text of the Clause and substitute the following re-numbered Sub-Clauses:

Insurance of Works 21.1
Without limiting his or the Contractor's obligations and responsibilities under Clause 20, the Employer will insure:

(a) the Works, together with materials and Plant for incorporation therein, to the full replacement cost (the term 'cost' in this context shall include profit), and
(b) an additional sum to cover any additional costs of and incidental to the rectification of loss or damage including professional fees and the cost of demolishing and removing any part of the Works and of removing debris of whatsoever nature.

Insurance of Contractor's Equipment 21.2
The Contractor shall, without limiting his obligations and responsibilities under Clause 20, insure the Contractor's Equipment and other things brought onto the site by the Contractor, for a sum sufficient to provide for their replacement at the Site.

Scope of Cover 21.3
The insurance in Sub-Clause 21.1 shall be in the joint names of the Contractor and the Employer and shall cover:

(a) the Employer and the Contractor against loss or damage as provided in the details of insurance annexed to these Conditions, from the start of work at the Site until

the date of issue of the relevant Taking-Over Certificate in respect of the Works or of any Section or part thereof as the case may be, and
(b) the Contractor for his liability:
  (i) during the Defects Liability Period for loss or damage arising from a cause occurring prior to the commencement of the Defects Liability Period, or
  (ii) occasioned by the Contractor in the course of any operations carried out by him for the purpose of complying with his obligations under Clauses 49 and 50.

Responsibility for Amounts not Recovered 21.4
Any amounts not insured or not recovered from the insurers shall be borne by the Employer or the Contractor in accordance with their responsibilities under Clause 20.

Clause 23
Delete the text of the Clause and substitute:

Third Party Insurance (including Employer's Property) 23.1
Without limiting his or the Contractor's obligations and responsibilities under Clause 22, the Employer will insure in the joint names of the Contractor and the Employer, against liabilities for death of or injury to any person (other than as provided in Clause 24) or loss of or damage to any property (other than the Works) arising out of the performance of the Contract, as provided in the details of insurance referred to in Sub-Clause 21.3.

Clause 25
Delete the text of the Clause and substitute:

Evidence and Terms of Insurances 25.1
The insurance policies to be arranged by the Employer pursuant to Clauses 21 and 23 shall be consistent with the general terms described in the Tender and copies of such policies shall when required be supplied by the Employer to the Contractor.

Adequacy of Insurances 25.2
The Employer shall notify the insurers of changes in the nature, extent or programme for execution of the Works and ensure the adequacy of the insurances at all times in accordance with the terms of the Contract and shall, when required, produce, to the Contractor the insurance policies in force and the receipts for payment of the premiums. No variations shall made to the insurances by the Employer without the prior approval of the Contractor.

Remedy on Employer's Failure to Insure 25.3
If and so far as the Employer fails to effect and keep in force any of the insurances referred to in Sub-Clause 25.1, then the Contractor may effect and keep in force any such insurance and pay any premium as may be necessary for that purpose and add the amount so paid to any monies due or to become due to the Contractor, or recover the same as a debt due from the Employer.

Compliance with Policy Conditions 25.4

In the event that the Contractor or the Employer fails to comply with conditions imposed by the insurance policies effected pursuant to the Contract, each shall indemnify the other against all losses and claims arising from such failure.

Clause 28

Sub-Clause 28.2 – Royalties

For a Contract comprising dredging and reclamation work and for any other Contract involving the dumping of materials the Sub-Clause may require to be varied.

> EXAMPLE
> Add second sentence as follows:
>
> The Contractor shall also be liable for all payments or compensation, if any, levied in relation to the dumping of part or all of any such materials.

It is sometimes the case on dredging contracts for the Employer to bear the costs of tonnage and other royalties, rent and other payments or compensation. If such conditions are to apply, Sub-Clause 28.2 should be varied either by adding wording or by deleting the existing wording and substituting new wording.

Clause 31

Where the particular requirements of other contractors are known within reasonable limits at the time of preparation of the Contract documents, details must be stated. The Specification is usually the appropriate place to do so but, exceptionally, some reference may be desirable in the Conditions. In that case, an additional Sub-Clause or Sub-Clauses could be added to this Clause.

Clause 34

It will generally be necessary to add a number of Sub-Clauses, to take account of the circumstances and locality of the Works, covering such matters as: permits and registration of expatriate employees; repatriation to place of recruitment; provision of temporary housing for employees; requirements in respect of accommodation for staff of Employer and Engineer; standards of accommodation to be provided; provision of access roads, hospital, school, power, water, drainage, fire services, refuse collection, communal buildings, shops, telephones; hours and conditions of working; rates of pay; compliance with labour legislation; maintenance of records of safety and health.

EXAMPLE SUB-CLAUSES (*to be numbered, as appropriate*)

Rates of Wages and Conditions of Labour 34.
The Contractor shall pay rates of wages and observe conditions of labour not less favourable than those established for the trade or industry where the work is carried out. In the absence of any rates of wages or conditions of labour so established, the Contractor shall pay rates of wages and observe conditions of labour which are not less favourable than the general level of wages and conditions observed by other employers whose general circumstances in the trade or industry in which the Contractor is engaged are similar.

Employment of Persons in the Service of Others 34.
The Contractor shall not recruit or attempt to recruit his staff and labour from amongst persons in the service of the Employer or the Engineer.

Repatriation of Labour 34.
The Contractor shall be responsible for the return to the place where they were recruited or to their domicile of all such persons as he recruited and employed for the purposes of or in connection with the Contract and shall maintain such persons as are to be so returned in suitable manner until they shall have left the Site or, in the case of persons who are not nationals of and have been recruited outside (*insert name of country*), shall have left (*insert name of country*).

Housing for Labour 34.
Save insofar as the Contract otherwise provides, the Contractor shall provide and maintain such accommodation and amenities as he may consider necessary for all his staff and labour, employed for the purposes of or in connection with the Contract, including all fencing, water supply (both for drinking and other purposes), electricity supply, sanitation, cookhouses, fire prevention and fire-fighting equipment, air conditioning, cookers, refrigerators, furniture and other requirements in connection with such accommodation or amenities. On completion of the Contract, unless otherwise agreed with the Employer, the temporary camps/housing provided by the Contractor shall be removed and the site reinstated to its original condition, all to the approval of the Engineer.

Accident Prevention Officer; Accidents 34.
The Contractor shall have on his staff at the Site an officer dealing only with questions regarding the safety and protection against accidents of all staff and labour. This officer shall be qualified for this work and shall have the authority to issue instructions and shall take protective measures to prevent accidents.

Health and Safety 34.
Due precautions shall be taken by the Contractor, and at his own cost, to ensure the safety of his staff and labour and, in collaboration with and to the requirements of the local health authorities, to ensure that medical staff, first aid equipment and stores, sick bay and suitable ambulance service are available at the camps, housing and on the Site at all times throughout the period of the Contract and that suitable arrangements are made for the prevention of epidemics and for all necessary welfare and hygiene requirements.

Appendix C

Measures against Insect and Pest Nuisance 34.

The Contractor shall at all times take the necessary precaution to protect all staff and labour employed on the Site from insect nuisance, rats and other pests and reduce the dangers to health and the general nuisance occasioned by the same. The Contractor shall provide his staff and labour with suitable prophylactics for the prevention of malaria and take steps to prevent the formation of stagnant pools of water. He shall comply with all the regulations of the local health authorities in these respects and shall in particular arrange to spray thoroughly with approved insecticide all buildings erected on the Site. Such treatment shall be carried out at least once a year or as instructed by the Engineer. The Contractor shall warn his staff and labour of the dangers of bilharzia and wild animals.

Epidemics 34.

In the event of any outbreak of illness of an epidemic nature, the Contractor shall comply with and carry out such regulations, orders and requirements as may be made by the Government, or the local medical or sanitary authorities, for the purpose of dealing with and overcoming the same.

Burial of the Dead 34.

The Contractor shall make all necessary arrangements for the transport, to any place as required for burial, of any of his expatriate employees or members of their families who may die in (insert name of country). The Contractor shall also be responsible, to the extent required by the local regulations, for making any arrangements with regard to burial of any of his local employees who may die while engaged upon the Works.

Supply of Foodstuffs 34.

The Contractor shall arrange for the provision of a sufficient supply of suitable food at reasonable prices for all his staff, labour and Subcontractors, for the purposes of or in connection with the Contract.

Supply of Water 34.

The Contractor shall, so far as is reasonably practical, having regard to local conditions, provide on the Site an adequate supply of drinking and other water for the use of his staff and labour.

Alcoholic Liquor or Drugs 34.

The Contractor shall not, otherwise than in accordance with the Statutes, Ordinances and Government Regulations or Orders for the time being in force, import, sell, give, barter or otherwise dispose of any alcoholic liquor or drugs, or permit or suffer any such importation, sale, gift, barter or disposal by his Subcontractors, agents, staff or labour.

Arms and Ammunition 34.

The Contractor shall not give, barter or otherwise dispose of to any person or persons, any arms or ammunition of any kind or permit or suffer the same as aforesaid.

Festivals and Religious Customs 34.

The Contractor shall in all dealings with his staff and labour have due regard to all recognised festivals, days of rest and religious or other customs.

Disorderly Conduct 34.
The Contractor shall at all times take all reasonable precautions to prevent any unlawful, riotous or disorderly conduct by or amongst his staff and labour and for the preservation of peace and protection of persons and property in the neighbourhood of the Works against the same.

Clause 35

Additional Sub-Clauses may be desirable to cover circumstances which require the maintenance of particular records or the provision of certain specific reports.

EXAMPLE SUB-CLAUSES (*to be numbered, as appropriate*)

Records of Safety and Health 35.
The Contractor shall maintain such records and make such reports concerning safety, health and welfare of persons and damage to property as the Engineer may from time to time prescribe.

Reporting of Accidents 35.
The Contractor shall report to the Engineer details of any accident as soon as possible after its occurrence. In the case of any fatality or serious accident, the Contractor shall, in addition, notify the Engineer immediately by the quickest available means.

Clause 40

For a Contract comprising dredging and some types of reclamation work the Clause may be varied.

Sub-Clause 40.1 – Suspension of Work

EXAMPLE
Delete paragraph (c) and renumber paragraph (d) as (c).

Sub-Clause 40.3 – Suspension Lasting more than 84 Days

EXAMPLE
In the first sentence delete the words ',(c) or (d)' and substitute 'or (c)'.

Clause 43

Sub-Clause 43.1 – Time for Completion

Where completion is stated to be by a date and not within a period of time, the Sub-Clause will require to be varied.

EXAMPLE
Delete the words, 'within the time . . . such extended time' and substitute 'by the date or dates stated in the Appendix to Tender for the whole of the Works or the Section (as the case may be) or such later date or dates'.

Clause 45

For a Contract located in an isolated area, where environmental restrictions do not apply, or where a Contract comprises work, such as dredging and reclamation, that may require continuous working, the Clause may be varied.

EXAMPLE
Delete Sub-Clause 45.1 and substitute:

Working Hours 45.1
Subject to any provision to the contrary contained in the Contract, the Contractor shall have the option to work continuously by day and by night and on locally recognised days of rest.

The Contractor's option may be further extended by substituting, in the place of the last three words:

'holidays or days of rest.'

Clause 47

Where it is desired to make provision for the payment of a bonus or bonuses for early completion, an additional Sub-Clause may be added.
In the case where a bonus is provided for early completion of the whole of the Works:

EXAMPLE SUB-CLAUSE

Bonus for Completion 47.3
If the Contractor achieves completion of the Works prior to the time prescribed by Clause 43, the Employer shall pay to the Contractor a sum of (insert figure) for every day which shall elapse between the date stated in the Taking-Over Certificate in respect of the Works issued in accordance with Clause 48 and the time prescribed in Clause 43.

or

In the case where bonuses are provided for early completion of Sections of the Works and details, other than the dates, are given in the Specification:

EXAMPLE SUB-CLAUSE

Bonus for Completion 47.3
Sections are required to be completed by the dates given in the Appendix to Tender in order that such Sections may be occupied and used by the Employer in advance of the completion of the whole of the Works.

Details of the work required to be executed to entitle the Contractor to bonus payments and the amount of the bonuses are stated in the Specification.

For the purposes of calculating bonus payments, the dates given in the Appendix to Tender for completion of Sections are fixed and, unless otherwise agreed, no other adjustments of the dates by reason of granting an extension of time pursuant to Clause 44 or any other Clause of these Conditions will be allowed.

Issue of certificates by the Engineer that the Sections were satisfactory and complete by the dates given on the certificates shall, subject to Clause 60, entitle the Contractor to the bonus payments calculated in accordance with the Specification.

Clause 48

Where it can be foreseen that, when the whole of the Works have been substantially completed, the Contractor may be prevented by reasons beyond his control from carrying out the Tests on Completion, an additional Sub-Clause may be added.

EXAMPLE SUB-CLAUSE

Prevention from Testing 48.5
If the Contractor is prevented from carrying out the Tests on Completion by a cause for which the Employer or the Engineer or other contractors employed by the Employer are responsible, the Employer shall be deemed to have taken over the Works on the date when the Tests on Completion would have been completed but for such prevention. The Engineer shall issue a Taking-Over Certificate accordingly. Provided always that the Works shall not be deemed to have been taken over if they are not substantially in accordance with the Contract.

If the Works are taken over under this Sub-Clause the Contractor shall nevertheless carry out the Tests on Completion during the Defects Liability Period. The Engineer shall require the Tests to be carried out by giving 14 days notice.

Any additional costs to which the Contractor may be put, in making the Tests on Completion during the Defects Liability Period, shall be added to the Contract Price.

Clause 49

For a Contract which includes a high proportion of Plant, an additional Sub-Clause may be necessary.

EXAMPLE SUB-CLAUSE

Extension of Defects Liability 49.5
The provisions of this Clause shall apply to all replacements or renewals of Plant carried out by the Contractor to remedy defects and damage as if the replacements and renewals had been taken over on the date they were completed. The Defects Liability Period for the Works shall be extended by a period equal to the period during which the Works cannot be used by reason of a defect or damage. If only part of the Works is affected the Defects Liability Period shall be extended only for that part. In neither case shall the Defects Liability Period be extended beyond 2 years from the date of taking over.

When progress in respect of Plant has been suspended under Clause 40, the Contractor's obligations under this Clause shall not apply to any defects occurring more than 3 years after the Time for Completion established on the date of the Letter of Acceptance.

For a Contract comprising dredging work an additional Sub-Clause may be added.

EXAMPLE SUB-CLAUSE

No Remedying of Defects of Dredging Work after Completion 49.5
Notwithstanding Sub-Clause 49.2, the Contractor shall have no responsibility for the remedying of defects, shrinkages or other faults in respect of dredging work after the date stated in the Taking-Over Certificate.

Clause 50

For a Contract comprising dredging work and where the second Example Sub-Clause 49.5 has been adopted, an additional Sub-Clause should be added.

EXAMPLE SUB-CLAUSE

No Responsibility for Cost of Searching of Dredging Work 50.2
Notwithstanding Sub-Clause 50.1, the Contractor shall have no responsibility to bear the cost of searching for any defect, shrinkage or other fault in respect of dredging work after the date stated in the Taking-Over Certificate.

Clause 51

Sub-Clause 51.1 – Variations

For a Contract comprising dredging and some types of reclamation work the Sub-Clause may require to be varied.

EXAMPLE
Add final sentence as follows:

Provided also that the Contractor shall be under no obligation to execute any variation which cannot be executed by the Contractor's Equipment being used or to be used on the Works.

Clause 52

Where provision is made in the Contract for payment in foreign currency, this Clause may be varied.

Sub-Clause 52.1 – Valuation of Variations

EXAMPLE
Add final sentence as follows:

The agreement, fixing or determination of any rates or prices as aforesaid shall include any foreign currency and the proportion thereof.

Sub-Clause 52.2 – Power of Engineer to Fix Rates

Add to first paragraph final sentence as follows:

The agreement or fixing of any rates or prices as aforesaid shall include any foreign currency and the proportion thereof.

Sub-Clause 52.3 – Variation Exceeding 15 per cent

Add final sentence as follows:

The adjustment or fixing of any sum as aforesaid shall have due regard to any foreign currency included in the Effective Contract Price and the proportion thereof.

Where it is required to place some limitation on the range of items for which the rates and prices may be subject to review, the Clause may be varied.

Sub-Clause 52.2 – Power of Engineer to Fix Rates

EXAMPLE
At the end of the first paragraph add:

Provided further that no change in the rate or price for any items contained in the Contract shall be considered unless such items accounts for an amount more than 2 per cent of the Contract Price, and the actual quantity of work executed under the item exceeds or falls short of the quantity set out in the Bill of Quantities by more than 25 per cent.

Clause 54

Where vesting of Contractor's Equipment, Temporary Works and materials in the Employer is required, additional Sub-Clauses may be added.

EXAMPLE WORDING AND SUB-CLAUSES
Sub-Clauses 54.2 and 54.3 shall be renumbered as 54.3 and 54.4 and Sub-Clauses 54.4 to 54.8 shall be renumbered as 54.6 to 54.10. Add additional Sub-Clauses as follows:

Vesting 54.2
All Contractor's Equipment, Temporary Works and materials owned by the Contractor, or by any company in which the Contractor has a controlling interest, shall, when on the Site, be deemed to be the property of the Employer. Provided always that the vesting of such property in the Employer shall not prejudice the right of the Contractor to the sole use of the said Contractor's Equipment, Temporary Works and materials for the purpose of the Works nor shall it affect the Contractor's responsibility to operate and maintain the same under the provisions of the Contract.

Revesting and Removal 54.5
Upon the removal, with the consent of the Engineer under Sub-Clauses 54.1, of any such Contractor's Equipment, Temporary Works or materials as have been deemed to have become the property of the Employer under Sub-Clause 54.2, the property therein shall be deemed to revest in the Contractor and, upon completion of the Works, the property in the remainder of such Contractor's Equipment, Temporary Works and materials shall, subject to Clause 63, be deemed to revest in the Contractor.

Clause 60

Additional Sub-Clauses may be necessary to cover certain other matters relating to payments.

Where payments are to be made in various currencies in predetermined proportions and calculated at fixed rates of exchange the following 3 Sub-Clauses, which should be taken together, may be added:

EXAMPLE SUB-CLAUSES (*to be numbered, as appropriate*)

Currency of Account and Rates of Exchange 60.
The currency of account shall be the (*insert name of currency*) and for the purposes of the Contract conversion between (*insert name of currency*) and other currencies stated in the Appendix to Tender shall be made in accordance with the Table of Exchange Rates in the Appendix to Tender. Conversion between the currencies stated in such Table other than the (*insert name of currency*) shall be made at rates of exchange determined by use of the relative rates of exchange between such currencies and the (*insert name of currency*) set out therein.

Payments to Contractor 60.
All payments to the Contractor by the Employer shall be made:

(a) in the case of payment(s) under Sub-Clause(s) 70.2 and (*insert number of any other applicable Clause*), in (*insert name of currency/ies*);
(b) in the case of payments for certain provisional sum items excluded from the Appendix to Tender, in the currencies and proportions applicable to these items at the time when the Engineer gives instructions for the work covered by these items to be carried out; and
(c) in any other case, including Increase or Decrease of Costs under Sub-Clause 70.1, in the currencies and proportions stated in the Appendix to Tender as applicable to such payment provided that the proportions of currencies stated in the Appendix to Tender may from time to time upon the application of either party be varied as may be agreed.

Payments to Employer 60.
All payments to the Employer by the Contractor including payments made by way of deduction or set-off shall be made:

(a) in the case of credit(s) under Sub-Clause(s) 70.2 and (*insert number of any other applicable Clause*) in (*insert name of currency/ies*);
(b) in the case of liquidated damages under Clause 47, in (*insert name of currency/ies*);
(c) in the case of reimbursement of any sum previously expended by the Employer, in the currency in which the sum was expended by the Employer; and
(d) in any other case, in such currency as may be agreed.

If the part payable in a particular currency of any sum payable to the Contractor is wholly or partly insufficient to satisfy by way of deduction or set-off a payment due to the Employer in that currency, in accordance with the provisions of this Sub-Clause, then the Employer may if he so desires make such deduction or set-off wholly or partly as the case may be from the balance of such sum payable in other currencies.

Where all payments are to be made in one currency the following Sub-Clause may be added:

EXAMPLE SUB-CLAUSE (*to be numbered, as appropriate*)

Currency of Account and Payments 60.
The currency of account shall be the (*insert name of currency*) and all payments made in accordance with the Contract shall be in (*insert name of currency*). Such (*insert name of currency*), other than for local costs, shall be fully convertible. The percentage of such payments attributed to local costs shall be as stated in the Appendix to Tender.

Where place of payment is to be defined the following Sub-Clause may be added:

EXAMPLE SUB-CLAUSE (*to be numbered, as appropriate*)

Place of Payment 60.
Payments to the Contractor by the Employer shall be made into a bank account nominated by the Contractor in the country of the currency of payment. Where payment is to be made in more that one currency separate bank accounts shall be nominated by the Contractor in the country of each currency and payments shall be made by the Employer accordingly.

Where provision is to be included for the advance payment the following Sub-Clause may be added:

EXAMPLE SUB-CLAUSE (*to be numbered, as appropriate*)

Advance Payment 60.
An advance payment of the amount stated in the Appendix to Tender shall, following the presentation by the Contractor to the Employer of an approved performance security in accordance with Sub-Clause 10.1 and a Guarantee in terms approved by the Employer for the full value of the advance payment, be certified by the engineer for payment to the Contractor. Such Guarantee shall be progressively reduced by the amount repaid by the Contractor as indicated in Interim Payment Certificates of the Engineer issued in accordance with this Clause. The advance payment shall not be subject to retention. The advance payment shall be repaid by way of reduction in Interim Payment Certificates commencing with the next certificate issued after the total certified value of the Permanent Works and any other items in the Bill of Quantities (excluding the deduction of retention) exceeds (*insert figure*) per cent of the sum stated in the Letter of Acceptance. The amount of the reduction in each Interim Payment Certificate shall be one (*insert fraction*) of the difference between the total value of the Permanent Works and any other items in the Bill of Quantities (excluding the deduction of retention) due for certification in such Interim Payment Certificate and the said value in the last preceding Interim Payment Certificate until the advance payment has been repaid in full. Provided that upon issue of a Taking-Over Certificate for the whole of the Works or upon the happening of any of the events specified in Sub-Clause 63.1 or termination under Clauses 65, 66 or 69, the whole of the balance then outstanding shall immediately become due and payable by the Contractor to the Employer.

Clause 67

Where it is considered desirable to add to Sub-Clauses 67.3 provisions with respect to the number of arbitrators, the place of arbitration and the language of arbitration, the following paragraphs may be added to Sub-Clause 67.3:

EXAMPLE
The arbitral tribunal shall consist of . . . (*a sole or three*) arbitrator(s).
The place of arbitration shall be . . . (*city and country*).
The language of the arbitration shall be . . .

It is desirable that the place of arbitration be situated in a state, other than that of the Employer or the Contractor, which has a modern and liberal arbitration law and which has ratified a bilateral or multilateral convention (such as the 1958 New York Convention on the Recognition and Enforcement of Foreign Arbitral Awards), or both, that would facilitate the enforcement of an arbitral award in the states of the parties to the Contract.

In the absence of stipulations as to the three above mentioned matters (number of arbitrators, place of arbitration and language of arbitration), the ICC will decide on the number of arbitrators (typically three in any substantial construction dispute) and on the place of arbitration. The arbitral tribunal will decide on the language of the arbitration if the parties cannot agree.

It may also be considered desirable in some cases for other parties to be joined into any arbitration between the Employer and the Contractor, thereby creating a multi-party arbitration. While this may be feasible, multi-party arbitration clauses require skilful draftmanship on a case-by-case basis. No satisfactory standard form of multi-party arbitration clause for international use has yet been developed.

Where it is decided that a settlement of dispute procedure, other than that of the International Chamber of Commerce (ICC), should be used the Clause may be varied.

Sub-Clause 67.3 – Arbitration

EXAMPLE
Following paragraph (b), delete the words 'shall be finally settled ... International Chamber of Commerce' and substitute 'shall be finally settled under the UNCITRAL Arbitration Rules as administered by (*insert name of administering authority*)'.

Where alternatives to the ICC are considered care should be taken to establish that the favoured alternative is appropriate for the circumstances of the Contract and that the wording of Clause 67 is checked and amended as may be necessary to avoid any ambiguity with the alternative. Care should be taken to define exactly how the arbitral tribunal is to be appointed and, where appropriate, an appointing authority should be designated.

Clause 68

Sub-Clause 68.2 – Notice to Employer and Engineer

For the purposes of this Sub-Clause the respective addresses are:

(a) The Employer (*insert address*)
(b) The Engineer (*insert address*)

The addresses should be inserted when the documents are being prepared prior to inviting tenders.

Clause 69

Sub-Clause 69.1 – Default of Employer

Where the Employer is a government it may be considered appropriate to vary the Sub-Clause.

> EXAMPLE
> Delete paragraph (c) and renumber paragraph (d) as (c).

Where the terms of the Sub-Clause, when read in conjunction with Sub-Clause 69.3, are in conflict with the law of the country the Sub-Clause may require to be varied.

> EXAMPLE
> Delete 'or' at the end of paragraph (c) and delete paragraph (d).

Clause 70

Three alternative methods of dealing with price adjustment are given below. The first alternative is suitable where a contract is of short duration and no price adjustment is to be made:

Sub-Clause 70.1 – Increase or Decrease in Cost

> EXAMPLE
> Delete the text of the Sub-Clause and substitute

Subject to Sub-Clause 70.2, the Contract Price shall not be subject to any adjustment in respect of rise or fall in the cost of labour, materials or any other matters affecting the cost of execution of the Contract.

Sub-Clause 70.2 – Subsequent Legislation

> EXAMPLE
> Delete the words ',other than under Sub-Clause 70.1,'.

The second alternative is suitable where price adjustment is to be made by establishing the difference in cost between the basic price and the current price of local labour and specified materials:

Sub-Clause 70.1 – Increase or Decrease in Cost

> EXAMPLE
> Delete the text of the Sub-Clause and substitute

Adjustments to the Contract Price shall be made in respect of rise or fall in the cost of local labour and specified materials as set out in this Sub-Clause.

(a) Local Workmen
   (i) For the purpose of this Sub-Clause:
   'Local Workmen' means skilled, semi-skilled and unskilled workmen of all trades engaged by the Contractor on the Site for the purpose of or in connection with the Contract or engaged full time by the Contractor off the Site for the purpose of or in connection with the contract (by way of illustration but not limitation: workmen engaged full time in any office, store, workshop or quarry);
   'Basic Rate' means the applicable basic minimum wage rate prevailing on the date 28 days prior to the latest date for submission of tenders by reason of any National or State Statute, Ordinance, Decree or other Law or any regulations or bye-law of any local or other duly constituted authority, or in order to conform with practice amongst good employers generally in the area where the Works are to be carried out; and
   'Current Rate' means the applicable basic minimum wage rate for Local Workmen prevailing on any date subsequent to the date 28 days prior to the latest date set for submission of tenders by reason of any National or State Statute, Ordinance, Decree or other Law or any regulation or bye-law of any local or other duly constituted authority, or in order to conform with practice amongst good employers generally in the area where the works are to be carried out.
   (ii) The adjustment to the Contract price under the terms of this Sub-Clause shall be calculated by multiplying the difference between the Basic and Current Rates for Local Workmen by:
   (a) the number of all hours actually worked, and
   (b) in respect of those hours worked at overtime rates, by the product of the number of said hours and the percentage addition required by the law to be paid by the Contractor for overtime.
   Such adjustment may be either an addition to or a deduction from the Contract Price.
   (iii) No other adjustment of the Contract price on account of fluctuation in the remuneration of Local Workmen shall be made.
(b) Specified Materials
   (i) For the purpose of this Sub-Clause:
   'Specified Materials' means the materials stated in Appendix (insert reference) to Tender required on the Site for the Execution and completion of the Works;
   'Basic Prices' means the current prices for the specified materials prevailing on the date 28 days prior to the latest date for submission of tenders; and
   'Current Prices' means the current prices for the specified materials prevailing at any date subsequent to the date 28 days prior to the latest date for submission of tenders.

(ii) The adjustment to the Contract Price under the terms of this Sub-Clause shall be calculated by applying the difference between the Basic and Current Prices to the quantity of the appropriate Specified Material which is delivered to the Site during the period for which the particular Current Price is effective. Such adjustment may be either an addition to or a deduction from the Contract Price.

(iii) The Contractor shall use due diligence to ensure that excessive wastage of the Specified Materials shall not occur. Any Specified Materials removed from the Site shall be clearly identified in the records required under paragraph (d) of this Sub-Clause.

(iv) The provisions of this Sub-Clause shall apply to fuels used in Contractor's Equipment engaged on the Site for the purposes of executing the Works, including vehicles owned by the Contractor (or hired by him under long term arrangements under which the Contractor is obligated to supply fuel) engaged in transporting any staff, labour, Contractor's Equipment, Temporary Works, Plant or materials to and from the Site. Such fuels shall be clearly identified in the records required under paragraph (d) of this Sub-Clause. The provisions of this Sub-Clause shall not apply to any fuels sold or supplied to any employee of the Contractor or to any person for use in any motor vehicle not being used for the purposes of the Contract.

(v) The Contractor shall at all times have regard to suitable markets and shall, whenever buying materials a variation in the cost of which would give rise to an adjustment of the Contract Price under this Sub-Clause, be diligent to buy or procure the same at the most economical prices as are consistent with the due performance by the Contractor of his obligations under the Contract.

If at any time there shall have been any lack of diligence, default or negligence on the part of the Contractor, whether in observing the above requirements or otherwise, then, for the purposes of adjusting the Contract Price pursuant hereto, no account shall be taken of any increase in cost which may be attributable to such lack of diligence, default or negligence and the amount by which any cost would have been decreased but for such lack of diligence, default or negligence shall be deducted from the Contract Price.

(vi) No other adjustment to the Contract Price on account of fluctuation in the cost of materials shall be made.

(c) Overheads and Profits Excluded

In determining the amount of any adjustment to the Contract Price pursuant to this Sub-Clause no account shall be taken of any overheads or profits.

(d) Notices and Records

The Contractor shall forthwith, upon the happening of any event which may or may be likely to give rise to adjustment of the Contract Price pursuant to this Sub-Clause, give notice thereof to the Engineer and the Contractor shall keep such books, accounts and other documents and records as are necessary to enable adjustment under this Sub-Clause to be made and shall, at the request of the Engineer, furnish any invoices, accounts, documents or records so kept and such other information as the Engineer may require.

(e) Adjustment after Date of Completion

Adjustment to the Contract Price, after the due date for completion of the whole of the Works pursuant to Clause 43, or after the date of completion of the whole of the Works certified pursuant to Clause 48, shall be made in accordance with Current Rates or Current Prices, as applicable, ruling at the due date for completion or the date stated in the Taking-Over Certificate, whichever is the earlier.

(f) Determination of Adjustment to Contract Price

The amount of any adjustment to the Contract Price pursuant to this Sub-Clause shall be determined by the Engineer in accordance with the foregoing rules.

EXAMPLE APPENDIX TO TENDER

For use in conjunction with the second alternative.

SPECIFIED MATERIALS

Material	Unit	Price and location	Transport to site	Price delivered to site	Remarks
Bitumen					
Diesel					
Petrol					
Lubricants					
Cement					
Reinforcing steel					
Explosives					

Notes:

1. The Contractor shall provide copies of quotations to substantiate all prices included in the above table.
2. All subsequent price substantiation shall be from the same source as original unless otherwise agreed by the Engineer.
3. The Contractor shall submit full explanation and provide substantiating documentation for the mode of transport to Site he proposes. Only the proposed documented mode of transport shall qualify for price adjustment.

(*Note*: Materials stated in the Appendix to Tender should be those of which substantial quantities are involved.)

The third alternative is suitable where price adjustment is to be made through the application of indices in a formula:

Sub-Clause 70.1 – Increase or Decrease in Cost

EXAMPLE

Delete the text of the Sub-Clause and substitute

(a) Adjustments to the Contract Price in respect of rise and fall in the cost of labour and materials and other matters affecting the cost of execution of the Works shall be calculated for each monthly statement pursuant to Sub-Clause 60.1, the Statement at Completion pursuant to Sub-Clause 60.5 and the Final Statement pursuant to Sub-Clause 60.6 in accordance with the provisions of this Sub-Clause if there shall be any changes in the following Index figures complied by (*insert details of source of indices*) and published by (*insert details of publication*):
  (i) the Index of the cost of Labour in (*insert name of country*),
  (ii) the Index of cost of (*insert other factor, as relevant*), or
  (iii) The Index of the cost of (*insert other factor, as relevant*).
(b) For the purpose of this Sub-Clause:
  (i) 'Base Index Figure' shall mean the index figure applicable on the date 28 days prior to the latest date for submission of tenders, and
  (ii) 'Current Index Figure' shall mean the index figure apllicable on the last day of the period to which the particular statement relates. Provided that in respect of any work the value of which is included in any such monthly statement (or Statement at Completion or Final Statement) and which was executed after the due date (or extended date) for completion of the whole of the Works, pursuant to Clause 43, the Currrent Index Figure shall be the index figure applicable on the aforesaid due date (or extended date) for completion of the whole of the Works.
  (iii) 'Effective Value' shall be the difference between:
    (a) The amount which is due to the Contractor under the provisions of Sub-Clauses 60.2, 60.5 or 60.8 (before deducting retention and excluding repayment of the advance payment) less any amounts for:
      — work executed under nominated Subcontracts
      — materials and Plant on the Site, as referred to in Sub-Clause 60.1(c)
      — dayworks, variations or any other items based on actual cost or current prices, and bonuses (if any)
      — adjustments under Clause 70,
    and
    (b) The amount calculated in accordance with (b) (iii) (a) of this Sub-Clause and included in the last preceding statement.
    (c) The adjustment to the Contract Price shall be calculated by multiplying the Effective Value by a Price Fluctuation Factor which shall be the net sum of the products obtained by multiplying each of the

proportions given in paragraph (d) of this Sub-Clause by the following fraction:

$$\frac{\text{Current Index Figure} - \text{Base Index Figure}}{\text{Base Index Figure}}$$

calculated using the relevant index figures.

(d) For the purpose of calculating the Price Fluctuation Factor, the proportions referred to in paragraph (c) of this Sub-Clause shall (irrespective of the actual constituents of the work) be as follows:

0. in respect of labour (and supervision) costs subject to adjustment by reference to the Index referred to in (a) (i) of this Sub-Clause;

0. in respect of by reference to the Index referred to in (a) (ii) of this Sub-Clause;

0. in respect of by reference to the Index referred to in (a) (iii) of this Sub-Clause;

0. in respect of all other costs which shall not be subject to any adjustment; 1.00 Total

(e) Where the value of an Index is not known at the time of calculation, the latest available value shall be used and any adjustment necessary shall be made in subsequent monthly statements.

(*Note*: The number of indices included under (a) of this Sub-Clause may be varied, if it is determined that a different number of factors should be separately identified, and in such case (d) of this Sub-Clause must be altered to be consistent.)

Clause 72

Sub-Clause 72.2 – Currency Proportions

Where it is decided that the rate or rates of exchange shall be established from a source other than the Central Bank of the country, the Sub-clause may be varied.

EXAMPLE
Delete the words from 'prevailing . . .' to the end of the sentence and substitute 'stated in the Appendix to Tender'.

Clause 73 onwards

Where circumstances require, additional clauses may be added.

EXAMPLE CLAUSES (*to be numbered, starting with Clause 73, as appropriate*).
Where the law applicable to the Contract does not cover bribery, the following example Clause may be added.

Bribes 1.
If the Contractor or any of his Subcontractors, agents or servants offers to give or agrees to offer or give to any person, any bribe, gift, gratuity or commission as an inducement or reward for doing or forbearing to do any action in relation to the Contract or any other contract with the Employer or for showing or forbearing to show favour or disfavour to any person in relation to the Contract or any other contract with the Employer, then the Employer may enter upon the Site and the Works and terminate the employment of the Contractor and the provisions of Clause 63 hereof shall apply as if such entry and termination had been made pursuant to that Clause.

Where circumstances require that particular confidentiality is observed, the following example Clause may be added.

Details to be Confidential 1.
The Contractor shall treat the details of the Contract as private and confidential, save insofar as may be necessary for the purposes thereof, and shall not publish or disclose the same or any particulars thereof in any trade or technical paper or elsewhere without the previous consent in writing of the Employer or the Engineer. If any dispute arises as to the necessity of any publication or disclosure for the purpose of the Contract the same shall be referred to the Employer whose determination shall be final.

Where the Contract is being financed wholly or in part by an international financial institution whose rules or policies require a restriction on the use of the funds provided, the following example Clause may be added.

Expenditure Restricted 1.
The Contractor shall not make any expenditures for the purpose of the Contract in the territories of any country which is not a member of (insert name of international financial institution) nor shall he make any expenditure for goods produced in or services supplied from such territories.

Where the Contractor may be a joint venture, the following example Clause may be added.

Joint and Several Liability 1.
If the Contractor is a joint venture of two or more persons, all such persons shall be jointly and severally bound to the Employer for the fulfilment of the terms of the Contract and shall designate one of such persons to act as leader with authority to bind the joint venture. The composition or the constitution of the joint venture shall not be altered without the prior consent of the Employer.

Part II – Conditions of Particular Application

Index	*Clause*
Access to Data	11.2
Accident Prevention Officer	34.
Accidents, Reporting of	35.
Additional Clauses	73.
Alcoholic Liquor	34.
Arbitration	67.
Arms and Ammunition	34.
Bonus for Completion	47.3
Boreholes and Exploratory Excavation	18.1
Burial of the Dead	34.
Cash Flow Estimate to be Submitted	14.3
Changes in Cost and Legislation	70.
Conditions of Labour and Rates of Wages	34.
Contract Agreement	9.1
Contractor's Equipment, Insurance of	21.1
Contractor's Representative, Language Ability of	15.2
Currency of Account and Rates of Exchange	60.
Currency Proportions	72.2
Data, Access to	11.2
Default of Employer	69.1
Disorderly Conduct	34.
Documents, Order of Precedence	5.2
Drugs	34.
Employer, Name and Address	1.1
Employment of Local Personnel	16.4
Employment of Persons in Service of Others	34.
Engineer, Name and Address	1.1
Engineer's Duties	2.1
Environment, Protection of	19.1
Epidemics	34.
Exploratory Excavation and Boreholes	18.1
Extension of Defects Liability	49.5
Facilities for Other Contractors	31.2
Festivals and Religious Customs	34.
Foodstuffs, Supply of	34.
Health and Safety of Staff and Labour	34.
Housing for Labour	34.

Inspection of Site	11.1
Insurance of the Works and Contractor's Equipment	21.1
Insurances Arranged by the Employer	21. 23 & 25
Interpreter to be made Available	15.2
Labour, Conditions of and Rates of Wages	34.
Labour, Health and Safety	34.
Labour, Housing for	34.
Labour, Rates of Wages and Conditions	34.
Labour, Repatriation of	34.
Labour, Special Provision for	34.
Language, Ability of Contractor's Representative	15.2
Language, Ability of Superintending Staff	16.3
Language, Ruling	5.1
Language/s	5.1
Law Applicable	5.1
Local Personnel, Employment of	16.4
Measures Against Insect and Pest Nuisance	34.
No Remedying of Defects in Dredging Work after Completion	49.5
No Responsibility for Cost of Searching of Dredging Work	50.2
Not Foreseeable Physical Obstructions or Conditions	12.2
Notice to Employer and Engineer	68.2
Other Contractors, Facilities for	31.2
Other Contractors, Opportunities for	31.1
Payments to Contractor	60.
Payments to Employer	60.
Performance, Example of Security Bond for	10.1
Performance Guarantee, Example	10.1
Performance Security	10.1
Performance Security, Source of	10.4
Physical Conditions or Obstructions, Not Foreseeable	12.2
Place of Payment	60.0
Power of Engineer to Fix Rates	52.2
Prevention from Testing	48.5
Programme to be Submitted	14.1
Protection of the Environment	19.1
Rates of Exchange	60.
Rates of Wages and Conditions of Labour	34.
Records of Safety and Health	35.

Religious Customs and Festivals	34.
Repatriation of Labour	34.
Reporting of Accidents	35.
Revesting and Removal of Contractor's Equipment, Temporary Works and Materials	54.5
Royalties	28.2
Safety, Security and Protection of the Environment	19.1
Security, Safety and Protection of the Environment	19.1
Site, Inspection of	11.1
Source of Performance Security	10.4
Staff, Health and Safety of	34.
Submission of Cash Flow Estimate	14.3
Submission of Programme	14.1
Superintending Staff, Language Ability of	16.3
Supply of Foodstuffs	34.
Supply of Water	34.
Surety Bond for Performance, Example of	10.1
Suspension Work	40.1
Testing, Prevention from	48.5
Time for Completion	43.1
Valuation of Variations	52.1
Variations	51.1
Variations Exceeding 15 per cent	52.3
Vesting of Contractor's Equipment, Temporary Works and Materials	54.2
Water, Supply of	34.
Working Hours	45.1
Works, Insurance of	21.1

Editorial Amendments In 1988 to Part II – Conditions of Particular Application

Following publication in 1987 of the Fourth Edition of the Conditions of Contract for Works of Civil Engineering Construction, a number of editorial amendments were agreed by FIDIC. The amendments have been incorporated during reprinting in 1988 and the list below clarifies the differences between the 1988 reprint and the original document.

Page 5 Clause 9. The words 'as stated in Sub-Clause 9.1 of Part I of these Conditions' have been added to the final line of the first paragraph.

Page 6 Sub-Clause 12.2. A comma has been moved from after the word 'words' to immediately before the word 'other'. The word 'Site' has been capitalised.

Page 7	Example performance guarantee. A comma previously appeared between the words 'and' and 'complete' in the third line of the paragraph beginning 'Whereas'.
	The fifth line of the paragraph beginning 'Now the Condition . . .' previously read '. . . default by the Contract . . .'
Page 12	Sub-Clause 21.3. (b) (ii). A full stop has been inserted following '50'.
Page 13	Sub-Clause 34. Repatriation of Labour. A comma has been inserted between the words 'country)' and 'shall'.
Page 14	Sub-Clause 34. Epidemics. The word 'Contractor' has been capitalised in the first line.
	Sub-Clause 34. Alcoholic Liquor or Drugs. The word 'Contractor' has been capitalised in the first line.
Page 15	Sub-Clause 40.3. This was previously incorrectly listed as 40.2.
Page 17	Sub-Clause 49.5. The last line of the first paragraph previously read '. . . extend beyond 730 days'.
Page 20	Sub-Clause 67.3. The word 'a' previously appeared before the bracket on the penultimate line of the Example.
Page 21	Sub-Clause 70.2. A comma has been moved from after the word 'words' to immediately before the word 'other'.

Further Amendments in 1992 to Part II – Conditions of Particular Application

The following amendments have been made to the 1988 reprint of the Fourth Edition. In addition, some minor changes in the use of punctuation (commas, semicolons, colons and stops), as well as the use of the words 'or' and 'and' have been introduced to attain uniformity in the style of all Clauses. These minor changes which improve the style, but which have no effect on the meaning of Clause, have not been listed below.

Contents	The words 'Index' and 'Editorial Amendments' have been added at the bottom of the page.
Page 1	INTRODUCTION. The words,' subject to minor modifications' have been added, and the word 'equally' changed to read 'also'.
Page 3	Sub-Clause 1.1. In the last sentence, the words 'for example the name of an International Financing Institution (IFI)' have been deleted.
Page 4	Clause 9. In the first paragraph the words 'of Part I' have been added. In the second paragraph '1' has been corrected to read 'I'.
Page 5	Clause 12. Sub-Clause 12.2. In the title the word 'Adverse' has been changed to read 'Not Foreseeable' (also amended in the Contents and the Index).

Page 10 Sub-Clause 21.1, sub-para (a). The words '(the term "cost" in this context shall include profit)' have been added.

Page 13 Example Sub-Clause for Supply of Foodstuffs. The words 'staff and labour, or his Subcontractors' have been changed to read 'staff, labour and Subcontractors'.

Page 19 Example Sub-Clause for Advance Payment. The words 'interim certificate' have been changed to read 'Interim Payment Certificate', in both the singular and plural.

Clause 67. The first four paragraphs of the commentary have been added.

Page 24 Sub-Clause 70.1. In the formula, the word 'Based' has been corrected to read 'Base'.

Page 25 Example Clause for Bribes. The word 'Sub-contractors' has been corrected to read 'Subcontractors'.

Example Clause for Details to be Confidential. In the sixth line, the words 'the decision of' have been deleted. In the seventh line, the word 'award' has been changed to read 'determination'.

Example Clause for Expenditure Restricted. In the commentary and text, the leading capital letters of the words 'International Financing Institutions' have been changed to small letters. In the commentary, the word 'Articles' has been changed to read 'rules or policies'. In the third line, the word 'not' has been corrected to read 'nor'.

Editorial Amendments (1988). In the last item, the words 'Line 21' have been corrected to read 'Page 21'.

References

Chapter 1 Background of the Red Book

1.1 'Statutes and By-Laws', a publication of FIDIC, Lausanne, Switzerland, October 1996.
1.2 'Bad Drafting Sires a Lawyer's Gift Horse', Max W. Abrahamson and 'A Charter to Riches for the Contractor', Thomas Akroyd, *New Civil Engineer*, 5 July 1973, London.
1.3 'The Modest Revision which Became a Torrent of Change', I. N. Duncan Wallace, QC, *New Civil Engineer*, 1 November 1973, London.
1.4 *The ICE Conditions of Contract Fifth Edition, A Commentary*, I. N. Duncan Wallace, Sweet & Maxwell, London, 1978. This book was published after the date of publication of the Third Edition of the Red Book. Comments made in this book in connection with the Red Book were linked to a previous book published in 1974 by Mr Duncan Wallace on that Form with the title *The International Civil Engineering Contract* and a subsequent supplement on the Third Edition published in 1980, see Reference 1.5 below.
1.5 *First Supplement to the International Civil Engineering Contract*, I. N. Duncan Wallace, QC, Sweet & Maxwell, London, 1980.
1.6 The applicable law of the contract is referred to by some as the 'proper' law of the contract and by others as the 'governing' law of the contract. These three terms are synonymous but as the wording of sub-clause 5.1(b) refers to 'the country or state the law of which shall apply to the contract . . .', the first of these terms will be used in the book.
1.7 See *Hudson's Building and Engineering Contracts*, Eleventh Edition, Volume I, pages 311 to 313, para. 2.133, where the case of *Neodox Ltd* v. *Swinton and Pendlebury BC* (1958) QBD (unreported) is quoted in respect of what is a reasonable time.
1.8 'Notes on Documents for Civil Engineering Contracts', a publication of FIDIC, Lausanne, Switzerland, March 1977, page 17.

Chapter 2 The Red Book is Based on a Domestic Contract

2.1 *Codigo de Hammurabi*, Edicion preparada por Federico Lara Peinado, Editora Nacional, Madrid, 1982. Hammurabi's Code contains the earliest available recorded rules of codified construction law and insurance.
2.2 *Private International Law*, Eleventh Edition, Cheshire and North, Butterworths, London, 1987, Chapter 18, page 447.
2.3 Article 1(3) of the UNCITRAL Model Law on International Commercial Arbitration as adopted by the United Nations Commission on International Trade Law, adopted on 21 June 1985.

2.4 *Applicable Law in International Commercial Arbitration*, Julian D. M. Lew, Oceana Publications, 1978, page 75.
2.5 This principle was accepted, for example, by the Member States of the European Economic Community in an international convention on 19 June 1980, in Rome, on 'The Law Applicable to Contractual Obligations'.
2.6 *Irish Conflicts of Law*, William Binchy, Butterworths, London, 1988, page 3.
2.7 *Law and Practice of International Commercial Arbitration*, Second Edition, Alan Redfern and Martin Hunter, Sweet & Maxwell, London, 1991, page 125.
2.8 *Applicable Law in International Commercial Arbitration* (see Reference 2.4 above), page 104.
2.9 For further information on this topic, see References 2.3 and 2.4 above and also *Amin Rasheed Shipping Corporation* v. *Kuwait Insurance Company* [1983] 3 WLR 241.
2.10 Delocalised arbitrations are those detached from the control of the law of the Seat where they are held. For a discussion on this topic, see *Law and Practice of International Commercial Arbitration* (Reference 2.7 above), page 81. See also a paper on 'Choice of Law Issues in International Arbitration' by Professor Michael Pryles, Singapore Conference on Current Legal Issues in International Commercial Litigation, October 1996.
2.11 The New York Convention: 1958, Convention on the Recognition and Enforcement of Foreign Arbitral Awards, New York, 10 June 1958. See also Commentary by Dr Albert Jan van den Berg, T.M.C. Asser Institute for International Law, The Hague, Netherlands.
2.12 There are recent moves to change this concept to the year 570 AD which is the year of Prophet Muhammed's birth.
2.13 *An Introduction to Comparative Law*, K. Zweigert and H. Kotz, Second Edition, translated by Tony Weir, Volume 1, Clarendon Press, Oxford, 1987, page 85.
2.14 *Major Legal Systems in the World Today*, Third Edition, Réne David and John E. C. Brierley, Stevens & Sons, London, 1985, page 94.
2.15 'British and French Statutory Drafting', Proceedings of the Franco-British Conference of 7 and 8 April 1986, edited by Sir William Dale, Institute of Advanced Legal Studies, University of London, page 1.
2.16 *Major Legal Systems in the World Today* (see Reference 2.14, above), page 120.
2.17 French Code Civil, Article 5, paragraph 70, Note 2.
2.18 *An Introduction to Comparative Law* (see Reference 2.13 above), Volume 1, page 188.
2.19 *Major Legal Systems in the World Today* (see Reference 2.14, above), page 316.
2.20 *Snell's Principles of Equity*, P. V. Baker and P. St. J. Langan, Sweet & Maxwell, London, 1982, page 5, and the judgment referred to in the quotation is from *Cawley & Co.* (1889) 42 ChD 209 at 236.
2.21 *Earl of Oxford's Case* (1615) 1 W & T 615, 21 Eng Rep 485, 487.
2.22 'British and French Statutory Drafting' (see Reference 2.15 above), page 17.
2.23 *Conway* v. *Rimmer* [1968] AC 910; 1 All ER 874.
2.24 'The Legislative Systems', a paper by Sir George Engle, QC, First Parliamentary Counsel, London, read at a Conference on Statutory Drafting, Institute of Advanced Legal Studies, University of London, April 1986, page 25.
2.25 'The Demands of the Legislator and the Lawyer and his Client', a paper by Norman Adamson QC, First Parliamentary Draftsman for Scotland, read at a Conference on Statutory Drafting, Institute of Advanced Legal Studies, University of London, April 1986, page 63.
2.26 *Morgan Grenfell (Local Authority Finance) Ltd* v. *Sunderland Borough Council and Seven Seas Dredging Ltd* (1990) 49 BLR 31.
2.27 Nudrat Majeed, 'Good Faith and Due Process: Lessons from the Shari'ah', *Arbitration International*, Volume 20, No. 1, 2004, page 101.

2.28   Quoted in the previous reference and referred to [1951] 1 ICLQ 247.
2.29   The Qur'anic verse is 4.115 which is translated as follows: 'If any one opposes the Messenger, after guidance has been made clear to him, and follows a path other than that of the believers, we shall leave him on his chosen path – we shall burn him in hell, an evil destination.'
2.30   *General Jurisprudential Introduction*, Mustapha Ahmed AlZarqa, second part, page 916, Dar AlQalam, Damascus.
2.31   *Masader al-haq in Islamic Jurisprudence*, T. III, page 41, A. Al Sanhoury, Dar Al Nahdha Al Arabiya, Cairo. (Dr Al Sanhoury is a famous Egyptian jurist who lived in the middle of the twentieth century.)
2.32   As in Reference 2.31, but page 49.
2.33   *Al-Majalla to Al Ahkam Al Adaliyyah* deals with the civil obligations of Islamic jurisprudence in which the questions which are of the most frequent occurrence have been collected together from reliable works and set out in a Code in the form of books. These books have been divided into chapters and the chapters into sections. The questions of detail which are applied in the courts are those questions which are set out in the chapters and sections.
2.34   'Good Faith and Due Process: Lessons from the Shari'iah', Nudrat Majeed, *Arbitration International*, Volume 20, No. 1 (2004), page 111, reference is also made to Shaybani with Sarakhsi's commentary, Volume IV, pages 60–61.
2.35   The holy Qur'an, Surah 2, Al Baqarah, v. 278.
2.36   The holy Qur'an, Surah 2, Al Baqarah, v. 279.
2.37   The holy Qur'an, Surah 2, Al Baqarah, v. 280.
2.38   The holy Qur'an, Surah 4, Al Nisa', v. 161.
2.39   *The Insurance Contract – Its nature and legality: A comparative study*, Dr Abdel Hadi Alsayed Muhammed Taqey AlHakeem, Al Halabi Legal Publications, page 323.
2.40   As an example of the last criteria, see reference to a contract of insurance in the text below.
2.41   The holy Qur'an, Surah 4, Al Nisa', v. 29.
2.42   The holy Qur'an, Surah 2, Al Baqarah, v. 188.
2.43   *The Insurance Contract – Its nature and legality: A comparative study*, Dr Abdel Hadi Al Sayed Mohamed Taqey Al Hakeem, Al Halabi Legal Publications, page 323.
2.44   *Commercial Law in Gulf States – The Islamic Legal Traditions*, Noel J. Coulson, Graham & Trotman, London, 1984, page 44.
2.45   See *Courtney & Fairbairn Ltd* v. *Tolaini Bros (Hotels) Ltd* [1975] 1 All ER 716, [1975] 1 WLR 297 and *May and Butcher* v. *R* [1934] 2 KB 17n.
2.46   It should be noted that Al-Majalla is representative of only one school of thought on Islamic jurisprudence, the *Hanafi* school.
2.47   The principle of good faith is prominent from the very beginning of a contract, i.e. from the offer and acceptance stage. The essential objective of the principle of offer and acceptance in contract law is to clearly prove that there was/is some form of agreement between the parties. The contract will be concluded successfully where each party carries out its own obligations within the contract to the satisfaction of the other party. Any kind of action by either party indicating an intention not to sign the contract will render the offer and acceptance null and void. The obligations of the parties to act in good faith therefore, originates from the time the substantive agreement comes into being. The significance of the intention and will of the parties is a persistent theme in the provisions of the Al-Majalla. See for example, articles 2, 3 and 44.
2.48   *Masadir al-haq*, Al Sanhuri, Volume 1, *Revival of Arabic Heritage* 1953–1954 (Beirut).
2.49   Qur'an IV: 135.

2.50 For a more in-depth discussion on this area, see 'Good Faith and Due Process: Lessons from the Shari'iah', Nudrat Majeed, *Arbitration International*, Volume 20, No. 1 [2004] page 108.

2.51 For some there is *gharar* when a number of elements are present but they differ as to what these elements are, for example, Sanhoury states that *gharar* exists when there is simultaneous ignorance and uncertainty regarding certain elements of the object for sale. See *Masader al-haq*, T. III, page 49, Nabil Saleh considers that the word *gharar* implies the three elements of risk, uncertainty and speculation.

Chapter 3 Legal Concepts Based on the Common Law System

3.1 Lord McNair, former President of the International Court of Justice, 'The General Principles of Law recognised by Civilised Nations' (1957) 33 BYIL 1 at page 7. See also *Amin Rasheed Shipping Corporation* v. *Kuwait Insurance Co.* [1983] 3 WLR 241 at 245.

3.2 *Clerk & Lindsell on Torts*, 18th Edition, Sweet & Maxwell, London, 2000, Section 1–01.

3.3 This is also the case under Irish Law, see for example the Hotel Proprietors Act 1963.

3.4 *Pollock on Contracts*, Sir Percy H. Winfield, Stevens, London, 1950, page 133.

3.5 As such, minors and those of unsound mind are incapable, in law, of creating a valid contract.

3.6 *Company Law in the Republic of Ireland*, Judge Ronan Keane, Butterworths, London, 1985, page 102.

3.7 In England, since 31 July 1990, by the Law of Property (Miscellaneous Provisions) Act 1989 and, so far as companies are concerned, by the Companies Act 1989, the formalities of affixing the seal to the signature in sealed contracts have been abolished. See page iv of *Construction Law Digest*, Volume 8, also page 10, BSP Professional Books, Oxford, 1990.

3.8 *Design Liability in the Construction Industry*, Fourth Edition, D. L. Cornes, BSP Professional Books, Oxford, 1994, Chapter 12.

3.9 Cheshire, Fifoot and Furmston's *Law of Contract*, Twelfth Edition, Butterworths, London, 1991, pages 128–130.

3.10 *Applegate* v. *Moss* [1971] 1 QB 406. The quotation is taken from *Law of Contract* (Reference 3.5 above). The last sentence in double quote marks is per Lord Denning MR at [1971] 1 QB 413.

3.11 *Wong Lai Ying and Others* v. *Chinachem Investment Co. Ltd* (1979) 13 BLR 81. See also Chapters 8 and 20 in *Law of Contract* (Reference 3.9 above).

3.12 See *Law of Contract*, Chesire, Fifott and Furmston, 12th Edition, Butterworths, London, pages 128–130.

3.13 *London Export Corporation Ltd* v. *Jubilee Coffee Roasting Co.* [1958] 1 WLR 661 at 675.

3.14 *Tai Hing Cotton Mill Ltd* v. *Liu Chong Hing Bank Ltd* [1986] 1 AC 80.

3.15 *The Moorcock* (1886–90) All ER 850.

3.16 *Trollope & Colls Ltd* v. *North-West Metropolitan Regional Hospital Board* [1973] 1 WLR 601 at 609, [1973] 2 All ER 260 at 268.

3.17 *Reigate* v. *Union Manufacturing Co. (Ramsbottom)* [1918] 1 KB 592 at 605.

3.18 *The Concept of Law*, H. L. A. Hart, Clarendon Law Series, Oxford University Press, Oxford, 1961.

3.19 See the German Civil Code, paragraph 241, where the building owner is entitled to demand performance from the contractor. See also 'Contractor's Liability for Design under German Construction Law', by Dr Christian Wiegand, a paper presented at a conference on 'The Liability of Contractors', Centre for Commercial Law Studies at Queen Mary College, London, 1986.

3.20 *The Oxford Companion to Law*, David M. Walker, Clarendon Press, Oxford, 1980.
3.21 *Hadley* v. *Baxendale* (1854) 9 Exch 341.
3.22 *Law of Contract* (see Reference 3.9 above), pages 595–598.
3.23 *Victoria Laundry (Windsor) Ltd* v. *Newman Industries Ltd* [1949] 2 KB 528; *The Heron II* [1969] 1 AC 350; and *H. Parsons (Livestock) Ltd* v. *Uttley Ingham & Co. Ltd* [1978] QB 791.
3.24 See page 595 of *Law of Contract* (Reference 3.9 above).
3.25 For a discussion on the terms 'penalty' and 'liquidated damages', see later in Section 17.5. See also Section 2.9, equity as a source of law.
3.26 *The Law of Contract*, Seventh Edition, G. H. Treitel, Stevens & Sons, London, 1987, pages 721 and 722.
3.27 *Appleby* v. *Myers* (1867), quoted in *Hudson's Building and Engineering Contracts*, Eleventh Edition, by I. N. Duncan Wallace, Sweet & Maxwell, London, 1994, Volume I, page 645, para. 4.251.
3.28 *Charon (Finchley) Ltd* v. *Singer Sewing Machines Ltd* (1968), 207 EG 140.
3.29 *Hudson's Building and Engineering Contracts* (see Reference 3.21 above). The words quoted and the reference to the case of *Charon* v. *Singer Machines Ltd*, appear in Volume I, page 508, para. 4.051.

Chapter 4 Drafting Principles

4.1 'Risk Management', Max W. Abrahamson, Appendix J of the discussion paper 'Construction, Insurance and Law' published by the International Federation of Consulting Engineers, FIDIC, Switzerland, March 1986.
4.2 *Construction Insurance*, Nael G. Bunni, Elsevier Applied Science Publishers, London, 1986, page 8.
4.3 *Communication for Professional Engineers*, Bill Scott, Thomas Telford Ltd, London, 1984, page 69.
4.4 *Monmouthshire County Council* v. *Costelloe and Kemple Ltd* (1965) 63 LGR 429. In this case Lord Justice Winn referred to clause 66 of the ICE Form in the following manner: '... I am very far from saying that I find this Clause 66 easily intelligible (and would add that I venture to think it might be reconsidered and possibly clarified by different wording) ...'.
4.5 It is reported in the *New Civil Engineer*, 30 April 1987, that during the hearing of amendments to the Animals (Scientific Procedures) Bill, in the House of Lords in England, the following was stated by one of their Lordships:

> 'My Lords, the difference between "may" and "shall" is one of the bugbears of the Statute law and of Parliamentary debate. I do not know how many hours I have spent in both Houses of Parliament debating the difference between "may" and "shall". I think we need an Act of Parliament to resolve this difficulty and to declare once and for all that "may" means "shall" ... anyway in this case "may" has been translated into "shall". Of course, we know that it meant "shall" all the time but I need not pursue it any further.'

To this statement he received the following response from another member of the House of Lords:

> 'My Lords, I should like to say this to the noble Lord. If the word "shall" is used, the judges will interpret it as "may". If the word "may" is used, the judges will interpret it as "shall".'

4.6 'Address to the Irish Branch of the Chartered Institute of Arbitrators', the late Hon. Mr Justice Niall McCarthy, Dublin, AGM of the Branch on 28 February 1985, *Arbitration Journal*, February 1986, page 61.

4.7 'Risk Assessment and Allocation – The Need for a Policy', John Barber, a paper read at a Conference on Construction Contract Policy, 14–16 September 1988, the Centre of Construction Law and Project Management, King's College, London.

4.8 Various articles and lectures at seminars and conferences on the subject of the standard forms as well as reference books have called for simpler and clearer wording. Amongst these are the following: 'Revisions to the Red Book for Civil Engineering Works: The Point of View of the Engineer', Humphrey Lloyd QC, *The International Construction Law Review* (ICLR), Volume 3, Part 5, October 1986; 'A Claims-Review-Board as a Way for an Amicable Settlement of Disputes, and Other Considerations on the Subject of Claims', G. Lodigiani, ICLR, Volume 3, Part 5, October 1986; 'FIDIC Conditions of Contract for Works of Civil Engineering Construction, Fourth Edition', Dr J. J. Goudsmit, a paper presented at a conference in London on these Conditions, October 1987.

Chapter 5 The Concept of a Trusted Independent Engineer

5.1 *FIDIC's Statutes and By-Laws*, a publication of FIDIC, Geneva, Switzerland, October 2002.

5.2 Article 2 of FIDIC's Statutes and By-Laws was most recently updated at the September 2004 conference in Copenhagen.

5.3 *Guide to the Use of Independent Consultants for Engineering Services*, Third Edition, publication of FIDIC, the International Federation of Consulting Engineers, Switzerland, 1980.

5.4 FIDIC's Statutes and By-Laws, a publication of FIDIC, Lausanne, Switzerland, October 1993 and 1994. See also: *Consulting Engineers 1913–1988: FIDIC over 75 Years*, Ragnar Widegren, FIDIC, Switzerland, 1988.

5.5 'Statutes and By-Laws', FIDIC, October 1996.

5.6 FIDIC *Policy Statement on Business Integrity*, a FIDIC publication, Geneva, Switzerland, or see also www.fidic.org

5.7 'Quality-Based Selection for the Procurement of Consulting Services', FIDIC, 1997.

5.8 *FIDIC Guidelines for the Selection of Consultants*, First Edition, FIDIC, 2003, page 1 of the Introduction. http://www1.fidic.org/resources/selection/qbs/qbs-defa.htm

5.9 *Risk and Insurance in Construction*, Nael G. Bunni, Second Edition, Spon Press, London, 2003, page 20.

5.10 The piece quoted was taken from a paper read at the 1983 FIDIC Conference in Florence by Mr Norman Westberg, a Finnish Consulting Engineer.

5.11 *FIDIC Guidelines for the Selection of Consultants*, First Edition, FIDIC, 2003, page 1 of Section 4. http://www.fidic.org/resources/selection/qbs/qbs-defa.htm

5.12 This is where the consultants are requested to submit their priced proposals in two sealed envelopes. One envelope will contain the proposal without the price included and the second will contain the prices proposed by the consultant. The owner/client should then examine the proposals and make their decision based on the first envelope only. Once they have chosen the best proposal, the second envelope will then be opened, in the presence of the consultancy firm chosen, and price negotiations will begin.

5.13 This is a system where points are assigned in the appraisal of qualification, ability, experience and price. FIDIC recommends that the weighted value of the price should be a maximum of 10%.

5.14 This system could be used when the owner/client will give a budget figure to the short list of firms that they have chosen, and the proposal is made with this budget in mind. The selection is then based on the best proposal submitted.

5.15 This is a system whereby those short-listed for the project are requested to submit their proposal including details regarding fees and any estimates of construction cost with designs.

5.16 This is where a small group of firms are short-listed to carry out the services required for the project based on their professional capabilities and are then asked to negotiate the price.

5.17 For a full list of the organisations that use the QBS system, see Section 6 of the *FIDIC Guidelines for the Selection of Consultants*, First Edition, FIDIC, 2003, http://www.fidic.org/resources/selection/qbs/qbs-defa.htm

5.18 'Making Effective Use of Consulting Engineers in Project Implementation', D. E. Cullivan, a paper presented at a seminar held in Abu Dhabi, November 1983. Report by FIDIC entitled Arab Funds/FIDIC Seminar, 1984, page 70.

5.19 'International Model Form of Agreement between Client and Consulting Engineer and International General Rules of Agreement between Client and Consulting Engineer'. Three separate documents were published by FIDIC designed to cover agreements relating to pre-investment studies: IGRA PI; design and supervision of construction of works: IGRA D&S; and project management services: IGRA PM, Switzerland. These documents were later replaced by FIDIC's White Book, as to which see Chapter 21.

5.20 'Recommended Tendering Procedures and Pre-Qualification of Contractors', J. J. de Greef, a paper presented at a seminar held in Jakarta, in 1984, on the subject of 'Consulting Engineering, a Development Resource'. Report published by FIDIC, Switzerland, 1985, page 137.

5.21 This dual role is confirmed by FIDIC in the 'Notes on Documents for Civil Engineering Contracts' published in 1977, page 7; and also in legal writing, see for example: 'Contractor's Claims under the FIDIC International Civil Engineering Contract', *International Business Lawyer*, June 1986, by Christopher R. Seppala; and 'Position and Function of the Engineer under FIDIC Civil Conditions of Contract', Conference paper by Geoffrey Hawker, June 1988.

Chapter 6   A Traditional Re-measurement Contract

6.1 It is reported in *Privatized infrastructure, the BOT approach* edited by C. Walker and A. J. Smith and published by Thomas Telford in 1995, that the acronym 'BOT' was first coined in the early 1980s by Turkey's late Prime Minister, Targut Ozal.

6.2 The cases which deal with the question of re-measurement and come to conflicting results mainly due to the differing provisions are *Aros Industries* v. *Electricity Commission of New South Wales* (1973) 2 NSWLR 186; *Mitsui Construction Co.* v. *The Attorney-General of Hong Kong* (1986) 2 Const LJ 133; and *Grinaker Construction* v. *Transvaal Provincial Administration* [1982] 1 SALR 78. See *Construction Contracts: Principles and Policies in Tort and Contract*, I. N. Duncan Wallace, Volume 2, Sweet & Maxwell, London, 1996, Chapter 24.

6.3 *FIDIC 4th, A Practical Legal Guide*, E. C. Corbett, Sweet & Maxwell, London, 1990, page 302.

6.4 *Construction Contracts: Principles and Policies in Tort and Contract*, (Reference 6.2 above), page 458.
6.5 The supply of the Schedule of make-up of rates and prices by the successful tenderer is a compulsory feature of the pre-contract procedures of the German VOB contract. See a paper by Professor Hermann Korbion entitled 'The Effects of Changed Conditions on the Contractor's Remuneration According to German Construction Law', presented at the First International Construction Law Conference held on 24 May 1982, at the University of Fribourg in Switzerland. The Schedule is also advocated by Mr Ian Duncan Wallace QC, as can be seen from his paper at the same conference under the title of 'Price Under Common Law Systems'. Both papers are published in *Selected Problems of Construction Law, International Approach*, University Press, Fribourg, Switzerland, Sweet & Maxwell, London, 1983.
6.6 Ian Duncan Wallace in *Construction Contracts: Principles and Policies in Tort and Contract* (Reference 6.2 above), Vol. 1, Chapter 26, argues against the validity of the real advantages referred to in paragraphs (a) and (b). Whilst the reasons given by him are attractive, the practical considerations of a practising engineer may be different.
6.7 *Civil Engineering Procedure*, Fourth Edition, the Institution of Civil Engineers, Thomas Telford, London, 1986.

Chapter 7 Sharing of Risks

7.1 Annual Reports of the International Chamber of Commerce, Paris; the section on the ICC International Court of Arbitration indicates a constant flow of international construction disputes. Litigation and arbitration cases around the world involving issues of professional negligence add to the list of disputes. Insurance and reinsurance loss statistics and reports complement this picture. See also *Collection of ICC Arbitral Awards 1974–1985*, Sigvard Jarvin and Yves Derains, ICC Publishing SA, ICC Publication No. 433, Paris, 1990; and *Collection of ICC Arbitral Awards 1986–1990*, Sigvard Jarvin, Yves Derains and Jean-Jacques Arnaldez, ICC Publishing SA, ICC Publication No. 514, Paris, 1994. See also Reference 8.1 below.
7.2 Figures published annually by Central Statistics Offices around the world indicate a high, if not the highest exposure at work to bodily or fatal injuries in construction. See also Health and Safety Statistics, HMSO UK and *Facts in Focus*, Statistics compiled by the Central Statistics Office, UK and published by Penguin Books in association with HMSO, UK.
7.3 Publications of the Munich Reinsurance Company and the Swiss Reinsurance Company are a valuable source of reference in this regard. These publications cover topics such as earthquakes, windstorms, flood and inundation, volcanic eruption and hailstorm. In 1978, the Munich Reinsurance Company published a world map of natural hazards which was updated in 1988. It indicates the intensity, frequency and reference period of various natural hazards (over 670 in number), catalogued in a chronological order and location, with the consequences in terms of loss of life and cost. These world-wide records go back in time to the tenth century. The map and the accompanying publications are extremely useful in risk management calculations and in any attempt at predicting future exposures through extrapolation from retro-spective exposure.
7.4 Construction (Design and Management) Regulations 1994; S.I. 1994 No. 3140.
7.5 In Arabic, the word 'Rizq' conveys a similar concept of receiving what destiny might dispose, a positive or a negative event.

7.6 'Risk Management and Insurance', Wolf-Rudiger Heilmann, a paper delivered at a Conference on Structural Failure, Product Liability and Technical Insurance, Technische Universität, Vienna, 1989, and published subsequently in *Forensic Engineering*, V. 2, Nos. 1/2, 1990, pages 119–134.

7.7 An extract from British Standard BS 4778: Part 3; Availability, reliability and maintainability terms. Section 3.1 Guide to concepts and related definitions: 1991, Quality Vocabulary. The British Standards Institution, Linford Wood, Milton Keynes, MKI4 6LE, UK, where complete copies of the standards can be obtained.

7.8 A wider definition and probably more correct is given by Australia/New Zealand Standard on Risk Management, AS/NZS 3951: 1995, which defines risk as inclusive of not only loss or damage, but also gain.

7.9 'Hyatt-Regency Walkway Collapse: Design Alternatives', George F. W. Hauck, A.S.C.E. Structural Engineering, Vol. 109, 5 May 1983.

7.10 Quoted from *Management of Engineering Risk*, Roger B. Keey, Centre for Advanced Engineering, University of Canterbury, New Zealand, April 2000.

7.11 *Engineering Ethics: Balancing Cost, Schedule and Risk*, R. L. B. Pinkus, L. J. Shuman, N. P. Hummon and H. Wolfe, Cambridge University Press, Cambridge, 1997.

7.12 *Practical risk management in the construction industry*, Leslie Edwards, Thomas Telford, London, 1995.

7.13 *Risk Management*, Max W. Abrahamson, [1983] ICLR 241, also published as Appendix J to a discussion paper *Construction, Insurance and Law* published by FIDIC – 1986, page 49; see also 'Defects: A summary and analysis of American Law', Justin Sweet, a paper published in *Selected Problems of Construction Law, International Approach*, Peter Gauch (Switzerland) and Justin Sweet (USA), University Press, Fribourg, Switzerland, Sweet & Maxwell, London, 1983, page 97.

7.14 As an example of the 'benefit' principle, although referred to as 'incentive' and not benefit, we can look at the NEC, which is commented on in the Grove Report, referred to below in reference 7.16, in the following manner: 'The NEC is said to have introduced a variant of the management standard which might be called the philosophy of incentive. The postulate is that risks should be placed on the party most in need of incentive (presumably already with the ability) to prevent and control them. This is thought to motivate people to play their part. An examination of the "compensation events" listed in the ECC does not, however demonstrate that this philosophy has been uniformly applied.'

7.15 'Price under Common Law System', I. N. Duncan Wallace; and 'Defects: A Summary and Analysis of American Law', Justin Sweet; pages 149 and 79 of *Selected Problems of Construction Law, International Approach*, Peter Gauch (Switzerland) and Justin Sweet (USA), University Press, Fribourg, Switzerland, Sweet & Maxwell, London, 1983, page 97.

7.16 *The Grove Report: Key Terms of 12 Leading Construction Contracts Are Compared and Evaluated*, published September 1998 and available on the website of Thelen Reid & Priest at www.constructionweblinks.com

7.17 A conference focusing on the report commissioned by the Government of Hong Kong SAR and prepared by Mr Jesse B. Grove III of Thelen Reid & Priest, New York, to carry out a 'fundamental review of the General Conditions of Contract, in particular the allocation and management of risk in the procurement and work projects . . .', for the report itself, see the website of Thelen Reid & Priest at: www.constructionweblinks.com

7.18 *The Grove Report*, Humphrey Lloyd [2001] 2 ICLR 302.

7.19 *Engineering Construction Risks*, Thompson and Perry, Science and Research Council, UK, 1992.
7.20 In paragraph 9.3 of this section of his report, Mr Grove quoted from an article by Mr Max Abrahamson, 'Risk Management', Max W. Abrahamson [1983] ICLR 241.
7.21 *Construction Insurance*, Nael G. Bunni, Elsevier Applied Science Publishers, London, 1986, Chapter 3, 'The Spectrum of Risks'.
7.22 'Construction, Insurance and Law', Nael G. Bunni, a paper delivered at a Conference on Structural Failure, Product Liability and Technical Insurance, Technische Universität, Vienna, 1989, and published subsequently in *Forensic Engineering*, Vol. 2, Nos. 1/2, 1990, page 163.
7.23 *FIDIC's New Suite of Contracts – Clauses 17 to 19: Risk, Responsibility, Liability, Indemnity and Force Majeure*, Nael G. Bunni, ICLR, Vol. 18, Part 3, July 2001.
7.24 See the Grove Report referred to in References 7.16 and 7.18 above.
7.25 The Institution of Civil Engineers (UK), *The Engineering and Construction Contract*, Second Edition (1995, reprinted with corrections May 1998) ('EEC' formerly the 'NEC').
7.26 The draftsman of the new suite of FIDIC's Forms of Contract, published in 1999, applied the same format of the 1995 Orange Book to these new forms and ended up with the same problem.
7.27 *FIDIC's New Suite of Contracts – Clauses 17 to 19: Risk, Responsibility, Liability, Indemnity and Force Majeure*, Nael G. Bunni, ICLR, Vol. 18, Part 3, July 2001.
7.28 The reply was made by Sir William Harris, Chairman of the Joint Contracts Committee (JCC) responsible for the revision of the ICE Form to its Fifth Edition, and Mr David Gardam, QC, Legal Adviser to the Committee.
7.29 'Clearing the Critics' Confusion', *New Civil Engineer*, 20 December 1973, London, page 33.
7.30 In 1991 the Sixth Edition of the ICE Form of Contract was published incorporating only some of the changes that were introduced by FIDIC in their 1992 reprint of the Red Book. The Seventh Edition of the ICE Form was published in 1999.
7.31 See also *Construction Contracts: Principles and Policies in Tort and Contract*, Vol. 1, I. N. Duncan Wallace, Sweet & Maxwell, London, 1986, chapter 27, paragraphs 27–34, page 474.
7.32 *Construction Insurance* (Reference 7.21 above), page 143.

Chapter 8 The Concepts in Practice

8.1 It is known that the International Court of Arbitration of the International Chamber of Commerce, whose Rules of Conciliation and Arbitration are specified in the Red Book, in 1988 received 304 requests for arbitration, involving parties from 86 different countries. In 1995, the number of cases went up to 427. Around 25% of these disputes were in the construction field. The number of requests for arbitration received by the ICC in the years 1982 to 1988 ranged from 50 to 80 per year: 'ICC Annual Reports', ICC Secretariat, Paris.
8.2 An article by Mr Zubair Iqbal, *Finance and Development*, June 1983.
8.3 'The World Bank Annual Report 1987', The World Bank, Washington, DC, 20433, 1987, page 8, and subsequent Reports for 1990 (page 13), 1994 and 1995.
8.4 'The Engineer's Approach to Claims', Nael G. Bunni, a paper delivered at a Conference on Civil Engineering Claims and Arbitration, London, Professional, Business and Industrial Management Studies, October 1975, in which a warning was made that the liability resultant from such responsibilities forms a heavy burden which could only produce conflict.

8.5 'The Changing Image of Consultants of the North Working in the South', T. A. Dabbagh, Engineering Adviser to the Kuwait Fund for Arab Economic Development, 1980.

8.6 D. E. Cullivan in a light-hearted note at a Seminar on the Consulting Engineer, a Development Resource, held in Jakarta in October 1984, told the following anecdote:

> '[A contractor colleague of his] was complaining about the horrible Conditions of Contract that he was being asked to sign which essentially made him responsible for Acts of God and of the Consulting Engineer, the two things being equivalent for him. Someone asked him: "If this is so horrible why do you insist on signing a Contract?" And he took a big sigh and said: "You know, Contractors have to be optimists. I am convinced that if a contract document said: 'The successful tenderer will be hanged by the neck as soon as the contract is signed, you would not lack for competitors!'" He continued: "They would think, well, perhaps I can negotiate my way out of it. Secondly, I am sure my lawyer will find a clause that will protect me. And failing that, maybe the rope will break."'

8.7 'Construction, Insurance and Law – A Discussion Paper', FIDIC, Switzerland, 1986, page 19.

8.8 *Construction Insurance*, Nael G. Bunni, Elsevier Applied Science Publishers, London, 1986, page 37.

8.9 Banwell Report, HMSO, London, 1964.

8.10 'Examination of Some Aspects of FIDIC Conditions of Contract (International) for Works of Civil Engineering Construction', M. A. Ibrahim and M. Y. Abdel A'Al. A paper presented at a seminar held in Abu Dhabi between the Arab Funds and FIDIC in November 1983. Report published by FIDIC in 1984, page 94.

8.11 *The International Civil Engineering Contract*, I. N. Duncan Wallace (a commentary on the second edition of the Red Book) Sweet & Maxwell, London, 1974. A supplement on the third edition in 1980.

8.12 'Problems of Applying the FIDIC Contract for Civil Works under the Civil Code System – A Comparison of the Legal Concepts Used by the FIDIC Contract with Those Used Under the Civil Code', Ali El Shalakany. This was a paper at a conference on FIDIC Conditions of Contract, Cairo, 1987; later published in *International Construction Law Review* (1989), page 266.

8.13 An example of such a rule is Article 198 of the Kuwaiti Civil Code which states:

> 'If, after entering into the contract, and before its final execution, general unforeseeable exceptional circumstances occur causing the performance of the obligation arising therefrom to be onerous, if not impossible threatening [the Contractor] with excessive losses, then, after balancing the interests of the two parties, the judge may contain the onerous obligation to within reasonable limits by either reducing the extent of the obligation or increasing its consideration. Any agreement to the contrary shall be null and void.'

This doctrine has also been developed in Europe, either by the courts as in Germany or through codification as in Switzerland (see Swiss Civil Code, Article 373, paragraph 2).

8.14 In this connection, see *Al Wasseett*, Professor Abdul Razak Al Sanhoori, Volume. 7; *Development of Kuwaiti Administrative Law*, Al-Magwari; and *Obligations of the Administration*, Dr Ibrahim Taha Al-Fayad.

8.15 The Egyptian *Conseil d'Etat* has applied the theory of physical obstructions in several cases but it refused compensation in others due to the absence of one or more of the conditions which must be satisfied if the theory is to apply. See Reference 8.12.

8.16 The Red Book provides specific periods for the submission of claims by both employer and contractor against each other, for example, see clauses 60 to 62, under the heading 'Certificates and Payments'. In effect, therefore, it provides for a period of limitation to apply in connection with known problems and defects. For hidden or latent defects, however, the applicable law of the contract imports into its provisions a separate and specific period of limitation, in most legal systems throughout the world.

8.17 'Investigate Don't Capitulate', Report of the Standing Committee on Professional Liability, 1984, FIDIC, Switzerland. See also subsequent annual reports of that Committee.

8.18 The legal rule in the United Arab Emirates is in Article 880 of the Civil Code which states:

> '1. If the intention of the contract was to construct buildings or other permanent construction works which were designed by the Engineer and to be executed by the Contract under the Engineer's supervision, they both jointly and severally warrant, to indemnify the Employer in respect of all occurrences of partial or total failure of what they had constructed or built and for all defects that might threaten the soundness and safety of the structure for a period of ten years unless a longer period is stipulated in the Contract. All this applies unless the period for which both parties of the Contract require these construction works to last is less than ten years.
> 2. The obligation to indemnify remains even if the damage or failure was due to a fault in the ground (the Site) or if the faulty construction was carried out with the approval of the Employer.
> 3. The period of ten years starts from the date of handing over of the Works.'

8.19 The Fourth Edition of the Red Book refers to 'release from performance'.

8.20 'EIC/FIDIC Questionnaire Survey: The use of the FIDIC Red Book', Final Report, June 1996, Department of Construction Management and Engineering, Reading University.

Chapter 9 The Revisions – Purposes and Consequences

9.1 *Guide to Use of FIDIC Conditions of Contract for Works of Civil Engineering Construction, Fourth Edition*, Fédération Internationale des Ingénieurs-Conseils, Switzerland, 1989, page 18. This publication contains 173 pages of commentary and the text of the Fourth Edition of the Red Book. (FIDIC, PO Box 86, CH 1000 Lausanne, 12-Chailly, Switzerland.)

9.2 Details of the advice sought and obtained are contained in the Guide referred to in Reference 9.1.

9.3 'Background and Overview', a paper presented by Tony Norris, a member of the drafting committee of the Fourth Edition and of the Civil Engineering Contracts Committee of FIDIC, at a Conference on the FIDIC Civil Conditions of Contract, Fourth Edition, Legal Studies and Services Ltd, London, June 1988.

9.4 'Background and Overview', Helge Sorensen, a paper presented at a two-day Seminar on the Revised FIDIC Conditions, Amsterdam, October 1987.

9.5 *Conditions of Contract for Electrical and Mechanical Works (Including Erection on Site)*, Third Edition, 1987.

9.6 In a paper presented to a seminar on the Fourth Edition of the FIDIC Form: Seminar on FIDIC conditions of Contract for Works of Civil Engineering Construction, October 1987, Dr J. J. Goudsmit wrote:

'Many changes and a number of them certainly improvements have been introduced. However, it remains questionable whether the opportunity to launch a new Red Book has been adequately used to simplify and clarify the contractual structure and the wording which up till now could be characterised as prolix and obscure.'

Chapter 10 Role of the Engineer

10.1 A standard form of contract between an employer and a consulting engineer is published by FIDIC under the title of 'Model Services Agreement between Client and Consultant', in two parts: Part I, Standard Conditions; and Part II, Conditions of Particular Application, 1990.

10.2 'Legal Liability in Contract Structures', Dr J. J. Goudsmit, *The Liability of Contractors*, Queen Mary College, University of London, Centre for Commercial Law Studies, Longman, page 18.

10.3 *Construction Insurance*, Nael G. Bunni, Elsevier Applied Science Publishers, London, 1986, page 145.

10.4 'New Defence Plan for the Engineer', Speaker's Corner, *New Civil Engineer*, 17 October 1985.

10.5 With respect to the reference to a quasi-judicial role, see Section 10.9.

10.6 *Moresk* v. *Hicks* [1966] 2 Lloyd's Rep 338.

10.7 *Design Liability in the Construction Industry*, D. L. Cornes, BSP Professional Books, Fourth Edition, Oxford, 1994, Chapter 4 at page 58.

10.8 The *Guide to Use of FIDIC Conditions of Contract for Works of Civil Engineering Construction, Fourth Edition* (see Reference 9.1), refers on page 56 to the design of temporary works as normally carried out by the contractor. The Guide adds that 'where this is the case, the Engineer may, depending upon the nature or importance of the Temporary Works, require information about their design'.

10.9 'Of judges and judging', a speech delivered to magistrates in 1964 by Lord Hailsham, the then Lord Chancellor, United Kingdom.

10.10 *Oldschool* v. *Gleeson* (1976) 4 BLR 103.

10.11 *Clayton* v. *Woodman* [1962] 1 WLR 585.

10.12 *East Ham* v. *Bernard Sunley* [1966] AC 406.

10.13 'FIDIC Policy Statement on the Role of the Consulting Engineer During Construction', a FIDIC Policy Statement, 1984, Lausanne, Switzerland. The Red Book is based on the idea that the best person to supervise a particular project is the one who had originally designed it. Clause 6 of the Form is drafted with the assumption that further drawings or instructions may be issued during the construction period by the engineer.

10.14 *Pacific Associates Inc* v. *Baxter* [1989] 2 All ER 159.

10.15 *Sutcliffe* v. *Thackrah* [1974] 1 All ER 319.

10.16 'Construction Contracts – Time for a Change', Clifford J. Evans, *Arbitration Journal*, August 1987.

Chapter 11 Responsibility and Liability of the Engineer

11.1 *The Oxford Companion to Law*, David M. Walker, Clarendon Press, Oxford, 1980.

11.2 'The Engineer's Liability to the Contractor: French Law', Christopher R. Seppala, a paper presented at the Conference of the International Bar Association's Section on Business Law Conference, Toronto, 1983.

11.3 *Bagot v. Stevens Scanlan & Co.* [1964] 3 WLR 1162; [1964] 3 All ER 577.
11.4 *Esso Petroleum Co. Ltd v. Mardon* [1976] QB 801; [1976] 2 All ER 5.
11.5 *Boorman v. Brown* (1844) 8 ER 1003.
11.6 *Nocton v. Lord Ashburton* [1914] AC 932.
11.7 *Midland Bank Trust Co. Ltd v. Hett, Stubbs & Kemp* [1978] 3 WLR 167.
11.8 *Finlay v. Murtagh* [1979] IR 249; and *Valdo Vulic v. Bohdam Bilinsky* (1982) NSW Supreme Court No. 177700/78.
11.9 *Rowe v. Turner Hopkins & Partners* [1980] NZLR 550 (New Zealand High Court).
11.10 *Hill Organisation Ltd v. Bernard Sunley & Son Ltd* (1983) 22 BLR 1.
11.11 *Tai Hing Cotton Mill Ltd v. Liu Chong Hing Bank Ltd* [1986] AC 80.
11.12 *Ernst and Whinney v. Willard Engineering (Dagenham) Ltd* (1987) 3 Const LJ 292; (1988) 40 BLR 67.
11.13 *Greater Nottingham Co-operative Society v. Cementation Piling & Foundations Ltd* [1989] QB 71; (1988) 41 BLR 43.
11.14 *Barclays Bank plc v. Fairclough Building Limited* [1995] QB 214.
11.15 *Lancashire & Cheshire Association of Baptist Churches Incorporated v. Howard & Seddon Partnership* [1993] 3 All ER 467.
11.16 *Gable House Estates v. The Halpern Partnership and Bovis Construction Ltd* 1995, Construction Law Digest, 12-CLD-03-01 (QBD).
11.17 *Holt v. Payne Skillington* [1995] Construction Law Digest, 13-CLD-06-01 (CA).
11.18 *Design Liability in the Construction Industry*, Fourth Edition, D. L. Cornes, BSP Professional Books, Oxford, 1994, page 60.
11.19 *Bolam v. Friern Hospital Management Committee* [1957] 2 All ER 118.
11.20 *Whitehouse v. Jordan* [1981] 1 WLR 246; 125 SJ 167; [1981] 1 All ER 267.
11.21 *Design Liability in the Construction Industry* (Reference 11.18 above), page 47.
11.22 *Medjuck & Budovitch Ltd v. Adi Ltd* 33 NBR 2nd 271 (80 Apr. 271, paragraph 110).
11.23 *Brickfield Properties v. Newton* [1971] 1 WLR 862.
11.24 *Moresk Cleaners Ltd v. Hicks* [1966] 2 Lloyd's Rep 338.
11.25 *The Liability of Contractors*, Edited by Humphrey Lloyd QC, and in particular the article by Max Abrahamson 'Contractors' Right over and against Architects, Engineers and Surveyors in respect of Liabilities incurred to the Employer', Centre for Commercial Law Studies, Queen Mary College, 1986, page 181.
11.26 A translation of Article 1792 is given in *Construction Insurance*, Nael G. Bunni, Elsevier Applied Science Publishers, London, 1986, page 156.
11.27 *Portsea Island Mutual Co-operative Society Ltd v. Michael Brasher Associates* (1989), see *Construction Law Digest* 7-CLD-10-05, 'The Duty to Take Care', 1989, BSP Professional Books, Oxford.
11.28 *Oldschool v. Gleeson (Construction) Ltd* (1976) 4 Build LR 103.
11.29 *Arenson v. Casson, Beckman, Rutley & Co.* [1975] 3 WLR 815.
11.30 *Pacific Associates v. Baxter* [1989] 2 All ER 159.
11.31 *D & F Estates Ltd v. Church Commissioners for England* [1989] AC 177. See also *Murphy v. Brentwood District Council* in Construction Law Digest 8-CLD-10-05, 1990, BSP Professional Books, Oxford.
11.32 *Donoghue v. Stevenson* [1932] AC 562.
11.33 *Greater Nottingham Co-operative Society v. Cementation Piling & Foundations Ltd* [1989] QB 71.
11.34 *Pacific Associates v. Baxter* [1989] 2 All ER 159. See also *Construction Law Digest* 6-CLD-06-01, 1989, BSP Professional Books, Oxford.
11.35 *Demers v. Dufresne Engineering Co. Ltd* (1979) 1 SCR 146 (Supreme Court of Canada).

11.36 See Reference 11.35 above.
11.37 *Al Wasseett in Interpretation of The Civil Code – First Volume, Contracts on Work*, A.A. Al Sanhoori, 1964, page 105.
11.38 *See Construction Law Digest* 8-CLD-10-03, 1990, where the decision in *Murphy* v. *Brentwood District Council* is discussed and where it is held that the case of *Dutton* v. *Bognor Regis District Council* is overruled and *Anns* v. *London Borough of Merton* is departed from.
11.39 As in Reference 11.38 above but with regard to complex structures. Reference is also made here to recent directives of the European Community.
11.40 Description of Civil Engineering in the Charter of 1828 of the Institution of Civil Engineers (UK).
11.41 *Rylands* v. *Fletcher* (1868) LR, 3 HL 330, where the defendant employed an independent contractor to construct a reservoir on his land. When the reservoir was filled, water flowed into a disused mine workings underneath, which communicated with and consequently flooded the claimant's mines. Despite the fact that there was no proof of negligence, it was held that a person who used his land in a non-natural way, as the defendant had done, for his own purposes, brought on his land, collected and kept there anything likely to do mischief if it escaped, must keep it in at his peril. Such a person was prima facie answerable for all damage which was the natural consequence of its escape, unless he excused himself by showing that the escape was due to the claimant's default, or was the consequence of *vis major*, or act of God: *The Oxford Companion to Law*, Professor David M. Walker. This principle was discussed in some detail in the more recent case of *Cambridge Water Co. Ltd* v. *Eastern Counties Leather plc* [1994] 1 All ER 53.
11.42 *Cagne* v. *Bertran* (1954) 43 Cal. 2d 481, 275 p. 2d15.
11.43 *Allied Properties* v. *Blume* (1972) 25 CA, 3d, 848.
11.44 *Xerox Corporation* v. *Turner Construction Company, et al.* (1973) GFIP Vol. IV, No. 4.
11.45 *Gravely* v. *The Providence Partnership* (1977) Federal Appellate Court, USA, GFIP Vol. VIII, No. 1.
11.46 *Whitehouse* v. *Jordan* [1981] 1 WLR 246; 125 SJ 167; [1981] 1 All ER 267.
11.47 *Medjuck & Budovitch Ltd* v. *Adi Ltd* 33 NBR 2nd 271 80 Apr. 271, paragraph 110.
11.48 See also the statement of Lord Denning in *Greaves (Contractors) Limited* v. *Baynham Meikle & Partners* below.
11.49 *Brickfield Properties* v. *Newton* [1971] 1 WLR 862 and also *Eckersley, T. E. and Others* v. *Binnie & Partners and Others* (1988) Con LR 1, CA.
11.50 See also the standard form of contract of sale under the United Nations Convention on Contracts for the International Sale of Goods. This convention has been ratified by many jurisdictions worldwide and applies to contracts of sale in the international field.
11.51 *Young & Marten* v. *McManus Childs* [1969] 1 AC 454; [1968] 3 WLR 630.
11.52 *Shanklin Pier Ltd* v. *Detel Products Ltd* (1951) 2 All ER 471.
11.53 *Greaves (Contractors) Limited* v. *Baynham Meikle & Partners* [1975] 1 WLR 1095.
11.54 *Independent Broadcasting Authority* v. *EMI Electronics and BICC Construction* (1980) 14 BLR 1.
11.55 *Norta Wallpapers (Ireland)* v. *Sisk and Sons (Dublin) Limited* (1978) IR 114.
11.56 *George Hawkins* v. *Chrysler (UK) Limited and Burne Associates* (1986) 38 BLR 36.
11.57 *Design Liability in the Construction Industry*, David L. Cornes, Fourth Edition, Blackwell Publishing, Oxford, 1994.
11.58 *Miller* v. *Cannon Hill Estates Limited* [1931] 2 KB 113.
11.59 *Hancock* v. *B. W. Brazier (Anerley) Limited* [1966] 2 All ER 901; [1966] 1 WLR 1317.
11.60 *Samuels* v. *Davis* [1943] 1 KB 526.

Chapter 12   The Employer's Obligations

12.1 *Hudson's Building and Engineering Contracts*, Eleventh Edition, I. N. Duncan Wallace, Sweet & Maxwell, London, 1994, Volume I, page 568, paras. 4.136 to 4.139.
12.2 Per Lord Justice Vaughan Williams in *Wells* v. *Army & Navy Co-operative Society* (1902) 86 LT 764. See *Hudson's Building and Engineering Contracts*, Eleventh Edition, Volume I, page 573, para. 4.146.
12.3 *AMF (International) Ltd* v. *Magnet Bowling Ltd* [1968] 1 WLR 1028.
12.4 *Hudson's Building and Engineering Contracts*, Eleventh Edition (Reference 12.1 above), Volume I, page 610, para. 4.202.

Chapter 13   The Contractor's Obligations

13.1 *Engineering Law and the ICE Contracts*, Fourth Edition, Max W. Abrahamson, Applied Science Publishers Ltd, London, 1975, clause 13.
13.2 *Hudson's Building and Engineering Contracts*, I. N. Duncan Wallace, Eleventh Edition, Sweet & Maxwell, London, 1994, Volume I, page 497.
13.3 Various legal cases are reported on this question in *Hudson's Building and Engineering Contracts* (Reference 13.2 above), Volume I, pages 497 to 508, which throw some light on whether a contractor should be paid or not.
13.4 *Francis* v. *Cockerell* (1870) LR 5 QB 501.
13.5 *G.H. Myers & Co.* v. *Brent Cross Service Co.* [1934] 1 KB 46 at page 55.
13.6 *Young & Marten Ltd* v. *McManus Childs* [1969] 1 AC 454.
13.7 Paragraph 633, Section 1, German Civil Code, Statutory Law Governing Contracts for Work.
13.8 *Verdingungsordnung für Bauleistungen*, Rules for all Public Construction Work, Paragraph 13, Section 8.
13.9 See, for example, Articles 875 to 878 inclusive of the Civil Code of the United Arab Emirates, 1985; and Articles 647 to 649 inclusive of the Libyan Civil Code 1954, Section 3 on Contracts for Work.
13.10 *Law of Contract* (see Reference 3.9 above), Chapter 16.
13.11 *Cottage Club Estates Ltd* v. *Woodside Estates Co. (Amersham) Ltd* [1928] 2 KB 463.
13.12 *Commercial Arbitration*, Second Edition, Sir Michael J. Mustill and Steward C. Boyd, Butterworths, London, 1989, page 138.

Chapter 14   Risks, Liabilities, Indemnities and Insurances

14.1 *Construction Insurance*, Nael G. Bunni, Elsevier Applied Science Publishers, London, 1986, Chapters 7 and 8.

14.2 The collapse event resulted in a dispute between the insured and the insurers which proceeded to arbitration and later to the Supreme Court of Queensland. The case became known as *Manufacturers' Mutual Insurance* v. *Queensland Government Railways* (1968) QWN 12. It is referred to in detail in *Construction Insurance* (Reference 14.1 above), page 193.

14.3 The words in quotation marks are quoted from the *Guide to Use of FIDIC Conditions of Contract for Works of Civil Engineering Construction, Fourth Edition*; see Reference 9.1 for a full reference.

14.4 See discussion of the mandatory requirement of the World Bank in connection with sub-clause 20.4 in Section 22.5.3 of the book.

14.5 After publication of the Fourth Edition, four conferences were organised in London, Paris, Amsterdam and Copenhagen, where major contributions were made on the changes made in the Red Book. These conferences were held in the second half of 1987.

14.6 A probable reason may be the fact that the equivalent clause in the Fifth Edition of the ICE Form does restrict the exclusion to property damage. The restriction is discussed by Max W. Abrahamson in his book *Engineering Law and the ICE Contracts*, Fourth Edition, Applied Science Publishers, London, 1975, page 95.

14.7 Speakers and commentators at the seminar organised by the Institution of Civil Engineers, London, in collaboration with FIDIC, as reported by Mr K. N. Drobig in the 'Proceedings of the Institution of Civil Engineers', Part I, August 1988, pages 821–836. Similar comments were also made at subsequent seminars in Paris, Amsterdam and Copenhagen.

14.8 See Chapter 9 and Reference 9.1 quoted above.

14.9 These definitions are taken from *Construction Insurance* (Reference 14.1 above), page 230.

Chapter 15   Performance and Other Securities

15.1 Whilst bonds and guarantees were originally issued in the main by banks, the security of an insurance or surety company is now equally acceptable in the majority of countries. The advantage for an enterprise to set up separate lines of credit for bonds and guarantees with surety companies, is that it would protect its lines of credit with banks for working capital purposes – which might otherwise be blocked at the very time when such working capital is most needed. Additionally, banks prefer to issue bonds in an onerous 'on demand' form and must therefore treat them as unpresented letters of credit.

15.2 Customs guarantees are issued to cover any customs duties that may become payable when imported goods, which are exempt from duty if re-exported within a specified time, are not re-exported within that time.

15.3 *Construction Insurance*, Nael G. Bunni, Elsevier Applied Science Publishers, London, 1986, page 160.

15.4 *Trade Industry* v. *Workington* [1937] AC 1 at page 17.

15.5 'Guarantees and Bonds in Construction Contracts', a paper by I. N. Duncan Wallace, reprinted in a book entitled *Construction Contracts: Principles and Policies in Tort and Contract*, Volume 1, Sweet & Maxwell, London, 1986.

15.6 *Mercers Co.* v. *New Hampshire Insurance Ltd* [1992] 2 Lloyd's Rep 365.

15.7 *Perar BV* v. *General Surety Guarantee* (1994) 66 BLR 72.

15.8 *Construction Contracts: Principles and Policies in Tort and Contract* (Reference 15.5 above), Volume 2, paragraph 19-01.

15.9 *Trafalgar House Construction Ltd* v. *General Surety and Guarantee Ltd* (1994) 66 BLR 42, CA, strongly criticised on both factual and textual grounds in 'Loose Cannons in the Court of Appeal: On Demand per Incuriam?' by I. N. Duncan Wallace in 1995, 10 Construction Law Journal 190. The case was later overruled by the House of Lords [1996] 1 AC 199.

15.10 These requirements are modified from and added to those suggested in 'Guarantees and Bonds in Construction Contracts' (Reference 15.5 above), page 308, paragraph 19–20.

15.11 *The Law and Practice of International Banking*, Graham A. Penn, Sweet & Maxwell, 1987, paragraph 12.23, page 282.

15.12 *Edward Owen Ltd* v. *Barclays Bank* [1978] QB 159, page 170.

15.13 *R.D. Harbottle (Mercantile) Ltd* v. *National Westminster Bank Ltd* [1978] QB 146 [1977] 2 All ER 862.

15.14 'Selecting Arbitrators for Construction Disputes', Guillermo Aguilar Alvarez, April 1990, (the then General Counsel of the ICC International Court of Arbitration, Paris).

15.15 ICC Uniform Demand Guarantees, ICC Publications No. 458, April 1992.

15.16 Guide to the ICC Uniform Rules for Demand Guarantees, by Professor Roy Goode, ICC Publication No. 510, October 1992.

15.17 ICC Uniform Rules for Contract Bonds, ICC Publication No. 524, September 1993.

15.18 The International Credit Insurance Association (ICIA) with its headquarters in Switzerland, brings together, on a world-wide basis, 42 member companies writing credit and/or guarantee insurance. The majority of the membership is comprised of companies specialising in these fields. They are located in 28 countries spread over all five continents.

15.19 *Construction Contracts: Principles and Policies in Tort and Contract* (Reference 15.8 above), paragraph 20–21.

15.20 *Trafalgar House* v. *General Surety* (1994) (Reference 15.9 above) and *Perar BV* v. *General Surety* (1994) (Reference 15.7 above).

15.21 'Bonds and Guarantees', D. W. Graham, a paper read at a seminar on FIDIC's Conditions of Contract for Electrical and Mechanical Work, Third Edition, London, 1987.

Chapter 16   Claims and Counterclaims

16.1 *The Oxford Companion to Law*, David M. Walker, Clarendon Press, Oxford, 1980, page 227.

16.2 *Hadley* v. *Baxendale* (1854) 9 Ex 341.

16.3 For example, *Craven-Ellis* v. *Canons Ltd* [1936] 2 KB 403.

16.4 For example, *Planché* v. *Colburn* [1831] 131 ER 305.

16.5 For example, *Fibrosa Spolka Akcyjna* v. *Fairbairn Lawson Combe Barbour Ltd* [1943] AC 32 p. 61.

16.6 *Hudson's Building and Engineering Contracts*, 11th Edition, Volume 1, page 144, paragraphs 1–263.

16.7 *Constable Hart Co. Ltd* v. *Peter Lind & Co. Ltd* [1978] 9 BLR 1.

16.8 On this topic, see two articles: 'Bones of Contention', by Nick Barrett, *New Civil Engineer*, 8 February 1996; and 'This is Your Lawyer Speaking', *Evening Standard*, 17 November 1994 based on a research by the International Financial Law Review.

16.9 See also the *Guide to Use of FIDIC Conditions of Contract for Works of Civil Engineering Construction, Fourth Edition* (Reference 9.1 above), page 112; and Chapter 6 on the meaning of a re-measurement contract.

16.10 *The International Civil Engineering Contract*, I. N. Duncan Wallace QC, Sweet & Maxwell, London, 1974, page 97.

16.11 Final Award in ICC Arbitration Case No. 5634 (1989). Relevant extracts reported in the ICC International Court of Arbitration Bulletin, Vol. 2, No. 1, May 1991, page 24.

16.12 *Tersons Ltd* v. *Stevenage Development Corporation* [1965] 1 QB 37.

16.13 *Mitsui Construction Co. Ltd* v. *Attorney-General of Hong Kong* (1986) 33 BLR, 1 (PC).

16.14 *The International Civil Engineering Contract*, I. N. Duncan Wallace QC, Sweet & Maxwell, London, 1974, page 43.

16.15 'Risks in Construction; Methods and Adverse Conditions', Max W. Abrahamson, King's College London, a five-day course on Civil Engineering Law and Arbitration, London, 1987. An interesting and useful analysis of the legal background to clause 12 is given in this lecture, published by King's College, page 78.

16.16 'Recommendations for Review of the Conditions of Contract (International) for the Works of Civil Engineering Construction', Dr Joachim E. Goedel, *International Construction Law Review*, Third Edition, March 1977.

16.17 'The Clause 12 Nightmare', Professor John Uff QC, *New Civil Engineer*, 6 July 1989, page 19.

16.18 *Consultant's Report on Review of General Conditions of Contract for Construction Works for the Government of the Hong Kong Special Administrative Region* or *The Grove Report*, Jesse B. Grove III, Thelen Reid & Priest LLP, New York, September 1998, section 6.3, which can be found on www.constructionweblinks.com/Resources/Industry_Reports__Newsletters/Nov_6_2000/grove_report.htm

16.19 'Clause 12 – Not So Much a Nightmare, More a Haze', Guy Cottam, *New Civil Engineer*, 17 August 1989, page 15.

16.20 *Construction Contracts: Principles and Policies in Tort and Contract*, Volume 1, I. N. Duncan Wallace QC, Sweet & Maxwell, London, 1986, page 382.

16.21 'Handling the Unexpected under the New Yellow Book', Geoffrey Hawker, Proceedings of the 1988 FIDIC Conference, Dublin, Workshop No. 8, September 1988.

16.22 'Price Under Common Law Systems', I. N. Duncan Wallace, QC, *Selected Problems of Construction Law, International Approach*, First International Construction Law Conference, 1982. Proceedings published in 1983 by University Press, Fribourg, Switzerland, page 169. See also Reference 16.19.

16.23 'Defects: A Summary and Analysis of American Law', Professor Justin Sweet, *Selected Problems of Construction Law, International Approach* (Reference 16.22 above), page 90.

16.24 *Ruxley Electronics and Construction Ltd* v. *Forsyth* [1994] 1 WLR 650, CA and [1995] 3 WLR 118, HL, reversing the Court of Appeal decision.

16.25 *Whittal Builders Co. Ltd* v. *Chester-le-Street District Council* [1996] 12 Const LJ 356; 11 Con LR 40.

16.26 The Hudson Formula, see *Construction Insurance* (Reference 11.1 above); the Emden Formula, see *Emden's Building Contracts and Practice*, Eighth Edition, Volume 2, Butterworths, London, page N/46; the Eichleay Formula, see *Building Contract Claims*, Third Edition, Powell-Smith and Sims, Blackwells, page 133.

16.27 *Property & Land Contractors Ltd* v. *Alfred McAlpine Homes Ltd* [1995] 47 Con LR 74.

16.28 *Norwest Holst Construction* v. *Cooperative Wholesale*, 17 February 1998 (unreported) but the subject of articles published at: www.clientplus.co.uk/clientplus/legal/asp/overheads.asp; and www.masons.com/php/page.php?page_id=ce09.doc.htm958825459

16.29 *Allied Maples Group Ltd* v. *Simmons & Simmons* [1995] 2 All ER 907; [1995] 1 WLR 1602.
16.30 *Hudson's Building and Engineering Contracts*, Eleventh Edition, I. N. Duncan Wallace, Sweet & Maxwell, Volume II, page 1086.
16.31 *Wharf Properties Ltd* v. *Eric Cumine Associates* (1991) 52 BLR 8.
16.32 *Ibid.*, per Lord Oliver, at page 21.
16.33 *Cervidone Construction Corp.* v. *US*, 931 F 2d 860 (Fed. Cir. 1991).
16.34 *McAlpine Humberoak Ltd* v. *McDermott International Inc* (1992) 58 BLR 1.
16.35 *J. Crosby & Sons Ltd* v. *Portland Urban District Council* (1967) 5 BLR 121.
16.36 *Leach* v. *London Borough of Merton* (1985) 32 BLR 68, at pages 102, 112.
16.37 Final Award in ICC Arbitration Case No. 5634 (1989), relevant extracts reported in the ICC International Court of Arbitration Bulletin, Vol. 2, No. 1, June 1991, page 23.
16.38 Building Law Reports, Volume 52, page 6.
16.39 *British Airways Pension Trustees* v. *Sir Robert McAlpine & Sons Ltd* (1994) 72 BLR 26.
16.40 *GMTC Tools & Equipment Ltd* v. *Yuasa Warwick Machinery Ltd* (1994) 73 BLR 102.
16.41 *John Doyle Construction Ltd* v. *Laing Management (Scotland) Ltd* (2004) BLM 21:7; CILL 2135, Court of Session, Inner House (Scotland).
16.42 Article entitled 'Considering global claims', 14 July 2004, published on www.contractjournal.com

Chapter 17 Delay in Completion and Claims for Extension of Time

17.1 If time were of the essence in construction contracts then non-performance by a certain date would give rise to a right not merely to damages but also to rescind (terminate) the contract – see definition of 'Essence of the Contract' in *The Oxford Companion to Law*, David M. Walker, Clarendon Press, 1980. Also, in *Carr* v. *J. A. Berriman Pty Ltd (1953) 27 ALJ 273* it was held that if the contractor's obligation to complete by a specified date is of the essence, then the employer is released from further obligation if the contractor is in breach and he can treat the contract as repudiated. The nature of construction contracts is such that some delay is highly probable, so a delay to the completion date typically gives rise to the right to levy liquidated damages but very rarely the right to terminate the contract.
17.2 *Peak Construction (Liverpool) Ltd* v. *McKinney Foundations Ltd* (1970) 1 BLR 111.
17.3 'Liquidated Damages and Extensions of Time in Construction Contracts', 2nd Edition, Brian Eggleston, Blackwell Publishing, Oxford, 1997.
17.4 *Gaymark Investments* v. *Walter Construction Group Ltd* [1999] NTSC 143 (Bailey J).
17.5 *Balfour Beatty Building Ltd* v. *Chestermount Properties Ltd* (1993) 62 BLR 12.
17.6 *Bilton* v. *Greater London Council* (1982) 20 BLR 1.
17.7 *Final Report on Construction Industry Arbitrations*, Nael G. Bunni and Humphrey Lloyd [2001] ICLR 644, Volume 18, Part 4, October 2001. Also available from www.iccwbo.org, search words 'construction industry'.
17.8 *The Society of Construction Law Delay and Disruption Protocol*, Society of Construction Law, Oxfordshire, October 2002, reprinted March 2003.
17.9 *Glenlion Construction* v. *The Guinness Trust* (1987) 39 BLR 89.
17.10 *Assoland Construction Pte Ltd* v. *Malayan Credit Properties Pte Ltd* [1993] 3 SLR 470.
17.11 *Consultant's Report on Review of General Conditions of Contract for Construction Works for the Government of the Hong Kong Special Administrative Region* or *The Grove Report*, Jesse B. Grove III, Thelen Reid & Priest LLP, New York, September 1998, section 11.6 of which can be

found on www.constructionweblinks.com/Resources/Industry_Reports__Newsletters/Nov_6_2000/grove_report.html

17.12 As provided, for example, in clause 2.3 of JCT 80.

17.13 As provided, for example, in clause 1.9 (in respect of the employer's requirements) and clause 4.7 of the 1999 Yellow Book.

17.14 As provided for, for example, in clause 13.7 of the 1999 Red Book.

17.15 As provided, for example, in clauses 35 and 36 of JCT 80.

17.16 As provided for, for example, in clause 8.5 of the 1999 Red Book.

17.17 *Abigroup Contractors Pty Ltd* v. *Peninsula Balmain Pty Ltd* [2001] NSWSC 752 (Barrett J).

17.18 Judge Humphry Lloyd QC in *Sindall Ltd* v. *Solland* (15 June 2001) TCC as reported on www.adjudication.co.uk/cases/sindall.htm

17.19 *City Inn* v. *Shepherd Construction* (2003) BLM 20:10, Court of Session, Inner House (Scotland).

17.20 See Reference 17.4 above.

17.21 *Turner Corporation Ltd* v. *Co-ordinated Industries Pty Ltd* (1994) 11 BCL 202 at 212.

17.22 *Turner Corporation Limited* v. *Austotel Pty Limited* [1994] 13 BCL 378 per Cole J at 384–385.

17.23 *Delay and Disruption in Construction Contracts*, Keith Pickavance, LLP Reference Publishing, London, 1997, p. 113.

17.24 *Royal Brompton Hospital NHS Trust* v. *Frederick A. Hammond and Others* [2000] EWHC Technology 39 (18 December 2000).

17.25 *GLC* v. *Cleveland Bridge & Eng. Co. Ltd* [1984] 34 BLR 50.

17.26 'The Practical Use of Critical Path Network Analysis on a Large Project', a paper read at a meeting of the Institution of Engineers in Ireland in 1990 by Paul Hackett, John Sisk and Son Ltd, the contractor responsible for the Square Towncentre at Tallaght, where the author's previous firm was responsible for the civil and structural engineering design and supervision.

17.27 *Walter Lawrence & Son Ltd* v. *Commercial Union Properties Ltd* (1984) 4 CLR 37; *Glenlion Construction Ltd* v. *Guinness Trust* (1987) CILL 360.

17.28 A study carried out by the Commonwealth Scientific and Industrial Research Organisation in Australia and reported in a paper by Terence M. Burke entitled, 'Delay under Australian Law', published in *Selected Problems of Construction Law: International Approach*, University Press, Fribourg, Switzerland, 1984.

17.29 *Henry Boot Construction (UK) Ltd* v. *Malmaison Hotel (Manchester) Ltd* (1999) 70 Con LR 32.

17.30 See Reference 17.24 above.

17.31 See Reference 17.5 above.

17.32 Keating on Building Contracts, Fifth Edition, Hon. Sir Anthony May, p. 583 (which in turn refers to *Glenlion Construction* v. *The Guinness Trust* (1987) 39 BLR 89 and *J.F. Finnegan* v. *Sheffield City Council* (1988) 43 BLR 124).

17.33 *Philips Hong Kong Ltd* v. *Attorney General of Hong Kong* (1993) 61 BLR 41.

17.34 *Law of Contract* (see ref 3.5 above), pages 620–625.

17.35 *Law* v. *Redditch Local Board* [1892] 1 QB 127 at 132.

17.36 *Robophone Facilities Ltd* v. *Blank* [1966] 1 WLR 1428 at 1447.

17.37 *Dunlop Pneumatic Tyre Co. Ltd* v. *New Garage and Motor Co. Ltd* [1915] AC 79 at 86.

17.38 See reference 17.19 above.

17.39 *Murray* v. *Leisureplay Plc* [2004] EWHC 1927 (QB) (5 August 2004).

17.40 *Temloc Ltd* v. *Errill Properties Ltd* (1987) CILL 376, as reported in a paper by Sir Patrick Garland, entitled 'Policy for Time, Conference on Construction Contract Policy: Improved Procedure and Practice', King's College London, September 1988.

808 References

17.41 The Society of Construction Law *Delay and Disruption Protocol*, October 2002, which can be downloaded (at no charge) at http://www.scl.org.uk/ and http://eotprotocol.com/
17.42 See also the *Final Report on Construction Arbitrations* published by ICC.

Chapter 18 Certificates and Payments

18.1 The SIA standard forms of contract (published by the Singapore Institute of Architects) have since 1987 offered the altrnative of fixed stage instalments in their measured contract forms with bills of quantities as well as in their lump sum forms.
18.2 *Hudson's Building and Engineering Contracts*, Eleventh Edition, I. N. Duncan Wallace, Sweet & Maxwell, Volume II, page 846.

Chapter 19 Disputes Settlement by Arbitration

19.1 Different nationalities mean different cultures, languages, standards, customs, etc., all intrinsic to possible misunderstandings and conflicts.
19.2 Teachings of Buddhism, Buddhist Promotions Foundations, Tokyo, Japan.
19.3 *The National Geographic Magazine*, June 1927, p.661,662, brought to life by Professor Park at the 2002 Freshfields Lecture 'Arbitration's Protean Nature: The Value of Rules and the Risks of Discretion'.
19.4 A Brehon was an arbitrator, an expositor of the law rather than a judge in the modern sense. Every King or chief of a substantial area had an official Brehon whose studies occupied 20 years of his life, and who had free land for his maintenance.
19.5 *Princes and Pirates: The Dublin Chamber of Commerce 1783–1983*, L. M. Cullen, Dublin Chamber of Commerce, Dublin, 1983.
19.6 Arbitration of international commercial disputes under the auspices of the ICC, Stephen R. Bond, the then Secretary General of the ICC Court of International Arbitration, International and ICC Arbitration, Conference Proceedings, Centre of Construction Law and Management, King's College London, 1990.
19.7 *Draft Report on Construction Industry Arbitrations* [Document 420/21-002] International Chamber of Commerce, Forum on Arbitration Issues and New Fields, Construction Arbitration Section.
19.8 International Court of Arbitration, Introduction to Arbitration, Advantages of Arbitration: www.iccwbo.org/court/english/arbitration/introduction.asp
19.9 See, in this connection, section 68 of the English Arbitration Act 1996.
19.10 *Northern Regional Health Authority* v. *Derek Crouch Construction Co. Ltd* [1984] QB 644; (1984) 26 BLR 1. However, this case has been overtaken by events in England since the introduction of the Courts and Legal Services Act 1990 which provides for a similar power of opening and revising an engineer's decision. However, in some other jurisdictions, this power may still lie solely with the arbitrator.
19.11 *The ICC in the context of International Arbitration*, J Gillis Wetter, Conference Proceedings, Centre of Construction Law and Management, King's College London, 1990, page 42. Mr Wetter also provided the quoted passage from Lord Justice Mustill in 'Arbitration: history and background', 6 *Journal Int Arb*, 2 June 1989, 43 at 49.

19.12 *Concord and Conflict in International Arbitration*, Rt. Hon. Sir Michael Kerr, the Keating Lecture, King's College London, October 1996, page 12.
19.13 See Reference 19.8 above.
19.14 Comment made in relation to clause 66 of the ICE Form, on which clause 67 of the Red Book is modelled, interpreting the case of *Monmouthshire County Council* v. *Costello & Kemple Ltd* (1965) 5 BLR 83 at pages 84 to 85.
19.15 *Halki Shipping Corporation* v. *Sopex Oils Ltd* [1997] EWCA Civ 3062 (19 December 1997) as reported on www.bailii.org/ew/cases/EWCA/Civ/1997/3062.html
19.16 *Ellerine Bros Pty Ltd* v. *Klinger* [1982] 1 WLR.
19.17 *Fastrack Contractors Ltd* v. *Morrison Construction Ltd* TCC (4 January 2000) as reported on www.adjudication.co.uk/cases/fastrack.htm
19.18 *Sindall Ltd* v. *Solland* TCC (15 June 2001) as reported on www.adjudication.co.uk/cases/sindall.htm
19.19 *Edmund Nuttall Ltd* v. *R.G. Carter Ltd* [2002] EWHC 400 (TCC) (21 March 2002) as reported on www.adjudication.co.uk/cases/nuttall.htm
19.20 *Hitec Power Protection BV* v. *MCI World Com Limited* [2002] EWHC 1953 (TCC) (15 August 2002) as reported on www.adjudication.co.uk/cases/hitec.htm
19.21 *Cowlin Construction Ltd* v. *CFW Architects (a firm)* [2002] EWHC 2914 (TCC) (15 November 2002) as reported on www.adjudication.co.uk/cases/cowlin.htm
19.22 *Beck Peppiatt Ltd* v. *Norwest Holst Construction Ltd* [2003] EWHC 822 (TCC) (20 March 2003) as reported on www.bailii.org/ew/cases/EWHC/TCC/2003/822.html
19.23 See Reference 19.18.
19.24 *Costain Ltd* v. *Wescol Steel Ltd* [2003] EWHC 312 (TCC) (24 January 2003) as reported on www.bailii.org/ew/cases/EWHC/TCC/2003/312.html
19.25 *Orange EBS Ltd* v. *ABB Ltd* [2003] EWHC 1187 (TCC) (22 May 2003) as reported on www.bailii.org/ew/cases/EWHC/TCC/2003/1187.html
19.26 *Dean and Dyball Construction Ltd* v. *Kenneth Grubb Associates Ltd* [2003] EWHC 2465 (TCC) (28 October 2003) as reported on www.bailii.org/ew/cases/EWHC/TCC/2003/2465.html
19.27 *London & Amsterdam Properties Ltd* v. *Waterman Partnership Ltd* [2003] EWHC 3059 (TCC) (18 December 2003) as reported on www.bailii.org/ew/cases/EWHC/TCC/2003/3059.html
19.28 This requirement is embodied in most systems of law as it is based on logic and reason. Under English law, see *Commercial Arbitration*, Second Edition, Sir M. J. Mustill and S. C. Boyd, Butterworths, London, 1989, pages 46–8. Under French law, see *La Notion d' Arbitrage*, Professor Jarrosson, Paris, 1987, page 35. Where the law in certain jurisdictions is silent on this question, such silence should not be interpreted against the necessity for a formulated dispute being in existence at the time when the arbitrator is appointed. For the meaning of dispute, see Section 19.7.
19.29 Most institutional rules respect the choice of the parties as to the number of arbitrators. Some, however, do not allow an *even* number as in the case of the rules of the UK Chartered Institute of Arbitrators, Article 3.1, which states that: 'Provided that the final number is uneven, the parties may agree on the number of arbitrators in the Tribunal.' Other institutions or organisations in this category include the Euro-Arab Chambers of Commerce; the United Nations Commission on International Trade Law Arbitration Rules; the Netherlands Arbitration Institute; and the Inter-American Commercial Arbitration Commission.

Where the institutional rules admit an even number of arbitrators, a mechanism to break a possible tie in the decision process must be provided, for instance, the Japan Commercial Arbitration Association grants a casting vote to the 'Chief Arbitrator' in Article 34.

19.30 In recent years the laws of many countries have upheld the effect of an arbitration agreement, for instance: in the United States, see the case of *Prima Pain* v. *Flood and Conklin*, US Supreme Court, 338 US 395 (1967); in France, the French *Cour de Cassation Cass Civ.*, May 1963; and in many countries in the Middle East, see *Arbitration in Private International Relations*, Professor Samia Rashed, Cairo, 1984, Volume 1, page 114.

19.31 *Heyman* v. *Darwins Ltd* [1942] AC 356 at 374.

19.32 Rules of the International Court of Arbitration of the International Chamber of Commerce, Paris, 1988 Edition, Article 11.

19.33 *Orion Compania Espanola de Seguros* v. *Belfort Maatschappij voor Algemene Verzekgringeen* [1962] 2 Lloyd's Rep 257.

19.34 *Commercial Arbitration*, Second Edition, Sir M. J. Mustill and S. C. Boyd, Butterworths, London, 1989, page 57.

19.35 To act as an *'amiable compositeur'* or to proceed *'ex aequo et bono'* is to disregard, within specific limits, strict legal rules or the contractual arrangements in the interest of arriving at an equitable determination of the dispute. The power to act as such is rarely conferred.

19.36 It is reported in *Law and Practice of International Commercial Arbitration*, Second Edition, Alan Redfern and Martin Hunter, Sweet & Maxwell, 1991, page 37, that to act as an *amiable compositeur* is expressly recognised in, amongst other countries, Argentina, Brazil, Chile, Egypt, France, Iraq, Lebanon, Mexico, Panama, Peru, Switzerland and Syria.

19.37 *Orion Compania Espanola de Seguros* v. *Belfort Maatschappij voor Algemene Verzekgringeen* [1962] 2 Lloyd's Rep 257 at 264 and confirmed in this respect by the decision in *Eagle Star Insurance Co. Ltd* v. *Yuval Insurance Co. Ltd* [1978] 1 Lloyd's Rep 357.

19.38 *Concord and Conflict in International Arbitration* (Reference 19.12 above), page 9.

19.39 *Commercial Arbitration* (Reference 19.34 above), page 219.

19.40 'Standard of Behaviour of Arbitrators', Fali S. Nariman, *Arbitration International*, Vol. 4, No. 4, October 1988, 311.

19.41 *Concord and Conflict in International Arbitration* (Reference 19.12 above), page 6.

19.42 A number of ICC awards were published in legal journals such as the *International Construction Law Review* (ICLR). Articles were also published and read at seminars and conferences on clause 67 and the interpretation of its wording. Some of these awards seem to have influenced the Committee in charge of the Fourth Edition of the Red Book. Amongst the most important of these are: 'The Pre-Arbitral Procedure for the Settlement of Disputes in the FIDIC (Civil Engineering) Conditions of Contract', Christopher R. Seppala, *ICLR*, Volume 3, Part 4, July 1986, page 330; 'ICC Court of Arbitration Case Notes', Sigvard Jarvin, *ICLR*, Volume 3, Part 5, October 1986, page 470; 'Revisions to the FIDIC Form of Civil Engineering Works: The Point of View of the Engineer', Humphrey Lloyd QC, *ICLR*, Volume 3, Part 5, October 1986, page 517; and 'Current Problems in International Construction Contracts', H. Andre-Dumont, Seminar Report, Seminar on International Construction Contracts organised by the Foundation for the Study of the Law and Practices of International Trade, Paris, February 1986, published in *ICLR*, Volume 3, Part 4, July 1986, page 413.

19.43 'Changes in the Procedure for the Resolution of Disputes', Christopher R. Seppala, a paper read at a conference in London and Paris on the new Fourth Edition of the Red Book entitled, International Construction Contracts FIDIC Conditions: The New Edition, September and October 1987.

19.44 'The Pre-Arbitral Procedure for the Settlement of Disputes in the FIDIC (Civil Engineering) Conditions of Contract', Christopher R. Seppala, *ICLR*, Volume 3, Part 4, July 1986, page 328.

19.45 *Engineering Law and the ICE Contracts*, Fourth Edition, Max W. Abrahamson, Applied Science Publishers, London, 1979, page 411.

19.46 *Paschen Contractors Inc* v. *John J. Calnan Co.* 13 Ill. App. 3d 485, 300 N.E. 2d 795 [App. Ct., 1st. Dist., 1st Div. (an intermediate appellate court of the State of Illinois) 1973]; *Methodist Church of Babylon* v. *Glen-Rich Const Corp*, 27 N.Y. 2d 357, 318 N.Y.S. 2d 297 [Ct. App. (the highest court of the State of New York) 1971].

19.47 This principle is accepted as can be seen from at least one reported arbitration case, ICC Arbitration Case No. 5428, reported in *Yearbook Commercial Arbitration*, Volume XIV (1989), published by Kluwer Law and Taxation, Netherlands. The Final Award in that case, which was rendered in 1988, dealt with this issue and as reported in the above publication, the following may be quoted:

> '[121 There is a world of difference between a "claim" as opposed to a "dispute" or "difference". The FIDIC scheme is clear. The Engineer is the port of first call if the Contractor has a claim during the performance of the contract. The claim may be for extension of time, extra expense, etc. Of whatever nature, the Contractor must give notice of it to the Engineer as soon as reasonably practical and he must, at the earliest that he can, quantify it and detail it. When that "claim" – of whatever description – goes to the Engineer, the Engineer may grant it in full, or in part or reject it altogether. This is the first tier of decision making by the Engineer. The Contractor may be happy with the result. That would be the end of the matter and there is no "dispute" or "difference". If, however, he is unhappy, he goes to the Engineer again. This time there is a "dispute" or "difference" and we get to the second tier of the FIDIC system where the Engineer acts (or should act) in a quasi-judicial role specifically under Clause 67.'

19.48 See, for example, the case of *Costain Ltd* v. *Wescol Steel Ltd* [2003] EWHC 312 (TCC) (24 January 2003) where the court, in referring to the judgment of *Halki Shipping Corporation* v. *Sopex Oils Ltd* [1997] EWCA Civ 3062, outlined the circumstances in which a dispute will be inferred. This has been summarised in Section 19.3 above.

19.49 Ibid.

19.50 In ICC Award No. 4840, an arbitral tribunal held that it had no jurisdiction over claims which were not submitted 'in the first place' to the engineer. This case was reported by Sigvard Jarvin in an article entitled 'ICC Court of Arbitration Case Note', *ICLR*, Volume 3, Part 3, April 1986.

19.51 *J. T. Mackley & Company Ltd* v. *Gosport Marina Ltd* [2002] EWHC 1315 (TCC) (3 July 2002) as reported on www.bailii.org/ew/cases/EWHC/TCC/2002/1315.html

19.52 'Notes on Documents for Civil Engineering Contracts', FIDIC, Lausanne, 1977, page 16.

19.53 See *Construction Contracts: Principles and Policies in Tort and Contract* (Reference 16.20 above), page 327.

19.54 *Engineering Law and the ICE Contracts* (Reference 19.45 above), page 290.

19.55 'ICC Court of Arbitration Case Notes', Sigvard Jarvin (the then General Counsel, ICC Court of Arbitration), *ICLR*, Volume 3, Part 5, October 1986, page 470. The author relates the findings of the arbitrators in ICC Arbitration Cases No. 4707 and No. 5029 at which legal experts appeared on both sides giving opinions as to the meaning of a few words in clause 67. The author comments that:

'It is obviously unsatisfactory that such a vital question as to how to present correctly a claim for arbitration under the terms of a standard and widely-used set of general conditions gives rise to different interpretations. I shall not discuss the learned opinions and arguments presented in favour of one or the other solution; the readers . . . have the benefit of the full quotes from the awards on these points.'

19.56 *Northern Regional Health Authority* v. *Derek Crouch Construction Co. Ltd* [1984] QB 644, (1984) 26 BLR 1. This case has been overtaken by events in England since the introduction of the Courts and Legal Services Act 1990 which provides for a similar power of opening and revising an engineer's decision. However, in some other jurisdictions, this power may still lie solely with the arbitrator.

19.57 This chart is adapted from one provided previously in 1987 for the Third Edition of the Conditions.

19.58 The ICC Rules of Arbitration (in English) can be viewed on and/or downloaded from www.iccwbo.org/court/english/arbitration/pdf_documents/rules/rules_arb_english.pdf

19.59 ICC Annual Reports, Publications Nos. 529 and 539, the International Chamber of Commerce, Paris, 1989.

19.60 Foreword to the Rules of Arbitration, International Chamber of Commerce, 1998.

19.61 The Main Objectives of the Revision, Yves Derains, as published in 'The New 1998 ICC Rules of Arbitration', *The ICC International Court of Arbitration Bulletin*.

19.62 Foreword to the Rules of Arbitration, International Chamber of Commerce, 1998.

19.63 *Guide to Arbitration*, a publication by the International Chamber of Commerce, ICC Publishing SA, Publication No. 382, Paris, September 1983. In this regard, see also the articles by Dominique Hascher on the scrutiny of draft awards published in the ICC International Court of Arbitration Bulletin, Vol. 6, No. 1, May 1995, and Vol. 7, No. 1, May 1996.

19.64 'International Commercial Arbitration Rules of the International Chamber of Commerce', Stephen R. Bond, Secretary General of the International Court of the ICC at the time; a paper delivered at the 1988 Annual Conference of the Chartered Institute of Arbitrators, Dublin, September 1988.

19.65 See also paragraph (c) on page 407.

19.66 As in Reference 19.6 above but page 29.

19.67 See for example the proposals made in an article by Professor Uff seeking to arrive at cost effective procedures: Cost-effective arbitration, John Uff, *Arbitration*, February 1993, page 39.

19.68 The ICC International Court of Arbitration Bulletin, Vol. 7, No. 1, May 1996.

19.69 'Disputes Provisions – Recent Developments at FIDIC and the World Bank', Robert Knutson, *International Construction Law Review*, Vol. 13, Part 2, April 1996.

19.70 *Bank Mellat* v. *GAA Development Construction Co.* [1988] 2 Lloyd's Rep 44.

19.71 'Post-conference review of ICC arbitration', Stephen R. Bond, Conference Proceedings, Centre of Construction Law and Management, King's College London, 1990, page 83.

19.72 'ICC United Kingdom: and its role in arbitration', David Sarre, Conference Proceedings, Centre of Construction Law and Management, King's College London, 1990, page 59.

19.73 *Guide to Arbitration* (Reference 19.63 above), Section 4.9, page 41.

19.74 'Terms of Reference Under the 1988 ICC Arbitration Rules – A Practical Guide', The ICC International Court of Arbitration Bulletin, Vol. 7, No. 1, May, 1996.

19.75 'Particular features of the ICC Rules of Arbitration', Humphrey Lloyd QC, Conference Proceedings, Centre of Construction Law and Management, King's College London, 1990, page 94.

19.76 'ICC Report on Dissenting and Separate Opinions', published in the ICC International Court of Arbitration Bulletin, Vol. 2, No. 1, May 1991. Other reports have also been issued by the ICC Commission on International Arbitration, such as those on 'Interim and Partial Awards' published in the ICC International Court of Arbitration Bulletin, Vol. 1, No. 2, December 1990 and on 'Status of Arbitrators', Vol. 7, No. 1, May 1996.

19.77 'A.D.R. Discussed', Hans van Houtte. A summary of an ICC seminar on the Settlement of International Commercial Disputes, the ICC International Court of Arbitration Bulletin, Vol. 7, No. 1, May 1996.

19.78 'Getting the best from ICC arbitration', Martin Harman, Conference Proceedings, Centre of Construction Law and Management, King's College London, 1990, page 147.

19.79 See also the comprehensive paper in connection with this topic: 'The Future of Arbitration (with particular reference to Construction disputed)', R. Fernyhough QC; a paper delivered to the Annual Conference of the Chartered Institute of Arbitrators, Maidstone, 13/15 June 1996.

19.80 'The Construction Dispute Resolution Group', Kenneth Severn, *Structural Engineer*, Vol. 69, No. 11, page 222.

19.81 *Concord and Conflict in International Arbitration* (Reference 19.12 above), page 7.

19.82 Report to the New Engineering Contract (NEC) committee, N. G. Bunni, September 1995.

19.83 In an interim report entitled 'Access to Justice', submitted by the Right Honourable Lord Woolf to the Lord Chancellor on the civil justice system in England and Wales, a number of the procedures developed in international arbitration were approved of and recommended for adoption. Some of these are incorporated in Section 19.11.

19.84 *Commercial Arbitration* (Reference 19.8 above), page 299. This passage is cited with approval by His Honour Judge Sir William Stabb QC in *Town & City* v. *Wiltshier Southern* (1988) 44 BLR 109 at 114.

19.85 'Cost-effective arbitration', John Uff, *Arbitration*, February 1993, page 33.

19.86 Ibid., page 32.

19.87 Final Report on Construction Industry Arbitrations, Nael G. Bunni and Humphrey Lloyd, *ICLR*, Vol. 18, Part 4, October 2001, page 644, and *ICC International Court of Arbitration Bulletin*, Vol. 12, No. 2, Fall 2001.

Chapter 20   Amicable Settlement Using Alternative Dispute Resolution

20.1 'Flip Flop Costs – A Tonic to Revive Arbitration', Richard D. S. Bloore, *Arbitration*, Volume 61, Number 2, May 1995.

20.2 'The Engineering and Construction Contract, 2nd Edition', an NEC document, Thomas Telford, London, 1995.

20.3 'Access to Justice', by the Right Honourable Lord Woolf, interim report to the Lord Chancellor on the civil justice system in England and Wales, June 1995.

20.4 'Alternative Dispute Resolution for Design Professionals', American Consulting Engineers' Council *Guidelines to Practice*, Volume 1, No. 7, Washington DC, 1988.

20.5 *Getting to Yes: Negotiating Agreements Without Giving In*, Roger Fisher and William Ury, Hutchinson & Co., London, 1981.

20.6 'Conciliation', John Tackaberry QC; a paper presented at the Conference on New Concepts in the Resolution of Disputes in International Construction Contracts, the Chartered Institute of Arbitrators, London, June 1989.

20.7 *Conciliation Procedure of the Institution of Civil Engineers*, 1994, London and *Conciliation Procedure of the Institution of Engineers of Ireland*, 1995.

20.8 The 1990 Standard General Conditions of Contract for Civil Engineering Works in South Africa include in their provisions a clause (clause 69) for mediation prior to arbitration, whereas the ICE 6th Edition refers to conciliation.

20.9 'Roles of a Mediator', Department of Education and Training of the American Arbitration Association, 140 West 51st Street, New York, NY 10020-1203, USA.

20.10 The National Construction Industry Arbitration Committee, referred to in abbreviated form as NCIAC, is a committee of the American Arbitration Association, whose address is given in Reference 20.9 above.

20.11 'Mediation', Peter H. Davies; a paper presented at a conference on New Concepts in the Resolution of Disputes in International Construction Contracts, the Chartered Institute of Arbitrators, London, June 1989.

20.12 'ADR: Alternative Dispute Resolution for the Construction Industry', a publication by the Association of Engineering Firms Practising in the Geosciences, Maryland, USA, 1988.

20.13 *Donovan Leisure Newton and Irvine ADR Practice Book*, edited by John H. Wilkinson, Wiley Law Publications, 1990, page 139.

20.14 *Martindale-Hubbell Dispute Resolution Directory*, Dispute Resolution Options. Published in co-operation with the American Arbitration Association, 1996.

20.15 'Zurich Mini-Trial', Rules of Procedure for the Zurich Mini-Trial, Zurich Chamber of Commerce, 5 October 1984.

20.16 'Mini-Trial Case Study', Jack K. Lemley, a paper presented at conference on New Concepts in the Resolution of Disputes in International Construction Contracts (Reference 20.11 above).

20.17 The Center for Public Resources, 680 Fifth Avenue, New York, NY, USA.

20.18 Housing Grants, Construction and Regeneration Act 1996, Chapter 53, HMSO, London.

20.19 Supplement to Fourth Edition 1987 of Conditions of Contract for civil engineering construction, Section A, 1996, FIDIC, Switzerland.

20.20 Definitions of the different ADR methods that are covered under the ICC ADR Rules can be found on pages 24 to 26 of the *Guide to ICC ADR* which is attached to the ADR Rules of 1 July 2001.

20.21 'ICC Pre-Arbitral Referee Procedure', ICC Publication No. 482, ICC Publishing SA, January 1990.

20.22 'The International Centre for Technical Expertise', ICC Publication No. 307, ICC Paris, 1977, containing the Role of the Centre and the Rules for Technical Expertise.

20.23 Under Article 1(A), it is stated that the Request should be made by 'any physical or legal person(s) or any court or tribunal (a "Person")'.

Chapter 21   FIDIC's Other Forms of Contract

21.1 'International Model Form of Agreement between Client and Consulting Engineer and International General Rules of Agreement between Client and Consulting Engineer'. The three separate documents, published by FIDIC, were designed to cover agreements relating to Pre-Investment Studies: IGRA PI; Design and Supervision of Construction of Works: IGRA D&S; and Project Management Services: IGRA PM.

21.2 *Construction Insurance*, Nael G. Bunni, Elsevier Applied Science Publishers, London, 1986.

21.3 *Guide to Use of FIDIC Conditions of Contract for Electrical and Mechanical Works, Third Edition*, Fédération Internationale des Ingénieurs Conseils, Switzerland, 1988. This publication contains 171 pages of commentary and the text of the third edition of the Yellow Book. FIDIC, PO Box 86, CH 1000 Lausanne, 12-Chailly, Switzerland.

21.4 Ibid.

21.5 See references to 'Due Consultation' in this book.

21.6 'The Liability and Insurance Clauses of the FIDIC E & M Form', Nael G. Bunni, *Construction Law Review*, July 1987.

21.7 The ICE Design and Construct Conditions of Contract, published by Thomas Telford Services Ltd for the joint sponsoring authorities: the Institution of Civil Engineers; the Association of Consulting Engineers; and the Federation of Civil Engineering Contractors. First edition, October 1992.

21.8 'FIDIC's Orange Book Conditions of Contract for Design – Build and Turnkey', Peter Booen; a paper presented at a seminar in London, The Study Group, February 1996.

21.9 *Guide to Use of FIDIC Conditions of Contract for Design-Build and Turnkey, First Edition*, FIDIC, Switzerland, 1996. This publication contains 163 pages of commentary and the text of the first edition of the Orange Book. FIDIC, PO Box 86, CH 1000 Lausanne, 12-Chailly, Switzerland.

21.10 'Conditions of Subcontract for Works of Civil Engineering Construction, for use in conjunction with the Red Book', Edition, 1994, FIDIC, PO Box 86, CH 1000 Lausanne, 12-Chailly, Switzerland.

21.11 'New FIDIC Publications: The Tendering Procedure and Subcontract', K. B. Norris; a paper presented at a seminar in Oxford, The Study Group, August 1995.

Chapter 22   The 1999 FIDIC Suite of Contracts

22.1 *FIDIC's Standard Forms of Contract – Principles and Scope of the Four New Books*, Christopher Wade, [2000] ICLR 1-229, Volume 17, Part 1, page 7.

22.2 It is indeed a pity that these new contract forms were not given new colours, which would have eliminated the necessity to refer to the Fourth Edition of the Red Book as the 1992 Red Book or as the Old Red Book. The drafting committee was advised to that effect by many correspondents, but appears to have chosen to disregard such advice.

22.3 For example, with the variations in the design function between the 1999 Red, Yellow and Silver Books, it could be argued that there should be different insurance requirements set out independently for each of them. In particular, the requirements in turnkey projects demand as a prerequisite the provision of additional insurance policies. The differences between these three forms are highlighted in Chapter 27.

22.4 See Annex C to the 1999 Red and Yellow Books: Example Form of Performance Security – Demand Guarantee.

22.5 See sub-clauses 16.1 and 16.2 of the 1999 Red and Yellow Books.

22.6 See sub-clause 4.1 of the Yellow Book and 4.1(c) of the Red Book and sub-clause 11.3 of both forms of contract in relation to the defects notification period.

22.7 See further discussion on the subject of fitness for purpose in Chapter 11.

22.8 See sub-clause 4.12 of the 1999 Red and Yellow Books.

22.9 See sub-clause 1.1.6.8 of the 1999 Red and Yellow Books.

22.10 *Donoghue* v. *Stevenson* [1932] AC 562.
22.11 The Guidance for the Preparation of Particular Conditions appears after the General Conditions in the book itself of each of the 1999 forms of contract.
22.12 See sub-clause 4.4(c) of the 1999 Red and Yellow Books.
22.13 See sub-clause 8.1 of the 1999 Red and Yellow Books.
22.14 See sub-clause 13.2 of the 1999 forms of contract.
22.15 See sub-clause 14.8 of the 1999 forms of contract.
22.16 See Chapter 26 for further discussion on the topic of dispute boards.
22.17 It is suggested that the intended capital value be around US$0.5 million.
22.18 See the first line of the Notes for Guidance of the Green Book.
22.19 As noted in the first paragraph (entitled 'General') of the Notes for Guidance of the Green Book.
22.20 See *FIDIC's Green Book – Clauses 6, 13 and 14. Risk, Responsibility, Liability, Indemnity, Insurance and Force Majeure*, Nael G. Bunni, ICLR [2002], Part 1.
22.21 The Third Edition of the Yellow Book and to some extent the Fourth Edition of the Red Book, both of which were first published in 1987, were the first forms of contract that recognised the natural flow of risk *to* responsibility *to* liability *to* indemnity *to* insurance. See in this connection, Chapter 7.
22.22 David M. Walker, *The Oxford Companion to Law*, Clarendon Press, Oxford, 1980.
22.23 Nael G. Bunni, *FIDIC's New Suite of Contracts – Clauses 17 to 19, Risk, Responsibility, Liability, Indemnity, Insurance and Force Majeure*, [2001] 18 ICLR 523, page 524.
22.24 See Mr Peter Booen's letter to the Editor of ICLR published [2001] 18 ICLR 714, 5th paragraph.
22.25 *FIDIC's New Suite of Contracts – Clauses 17 to 19, Risk, Responsibility, Liability, Indemnity, Insurance and* Force Majeure, Nael G. Bunni, [2001] 18 ICLR, page 525, last paragraph.
22.26 There are no case precedents to support this view and the courts will not imply an allocation of risks. As Seymore J said in the English case of *Carillion Construction Ltd.* v. *Farebrother and Others* [2002] EWHC 216 (TCC): 'Where the parties are in a contractual relationship in respect of a particular subject-matter it is, in my judgment, to be presumed that they have, by their contract, made that provision for the allocation of responsibilities and risks in relation to that subject-matter which they consider appropriate. It is not for the Court to disrupt the balance as struck by the parties by their contract and substitute some different allocation of responsibilities or risk. To do so would not be fair, just or reasonable.'
22.27 See the first paragraph of the Notes for Guidance of the Green Book.
22.28 *FIDIC's New Suite of Contracts – Clauses 17 to 19, Risk, Responsibility, Liability, Indemnity, Insurance and* Force Majeure, Nael G. Bunni, [2001] 18 ICLR 523, page 528, where a similar proposal is made in connection with FIDIC's three major forms, but referring to the special risks as exceptional risks.
22.29 *'Risk and Insurance in Construction,'* Second Edition, Nael G. Bunni, Spon Press, London, 2003, page 187,188. See also for example clause 13 in the Fourth Edition of the Red Book and clause 19.7 of the 1999 Red Book.
22.30 *Raineri* v. *Miles* [1981] AC 1050, 1086.
22.31 *Force Majeure and Frustration of Contract*, edited by E McKendrick, Lloyd's of London Press, Second Edition, 1995, page 43.
22.32 *Force Majeure and Frustration of Contract*, edited by E McKendrick, Lloyd's of London Press, Second Edition, 1995, page 39.

Chapter 23   The 1999 Red Book

23.1   Peter L. Booen, *The Three Major New FIDIC Books*, [2000] ICLR 1-229, page 24.
23.2   Some of these problems have been isolated and resolved by amendments incorporated in the newly proposed conditions of contract in Ireland, which are based on the 1999 Red Book.
23.3   *The FIDIC Contracts Guide* (with detailed Guidance on using the First Editions of FIDIC's Conditions of Contract for Construction; Conditions of Contract for Plant and Design-Build; and Conditions of Contract for EPC/Turnkey Projects), published by FIDIC in 2000, Switzerland.
23.4   *EIC Contractor's Guide to the FIDIC Conditions of Contract for Construction*, March 2002, reprint March 2003 with editorial amendments, page 6.
23.5   As in Reference 23.3, but page 71.
23.6   See generally Peter L. Booen, The Three Major New FIDIC Books, [2000] ICLR 1-229, page 29.
23.7   See page 8 of the *EIC Contractor's Guide to the FIDIC Conditions of Contract for Construction*, March 2002, reprint March 2003 with editorial amendments.
23.8   See Section 10.3.1 in Chapter 10.
23.9   *Reader's Digest Oxford Complete Word-finder*, published by Reader's Digest in 1993.
23.10   *Semco Salvage Marine Pte Ltd* v. *Lancer Navigation Ltd* [1996] 1 Lloyd's Rep. 449, 457, 459, CA.
23.11   See further details of the principle of fitness for purpose in Chapter 11, Section 11.7.
23.12   See sub-clause 1.1.2.6 in this connection.
23.13   *EIC Contractor's Guide to the FIDIC Conditions of Contract for Construction*, March 2002, reprint March 2003 with editorial amendments, page 12.
23.14   The risks in paragraph (h) of sub-clause 17.3 are: 'any operation of the forces of nature which is unforeseeable or against which an experienced contractor could not reasonably have been expected to have taken adequate preventative precautions'.
23.15   For example, the Form of Contract for Civil Engineering Construction of the Institution of Engineers of Ireland.
23.16   Sub-clause 17.1(b)(ii) of the new suite of contracts provides that 'The Contractor shall indemnify and hold harmless the Employer, the Employer's Personnel, and their respective agents, against and from all claims, damages, losses and expenses (including legal fees and expenses) in respect of: . . . damage to or loss of any property, real or personal (other than the Works), to the extent that such damage or loss: . . . is attributable to any negligence, wilful act or breach of the Contract by the Contractor . . .'
23.17   In his article entitled FIDIC's New Standard Forms of Contract – Force Majeure, Claims, Disputes and Other Clauses' [2000] 17 ICLR 235, Mr C. Seppala explains on page 238 that FIDIC adopted this change in line with the policy in the major UK and other standard forms.
23.18   Ibid.
23.19   Clause 22 of the ICE form, whether the Fifth Edition, which is referred to in the referenced footnote in *Hudson's Building and Engineering Contracts (1995)*, or the Sixth or the Seventh Editions, does not refer to negligence by the contractor and does not distinguish between the indemnity required for property damage as against that for bodily injury, disease or death of any person.
23.20   As Reference 23.13 above, but page 22.
23.21   As defined in the first paragraph of sub-clause 18.1.
23.22   See the second line of the third paragraph of clause 18.

23.23 See Reference 23.14 above.

23.24 The reference to financing charges is made in the Guide.

Chapter 24 The 1999 Yellow Book

24.1 'Design Build: Challenges and Opportunities', by Richard Kell, President, FIDIC, FIDIC2004\FIDICORI\DesignBuild_Paper.RAK.doc

24.2 Paragraph 6.4.2 of 'Legal And Procurement Practices in Building And Construction', First Domain Report, August 2002 by the International Council for Research and Innovation in Building and Construction, Thematic Network PeBBu: Performance Based Building.

24.3 *EIC Contractor's Guide to the FIDIC Conditions of Contract for Plant and Design Build (The New Yellow Book)*, The International Construction Law Review [2003] *ICLR* Volume 20, Part 3, July 2003, page 332.

24.4 This problem also exists in the Red Book, as discussed in Chapter 17, section 17.6.

24.5 Hands-on experience of the author as DAB on various recent projects under the 1999 Yellow Book.

24.6 See Chapter 11.

24.7 See section 24.1 above in respect of comments made by the European International Contractors (EIC) and in particular page 343 of the article in Reference 24.3 above, where they go so far as stating that 'the Yellow Book contains no obligation on the Employer to provide a definition of the intended purpose'.

24.8 See the definition of 'Employer's Requirements' in sub-clause 1.1.1.5 of the 1999 Yellow Book.

24.9 *FIDIC Contracts Guide*, a publication by FIDIC, Geneva, Switzerland.

24.10 See the *FIDIC Contract's Guide*, page 171.

24.11 Under the 1999 Red Book, sub-clause 8.3(b) states that the programme should include each of the stages for work by each nominated sub-contractor.

Chapter 25 The 1999 Silver Book

25.1 Christopher Wade, 'The Silver Book: The Reality', *ICLR*, Volume 18, Part 3, July 2001, page 500.

25.2 See the second paragraph of sub-clause 5.1.

25.3 *EIC Contractor's Guide to the FIDIC Conditions of Contract for EPC Turnkey Projects*, pages 17 to 21.

25.4 A comparison between sub-clause 17.3 'Employer's Risks' in the Silver Book and that in the other two Books would reveal that these risks have been shifted to the contractor.

25.5 See page 113 of the *FIDIC's Contracts Guide*.

25.6 'The Three Major New FIDIC Books', Peter L. Booen, [2000] *ICLR* 1-229, Volume 17, Part 1, January 2000, page 37.

Chapter 26 Dispute Boards

26.1 *Final Report on Construction Industry Arbitrations*, dated September 2001, ICC Commission on International Arbitration, Forum on 'Arbitration and New Fields', Construction Arbitration Section.

26.2 Hands on experience of the author and from the following articles at the Institution of Engineers of Ireland's training and assessment course on Dispute Adjudication Board, held on 19–22 January 2004, Dublin: 'Setting up the DAB' by Peter H. Chapman; and 'The Role of the DAB' by Gordon L. Jaynes. Also, reference is made to 'DRB's in Practice – Keys to Success' by Robert J. Smith, a presentation at the World Bank on 9 October 2002, in Washington.

26.3 'Periodical Meetings of the DAB' by Michael Mortimer Hawkins, a presentation at the Institution of Engineers of Ireland's training and assessment course on Dispute Adjudication Board, held on 19–22 January 2004, Dublin.

26.4 'Dispute Boards', a presentation by Peter H. J. Chapman, delivered at a course on Dispute Boards organised by the Institution of Engineers of Ireland in January 2004.

26.5 *Better Contracting/or Underground Construction*, a report published in 1974 as a result of the study sponsored by the US National Committee on Tunneling Technology in 1972.

26.6 See Section 1 – Chapter 1 of the DRBF Manual of May 2004. The Dispute Resolution Board Foundation (DRBF) was established in 1996 to promote use of the process and serve as a technical clearinghouse for owners, contractors and board members in order to improve the dispute resolution process. The DRBF provides DRB information and educational opportunities for all parties involved in construction disputes. The Foundation's website is www.drb.org

26.7 See the bar charts in Appendix A of the DRBF Manual referred to in the previous reference.

26.8 'A "Claims Review Board" as a Way for an Amicable Settlement of Disputes, and Other Considerations on the Subject of Claims', G. Lodigiani, *ICLR*, Volume 3, Part 5, October 1986, page 498.

26.9 World Construction Review/Outlook 2003/4, Davis Langdon PKS, www.davislangdon.com

26.10 As explained earlier in Chapters 8, 10 and particularly Chapter 19, Section 19.8.

26.11 See Chapter 20 regarding the ICC ADR Rules.

26.12 The ICC Dispute Board Rules were published in September 2004.

26.13 In at least two projects where the author acted as a single DRB, no disputes occurred and when asked at the end of the construction period why this happened, both parties replied that they knew what my answer would be.

26.14 As in Reference 26.6 above.

26.15 See Section 19.8 above.

26.16 See Section 20.8 above. It should be noted that some commentators make the mistaken conclusion that the Dispute Adjudication Board has some connection with Statutory Adjudication in the UK. This statute came into being in 1996 under the title The Housing Grants, Construction and Regeneration Act 1996.

26.17 *ICC Rules for Dispute Boards* published in November 2004, Document 420/504.

26.18 The World Bank recommends US$50 million in value as a cut-off limit above which a board of three members should be used.

26.19 Hong Kong Airport had a dispute review group of six members plus a convenor convening the range of expertise required. The convenor selected one or three members of the group depending on the nature and complexity of the dispute.

26.20 US$50 million is the cut-off limit for one member board under the World Bank form of contract, whereas US$20 billion is the estimated cost of a project such as the Hong Kong Airport.

26.21 See the example sub-clause for the pre-arbitral decisions by the engineer on page 20 of the Guidance Notes of the 1999 FIDIC Red Book.

26.22 See pages 63 to 66 of the 1999 Red Book; pages 64 to 67 of the 1999 Yellow Book; and pages 57 to 60 of the 1999 Silver Book.
26.23 See pages 19 to 21 of the 1999 Red Book; and pages 20 to 22 of the 1999 Yellow and Silver Books.
26.24 See Condition 3 of The General Conditions of Dispute Adjudication Agreement, page 63 of the 1999 Red Book, and also Rule 5 of The Procedural Rules as annexed to the Appendix, page 67 of the 1999 Red Book.
26.25 See Chapter 5, Sections 5.1.1 and 5.4, Chapter 9, Section 9.3.4, and Chapter 19, Section 19.10, for the meaning of independence and impartiality.
26.26 The General Conditions of Dispute Adjudication Agreement, pages 63 to 66 of the 1999 Red Book. See also Section 26.9 below.
26.27 It should be noted that these obligations differ slightly between the 1999 Red, Yellow and Silver books.
26.28 See *Hitec Power Protection BV v. MCI Worldcom Ltd* [2002] EWHC 1953, reported in 5 BLR 83, at page 89.
26.29 Chapter 4 of the Manual of The Dispute Resolution Board Foundation (DRBF), May 2004.
26.30 Per Lord Justice Donaldson, in 1981, then Master of the Rolls, in *Bremen Handelsgesellschaft GmbH v. Westzucker GmbH (No. 2); Westzucker GmbH v. Burge GmbG* [1981] 2 Lloyd's Rep 130, CA.
26.31 *The Canterbury Tales* (c. 1387), Melibee, 2240.

Chapter 27 A Precise Record of the Alterations, Omissions and Additions in the 1999 Yellow and Silver Books as compared with the 1999 Red Book

27.1. *The FIDIC Contracts Guide – with Detailed Guidance on Using the First Editions Of FIDIC'S Conditions of Contract for Construction, Conditions of Contract for Plant and Design-Build, and Conditions of Contract for EPC/TURNKEY Projects*, Fédération Internationale des Ingénieurs-Conseils, Switzerland, 2000.

Table of Cases

The following abbreviations of reports are used:

AC, App	Law Reports, Appeal Cases
ALJ	Australian Law Journal
All ER	All England Law Reports
Beav	Beavan's Rolls Court Reports
BLM	Building Law Monthly
BLR	Building Law Reports
CA	Court of Appeal
Cal	California Reports
ChD	Law Reports, Chancery Cases
CILL	Construction Industry Law Letter
CLD	Construction Law Digest
CLR	Construction Law Reports
Con LR	Construction Law Reports
Const LJ	Construction Law Journal
EG	Estates Gazette
ER	English Reports
Exch	Law Reports, Exchequer Cases
EWCA	England and Wales Court of Appeal
EWHC	England and Wales High Court
Fed Cir	Federal Circuit Bar Journal
HL	House of Lords
ICLQ	International and Comparative Law Quarterly
IR	Irish Reports
KB	Law Reports, King's Bench
Lloyd's Rep	Lloyds' Reports
LR	Law Reports
LT	Law Times Reports
NBR	New Brunswick Reports
NE	North Eastern Reports
NSWLR	New South Wales Law Reports
NSWSC	New South Wales Supreme Court
NTSC	Northern Territory Supreme Court
NY	New York Court of Appeal
NYS	New York State Court Reports

NZLR	New Zealand Law Reports
QB	Law Reports, Queen's Bench
QWN	Queensland Weekly Notes
SALR	South African Law Reports
SCR	Supreme Court Reports
SJ	Solicitors' Journal
SLR	Student Law Review
US	United States: Supreme Court Reports
WLR	Weekly Law Reports

Abigroup Contractors Pty Ltd v. Peninsula Balmain Pty Ltd [2001] NSWSC 752 348
Allied Maples Group Ltd v. Simmons & Simmons [1995] 2 All ER 907; [1995]
  1 WLR 1602 335
Allied Properties v. Blume [1972] 25 CA, 3d, 848 202
AMF (International) Limited v. Magnet Bowling Ltd [1968] 1 WLR 1028 .... Ref. 12.3
Amin Rasheed Shipping Corporation v. Kuwait Insurance Company [1983]
  3 WLR 241 .... Refs 2.9 and 3.1
Appleby v. Myers [1867] unreported 68
Applegate v. Moss [1971] 1 QB 406 .... 57, 58
Arenson v. Casson, Beckman, Rutley & Co. [1975] 3 WLR 815 194
Aros Industries v. Electricity Commission of New South Wales (1973)
  2 NSWLR 186 .... Ref. 6.2
Assoland Construction Pte Ltd v. Malayan Credit Properties Pte Ltd [1993] 3 SLR 470 346

Bagot v. Stevens Scanlan & Co. [1964] 3 WLR 1162; [1964] 3 All ER 577 187
Balfour Beatty Building Ltd v. Chestermount Properties Ltd (1993) 62 BLR 12 .... 344, 365
Bank Mellat v. GAA Development Construction Co. [1988] 2 Lloyd's Rep. 44 ...... Ref. 19.70
Barclays Bank Plc v. Fairclough Building Limited [1995] QB 214 188
Beck Peppiatt Ltd v. Norwest Holst Construction Ltd [2003] EWHC 822 (TCC) 391
Bilton v. Greater London Council (1982) 20 BLR 1 .... Ref. 17.6
Bolam v. Friern Hospital Management Committee [1957] 2 All ER 118 190
Boorman v. Brown [1844] 8 ER 1003 .... Ref. 11.5
Bremen Handelsgesellschaft GmbH v. Westzucker GmbH (No. 2) [1981]
  2 Lloyd's Rep. 130, CA .... Ref. 26.30
Brickfield Properties v. Newton [1971] 1 WLR 862 .... Refs 11.23 and 11.49
British Airways Pension Trustees v. Sir Robert McAlpine & Sons Ltd (1994) 72 BLR 26 339

Cagne v. Bertran [1954] 43 Cal. 2d 481, 275, p.2d15 202
Cambridge Water Co. Ltd v. Eastern Counties Leather Plc [1994] 1 All ER 53 ...... Ref. 11.41
Carillion Construction Ltd v. Farebrother and Others [2002] EWHC 216 (TCC) ..... Ref. 22.26
Carr v. J. A. Berriman Pty Ltd [1953] 27 ALJ 273 .... Ref. 17.1
Cawley & Co. [1889] 42 ChD 209 .... Ref. 2.20
Cervidone Construction Corp. v. US 931 F 2d 860 (Fed Cir 1991) 337
Charon v. Singer Sewing Machines Ltd [1968] 207 EG 140 68
City Inn v. Shepherd Construction [2003] BLM 20:10, Court of Session, Inner House
  (Scotland) 349, 373
Clayton v. Woodman [1962] 1 WLR 585 177

Constable Hart & Co. Ltd v. Peter Lind & Co. Ltd [1978] 9 BLR 1 297
Conway v. Rimmer [1968] AC 910; [1968] 1 All ER 874 .... Ref. 2.23
Costain Ltd v. Wescol Steel Ltd [2003] EWHC 312 (TCC) 392
Cottage Club Estates Ltd v. Woodside Estates Co. (Amersham) Ltd [1928] 2 KB 463 238
Courtney & Fairbairn Ltd v. Tolaini Bros (Hotels) Ltd [1975] 1 All ER 716; [1975]
 1 WLR 297 .... Ref. 2.44
Cowlin Construction Ltd v. CFW Architects [2002] EWHC 2914 (TCC) 391
Craven Ellis v. Canons Ltd [1936] 2 KB 403 .... Ref. 16.3
Crosby J. & Sons Ltd v. Portland Urban District Council (1967) 5 BLR 121 337

Dean and Dyball Construction Ltd v. Kenneth Grubb Associates Ltd [2003]
 EWHC 2465 (TCC) 392
Demers v. Dufresne Engineering Co. Ltd (1979) 1 SCR 146 (Supreme Court of Canada) 198
D & F Estates Ltd v. Church Commissioners for England [1989] AC 177 195
Donoghue v. Stephenson [1932] AC 562 195
Dunlop Pneumatic Tyre Co. Ltd v. New Garage & Motor Co. [1915] AC 79 .... 372, 373

Eagle Star Insurance Co. Ltd v. Yuval Insurance Co. Ltd [1978] 1 Lloyd's Rep. 357 ... Ref. 19.37
Earl of Oxford's Case [1615] 1 W&T 615, 21 Eng. Rep. 485 .... Ref. 2.21
East Ham v. Bernard Sunley & Sons [1966] AC 406 177
Eckersley T. E. and Others v. Binnie & Partners and Others (1988) Con LR 1, CA ... Ref. 11.49
Edmund Nuttall Ltd v. R. G. Carter Ltd [2002] EWHC 400 (TCC) 391
Edward Owen Ltd v. Barclays Bank [1978] QB 159 .... 277, 282
Ellerine Bros Pty Ltd v. Klinger [1982] 1 WLR 390
Ernst and Whinney v. Willard Engineering (Dagenham) Ltd (1987) 3 Const LJ 292;
 (1988) 40 BLR 67 188
Esso Petroleum Co. Ltd v. Mardon [1976] QB 801; [1976] 2 All ER 5 187

Fastrack Contractors Ltd v. Morrison Construction Ltd [2000] BLR 168 .... 390, 392
Fibrosa Spolka Akcyjna v. Fairbairn Lawson Combe Barbour Ltd [1943] AC 32 ...... Ref. 16.5
Finlay v. Murtagh [1979] IR 249 .... Ref. 11.8
Finnegan J. F. v. Sheffield City Council (1988) 43 BLR 124 .... Ref. 17.32
Francis v. Cockerell (1870) LR 5 QB 501 .... Ref. 13.4

Gable House Estates v. The Halpern Partnership and Bovis Construction Ltd (1995)
 12-CLD-03-01 (QBD) 189
Gaymark Investments v. Walter Construction Group Ltd [1999] NTSC 143 ...... 344, 350, 367
George Hawkins v. Chrysler (UK) Limited and Burne Associates (1986) 38 BLR 36 204
GLC v. Cleveland Bridge & Engineering Co. [1984] 34 BLR 50 355
Glenlion Construction Ltd v. The Guinness Trust (1987) 39 BLR 89 .... 346, 363
GMTC Tools & Equipment Ltd v. Yuasa Warwick Machinery Ltd (1994) 73 BLR 102 339
Graveley v. The Providence Partnership (1977) Federal Appellate Court, USA, GFIP
 Vol. VIII, No. 1 202
Greater Nottingham Co-operative Society v. Cementation Piling & Foundations Ltd
 [1989] QB 71; (1988) 41 BLR 43 .... 188, 196
Greaves (Contractors) Limited v. Baynham Meikle & Partners [1975] 1 WLR 1095 .... 203, 204
Grinaker Construction v. Transvaal Provincial Administration [1982] 1 SALR 78 .... Ref. 6.2

Hadley v. Baxendale (1854) 9 Exch 341 . . . . 66, 295
Halki Shipping Corporation v. Sopex Oils Ltd [1997] EWCA Civ 3062 . . . . 390, 392
Hancock v. B. W. Brazier (Anerley) Limited [1966] 2 All ER 901; [1966] 1 WLR 1317 205
Harbottle R. D. (Mercantile) Ltd v. National Westminster Bank Ltd [1978] QB 146;
    [1977] 2 All ER 862 278
Henry Boot Construction (UK) Ltd v. Malmaison Hotel (Manchester) Ltd (1999)
    70 Con LR 32 365
Heyman v. Darwins Ltd [1942] AC 356 . . . . Ref. 19.31
Hill Organisation Ltd v. Bernard Sunley & Son Ltd (1983) 22 BLR 1 . . . . Ref. 11.10
Hitec Power Protection BV v. MCI World Com Limited [2002] EWHC 1953; 5 BLR 83 391
Holt v. Payne Skillington (1995) 13-CLD-06-01, CA 189

Independent Broadcasting Authority v. EMI Electronics and BICC Construction (1980)
    14 BLR 1 . . . . 204, 205

John Doyle Construction Limited v. Laing Management (Scotland) Limited (2004)
    BLM 21:7; CILL 2135, Court of Session, Inner House (Scotland) 339

Lancashire & Cheshire Association of Baptist Churches Incorporated v. Howard & Seddon
    Partnership [1993] 3 All ER 467 189
Law v. Redditch Local Board [1892] 1 QB 127 . . . . Ref. 17.35
Leach v. London Borough of Merton [1985] 32 BLR 68 . . . . 338, 348
London & Amsterdam Properties Ltd v. Waterman Partnership Ltd [2003] EWHC 3059
    (TCC) 392
London Export Corporation Ltd v. Jubilee Coffee Roasting Co. [1958] 1 WLR 661 . . . . Ref. 3.13

Mackley J. T. & Company Ltd v. Gosport Marina Ltd [2002] EWHC 1315 (TCC) 404
Manufacturers' Mutual Insurance v. Queensland Government Railways (1968)
    QWN 12 . . . . Ref. 14.2
May and Butcher v. R [1934] 2 KB 17n . . . . Ref. 2.45
McAlpine Humberoak Ltd v. McDermott International Inc. [1992] 58 BLR 1 337
Medjuck & Budovitch Ltd v. Adi Ltd 33 NBR 2nd 271 80 Apr. 271, paragraph 210 203
Mercers Co. v. New Hampshire Insurance Co. Ltd [1992] 2 Lloyd's Rep. 365 273
Methodist Church of Babylon v. Glen-Rich Const Corp (1971) 27 NY 2d 357, 318 NYS
    2d 297 . . . . Ref. 19.46
Midland Bank Trust Co. Ltd v. Hett, Stubbs & Kemp [1978] 3 WLR 167 187
Miller v. Cannon Hill Estates Limited [1931] 2 KB 113 205
Mitsui Construction Co. v. The Attorney General of Hong Kong (1986) 33 BLR 1 (PC) 310
Monmouthshire County Council v. Costello & Kemple Ltd [1965] 5 BLR 83 389
Moresk Cleaners Ltd v. Hicks [1966] 2 Lloyd's Rep 338 . . . . 161, 191
Morgan Grenfell (Local Authority Finance) Ltd v. Sunderland Borough Council and
    Seven Seas Dredging Ltd [1990] 49 BLR 31 40
Murphy v. Brentwood District Council (1990) 8-CLD-10-05 . . . . Refs 11.31 and 11.38
Murray v. Leisureplay Plc [2004] EWHC 1927 (QBD) 373
Myers G. H. & Co. v. Brent Cross Service Co. [1934] 1 KB 46 . . . . Ref. 13.5

Neodox Ltd v. Swinton & Pendlebury BC (1958) unreported (QBD) . . . . Ref. 1.7
Nocton v. Lord Ashburton [1914] AC 932 . . . . Ref. 11.6

Norta Wallpapers (Ireland) *v.* Sisk & Sons (Dublin) Ltd [1978] IR 114 204
Northern Regional Health Authority *v.* Derek Crouch Construction Co. Ltd [1984] QB 644;
    (1984) 26 BLR 1 . . . . Refs 19.10 and 19.56
Norwest Holst Construction *v.* Cooperative Wholesale, 17 February 1998, unreported 335

Oldschool *v.* Gleeson (Construction) Ltd (1976) 4 BLR 103 . . . . 175, 194
Orange EBS Ltd *v.* ABB Ltd [2003] EWHC 1187 (TCC) 392
Orion Compania Espanola de Seguros *v.* Belfort Maatschappij voor Algemene
    Verzekgringen [1962] 2 Lloyd's Rep 257 397

Pacific Associates Inc. *v.* Baxter [1989] 2 All ER 159 . . . . 180, 195, 196
Parkin *v.* Thorold [1852] 16 Beav 59 37
Parsons H. (Livestock) Ltd *v.* Uttley Ingham & Co. Ltd [1978] QB 791 . . . . Ref. 3.23
Paschen Contractors Inc. *v.* John J. Calnan Co. (1973) 13 Ill. App. 3d 485, 300 NE
    2d 795 . . . . Ref. 19.46
Peak Construction (Liverpool) Ltd *v.* McKinney Foundations Ltd (1970) 1 BLR 111 343
Perar BV *v.* General Surety Guarantee (1994) 66 BLR 72 . . . . Refs 15.7 and 15.20
Petroleum Development (Trucial Costs) Ltd *v.* Shaikh of Abu Dhabi [1951] 1 ICLQ 247 41
Phillips Hong Kong Ltd *v.* Attorney General of Hong Kong [1993] 61 BLR 41 370
Planché *v.* Colburn [1831] 131 ER 305 . . . . Ref. 16.4
Portsea Island Mutual Cooperative Society Ltd *v.* Michael Brasher Associates (1989)
    7-CLD-10-05 193
Prima Pain *v.* Flood and Conklin (1967) 338 US 395 . . . . Ref. 19.30
Property & Land Contractors Ltd *v.* Alfred McAlpine Homes Ltd [1995] 47 Con LR 74 335

Raineri *v.* Miles [1981] AC 1050 . . . . Ref. 22.30
Reigate *v.* Union Manufacturing Co. (Ramsbottom) [1918] 1 KB 592 . . . . Ref. 3.17
Robophone Facilities Ltd *v.* Blank [1966] 1 WLR 1428 . . . . Ref. 17.36
Rowe *v.* Turner Hopkins & Partners [1980] NZLR 550 (New Zealand High Court) . . . Ref. 11.9
Royal Brompton Hospital NHS Trust *v.* Frederick A. Hammond and Others [2000]
    EWHC Technology 39 (18 December 2000) 354, 365
Ruxley Electronics & Construction Ltd *v.* Forsyth [1994] 1 WLR 650, CA; [1995] 3 WLR
    118 HL 317
Rylands *v.* Fletcher [1868] LR, 3 HL 330 201

Samuels *v.* Davis [1943] 1 KB 526 205
Semco Salvage Marine Pte Ltd *v.* Lancer Navigation Ltd [1996] 1 Lloyd's Rep 449, CA 524
Shanklin Pier Ltd *v.* Detel Products Ltd [1951] 2 All ER 471 204
Sindall Ltd *v.* Solland (15 June 2001) (TCC) www.adjudication.co.uk 348, 390, 391
Sutcliffe *v.* Thackrah [1974] 1 All ER 319 180

Tai Hing Cotton Mill Ltd *v.* Liu Chong Hing Bank [1986] 1 AC 80 . . . . 60, 188
Temloc Ltd *v.* Errill Properties Ltd [1987] CILL 376 . . . . Ref. 17.40
Tersons Ltd *v.* Stevenage Development Corporation [1965] 1 QB 37 . . . . Ref. 16.12
The Heron II [1969] 1 AC 350 . . . . Ref. 3.23
The Moorcock Case [1886–1890] All ER 850 60
Town & City *v.* Wiltshire Southern (1988) 44 BLR 109 . . . . Ref. 19.84
Trade Industry *v.* Workington [1937] AC 1 . . . . Ref. 15.4

Trafalgar House Construction (Regions) Ltd *v.* General Surety & Guarantee Ltd [1994]
  66 BLR 42 274
Trollope & Colls Ltd *v.* Northwest Metropolitan Regional Hospital Board [1973] 2 All
  ER 260 .... 60, 61
Turner Corporation Ltd *v.* Austotel Pty Ltd (1994) 13 BCL 378 351
Turner Corporation Ltd *v.* Coordinated Industries Pty Ltd [1994] 11 BCL 202 350

Valdo Vulic *v.* Bohdam Bilinsky [1982] NSW Supreme Court No. 177700/78 .... Ref. 11.8
Victoria Laundry (Windsor) Ltd *v.* Newman Industries Ltd [1949] 2 KB 528 .... Ref. 3.23
Viking Grain Storage Limited *v.* T. H. White Installations Limited and Another (1985)
  33 BLR 103 206

Walter Lawrence & Son Ltd *v.* Commercial Union Properties Ltd [1984] 4 CLR 37 363
Wells *v.* Army & Navy Co-operative Society (1902) 86 LT 764 .... Ref. 12.2
Westzucker GmbH *v.* Burge GmbH [1981] 2 Lloyd's Rep 130, CA .... Ref. 26.30
Wharf Properties Ltd *v.* Eric Cumine Associates [1991] 52 BLR 8 .... 336, 339
Whitehouse *v.* Jordan [1981] 1 WLR 246; 125 SJ 167; [1981] 1 All ER 267 .... 190, 203
Whittal Builders Co. Ltd *v.* Alfred McAlpine Homes Ltd (1996) 12 Const LJ 356; (1996)
  11 Con LR 40 .... Ref. 16.25
Wong Lai Ying and Others *v.* Chinachem Investment Co. Ltd (1979) 13 BLR 81 ...... Ref. 3.11

Xerox Corporation *v.* Turner Construction Co. and Others [1973] GFIP, Vol. IV, No. 4 202

Young & Marten *v.* McManus Childs [1969] 1 AC 454; [1968] 3 WLR 630 .... 204, 231

Index

ACE Form of Contract, 4, 10, 11, 17, 49, 70, 531
  compared with the FIDIC Form, 6
  compared with the ICE Form, 4
  origin, 4
acceleration, 230, 350, 369
  cost of, 369
acceptance
  by engineer, 139
  letter of
    definition, 224
  obligations following, 223, 224, 263, 286
  obligations prior to, 223, 262, 286
  of tender, 158
access to site, 212
accident or injury to workmen, 508
  insurance against, 261, 508, 518
  prevention of, 763
  to employees, 261
act
  judicature, 28, 34
  limitation, 57, 189
  of foreign enemies, 246, 248
  of God, 103, 367
  of parliament, 35, 38
adverse physical conditions, *see* claims
agent, 62, 260, 261, 266
  engineer as employer's, 82, 126, 155–6, 162–3, 179, 186, 195, 209, 213, 241
agreement, 4, 6, 14, 49, 53, 55
  ad hoc, 394
  arbitration, 22, 31, 289, 294, 394–5
  as a source of law, 395
  client/consultant, 84, 158, 463
  concession, 85
  definition, 55

enforcement, 397
existing, 393
in amicable settlement, *see* amicable settlement
in the Red Book, 396, 399
standard forms of, 3, 6, 17, 18, 51, 68, 70, 71, 81, 84
validity, 395
alternative dispute resolution, 14, 151, 152, 385, 408, 437, 438
adjudication, 453
  board, *see* Dispute Board
advantages of, 440
conciliation, 445
  advantages, 446
  attendance at, 448
  characteristics, 446
  conciliator, 447
  definition, 446
  process, 448
  rules, 445
  when to conciliate, 446
dispute boards, *see* Dispute Board
mediation, 443
methods of, 439
mini trial, 451
negotiation, 441, 442
negotiators, 441
pre-arbitral referee, 457
amicable settlement, 150, 221, 439
  agreement, 440, 442, 443, 445, 450, 457
  procedure for, 407
Appendix to Tender, *see* tender
applicable law, 8, 10, 18, 21, 22, 30, 50, 117, 186
  claims under, 296, 300, 320, 336
  imported provisions, 118, 119

applicable law (cont'd)
  indemnity, 135
  in international construction, 19, 22, 24
  of contract, 114, 228
  selection of, 20, 21
approval, 140, 147, 160
  by the employer, 135, 164, 209, 213
    in an emergency, 164
    of engineer's duty and authority, 135, 163, 164, 167, 168, 210
    of insurance policies, 214
    of securities, 213, 286
  by the engineer, 364
    of breakdown of lump sum items, 148, 225, 235
    of contractor's representative, 167, 168
    of design, 138, 234
    of drawings etc., 92, 160, 171
    of programme, 140
    of receipts and vouchers, 235
  definition, 134
  in Orange Book, 477
  in Yellow Book, 465–71
  of the Fourth Edition of Red Book, 128
arbitration, 109, 110, 112, 113, 115, 118, 144, 149, 150–3, 162, 180–6, 288, 385
  advantages of 294, 375, 387, 425
  ad hoc, 394, 422
  agreement, 19, 31, 199, 296, 298, 301, 377–9, 393–9, 417
  and role of national courts, 19, 389, 397
  and ICC Court, see ICC Rules
  applicable law, see applicable law
  as a source of law, 396
  award, 50, 280, 389, 397, 446
  background, 385
  cost, 426, 428
  criticism, 424–37
  custom, 35, 396
  definition of, 393
  delay, 426
  disputes settlement by, 385
  failure to comply with engineer's decision, 409, 419, 512, 539
  failure to give notice of intention to commence, 150, 406
  ICC awards, see ICC Rules
  initiation, 394, 409
  intention to commence, 150, 214, 237, 239, 406–8, 414, 439
  international commercial, 19, 21–3, 31
    definition, 19
  losing favour, 425
  notice to commence, 237, 239, 406
    functions, 406
  place of, 19, 22, 23, 394, 416, 417, 418, 421, 773, 774, see also seat
  prior to substantial completion, 409
  procedural law, 22
  procedure under clause 13, 150, 180, 197, 399, 403, 408
  reference to, during progress of works, 10
  remedies, 427
  right to, 13, 150, 238
  seat of, 21, 23, 394, 416, 423, 772
  sources of law, 395
  UNCITRAL Rules, 20, 445
arbitrator, 72, 388, 394, 397
  appointment of, 7, 23, 388, 395, 415, 417
  award of, 389, 394, 420, 421
  duties of, 398
  independence of, 398, 415
  jurisdiction, 393, 394, 395
  legal or technical, 398
  number to be appointed, 408
  power to open up, review and revise, 377, 388, 409
  qualifications of, 398, 399
  rules for appointment, 420
  sole, 408, 417, 429
  technical or legal, 398, 399
  under ICC Rules, 414, 415, 416, 418, 420
assignment, 10, 23, 58, 209, 238
award
  enforcement of arbitral, 23, 31, 376, 388
  of contract, 213, 254, 582
  under ICC rules, see ICC Rules

benefit, 160, 173, 257, 267
  principle of, 154, 201
bill of quantities
  contract documentation, 56, 80
  contracts with, 89

estimates only, 314, 547, 578
measurement of quantities, 303, 304, 305
preliminary items, 301
use of, 15, 17, 87, 89, 90, 105, 300, 303, 304, 310, 318
bond/guarantee, 125, 232, 269, 273, *see also* performance
  advanced payment guarantees, 290
  bid, 272, 290
  characteristics, 272
  company suretyship, 291
  compatibility with applicable law, 285
  contract bonds, 281
  for 'proper' execution, 272
  maintenance or defects liability, 272, 291
  notice, 108
  on demand, 107, 270, 273–7
    ICC Uniform Rules, 280
    under the 1999 forms of contract, 491, 527
  performance, 10, 102, 136, 138, 217, 218, 223, 228, 230, 270, 271, 272, 273
  retention, 291
bonus, 6, 346, 376
boreholes, 325
breach
  claim for, 148, 279, 299, 300, 332
  failure to insure, 10, 134
  guarantee against, 276
  not liable for, 227
  of contract, 8, 9, 53, 57, 120, 142, 186, 189, 212, 221, 228, 229, 234, 237, 256, 262, 331, 346
  of duty, 54, 185, 194, 197
  of statutory duty, 54, 200
  remedies for breach of contract, 51, 62, 67

CAR policy, 243, 248, 253, 257, 265
cash-flow estimates, 224, 234
certificate, 6, 9, 15, 17, 146, 110, 130, 141, 287, 364, 450
  defects liability, 139, 225, 227, 240, 286, 289, 319, 381
  effect of taking-over, 289, 367
  final, 11, 12, 13, 130, 139, 377, 381, 511
  interim 11, 12, 13, 130, 139, 273, 377, 510

  late certification, 384
  non-payment or failure to make payment, 151, 378, 384
  of valuation at date of termination, 287, 291, 364, 384
  payment, 11, 12, 13, 130, 218, 521, 522, 526, 528, 529, 537, 548, 588
  taking-over, 169–73, 179, 219, 230, 240, 245, 256, 291, 302, 318, 327, 370, 375, 481
    under 1992 Red Book, 379
    under 1999 Red Book, 545
    under 1999 Silver Book, 594
    under 1999 Yellow Book, 564, 571, 579
civil war, 247, 249, 266, 267, 323
claims and counterclaims, 293
  adverse physical obstructions or conditions, 5, 132, 139, 297, 299, 309–15, 318
  acceleration, 230, 369
  based on grounds of breach, 298–301, 316, 320, 332
  based on provisions of the contract, 327, 405
  categories, 294–7, 342
  compliance with statutes, 316
  concepts of, 234
  definition of, 293, 294, 402
  delay *see* delay
  dispute as opposed to, 389–93
  disruption, 160, 269, 334, 366, 368, 375, 430, 433
  employer's risks, 315
  ex gratia claim, 297
  extension of time, 343, 347
  failure to commence, 320
  failure to perform obligations, 187, 320, 326, 370, 540, 598
  fluctuations of cost, 316
  for damage to bridges, 171, 214, 234, 326
  for defects, 317
  global approach to, 270, 336
  heads of, 333, 367
  intention to, 145, 172, 275, 287, 301, 328–31, 430
  legal basis, 293
  measurement changes, 299, 304
  nature of, 293

claims and counterclaims (*cont'd*)
  notice of, 330, 341, 483, 525
  procedure, 293, 298, 328
  prolongation, 334, 366, 375, 553
  quantum, 333, 342
  quantum meruit, 65, 296
  records, 332
  remedy, 295, 296, 327
  settlement of, 150, 438, 483, 606
  specified events, 293
    antiquities, 326
    boreholes, 325
    incorrect data, 325
    interference, 326
    payment of fees, 326
    provision of facilities, 326
    royalties, 326
    tests, 326
    uncovering of work, 327
  types of, 298, 366
  under performance security, 139, 163, 228, 281, 282, 285, 292, 293
    ECGD, 289
    ICC URCB Rules, 280
    notice requirements, 279
  under tort, 296
  valuation of, 301
  variations, 298, 299, 305, 317, 501, 546, 556, 571, 595
  varied work, 235, 301
code, 17, 27, 29
  civil, 26, 28, 29, 33, 44, 84, 118, 147, 149, 192, 198
    Egyptian, 47, 85, 115
    French, 22, 26, 28, 29, 41, 86, 116, 120, 145, 147, 149, 154, 189, 192, 195, 198, 202
    German, 26, 51, 147, 181, 193
    Kuwaiti, 86, 116
    Libyan, 181
    Quebec, 154, 198
    Swiss, 30, 32
    UAE, 86, 116, 155, 181, 198, 202
  FIDIC's code of ethics, 75–7, 82
  Hammurabi, 17 (ref. 2.1)
  of practice, 6, 61, 65, 81, 179
    of a consulting engineer, 6, 81

commence
  arbitration, 150, 214, 237, 239, 404, 406, 407, 408, 439, 628
    under the Yellow and Orange Books, 454
  failure to, 299, 320
  notice to, 146, 211, 224, 225, 344, 406
commencement
  date, 145, 212, 225, 256, 262, 344, 411, 466, 493
  date under Green Book, 501
  date under 1999 Red Book, 529
  date under 1987 Yellow Book, 466
  date under 1999 Yellow Book, 553, 557, 569
  meaning of, 225
  of works, 228, 362, 509, 516, 560, 563
commotion, 266
completion, 16, 48, 50, 57, 61, 220, 221, 223–7, 229, 250, 251, 272, 273, 277, 288, 314
  certificate, 9, 119, 245, 292
  practical, 57, 191, 194, 348, 350, 351
  substantial, 172, 223, 224, 240, 300, 509, 516
    see also taking-over certificate
  tests on, 130, 236, 237, 345, 379, 469, 524, 544, 545, 556, 564, 573–5, 593, 594
    definition, 130
  time for, 146, 225, 230, 294, 343–7, 355, 358
    definition, 345
conciliation, *see* alternative dispute resolution
concurrent liability, 53, 187, 188, 189, 190
Conditions of Particular Application, Part II, 4, 7, 11, 51, 208, 466, 473, 752
consideration, 45, 55–7
consultation, 107, 135, 142, 166, 218, 401, 406, 481
  due, 135, 219, 220, 240
  and negotiation, 441
  clause 6.4, 137, 167, 320, 345
  clause 12.2, 171, 311
  clause 14.1, 140, 221
  clause 30.3, 171, 221
  clause 30.4, 171
  clause 36.5, 168
  clause 37.4, 168, 318
  clause 38.2, 174, 332
  clause 39.1, 171
  clause 39.2, 171
  clause 40.2, 145, 169, 171

clause 42, 174, 328
clause 42.2, 171
clause 44, 144, 174, 213, 345–6, 358
clause 44.1, 171
clause 44.3, 172
clause 46, 170, 174
clause 46.1, 169, 172
clause 49.4, 169
clause 50.1, 169, 319
clause 52.2, 301
clause 64.1, 170
clause 65.5, 173
clause 65.8, 173
clause 67, 403, 410
clause 69.4, 173
clause 70.1, 170
clause 70.2, 170, 317
duty to, 220, 240
meaning, 220
prior to making decisions, 196
under the Yellow Book, 444, 448
requirement for, 135
contingency, 99, 106, 279, 306, 346, 581, 587
contract, 8, 26, 37, 40, 47, 53, 54, 78, see also agreement
build-operate-transfer type, 85, 473
choice of, 83
construction contract, 3, 8, 44, 58
characteristics, 93
inherent characteristics, 93, 109
responsibility to complete, 68
under 1987 Yellow Book, 463
contents of, 48, 59
cost reimbursable, 87, 90
design and build, 206, 207, 274, 282, 463, 471, 475, 552, 554, 567
electrical and mechanical, 15, 126, 127, 463, 464, 465
execution of, 188
general principles, 54
intent, 45, 55
international, 11, 19, 23, 466
legal capacity, 56
lump sum, 87, 90, 91, 476
management, 86
performance of, 58
specific performance, 64, 268

pre-requisites of a contract, 55
privity of, 58
re-measurement contract, 87
schedule of rates, 90
standard forms, 3, 17, 81, 112, see also agreement
contractor, 85
care and skill, 231
default by, 106, 117, 133, 171, see also default
design by, 86, 115, 116, 137, 232, 234, 238, 496, 560, 572
determination of engineer in favour of, 131, 132, 139
duties of, 69, 225
equipment, 130, 133, 144, 148, 173, 233, 234, 235, 249, 378, 567–8, 510, 573, 595
insurance of, 252, 253, 256, 507, 518, 521
re-export of, 513
removal of, 250, 252, 512, 515, 518, 666, 668
to remain on site, 285
transport of, 514
valuation of, 384
excused from completion, 227
further information, 240
obligations of, 67, 222–5
administrative, 236
after substantial completion, 240
care of works, 239, 380, 507
implied, 231
risk sharing, 110
to acquire patent right, 229
to administer, 236–40
to commence, 328
to complete, 8, 68, 139, 225, 229, 230, 238, 240, 343, 375, 499, 544
to give notice, 236, 237
to limit interference, 326
to make payments, 316, 331
to proceed, 239
to provide indemnities, 225, 230, 222, 233, 259, 260
to provide insurance, 222, 223, 260, 264, 481, 507
to provide securities, 222, 233, 284, 285

contractor (*cont'd*)
  to remedy defects, 240, 246, 247, 253, 315, 321
  to satisfy himself as to correctness and sufficiency of tender, 223, 314
  to supply information, 234–6
    for cash flow, 221, 234
    for design, 230
    for the site, 222
    on costs for purpose of clause 63, 239
    on nominated sub-contractors, 236
    on payment, 236
    re draft final statement, 240
    re statement of completion, 240
  to strengthen and/or improve routes to site, 240, 263, 331, 332
convention
  EEC Convention on the Law Applicable to Contractual Obligations, 21, 22, 23
  international, 19, 28, 30, 31, 39, 151, 409
  1927 Geneva Convention, 31
  1958 New York Convention, 23, 30, 31, 387
correction of certificates, 511, 517
  cost, 7–11, 49, 69, 89, 90, 94–8, 128–32, 145, 146, 150, 163, 182, 186, 217, 225, 232, 251–5, 320, 359, 360
  definition, 131
  extra, 8, 9, 252, 338, 385, 401
  increased or decreased, 133, 173, 219, 236, 250, 512, 518
  prime, 10
courts
  hierarchy under common law, 36
  International Court of the ICC, 23, 395, 415, 423
    appointment of arbitrators, 415
    functions, 415, 416
    independence, 418
    procedure, 416
    publications, 423
    scrutiny of awards, 416
  London Court of International Arbitration, 23
  Permanent Court of Arbitration, The Hague, 476
crime, 54, 185, 200

critical
  activity, 362, 366–8
  delay, 95, 363, 366, 368
  event, 362
  path, 346, 359, 361, 362, 365, 367
currency, 219, 299, 316
  fluctuations, 237, 253, 255, 316
  payment in foreign, 258, 288
  proportions, 317
  restrictions, 7, 51, 133, 212, 214, 215, 512, 517
custom, 8, 28, 30, 35, 39, 41, 43, 44, 47, 59, 60, 387, 388, 396, 550

damage to works, 245
  damage to third party property, 259
  duty of contractor, 245, 246, 256
  rectify, 245, 248
damages, 8, 34, 57, 62, 65, 67, 259, 260, 266, 325
  definition, 370
  liquidated, 4, 37, 67, 117, 134, 285
  meaning, 62
  under sub-contract, 480
  under Yellow Book, 469
daywork, 132, 134, 235, 303, 377
decision
  judicial, 24, 27, 29, 199, 399
  of dispute board, *see* Dispute Board
  of employer, 208, 221, 259
  of engineer, 136, 150, 158–60, 403, 409
    re sub-clause 60.6, 149
    under clause 67, 173, 175, 237, 403–4, 409
    under 1987 Yellow Book, 467, 470
  of Dispute Adjudication Board, *see* Dispute Board
  of Dispute Review Board, *see* Dispute Board
deductibles, 259, 262, 534
default by contractor, 117, 133, 147, 171, 175, 211, 258, 281, 323–5
  consequence, 270, 323
  definition, 324
  in carrying out an instruction of engineer, 318–19
  rate of progress, 229
  re Dispute Boards, *see* Dispute Board
  securities, 139, 214, 284

suspension, 237, 364
termination, 214, 229, 324
  payment under clause 65.8, 325
default by employer, 5, 130, 135, 138, 229, 325
  non-payment, 151, 378
  payment on termination, 325, 384
defects, 192, 195, 232, 240, 299
  cost of remedying, 132, 194, 258, 319
  definition, 317
  latent, 58, 119, 272, 275, 286
  notice of, 138
  obligation to remedy, 317
  remedying, 50, 162, 168, 169, 217, 223, 227, 240, 260, 310, 318, 319
defects liability period, 217, 237, 240, 245, 246, 253, 256, 272, 291, 318–19, 325, 381, 476
delay, 253, 298, 320, 343, 344, 345
  antecedent, 216
  by employer, 145, 219, 334, 335, 344
  cost of, 319, 320, 323, 358
  critical, 95, 333, 366
  definition, 362, 363
  due to late issue of drawings or instruction, 115, 320, 343
  extension of time for, 359
  in arbitration, 425, 426, 438
  in completion, 359
  non-critical, 333, 363, 368
  notice of, 146, 237, 320
  network analysis for, 359, 363
  risk, 100
  non-performance, 284, 286
  special risks, and, determinations, 156, 318, 506, 513, 514, 522, 526, 583
  under Orange Book, 472
  under Yellow Book, 466, 468
design, 156, 162, 216
  by contractor, 126, 137, 222, 234, 238
  by engineer, 141, 155, 157, 210
  defective, 141, 142, 256, 257, 263
  of temporary works, 127, 137
direct negotiation, 441–3
dispute, 12–14, 112, 115
  adjudication, 135, 400, 452, 453, see also Dispute Board
    under Orange Book, 475

alternative dispute resolution, see alternative dispute resolution
amicable settlement of, 221, 239, 400
  procedure under clause 67, 407
between employer and contractor, 300
creating a dispute when one does not exist, 402
definition, 402
denial of existence of, 393, 400, 401
existence of, 393, 400, 401
or difference, 402
procedure when a dispute arises, 403
referred to arbitration, 214, 237, 393, 394
  under Orange Book, 475
  under sub-contract, 483
  under 1987 Yellow Book, 470
referred to Dispute Board, 403
referred to engineer, 399
settlement of
  by alternative methods, 439
    under Orange Book, 474, 475
    under sub-contract, 483
    under 1987 Yellow Book, 469, 470
  by arbitration, 387
Dispute Board
  advantages, 600
  background, 602
  chairman, 615
  cost of, 631
  decision
    binding effect, 618
  delay in appointment, 617
  for an opinion, 618, 625
  hearings, 628
    procedure, 622
    to decision, 629
  impartiality and independence, 620
  indemnity, 619
  liability, 619
  meetings, 623
    agenda, 621, 624
  obligations, 613
    following a hearing, 629
    of the board, 620
    of the parties, 617
  power of, 622

Dispute Board (cont'd)
  referrals to, 625
    for a decision, 626
  remuneration of, 629, 630
  role of, 612
  selection of members, 616
  site visits, 621, 623
    attendance, 619
  types of
    Combined Dispute Board, 607
    Dispute Adjudication Board (DAB), 606
    Dispute Review Boards (DRB), 604
  varieties, 607
disruption, 8, 160, 269, 334, 363, 366, 368, 375, 430, 433
  cost of, 8, 334
  definition, 365
drafting principles, 70, 113, 464
drawings
  delay in supplying, 137, 219
  failure by contractor to submit, 167, 213
  failure by engineer to submit, 127, 167, 237
  return of, 381
  supply, 155, 158, 160, 161, 193
duty of care, 54, 190
  engineer's, 193–8
  requirements, 193
  standard of, 190

ECGD, 284
EIC/FIDIC survey of 1996, 120
employer
  default by, 145, 229, 237, 325
  not to interfere, 215
  obligations of, 208–21
    appointment of engineer, 209
    definition of requirements, 209
    possession of and access to site, 211–13
    to appoint nominated sub-contractors, 217
    to comply with insurance conditions, 214, 221
    to consult, 219
    to give notice, 209, 213, 214
    to make payment, 218
    to nominate, 209, 217
    to permit construction of the whole of the works, 218, 219
    to provide all available data, 209, 214
    to provide instructions, 213–15
    to provide permissions, 214
    to resolve disputes through amicable settlement, 221
    to supply material and workmen, 216
  rescission by, 261, 328
  risks of, 104, 141, 142, 168, 233, 245–51, 254, 257, 259, 294, 295, 315
  role of, 478, 492, 497, 498, 501, 507, 512, 518, 524, 531, 534, 582, 586, 587
  termination of contract by, 250, 251
engineer, 73
  appointment of, 209
  assistant of, 166
    appointment of, 165, 167
    authority of, 166, 167
    instructions given by, 165
  authority of, 162, *see also* duties
    delegation of, 138, 166
  decision of, 150, 404
    conditions precedent to arbitration, 182, 404
    failure to comply with, 151, 409
    failure to give notice of, 402, 404
    in relation to clause 60.6, 149, 240, 381
    reasons for, 181
    re-opened, reviewed and revised, 181, 388
    under clause 67, 150, 175, 214, 404
    under Yellow Book, 445, 449
  definition of, 74, 129, 163
    under the Yellow Book, 75, 129
  dispute referred to, 183, 185
  duty or duties, 8, 162
    of care, 190, 194
    of design, 194
    passive duties, 173–5
    proactive duties, 167–8
    reactive duties, 170–3
    standard of, 190
    to consult, *see* consultation, due
    to resolve disputes until end of limitation period, 364
    under the Yellow Book, 446

identity, 125
impartiality of, 82, 165
   under the Yellow Book, 445
independence of, 73
liability of, 184
   concurrent, 189, 196
   decennial, 192
   in negligence under common law, 193
   in negligence under Romano-Germanic law, 198
   joint and several, 119, 198
   knock-on, 191
notice by, 169–77
remuneration of, 74, 76, 77
replacement of, 210
   under the Yellow Book, 446
representative of, 4, 159, 160, 165, 167, 169
   appointment of, 165
   responsibility of, 165
responsibility of, 184–7
   towards contractor, 192
   towards employer, 187–9
   towards society, 200
   towards third parties, 199
role of, 14, 81, 82, 175
   as adjudicator, 14, 82, 180–3
   as agent, 82, 162
   as certifier, 9, 82, 178–80
   as designer, 155, 157
   as quasi-arbitrator, 14, 82, 180–2
   as supervisor, 155, 175–8
   as witness, 409
   criticism of, 149, 186, 383, 387, 388
services provided by, 77
supervision by, 155
trust, 114
under the Yellow Book, 445
equity, 25, 34, 35, 37, 52
   Rules of, 34, 37
exclusion clauses, 68
expense, 143
Export Credit Guarantee Department, *see* ECGD
extension of time, 12, 137, 343–58
   calculation of, 137, 358
   claim for, 350, 557
   clause 6.4, 167, 320
   clause 12, 137, 171
   clause 27, 143, 17
   clause 36.5, 144, 168
   clause 40, 145, 168
   clause 42, 146, 171
   clause 44, 345–7, 358
   clause 46, 147
   clause 48, 172
   clause 69, 151
   duty to consult relating to, 219
   employer to be kept informed, 133
   notice for, 235
   relevant clauses, 353
   under Orange Book, 453
   under Yellow Book, 448
extension of the time for completion, 146, 346, 348

failure
   by contractor to
      carry out instructions, 131, 133, 510, 514
      commence, 320, 328, 329
      complete, 295, 344, 370, 374
      comply with clause 53, 148
      comply with Dispute Board's decision, 537, 538, 549
      comply with engineer's decision, 151, 318, 549
      comply with insurance conditions, 143
      insure, 98, 105, 131, 141, 508
      notify, 332, 350
      pay, 400
      perform his obligations, 132, 147, 173, 231, 233, 400
      proceed with work, 5, 357
      submit drawings, 137, 167, 171, 506, 513
   by employer to
      comply with Adjudication Board's decision, 519
      comply with Dispute Board's decision, 538
      comply with engineer's decision, 151, 549
      give possession, 132, 211, 219, 323, 353, 509
      other obligations, 326
      pay, 179, 218, 384

failure (cont'd)
  by engineer to
    approve drawings, 401
    give notice of decision, 150
    issue drawings or instructions, 8, 120, 127, 301, 566
FIDIC, 6, 14, 73–8, 119, 125
  Code of Ethics, 76, 82
  Codes of Practice, 7, 61, 65
  Contracts Committee, 463, 487, 581
  definition of, 6
  Guide to the red book, *see* Red Book
  Guide to the use of independent consultants, 77
  International General Rules for Agreement, 463
  Notes to third edition, 10
  Professional Liability Committee, 119
  publications, 483
  statutes, 73–8
FIDIC 1992 Fourth Edition, *see* Red Book, 1992
FIDIC 1999 Books
  Green Book, 488, 494, 502
  1999 Red Book, 488, 503–50
  1999 Silver Book, 488, 581–98
  1999 Yellow Book, 488, 551–80
final certificate, *see* certificate
final statement, 148, 149, 170, 219, 328, 332, 511, 517, 597, 643
  draft, 173, 240, 381
fitness for purpose, 190, 202–4, 206, 207, 232, 476, 526, 554, 555, 565–7, 571, 575, 582
fluctuations, 301, 320
  calculation of, 255, 321
  claim in respect of, 237, 301
  cost, 316, 378
  currency, 299, 316
  labour and material, 316
force majeure, 199, 466, 470, 477, 492, 498–500, 511, 518, 528, 530–5, 586
forces of nature, 5, 141, 142, 246, 248, 315, 587
foreign currency, *see* currency
fossils, 132, 134, 143, 171, 219, 234, 324, 353, 478, 515, 524, 540, 544
frustration, 59, 118–20, 499, 500, 530

guarantee, *see* securities and guarantees
  advance payments, 272, 290, 547
  export credit, 284
  payment, 272, 273
  performance, 270, 277–9
  unconditional, 271, 287, 288, 291, 491, 588, 641, 715, 786

hazard, 93, 95–8, 105, 107
  analysis, 100
  classification, 98
  consequences, 98, 100
  definition, 94, 266
  identification of, 100, 310
  meaning, 95
hostilities, 246, 248, 266
  definition, 266

ICC Rules of Arbitration, 407, 414
ICC Rules of Conciliation and Arbitration, 7, 280, 408, 414, 471, 483
  advantages, 422
  articles, 420
  award under, 280, 422, 423
  conciliation, 445
  criticism, 424
  party autonomy, 423
  procedure, 416, 431
  terms of reference, 430
    advantages, 419
ICE Form of Contract, 4, 7, 9, 17, 72, 73, 156, 259, 532
  changes made in devising the Red Book, 113
  role of the engineer, 156, 158
impartiality
  definition, 165
  requirement for, 177, 182, 185, 186
implied terms, 58
impossibility, 325, 338, 535
  frustration, *see* frustration
  physical or legal, 227, 228, 229, 230, 318
indemnity, 175, 195, 206, 233, 248, 259, 260, 266, 480
  and insurance, 111, 241, 242, 244, 269
  limit of, 261, 262
  by contractor, 144, 233, 234, 259, 260, 326, 478
  by employer, 5, 135, 259, 326

definition, 266
  for damage to persons and property, 259, 507
  for infringement of patent rights, 144, 233, 508
  to Dispute Adjudication and Review Board Members, *see* Dispute Board
injunctions, 34, 61, 64
inspection, 166, 509, 513
  as an advantage to arbitration, 376
  delegate, 168
  of insurance policies, 266
  of material, plant, etc., 133, 232
  of records, 172, 305, 329
  of site, 310, 506, 514
  of work, 155, 159
  under the Orange Book, 457
instructions, 8, 133, 134, 167
  by employer, 209, 592
    as and when required, 213
  by engineer, 136, 156, 160, 162, 167, 168, 171, 174, 192, 232, 379, 523
    disputed under Yellow Book, 466, 467, 468
    during Defects Liability Period, 323, 510
    for dealing with fossils, etc., 134, 171, 326
    for variations, 147, 368
    late, 320, 566
    leading to clause, 67, 184, 393, 401
    on records, 167, 329
    relating to certificates, 366
    relating to material or plant, 230, 232
      not in accordance with contract, 232
      failure to carry out, 214, 300
    relating to programme, 178
    reviewed by arbitrator, 391, 392
    to delegate, 589, 590
    to search, 323
  in writing, 379, 523, 566
    relating to securities, 286
    to omit work, 147, 301
    to tenderers, 196, 568, 591
    under the sub-contract form, 460
insurance, 111, 233, 241, 252–66
  adequacy of, 143, 252, 263
  against defective design, 142
  against defective materials, 9, 142, 257
  against defective workmanship, 142
  against unfair calling of securities, 284, 287
  approval of, 533
  average clause, 206, 253
  claim settlement, 253, 262
  conditions, 143, 262
    arranged by employer, 213, 248, 258
    failure to comply, 143, 263
  contractors' all risk policy, 247, 248, 253, 256, 257, 262, 263
  cross liability, 261
  decennial, 119, 191, 192, 198
  deductible, 108, 259, 262, 534
  definition, 266
  engineer's duty relating to, 160, 174, 177
  evidence of, 223, 224, 262
    notice, 236, 265
  exclusions, 142, 252, 258, 292, 532
  failure to keep, 213, 266
  failure to provide, 508
  indemnity and, *see* indemnity
  inherent characteristics of construction, 111
  injury to workmen, 261, 508
  joint, 5, 143, 220, 256, 260
  limit of indemnity, *see* indemnity
  notice of change, 140, 213
  of works, 142, 537
  period of, 252, 253, 256, 260–5, 480
  policies, 481, 501
    contractors' all risk, 248, 258
    employer's liability, 264
    production of, 262
    professional indemnity, 206, 207, 257, 476
    third party, 5, 9, 143, 220, 260, 507, 518
  purpose of, 255
  requirements, 143, 244, 252, 260, 262, 533
  scope of cover, 257, 261
  sum insured, 253, 257, 262
    definition, 142
    full replacement cost, 131, 252, 262
  terms of insurance policies, 221, 225, 262
    agreement on, 223
  to be provided, 243, 253, 263, 481, 500
  under-insured, 253
  under the Orange Book, 480
  under the sub-contract form, 479
  under the Yellow Book, 466
  unfair calling of securities, 284, 287

insurrection, 247, 249, 267
interest, 148, 179, 218, 336, 343, 371, 378, 384, 530, 545
   under the sub-contract form, 482
international
   arbitration, 93, 388, 389, 397, 427, 538, *see also* arbitration
   contracts, 16, 18, 71, *see* contract
items of interest discovered on site, 331

joint and several, 198
joint insurance, *see* insurance

knock-on-liability, 191

language, 71, 72
Latent Damage Act, 57
law
   administrative law, 26
   applicable law of contract, *see* applicable law
   areas of laws affecting construction, 32, 33, 40
   common, 24, 25, 32–40, 52, 61, 70, 193, 198
     Red Book and the, 54
   conflict, 51
   express choice of applicable law, 20
   inferred choice of applicable law, 20–2
   Islamic, 24, 25, 41–8
   governing enforcement of awards, 22
   governing procedure, 22
   procedural, 52
   Roman, 24
   Romano-Germanic, 24–33, 193, 198
   socialist, 24, 25, 41
   sources, 28, 35, 42
   sources in arbitration, 396
   substantive, 52
legal capacity, 56
legal impossibility, 228
legal systems
   absence of, 116
   common law, 24, 35, 52, 193, 198
   contemporary, 24
   diversity of, 17–19
   Islamic Law, 24, 25, 41
   Romano-Germanic, 24, 25, 27, 193, 198
   socialist laws, 24, 25, 41

legality of objectives, 55, 56
legislation, 28, 35
   primary and subordinate, 35
letter of acceptance, 163, 169, 223, 224, 325
   contractor's obligations and, 224, 225, 233
   definition, 224
   employer's obligations, 219
   insurance and, 224, 225, 262
   securities and, 224, 233, 286, 290
letter of credit, 270, 276
   standby, 271
liability, 108, 111, 241, 249, 251
   concurrent, 53, 187, 189
   contractor's, 229
     for fitness for purpose, 325
     statutory, 233
   decennial, 119, 191, 192, 198
   defects, 194
   design, 115, 194
   difference as to whether in tort or contract, 189
   engineer's, 156, 184, 187, 189
     criminal, 200
     in common law, 193
     in Romano-Germanic system, 198
     skill and care, 157, 190
     statutory, 207
     towards contractor, 193
     towards employer, 187
     towards third parties, 193
   employer's, 159, 245
   joint and several, 118, 198
   knock-on-liability, 191
   limitation of, 70, 119
   period, *see* defects liability period
   retained, 135
   to third parties, 142, 198, 437, 438
     under contract, 187, 188, 193
   tort, 52, 54, 81–2, 187, 189, 194
libel, 52
limitation
   Act, 46, 57–8, 189
   periods, 57, 58, 119, 189, 377
   statutes of, 319
liquidated damages, 67, 370
   definition, 67, 109
   for delay, 117, 134

limit of, 285
penalty and, 37, 278, 370–1
reduction of, 376
right to deduct, 375, 376, 378
under the Yellow Book, 469
lump sum, 87, 472
contract, 87, 90, 472
items, 235
breakdown of, 169, 225
Orange Book, 472, 476
Supplement of November 1996, 14, 91
basis of payment, 91

maintenance
bond, 272, 282, 291
certificate, 7, 159
manuals, 137, 160, 171, 477
measurement, 87, 303, 304
duty of engineer, 158
give notice, 304
of risk, 98
particulars of, 235
principles of, 304
re-measurement, 15, 83, 87, 88
standard method, 5, 158
valuation by, 302, 304
under the 1992 Red Book, 83, 87, 113, 377
under the Yellow Book, 87
with bill of quantities, 89
mediation, 443
advantages, 444
rules, 445
where unsuitable, 444
mediator, 444
as negotiator, 444
role, 444
methods of construction, 158
responsibility of contractor, 215
mini-trial, 451, 452
development of, 451
procedure, 451
rules, 452

negligence
claim in, 193
criminal, 52

definition, 54
in certifying, 194
onus of proof, 233
pre-requisites, 193
New York Convention of 1958, *see* convention
nominated sub-contractor, 132, 133, 159, 179, 219
notice
by contractor, *see* contractor
by employer, *see* employer
by engineer, *see* engineer
delaying events, 147
intention to claim, 145, 274, 301
intention to commence arbitration, 150, 214, 237, 239, 406
of claim, 64
of defects, 138
relating to securities, 107, 275, 286, 287
to commence, *see* commence
failure to give notice of intention to commence, 407
novation, 238
nuisance, 53

obligations
of contractor, *see* contractor
of employer, *see* employer
of engineer, *see* engineer
unfulfilled, 317
obstruction, 306
unforeseen, 139, 492
operation and maintenance manuals, *see* maintenance
Orange Book, 471
background, 473
employer's requirements, 475
fitness for purpose, 476
risk, 477
Ouzel Galley, 385

patent right, 144, 229, 233
infringement of, 233
payment, 12, 14, 132, 148, 159, 160, 162, 172, 377
certificate, *see* certificate
contractor to make, 168, 240
of royalties, 236

payment (cont'd)
  employer to make, 209, 211, 218, 219
  failing to make, 151, 320, 381
  guarantees, 272, 290
  if contract is terminated, 133, 219, 250
  if performance is released, 133
  in foreign currency, 252, 258
  of certificates, see certificate
  of claims, 132, 218, 237, 295, 301
    basis, 87, 91, 248, 301, 332
  to Dispute Board members, see Dispute Boards
  to nominated sub-contractors, 132, 179, 218, 236
  under the Orange Book, 474, 475
  under the sub-contract form, 479, 481–3
  under the Yellow Book, 465, 471
penalties, 35, 64, 117, 325
  and liquidated damages, 117, 278
performance
  bond, see securities
  of administrative functions, 222, 236
  of sub-contractor, 238
  proper, 224, 233, 271, 284, 288
  release from, 118, 219, 319, 324, 325
    payment in the event of, 132, 133
  securities and guarantees, see securities
physical and legal impossibility, 229, 318
  meaning, 227
physical obstruction or conditions, 118, 139, 218, 309, 314, 353, 492
  meaning, 139
  notice of, 171, 237
  Pacific Associates case, 196
  recovery as only costs, 131
plant, 126, 130, 133, 137, 144
  definition, 130
pre-arbitral referee procedure, 457
priority of contract documents, 136
procedure
  applicable arbitration rules of, 408
  arbitration, 388, 395, 425, 427
    under the Orange Book, 475
    under the sub-contract form, 482
    under the 1987 Yellow Book, 468, 470
  for amicable settlement, 407, 440, 445, 483
    conciliation, 445, 448

  for arbitration under clause 67, 403, 405, 408, 412
  for claims, 328
    charts, 330, 331
  for decision of Dispute Boards, see Dispute Boards
  law governing, 22
  mini-trial, 451
  pre-arbitral reference, 457
  under ICC rules
    for ADR, 445, 453–7
    for arbitration, 408, 416
    for conciliation, 414, 420
programme, 216, 224, 358, 364
  by contractor, 234
    duties, 224
  charts, 356, 357
  network, 358, 362
programming, 358
progress of the works, 162, 178, 209, 332, 352, 364, 377
  acceleration of, 369
  charts, 356, 357
  not to interfere with, 215
  supervision of, 159, 169
  suspend, see suspension
project network analysis, 362
provisional items, 158
provisional sum, 161, 170, 175, 220, 236, 304, 314, 325
  meaning, 220

quality
  of materials, 230, 232, 318, 508, 514, 515
  of plant, 232, 318, 508, 514, 515
  of workmanship, 232, 318, 508, 514, 515
quantities, 87–90, 303, 309, 456
  adjustment of estimated, 303, 304
  bill of, see bill of quantities
  increase or decrease, 300
    in excess of 15%, 303
  measurement of, 304
  re-measurement contract, 15, 83, 87, 91
  variations and, 305
quantum and claims, 333, 337, 417
quantum meruit, 65, 295
Qu'ran, 25, 42, 44, 46, 47

rebellion, 247, 249
  definition, 267
rectification of
  contract, 64
  damage or loss, 142, 168, 174, 246, 252, 760
  defects, *see* defects
Red Book
  1957 First Edition, 6, 9
    based on a domestic contract, 17
    concepts in practice, 112
    concepts of, 6
    on a lump sum basis, 14
    origins, 6
  1967 Second Edition, 7–10
  1977 Third Edition, 7–11, 15, 125, 155
  1987 Fourth Edition, 11–13, 125, 153
    amendment to, 11
    guide, 128
    insurance requirements, 143, 252, 260, 262, 533
    performance securities, 272, 284, 288, 290
    principal changes from the third, 125–52
    reprint, 1992, 140, 149, 152
  sub-contract form, 84, 460
  Supplement of 1996, 14, 91, 151, 409, 411
  1999 Edition, 15, 487, 503, 613, 615, 627
reinsurance, 108, 259
remedies in law, 62
  chart, 62
  damages, 8, 34, 62, 65
    action for, 52, 54, 57
  equitable, 37, 62, 64
  for breach of contract, 8, 62, 67
  injunctions, 64
  punitive, 67
  specific performance, 64
removal of contractor's equipment, 173, 250, 251, 252, 512, 518
repudiation, 59, 280, 295, 395
rescission, 58, 64, 323, 324
responsibility
  and liability, 15, 108, 112, 152, 184, 193, 241
  definition, 184
  delegation of, 191
  for variations, 301
  of contractor, 138, 170, 175, 176, 311, 368, 541, 542

  of employer, 169, 248, 260, 261, 316, 340, 367
  of engineer, 136, 184, 312
  passing of, 105, 314, 380
  shared, 116
  to complete, 68
retention
  bond, 282, 291
  monies, 285, 291, 378, 380, 482, 511, 578
  under sub-contract form, 482
riot, 268
risk, 94, 95, 111, 268
  allocation, 5, 86, 100–8, 115, 478, 492, 582
    principles of, 105
  assessment, 98, 587
  assumption of, 54
  avoidance of, 115
  contractor's, 233, 245–8, 259, 478, 531
  definition, 94
  delayed completion, 103
  design, 161, 247
  employer's, 229, 233, 246, 345
    under the 1999 Red Book, 104
  evaluation, 100
  financial, 85, 113, 158, 244, 478, 491
  forces of nature, 248
  improper execution, 284
  management, 99
  meaning of, 105
  of damage or injury, 54, 108
  of war, 246
  quantification, 100
  sharing, 93
  significance of, 100, 105, 241
  special, 248
  uninsurable, 103, 257
role
  of engineer, 14, 81, 114, 155
  of judge, 28
royalties
  liability to pay, 239

safety, 94, 141, 198, 211, 220
securities and guarantees
  forms of, 270
    accessory suretyship, 271, 281
    conditional guarantee, 270

securities and guarantees (*cont'd*)
    demand guarantee, 271
    documentary credit, 271
    suretyship guarantee, 271
  notice, 275, 286, 287
  period of validity, 272, 275, 286
  types of
    bid bonds, 272
    concession bonds, 272
    contract bonds, 271
      features of ICC uniform rules, 281
      ICC uniform rules, 280
    customs bonds, 271
    demand guarantees, 276
      features of ICC uniform rules, 281
      ICC uniform rules, 280
    fidelity bonds, 272
    financial bonds, 272
    maintenance bonds, 272
    payment guarantees, 272
    performance bonds, 273
      requirements, 274
  under the 1992 Red Book, 284
    delayed completion, 284
    examples of, 288
    failure to complete the works, 284
    improper execution, 284
  under the sub-contract, 482
  under the Yellow Book, 287, 469
  unfair calling of, 284
Shari'ah, 25, 42
site, 209, 211, 223
  accepted as contractor finds it, 212
  access to 174, 212, 213, 239
    by owners of properties forming part of the site, 212
  clearance of, 168, 240
  damage to, 259
  definition, 209, 212
  possession of, 171, 174, 209, 211
    failure to give, 219, 323
    the whole site, 212
slander, 53
sources of law, 19, 25, 28, 35
  in arbitration, 394
special risks, *see* risk
statement on completion, 173, 240, 332, 381

statute, 60
strike, 268
supervision, 168, 175, 178, 191, 194, 198
suspension, 173, 175, 237, 299, 320
  charts, 321, 322
  consequences of, 219, 323
  contractor's entitlement to, 132, 151, 219
  engineer's determination following, 132
  more than 84 days, 145, 229
  power of engineer, 320
  war, *see* war

tender, 223
  acceptance, 224, 364
  Appendix to Tender
    completion period, 225, 230, 327, 343
    details of listed items, 210
    Dispute Boards, *see* Dispute Boards
    insurance limits, 260
    interest, 384
    limit of liquidated damages, 327, 344, 370, 372
    minimum amount of interim certificate, 218, 377
    percentage of the invoice value of listed materials, 377
    rate of interest, 218
    retention percentage, 285, 291, 327, 378
    security sum, 224, 233, 285, 288
    site, 326
    in the Orange Book, 474, 475
    in the sub-contract form, 480
    in the 1996 Supplement, 400, 409
  competitive, 157, 217, 313
  form of, 91
    Yellow Book, 465
  stage, 211, 224, 376, 475
  sufficiency of, 311
termination, 118, 150, 299, 320, 324
  arbitration and, 388
  assignment after, 239
  by contractor, 229
  by employer, 324, 328
  of a security, 274, 280, 286

payment on or after, 133, 219, 325
under clause 65, 250
valuation at, 133, 179, 324, 325, 377, 384
testing, 145, 168, 232, 318
costs of, 325
from time to time, 318
on completion, 178, 237, 379, 380
definition, 130
under the Orange Book, 475
third party insurance, *see* insurance
Time for Completion, 169, 222, 225, 229, 230, 234, 353, 359, 379
claim, 299, 327, 365
definition, 343
delay to the, 362, 363, 366, 368
extension of, 168, 215, 353, 466
under the Yellow Book, 468, 469
tort, 56, 296
cause of action, 57
damages in, 67
liability in, 187–99
concurrency, 196
meaning, 52
period of limitation, 119
trespass, 53
trust, 73, 112, 114

UNCITRAL Arbitration Rules, 20, 502
UNCITRAL Conciliation Rules, 445
UNCITRAL Model Law, 19

variations, 9, 88, 92, 132, 147, 172, 186, 229, 298, 299, 306, 317, 501, 546, 556, 571, 595
exceeding 15%, 132, 169, 510
instruction for, *see* instructions
nature of, 299
orders, 306–9
reasons for, 299
records, 299
responsibility for, 302
valuation of, 174, 302, 510

war
civil, 247, 249, 266, 267, 323
definition, 266
definition, 268
risk of, 246, 250
suspension of works and, 323
termination and, 250
White Book, 84, 463, 499
works
meaning, 130, 157
progress of, 8, 10, 169, 172, 209, 215, 501, 543, 547, 592, 607
World Bank, 85, 112, 113, 138, 248, 608, 631

Yellow Book 1987
essential features, 465
format, 466
performance security, 469

Zurich Chamber of Commerce, 451, 452